Sustainability Sciences in Asia and Africa

Agroecosystem Dynamics and Sustainable Practices

Series Editors

Kamel A. Abd-Elsalam, Plant Pathology Research Institute, Agricultural Research Center, Giza, Egypt

Olubukola O. Babalola, Faculty of Natural & Agricultural Sciences, North-West University, Mmabatho, South Africa

This book series publishes research associated with the prevailing opportunities and challenges faced by agroecosystems in Asia and Africa. It also explores the future scope of agroecological techniques and offers guidance on sustainable management of these systems. The series discusses important areas such as sustainable agricultural production, soil health, water resource management, biodiversity conservation, animal health and productivity, insect management strategies, genome editing and climate change mitigation. Further, it includes the challenges faced by African agriculture in arid and semi-arid regions, particularly in response to climate change. It highlights issues such as water shortages, soil degradation, desertification, and the need for sustainable farming techniques.

Sustainable agroecosystem maintenance entails strategies that improve overall environmental health and economic viability while limiting negative environmental impacts. The series takes a multidisciplinary approach to investigate creative ways for boosting sustainable agriculture, increasing food security, and improving lives for people in global south. This book series acts as a global publishing platform to collate and disseminate knowledge on fostering resilient and productive agroecosystems, thus enhancing accessibility, and fostering informed decision-making.

Arafat Abdel Hamed Abdel Latef
Editor

Harnessing Vitamins and Organic Acids for Plant Stress Resilience

Springer

Editor
Arafat Abdel Hamed Abdel Latef
Department of Botany and Microbiology
Faculty of Science
Qena University
Qena, Egypt

ISSN 2730-6771 ISSN 2730-678X (electronic)
Sustainability Sciences in Asia and Africa
ISSN 3059-247X ISSN 3059-2488 (electronic)
Agroecosystem Dynamics and Sustainable Practices
ISBN 978-981-95-8592-2 ISBN 978-981-95-8593-9 (eBook)
https://doi.org/10.1007/978-981-95-8593-9

This Springer imprint is published by the registered company Springer Nature Singapore Pte Ltd.
The registered company address is: 152 Beach Road, #21-01/04 Gateway East, Singapore 189721, Singapore

Preface

The search for viable strategies to strengthen plant defense is now more important than ever because of the continuous threats posed to crops by drought, salinity, diseases, and other environmental challenges. This pressing need is addressed in this book, *Harnessing Vitamins and Organic Acids for Plant Stress Resilience*, which concentrates on a revolutionary and environmentally friendly method: the exogenous application of vitamins and organic acids to actively safeguard plants and increase their inherent resilience.

The main topic discussed in this book is how the application of compounds such as thiamin, ascorbic acid, or humic acid can proactively initiate defense mechanisms, strengthen antioxidant systems, and modify metabolism to aid plants in coping with both abiotic and biotic stressors. Every chapter is devoted to a certain compound, carefully describing the scientific basis for its protective function.

This compilation brings together cutting-edge studies showing that these uses may greatly lessen the harmful consequences of stress, resulting in higher yields and healthier plants. This book is intended to serve as an indispensable resource for students, agricultural professionals, and scholars. It offers a novel set of tools for managing sustainable crops and seeks to connect laboratory research with real-world applications. We may unlock a new era in agriculture that is resilient, productive, and in tune with nature by utilizing the potential of vitamins and organic acids.

Qena, Egypt — Arafat Abdel Hamed Abdel Latef

Contents

Editor and Contributors

About the Editor

Arafat Abdel Hamed Abdel Latef is a Professor at the Department of Botany and Microbiology, Faculty of Science, Qena University (formerly South Valley University), Egypt. He has published 100 research articles in peer-reviewed journals and 40 book chapters in international books. He served as editor of ten published international books. He serves as an editorial member of several reputed journals. Furthermore, he is an expert reviewer for over 150 international journals. He was one of the top 2% of scientists in the world, according to Stanford University, in 2020, 2021, 2022, 2023, 2024, and 2025. He has received several awards, such as the Springer Nature Editor of Distinction Award 2025, the Egypt State Award of Excellence in Agricultural Sciences 2021, the South Valley University Award of Scientific Excellence 2019, the Top Peer Reviewer Award 2019 Powered by Web of Science and Publons, and the Qena University / South Valley University Award of Scientific Publication for the years from 2010 to 2025.

Contributors

Yara T. AbdElaziz Botany and Microbiology Department, Faculty of Science, Assiut University, Assiut, Egypt

Aouabe Abdelhamid Centre d'Agrobiotechnologie et Bioingénierie, Unité de Recherche Labellisée CNRST (Centre AgroBiotech-URL-CNRST-05), Cadi Ayyad University, Marrakesh, Morocco

Zainul Abideen Dr. Muhammad Ajmal Khan Institute of Sustainable Halophyte Utilization, University of Karachi, Karachi, Pakistan

Department of Applied Biology, University of Sharjah, Sharjah, United Arab Emirates

Jaber Mehdiniya Afra Department of Agricultural Sciences, National University of Skills (NUS), Tehran, Iran

Reza Afsharianzadeh Department of Agronomy, Faculty of Agriculture, Science and Research Branch, Islamic Azad University, Tehran, Iran

Affan Afzal Department of Microbiology, Faculty of Veterinary and Animal Sciences, The Islamia University of Bahawalpur, Bahawalpur, Pakistan

Mohamed Ait-El-Mokhtar Centre d'Agrobiotechnologie et Bioingénierie, Unité de Recherche Labellisée CNRST (Centre AgroBiotech-URL-CNRST-05), Cadi Ayyad University, Marrakesh, Morocco

Laboratory of Biotechnology, Agri-food, Materials, and Environment (LBAME), Department of Biology, Faculty of Science and Techniques Mohammedia, Hassan II University of Casablanca, Mohammedia, Morocco

Laboratory of Biochemistry, Environment & Agri-Food URAC 36, Department of Biology, Faculty of Science and Techniques—Mohammedia, Hassan II University of Casablanca, Mohammedia, Morocco

Hafsa Akter Department of Horticulture, Sher-e-Bangla Agricultural University, Dhaka, Bangladesh

Baber Ali School of Science, Western Sydney University, Penrith, NSW, Australia

Md. Arfan Ali Department of Horticulture, Sher-e-Bangla Agricultural University, Dhaka, Bangladesh

Mohsin Ali State Key Laboratory of Microbial Resources, Institute of Microbiology, Chinese Academy of Sciences, Beijing, China

University of Chinese Academy of Sciences, Beijing, China

Saeideh Alipoor Department of Agronomy and Plant Breeding Sciences, College of Aburaihan, University of Tehran, Pakdasht, Iran

Sevda Alizadeh Department of Biology, Faculty of Science, Urmia University, Urmia, Iran

Zahra Alizadeh Department of Plant, Cell and Molecular Biology, Faculty of Natural Sciences, University of Tabriz, Tabriz, Iran

Abdelhamid Aouabe Centre d'Agrobiotechnologie et Bioingénierie, Unité de Recherche Labellisée CNRST (Centre AgroBiotech-URL-CNRST-05), Cadi Ayyad University, Marrakesh, Morocco

Faculty of Sciences Semlalia Marrakech, Laboratory of Excellence in Agrobiotechnology and Bioengineering, AgroBiotech Center, CNRST-Labeled Research Unit (URL05-CNRST), (A)biotic Stress Physiology, Cadi Ayyad University, Marrakech, Morocco

Md. Riyadh Arefin Department of Crop Botany, Bangladesh Agricultural University, Mymensingh, Bangladesh

Botany Division, Bangladesh Tea Research Institute, Moulvibazar, Bangladesh

Berrin Arslan Faculty of Agriculture, Department of Plant Protection, Siirt University, Siirt, Turkey

Ali Bandehagh School of Molecular Sciences, Australian Plant Phenomics Network, ARC Training Centre in Predictive Breeding for Agricultural Futures and UWA Institute of Agriculture, The University of Western Australia, Crawley, Boorloo, WA, Australia

Behzat Baran Plant Protection Research Institute, Sur, Diyarbakır, Turkey

Marouane Baslam Centre d'Agrobiotechnologie et Bioingénierie, Unité de Recherche Labellisée CNRST (Centre AgroBiotech-URL-CNRST-05), Cadi Ayyad University, Marrakesh, Morocco

GrowSmart, Amsterdam, The Netherlands

Faculty of Sciences Semlalia Marrakech, Laboratory of Excellence in Agrobiotechnology and Bioengineering, AgroBiotech Center, CNRST-Labeled Research Unit (URL05-CNRST), (A)biotic Stress Physiology, Cadi Ayyad University, Marrakech, Morocco

Raja Ben-Laouane Laboratory of Environment and Health, Department of Biology, Faculty of Science and Techniques, Moulay Ismail University, Errachidia, Morocco

Centre d'Agrobiotechnologie et Bioingénierie, Unité de Recherche Labellisée CNRST (Centre AgroBiotech-URL-CNRST-05), Cadi Ayyad University, Marrakesh, Morocco

FSTE-FSM Joint Laboratory: NRHE-UMI, Bioresources, Environment and Health Research Team, Moulay Ismail University, Meknes, Morocco

Laboratory of Environment and Health, Department of Biology, Faculty of Science and Techniques, Errachidia, Morocco

Center of Agrobiotechnology and Bioengineering, Research Unit labelled CNRST (Centre AgroBiotech-URL-7 CNRST-05), Cadi Ayyad University, Marrakesh, Morocco

Laboratory of Environment and Health, Department of Biology, Faculty of Science and Technology of Errachidia, Moulay Ismail University of Meknes, Meknes, Morocco

Abderrahim Boutasknit Center of Agrobiotechnology and Bioengineering, Research Unit labelled CNRST (Centre AgroBiotech-URL-7 CNRST-05), Cadi Ayyad University, Marrakesh, Morocco

Laboratory of Biology, Geosciences, Physics and Environment (OLMAN-LBGPE), Pluridisciplinary Faculty of Nador, University Mohamed Premier, Nador, Morocco
Faculty of Sciences Semlalia Marrakech, Laboratory of Excellence in Agrobiotechnology and Bioengineering, AgroBiotech Center, CNRST-Labeled Research Unit (URL05-CNRST), (A)biotic Stress Physiology, Cadi Ayyad University, Marrakech, Morocco

Department of Biology, Multidisciplinary Faculty of Nador, Mohamed First University—Oujda, Nador, Morocco

Masoud Chamani Department of Plant Protection, Faculty of Agriculture and Natural Resources, University of Mohaghegh Ardabili, Ardabil, Iran

Shyam Sundar Das Department of Agronomy, Dr Rajendra Prasad Central Agricultural University, Pusa, Samastipur, Bihar, India

Srilekha Das Department of Agronomy, Dr Rajendra Prasad Central Agricultural University, Pusa, Samastipur, Bihar, India

Durjoy Kumar Datta Department of Horticulture, Sher-e-Bangla Agricultural University, Dhaka, Bangladesh

Sumanta Daw Department of Microbiology, Maulana Abul Kalam Azad University of Technology, Kolkata, West Bengal, India

Mona F. A. Dawood Botany and Microbiology Department, Faculty of Science, Assiut University, Assiut, Egypt

Mona Gergis Dawood Botany Department, Agriculture and Biological Institute, National Research Centre, Giza, Egypt

Zahra Dehghanian Department of Biotechnology, Faculty of Agriculture, Azarbaijan Shahid Madani University, Tabriz, Iran

Murat Dikilitas Faculty of Agriculture, Department of Plant Protection, Harran University, S. Urfa, Turkey

Aicha Nait Douch Centre d'Agrobiotechnologie et Bioingénierie, Unité de Recherche Labellisée CNRST (Centre AgroBiotech-URL-CNRST-05), Cadi Ayyad University, Marrakesh, Morocco

Mamdouh Eissa Department of Soils and Water, Faculty of Agriculture, Assiut University, Assiut, Egypt

Nizar El Mazouni Center of Agrobiotechnology and Bioengineering, Research Unit labelled CNRST (Centre AgroBiotech-URL-7 CNRST-05), Cadi Ayyad University, Marrakesh, Morocco

Faculty of Sciences Semlalia Marrakech, Laboratory of Excellence in Agrobiotechnology and Bioengineering, AgroBiotech Center, CNRST-Labeled Research Unit (URL05-CNRST), (A)biotic Stress Physiology, Cadi Ayyad University, Marrakech, Morocco

Farid Errouh Center of Agrobiotechnology and Bioengineering, Research Unit labelled CNRST (Centre AgroBiotech-URL-7 CNRST-05), Cadi Ayyad University, Marrakesh, Morocco

Faculty of Sciences Semlalia Marrakech, Laboratory of Excellence in Agrobiotechnology and Bioengineering, AgroBiotech Center, CNRST-Labeled Research Unit (URL05-CNRST), (A)biotic Stress Physiology, Cadi Ayyad University, Marrakech, Morocco

Boujemaa Fassih Centre d'Agrobiotechnologie et Bioingénierie, Unité de Recherche Labellisée CNRST (Centre AgroBiotech-URL-CNRST-05), Cadi Ayyad University, Marrakesh, Morocco

Jannatul Ferdous Department of Horticulture, Sher-e-Bangla Agricultural University, Dhaka, Bangladesh

Robabeh Hosseini Ghavamabad Centre for Crop and Disease Management, School of Molecular and Life Sciences, Curtin University, Perth, WA, Australia

Hansa Gul Department of Zoology, University of Mianwali, Mianwali, Pakistan

Sohel Hossen Department of Horticulture, Sher-e-Bangla Agricultural University, Dhaka, Bangladesh

Shabab Hussain Department of Chemical, Biological, Pharmaceutical and Environmental Sciences, University of Messina, Messina, Italy

Nafiseh Jalali Agroecology, Shirvan Institute of Higher Education, Ministry of Science, Research and Technology, Shirvan, Iran

Iman Janah Centre d'Agrobiotechnologie et Bioingénierie, Unité de Recherche Labellisée CNRST (Centre AgroBiotech-URL-CNRST-05), Cadi Ayyad University, Marrakesh, Morocco

Alevcan Kaplan Department of Crop and Animal Production, Sason Vocational School, Batman University, Batman, Turkey

Muhammad Nauman Khan Department of Botany, University of Chakwal, Punjab, Pakistan

Arafat Abdel Hamed Abdel Latef Department of Botany and Microbiology, Faculty of Science, Qena University, Qena, Egypt

Abdelilah Meddich Centre d'Agrobiotechnologie et Bioingénierie, Unité de Recherche Labellisée CNRST (Centre AgroBiotech-URL-CNRST-05), Cadi Ayyad University, Marrakesh, Morocco

African Sustainable Agriculture Research Institute (ASARI), University Mohamed VI Polytechnic (UM6P), Morocco

Faculty of Sciences Semlalia Marrakech, Laboratory of Excellence in Agrobiotechnology and Bioengineering, AgroBiotech Center, CNRST-Labeled Research Unit (URL05-CNRST), (A)biotic Stress Physiology, Cadi Ayyad University, Marrakech, Morocco

Mohamed Imran Mir Department of Plant Sciences, School of Life Sciences, University of Hyderabad, Hyderabad, Telangana, India

Sina Siavash Moghaddam Department of Plant Production and Genetics, Faculty of Agricultural Sciences, University of Guilan, Rasht, Iran

Murad Muhammad State Key Laboratory of Desert and Oasis Ecology, Key Laboratory of Ecological Safety and Sustainable Development in Arid Lands, Xinjiang Institute of Ecology and Geography, Chinese Academy of Sciences, Beijing, China
University of Chinese Academy of Sciences, Beijing, China

Redouane Ouhaddou Centre d'Agrobiotechnologie et Bioingénierie, Unité de Recherche Labellisée CNRST (Centre AgroBiotech-URL-CNRST-05), Cadi Ayyad University, Marrakesh, Morocco

Khursheda Parvin Department of Horticulture, Sher-e-Bangla Agricultural University, Dhaka, Bangladesh

Latifeh Pourakbar Department of Biology, Faculty of Science, Urmia University, Urmia, Iran

Biswajit Pramanick Department of Agronomy, Dr Rajendra Prasad Central Agricultural University, Pusa, Samastipur, Bihar, India

Abdul Razzaq Qarshi Herb Research Center, Qarshi Industries (Pvt.) Ltd., Hattar, Haripur, Pakistan
Qarshi University, Lahore, Punjab, Pakistan

Saleha Saeed Department of Botany, University of Peshawar, Peshawar, Pakistan

A. K. M. Golam Sarwar Department of Crop Botany, Bangladesh Agricultural University, Mymensingh, Bangladesh

Seyede Roghie Ghadirnezhad Shiade Department of Agronomy, Sari Agricultural Science and Natural Resources University, Sari, Iran

Mahmud Hossain Sumon Department of Soil Science, Bangladesh Agricultural University, Mymensingh, Bangladesh

Nicolas L. Taylor School of Molecular Sciences, Australian Plant Phenomics Network, ARC Training Centre in Predictive Breeding for Agricultural Futures and UWA Institute of Agriculture, The University of Western Australia, Crawley, Boorloo, WA, Australia

Samar G. Thabet Department of Botany, Faculty of Science, Fayoum University, Fayoum, Egypt

Sami Ullah Department of Botany, University of Peshawar, Peshawar, Pakistan

Rajeev Kumar Yadav Department of Horticulture, Dr Rajendra Prasad Central Agricultural University, Pusa, Samastipur, Bihar, India

Tabassum Yaseen Department of Botany, Bacha Khan University Charsadda, Khyber Pakhtunkhwa, Pakistan

Afshar Zafarani Department of Plant Production and Genetics, Faculty of Agriculture and Natural Resources, University of Mohaghegh Ardabili, Ardabil, Iran

Btissam Zoubi Laboratory of Biotechnology, Environment, Agri-Food, and Health, Faculty of Sciences Dhar El Mahraz, Sidi Mohammed Ben Abdellah University, Fez, Morocco

Center of Agrobiotechnology and Bioengineering, Research Unit Labeled CNRST, Cadi Ayyad University, Marrakech, Morocco

Chapter 1
Thiamine: A Shield Against Plant Stress

Zahra Dehghanian, Robabeh Hosseini Ghavamabad, Masoud Chamani, Ali Bandehagh, and Nicolas L. Taylor

Abstract Thiamine (vitamin B1) is an essential micronutrient required for cellular metabolism and energy production in all living organisms. Its biologically active form, thiamine pyrophosphate (TPP), serves as a critical cofactor for several key enzymatic processes, including those involved in glycolysis, the oxidative pentose phosphate pathway, and the tricarboxylic acid (TCA) cycle. Through these pathways, TPP facilitates the conversion of carbohydrates, lipids, and proteins into usable energy. Beyond its metabolic importance, recent research has highlighted thiamine's role in enhancing plant immunity by activating systemic acquired resistance (SAR) and strengthening defence responses against both biotic and abiotic stressors. Multiple studies have shown that the expression of genes involved in thiamine biosynthesis increases in plants exposed to various stresses. Investigations of gene expression, thiamine and intermediate metabolite accumulation, and enzyme activity complementation have all supported the protective role of thiamine in plants. This chapter will examine thiamine's function as a stress-responsive signalling

Z. Dehghanian
Department of Biotechnology, Faculty of Agriculture, Azarbaijan Shahid Madani University, Tabriz, Iran
e-mail: zahra.dehghanian@azaruniv.ac.ir

R. H. Ghavamabad
Centre for Crop and Disease Management, School of Molecular and Life Sciences, Curtin University, Perth, WA, Australia
e-mail: robabeh.hosseini@postgrad.curtin.edu.au

M. Chamani
Department of Plant Protection, Faculty of Agriculture and Natural Resources, University of Mohaghegh Ardabili, Ardabil, Iran

A. Bandehagh (✉) · N. L. Taylor (✉)
School of Molecular Sciences, Australian Plant Phenomics Network, ARC Training Centre in Predictive Breeding for Agricultural Futures and UWA Institute of Agriculture, The University of Western Australia, Crawley, Boorloo, WA, Australia
e-mail: ali.bandehagh@uwa.edu.au; nicolas.taylor@uwa.edu.au

A. A. H. Abdel Latef (ed.), *Harnessing Vitamins and Organic Acids for Plant Stress Resilience*, Sustainability Sciences in Asia and Africa,
https://doi.org/10.1007/978-981-95-8593-9_1

molecule, outline its biosynthetic pathway, and discuss its regulatory mechanisms. Additionally, the chapter will briefly address the practical implications of these findings, reinforcing thiamine's proposed role in plant stress resilience.

Keywords Thiamine · Thiamine pyrophosphate (TPP) · Tricarboxylic acid (TCA) cycle · Biotic and abiotic stress · Gene regulation · Plant stress protection

1.1 Introduction

As immobile organisms, plants are continually exposed to a variety of harmful biotic and abiotic stresses throughout their life cycle (Bandehagh et al. 2023; Afzal et al. 2026; Barman et al. 2026). In crop species, stressors such as environmental fluctuations, climate change, and pathogen attacks can significantly reduce yield and productivity. Modern research is increasingly focused on developing strategies to mitigate the effects of these stresses in ways that are both sustainable and environmentally friendly (Abdel Latef et al. 2026). In this context, the study of plant-microbe interactions, particularly the use of biocontrol agents, has attracted considerable attention. These interactions have been shown to alleviate stress in host plants, offering promising opportunities to enhance plant resilience (Boivin et al. 2016). Furthermore, research has identified a feedback regulatory mechanism through which exposure to stress can promote increased tolerance and even induce resistance in plants (Goyer 2010).

Thiamine (vitamin B_1) has emerged as a valuable compound in the field of plant stress physiology. Traditionally recognised for its role as a coenzyme in key metabolic pathways, such as the pyruvate dehydrogenase (PDH) complex, α-KG dehydrogenase (α-KGDH), and transketolase (TK) reactions, thiamine also exhibits broader regulatory functions that extend beyond primary metabolism (Subki et al. 2018). Studies have shown that thiamine enhances plant resilience to both abiotic and biotic stressors through multiple mechanisms. These include regulating reactive oxygen species (ROS) scavenging, activating key defence-related genes, and modulating hormone signalling pathways (Wolak et al. 2015). The exogenous application of thiamine, as well as its increased biosynthesis in response to stress, has been linked to improved plant performance under adverse conditions, including drought, salinity, and oxidative stress. This enhancement is often attributed to the stimulation of antioxidant enzyme activity and the activation of osmoprotective responses (Kaya et al. 2015). In addition, thiamine has been implicated in the activation of systemic acquired resistance (SAR) and other induced resistance mechanisms, thereby reinforcing plant defences against a wide range of pathogens (Boubakri et al. 2016). Given the growing prevalence and severity of environmental stresses driven by climate change, there is an urgent need for sustainable approaches to crop protection. This chapter aims to explore the potential of thiamine to enhance plant resistance to both biotic and abiotic stresses by elucidating its underlying mechanisms, including the detoxification of ROS, the activation of defence-related gene expression, and the regulation of hormonal signalling pathways. It aims to contribute to the development of environmentally friendly and sustainable strategies for enhancing crop resilience and productivity in the face of escalating environmental challenges.

1.2 Biochemistry and Physiological Role of Thiamine in Plants

1.2.1 Biosynthesis, Transport, and Distribution in Plant Tissues

The thiamine biosynthetic pathway in plants exhibits a high degree of conservation with the corresponding pathways characterised in bacteria and yeast (Li et al. 2010). The pathway begins with the conversion of the precursor 5-aminoimidazole ribonucleotide (AIR) into the pyrimidine moiety of thiamine, specifically, hydroxymethylpyrimidine phosphate (HMP-P), a reaction catalysed by the enzyme encoded by the *THIC* gene (HMP synthase) (Fig. 1.1). In a parallel branch, the thiazole moiety is synthesised from nicotinamide adenine dinucleotide (NAD), glycine, and a sulphur donor, forming hydroxyethylthiazole phosphate (HET-P) through the action of (HET-P) synthase, encoded by *THI4*. The two moieties, pyrimidine and thiazole, are subsequently condensed by the bifunctional enzyme *TH1* to produce thiamine monophosphate (TMP). TMP is then dephosphorylated by thiamine monophosphate phosphatase (*TH2*) to yield free thiamine (Mimura et al. 2016). All biosynthetic steps take place in the chloroplast, except for the final activation step, in which thiamine is phosphorylated into its biologically active form, thiamine pyrophosphate (TPP), by thiamine pyrophosphokinase (TPK) in the cytosol. Since TPP exerts its primary biological functions in mitochondria, it must subsequently be transported into the mitochondria or chloroplasts to become functionally active (Goyer 2017). Transcripts of *THI4*, *THIC*, *TH1*, and *TPK* have been identified in various plant tissues, including leaves, roots, and stems. Notably, transcript levels are highest in leaves, supporting the view that thiamine biosynthesis predominantly occurs in chloroplasts. Moreover, the regulation of thiamine biosynthesis appears to be tissue-specific, suggesting distinct control mechanisms across plant organs (Yusof 2019). Investigations into the expression of *THI4* and *THIC* found no detectable amplification in the roots of oil palm seedlings (Yee et al. 2016), further highlighting organ-specific expression patterns.

The source of thiamine in plants varies across different developmental stages, drawing from seed storage reserves, de novo biosynthesis, and uptake from the soil. During seed maturation, thiamine gradually accumulates in the germ, in parallel with an increase in total soluble protein content (Phosaeng and Phraprasert 2021). In

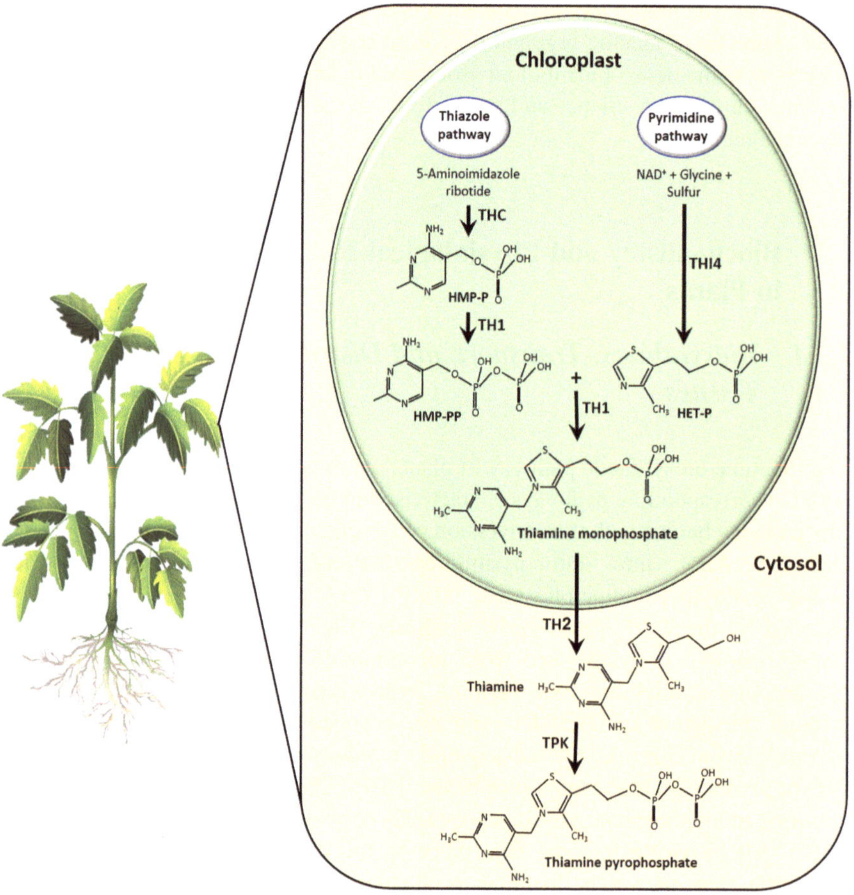

Fig. 1.1 The thiamine biosynthesis pathway in plants involves several key intermediates and enzymes. HMP-P; hydroxymethylpyrimidine phosphate, HMP-PP; hydroxymethylpyrimidine pyrophosphate, HET-P; hydroxyethylthiazole phosphate. The enzymatic steps in this pathway are catalysed by THIC (hydroxymethylpyrimidine synthase), THI4 (thiazole biosynthetic protein), TH1 (bifunctional enzyme with hydroxymethylpyrimidine phosphate kinase and thiamine phosphate synthase activity), TH2 (thiamine monophosphate phosphatase), and TPK (thiamine pyrophosphate kinase). (Figure adapted from Kamarudin et al. (2017))

seeds, thiamine is primarily stored in its unphosphorylated form, while phosphorylated derivatives represent only about 5% of the total thiamine content, even in mature seeds. Long-term storage is facilitated by specialised thiamine-binding proteins (TBPs), which have been identified in various plant species (Adamek-Świerczyńska et al. 2000; Phosaeng and Phraprasert 2021; RąPala-Kozik 1995). The molecular mechanisms underlying the interaction between thiamine and TBPs have been extensively investigated (Gołda et al. 2004; Rapala-Kozik et al. 1999), and TBPs are proposed to be specialised isoforms of primary seed storage globulins (Adamek-Świerczyńska and Kozik 2002; Radoviü et al. 2012). Studies assessing total seed thiamine levels suggest that germinating seedlings primarily rely on pre-stored thiamine reserves, as thiamine content either remains stable (Ghani et al. 2016) or slightly declines during early germination (Gołda et al. 2004).

Depending on the plant species, de novo thiamine biosynthesis typically begins 2–4 days after seed imbibition (Gołda et al. 2004). In cereal seeds, TPK activity has been shown to increase progressively throughout germination and seedling development (Gołda et al. 2004; Goyer 2017). Whilst thiamine biosynthesis occurs in the chloroplasts, thiamine diphosphate (TDP) formation takes place in the cytosol (Subki et al. 2018). TDP functions as a cofactor in multiple cellular compartments, suggesting the presence of specialised transporters for thiamine, TMP, or TDP. However, such transporters have yet to be fully characterised. TMP synthesised de novo in the chloroplast may be dephosphorylated by nonspecific phosphatases (Subki et al. 2018), although the exact subcellular site of this reaction remains unresolved. While thiamine is pyrophosphorylated in the cytosol, the resulting TDP is essential for mitochondrial functions, implying the existence of a TDP transporter in the inner mitochondrial membrane. Indeed, mitochondrial TDP transporters have been identified in humans, yeast, and *Drosophila melanogaster* (Curcio et al. 2020) and homologues have been found in *Arabidopsis* (Millar and Heazlewood 2003). Putative transporters may also facilitate thiamine uptake from seed reserves or the rhizosphere. Given that the complete de novo synthesis of TMP is confined to chloroplasts, green tissues serve as the primary sites of thiamine biosynthesis, from which the vitamin is transported to sink tissues with higher demand, such as roots. Consistently, genes encoding HMP-P synthase, HET-P synthase, and HMP-P kinase/TMP synthase are predominantly, if not exclusively, expressed in leaves (Subki et al. 2018). By contrast, TPK is expressed throughout all plant tissues at varying abundance (Tunc-Ozdemir et al. 2009), ensuring that both endogenous and exogenous thiamine can be efficiently converted into TDP.

Another source of thiamine is root absorption from the soil (Kaya et al. 2015), as most roots lack the capacity for thiamine biosynthesis. This uptake appears to occur independently of metabolic energy and is likely facilitated by a passive, transpiration-driven mechanism. Once absorbed, thiamine is transported to other organs via the xylem (Mozafar 2018). Additionally, thiamine and its phosphorylated derivatives can be delivered to seedlings through foliar application and eventually, they become evenly distributed throughout plant tissues (Kaya et al. 2015). This translocation likely occurs via the phloem and may follow a strictly basipetal direction, as observed in tomato petioles (Martinis et al. 2016), or proceed bidirectionally, both acropetally and basipetally, as shown in other species (Ghaffar et al. 2019). Recent studies have demonstrated that foliar application of TMP and TDP can induce disease resistance (Boubakri et al. 2016) and rescue the phenotypes of *Arabidopsis* TPK double mutants (Ajjawi et al. 2007a, b). These findings support the involvement of the vascular system in distributing thiamine phosphates via an apoplastic route. Moreover, leaves that emerge following thiamine treatment are capable of accumulating the vitamin, suggesting potential remobilisation from older tissues (Urooj et al. 2024). The transport of thiamine through the phloem from leaves to developing kernels has been reported in crops such as maize, wheat, and rice

(Mozafar 2018). In maize, thiamine content in the embryo is more than tenfold higher than that in the endosperm (Woodward 2010).

1.2.2 Thiamine's Role in Central Metabolic Pathways

Thiamine and its vitamers perform multiple essential functions within plant cells, with TDP being the most extensively studied due to its role as a coenzyme for several key enzymes in central metabolism. These include enzymes involved in photosynthesis, such as TK, and those central to respiration, such as PDH and α-KGDH, which operate within the Calvin cycle in chloroplasts and the tricarboxylic acid (TCA) cycle in mitochondria, respectively. The intracellular and organellar concentrations of TDP modulate metabolic flux through these critical pathways.

Among TCA cycle enzymes, α-KGDH exhibits one of the highest flux control coefficients (Araujo et al. 2012) and reduced TDP levels have been associated with decreased flux through the TCA cycle, primarily due to diminished α-KGDH activity (Joshi et al. 2019). Conversely, elevated TDP concentrations enhance metabolic flux and overall respiration rates (Rosado-Souza et al. 2019). Additionally, TDP modulation affects the abundance of several metabolites, including photosynthetic pigments (Noordally et al. 2020).

Emerging evidence highlights the importance of maintaining TDP homeostasis (Fig. 1.2). TDP concentrations within organelles appear to be rate-limiting, contributing to the reprogramming of metabolic fluxes in central pathways (Joshi et al. 2019). Regulation of TDP levels involves multiple mechanisms, including a thiamine riboswitch, daily oscillations in biosynthesis and transport, and control by the circadian clock (Noordally et al. 2020). This multilayered regulation underscores the crucial role of TDP in plant growth and metabolism, suggesting a potential involvement in regulating diel metabolite rhythms. These rhythms reflect the need to adjust plant metabolism in response to light–dark cycles, a necessity imposed by its sessile lifestyle (Kim et al. 2017).

Disruption of TDP regulation, such as riboswitch malfunction, impairs adaptation to changing photoperiods, resulting in severe growth defects (Rosado-Souza et al. 2019). Interestingly, overexpression of TK has been shown to sequester intracellular TDP, thereby restricting its availability to other TDP-dependent enzymes and resulting in stunted growth and chlorophyll deficiency (Khozaei et al. 2015). Therefore, precise control of cellular TDP levels is vital for maintaining metabolic homeostasis, photoperiodic adaptation, and plant health. This complexity also opens the door to potential metabolic engineering strategies. Fine-tuning TDP levels may influence photosynthesis, carbon assimilation, and the accumulation of sugars, amino acids, and lipids, traits with implications for crop yield improvement.

Given predictions that global food production must increase by 40–70% to avoid a Malthusian crisis (Rosado-Souza et al. 2019), innovative, sustainable agricultural strategies are urgently needed. The effects of climate change and unsustainable practices on biodiversity, soil, and water resources intensify this urgency. For decades, plant science has focused on increasing photosynthetic efficiency and

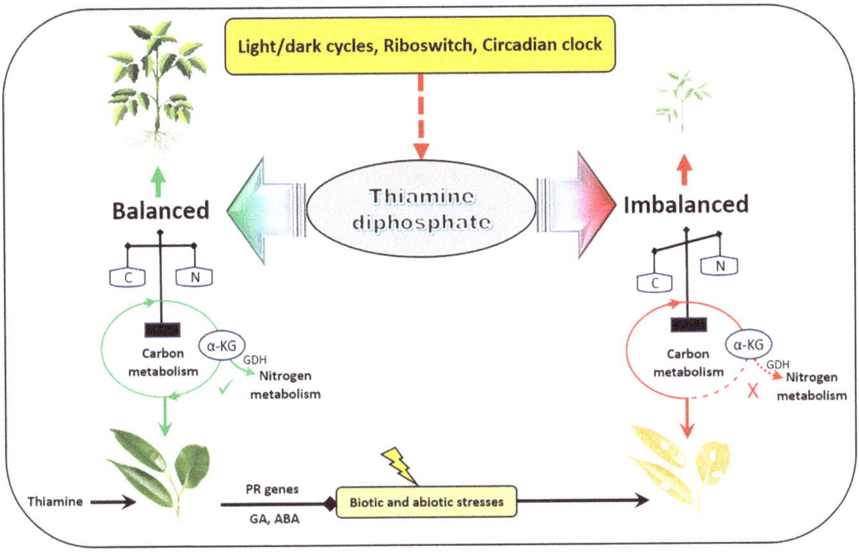

Fig. 1.2 Thiamine diphosphate (TDP) concentrations are tightly regulated by the circadian clock, light–dark cycles, and a riboswitch mechanism. Maintaining TDP homeostasis is essential and may be linked to the coordination of carbon (C) and nitrogen (N) metabolism, potentially mediated via α-ketoglutarate (α-KG) levels. α-KG serves as a key metabolic hub, being processed by α-ketoglutarate dehydrogenase (α-KGDH; not shown) in the TCA cycle and by glutamate dehydrogenase (GDH) in nitrogen assimilation. While thiamine derivatives such as adenosine thiamine triphosphate (ATTP) and thiamine triphosphate (TTP) have been shown to allosterically activate GDH in mammalian systems, their roles in plants remain to be elucidated. Exogenous thiamine application primes defence responses against biotic stress by upregulating pathogenesis-related (PR) genes and salicylic acid (SA) signalling, while also enhancing abiotic stress tolerance through abscisic acid (ABA)-mediated pathways. Disruption of the riboswitch (indicated by the red line) leads to TDP imbalance, compromising carbon–nitrogen homeostasis, plant health, and crop productivity (Fitzpatrick and Chapman 2020)

improving carbon allocation to storage organs like grains and tubers (Simkin et al. 2019; Smith et al. 2018). Research has also aimed to reduce respiratory energy losses to redirect resources toward biosynthesis and growth (Amthor et al. 2019).

Within this context, engineering more stable, non-suicidal variants of thiamine biosynthetic enzymes has been proposed to reduce the energetic cost of thiamine production and enhance crop yield potential (Sun et al. 2019). Thiamine plays a central role in both photosynthetic carbon assimilation and mitochondrial energy production, positioning it as a promising target for yield improvement.

However, optimising yield also requires coordination between carbon and nitrogen metabolism. The metabolite α-KG, also known as 2-oxoglutarate, lies at the intersection of these two pathways (Fig. 1.2). While the regulatory mechanisms governing carbon–nitrogen coordination in plants remain partially understood, α-KG is considered a key metabolic regulator, particularly in well-studied bacterial systems (Huergo and Dixon 2015). In plants, α-KG may be channelled through the TCA cycle via α-KGDH or diverted into nitrogen assimilation pathways through glutamate dehydrogenase (GDH). In mammals, thiamine derivatives such as thiamine triphosphate (TTP) and adenosine thiamine triphosphate (ATTP) can allosterically activate GDH, suggesting a potential regulatory role in the fate of α-KG (Aleshin et al. 2020).

In addition to its central metabolic functions, thiamine is increasingly recognised for its role in plant responses to abiotic and biotic stress. Abiotic stress conditions, including salinity, temperature extremes, osmotic stress, and oxidative stress, trigger increased expression of thiamine biosynthetic genes (Tunc-Ozdemir et al. 2009). In *Arabidopsis*, the transcripts of *THIC* and *THI1* increase rapidly in response to stress, likely as part of an abscisic acid (ABA)-regulated early stress response (Rapala-Kozik et al. 2012). This upregulation is thought to elevate TDP availability, supporting the activity of TDP-dependent enzymes critical to stress defence. Although direct evidence is limited, thiamine and its derivatives have been proposed to function as antioxidants, scavenging ROS in plants (Rosado-Souza et al. 2020). In contrast, the exogenous application of thiamine has been shown to enhance resistance to pathogens by inducing the expression of pathogenesis-related (PR) genes. These protective effects can last up to 15 days and are associated with primed SAR, mediated by elevated salicylic acid (SA) levels (Westman et al. 2019) (Fig. 1.2). Such responses have been observed across various crops, including wheat, peas, barley, oats, and millet, where thiamine treatment enhances resistance to both herbivores and fungal pathogens (Hamada et al. 2018). Conversely, rice plants with reduced thiamine levels are more susceptible to bacterial blight infections (Bahuguna et al. 2012).

Thiamine has also been linked to the establishment and maintenance of beneficial symbiotic relationships, such as root nodulation and arbuscular mycorrhizal associations with endophytic fungi, both of which enhance nutrient acquisition, particularly of nitrogen and phosphorus (Nagae et al. 2016). These findings position thiamine as a vital regulator of plant growth, stress adaptation, and productivity.

In summary, enhancing thiamine levels in plants holds significant potential for enhancing stress resistance, improving yield stability, and increasing metabolic resilience in the face of climate change.

1.3 Metabolic Interactions

1.3.1 Synergistic or Antagonistic Effects of Thiamine with Other Vitamins, Organic Acids, or Micronutrients

Thiamine plays a central role in plant stress responses and exhibits synergistic effects when applied in combination with ascorbate, riboflavin, or α-KG. These combinatory treatments enhance antioxidant defences, support tissue regeneration, and strengthen metabolic resilience, underscoring their potential application in sustainable agricultural practices.

1.3.1.1 Thiamine and Ascorbate (Vitamin C)

Beyond its direct involvement in organic acid metabolism, thiamine plays a critical role in vitamin C-mediated antioxidant defence mechanisms. This functional synergy is particularly evident under environmental stress conditions such as high light intensity, salinity, and low temperatures. Thiamine contributes to NADPH production via its role in the pentose phosphate pathway, specifically through the TDP-dependent enzyme TK. NADPH is essential for the regeneration of reduced ascorbate within the ascorbate–glutathione (ASC–GSH) cycle, which is responsible for detoxifying ROS (Rosado-Souza et al. 2020). Experimental evidence from *Arabidopsis* indicates that thiamine supplementation enhances redox homeostasis and reduces oxidative stress markers, highlighting the importance of this metabolic interaction (Tunc-Ozdemir et al. 2009).

1.3.1.2 With Other B Vitamins

Thiamine's role as a coenzyme in central metabolism is complemented by other B vitamins, notably riboflavin (vitamin B_2) and pyridoxine (vitamin B_6). Riboflavin supports flavoproteins involved in key redox reactions, thereby potentially enhancing the activity of thiamine-dependent enzymes in both the TCA cycle and the pentose phosphate pathway. Pyridoxine, as a coenzyme in amino acid metabolism, further contributes to cellular metabolic balance under stress conditions. The combined application of B vitamins has been shown to promote plant growth and enhance tolerance to environmental stressors, indicating a synergistic interaction that optimises metabolic efficiency and improves overall plant performance (Roje 2007).

1.3.1.3 Thiamine and Organic Acid Metabolism

TDP functions as a crucial coenzyme for key enzymes in the TCA cycle, including PDH and α-KGDH. Through these roles, TDP facilitates the conversion of organic acids, such as pyruvate and α-KG, into TCA cycle intermediates, thereby supporting mitochondrial respiration and energy production in plants. A reduction in thiamine levels can significantly disrupt organic acid metabolism by decreasing the activity of these enzymes, impairing TCA cycle flux. This may lead to reduced production of essential organic acids required for energy generation and for maintaining carbon and nitrogen balance. Consequently, such disruptions can impair respiration, cause the accumulation or depletion of specific organic acids, and interfere with related pathways such as photosynthesis and amino acid biosynthesis (Fitzpatrick and Chapman 2020). In *Arabidopsis* mutants deficient in thiamine, significant alterations in organic acid and amino acid profiles have been reported, including elevated malate and decreased fumarate levels, indicative of disrupted TCA cycle function likely due to impaired PDH activity. Thiamine supplementation partially restores these metabolic imbalances, underscoring its essential role in maintaining organic acid flux and overall metabolic homeostasis (Chung et al. 2024).

1.3.1.4 Interactions with Micronutrients

Although the interaction of thiamine with other vitamins and organic acids has been extensively studied, its relationships with micronutrients such as iron, zinc, and manganese remain relatively underexplored. Given thiamine's pivotal role in carbon and nitrogen metabolism, particularly within the pentose phosphate pathway and amino acid biosynthesis, it may indirectly influence the uptake, utilisation, or functional efficiency of these essential micronutrients (Subki et al. 2018). For instance, iron and zinc serve as critical cofactors for several enzymes involved in these metabolic pathways, and thiamine-dependent processes may modulate their activity or bioavailability (Tripathi et al. 2015). Nevertheless, the current body of literature lacks direct experimental evidence of synergistic or antagonistic interactions between thiamine and micronutrients under abiotic or biotic stress conditions, highlighting a significant gap and promising avenue for future research.

1.4 Thiamine and Abiotic Stress Tolerance: Mechanisms and Applications

In higher plants, a regulatory link between thiamine biosynthesis and thiamine-dependent metabolic pathways may become particularly prominent under abiotic or biotic stress conditions. Several key stress-responsive metabolic processes, including glycolysis, the TCA cycle, and photosynthesis, rely on TDP-dependent enzymes. Common stress-induced responses involving these pathways include the accumulation of soluble sugars, which fulfil multiple regulatory and osmoprotective functions (Handa et al. 2018); increased energy production to support the synthesis of stress-related proteins and osmolytes (Liang et al. 2016); and the activation of antioxidant systems that depend on NADPH availability (Valderrama et al. 2006). Despite the centrality of these pathways, only a limited number of TDP-dependent enzymes are transcriptionally or post-transcriptionally regulated in response to stress. Among them, pyruvate decarboxylase (PDC) is the most extensively characterised, particularly for its role in redirecting glycolytic intermediates toward ethanolic fermentation. PDC expression is significantly upregulated under various stress conditions, including anoxia, low temperature, mannitol-induced osmotic stress, high salinity, ABA treatment, ROS accumulation, and mechanical wounding (Chirkova and Yemelyanov 2018; Juri 2017).

1.4.1 Mechanisms of Stress Mitigation

1.4.1.1 Function in Drought Tolerance

Drought stress is a major abiotic factor that limits plant growth and productivity (Danish et al. 2020a; Faozi and Fajrin 2024; Umer et al. 2026). Under drought conditions, plants experience a range of physiological disturbances, including impaired water uptake, reduced photosynthetic efficiency, elevated respiration rates, and stomatal closure. These physiological disruptions lead to visible symptoms such as leaf yellowing, curling, premature senescence, and wilting (Bhusal et al. 2020). Drought stress is also closely associated with oxidative stress, as water deficits stimulate the overproduction of ROS, which can damage cellular structures and interfere with essential metabolic functions (Jabeen et al. 2021). In response, plants enhance their antioxidant defence systems by upregulating enzymatic antioxidants, including superoxide dismutase (SOD), peroxidase (POD), and catalase (CAT), as well as accumulating non-enzymatic antioxidants such as phenolic compounds and vitamins, thereby maintaining physiological balance and supporting growth under adverse conditions (Jabeen et al. 2021).

Thiamine, a key component of the vitamin B complex, plays a crucial role in mitigating drought-induced damage. Under drought and other stress conditions, endogenous thiamine biosynthesis may be insufficient, and exogenous application has been shown to improve plant growth, yield, and photosynthetic performance (Polash et al. 2019).

Thiamine treatment has been particularly effective in enhancing the growth and biochemical traits of maize under environmental stress, likely due to its role in activating antioxidant defences, mediated in part by TK activity within the glycolytic pathway (El-Bassiouny et al. 2023). In tomato plants, foliar application of thiamine mitigated the adverse effects of drought, improving parameters such as plant height, stem diameter, leaf number, and both fruit quantity and weight under moderate and severe drought conditions (Zamanipour 2021). The highest efficacy was observed with a 100 mg L^{-1} thiamine treatment under well-watered conditions, although the benefits were comparatively reduced under severe drought.

Thiamine has also been shown to support the biosynthesis and retention of photosynthetic pigments, promote the accumulation of osmoprotectants such as proline and glycine betaine, and reduce membrane damage by lowering malondialdehyde (MDA) and hydrogen peroxide (H_2O_2) levels (Jabeen et al. 2022). Foliar thiamine application significantly increased antioxidant enzyme activities (SOD, POD, CAT), boosted ascorbic acid levels, and enhanced membrane stability under drought (Hamada and Al-Hakimi 2009). Similar findings were reported in mung bean, where thiamine-treated plants displayed improved antioxidant defences, osmolyte accumulation, and yield compared to untreated controls (Jabeen et al. 2021). Moreover, in rice, thiamine treatment under drought stress improved physiological traits such as chlorophyll content and relative water content, while enhancing antioxidant activity and overall drought tolerance (Danish et al. 2020b).

1.4.1.2 Effects Under Salinity Stress

Soil salinity negatively impacts plant growth and productivity by increasing the levels of ROS, such as H_2O_2, superoxide, and hydroxyl radicals (Ashraf and Akram 2009; Azooz et al. 2026). These reactive molecules can damage vital cellular components, including proteins and lipids (Golldack et al. 2014; Noctor et al. 2014). To counteract this oxidative stress, plants enhance their antioxidant defence systems by increasing the activity of key enzymes, such as SOD, CAT, and POD (Kachout et al. 2013). Several plant growth regulators play a role in modulating plant growth and development under stress conditions, including those caused by salinity. Thiamine is among these regulatory compounds, playing a crucial role in the growth and differentiation of particular plant species (Rapala-Kozik et al. 2012). Evidence from in vitro studies also suggests that thiamine may act directly as an antioxidant (Tunc-Ozdemir et al. 2009). Furthermore, seed priming or exogenous thiamine application under salt stress has been shown to increase intracellular thiamine levels and enhance plant tolerance to salinity (Kaya et al. 2015).

In maize (*Zea mays L.*) cultivars subjected to salt stress, thiamine application improved photosynthetic pigment content, reduced membrane permeability, lowered MDA and H_2O_2 abundance, and modulated the activity of key antioxidant enzymes under saline conditions (Kaya et al. 2015). Rapala-Kozik et al. (2012) observed that osmotic and salt stress in *Arabidopsis* led to the upregulation of genes involved in thiamine biosynthesis and those encoding TDP-dependent enzymes. Additionally, TDP and its associated enzymes, α-KGDH, PDH, and TK were simultaneously upregulated, reflecting their integral roles in key metabolic pathways that support plant responses to extreme environmental conditions. Furthermore, Rahman et al. (2017) reported that thiamine biosynthesis was enhanced under salinity stress due to the increased expression of its biosynthetic genes. In a separate study, Wouyou et al. (2017) observed that salt stress led to increased accumulation of thiamine and other vitamins in *Amaranthus* leaves. Tuna et al. (2013) found that thiamine application in maize mitigated the adverse effects of salinity by enhancing antioxidant enzyme activities. Moreover, thiamine treatment alleviated the alterations in protein metabolism caused by salt stress in plants (Xie et al. 2020). Exogenous application of thiamine through foliar spraying has been shown to improve salt stress tolerance in maize and sunflower (Hamada and Al-Hakimi 2009).

El-Metwally and Sadak (2019) reported that faba bean (*Vicia faba* L.) plants exhibited improved salt tolerance associated with increased levels of proline, free amino acids, and total carbohydrates. Additionally, the study documented enhancements in leaf area index, yield, and its components, as well as the chemical composition of the harvested seeds. Foliar application of thiamine in maize cultivars promoted tolerance to salinity-induced oxidative stress (Kaya et al. 2015).

Urooj et al. (2024) investigated the effect of thiamine seed priming on the growth and physiological responses of wheat (*Triticum aestivum* L.) under salinity stress. Their results showed that thiamine priming significantly reduced salinity-induced oxidative damage by decreasing ROS production and lipid peroxidation. Additionally, thiamine enhanced osmotic regulation by promoting the accumulation of osmolytes, thereby improving water retention and cellular hydration. The activities of crucial antioxidant enzymes, such as SOD, CAT, and ascorbate peroxidase (APX), were significantly elevated, facilitating ROS detoxification and protection from oxidative stress. These results highlight thiamine seed priming as a valuable approach to bolster wheat tolerance to salinity, with significant potential for improving crop resilience in saline conditions (Urooj et al. 2024). Similarly, Hussein and Orabi (2025) demonstrated that exogenous thiamine application enhanced growth and dry weight in grain sorghum under salt stress by stimulating root development, reducing oxidative damage through decreased O_2^- and MDA levels, and elevating antioxidant enzyme activities, thereby improving salinity tolerance.

1.4.1.3 Response to Temperature Extremes (Heat and Cold)

Plants are constantly exposed to environmental fluctuations, among which temperature extremes—both low and high—pose major challenges to their survival and productivity. These temperature stresses trigger physiological and biochemical changes that can disrupt cellular functions at multiple levels (Suzuki et al. 2014). Temperature extremes, both cold and heat stress, can significantly impair plant growth, development, and yield. Cold stress may damage cellular membranes, inhibit photosynthesis, and lead to oxidative stress through the accumulation of ROS (Manasa et al. 2022). In contrast, heat stress can result in protein denaturation, increased membrane permeability, and disruption of metabolic homeostasis (Wani and Kumar 2020). To overcome these challenges, plants have evolved sophisticated defence strategies, with thiamine playing a pivotal role in these adaptive responses.

Thiamine in Plant Response to Cold Stress

Exposure to low-temperature stress induces a complex array of physiological and biochemical adaptations in plants, characterised by the synthesis of protective metabolites and the enhanced expression of genes associated with stress tolerance (Raza et al. 2023). Emerging evidence suggests that thiamine plays a crucial role in these adaptive mechanisms (Li et al. 2022). In *Arabidopsis thaliana*, exposure to cold stress at 4 °C for 24 h led to a 1.4-fold increase in total thiamine abundance, encompassing free thiamine, TMP, and TPP (Rapala-Kozik et al. 2012). This increase suggests that thiamine biosynthesis is upregulated during cold stress, potentially mitigating oxidative damage and maintaining metabolic homeostasis (Li 2014). Additional evidence is provided by research on *Vitis vinifera* (grapevine), where transcriptomic analyses using machine learning approaches identified upregulation of key thiamine biosynthesis genes, including *THIAMINE THIAZOLE SYNTHASE 1* and *2* (*THI1-1*, *THI1-2*), *PHOSPHOMETHYLPYRIMIDINE SYNTHASE* (*THIC*), and *1-DEOXY-D-XYLULOSE-5-PHOSPHATE SYNTHASE* (*DXS*), in response to cold acclimation (4 °C) and freeze-shock exposure (4 °C to −3 °C) (Konecny et al. 2023). Comparable induction of thiamine biosynthesis genes has been observed in various crop species, including *Triticum aestivum* and rice

(*Oryza sativa*), where cold stress stimulates the expression of *THI1* and *THIC* genes, thereby enhancing stress resilience (Fitzpatrick and Chapman 2020). In maize, both the transcription of thiamine biosynthesis genes and the activity of TPK increase in response to osmotic and oxidative stresses, suggesting a conserved protective function (Subki et al. 2018).

Thiamine in Plant Response to Heat Stress

Heat stress inflicts significant cellular damage in plants, including protein denaturation and aggregation, increased membrane lipid fluidity, enzymatic inactivation within chloroplasts and mitochondria, suppression of protein synthesis, and compromised membrane integrity (Zahra et al. 2023). These primary and secondary injuries provoke elevated ROS production, disruption of cellular homeostasis, and may culminate in cell death under severe or prolonged stress conditions (Ghosh et al. 2018). Additionally, heat stress disrupts microtubule organisation, thereby impairing cell division and growth processes (Parker et al. 2014). Plants counteract such stress at the molecular level by modulating gene expression and accumulating stress-responsive proteins, particularly heat shock proteins (HSPs), which function as molecular chaperones to safeguard cellular proteins and preserve essential functions, including photosynthesis and membrane stability (ul Haq et al. 2019). During heat exposure, thiamine biosynthesis-related genes such as thiamine thiazole synthase (*THI1/THI4*) and phosphomethylpyrimidine synthase (*THIC*) exhibit transient upregulation across various plant species. This response is believed to be linked to thiamine's protective role in DNA stabilisation and repair under stress conditions (Xu and Huang 2018). Nevertheless, extended or intense stress may lead to the repression of certain thiamine biosynthetic genes, reflecting a nuanced regulatory mechanism influenced by the severity and duration of stress.

These combined biochemical and physiological adaptations contribute synergistically to enhanced stress resilience, supporting continued growth and development during exposure to elevated temperatures (Jabeen et al. 2022). In *Oryza sativa*, exposure to heat stress triggers a transient increase in the expression of thiamine biosynthetic genes, including *THI1* and *THIC*, which is associated with improved mechanisms for DNA protection and repair (Faustino et al. 2024). The application of exogenous thiamine has been shown to reduce ROS accumulation and lower lipid peroxidation, thereby enhancing heat stress tolerance in rice seedlings (Rosyida and Kristanto 2022). Similarly, *Zea mays* seedlings subjected to high temperatures exhibit increased activities of key antioxidant enzymes and improved membrane integrity following thiamine treatment, effectively minimising oxidative damage and supporting sustained growth under heat stress conditions (de Queiroz 2024).

1.4.2 Heavy Metal Stress Mitigation

The extraordinary environmental accumulation and trophic-level amplification of heavy metals represent a critical and pervasive threat to the biotic integrity of ecosystems, affecting both autotrophic and heterotrophic organisms (Buzea and Pacheco 2020). At toxic concentrations, heavy metals can interact with essential cellular components, such as nuclear proteins and DNA, resulting in a significant increase in the production of ROS (Mansoor et al. 2023). This can lead to profound morphological, metabolic, and physiological disturbances in plants, including symptoms such as shoot chlorosis, enhanced lipid peroxidation, and accelerated protein degradation (Kumari et al. 2022).

A study examining the impact of thiamine on *Brassica napus* (canola) plants exposed to cadmium toxicity demonstrated that thiamine application significantly reduced oxidative stress indicators, such as MDA and H_2O_2, along with decreasing cadmium accumulation in leaf tissues. Furthermore, it enhanced the activity of antioxidant enzymes and increased non-enzymatic antioxidant levels, reflecting an overall reinforcement of the defence mechanisms against oxidative damage (Sanjari et al. 2019).

Exposure to elevated arsenic levels in maize triggers oxidative stress markers such as MDA, H_2O_2, and proline, along with increased activities of antioxidant enzymes including CAT, SOD, POD, and APX. In a separate study, exogenous thiamine application significantly improved multiple morphophysiological and biochemical attributes in lentil plants subjected to lead (Pb) toxicity, while concurrently alleviating oxidative damage by decreasing H_2O_2 accumulation (Bouhadi et al. 2024). In *Lens culinaris* (lentil) plants subjected to Pb toxicity, exogenous thiamine application significantly improved morphophysiological and biochemical attributes while concurrently alleviating oxidative damage by decreasing H_2O_2 accumulation and boosting antioxidant enzyme activities (Bouhadi et al. 2024).

1.5 Thiamine and Biotic Stress Tolerance: Mechanisms and Applications

Thiamine is increasingly recognised as an essential molecule in boosting plant resilience against biotic stresses, such as pathogens and pests. Applying thiamine externally has been shown to enhance resistance to various pathogens by upregulating PR genes and activating SAR via SA signalling (Ahn et al. 2005). Also, thiamine affects plant–microbe relationships, including symbioses with nitrogen-fixing bacteria and mycorrhizal fungi, which can enhance nutrient uptake and overall plant resilience (Ahn et al. 2007; Deryło and Skorupska 1993).

As global agricultural systems grapple with rising pathogen pressures and climate-induced stresses, understanding the mechanisms by which thiamine confers biotic stress tolerance becomes essential for developing sustainable crop management strategies. As a broad-spectrum resistance inducer, thiamine protects plants against a range of bacterial, viral, and fungal diseases (Nie and Xu 2016).

Thiamine enhances plant defence by priming the immune system, specifically through the upregulation of PR genes. These genes encode antimicrobial proteins that enable plants to mount a faster and stronger defence response upon pathogen attack (Ahn et al. 2005; Westman et al. 2019). This mechanism is facilitated by several calcium ions and SA-mediated signalling pathways, highlighting the complexity of thiamine's mode of action. This signalling mobilisation includes the upregulation of protein kinase C activity, a crucial component of plant defence responses (Ahn et al. 2005).

For instance, research on wheat, peas, barley, oats, and millet has shown that thiamine treatment boosts resistance to fungal pathogens and herbivorous insects, with protective effects lasting up to 15 days after application (Hamada and Jonsson 2013; Pushpalatha et al. 2011). Thiamine affects the SA signalling pathway, a crucial regulator of systemic immunity in plants. Activation of SA-dependent pathways results in the build-up of defence-related compounds and the strengthening of cell walls, thereby reducing pathogen entry. Evidence of this includes the fact that thiamine application has been demonstrated to increase the expression of defence-associated genes and the accumulation of phenolic compounds, which are poisonous to many pathogens (Vidal et al. 1997).

In addition to systemic signalling, thiamine biosynthesis genes, particularly *THIC* and *THI1*, are rapidly induced under biotic stress conditions, indicating a role for endogenous thiamine in stress responses (Li et al. 2022). The accumulation of thiamine vitamers, including TTP and TDP, contributes to maintaining the metabolic fluxes required for energy production and defence metabolism during stress (Hofmann et al. 2020).

Thiamine biosynthesis in plants is a highly compartmentalised process involving the plastids, cytosol, and mitochondria, where distinct enzymatic steps contribute to the assembly and activation of the vitamin. The biosynthetic pathway is divided into the synthesis of two primary structural moieties, the pyrimidine and thiazole rings, which are subsequently condensed to form thiamine monophosphate (TMP), followed by phosphorylation to the active coenzyme thiamine diphosphate (TDP). The key genes involved in this process include *THIC*, *THI1*, *THI4*, *HETK*, *TDPK*, and *TPC*, each playing specific roles in precursor formation, condensation, and activation (Goyer 2010; Pourcel et al. 2013; Rapala-Kozik et al. 2012). A schematic overview of this biosynthetic pathway, along with its cellular compartmentalisation and enzymatic steps, is presented in Fig. 1.3.

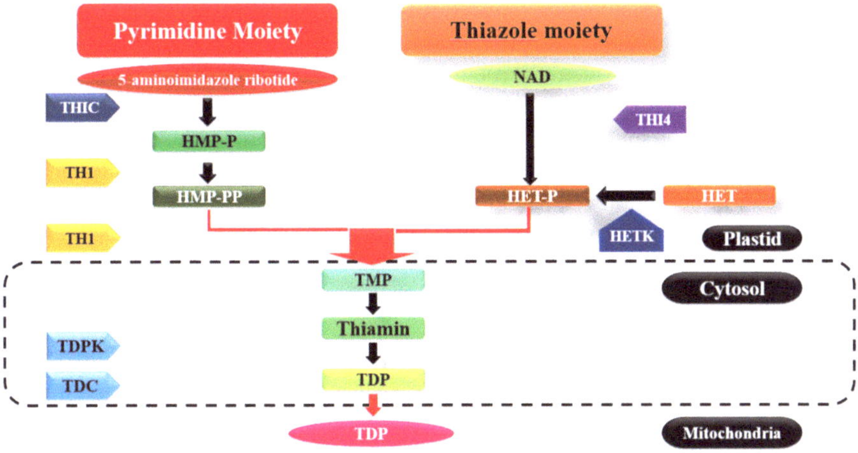

Fig. 1.3 Schematic representation of the thiamine (vitamin B1) biosynthetic pathway in plants. This pathway involves the biosynthesis of two key molecular parts: the pyrimidine and thiazole rings. The pyrimidine moiety is synthesised from 5-aminoimidazole ribotide via the action of the THIC gene, producing HMP-P (4-amino-2-methyl-5-hydroxymethylpyrimidine phosphate). The enzyme encoded by *TH1I* then converts HMP-P to HMP-PP, its pyrophosphate form. Concurrently, the thiazole moiety is synthesised from NAD through the activity of the THI4 gene, resulting in HET-P (4-methyl-5-β-hydroxyethylthiazole phosphate). HETK can also phosphorylate HET to HET-P. The two moieties, HMP-PP and HET-P, are subsequently condensed by the *TH1I* enzyme to form thiamine monophosphate (TMP) in either the plastid or cytosol. TMP is then dephosphorylated to yield free thiamine, which is phosphorylated by thiamine diphosphokinase (TDPK) to produce the active coenzyme thiamine diphosphate (TDP). TDP is transported into the mitochondria, where it functions as a vital cofactor for several key metabolic enzymes. This pathway highlights the compartmentalisation of thiamine biosynthesis and activation within plant cells, involving plastids, the cytosol, and mitochondria. It also underscores the essential roles of the core biosynthetic genes, *THIC*, *THI1*, *THI4*, *HETK*, *TDPK*, and *TPC*, in regulating thiamine homeostasis. (This figure is adapted from Guan et al. (2014))

1.5.1 Mechanisms of Biotic Stress Mitigation

1.5.1.1 Effects Against Fungal, Bacterial, and Viral Pathogens

Thiamine-treated plants exhibit enhanced resistance to a wide range of fungal, bacterial, and viral infections, a phenomenon observed in various plant species, including *Oryza sativa Arabidopsis thaliana*, and several vegetable crops, underscoring thiamine's broad applicability (Ahn et al. 2005). Thiamine effectively suppresses the growth and spore formation of *Phytophthora nicotianae* while enhancing tobacco plant defences, resulting in significantly reduced disease severity (Suohui et al. 2022). In *Oryza sativa* L., the activation of resistance mechanisms significantly enhances defence against diseases such as sheath blight, caused by *Rhizoctonia solani*, and bacterial leaf blight, caused by *Xanthomonas oryzae* pv. *oryzae*. Similarly, thiamine-induced resistance can protect soybean crops from charcoal rot disease (Abdel-Monaim 2011; Ahn et al. 2005; Bahuguna et al. 2012).

In the case of soybean charcoal rot, thiamine not only reduced disease severity but also enhanced several physiological and biochemical defence responses. It significantly increased the number of nodules per plant and improved both fresh and dry nodule weights, indicating enhanced symbiotic performance and overall plant vigour. At the molecular level, thiamine significantly increased the activity of key defence-related enzymes, including POD, polyphenol oxidase (PPO), phenylalanine ammonia-lyase (PAL), and chitinase. These responses suggest that thiamine induces systemic resistance, strengthening plant defences against pathogen attack (Abdel-Monaim 2011).

For fungal pathogens such as *Rhizoctonia solani*, the causal agent of sheath blight, thiamine-induced defence responses reduce disease severity by reinforcing cellular defences and activating key signalling pathways. In rice infected with *Xanthomonas oryzae* pv. *oryzae*, thiamine enhances resistance primarily through SAR mediated by SA-dependent signalling. This priming effect upregulates defence-related genes, particularly *PR-1*, which plays a crucial role in limiting bacterial growth and proliferation. The effectiveness of thiamine-mediated resistance depends on the integrity of SA and calcium signalling pathways, reflecting the complexity of the plant immune response. Notably, thiamine does not possess direct antibacterial activity; rather, it strengthens the plant's innate immune system, thereby improving its capacity to respond swiftly and effectively to pathogen invasion (Ahn et al. 2005).

Thiamine acts as a priming agent in rice, enhancing innate immune responses and increasing resistance to a broad spectrum of biotic stresses caused by fungal, bacterial, and viral pathogens. Thiamine treatment stimulates the activity of key defence-related enzymes such as PAL and SOD, which are crucial for phenolic compound biosynthesis and ROS detoxification, respectively. Additionally, thiamine induces the accumulation of H_2O_2, a signalling molecule involved in defence responses, and increases the total phenolic content, compounds known for their antimicrobial properties. These coordinated biochemical and molecular changes reinforce SAR, priming the plant for more rapid and robust activation of defence mechanisms upon pathogen challenge. This heightened state of readiness enables rice plants to effectively limit pathogen invasion and disease progression (Bahuguna et al. 2012).

Thiamine also enhances maize resistance to viral infection by activating the mitogen-activated protein kinase (MAPK) signalling pathway and promoting lignin synthesis in cell walls, both of which contribute to strengthened antiviral defences. Maize chlorotic mottle virus (MCMV) disrupts this defence via its P31 protein, which interacts with the key thiamine biosynthesis enzyme *ZmTHIC*, preventing its localisation to the chloroplast and thereby reducing thiamine production. Silencing *ZmTHIC* leads to decreased thiamine levels and increased viral accumulation, highlighting the critical role of thiamine biosynthesis in antiviral immunity (Sun et al. 2025).

In *Hordeum vulgare* (barley), thiamine also enhances the immune system, improving resistance to viral pathogens such as Barley yellow dwarf virus and supporting overall plant health (Han et al. 2024). Thiamine has likewise been shown to increase potato resistance against *Potato virus Y* (PVY), a major yield-limiting pathogen worldwide and in trials with 'Ranger Russet' potatoes, the highest concentration of thiamine (100 mM) significantly reduced viral levels across all sampling dates and leaf samples, with repeated applications further decreasing virus accumulation. These findings highlight thiamine's potential as an effective natural agent to enhance plant immunity and manage viral diseases in potatoes (Vinchesi et al. 2017).

Another benefit of thiamine in the context of viral disease management is its role in improving plant tissue culture and mitigating physiological disorders. In watermelon, the addition of thiamine HCl (50 mg/L) to culture media effectively alleviates hyperhydricity, a common problem that impairs healthy in vitro growth (Huang et al. 2011). This improvement in tissue culture quality facilitates the use of transgenic strategies, such as expressing the *nucleocapsid (N)* gene of *Watermelon silver mottle virus* (WSMoV), to reduce viral infection. This is especially valuable given the limited availability of natural resistance sources. Thus, thiamine acts both as a growth enhancer in tissue culture and as a natural elicitor of antiviral resistance (Huang et al. 2016).

1.5.1.2 Activation of SAR by Thiamine

Thiamine significantly contributes to the activation of SAR in plants, functioning as an elicitor that primes the plant's immune system. This priming enhances the expression of defence-related genes, including PR proteins such as *PR-1*, a key marker of SAR. Thiamine treatment activates SAR primarily through a (SA)-dependent signalling pathway. This dependence is supported by findings showing a significant reduction in the expression of defence-related genes in SA-deficient *Arabidopsis thaliana* plants following thiamine application (Ahn et al. 2005). In addition to SA signalling, calcium signalling is also involved. Thiamine treatment induces the accumulation of defence enzymes, such as PAL, and elevates the levels of secondary metabolites and ROS, including H_2O_2. It also stimulates other biochemical changes that collectively reinforce defence responses. The biosynthesis of thiamine itself is upregulated during pathogen attack, and mutants deficient in thiamine biosynthesis display increased susceptibility to pathogens and fail to develop effective SAR. Notably, exogenous application of thiamine can restore SAR in these mutants. Highlighting its dual role as both a metabolic cofactor and a natural inducer of SAR, thiamine emerges as a promising agent for enhancing plant immunity against a broad spectrum of pathogens (Goyer 2010).

Thiamine foliar application primes plant immunity and has been shown to control various diseases effectively; however, its efficacy against potato pathogens has been less frequently explored. Recent research indicates that foliar application of thiamine reduced early blight lesion size in potato by approximately 33%, although the protective effect was both temporally and spatially limited. Transcriptomic analysis revealed that thiamine influenced the expression of 308 genes involved in multiple pathways, including the biosynthesis of SA, secondary metabolites, and fatty acids. Additionally, thiamine delayed the downregulation of photosynthesis-related genes in plants inoculated with *Alternaria solani*. Metabolite profiling suggests that thiamine primes plant defences by altering primary metabolic processes (Berrian et al. 2025).

1.5.1.3 Modulation of Defence-Related Genes by Thiamine

Thiamine treatment induces the transient expression of PR genes in rice and other plants (Bahuguna et al. 2012). The transient expression of defence-related genes is a crucial component of the plant's immune response. Thiamine treatment has been shown to enhance this response by inducing a more rapid and robust expression of PR genes. In *Arabidopsis thaliana* Col-0, thiamine specifically activated genes such as *PR-1*, *PBZ1*, and *POX22.3*, which are markers of the SA-dependent pathway, without inducing *PDF1.2*, a marker associated with the jasmonic acid/ethylene (JA/ET)-dependent pathway. This selective activation highlights a finely tuned defence signalling mechanism that enables the plant to tailor its response to specific stress stimuli (Ahn et al. 2005).

Treatment with benzothiadiazole (BTH), a functional analogue of SA, has been shown to enhance thiamine metabolism and activate immune responses in tomato (*Solanum lycopersicum*) plants. This process is characterised by the upregulation of genes involved in cellular detoxification and systemic acquired resistance, emphasising thiamine's pivotal role in modulating defence-related gene expression (Frąckowiak et al. 2022).

In cotton, exposure to biotic stressors such as the pathogenic fungus *Verticillium dahliae* rapidly induces the expression of key thiamine biosynthetic genes, including *GhTHIC* and *GhTHI1*. This upregulation is part of a broader defence response that also involves calcium signalling and the activation of downstream stress-responsive genes. Notably, exogenous application of thiamine has been shown to enhance cotton's resistance to biotic stress by boosting these defence pathways (Li et al. 2022). Hormonal cross-talk between the SA and JA pathways is pivotal in regulating plant defence gene expression under biotic stress, enabling the prioritisation of responses against specific pathogens or herbivores. SA often suppresses JA-responsive genes, such as *PDF1.2* and *VSP2*, to favour resistance against biotrophic pathogens, mediated by key regulators including *NPR1*, *TGA*, and *WRKY* transcription factors (Pieterse et al. 2012). Ethylene enhances JA-driven *PDF1.2* expression to support defence against necrotrophs, while inhibiting the JA–MYC branch, thereby influencing herbivore resistance. Thiamine may enhance SA–JA cross-talk by priming SA-dependent PR gene expression and thereby strengthening SAR (Boubakri et al. 2012).

Recent studies have identified thiamine thiazole synthase (*BcTHI4*) as a key enzyme in thiamine biosynthesis that can be targeted to control plant pathogenic fungi such as *Botrytis cinerea*, a pathogen notorious for multidrug resistance. The compound Z24 has been shown to inhibit *BcTHI4*, disrupting thiamine production and thereby suppressing fungal growth. This inhibition not only impacts enzyme function but also induces alternative splicing of the *BcTHI4* gene, suggesting a complex regulatory response. These findings highlight thiamine biosynthesis as a promising target for developing new antifungal agents, emphasising the protective effect of thiamine-related pathways in plant immunity and disease management (Zhao et al. 2025).

Notably, thiamine-induced gene expression is not restricted to defence-specific genes but also includes those involved in energy metabolism and signal transduction, such as MAPK genes, reflecting the broader role of thiamine in integrating stress responses with cellular energy status (Goyer 2010; Pieterse et al. 2012). Collectively, the ability of thiamine to modulate a wide array of defence-related genes underscores its potential as a molecular tool to enhance plant immunity in an environmentally sustainable manner.

1.5.2 Exogenous Application Strategies

1.5.2.1 Practical Applications in Plant–Pathogen Interaction Studies

Applying thiamine exogenously is a viable method for improving plant resilience to biotic stressors. Multiple studies have demonstrated that the exogenous application of thiamine induces SAR in various plant species, including rice, *Arabidopsis thaliana*, tobacco, and cucumber, leading to the rapid and elevated expression of defence-related mRNAs in response to virulent pathogens (Ahn et al. 2005). Studies on tomato plants have shown that thiamine can induce resistance against *Rhizoctonia solani*, a destructive necrotrophic fungus (Kheyri et al. 2022). In *Arabidopsis thaliana*, thiamine treatment followed by pathogen challenge leads to H_2O_2 accumulation and PR1/PAL1 transcription (Ahn et al. 2007). This approach facilitates the investigation of the effects of thiamine priming on defence responses, providing insights into the underlying mechanisms. Research on *Elaeis guineensis* (Oil palm) shows that thiamine biosynthesis genes are upregulated during interactions with *Hendersonia toruloidea* (Kamarudin et al. 2017). Thiamine can be used as an environmentally safe vitamin to protect tomato plants, making it a viable option for sustainable agriculture (Kheyri et al. 2022).

1.5.2.2 Case Studies: Thiamine in Disease Resistance of Crops and Model Plants

Thiamine has a range of multifaceted functions in enhancing plant defences against pests, pathogens, and other biotic stressors (Table 1.1). In the following sections, we provide a detailed examination of these roles, exploring the underlying physiological, biochemical, and molecular mechanisms through which thiamine contributes to plant immunity and resistance.

Table 1.1 Case-based integration of thiamine (vitamin B_1) into plant disease and pest management strategies

Crop/pathosystem	Target pathogen/pest	Thiamine function	Integration strategy	Key references
Rice (*Oryza sativa*)	*Rhizoctonia solani* (Sheath blight)	Enhances systemic resistance via antioxidant activation and phenolic pathway stimulation	Foliar application during early disease pressure as a replacement/ reducer of fungicide use	Bahuguna et al. (2012)
Soybean (*Glycine max*)	*Macrophomina phaseolina* (Charcoal rot)	Induces lignin, phenol accumulation, activates defence enzymes	Seed priming or early foliar spray in IPDM programs to reduce rot severity	Abdel-Monaim (2011)
Grapevine (*Vitis vinifera*, cv. *Chardonnay*)	*Plasmopara viticola* (Downy mildew)	Modulates phenylpropanoid metabolism, increases stilbenes, flavonoids, and antioxidants	Precision foliar application pre-infection; complements resistant varieties or fungicide reduction	Boubakri et al. (2013)
Carnation (*Dianthus caryophyllus*)	*Fusarium oxysporum* f.sp. *dianthi*	Induces specialised flavonoids and antifungal metabolites	Foliar spray on resistant cultivars; integration with conventional chemical programs	Pérez-Mora et al. (2025)
Wheat (*Triticum aestivum*), Barley (*Hordeum vulgare*), Oat (*Avena sativa*), Pea (*Pisum sativum*)	Aphids: *Ropalosiphum padi* *Acyrtosiphon pisum* *Myzus. persicae*	Suppresses fecundity and host acceptance, acts as an antifeedant	Thiamine seed priming as a biopesticide alternative in pest management programs	Hamada et al. (2018)
Tomato (*Solanum lycopersicum*)	*Pseudomonas syringae* pv. *tomato* DC3000	Pathogen requires thiamine for virulence; biosynthesis mutants are avirulent unless thiamine is provided	Diagnostic monitoring required; avoid indiscriminate thiamine use in high-risk zones	Liu et al. (2022)
Multiple hosts (e.g. cotton, tomato)	*Verticillium dahliae* (Wilt)	Requires VdThit transporter for virulence; deletion reduces infectivity; restored by exogenous thiamine	Target thiamine uptake in pathogens; consider use of biosynthesis inhibitors as a companion strategy	Liu et al. (2022)

Dual Role in Pathogen Ecology

Although thiamine is widely recognised for its role in enhancing plant resistance and alleviating biotic stress, its application does not always guarantee disease reduction or effective stress control. The effects of thiamine can be complex and context-dependent, and in some cases, it may even support pathogen survival under certain conditions. For example, a case study involving the bacterial pathogen *Pseudomonas syringae* pv. Tomato DC3000 (*Pst*DC3000) showed that thiamine biosynthesis is essential for the pathogen's growth, stress tolerance, and virulence. Mutants lacking key thiamine biosynthesis genes exhibited reduced growth and virulence in tomato plants, but these deficiencies were restored when exogenous thiamine was supplied. This finding suggests that while limiting thiamine biosynthesis can reduce infection, the external availability of thiamine might inadvertently support pathogen survival if not carefully managed. Additionally, mutations that impair stress tolerance also led to decreased colonisation, further emphasising thiamine's involvement in multiple pathways that influence bacterial fitness. Therefore, although targeting thiamine biosynthesis presents promising opportunities for disease control, its dual roles require careful consideration to avoid unintentionally promoting pathogen growth (Liu et al. 2022).

A similar pattern is observed with the fungus *Verticillium dahliae*, a major wilt-causing pathogen (Qi et al. 2016). The thiamine transporter protein VdThit is essential for fungal growth, reproduction, and pathogenicity. Deletion of the VdThit gene impairs fungal development and virulence, effects that can be partially reversed by exogenous thiamine supplementation. However, the reliance of *V. dahliae* on thiamine uptake also reveals a vulnerability that could be exploited for disease control. In this dual-faceted role, thiamine protects plants by boosting their defence mechanisms, yet it can also facilitate pathogen virulence when utilised by organisms such as *V. dahliae*. This underscores the importance of balanced management strategies in crop protection to maximise plant benefits while minimising the risks of supporting pathogen growth (Zhang et al. 2022).

Metabolic and Structural Fortification

Charcoal rot, caused by *Macrophomina phaseolina*, is a significant disease that severely impacts soybean crops worldwide. Research demonstrates that thiamine is highly effective in inducing systemic resistance in soybeans, leading to reduced disease severity and improved plant growth under both greenhouse and field conditions. Thiamine treatment enhances the activity of key defence enzymes and increases the accumulation of phenolic compounds and lignin, which are crucial for strengthening cell walls and preventing pathogen invasion. This metabolic and structural fortification enhances the plant's natural defence mechanisms, making it more resilient to infection (Abdel-Monaim 2011).

Significantly, the benefits of thiamine are not limited to soybeans. When applied exogenously, thiamine has been shown to enhance similar metabolic and structural defences in a variety of plant species. It stimulates the synthesis of phenolic compounds and lignin while also improving the activity of antioxidant enzymes, including PODs and PPO. These enzymes help neutralise ROS generated during pathogen attack, further supporting plant health. As a result, thiamine not only improves resistance to infection but also promotes overall plant vigour and recovery following stress. Such broad-spectrum protective effects have been observed in crops such as rice, cucumber, and *Arabidopsis thaliana*, underscoring thiamine's versatile and valuable role in plant defence (Ahn et al. 2005; Bahuguna et al. 2012).

Activation of Defensive Genes

Thiamine has emerged as a promising activator of plant defence against sheath blight in rice, a disease caused by *Rhizoctonia solani*. Research shows that thiamine enhances systemic resistance by stimulating antioxidant activity, increasing the accumulation of phenolic compounds, and upregulating defence-related genes and enzymes. These changes collectively enhance the plant's ability to resist infection. Beyond its protective effects, thiamine also helps maintain essential physiological functions in rice plants under pathogen stress, supporting overall plant health and resilience. This means that, even when under attack, rice plants treated with thiamine are better able to sustain growth and productivity. From an integrated disease management (IDM) perspective, thiamine represents a sustainable and eco-friendly alternative to traditional chemical fungicides. Its dual role, both in boosting innate immunity and in preserving physiological performance, makes it a valuable tool for developing holistic, low-input strategies for rice disease control (Bahuguna et al. 2012).

Insect Antifeedant and Reproductive Suppression

Thiamine has been shown to activate plant defences and reduce pest infestations, including those caused by pathogens, nematodes, and aphids. For example, a study (Hamada et al. 2018) investigating thiamine seed treatments in wheat, barley, oat, and pea found that thiamine significantly decreased both reproduction and host acceptance of specialist aphids, bird cherry-oat aphid (*Rhopalosiphum padi* L.) and pea aphid (*Acyrthosiphon pisum* Harris), as well as the generalist green peach aphid. Key reproductive parameters, such as fecundity, reproductive lifespan, and intrinsic growth rate, were notably reduced, particularly in the bird cherry-oat aphid. In contrast, the biocontrol bacterium *Pseudomonas chlororaphis* MA 342 had minimal impact.

These findings highlight thiamine's potential as an effective seed treatment to enhance plant resistance and manage aphid pests by mitigating biotic stress across different aphid species. Thiamine seed priming offers a promising, broad-spectrum strategy to reduce aphid populations and associated biotic stress, thereby supporting sustainable crop protection efforts (Hamada et al. 2018).

Enhancement of Antioxidant Capacity

Thiamine has been shown to induce resistance against *Plasmopara viticola*, the causative agent of downy mildew in the grapevine cultivar Chardonnay (Boubakri et al. 2013). This resistance is associated with the activation of the phenylpropanoid pathway, resulting in increased expression of related genes and the accumulation of secondary metabolites, including stilbenes, phenolics, flavonoids, and lignin. Thiamine treatment also enhanced the total antioxidant capacity of grapevine plants by 3.5-fold compared to the control, with compounds such as caftaric acid contributing significantly to this effect.

As the plant's metabolism shifts in response to thiamine, these secondary metabolites play a crucial role in defence. Microscopic analysis revealed that the accumulation of stilbenes coincided with altered fungal growth in treated plants, highlighting stilbenes as key components of thiamine-induced resistance. This finding demonstrates how thiamine not only boosts the plant's antioxidant defences but also directly impacts pathogen development.

This study presents the first evidence that thiamine influences secondary metabolism in grapevine, enhancing defence against *P. viticola*. It offers a new understanding of how vitamins mediate disease resistance. Thiamine's ability to increase antioxidant metabolites and phenylpropanoid-derived compounds highlights its potential as a natural inducer of disease resistance, supporting sustainable vineyard protection approaches (Boubakri et al. 2013).

Induced Resistance via Priming

In carnations infected with *Fusarium oxysporum* sp. *dianthi*, foliar application of thiamine significantly reduces disease incidence and severity. This effect is primarily achieved by increasing the abundance of glycosylated, malonyl glycosylated, and methylated flavonoids, as well as anthranilic acid derivatives, all of which are associated with enhanced antioxidant and antifungal activities. The priming effect of thiamine is especially pronounced in resistant cultivars, highlighting its potential as a bio-stimulant to enhance plant immunity (Pérez-Mora et al. 2025). The priming effect of thiamine has been widely documented in both crop species and model plants, where it enhances disease resistance by preparing the immune system for a more rapid and robust response upon pathogen attack. In crops such as rice, wheat, and tomato, as well as in model plants like *Arabidopsis thaliana*, exogenous thiamine application has been shown to activate a range of defence mechanisms. These include the upregulation of PR genes, increased production of ROS, and the accumulation of secondary metabolites such as phenolics and flavonoids. This primed state enables plants to respond more efficiently to subsequent infections, resulting in reduced disease severity and improved overall plant health. For instance, Ahn et al. (2005) showed that a thiamine treatment in *Arabidopsis thaliana* led to the rapid induction of defence-related mRNAs and enhanced resistance to bacterial and fungal pathogens. Similarly, Bahuguna et al. (2012) reported that thiamine application in rice not only reduced disease incidence but also improved physiological performance under pathogen stress. These findings underscore the broad-spectrum potential of thiamine as a priming agent for disease resistance in both economically important crops and model plant systems.

1.6 Thiamine Uptake, Bioavailability and Regulatory Considerations in Agriculture

1.6.1 Uptake and Bioavailability of Thiamine

1.6.1.1 Efficiency of Thiamine Absorption and Transport in Plant Tissues

Thiamine uptake in plants is primarily facilitated by energy-dependent transporter systems located in the plasma membranes of root and foliar cells. High-affinity transporters, such as those in the SLC19 family, become especially active under thiamine-deficient conditions, enabling efficient acquisition even at low external concentrations (Rapala-Kozik et al. 2012). Once internalised, thiamine is rapidly phosphorylated to its biologically active forms, TMP and TPP, which serve as vital cofactors in central metabolic pathways, including the Calvin cycle, the TCA cycle, and the pentose phosphate pathway (Goyer 2010).

Beyond its metabolic role, exogenous thiamine application can induce SAR in plants, signalling defence activation in both local and distant tissues (Ahn et al. 2005). The persistence of this protective effect for over 15 days suggests that thiamine or its associated signalling components are systemically mobile via the plant vasculature (Ahn et al. 2005; Shah and Zeier 2013).

Despite being water-soluble, thiamine exhibits limited mobility in the phloem, which may restrict its distribution to distal tissues. This limitation may require targeted application strategies or repeated dosing to ensure adequate systemic coverage (Liu et al. 2022). In contrast, root-applied thiamine can be efficiently translocated via the xylem, especially under high transpiration rates. The cohesion-tension mechanism and intact vascular connectivity facilitate the upward transport of dissolved compounds through xylem conduits (Else et al. 2006). Environmental variables, such as soil pH, microbial activity, nutrient interactions (notably with nitrogen and phosphorus), and abiotic stresses like drought or salinity, also modulate thiamine uptake and mobility. Thiamine availability in the rhizosphere is shaped by soil conditions, while thiamine-producing rhizobacteria can supplement plant requirements. Under drought or saline conditions, exogenous thiamine enhances antioxidant defences and improves nutrient acquisition by stabilising membrane integrity and transporter activity (Fitzpatrick and Chapman 2020; Ma et al. 2020; Urooj et al. 2024). Plant species and developmental stage significantly influence thiamine responsiveness, with actively dividing tissues, such as meristems and seedlings, exhibiting a heightened dependence. In *Zea mays* and *Arabidopsis thaliana*, apical meristems and root tips require external thiamine for proper development (Mozafar and Oertli 1993; Woodward 2010), and early-stage applications in crops like cotton and rice enhance stress tolerance and redox balance (Li et al. 2022; Mozafar and Oertli 1992).

Microbial interactions further contribute to thiamine homeostasis. Endophytic bacteria, such as *Pseudomonas aeruginosa* and *Burkholderia cepacia,* have been shown to upregulate plant thiamine biosynthetic genes (*THIC* and *THI4*) and enhance thiamine accumulation in oil palm seedlings (Johari et al. 2024). Similarly, colonisation by the endophytic fungus *Hendersonia toruloidea* increases thiamine levels in host tissues (Kamarudin et al. 2017). In wheat, the combined application of endophytes and thiamine enhances drought resilience, suggesting a synergistic role in stress adaptation (Amjad et al. 2021). Although there is little direct evidence that thiamine transporter regulation occurs during stress, the activation of thiamine biosynthesis genes in adverse environments suggests that internal redistribution mechanisms are coordinated.

1.6.1.2 Environmental and Physiological Factors Influencing Thiamine Uptake and Effectiveness

The uptake and efficacy of thiamine in plants are shaped by a complex interplay of environmental and physiological factors that can either facilitate or hinder its role as a biostimulant and resistance inducer. Environmental variables such as soil pH, organic matter content, and microbial diversity directly influence the stability and bioavailability of thiamine. It tends to remain more stable under mildly acidic conditions (pH 4–5), while its degradation is accelerated in neutral to alkaline soils (pH $\geq$6), potentially limiting its effectiveness in the rhizosphere (Voelker et al. 2021). On the other hand, specific plant growth-promoting rhizobacteria (PGPR) and microbial endophytes are capable of producing and releasing thiamine or its biosynthetic intermediates, increasing their local concentration and promoting uptake, particularly under nutrient-deficient or stressful environmental conditions (Bharti et al. 2016; Kamarudin et al. 2017).

Abiotic stresses such as drought, high salinity, and temperature extremes can negatively impact thiamine uptake by damaging membrane integrity or impairing the activity of transporter proteins. While thiamine plays a role in inducing SAR, its signalling efficiency may be influenced by external conditions, including humidity and the functionality of the plant's vascular system (Shah and Zeier 2013). In response to oxidative stress, plants often increase endogenous thiamine biosynthesis and may upregulate the expression of transporter genes, indicating that thiamine metabolism is actively regulated during stress adaptation (Tunc-Ozdemir et al. 2009).

Younger, actively growing tissues generally exhibit higher rates of thiamine uptake, which is associated with increased metabolic activity and the elevated expression of transporters (Strobbe et al. 2021). Application through the roots allows for systemic distribution of thiamine via xylem transport, particularly under high transpiration rates. In contrast, foliar application faces challenges such as limited phloem mobility and reduced penetration across leaf surfaces (Goyer 2010).

Nutrient interactions, particularly with nitrogen and phosphorus, can influence the composition of root exudates, which in turn alter the rhizosphere microbial communities and thiamine availability (Rapala-Kozik et al. 2012). Regulation of thiamine metabolism in plants under abiotic stresses, including drought, salinity, and extreme temperatures, appears to follow a conserved adaptive strategy also observed in microorganisms (Fitzpatrick and Chapman 2020; Rapala-Kozik et al. 2012). Exogenous thiamine treatment has been shown to reduce the damage caused by such stresses by supporting biomass production, enhancing photosynthetic performance, and strengthening the antioxidant defence system, all of which contribute to improved plant growth and yield (Sayed and Gadallah 2002; Tunc-Ozdemir et al. 2009). These diverse physiological effects underscore the need to tailor thiamine application protocols to species, developmental stage, and environmental context to realise its agronomic potential fully.

1.6.2 Regulatory and Safety Considerations

1.6.2.1 Regulatory Limitations on Thiamine Application in Agriculture

Although thiamine is increasingly recognised for its role in enhancing plant resilience and stress tolerance, its application in agriculture remains limited due to regulatory ambiguity and inconsistent classification across jurisdictions. In the European Union, thiamine is not currently approved as an active substance under Regulation (EC) No 1107/2009, which governs plant protection products, thereby restricting its direct use as a pesticide or defence elicitor. However, under Regulation (EU) 2019/1009, thiamine may be included in the category of plant biostimulants, provided that adequate evidence is presented regarding its agronomic functionality, safety, and efficacy (Du Jardin 2015; Ricci et al. 2019). In the United States, although thiamine holds GRAS (Generally Recognised As Safe) status for food use, it is not registered as an active ingredient under the Federal Insecticide, Fungicide, and Rodenticide Act (FIFRA). Furthermore, regulatory frameworks for biostimulants are still under development in many jurisdictions (Yakhin et al. 2017). These limitations underscore the need for harmonised international standards, clearer functional classifications, and evidence-based regulatory protocols that acknowledge the low-risk profile and multifunctional nature of thiamine in integrated crop management strategies.

Despite its favourable safety characteristics, thiamine often falls into a regulatory grey area, primarily due to the lack of clear distinctions between conventional agrochemicals and biostimulants in many countries. Full commercial approval may require additional data regarding its mode of action, environmental fate, and formulation stability (Du Jardin 2015). The absence of globally harmonised definitions and testing protocols further complicates its broader adoption and hampers international trade (Yakhin et al. 2017).

While existing literature emphasises thiamine's effectiveness in promoting plant growth and enhancing stress resistance (Amjad et al. 2021; Tunc-Ozdemir et al. 2009), relatively little attention has been given to the legal and regulatory frameworks surrounding its agricultural use (Du Jardin 2015, Ricci et al. 2019). To support compliant and effective deployment, further research is essential to bridge these regulatory gaps and to establish standardised guidelines tailored to thiamine's unique properties.

1.6.2.2 Environmental and Safety Concerns Associated with Thiamine Use as an Agrochemical

Thiamine is increasingly recognised as an environmentally benign compound with significant potential in sustainable agriculture, particularly for inducing plant defence responses (Kheyri et al. 2022). Its exogenous application has demonstrated efficacy against soilborne pathogens, such as *Rhizoctonia solani,* in tomato, likely through the activation of defence-related genes and modulation of oxidative signalling pathways. As a naturally occurring metabolite, thiamine serves dual functions, as a cofactor in central metabolic processes and as a signalling molecule involved in redox homeostasis and SAR (Gonzalez-Bosch 2018; Tunc-Ozdemir et al. 2009). These properties align with the increasing emphasis on eco-compatible strategies for plant protection. However, its field application warrants careful environmental and safety assessment. Excessive use may disrupt soil nutrient dynamics and alter microbial community structure, potentially leading to adverse impacts on soil health and biodiversity (Berg 2009; Bharti et al. 2016). A deeper understanding of dose-dependent effects and long-term interactions with soil microbiota is crucial for ensuring sustainable use.

Thiamine is generally considered safe and non-toxic to humans and animals; however, overapplication could result in residue accumulation in edible plant tissues, raising potential food safety concerns (Du Jardin 2015). Although its high water solubility and microbial degradability contribute to low environmental persistence (Fitzpatrick and Chapman 2020), abiotic stresses such as drought and salinity may reduce thiamine uptake and limit its effectiveness (Tunc-Ozdemir et al. 2009). Furthermore, potential synergistic or antagonistic interactions with other agrochemicals could influence both its efficacy and safety (Yakhin et al. 2017). Therefore, the development of comprehensive guidelines, covering application rates, timing, and residue monitoring, is essential to optimise its benefits while minimising associated risks.

1.7 Molecular and Omics Perspectives

1.7.1 Transcriptomic and Proteomic Responses to Thiamine

Thiamine functions as a multifaceted molecule in plants, with transcriptional and translational controls for its biosynthesis and metabolism playing a central role in plant adaptation to environmental stresses. Consequently, thiamine metabolism has emerged as a promising target for strategies aimed at enhancing stress tolerance in crops (Strobbe et al. 2021). Transcriptomic analyses have shown that genes encoding key enzymes involved in thiamine biosynthesis, including *THIC* (phosphomethylpyrimidine synthase), *THI1/THI4* (thiazole synthase), *TH1* (bifunctional hydroxyethylthiazole kinase/thiamine phosphate synthase), *TH2* (thiamine phosphate phosphatase), and *TPK* (thiamine pyrophosphate kinase), are dynamically regulated in response to stressors such as drought, salinity, cold, and oxidative conditions. For instance, in maize seedlings, *TPK* activity increases significantly under water deficit, salinity, and oxidative stress, along with elevated levels of TDP hydrolase enzymes (Rapala-Kozik et al. 2008). In creeping bentgrass, the expression of *THI1* and *THIC* is transiently induced under osmotic stress but declines under prolonged heat and drought exposure, suggesting a time-sensitive regulatory mechanism (Fan 2024).

Exposure to increasing concentrations of polyethylene glycol (PEG) and NaCl, used to simulate osmotic and salt stress, respectively, results in significant increases in total thiamine content in maize seedlings (Tunc-Ozdemir et al. 2009). In *Vitis vinifera* subjected to cold stress, transcript levels of *THI1-1*, *THI1-2*, *THIC*, and *DXS* are markedly upregulated, underscoring the importance of thiamine biosynthesis during cold acclimation (Konecny et al. 2023). In *Arabidopsis thaliana*, the genes *THI4*, *THIC*, *TH1*, and *TPK* respond to cold, osmotic, and salt stresses, with *TPK* showing the strongest induction (Tunc-Ozdemir et al. 2009). Similarly, in oil palm, expression of *THI1/THI4* and *THIC* increases under oxidative, salt, and osmotic stresses, as well as during early infection stages with *Ganoderma boninense*. However, their transcript levels decline under prolonged stress, indicating possible feedback regulation or adaptive signalling (Yusof 2019). In cotton (*Gossypium hirsutum*), the major thiamine biosynthetic genes *GhTHIC* and *GhTHI1* are significantly upregulated in the early phases of stress. Thiamine is essential for cotton growth and development, and its deficiency can be alleviated through external supplementation. Furthermore, exogenous thiamine enhances cotton stress tolerance by improving calcium-mediated signalling and activating downstream stress-responsive genes (Li et al. 2022). A notable example of transcriptional regulation in thiamine metabolism involves the *THIC* gene, whose expression is controlled by environmental cues such as light and circadian rhythms, as well as by riboswitch-mediated alternative splicing in the 3′ untranslated region (3′-UTR), affecting mRNA stability and enzyme production (Yadav et al. 2015). This post-transcriptional mechanism enables plants to fine-tune thiamine levels in response to metabolic demands and environmental conditions (Llavero Pasquina 2020).

Proteomic analyses further corroborate transcriptomic findings by demonstrating changes in protein abundance and activity in response to thiamine application or environmental stress (Rapala-Kozik et al. 2012). For example, thiamine-dependent enzymes such as TK, which are essential for the pentose phosphate pathway and maintaining redox balance, are upregulated during drought and salinity stress, contributing to enhanced metabolic adaptation and antioxidant capacity (Yusof 2019). Moreover, external thiamine application increases the accumulation of defence-related proteins, including PR proteins and antioxidant enzymes, thereby reinforcing SAR and improving pathogen defence.

In *A. thaliana*, thiamine treatment triggers the expression of defence-related genes such as *PR-1* and PAL (Yusof 2019). Likewise, in soybeans and grapevines, thiamine treatment enhances the activity of defence enzymes, such as POD, PPO, and chitinase (Rapala-Kozik et al. 2012). Extensive transcriptomic and proteomic studies across different plant species have confirmed that thiamine metabolism is intricately linked with stress adaptation mechanisms (Zheng et al. 2022). The synchronised upregulation of biosynthetic genes and their protein products results in increased thiamine levels, supporting stress-responsive metabolic reprogramming and defence priming. However, overexpression of these biosynthetic genes does not always translate into improved disease resistance, suggesting that additional regulatory mechanisms, beyond vitamin accumulation, are necessary for fully effective defence responses (Li et al. 2022).

Recent studies have shown that thiamine treatment can induce SAR in *Oryza sativa*, *Arabidopsis*, tobacco, and cucumber. Ahn et al. (2007) reported that plants pre-treated with thiamine exhibited faster and more robust accumulation of PR mRNAs after pathogen exposure, indicating that thiamine primes the plant's immune system for a heightened defensive response. Supporting earlier findings, *PR-1* expression was not observed in *Arabidopsis* plants expressing *NahG* or in *npr1* mutants, both of which are impaired in SA signalling, confirming the SA dependence of thiamine-mediated defence activation (Suohui et al. 2022). Additionally, thiamine-induced SAR involves calcium (Ca^{2+})-dependent signalling and requires H_2O_2 as a central signalling molecule (Berrian 2023). In *Arabidopsis*, thiamine treatment followed by pathogen challenge results in H_2O_2 accumulation, callose deposition, and the activation of defence genes, such as *PR-1* and *PAL1*. When H_2O_2 is removed by CAT treatment, these responses and SAR are significantly diminished, demonstrating that thiamine-induced defence priming is critically dependent on H_2O_2 signalling (Tao et al. 2021).

1.7.2 Hormonal Cross-Talk and Stress Signalling Pathways

Thiamine, primarily in its active form as TDP, functions as an essential coenzyme for several key metabolic enzymes involved in photosynthesis and respiration, including TK and α-KGDH (Yazdani 2015). These enzymes are crucial for sustaining carbon metabolism and maintaining cellular energy homeostasis, both of which are significantly affected under environmental stress conditions (Fisk 2015). Abiotic stressors induce the upregulation of thiamine biosynthesis genes, particularly *THIC* and *THI1* (Dong et al. 2015). This response is often mediated by the stress hormone ABA, highlighting thiamine's involvement in early stress signalling and metabolic adjustment (Li et al. 2022). As such, thiamine functions as a metabolic regulator that enhances plant resilience by supporting critical biochemical pathways during adverse environmental conditions (Subki et al. 2018).

The exogenous application of thiamine has been demonstrated to enhance plant resistance to pathogenic infections by stimulating the expression of PR genes and activating SAR pathways. This priming of immune responses is primarily regulated through the SA signalling cascade, a central hormonal pathway involved in plant immunity (Berrian et al. 2025). The interaction between thiamine and SA-dependent pathways emphasises thiamine's role in modulating hormonal defence responses, thereby strengthening plant resistance to biotic stressors (Dong et al. 2015). Thiamine metabolism also intersects with other stress-responsive processes, including the regulation of circadian rhythms, the biosynthesis of phenylpropanoids and flavonoids, and the synthesis of cytokinins (e.g. zeatin), all of which are tightly integrated within broader hormonal signalling networks (Li et al. 2022). In cotton (*Gossypium hirsutum*), the thiamine biosynthesis genes *GhTHIC* and *GhTHI1* are rapidly upregulated in response to both abiotic and biotic stress conditions (Li et al. 2022). Additionally, exogenous thiamine treatment enhances stress tolerance by modulating calcium-mediated signalling pathways and activating downstream stress-responsive genes, supporting its role as a signalling molecule involved in coordinating growth and defence responses (Yusof 2019).

Although thiamine is not conventionally classified as a phytohormone, it both influences and is influenced by major hormonal pathways such as ABA and SA during stress adaptation (Strobbe et al. 2021). Phytohormones, including auxin, gibberellin, cytokinin, ethylene, and salicylic acid, interact through complex cross-regulatory networks that orchestrate plant responses to environmental challenges (González Ortega-Villaizán et al. 2024). Thiamine's regulation of metabolic and signalling pathways positions it as a central integrator within the plant's hormonal network, coordinating physiological functions and defence mechanisms to enhance survival under stress conditions (Rosado-Souza et al. 2020).

1.8 Limitations and Knowledge Gaps in Thiamine Application for Plant Stress Tolerance

Despite growing evidence that thiamine functions as a biostimulant, enhancing plant resilience under various stress conditions, its practical application in agriculture remains limited due to several unresolved scientific and technical challenges.

One major limitation lies in the incomplete understanding of thiamine's molecular mechanisms of action in plant stress physiology. While thiamine is known to participate in central metabolic pathways, including the pentose phosphate pathway and the TCA cycle, as a cofactor for enzymes such as TK and PDH, its role within the signalling networks that govern stress responses remains poorly defined (Goyer 2010; Strobbe et al. 2021; Subki et al. 2018). Recent transcriptomic studies have shown that thiamine modulates the expression of stress-responsive genes involved in antioxidant defence, osmoprotectant biosynthesis, and hormonal signalling, thereby contributing to enhanced drought tolerance in species such as *Pterocarya stenoptera* (Zhang et al. 2023). However, the precise regulatory circuits and downstream targets of thiamine-mediated signalling remain elusive.

Another critical knowledge gap concerns the species-specific and context-dependent nature of thiamine responses. Experimental evidence indicates that the effectiveness of thiamine treatment varies considerably across plant taxa, developmental stages, and types of abiotic or biotic stressors. For example, exogenous thiamine improved drought tolerance in maize and salt tolerance in both sunflower and maize, but showed limited or no effects in other species under similar conditions (Balmer et al. 2013; Hamada and Al-Hakimi 2009). This variability presents a significant challenge to the development of standardised and broadly applicable treatment protocols.

In addition, optimal dosage, timing, and delivery methods for thiamine application have not been established. Potential interactions with soil microbiota, nutrient regimes, and other agrochemicals remain poorly understood and insufficiently explored. Long-term field studies evaluating the agronomic, ecological, and economic impacts of thiamine supplementation are also notably lacking. Addressing these limitations will require an integrative research strategy that combines functional genomics, metabolomics, and systems biology. Furthermore, well-designed, multilocational field trials will be essential to generate reproducible data and establish scalable, evidence-based application guidelines.

1.8.1 Gaps in Understanding Application Effects Across Environmental Conditions

Environmental variables significantly influence the effectiveness of thiamine application in enhancing plant stress tolerance; however, these interactions remain insufficiently understood. Soil properties, such as pH, organic matter content, and moisture availability, can affect thiamine uptake and translocation in plants, thereby influencing their physiological performance under stress conditions (Goyer 2010; Tunc-Ozdemir et al. 2009). For instance, drought-induced changes in root permeability and water potential may impede thiamine absorption, thereby reducing its protective effects (Sayed and Gadallah 2002; Urooj et al. 2024).

Abiotic stressors may also impact the expression of thiamine biosynthetic and transporter genes. However, these interactions have been examined in only a limited number of model or crop species. While exogenous thiamine has demonstrated positive effects on stress tolerance in various plants, its efficacy is highly context-dependent, making it difficult to generalise results across different environmental conditions (Rosyida and Kristanto 2022; Sayed and Gadallah 2002; Zhang et al. 2023).

Emerging evidence suggests that thiamine functions not only as a metabolic cofactor but also plays a role in regulating plant stress signalling and development. Nonetheless, its integration into broader environmental response networks, such as those influenced by light intensity or atmospheric CO_2 concentration, remains largely unexplored (Goyer 2010; Rapala-Kozik et al. 2012). These knowledge gaps present a significant challenge in designing robust and reproducible thiamine-based treatment strategies across diverse agroecological settings.

Furthermore, the potential synergistic or antagonistic interactions between thiamine and other agricultural inputs, such as plant growth regulators, fertilisers, or biostimulants, remain poorly understood. Without a clearer understanding of how thiamine operates within complex soil–plant–environment systems, the development of standardised and effective application protocols remains out of reach.

1.8.2 Other Key Knowledge Gaps

In addition to environmental and molecular uncertainties, several other challenges continue to limit the practical application of thiamine in agriculture. While species-specific responses have been addressed, the influence of plant developmental stage on thiamine sensitivity remains insufficiently understood and warrants further investigation. Similarly, the long-term consequences of repeated thiamine application on soil health and rhizosphere microbial communities are largely undocumented (Tunc-Ozdemir et al. 2009).

1.9 Future Directions and Agricultural Applications of Thiamine

Future efforts should prioritise the development of optimised thiamine application strategies tailored to specific crop species, developmental stages, and prevailing environmental conditions (Yakhin et al. 2017). Advances in nanotechnology, such as controlled-release formulations, may improve the bioavailability of thiamine and reduce its degradation under field conditions (Pandey et al. 2023).

Combining thiamine with existing disease management practices, including the use of biocontrol agents and PGPR, may help reduce reliance on synthetic agrochemicals while enhancing plant resistance to pathogens. Furthermore, elucidating its interactions with hormonal signalling pathways and epigenetic mechanisms could support targeted breeding strategies aimed at increasing endogenous thiamine-mediated resilience (Goyer 2010). Large-scale field trials conducted across various agroecological regions are necessary to validate laboratory findings and refine practical guidelines for agricultural implementation. Precision agriculture technologies could also facilitate the use of thiamine by enabling timely, site-specific applications.

In the context of climate-resilient farming, thiamine aligns well with organic and low-input systems, owing to its low toxicity, environmental compatibility, and synergistic interactions with biostimulants and beneficial microorganisms (Tunc-Ozdemir et al. 2009; Yakhin et al. 2017). Integrating thiamine into broader stress-mitigation strategies, such as integrated pest management (IPM), nutrient optimisation, and molecular breeding, provides a comprehensive approach to enhancing plant performance under complex stress scenarios.

From a biotechnological standpoint, enhancing thiamine biosynthesis through the overexpression of genes such as *THIC* and *THI1* has demonstrated promising results in *Arabidopsis*, including improved tolerance to drought, salinity, and oxidative stress (Abidin et al. 2016; Chen et al. 2019; Pourcel et al. 2013). Tools like CRISPR/Cas genome editing further enable the precise manipulation of thiamine-related genes without introducing foreign DNA, thereby supporting the development of climate-resilient crop varieties. However, future research should address tissue-specific expression, the metabolic trade-offs of gene overexpression, and system-wide regulatory effects to ensure stable and effective agronomic outcomes.

1.10 Conclusion

Thiamine stands out as a critical micronutrient in plant biology, with functions that extend far beyond its traditional role as a coenzyme in central metabolic processes such as glycolysis, the pentose phosphate pathway, and the TCA cycle. The evidence presented in this chapter underscores the indispensable contributions thiamine makes not only to primary metabolism but also to the regulation of plant responses under a wide range of stress conditions. Recent advances in plant science have revealed the role of thiamine as a signalling molecule capable of orchestrating complex defence mechanisms against both biotic and abiotic stressors. It plays a vital role in the detoxification of ROS, the activation of antioxidant enzymes, and the modulation of gene expression related to defence and hormonal signalling, particularly involving SA and ABA. Both exogenous application and endogenous induction of thiamine have been consistently linked with enhanced plant resilience, improved growth, and increased yield under adverse conditions such as drought, salinity, and pathogen pressure. The biosynthesis and transport of thiamine in plants are governed by a sophisticated regulatory network designed to ensure its availability in metabolically active tissues. The interplay among de novo synthesis, seed reserves, and root uptake highlights the evolutionary importance of maintaining thiamine homeostasis for plant survival and adaptation. Despite these encouraging findings, several key challenges remain. The lack of standardised protocols for thiamine application, limited understanding of its underlying molecular mechanisms, and a scarcity of long-term field data constrain its broader adoption. Additionally, evaluating the long-term ecological effects and establishing clear regulatory frameworks are crucial to support the safe and effective use of this technology in agricultural systems.

In summary, thiamine offers considerable potential as a sustainable tool for improving crop resilience in the face of mounting environmental challenges. Its dual function as a metabolic cofactor and stress signalling regulator makes it a promising candidate for future research and innovation in plant physiology and agricultural biotechnology. The insights discussed in this chapter lay the groundwork for continued exploration of thiamine's diverse roles and its integration into next-generation crop management strategies.

References

Abdel Latef AAH, Hameed M, Khan Z, Khan S, Imran A, Fatima N (2026) Foliar stimulants: A sustainable strategy for enhancing plant resilience to abiotic and biotic stresses. Physiologia Plantarum 178(1):e70741

Abdel-Monaim MF (2011) Role of riboflavin and thiamine in induced resistance against charcoal rot disease of soybean. Egypt J Phytopathol 39:1–23

Abidin AAZ, Wong S, Rahman NSA, Idris ZHC, Balia Yusof Z (2016) Osmotic, oxidative and salinity stresses upregulate the expressions of thiamine (Vitamin B1) biosynthesis genes (*THIC* and *THI1/THI4*) in oil palm (*Elaies guineensis*). J Oil Palm Res 28:308–319

Adamek-Świerczyńska S, Kozik A (2002) Multiple thiamine-binding proteins of legume seeds. Thiamine-binding vicilin of Vicia faba versus thiamine-binding albumin of Pisum sativum. Plant Physiol Biochem 40:735–741

Adamek-Świerczyńska S, Rąpała-Kozik M, Kozik A (2000) Purification and preliminary characterisation of a thiamine-binding protein from maize seeds. J Plant Physiol 156:635–639

Afzal A, Ali B, Samarasinghe KGBA, Khan MN, Javed MA, Saleem A (2026) Overview of Biotic Stress and Its Agricultural Impact. In: Abdel Latef AAH (eds) Chitosan Nanoparticles for Plant Resilience Under Abiotic and Biotic Stressors. Smart Nanomaterials Technology. Springer, Singapore. https://doi.org/10.1007/978-981-95-6422-4_3

Ahn I-P, Kim S, Lee Y-H (2005) Vitamin B1 functions as an activator of plant disease resistance. Plant Physiol 138:1505–1515

Ahn I-P, Kim S, Lee Y-H, Suh S-C (2007) Vitamin B1-induced priming is dependent on hydrogen peroxide and the NPR1 gene in Arabidopsis. Plant Physiol 143:838–848

Ajjawi I, Rodriguez Milla MA, Cushman J, Shintani DK (2007a) Thiamin pyrophosphokinase is required for thiamin cofactor activation in Arabidopsis. Plant Mol Biol 65:151–162

Ajjawi I, Tsegaye Y, Shintani D (2007b) Determination of the genetic, molecular, and biochemical basis of the Arabidopsis thaliana thiamin auxotroph th1. Arch Biochem Biophys 459:107–114

Aleshin V, Mezhenska O, Parkhomenko Y, Kaehne T, Bunik V (2020) Thiamine mono-and diphosphate phosphatases in bovine brain synaptosomes. Biochem Mosc 85:378–386

Amjad SF, Mansoora N, Yaseen S, Kamal A, Butt B, Matloob H, Alamri SA, Alrumman SA, Eid EM, Shahbaz M (2021) Combined use of endophytic bacteria and pre-sowing treatment of thiamine mitigates the adverse effects of drought stress in wheat (*Triticum aestivum* L.) cultivars. Sustainability 13:6582

Amthor JS, Bar-Even A, Hanson AD, Millar AH, Stitt M, Sweetlove LJ, Tyerman SD (2019) Engineering strategies to boost crop productivity by cutting respiratory carbon loss. Plant Cell 31:297–314

Araujo WL, Nunes-Nesi A, Nikoloski Z, Sweetlove LJ, Fernie AR (2012) Metabolic control and regulation of the tricarboxylic acid cycle in photosynthetic and heterotrophic plant tissues. Plant Cell Environ 35:1–21

Ashraf M, Akram NA (2009) Improving salinity tolerance of plants through conventional breeding and genetic engineering: an analytical comparison. Biotechnol Adv 27:744–752

Azooz MM, Abdel Latef AAH, Abdelmajeed WG, et al. (2026) Biochar ameliorates salt stress in maize by modulating physiological responses and expression of SOS1 and NHX2 genes. Journal of Soil Science and Plant Nutrition. https://doi.org/10.1007/s42729-026-03123-w

Bahuguna RN, Joshi R, Shukla A, Pandey M, Kumar J (2012) Thiamine primed defense provides reliable alternative to systemic fungicide carbendazim against sheath blight disease in rice (*Oryza sativa* L.). Plant Physiol Biochem 57:159–167

Balmer D, de Papajewski DV, Planchamp C, Glauser G, Mauch-Mani B (2013) Induced resistance in maize is based on organ-specific defence responses. Plant J 74:213–225

Bandehagh A, Dehghanian Z, Habibi K, Abdel Latef AAH (2023) The Role of Secondary Metabolites in Thymus vulgaris under Abiotic Stress. In: Latef AAH (Ed.). (2023). Medicinal Plant Responses to Stressful Conditions (1st ed.). CRC Press. https://www.taylorfrancis.com/chapters/edit/10.1201/9781003242963-22

Barman M, Das S, Kundu R, Mukherjee S, Singh SK, Abdel Latef AA (2026) Overview of Abiotic Stress and Its Agricultural Impact. In: Abdel Latef AAH (eds) Chitosan Nanoparticles for Plant Resilience Under Abiotic and Biotic Stressors. Smart Nanomaterials Technology. Springer, Singapore. https://doi.org/10.1007/978-981-95-6422-4_2

Berg G (2009) Plant–microbe interactions promoting plant growth and health: perspectives for controlled use of microorganisms in agriculture. Appl Microbiol Biotechnol 84:11–18

Berrian TW (2023) Thiamin as an immunity inducer against diseases of potato: investigation of its effectiveness and molecular mechanisms, Oregon State University

Berrian TW, Fabian ML, Rogan CJ, Anderson JC, Clarke CR, Goyer AJ (2025) Investigation of the effectiveness and molecular mechanisms of thiamin priming to control early blight disease in potato. Phytopathology 115:234–246

Bharti N, Pandey SS, Barnawal D, Patel VK, Kalra A (2016) Plant growth promoting rhizobacteria Dietzia natronolimnaea modulates the expression of stress responsive genes providing protection of wheat from salinity stress. Sci Rep 6:34768

Bhusal N, Kim HS, Han S-G, Yoon T-M (2020) Photosynthetic traits and plant–water relations of two apple cultivars grown as bi-leader trees under long-term waterlogging conditions. Environ Exp Bot 176:104111

Boivin S, Fonouni-Farde C, Frugier F (2016) How auxin and cytokinin phytohormones modulate root microbe interactions. Front Plant Sci 7:1240

Boubakri H, Wahab MA, Chong J, Bertsch C, Mliki A, Soustre-Gacougnolle I (2012) Thiamine induced resistance to Plasmopara viticola in grapevine and elicited host–defense responses, including HR like-cell death. Plant Physiol Biochem 57:120–133

Boubakri H, Poutaraud A, Wahab MA, Clayeux C, Baltenweck-Guyot R, Steyer D, Marcic C, Mliki A, Soustre-Gacougnolle I (2013) Thiamine modulates metabolism of the phenylpropanoid pathway leading to enhanced resistance to Plasmopara viticola in grapevine. BMC Plant Biol 13:1–15

Boubakri H, Gargouri M, Mliki A, Brini F, Chong J, Jbara M (2016) Vitamins for enhancing plant resistance. Planta 244:529–543

Bouhadi M, El Kouali M, Samir K, Elbouhmadi K, Talbi M, Fougrach H (2024) Exogenous application of thiamine and nicotinic acid improves tolerance and morpho-physiological parameters of lens culinaris under lead (Pb) exposure. J Plant Growth Regul 43:4185–4198

Buzea C, Pacheco I (2020) Heavy metals: definition, toxicity, and uptake in plants. In: Cellular and molecular phytotoxicity of heavy metals. Springer, pp 1–17. https://doi.org/10.1007/978-3-030-45975-8_1

Chen K, Wang Y, Zhang R, Zhang H, Gao C (2019) CRISPR/Cas genome editing and precision plant breeding in agriculture. Annu Rev Plant Biol 70:667–697

Chirkova T, Yemelyanov V (2018) The study of plant adaptation to oxygen deficiency in Saint Petersburg University. Biol Commun 63:17–31

Chung YH, Chen TC, Yang WJ, Chen SZ, Chang JM, Hsieh WY, Hsieh MH (2024) Ectopic expression of a bacterial thiamin monophosphate kinase enhances vitamin B1 biosynthesis in plants. Plant J 117:1330–1343

Curcio R, Lunetti P, Zara V, Ferramosca A, Marra F, Fiermonte G, Cappello AR, De Leonardis F, Capobianco L, Dolce V (2020) Drosophila melanogaster mitochondrial carriers: similarities and differences with the human carriers. Int J Mol Sci 21:6052

Danish S, Zafar-ul-Hye M, Fahad S, Saud S, Brtnicky M, Hammerschmiedt T, Datta R (2020a) Drought stress alleviation by ACC deaminase producing Achromobacter xylosoxidans and Enterobacter cloacae, with and without timber waste biochar in maize. Sustainability 12:6286

Danish S, Zafar-Ul-Hye M, Hussain S, Riaz M, Qayyum MF (2020b) Mitigation of drought stress in maize through inoculation with drought tolerant ACC deaminase containing PGPR under axenic conditions. Pak J Bot 52:49–60

de Queiroz AR (2024) Optimizing plant growth and photoprotection through exogenous antioxidants. PhD thesis, University of Nebraska–Lincoln

Deryło M, Skorupska A (1993) Enhancement of symbiotic nitrogen fixation by vitamin-secreting fluorescent Pseudomonas. Plant Soil 154:211–217

Dong W, Stockwell VO, Goyer A (2015) Enhancement of thiamin content in Arabidopsis thaliana by metabolic engineering. Plant Cell Physiol 56:2285–2296

Du Jardin P (2015) Plant biostimulants: definition, concept, main categories and regulation. Sci Hortic 196:3–14

El-Bassiouny H, Sadak MS, Mahfouz S, El-Enany M, Elewa T (2023) Use of thiamine, pyridoxine and bio stimulant for better yield of wheat plants under water stress: growth, osmoregulations, antioxidantive defence and protein pattern. Egypt J Chem 66:407–424

El-Metwally IM, Sadak MS (2019) Physiological role of thiamine and weed control treatments on faba bean and associated weeds grown under salt affected soil. Bull Natl Res Cent 43:1–16

Else MA, Taylor JM, Atkinson CJ (2006) Anti-transpirant activity in xylem sap from flooded tomato (*Lycopersicon esculentum* Mill.) plants is not due to pH-mediated redistributions of root- or shoot-sourced ABA. J Exp Bot 57:3349–3357

Fan Q (2024) Understanding heat stress mechanisms in creeping bentgrass: from physiology to molecular biology with a focus on protein metabolism. PhD thesis, University of Georgia. Available through ProQuest Dissertations & Theses (No. 31633023)

Faozi K, Fajrin PP (2024) Assessing the air salinity on agro-physiological response of Brassica oleracea var. capitata and Brassica oleracea var. botrytis. J Appl Nat Sci 16:77

Faustino M, Lourenço T, Strobbe S, Cao D, Fonseca A, Rocha I, Van Der Straeten D, Oliveira MM (2024) OsTH1 is a key player in thiamin biosynthesis in rice. Sci Rep 14:13591

Fisk S (2015) The effect of increased plastid transketolase activity on thiamine metabolism in transgenic tobacco plants. PhD thesis, University of Essex

Fitzpatrick TB, Chapman LM (2020) The importance of thiamine (vitamin B1) in plant health: from crop yield to biofortification. J Biol Chem 295:12002–12013

Frąckowiak P, Wrzesińska B, Wieczorek P, Sanchez-Bel P, Kunz L, Dittmann A, Obrępalska-Stęplowska A (2022) Deciphering of benzothiadiazole (BTH)-induced response of tomato (*Solanum lycopersicum* L.) and its effect on early response to virus infection through the multi-omics approach. Plant Soil 481:511–534

Ghaffar A, Akram NA, Ashraf M, Ashraf Y, Sadiq M (2019) Thiamin-induced variations in oxidative defense processes in white clover (Trifolium repens L.) under water deficit stress. Turk J Bot 43:58–66

Ghani M, Kulkarni KP, Song JT, Shannon JG, Lee J-D (2016) Soybean sprouts: a review of nutrient composition, health benefits and genetic variation. Plant Breed Biotechnol 4:398–412

Ghosh N, Das A, Chaffee S, Roy S, Sen CK (2018) Reactive oxygen species, oxidative damage and cell death, immunity and inflammation in health and disease. Academic Press, pp 45–55

Gołda A, Szyniarowski P, Ostrowska K, Kozik A, Rąpała-Kozik M (2004) Thiamine binding and metabolism in germinating seeds of selected cereals and legumes. Plant Physiol Biochem 42: 187–195

Golldack D, Li C, Mohan H, Probst N (2014) Tolerance to drought and salt stress in plants: unraveling the signaling networks. Front Plant Sci 5:151

González Ortega-Villaizán A, King E, Patel MK, Pollmann S (2024) Plant hormone crosstalk under abiotic stress conditions. In: Progress in botany, vol 85. Springer, pp 167–194. https://doi.org/10.1007/124_2024_80

Gonzalez-Bosch C (2018) Priming plant resistance by activation of redox-sensitive genes. Free Radic Biol Med 122:171–180

Goyer A (2010) Thiamine in plants: aspects of its metabolism and functions. Phytochemistry 71:1615–1624

Goyer A (2017) Thiamin biofortification of crops. Curr Opin Biotechnol 44:1–7

Guan J-C, Hasnain G, Garrett TJ, Chase CD, Gregory J, Hanson AD, McCarty DR (2014) Divisions of labor in the thiamin biosynthetic pathway among organs of maize. Front Plant Sci 5:370

Hamada A, Al-Hakimi A (2009) Exogenous ascorbic acid or thiamine increases the resistance of sunflower and maize plants to salt stress. Acta Agron Hung 57:335–347

Hamada AM, Jonsson LM (2013) Thiamine treatments alleviate aphid infestations in barley and pea. Phytochemistry 94:135–141

Hamada A, Fatehi J, Jonsson L (2018) Seed treatments with thiamine reduce the performance of generalist and specialist aphids on crop plants. Bull Entomol Res 108:84–92

Han X, Yang X, Chen S, Wang H, Liu X, Wang D, Yang J, Chen L, Sun B, Li H (2024) Barley yellow dwarf virus-GAV 17K protein disrupts thiamine biosynthesis to facilitate viral infection in plants. Plant J 119:432–444

Handa N, Kohli SK, Kaur R, Sharma A, Kumar V, Thukral AK, Arora S, Bhardwaj R (2018) Role of compatible solutes in enhancing antioxidative defense in plants exposed to metal toxicity. In: Plants under metal and metalloid stress: responses, tolerance and remediation, pp 207–228. https://doi.org/10.1007/978-981-13-2242-6_7

Hofmann M, Loubéry S, Fitzpatrick TB (2020) On the nature of thiamine triphosphate in Arabidopsis. Plant direct 4:e00258

Huang WK, Ji HL, Gheysen G, Kyndt T (2016) Thiamine-induced priming against root-knot nematode infection in rice involves lignification and hydrogen peroxide generation. Mol Plant Pathol 17:614–624

Huang YC, Chiang CH, Li CM et al. (2011) Transgenic watermelon lines expressing the nucleocapsid gene of Watermelon silver mottle virus and the role of thiamine in reducing hyperhydricity in regenerated shoots. Plant Cell Tiss Organ Cult 106:21–29. https://doi.org/10.1007/s11240-010-9889-z

Huergo LF, Dixon R (2015) The emergence of 2-oxoglutarate as a master regulator metabolite. Microbiol Mol Biol Rev 79:419–435

Hussein M, Orabi S (2025) Response of grain sorghum to alpha-tochopherol and thiamine under salinity. Egypt J Agron 47:1–13

Jabeen M, Akram NA, Ashraf M, Alyemeni MN, Ahmad P (2021) Thiamin stimulates growth and secondary metabolites in turnip (*Brassica rapa L.*) leaf and root under drought stress. Physiol Plant 172:1399–1411

Jabeen M, Akram NA, Ashraf M, Tyagi A, El-Sheikh MA, Ahmad P (2022) Thiamin stimulates growth, yield quality and key biochemical processes of cauliflower (*Brassica oleracea* L. var. Botrytis) under arid conditions. PLoS One 17:e0266372

Johari NAF, Abidin AAZ, Ismail NFN, Yusof ZNB (2024) Endophytic bacteria induce thiamine (vitamin B1) production in oil palm (*Elaeis guineensis*). Trop Life Sci Res 35:1

Joshi J, Folz JS, Gregory JF III, McCarty DR, Fiehn O, Hanson AD (2019) Rethinking the PDH bypass and GABA shunt as thiamin-deficiency workarounds. Plant Physiol 181:389–393

Juri NMBM (2017) Transcriptomic analysis of Malaysian rice seedlings (*Oryza sativa* l. Ssp. Indica) in early responses to salt-shock, Masters thesis, Universiti Putra Malaysia

Kachout SS, Hamza KJ, Bouraoui NK, Leclerc JC, Ouerghi Z (2013) Salt-induced changes in antioxidative enzyme activities in shoot tissues of two Atriplex varieties. Not Bot Horti Agrobot Cluj Na 41:115–121

Kamarudin AN, Lai KS, Lamasudin DU, Idris AS, Balia Yusof ZN (2017) Enhancement of thiamine biosynthesis in oil palm seedlings by colonization of endophytic fungus Hendersonia toruloidea. Front Plant Sci 8:1799
Kaya C, Ashraf M, Sonmez O, Tuna AL, Polat T, Aydemir S (2015) Exogenous application of thiamin promotes growth and antioxidative defense system at initial phases of development in salt-stressed plants of two maize cultivars differing in salinity tolerance. Acta Physiol Plant 37: 1–12
Kheyri F, Taheri P, Jafarinejad-Farsangi S (2022) Thiamine and Piriformospora indica induce bean resistance against Rhizoctonia solani: the role of polyamines in association with iron and reactive oxygen species. Biol Control 172:104955
Khozaei M, Fisk S, Lawson T, Gibon Y, Sulpice R, Stitt M, Lefebvre SC, Raines CA (2015) Overexpression of plastid transketolase in tobacco results in a thiamine auxotrophic phenotype. Plant Cell 27:432–447
Kim JA, Kim H-S, Choi S-H, Jang J-Y, Jeong M-J, Lee SI (2017) The importance of the circadian clock in regulating plant metabolism. Int J Mol Sci 18:2680
Konecny T, Nikoghosyan M, Binder H (2023) Machine learning extracts marks of thiamine's role in cold acclimation in the transcriptome of Vitis vinifera. Front Plant Sci 14:1303542
Kumari A, Sharma B, Singh BN, Hidangmayum A, Jatav HS, Chandra K, Singhal RK, Sathyanarayana E, Patra A, Mohapatra KK (2022) Physiological mechanisms and adaptation strategies of plants under nutrient deficiency and toxicity conditions. In: Plant perspectives to global climate changes. Academic Press, pp 173–194
Li K-T (2014) Metabolic engineering of Vitamin B1 and Vitamin B6 in staple crops. ETH Zurich, Zurich
Li M, Petteys BJ, McClure JM, Valsakumar V, Bekiranov S, Frank EL, Smith JS (2010) Thiamine biosynthesis in Saccharomyces cerevisiae is regulated by the NAD+-dependent histone deacetylase Hst1. Mol Cell Biol 30:3329–3341
Li W, Mi X, Jin X, Zhang D, Zhu G, Shang X, Zhang D, Guo W (2022) Thiamine functions as a key activator for modulating plant health and broad-spectrum tolerance in cotton. Plant J 111:374–390
Liang C, Meng Z, Meng Z, Malik W, Yan R, Lwin KM, Lin F, Wang Y, Sun G, Zhou T (2016) GhABF2, a bZIP transcription factor, confers drought and salinity tolerance in cotton (*Gossypium hirsutum* L.). Sci Rep 6:35040
Liu J, Zhang X, Deng S, Wang H, Zhao Y (2022) Thiamine is required for virulence and survival of pseudomonas syringae pv. tomato DC3000 on tomatoes. Front Microbiol 13:903258
Llavero Pasquina M (2020) Molecular biology of B vitamin metabolism genes and their regulation in Phaeodactylum tricornutum. https://doi.org/10.17863/CAM.61337
Ma Y, Dias MC, Freitas H (2020) Drought and salinity stress responses and microbe-induced tolerance in plants. Front Plant Sci 11:591911
Manasa SL, Panigrahy M, Panigrahi KC, Rout GR (2022) Overview of cold stress regulation in plants. Bot Rev 88:359–387
Mansoor S, Ali A, Kour N, Bornhorst J, AlHarbi K, Rinklebe J, Abd El Moneim D, Ahmad P, Chung YS (2023) Heavy metal induced oxidative stress mitigation and ROS scavenging in plants. Plants 12:3003
Martinis J, Gas-Pascual E, Szydlowski N, Crèvecoeur M, Gisler A, Bürkle L, Fitzpatrick TB (2016) Long-distance transport of thiamine (vitamin B1) is concomitant with that of polyamines. Plant Physiol 171:542–553
Millar AH, Heazlewood JL (2003) Genomic and proteomic analysis of mitochondrial carrier proteins in Arabidopsis. Plant Physiol 131:443–453
Mimura M, Zallot R, Niehaus TD, Hasnain G, Gidda SK, Nguyen TN, Anderson EM, Mullen RT, Brown G, Yakunin AF (2016) Arabidopsis TH2 encodes the orphan enzyme thiamin monophosphate phosphatase. Plant Cell 28:2683–2696
Mozafar A (2018) Plant vitamins. CRC Press, Boca Raton

Mozafar A, Oertli J (1992) Uptake and transport of Thiamin (vitamin B 1) by barley and soybean. J Plant Physiol 139:436–442

Mozafar A, Oertli J (1993) Thiamin (vitamin B1): translocation and metabolism by soybean seedling. J Plant Physiol 142:438–445

Nagae M, Parniske M, Kawaguchi M, Takeda N (2016) The thiamine biosynthesis gene THI1 promotes nodule growth and seed maturation. Plant Physiol 172:2033–2043

Nie S, Xu H (2016) Riboflavin-induced disease resistance requires the mitogen-activated protein kinases 3 and 6 in Arabidopsis thaliana. PLoS One 11:e0153175

Noctor G, Mhamdi A, Foyer CH (2014) The roles of reactive oxygen metabolism in drought: not so cut and dried. Plant Physiol 164:1636–1648

Noordally ZB, Trichtinger C, Dalvit I, Hofmann M, Roux C, Zamboni N, Pourcel L, Gas-Pascual E, Gisler A, Fitzpatrick TB (2020) The coenzyme thiamine diphosphate displays a daily rhythm in the Arabidopsis nucleus. Commun Biol 3:209

Pandey G, Tripathi S, Bajpai S, Kamboj M (2023) Approaches, challenges, and prospects of nanotechnology for sustainable agriculture. In: Agricultural and environmental nanotechnology: novel technologies and their ecological impact. Springer, Singapore, pp 83–103

Parker AL, Kavallaris M, McCarroll JA (2014) Microtubules and their role in cellular stress in cancer. Front Oncol 4:153

Pérez-Mora W, Coy-Barrera E, Melgarejo LM, Ardila HD (2025) Metabolite profiling-based analysis reveals up-regulated shikimate derivatives in carnation roots associated with the thiamine-induced resistance against Fusarium oxysporum f. sp. dianthi. Physiol Mol Plant Pathol 136:102535

Phosaeng M, Phraprasert P (2021) Thiamine biosynthesis during grain development of rice cultivars. Burapha Sci J 26(3):1826–1835

Pieterse CM, Van der Does D, Zamioudis C, Leon-Reyes A, Van Wees SC (2012) Hormonal modulation of plant immunity. Annu Rev Cell Dev Biol 28:489–521

Polash MAS, Sakil MA, Hossain MA (2019) Plants responses and their physiological and biochemical defense mechanisms against salinity: a review. Trop Plant Res 6:250–274

Pourcel L, Moulin M, Fitzpatrick TB (2013) Examining strategies to facilitate vitamin B1 biofortification of plants by genetic engineering. Front Plant Sci 4:160

Pushpalatha H, Sudisha J, Geetha N, Amruthesh K, Shekar Shetty H (2011) Thiamine seed treatment enhances LOX expression, promotes growth and induces downy mildew disease resistance in pearl millet. Biol Plant 55:522–527

Qi X, Su X, Guo H, Qi J, Cheng H (2016) VdThit, a thiamine transport protein, is required for pathogenicity of the vascular pathogen Verticillium dahliae. Mol Plant-Microbe Interact 29: 545–559

Radović S, Miljuš-Djukić J, Samardžić J, Banović B, Nikolić D, Milisavljević M, Timotijević G (2012) Buckwheat as a model plant in molecular biology. Eur J Plant Sci Biotech 6(2):11–16

Rahman NSA, Ghazali NS, Yusof ZNB (2017) Changes in the expressions of thiamine biosynthesis genes (*THIC* and *THI1/THI4*) in oil palm (*Elaies guineensis*) as response to salinity stress. Res Crops 18:634

RąPala-Kozik M (1995) Protein-attributable thiamine-binding activity in gymnosperm seeds. J Plant Physiol 146:760–762

Rapala-Kozik M, Chernikevich IP, Kozik A (1999) Ligand–protein interaction in plant seed thiamine-binding proteins. Binding of various thiamine analogues to the sepharose-immobilized buckwheat-seed protein. J Protein Chem 18:721–728

Rapala-Kozik M, Kowalska E, Ostrowska K (2008) Modulation of thiamine metabolism in Zea mays seedlings under conditions of abiotic stress. J Exp Bot 59:4133–4143

Rapala-Kozik M, Wolak N, Kujda M, Banas AK (2012) The upregulation of thiamine (vitamin B 1) biosynthesis in Arabidopsis thaliana seedlings under salt and osmotic stress conditions is mediated by abscisic acid at the early stages of this stress response. BMC Plant Biol 12:1–14

Raza A, Charagh S, Najafi-Kakavand S, Abbas S, Shoaib Y, Anwar S, Sharifi S, Lu G, Siddique KH (2023) Role of phytohormones in regulating cold stress tolerance: physiological and molecular approaches for developing cold-smart crop plants. Plant Stress 8:100152

Ricci M, Tilbury L, Daridon B, Sukalac K (2019) General principles to justify plant biostimulant claims. Front Plant Sci 10:494

Roje S (2007) Vitamin B biosynthesis in plants. Phytochemistry 68:1904–1921

Rosado-Souza L, Proost S, Moulin M, Bergmann S, Bocobza SE, Aharoni A, Fitzpatrick TB, Mutwil M, Fernie AR, Obata T (2019) Appropriate thiamin pyrophosphate levels are required for acclimation to changes in photoperiod. Plant Physiol 180:185–197

Rosado-Souza L, Fernie AR, Aarabi F (2020) Ascorbate and thiamin: metabolic modulators in plant acclimation responses. Plants 9:101

Rosyida R, Kristanto BA (2022) The role of antioxidant compounds in plant heat tolerance. Open Access Res J Life Sci 4:11–15

Sanjari S, Keramat B, Nadernejad N, Mozafari H (2019) Ameliorative effects of 24-epibrassinolide and thiamine on excess cadmium-induced oxidative stress in Canola (*Brassica napus* L.) plants. J Plant Interact 14:359–368

Sayed S, Gadallah M (2002) Effects of shoot and root application of thiamin on salt-stressed sunflower plants. Plant Growth Regul 36:71–80

Shah J, Zeier J (2013) Long-distance communication and signal amplification in systemic acquired resistance. Front Plant Sci 4:30

Simkin AJ, López-Calcagno PE, Raines CA (2019) Feeding the world: improving photosynthetic efficiency for sustainable crop production. J Exp Bot 70:1119–1140

Smith MR, Rao IM, Merchant A (2018) Source-sink relationships in crop plants and their influence on yield development and nutritional quality. Front Plant Sci 9:1889

Strobbe S, Verstraete J, Stove C, Van Der Straeten D (2021) Metabolic engineering provides insight into the regulation of thiamin biosynthesis in plants. Plant Physiol 186:1832–1847

Subki A, Abidin AAZ, Yusof ZNB (2018) The role of thiamine in plants and current perspectives in crop improvement. In: B group vitamins-current uses and perspectives, vol 5, pp 33–44. https://doi.org/10.5772/intechopen.79350

Sun J, Sigler CL, Beaudoin GA, Joshi J, Patterson JA, Cho KH, Ralat MA, Gregory JF III, Clark DG, Deng Z (2019) Parts-prospecting for a high-efficiency thiamin thiazole biosynthesis pathway. Plant Physiol 179:958–968

Sun S, Han X, Zhang Y, Xie L, Li H, Fan Z, Wang Q, Jia S, Yang X, Yuan H (2025) Exogenous thiamine regulates multiple plant immunity-associated pathways to inhibit MCMV infection. J Agric Food Chem 73(23):14353–14362

Suohui T, Yanping C, Shuhui Z, Zhihua L, Honggang J, Jun L, Tao L (2022) Thiamine induces resistance in tobacco against black shank. Australas Plant Pathol 51:231–243

Suzuki N, Rivero RM, Shulaev V, Blumwald E, Mittler R (2014) Abiotic and biotic stress combinations. New Phytol 203:32–43

Tao L, Suohui T, Yanping C, Shuhui Z, Jian M, Zhijiang W, Xianwen Y (2021) Priming effect of thiamine on the enhancement of induced resistance to the plant disease phytophthora nicotianae in tobacco. https://doi.org/10.21203/rs.3.rs-279029/v1

Tripathi DK, Singh S, Singh S, Mishra S, Chauhan D, Dubey N (2015) Micronutrients and their diverse role in agricultural crops: advances and future prospective. Acta Physiol Plant 37:1–14

Tuna AL, Kaya C, Altunlu H, Ashraf M (2013) Mitigation effects of non-enzymatic antioxidants in maize (‘*Zea mays*’ L.) plants under salinity stress. Aust J Crop Sci 7:1181–1188

Tunc-Ozdemir M, Miller G, Song L, Kim J, Sodek A, Koussevitzky S, Misra AN, Mittler R, Shintani D (2009) Thiamin confers enhanced tolerance to oxidative stress in Arabidopsis. Plant Physiol 151:421–432

ul Haq S, Khan A, Ali M, Khattak AM, Gai W-X, Zhang H-X, Wei A-M, Gong Z-H (2019) Heat shock proteins: dynamic biomolecules to counter plant biotic and abiotic stresses. Int J Mol Sci 20:5321

Umer S, Akram NA, Ashraf M, Abdel Latef AAH (2026) Pre-Sowing application of methionine and proline mitigates drought stress in two canola (Brassica napus L.) cultivars. Russ J Plant Physiol 73:48

Urooj S, Rasheed R, Ashraf MA, Ali S, Hussain I (2024) Thiamine regulated osmolyte accumulation, nutrient acquisition, and ROS metabolism to lessen salinity effects on wheat (*Triticum aestivum* L.). J Soil Sci Plant Nutr 24:3560–3578

Valderrama R, Corpas FJ, Carreras A, Gómez-Rodríguez MV, Chaki M, Pedrajas JR, Fernandez-Ocana A, Del Rio LA, Barroso JB (2006) The dehydrogenase-mediated recycling of NADPH is a key antioxidant system against salt-induced oxidative stress in olive plants. Plant Cell Environ 29:1449–1459

Vidal S, de León IP, Denecke J, Palva ET (1997) Salicylic acid and the plant pathogen Erwinia carotovora induce defense genes via antagonistic pathways. Plant J 11:115–123

Vinchesi AC, Rondon SI, Goyer A (2017) Priming potato with thiamin to control potato virus Y. Am J Potato Res 94:120–128

Voelker AL, Taylor LS, Mauer LJ (2021) Effect of pH and concentration on the chemical stability and reaction kinetics of thiamine mononitrate and thiamine chloride hydrochloride in solution. BMC Chem 15:1–14

Wani SH, Kumar V (2020) Heat stress tolerance in plants: physiological, molecular and genetic perspectives. John Wiley & Sons, Hoboken

Westman SM, Kloth KJ, Hanson J, Ohlsson AB, Albrectsen BR (2019) Defence priming in Arabidopsis–a meta-analysis. Sci Rep 9:13309

Wolak N, Tomasi M, Kozik A, Rapala-Kozik M (2015) Characterization of thiamine uptake and utilization in Candida spp. subjected to oxidative stress. Acta Biochim Pol 62:445–455

Woodward J (2010) A maize thiamine auxotroph is defective in meristem maintenance and leaf blade development. PhD thesis, Cornell University

Wouyou AD, Ahissou EA, Gandonou CB, Komlan FA, Houngbeme A, Gbaguidi FA, Ahissou H, Lagnika L, Zanklan SA, Lutts S (2017) Salinity increased vitamins concentration in Amaranthus cruentus leaves. Afr J Biotechnol 16:2106–2111

Xie Z, Wang C, Zhu S, Wang W, Xu J, Zhao X (2020) Characterizing the metabolites related to rice salt tolerance with introgression lines exhibiting contrasting performances in response to saline conditions. Plant Growth Regul 92:157–167

Xu Y, Huang B (2018) Comparative transcriptomic analysis reveals common molecular factors responsive to heat and drought stress in Agrostis stolonifera. Sci Rep 8:15181

Yadav S, Swati D, Chandrasekharan H (2015) Thiamine pyrophosphate riboswitch in some representative plant species: a bioinformatics study. J Comput Biol 22:1–9

Yakhin OI, Lubyanov AA, Yakhin IA, Brown PH (2017) Biostimulants in plant science: a global perspective. Front Plant Sci 7:2049

Yazdani M (2015) Genetic engineering of plant seeds to increase thiamin (vitamin B1) content. University of Nevada, Reno

Yee WS, Aziz SDA, Yusof ZNB (2016) Osmotic stress upregulates the transcription of thiamine (vitamin B1) biosynthesis genes (*THIC* and *THI4*) in oil palm (Elaies guineensis). Afr J Biotechnol 15:1566–1574

Yusof ZNB (2019) Thiamine and its role in protection against stress in plants (enhancement in thiamine content for nutritional quality improvement). In: Nutritional quality improvement in plants. Springer, Cham, pp 177–186

Zahra N, Hafeez MB, Ghaffar A, Kausar A, Al Zeidi M, Siddique KH, Farooq M (2023) Plant photosynthesis under heat stress: effects and management. Environ Exp Bot 206:105178

Zamanipour M (2021) Effects of pyridoxine, thiamine and folic acid on growth, reproductive and biochemical characteristics of Delphus tomato. J Hortic Sci 35:283–300

Zhang Y, Zhou J, Zhao L, Feng Z, Wei F, Bai H, Feng H, Zhu H (2022) A review of the pathogenicity mechanism of Verticillium dahliae in cotton. J Cotton Res 5:3

Zhang W, Wang S-C, Li Y (2023) Molecular mechanism of thiamine in mitigating drought stress in Chinese wingnut (Pterocarya stenoptera): insights from transcriptomics. Ecotoxicol Environ Saf 263:115307

Zhao W-B, An J-X, Jin Y-R, Jing C-X, Zhang S-Y, Liang H-J, Dai T-L, Luo X-F, Zhang B-Q, Zhang Z-J (2025) Indoloquinoline alkaloid neocryptolepine derivative inhibits Botrytis cinerea by targeting thiamine thiazole synthase. Sci Adv 11:eadq5329

Zheng L, Wu W, Chen Q, Zhang G, Gao F, Zhou Y (2022) Integrated transcriptomics, proteomics, and metabolomics identified biological processes and metabolic pathways involved in heat stress response in jojoba. Ind Crop Prod 183:114946

Chapter 2
Riboflavin: Enhancing Plant Resilience Under Stress Conditions

Murat Dikilitas, Behzat Baran, Berrin Arslan, and Arafat Abdel Hamed Abdel Latef

Abstract Plants are constantly exposed to abiotic (drought, high temperatures, salinity, chilling, and air pollution) and biotic (pathogens, insects, and other animals) stressors, or to both, either simultaneously or sequentially. Several biochemical, molecular, and physiological strategies have been proposed to improve crop plants' performance under stress conditions. The main goal is either to remove the stressor from around the plants or to treat the affected parts, or to keep the plants as resilient as possible from seed to seed to the next generation by boosting immunity. Eliminating stressors from around the plants or stressed parts is not always straightforward. The simplest way to protect plants from stress is to prevent it. Since plants lack a closed circulatory system like animals or humans, their immunity may not reach the same levels as in other organisms. However, recently, numerous studies and publications have emerged regarding induced resistance or immunity in plants. Several key molecules that enhance plant resilience to single or multiple stressors and that trigger biochemical and molecular pathways have been proposed. In this chapter, we examine the role of riboflavin and its potential applications, alongside other emerging compounds, to enhance plant resistance or tolerance to abiotic, biotic, or combined stresses, to promote sustainable agriculture.

Keywords Riboflavin · Combined stress · Quality · Drought · Pathogen

M. Dikilitas (✉)
Faculty of Agriculture, Department of Plant Protection, Harran University, S. Urfa, Turkey

B. Baran
Plant Protection Research Institute, Sur, Diyarbakır, Turkey

B. Arslan
Faculty of Agriculture, Department of Plant Protection, Siirt University, Siirt, Turkey

A. A. H. Abdel Latef
Department of Botany and Microbiology, Faculty of Science, Qena University, Qena, Egypt

A. A. H. Abdel Latef (ed.), *Harnessing Vitamins and Organic Acids for Plant Stress Resilience*, Sustainability Sciences in Asia and Africa,
https://doi.org/10.1007/978-981-95-8593-9_2

2.1 Introduction

Not long ago, humanity faced a series of challenges, including the COVID-19 pandemic, unprecedented floods, droughts, and insect plagues in many regions of the world, primarily in the Northern Hemisphere. These challenges should be taken seriously; otherwise, the future will not look very bright if we cannot implement significant measures. Therefore, we need to explore new sources or pathways in agricultural sciences. Although conventional breeding, biotechnology, and agronomic advancements have made major contributions to increasing agricultural yields, these improvements are not entirely sufficient given the projected global population growth. Increased abiotic and biotic stress factors, as well as their potential combinations, have also contributed to malnutrition and have overshadowed major agricultural progress. We are hopeful about the potential increase in yield to support the growing population; however, research on biochemical and molecular pathways to improve crop quality remains limited and elusive. Thus, every perspective should be considered and discussed to boost both crop yield and quality. So far, agricultural research has mostly focused on crop yield, but a new concept called "hidden hunger" has gained attention scientifically, although crop production has been maintained in many parts of the world. The concept of hidden hunger would become more significant in agricultural studies if crop quality were to decline or stay low. We rely on plants for specific vitamins and metabolites; therefore, we must grow them to their full potential in nutritional value, even under stress conditions. A large portion of the global population suffers from hidden hunger. Hidden hunger has been linked to long-term health and economic losses (Jiang et al. 2021). Recently, a new approach has been introduced as an alternative to genetic engineering to improve crop quality. It is biofortification, a concept that involves increasing the bioavailability and levels of nutrients in crops to enhance human health through metabolic engineering. This has been proven to be a cost-effective approach to minimize the impact of vitamin deficiency in human and animal health (Zikankuba et al. 2019; Jiang et al. 2021; Kumar et al. 2023). This new phenomenon is called micronutrients, a crucial chemical substance required in trace amounts for growth and development. A deficiency of micronutrients would not only exhibit the symptoms associated with a specific substance, but it could also lead to detrimental effects. It is essential to note that, in this context, a high dose of micronutrients or vitamins can also be harmful to living organisms due to cross-signaling or overlapping of antioxidant- or oxidant-related responses.

Abiotic and biotic stress factors are not only two significant crop yield-limiting stressors, but their combination can also be considered as a third group of stress that exerts more devastating effects on crop plants. Since plants are immobile organisms, they have developed advanced mechanisms to cope with stress factors. Its growth and stress resistance/tolerance depend entirely on balanced oxidative metabolism with a powerful antioxidant capacity that helps maintain homeostasis. When stress results in an imbalance, reactive oxygen species (ROS) begin to accumulate in plant cells, and these cells must then rebalance their metabolism and antioxidant systems through the synthesis of stress-metabolites and antioxidant enzymes. If the rebalancing of homeostasis cannot be achieved, the exogenous application of antioxidants should be considered to improve both plant growth and resilience to stress (Dikilitas et al. 2021; Rodrigues de Queiroz et al. 2023). So, plants can adjust their metabolism to the changing stress conditions. To combat stress, plants activate a complex defense mechanism triggered by signaling molecules that synthesize low concentrations of ROS or detoxify high concentrations of ROS via antioxidant enzymes, followed by the expression of defense-related genes and hormonal signaling (Ahammed et al. 2024). Plant metabolism can be affected in three ways: photosynthesis, respiration, and stress management through oxidative respiration. All these main metabolisms are interconnected. Therefore, any changes in these pathways would affect all processes carried out in cells. The exogenous application of antioxidants has a promising role in improving plant health and resistance. This method of application has gained an excellent reputation and is likely to play a significant role in the future of agriculture. Since application and possible optimization are easier than genetic engineering in the short term, it could be widely used soon. Several antioxidants have been recently introduced into agriculture. Their common pathway is involved in chemical reduction and activation in antioxidant mechanisms. These antioxidants are glutathione, ascorbate, melatonin, tocopherol, thiamine, flavins, chitosan, and proline. From these antioxidants, riboflavin serves as the backbone of the structural elements in other molecules (Fig. 2.1).

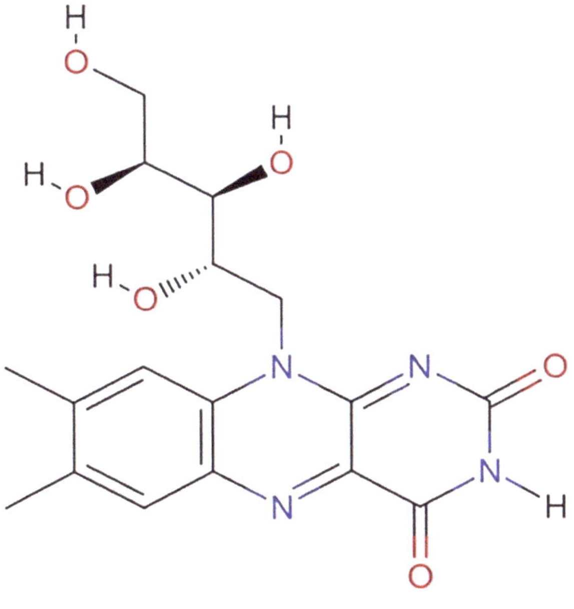

Fig. 2.1 Chemical structure of riboflavin (https://pubchem.ncbi.nlm.nih.gov/compound/Riboflavin). Molecular formula ($C_{17}H_{20}N_4O_6$)

Riboflavin, a water-soluble vitamin, has gained significant attention as a signaling molecule that enhances stress tolerance or resistance in recent years (Tripathi et al. 2024). In addition to increasing defense mechanisms against fungal, bacterial, and viral diseases, riboflavin may also induce tolerance to abiotic stress and enable growth under stress conditions. It is also active in the breakdown of carbohydrate, protein, and lipid molecules. Therefore, its role extends beyond being a mere coenzyme involved in systemic acquired resistance (SAR) or induced systemic resistance (ISR), as it also regulates oxidative stress in plants (Zhao et al. 2024).

However, the metabolism and regulatory mechanism of B vitamins during growth and seed protein development remain largely unclear. Zhou et al. (2023) investigated the development of six B vitamins during the growth of *Areca catechu* nuts via targeted metabolomics. They identified 88 structural genes related to the B vitamin biosynthesis pathway using RNA-seq. The authors showed that transcription factors such as *AcbZIP21*, *AcMYB84*, and *AcARF32* regulated thiamine and riboflavin accumulation in areca nuts.

Riboflavin was first discovered in 1879 from milk whey, exhibiting a yellow, fluorescent substance (Ruchala et al. 2025). Riboflavin is chemically synthesized from carbohydrate-based precursors such as D-glucose and D-ribose, followed by the hydrolysis of starch or microbial fermentation (Zhang et al. 2022; Ruchala et al. 2025).

B vitamins are a class of water-soluble vitamins participating as coenzymes or cofactors in energy metabolism (Fitzpatrick et al. 2012; Zhou et al. 2023; Suzuki et al. 2020). They are essential nutrients for the growth of all organisms. Therefore, their presence not only affects plant health but also significantly impacts human and animal health, which is dependent on the level of B vitamins. They could be classified as thiamine (vitamin B1), riboflavin (vitamin B2), niacin, nicotinamide, and nicotinamide riboside (vitamin B3), pantothenic acid (vitamin B5), pyridoxine, pyridoxal, and pyridoxamine (vitamin B6), biotin (vitamin B7), folic acid (vitamin B9), and cobalamin (vitamin B12) (Fitzpatrick et al. 2012; Kraemer et al. 2012; Zhou et al. 2023). All these vitamins have significant roles in human metabolic health as well as in plant metabolism. Deficiency of B vitamins has been associated with various health disorders, including heart disease (vitamin B1), dermatitis, tongue inflammation (vitamin B2), and neuropathy (vitamin B6) (Zhou et al. 2023). Although B vitamins play significant roles in human and animal health, most of them cannot be de novo biosynthesized in human and animal cells; therefore, they must be obtained from plant foods (Hanna et al. 2022). Thus, the contents of B vitamins in plants are reflected in both human and animal health. Their deficiency in plants could also be reflected in human and animal metabolisms.

B vitamin accumulation and biosynthesis respond to a variety of environmental changes and stresses caused by abiotic and biotic stresses in plants (Rosado-Souza et al. 2020; Zhou et al. 2023). For example, a decrease in vitamin B1, vitamin B2, vitamin B5, and vitamin B9 was associated with high CO_2 levels in rice plants (Zhou et al. 2011). On the other hand, an increase in biosynthesis of vitamin B1 and vitamin B6 and related gene expressions was associated with drought and salt stress in plants

(Bagri et al. 2018; Zhou et al. 2023). There were also reports of vitamin B synthesis and gene expressions in plants exposed to cold, heat stress, etc. (Xiang et al. 2020; Zhou et al. 2023). Therefore, we may think that if the application of antioxidants or B vitamins yields successful and promising results, we could improve the conditions of plants and maintain, or even enhance, their tolerance or resistance mechanisms. Numerous studies have demonstrated that the exogenous application of vitamin B1 and vitamin B2 can improve the disease resistance of *Arabidopsis*, tobacco, tomato, and rice (Zhou et al. 2015, 2016). Therefore, changes in B vitamin levels reflect the responses of plants to diseases, abiotic stressors, and even their combinations. Hence, precise measurement of B vitamin levels in human and plant cells is very crucial to determine the tolerance or resistance level of the organisms. The use of high-performance liquid chromatography (HPLC) and high-performance liquid chromatography-tandem mass spectrometry (HPLC-MS) has enabled us to precisely determine the levels of B vitamins.

In recent years, many crucial data and scientific approaches have been presented regarding biosynthetic genes, transporters, transcription factors, metabolomics, and metagenomics of B vitamin metabolism (Jiang et al. 2021; Zhou et al. 2023). For example, the molecular pathway of pyrimidine synthesis connected to the vitamin B1 biosynthesis of *Arabidopsis* (Kong et al. 2008; Zhou et al. 2023), thiamine biosynthesis (Noordally et al. 2020; Zhou et al. 2023), a gene, *AtPyrR*, encoding a protein involved in vitamin B2 (riboflavin) and flavin adenine dinucleotide (FAD) biosynthesis in *Arabidopsis* (Hasnain et al. 2013; Zhou et al. 2023) has been revealed. Many molecular and biochemical pathways are nearing completion soon. Altogether, research on the synthesis, regulation, and transport of B vitamins would enable us to enhance plant metabolism, as well as contribute to the health of animals and humans.

Recently, significant progress in metabolomics, particularly in the field of B vitamins, has been made. However, most studies have focused on the identification of B vitamin-related enzymes and transporters (Jiang et al. 2021; Zhou et al. 2023); there is little information about transcription factors (TFs) in B vitamin metabolism. It is often referred to as a growth and development vitamin that controls the defensive antioxidant system in plants (Ashoori and Saedisomeolia 2014; Nizar et al. 2023).

Riboflavin is widely known as a photosensitizer, playing a significant role in electron transfer processes during photosynthesis. Therefore, it is a vital component of photosynthesis along with the role of energy metabolism (Hasanuzzaman et al. 2020; Nizar et al. 2023). During light-dependent reactions, it facilitates the transfer of electrons to produce ATP and NADPH, which are crucial energy carriers for the dark reactions of photosynthesis. It is also a vital component of carotenoids, responsible for capturing light energy during stress.

Several studies have demonstrated that riboflavin enhances antioxidant accumulation while reducing oxidant levels in plants (Nizar et al. 2023). For example, Deng et al. (2014) and Nizar et al. (2023) reported that low and moderate levels of riboflavin application conferred drought tolerance on the plants. It has been shown

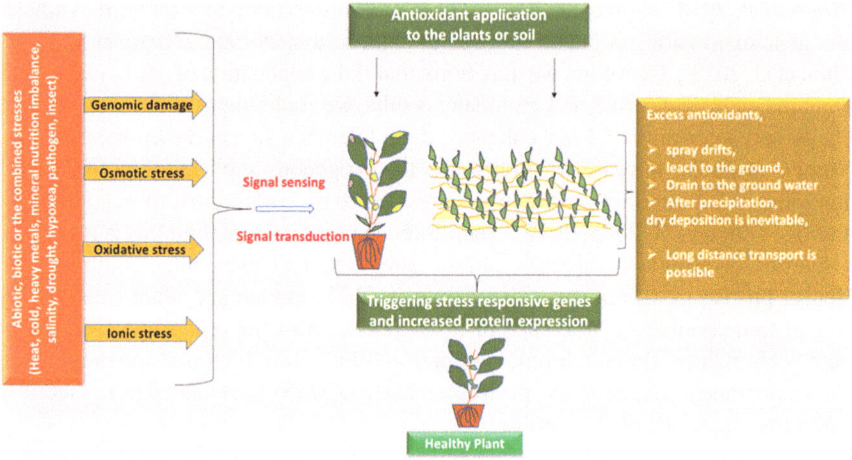

Fig. 2.2 Antioxidant application under stress conditions

that riboflavin regulates the expression of various genes involved in plant growth, development, and signal transduction pathways that trigger molecular defense mechanisms.

Although we have mentioned the positive aspects of antioxidant applications and mechanisms in various contexts, we must also acknowledge the potential adverse effects of antioxidants. When antioxidants are applied to the soil, there is always a risk that the long half-life of the antioxidants may pose a potential danger to the future of agriculture (Fig. 2.2). Due to the application of antioxidants, contamination of the air, soil, and water can potentially pollute the environment. Applications by spraying may drift away from the application point or target plants due to wind, irrigation water, underground water movement, leaching, or chemical application devices. They may bind to soil particles and accumulate there, and their concentrations can remain high over a long period of time if they are not decomposed or converted into inactive chemicals. Contamination of antioxidants primarily occurs near the field of application; however, long-distance contamination should not be overlooked. Although there are no significant reports of contamination or adverse side effects associated with antioxidant applications, we should remain vigilant and prepare ourselves for any potential unwanted consequences.

The exogenous application of antioxidants may pose potential pollution risks to both aquatic and terrestrial ecosystems. Uncontrolled use of antioxidants could contaminate soil, air, and water. Applications that are sprayed in windy weather may cause the compound to drift away from the target plants or areas. Soil contamination may occur when antioxidant compounds bind to the soil particles at high concentrations. In the long run, we may have to tackle the pollution caused by antioxidants.

2.2 Riboflavin Biosynthesis and Transport in Plants and Its Main Functions in Organisms

Riboflavin is biosynthesized de novo in microorganisms and plants. Humans and animals obtain vitamins via dietary supplements. Riboflavin is synthesized in plastids from guanosine triphosphate (GTP) and ribulose-5-phosphate in plant cells (Yang et al. 2021). Although many enzymes have been involved, two key enzymes, riboflavin synthase and lumazine synthase, play significant roles. Followed by phosphorylation, it is converted into flavin mononucleotide (FMN) by riboflavin kinase and further adenylated to flavin adenine dinucleotide (FAD) by FAD synthase. The active forms of riboflavin serve as cofactors in redox reactions, photosynthesis, and respiration (Zhang et al. 2023). It is transported within the plant tissues through a series of specialized riboflavin transporters, which are upregulated under stress conditions (Namba et al. 2023; Hayat et al. 2025).

Riboflavin, or vitamin B2, is an indispensable compound for all living organisms in terms of growth and metabolism. It is responsible for the generation of ROS and NADPH oxidase (for oxygen production, FAD required), and fatty acyl CoA dehydrogenase (for H_2O_2 production, FAD required) (Schramm et al. 2014). Therefore, riboflavin deficiency would seriously affect the FAD-containing enzymes, metabolism and pathways. Over the past decades, our knowledge about plant vitamin biosynthesis has increased significantly with the progress and development of metabolomics studies. Enzymes and regulatory factors involved in vitamin biosynthesis are now firmly established (Fig. 2.3).

The vitamin is biosynthesized from one molecule of Guanozin-5′-trifosfat (GTP) and two molecules of ribulose phosphate, regulated by enzymes. GTP

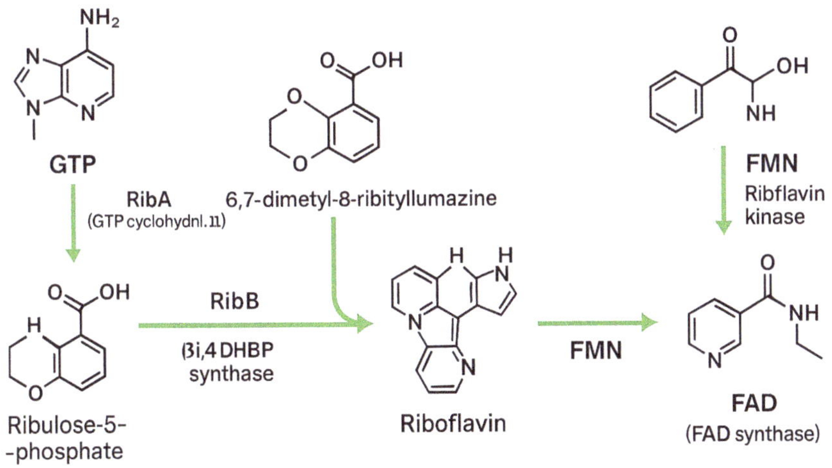

Fig. 2.3 Overview of riboflavin biosynthesis. The illustration for riboflavin biosynthesis is simplified from the source of KEGG Pathway (https://www.kegg.jp/pathway/ko00740)

Cyclohydrolase II converts GTP into a pyrimidine precursor. Lumazine synthase enzyme forms lumazine, an immediate precursor to riboflavin. Riboflavin synthase enzyme joins two lumazine molecules to form riboflavin. GTP and ribulose-5-phosphate are converted into riboflavin in plant cells. Followed by phosphorylation, it is then converted into FMN and FAD. Since B vitamins have significant roles in plant and human metabolisms and their deficiencies have non-repairable negative impacts on organisms, we, therefore, summarized their functions in plants and human metabolisms (Table 2.1).

2.3 Role of Riboflavin in Abiotic/Biotic Stress Tolerance/Resistance

The importance of riboflavin in plants has received little attention; however, as the mechanism of riboflavin is understood in cell metabolism, we can use it for the remediation and improvement of crop plants cultivated under abiotic and biotic stress conditions. Recent studies have shown that riboflavin can be used like other antioxidants for the remediation of crop plants. Exogenous plant antioxidant systems help modulate ROS signaling, mitigate ROS damage, and improve the conditions of plants. In general, antioxidants possess reductants and enzymes that can detoxify ROS. Since antioxidants provide natural stress resistance, the interest in harnessing antioxidant use to increase crop yield and resistance has been growing recently. Most vitamins synthesized by plants are linked to plant nitrogen metabolism. In this section, we summarize the responses of crop plants treated with riboflavin under major stress factors.

2.3.1 Drought Stress

Under drought conditions, the water deficit leads to stomatal closure, reduced CO_2 assimilation, and increased ROS accumulation (Li et al. 2023; Dawood et al. 2025; Fatema et al. 2026; Jameel et al. 2026; Umer et al. 2026). The application of riboflavin has been shown to reduce membrane lipid peroxidation and enhance the activities of antioxidant enzymes, including superoxide dismutase (SOD), catalase (CAT), peroxidase (POD), ascorbate peroxidase (APX), and glutathione reductase (GR), in plants under drought stress. The application of riboflavin has also been reported to improve stomatal regulation and water-use efficiency. Maintaining chlorophyll content and photosynthetic efficiency, as well as the modulation of abscisic acid (ABA) through drought-responsive genes, has also been regulated with the application of riboflavin. Deng et al. (2014) reported that 4-week-old tobacco plants treated with different concentrations of riboflavin (0-, 4-, 20-, 100-, and 500 μM) for 5 days exhibited low and moderate levels of riboflavin, which

Table 2.1 Functions of B vitamins in human and plant metabolism

Vitamins	Biological functions in humans	Deficiency symptoms in humans	Biological functions in plants	Deficiency symptoms in plants	Etiology of vitamin deficiency in humans/plants	References
Provitamin A and other carotenoids	Involved in antioxidant mechanisms	Night blindness, oxidative stress	Precursor of retinol, involved in photoprotection and stress signaling	Reduced chlorophyll, impaired photoprotection	Poor dietary intake, oxidative stress, and environmental factors	Jiang et al. (2021)
Vitamin B1 (Thiamine)	Cofactor in carbohydrate metabolism; nervous system regulation	Beriberi, Wernicke's encephalopathy, neurodegeneration	Involved in carbon metabolism and energy production	Growth retardation, chlorosis	Malnutrition, heavy metal stress, impaired uptake	Singh (2023)
Vitamin B2 (Riboflavin)	Involved in the antioxidant mechanism, and antioxidant defense	Neurodegeneration, dermatitis	Involved in the glutathione cycle, cofactor in redox enzymes	Reduced growth, oxidative stress	UV exposure, soil nutrient depletion	Singh (2023)
Vitamin B3 (Niacin)	Precursor of NAD/NADP; essential for metabolism	Pellagra: dermatitis, diarrhea, dementia	Required for NAD $(P)^+$ biosynthesis	Reduced photosynthesis, oxidative stress	Deficient tryptophan metabolism, poor nutrition	Wang et al. (2022)
Vitamin B5 (Pantothenic acid)	Essential for Coenzyme A synthesis	Fatigue, neurodegeneration	CoA-dependent enzyme function	Chlorosis, stunted growth	Stress-induced degradation, soil deficiency	Mignani et al. (2021)
Vitamin B6 (Pyridoxine)	Cofactor in amino acid metabolism and neurotransmitter synthesis	Neuropathy, cardiovascular disorders	Amino acid metabolism; stress defense	Delayed development, oxidative damage	Inhibited biosynthesis during stress	Chen et al. (2021)
Vitamin B7 (Biotin)	Cofactor in carboxylation reactions	Dermatitis, conjunctivitis	Supports fatty acid synthesis, CO_2 fixation	Abnormal growth, leaf necrosis	Low microbial availability, nutrient-poor soils	Jiang et al. (2021)
Vitamin B9 (Folate)	DNA synthesis, methylation, and cell division	Anemia, neural tube defects	Cell division and DNA synthesis	Impaired meristem growth, foliar damage	UV radiation, enzyme mutations	Chen et al. (2021)

(continued)

Table 2.1 (continued)

Vitamins	Biological functions in humans	Deficiency symptoms in humans	Biological functions in plants	Deficiency symptoms in plants	Etiology of vitamin deficiency in humans/plants	References
Vitamin C (Ascorbic acid)	Antioxidant, reactive oxygen species (ROS) scavenger, immune enhancer	Scurvy, immune suppression	ROS scavenger, hormone biosynthesis	Leaf curling, oxidative lesions	Oxidative load, limited recycling	Jiang et al. (2021)
Vitamin E (Tocopherol)	Lipid-soluble antioxidant protects membranes	Oxidative stress, cardiovascular issues	Protects chloroplast membranes from ROS	Photoinhibition, reduced seed viability	Stress-induced peroxidation, low uptake	Rahman et al. (2022)

Stress reacts by affecting vitamin accumulation. Understanding the mechanism will significantly benefit the biofortification of organisms with vitamins. This table was prepared with the help of the following references: Chen et al. (2021), Jiang et al. (2021), Mignani et al. (2021), Rahman et al. (2022), Wang et al. (2022), and Singh (2023) enhanced their drought tolerance. In contrast, the highest concentration (500 μM) of riboflavin impaired metabolism and reduced drought tolerance. The low and moderate doses of riboflavin induced ROS synthesis and triggered defense mechanisms via increasing signaling pathways. The highest dose of riboflavin, on the other hand, resulted in the accumulation of ROS, and malondialdehyde (MDA), along with an increase in antioxidant enzymes. Castro-Cegrí et al. (2024) also reported that application of riboflavin protected and extended postharvest shelf life of zucchini fruits (*Cucurbita pepo* L.) by increasing antioxidant levels. They reported that riboflavin enabled metabolic reprogramming with increasing terpenoids and phenolic compounds during cold storage and regulated the biosynthesis of alkanes, phytohormones, and vitamins.

In general, treatment with ROS-producing chemicals at low doses can always induce tolerance or resistance mechanisms in plants. For example, treating plants with low doses of toxic substances, such as paraquat—a type of herbicide, induced tolerance in plants. The principle lies at the heart of the cell's immunity mechanism and the activation of the defense response. As Nietzsche said (a German philosopher), "was mich nicht umbringt macht mich stärker," that is, "what doesn't kill me makes me stronger," which was scientifically proven. Deng et al. (2014) stated that low levels of riboflavin activated the antioxidant defense mechanism (priming effect) through reduced levels of ROS production. However, high levels of riboflavin can generate excessive levels of ROS, which can impair the antioxidant defense system. As a result of that, toxic substances can accumulate and impair the whole metabolism.

In general, as a defense principle, abiotic stress tolerance is preferred over biotic stress resistance in plants (Fujita et al. 2006; Deng et al. 2014). This may be due to the higher expression of signals in response to drought stress compared to the sensing of stress signals by the infecting pathogens. The same case can be observed with those of abiotic stress elicitors when compared to those of biotic stress elicitors. For example, Fujita et al. (2006) and Deng et al. (2014) stated that high doses of riboflavin may increase the synthesis of ABA, which is involved in stomatal functions, primarily playing a role in stomatal closure. This may prevent photosynthesis, thereby affecting antioxidant metabolism, which could result in an increased accumulation of ROS. A high accumulation of ABA can also suppress pathogen resistance signaling pathways (ethylene synthesis) (Fujita et al. 2006; Deng et al. 2014). This may be due to the synthesis of proteins associated with both abiotic and biotic pathways, in which only one pathway is preferred to conserve energy and combat stress mechanisms. Therefore, high concentrations of riboflavin should be avoided in protecting plants from pathogens under drought or other abiotic stress conditions, as this may lead to more devastating consequences. Similarly, Mostofa et al. (2018) stated that methylglyoxal (MG) could induce signaling cascades through Ca^{2+}, ROS, K^+, and ABA; however, an overlap between MG-responsive and stress-responsive signaling events might exist in plants at the same time. The authors stated that once MG was accumulated in cells, it provoked detrimental effects by generating ROS and inactivating antioxidant systems. Although plants are equipped with detoxifying systems to protect crucial cellular organelles from

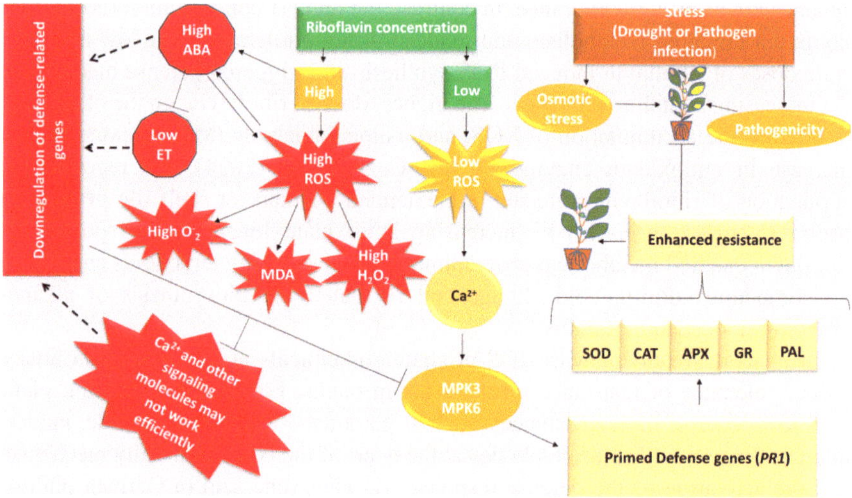

Fig. 2.4 Riboflavin-induced activation of the plant defense mechanism. "-----" dotted line represents suppressed or downregulated genes. "T" bar or perpendicular line represents an inhibited step

high doses of antioxidants, overlapping stress mechanisms with eliciting response mechanisms could be detrimental to the cell. The mechanism is briefly illustrated in Fig. 2.4. Application of riboflavin should be evaluated at different stress conditions, such as single or combined stress factor conditions (abiotic and biotic), which could come out simultaneously or consecutively. Combined stresses are commonly observed in many areas of the world; therefore, elucidation of B vitamins under these circumstances is crucially important and revealing the role of riboflavin under these conditions would enable us to find the right pathways responsible for both abiotic stress tolerance and biotic stress resistance mechanisms.

Riboflavin triggers defense systems through ROS and Ca^{2+} dependent signaling pathways that activate defense-related enzymes. When riboflavin is applied at low concentrations, it leads to low accumulation of ROS, which act as primers and activate the defense system. When riboflavin is applied at high concentrations, it may lead to a high accumulation of ROS, MDA, and ABA, which can cause toxicity in cells and impair plant defense systems. Signaling molecules may not sense the signal due to a high level of stress, and defense mechanisms may be impaired. Ethylene (ET) synthesis, which responds to pathogenicity, may also be interrupted. The application of riboflavin requires further elucidation under the combined stress of abiotic and biotic stressors, which have become increasingly prevalent in recent decades. Application time and concentration may help reveal the mode of action of riboflavin under complex stress conditions. ROS produced upon stress and the ROS produced upon application of riboflavin should be distinguished and evaluated to reveal the mechanism. The pathway was illustrated with the help of references previously studied (Deng et al. 2014; Nie and Xu 2016; Czarnocka and Karpiński 2018; Zou et al. 2025). The production and scavenging of ROS also depend on the type of stimulus. ROS can be generated and eliminated in various cellular compartments, including the apoplast, plasma membrane, chloroplasts, mitochondria, and endoplasmic reticulum. Different ROS have various half-life times, and they can affect different components such as lipids, proteins, and nucleic acids.

Recently, Abdelhalim et al. (2023) made a foliar application of 6 mM ascorbic acid or 2 mM vitamin B2 to wheat plants to reduce evapotranspiration and increase crop yield. The foliar applications of both compounds resulted in a 20% reduction in irrigation water with no significant yield reductions observed. Similarly, Ali et al. (2024) reported that vitamin B6 and riboflavin played a crucial role in drought tolerance, and their priming application significantly enhanced stress tolerance during rapeseed germination.

2.3.2 *Salinity Stress*

Salinity disrupts ionic and nutrient balances, osmotic pressure, and results in toxicity in plant cells. It triggers the accumulation of ROS, leading to impaired plant growth (Azooz et al. 2026). Riboflavin has been shown to improve salinity stress by enhancing ion homeostasis, followed by modulation of Na^+/K^+ transporters (Jiadkong and Ueda 2024) and improving membrane integrity. It also contributes to the activation of mitogen-activated protein kinase (MAPK) signaling pathways, proline and glycine betaine synthesis (Jagodzik et al. 2018). The involvement of riboflavin biosynthesis in response to salinity stress has been largely unknown; however, recent studies have demonstrated that riboflavin can be utilized to enhance the conditions of crop plants. For example, Jiadkong et al. (2024) showed that higher biomass accumulation and lower concentrations of MDA and H_2O_2 in riboflavin-pretreated rice seedlings under salinity stress were achieved. The authors achieved the scavenging of H_2O_2 with increasing riboflavin concentration. They stated that riboflavin functioned as a non-enzymatic antioxidant in ROS detoxification.

Riboflavin pretreated seedlings accumulated more Na^+ in the roots than in the leaf blades due to upregulation of *OsHKT2;1*, *OsNHX1*, and *OsHKT1;4* genes in the roots and leaf sheaths, respectively, and the higher Na^+ accumulation in the roots limited translocation of Na^+ into leaf blades and leaves, thus maintaining a lower Na^+/K^+ balance. Some workers partly revealed the mechanism of riboflavin. Rodríguez-Celma et al. (2011) and Jiadkong et al. (2024) stated that *Medicago truncatula* roots accumulated riboflavin under iron deficiency and it was translocated to the shoots. When accumulated there, it acted as a non-enzymatic antioxidant that minimized ROS overaccumulation and MDA under stress. Jiadkong et al. (2024) also revealed that 54 genes were upregulated, and 9 genes were downregulated via transcriptome analysis by RNA-seq in the roots of riboflavin-pretreated rice seedlings. Exogenous riboflavin application was made indirectly to mitigate the stress of salinity on the growth and metabolic status of *Chrorella sorokiniana* when co-cultured with riboflavin-producing bacterium *Azospirillum brasilense* (Palacios et al. 2021). The microalgae and *A. brasilense* co-cultures at various salinity levels

(0–3% NaCl) exhibited growth promotion and it was found to be more cost-effective. The authors finally suggested that this method of application could be used where irrigation water contains high salinity. Due to the continuous synthesis of riboflavin by *A. brasilense* in soils, growers may not need to make frequent applications to the soil under irrigation with saline water.

2.3.3 Temperature Stress

As is commonly known, high or low temperature stress causes protein denaturation and ROS accumulation (Zhou et al. 2025; Ahmad et al. 2026). With the increase in CO_2 concentrations in the air, plant metabolism appears to be deteriorating as temperatures escalate to remarkably higher levels. In recent years, quite a few reports showed that metabolite and gene expression analyses from several species were downregulated regarding vitamin synthesis under high temperatures (Dong et al. 2018; Loladze et al. 2019; Jiang et al. 2021). High temperature stress does not only come along with drought stress, but also high light intensity accompanies it. Therefore, we should be aware of the sterility caused by light. Several reports have suggested that riboflavin stabilizes the redox status, protects photosystem II efficiency, and enhances heat shock proteins (HSPs), including HSP70 and HSP90, as well as their related genes (Du et al. 2021).

2.3.4 Heavy Metal Stress

Exposure to heavy metals such as mercury (Hg), lead (Pb), arsenic (As), and cadmium (Cd) can cause significant oxidative stress in plants. This is due to the production of ROS, which can result in protein oxidation, DNA damage, and lipid peroxidation (Hasanuzzaman et al. 2020; Nizar et al. 2023; Abdel Latef 2024). Riboflavin plays a key role in mitigating heavy metal toxicity by acting as a cofactor and redox mediator in antioxidant enzyme systems.

The enhancement of the antioxidant defense system by riboflavin involves the upregulation of enzymes such as CAT, SOD, GR, and POD, which detoxify ROS and protect cellular organelles from oxidative damage (Ashoori and Saedisomeolia 2014; Rodrigues de Queiroz et al. 2023). Nizar et al. (2023) reported that the administration of vitamin B2 alleviated cadmium-induced oxidative stress in *Silybum marianum* by increasing the content of vitamins and essential nutrients, thereby improving the accumulation of silymarin under Cd stress. Similarly, riboflavin supplementation reduces Cd absorption by enhancing cell wall binding and chelation through organic acids and thiol-containing peptides, such as phytochelatins and glutathione (Seregin and Kozhevnikova 2023). Riboflavin can affect metal transporters and genes related to detoxification. Studies have demonstrated that riboflavin initiates signaling via ROS- and Ca^{2+}-dependent pathways, activating stress–responsive transcription factors such as DREB, WRKY, and NAC family members. This regulates gene expression related to heavy metal detoxification (Jagodzik et al. 2018; Namba et al. 2023). In *Arabidopsis thaliana*, riboflavin has been shown to maintain iron homeostasis under iron deficiency conditions by promoting flavin accumulation and transporting it from roots to shoots. This suggests a potential role in micronutrient regulation under metal stress (Rodríguez-Celma et al. 2011). Furthermore, combining riboflavin with other vitamins (e.g., thiamine or pyridoxine) has been shown to enhance the detoxification capacity of plants synergistically under heavy metal contamination (Ahmed and Sattar 2024). This combination increased chlorophyll retention, improved antioxidant potential, and stabilized membrane integrity.

In conclusion, riboflavin alleviates heavy metal stress by regulating antioxidant ion transporters, enhancing enzyme activity, stimulating chelation mechanisms, and activating stress-related signaling networks. Therefore, exogenously applying riboflavin may be an effective strategy for increasing plant tolerance to heavy metal contamination in soils, particularly under combined abiotic stress conditions (Nizar et al. 2023; Hasanuzzaman et al. 2020; Zhang et al. 2023).

2.3.5 Pathogen Resistance

Riboflavin functions as an abiotic elicitor that induces ISR or SAR, protecting plants against a wide range of abiotic and biotic factors such as pathogens (Kamle et al. 2020). It activates defense-related genes, enhances the accumulation of salicylic acid (SA) and H_2O_2, and stimulates callose deposition and lignification at infection sites through the phenylpropanoid pathway. Nie and Xu (2016) stated that thiamine functioned as a plant activator and conferred SAR through priming, which required H_2O_2 and signaling proteins. They also stated that treatments with riboflavin protected tobacco and *Arabidopsis* from fungal and bacterial infections without inhibiting the growth of pathogens. Riboflavin, like thiamine, initiates signal transduction. There are also reports of inhibitory effects against various fungal pathogens, such as *Fusarium* spp. and *Botrytis* spp., as well as viruses, via the inhibition of mycelial growth and sporulation through pathogenesis-related (PR) proteins and metabolites. Pyridoxine, folic acid, riboflavin, niacin, and D-biotin were used to treat pearl millet seeds to induce resistance against downy mildew disease caused by *Sclerospora graminicola* (Pushpalatha et al. 2007). The applied vitamins significantly protected the plant from the disease agent and increased the yield performance compared to that of the control. Again, Dong and Beer (2000) and Pushpalatha et al. (2007) reported that riboflavin was used to protect *Arabidopsis thaliana* from *Peronospora parasitica* by foliar spray. The authors stated that foliar spray of riboflavin triggered signal transduction and led to increases in peroxidation, antioxidant defense enzymes, and upregulated pathogen-related (PR) genes involved in PR protein production. It was also stated that the application of vitamins, even at higher

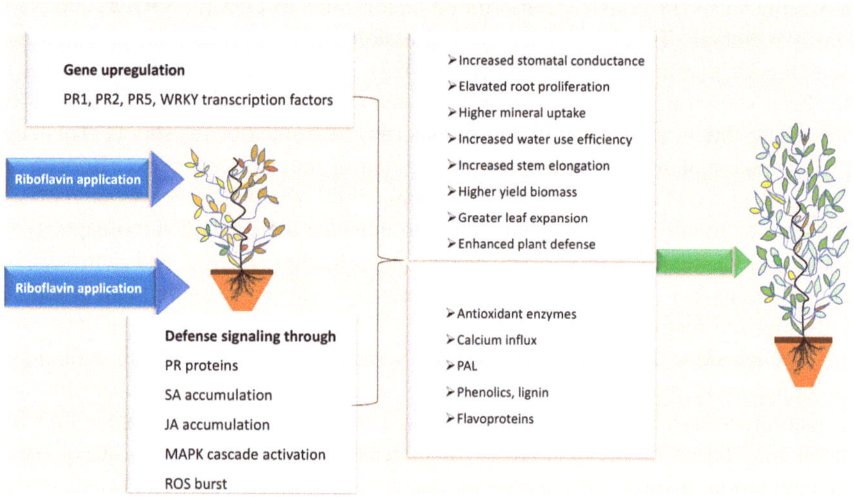

Fig. 2.5 Riboflavin mechanisms in plants

doses, was not phytotoxic to plants. Enhancing the efficacy of seed priming through vitamin-based formulations may help increase SAR and ISR in pearl millet seeds, ultimately improving yield and growth. Nie and Xu (2016) stated that riboflavin induced the transient expression of the *PR1* gene; however, following infection with *Pseudomonas syringae* pv. DC3000 inoculation resulted in stronger expression of the *PR1* gene transcription. The transcription levels of mitogen-activated protein kinase enzymes, MPK3 and MPK6, were primed following riboflavin treatment. They stated that these enzymes were responsible for enhanced defense gene expression (Fig. 2.5). Exogenous riboflavin activates SA pathways, enhances H_2O_2 accumulation, and promotes expression of PR genes, leading to SAR. Deposition of callose, lignin, and phenolic substances is evident to give strength to the cell wall, followed by the application of riboflavin. When combined with beneficial rhizobacteria, riboflavin enhances the ISR pathway by interacting with jasmonic acid (JA) and ethylene (ET) and promotes defense-related gene expression without triggering hypersensitive responses. Taheri and Tarighi (2010) demonstrated that the application of riboflavin primed the expression of the lipoxygenase (*LOX*) gene in the octadecanoid pathway and enhanced lignification in rice plants infected with *Rhizoctonia solani*.

The beneficial effects of riboflavin were also tested on crop quality. Wu et al. (2024) applied riboflavin to two different genotypes of fragrant rice at different (booting and flowering) growth stages at a 20 ppm concentration. The results revealed that the yield increase was significantly higher in the treated groups when compared to the non-treated control group. Spraying riboflavin at the flowering stage had a greater impact on yield and quality parameters. Riboflavin application significantly increased chlorophyll content, resulting in increased total starch, soluble sugars, and sucrose accumulation in leaves, which was subsequently transported to the seeds. The level of sucrose synthase and sucrose phosphate synthase also increased. This led to an increase in the aromatic substances in rice grains. As a common physiological mechanism, the increased carbohydrate content in leaves facilitates the transport of soluble sugars from leaves to grains, resulting in a reduction in soluble sugar content and an increase in starch content. This phenomenon increases the aromatic substances in cells.

Riboflavin has also been reported to enhance the immune system in humans, particularly in response to pathogen infections caused by *Staphylococcus aureus* and *Listeria monocytogenes* (Munteanu and Schwartz 2023).

2.4 Molecular Mechanisms and Signaling

Riboflavin regulates ROS signaling. It acts as a signaling molecule when the concentration is low; however, when applied at high concentrations, it leads to the accumulation of ROS. It influences NADPH oxidase activity and modulates gene expression via ROS-responsive transcription factors. It also regulates hormonal interactions via enhancing SA signaling involved in the SAR pathway. It activates JA hormones during the ISR pathway and regulates ABA signaling under water stress or drought conditions. There are also reports suggesting that riboflavin may influence chromatin remodeling and DNA methylation (Abdul et al. 2017). The riboflavin mechanisms and signaling network were presented in Figs. 2.5 and 2.6.

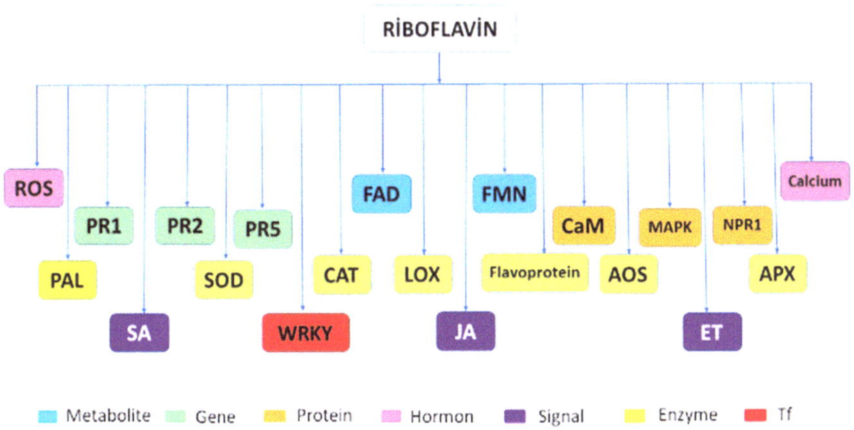

Fig. 2.6 Riboflavin signaling network in plants. Riboflavin triggers various pathways in plant cells. (This figure was prepared with the help of multiple references Song and Wang (2003), Zhang et al. (2009), Mao et al. (2023), and Ruchala et al. (2025))

2.5 Application of Riboflavin in Crop Management

Several studies have demonstrated that riboflavin application can enhance plant resistance and tolerance to biotic and abiotic stress factors. Riboflavin can be applied through foliar spray to seeds to improve germination rates under water and salinity stress, enhance root development, and improve seed quality parameters. It also reduces disease incidence and yield loss.

Riboflavin can be integrated with other agents such as seaweed extracts, humic substances, and microbial inoculants. It is possible to encapsulate riboflavin for controlled release and enhanced uptake.

Riboflavin has been used to remediate the condition of plants under climate change safely due to its non-toxic properties. Several studies have shown that the growth and development of wheat, rice, tomato, and soybean increased with riboflavin applications, especially under low-input farming systems. Vitamin applications on plants, including vitamin B2, are summarized in Table 2.2.

Table 2.2 Application of B vitamins to plants

B vitamin application	Plant species	Stress type	Mode of action	References
Foliar thiamine (B1) application	*Vicia faba* (faba bean)	Salinity stress	Enhances photosynthetic pigments, soluble sugars/proteins, antioxidant capacity	Ahmed and Sattar (2024)
Foliar thiamine (B1) + IAA	*Zea mays* (maize)	Arsenic stress	Modulates physiological and biochemical traits	Atif et al. (2022)
Foliar riboflavin (B2) application	*Oryza sativa* (rice)	Salinity stress	Enhances salinity tolerance via biomass increase, ion homeostasis, ROS detox	Jiadkong et al. (2024)
Foliar/exogenous riboflavin	Tobacco (*Nicotiana* spp.)	Drought stress	Boosts drought tolerance through ROS modulation	Deng et al. (2014)
Exogenous pyridoxine (B6)	*Triticum aestivum* (wheat)	Salinity stress	Enhance growth and antioxidant enzyme activity	Wang et al. (2021)
Foliar thiamine/B6 combination	*Vicia faba* (faba bean)	Salinity stress	Improves ion uptake, pigments, sugars, proteins, antioxidant capacity	Ahmed and Sattar (2024)
Seed priming with riboflavin	*Lactuca sativa* (lettuce)	Saline irrigation	Enhances plant health under saline irrigation	Rico-García et al. (2023)
Exogenous vitamin B12 (cobalamin)	*Phaseolus vulgaris* (common bean)	Salinity stress	Increases chlorophyll, CAT, POD activities under NaCl	Keshavarz and Moghadam (2017)

2.6 Conclusion and Future Prospects

Global crop production is expected to face severe challenges in the future due to the combined effects of abiotic and biotic stress factors. This may result in more than 50% crop loss. Therefore, agriculture is a crucial sector that requires development in light of recent findings and emerging approaches. Promoting plant development is one of the most effective means of combating stress. This depends on multiple factors, including mineral nutrition, irrigation, and environmental conditions. The demand for food is on a constant rise due to the increase in global population, along with principal abiotic stresses such as drought and high temperature. To meet this escalating demand, agricultural production must increase by 60 to 70% by 2050 (Chieb and Gachomo 2023; Hossain et al. 2023; Bouremani et al. 2024). Riboflavin is not only a vitamin but also a potent inducer of plant immunity for stress tolerance/ resistance. Therefore, it represents a promising natural molecule for a wide range of stresses. Future work should focus on signaling networks, metabolomics, and genomics involving riboflavin to elucidate the detailed characteristics of this compound. A combination of riboflavin with other nanotechnology products would increase the potential use and capacity of riboflavin. Genetic modifications or enhancements should involve the study of riboflavin followed by the release of a detailed mechanism. Elucidating riboflavin pathways for enhanced stress tolerance and adaptation in a large-scale framework for crops, as well as the preparation of riboflavin-based formulas, would make a significant contribution to this sector.

Agronomic technologies such as molecular breeding, metabolic engineering, and biochemical fortification have both advantages and disadvantages. It is essential to recognize that micronutrient and vitamin deficiencies cannot be separated. Therefore, our approach should encompass a comprehensive view of all metabolic pathways in combination.

Since riboflavin is naturally synthesized by plants and most microorganisms (*Bacillus subtilis*, *Ashbya gossypii*, *Candida famata*) (Ruchala et al. 2025), we can grow them for industrial applications and apply CRISPR/Cas9 genome editing in these microorganisms to increase riboflavin synthesis. To date, several genes have been identified as playing a role in the production of riboflavin.

Since human beings and animals totally depend on plants for B vitamin uptake, fortification of plants with riboflavin contents under stress conditions would be a sustainable approach. Otherwise, increasing crop yield would be successful without considering the quality parameters for sustainable agriculture and population health. There is a great concern about ageing populations that show symptoms of vitamin deficiencies; therefore, biofortification of the organisms with vitamins would ease metabolic disorders. Future farming should concentrate on functional food and micronutrients for the next generations. Therefore, pathways regarding vitamins should be revealed under normal and stressful conditions.

References

Abdelhalim AK, Sallam AA, Waleed F, Merah O, Abdelhamid MT (2023) Riboflavin or ascorbic acid enhances wheat drought tolerance by improving both yield and water productivity. https://doi.org/10.20944/preprints202312.1944.v1

Abdel Latef AAH (2024). Biochar in Mitigating Abiotic Stress in Plants. Academic Press. Elsevier Inc. pp. 1–375. https://doi.org/10.1016/C2023-0-01389-9

Abdul QA, Yu BP, Chung HY, Jung HA, Choi JS (2017) Epigenetic modifications of gene expression by lifestyle and environment. Arch Pharm Res 40(11):1219–1237. https://doi.org/10.1007/s12272-017-0973-3

Ahammed GJ, Li Z, Chen J, Dong Y, Qu K, Guo T et al (2024) Reactive oxygen species signaling in melatonin-mediated plant stress response. Plant Physiol Biochem 207:108398. https://doi.org/10.1016/j.plaphy.2024.108398

Ahmad N, Ahmad S, Islam MS, Khan MN, Abdel Latef AAH (2026) Chitosan nanoparticles under temperature extremes. In: Abdel Latef AAH (eds) Chitosan nanoparticles for plant resilience under abiotic and biotic stressors. Smart Nanomaterials Technology. Springer, Singapore. https://doi.org/10.1007/978-981-95-6422-4_6

Ahmed EZ, Sattar AMAE (2024) Improvement of Vicia faba plant tolerance under salinity stress by the application of thiamine and pyridoxine vitamins. Sci Rep 14(1):22367. https://doi.org/10.1038/s41598-024-72511-y

Ali M, Zhao X, Hussain S, Li X, Li F, Xinhua Z (2024) Physiological and transcriptomic analysis revealed the quality regulatory mechanism of pyridoxine (vitamin B6) in fresh-cut apple. Postharvest Biol Technol 217:113098. https://doi.org/10.1016/j.postharvbio.2024.113098

Ashoori M, Saedisomeolia A (2014) Riboflavin (vitamin B2) and oxidative stress: a review. Br J Nutr 111(11):1985–1991. https://doi.org/10.1017/S0007114514000178

Atif M, Perveen S, Parveen A, Mahmood S, Saeed M, Zafar S (2022) Thiamine and indole-3-acetic acid induced modulations in physiological and biochemical characteristics of maize (Zea mays L.) under arsenic stress. Sustainability 14(20):13288. https://doi.org/10.3390/su142013288

Azooz MM, Abdel Latef AAH, Abdelmajeed WG et al. (2026) Biochar ameliorates salt stress in maize by modulating physiological responses and expression of SOS1 and NHX2 genes. Journal of Soil Science and Plant Nutrition. https://doi.org/10.1007/s42729-026-03123-w

Bagri DS, Upadhyaya DC, Kumar A, Upadhyaya CP (2018) Overexpression of PDX-II gene in potato (*Solanum tuberosum* L.) leads to the enhanced accumulation of vitamin B6 in tuber tissues and tolerance to abiotic stresses. Plant Sci 272:267–275. https://doi.org/10.1016/j.plantsci.2018.04.024

Bouremani N, Cherif-Silini H, Silini A, Rabhi NEH, Bouket AC, Belbahri L (2024) Osmotolerant plant growth promoting bacteria mitigate adverse effects of drought stress on wheat growth. AIMS Microbiol 10(3):507. https://doi.org/10.3934/microbiol.2024025

Castro-Cegrí A, García A, Garrido D, Palma F (2024) Riboflavin improves postharvest cold tolerance in zucchini fruit inducing non-enzymatic antioxidant response and phenolic metabolism. Plant Physiol Biochem 217:109270. https://doi.org/10.1016/j.plaphy.2024.109270

Chen O, Mah E, Dioum E, Marwaha A, Shanmugam S, Malleshi N et al (2021) The role of oat nutrients in the immune system: a narrative review. Nutrients 13(4):1048. https://doi.org/10.3390/nu13041048

Chieb M, Gachomo EW (2023) The role of plant growth promoting rhizobacteria in plant drought stress responses. BMC Plant Biol 23(1):407. https://doi.org/10.1186/s12870-023-04403-8

Czarnocka W, Karpiński S (2018) Friend or foe? Reactive oxygen species production, scavenging and signaling in plant response to environmental stresses. Free Radic Biol Med 122:4–20. https://doi.org/10.1016/j.freeradbiomed.2018.01.011

Dawood MFA, Tahjib-Ul-Arif MD, Shirazy BJ, Abdel Latef AAH (2025) Unraveling the role of κ-carrageenan on the combined effect of drought and chromium stress in wheat (Triticum aestivium L.). Plant Physiology and Biochemistry 223:J109781

Deng B, Jin X, Yang Y, Lin Z, Zhang Y (2014) The regulatory role of riboflavin in the drought tolerance of tobacco plants depends on ROS production. Plant Growth Regul 72(3):269–277. https://doi.org/10.1007/s10725-013-9858-8

Dikilitas M, Simsek E, Karakas S, Abdel Latef AAH (2021) Abiotic stresses and their interactions with each other on plant growth, development and defense mechanisms. In: Abdel Latef AA (ed.) Organic Solutes, Oxidative Stress, and Antioxidant Enzymes under Abiotic Stressors. CRC Press/Taylor & Francis Group. https://doi.org/10.1201/9781003022879-1

Dong H, Beer SV (2000) Riboflavin induces disease resistance in plants by activating a novel signal transduction pathway. Phytopathology 90(8):801–811. https://doi.org/10.1094/PHYTO.2000.90.8.801

Dong J, Gruda N, Lam SK, Li X, Duan Z (2018) Effects of elevated CO2 on nutritional quality of vegetables: a review. Front Plant Sci 9:924. https://doi.org/10.3389/fpls.2018.00924

Du W, Ruan C, Li J, Li H, Ding J, Zhao S, Jiang X (2021) Quantitative proteomic analysis of *Xanthoceras sorbifolium* Bunge seedlings in response to drought and heat stress. Plant Physiol Biochem 160:8–17. https://doi.org/10.1016/j.plaphy.2021.01.002

Fatima A, Akram NA, Talha M, Younis M, Ashraf M, Abdel Latef AAH (2026) Comparative Effects of Externally Applied Glycinebetaine and Neem (Azadirachta indica) Leaf Extract on Plant Growth, Physio-Biochemical Characteristics and Yield Production of Canola (Brassica napus L.) Subjected to Drought Stress. Russ J Plant Physiol 73:18

Fitzpatrick TB, Basset GJ, Borel P, Carrari F, DellaPenna D, Fraser PD et al (2012) Vitamin deficiencies in humans: can plant science help? Plant Cell 24(2):395–414. https://doi.org/10.1105/tpc.111.093120

Fujita M, Fujita Y, Noutoshi Y, Takahashi F, Narusaka Y, Yamaguchi-Shinozaki K, Shinozaki K (2006) Crosstalk between abiotic and biotic stress responses: a current view from the points of convergence in the stress signaling networks. Curr Opin Plant Biol 9(4):436–442. https://doi.org/10.1016/j.pbi.2006.05.014

Hanna M, Jaqua E, Nguyen V, Clay JB (2022) Vitamins: functions and uses in medicine. Perm J 26(2):89–97. https://doi.org/10.7812/TPP/21.204

Hasanuzzaman M, Bhuyan MB, Parvin K, Bhuiyan TF, Anee TI, Nahar K et al (2020) Regulation of ROS metabolism in plants under environmental stress: a review of recent experimental evidence. Int J Mol Sci 21(22):8695. https://doi.org/10.3390/ijms21228695

Hasnain G, Frelin O, Roje S, Ellens KW, Ali K, Guan JC et al (2013) Identification and characterization of the missing pyrimidine reductase in the plant riboflavin biosynthesis pathway. Plant Physiol 161(1):48–56. https://doi.org/10.1104/pp.112.208488

Hayat M, Wang Z, Liu X, Gul Z, Bai Q, Ali S (2025) Transcriptomic and metabolomic analysis of fludioxonil-induced stress response and resistance in the poplar leaf blight fungus (*Alternaria alternata*). J For Res 36(1):60. https://doi.org/10.1007/s11676-025-01854-7

Hossain A, Hassan Z, Sohag MH, Khan M (2023) Impact of the endophytic and rhizospheric bacteria on crop development: prospects for advancing climate-smart agriculture. J Crop Sci Biotechnol 26(4):405–431. https://doi.org/10.1007/s12892-023-00195-3

Jagodzik P, Tajdel-Zielinska M, Ciesla A, Marczak M, Ludwikow A (2018) Mitogen-activated protein kinase cascades in plant hormone signaling. Front Plant Sci 9:1387. https://doi.org/10.3389/fpls.2018.01387

Jameel A, Akram NA, Bashir R, Ashraf M, Abdel Latef AA (2026) Salicylic acid enhances yield and alleviates drought-induced oxidative damage in two carrot cultivars with contrasting tolerance. Russ J Plant Physiol 73:23

Jiadkong K, Ueda A (2024) Effects of riboflavin application on rice growth under salinized soil conditions. Cereal Res Commun 52(4):1565–1577. https://doi.org/10.1007/s42976-024-00504-8

Jiadkong K, Fauzia AN, Yamaguchi N, Ueda A (2024) Exogenous riboflavin (vitamin B2) application enhances salinity tolerance through the activation of its biosynthesis in rice seedlings under salinity stress. Plant Sci 339:111929. https://doi.org/10.1016/j.plantsci.2023.111929

Jiang L, Strobbe S, Van Der Straeten D, Zhang C (2021) Regulation of plant vitamin metabolism: backbone of biofortification for the alleviation of hidden hunger. Mol Plant 14(1):40–60. https://doi.org/10.1016/j.molp.2020.11.019

Kamle M, Borah R, Bora H, Jaiswal AK, Singh RK, Kumar P (2020) Systemic acquired resistance (SAR) and induced systemic resistance (ISR): role and mechanism of action against phytopathogens. In: Fungal biotechnology and bioengineering. Springer International Publishing, Cham, pp 457–470. https://doi.org/10.1007/978-3-030-41870-0_20

Keshavarz H, Moghadam RSG (2017) Seed priming with cobalamin (vitamin B12) provides significant protection against salinity stress in the common bean. Rhizosphere 3:143–149. https://doi.org/10.1016/j.rhisph.2017.04.010

Kong D, Zhu Y, Wu H, Cheng X, Liang H, Ling HQ (2008) AtTHIC, a gene involved in thiamine biosynthesis in Arabidopsis thaliana. Cell Res 18(5):566–576. https://doi.org/10.1038/cr.2008.35

Kraemer K, Semba RD, Eggersdorfer M, Schaumberg DA (2012) Introduction: the diverse and essential biological functions of vitamins. Ann Nutr Metab 61(3):185–191. https://doi.org/10.1159/000343103

Kumar A, P Nirmal, Kumar M, Jose A, Tomer V, Oz E et al (2023) Major phytochemicals: recent advances in health benefits and extraction method. Molecules 28(2):887. https://doi.org/10.3390/molecules28020887

Li XC, Chang C, Pei ZM (2023) Reactive oxygen species in drought-induced stomatal closure: the potential roles of NPR1. Plants 12(18):3194. https://doi.org/10.3390/plants12183194

Loladze I, Nolan JM, Ziska LH, Knobbe AR (2019) Rising atmospheric CO2 lowers concentrations of plant carotenoids essential to human health: a meta-analysis. Mol Nutr Food Res 63(15):1801047. https://doi.org/10.1002/mnfr.201801047

Mao L, Zhu R, Yi K, Wang X, Sun J (2023) Transcriptome analysis of *Vicia villosa* in response to low phosphorus stress at seedling stage. Agronomy 13(7):1665. https://doi.org/10.3390/agronomy13071665

Mignani L, Gnutti B, Zizioli D, Finazzi D (2021) Coenzyme A biochemistry: from neurodevelopment to neurodegeneration. Brain Sci 11(8):1031. https://doi.org/10.3390/brainsci11081031

Mostofa MG, Ghosh A, Li ZG, Siddiqui MN, Fujita M, Tran LSP (2018) Methylglyoxal–a signaling molecule in plant abiotic stress responses. Free Radic Biol Med 122:96–109. https://doi.org/10.1016/j.freeradbiomed.2018.03.009

Munteanu C, Schwartz B (2023) B vitamins, glucoronolactone and the immune system: bioavailability, doses and efficiency. Nutrients 16(1):24. https://doi.org/10.3390/nu16010024

Namba J, Harada M, Toda Y, Maruta T, Ishikawa T, Shigeoka S et al (2023) AtDREB2G is a novel regulator of riboflavin biosynthesis under low-temperature stress and abscisic acid treatment in Arabidopsis thaliana. bioRxiv. https://doi.org/10.1101/2023.09.06.556598

Nie S, Xu H (2016) Riboflavin-induced disease resistance requires the mitogen-activated protein kinases 3 and 6 in Arabidopsis thaliana. PLoS One 11(4):e0153175. https://doi.org/10.1371/journal.pone.0153175

Nizar M, Shaukat K, Samad A, Zahra N, Saleem I, Raza A (2023) Impact of exogenous salicylic acid and hydrogen peroxide on vitamins, essential nutrients and silymarin content of milk thistle grown at two different altitudes of balochistan under cadmium stress. J Agric Food Res 13:100643. https://doi.org/10.1016/j.jafr.2023.100643

Noordally ZB, Trichtinger C, Dalvit I, Hofmann M, Roux C, Zamboni N et al (2020) The coenzyme thiamine diphosphate displays a daily rhythm in the Arabidopsis nucleus. Commun Biol 3(1):209. https://doi.org/10.1038/s42003-020-0927-z

Palacios OA, López BR, Palacios-Espinosa A, Hernández-Sandoval FE, de-Bashan LE (2021) The immediate effect of riboflavin and lumichrome on the mitigation of saline stress in the microalga *Chlorella sorokiniana* by the plant-growth-promoting bacterium *Azospirillum brasilense*. Algal Res 58:102424. https://doi.org/10.1016/j.algal.2021.102424

Pushpalatha HG, Mythrashree SR, Shetty R, Geetha NP, Sharathchandra RG, Amruthesh KN, Shetty HS (2007) Ability of vitamins to induce downy mildew disease resistance and growth promotion in pearl millet. Crop Prot 26(11):1674–1681. https://doi.org/10.1016/j.cropro.2007.02.012

Rahman MM, Islam F, -Or-Rashid MH, Mamun AA, Rahaman MS, Islam MM et al (2022) The gut microbiota (microbiome) in cardiovascular disease and its therapeutic regulation. Front Cell Infect Microbiol 12:903570. https://doi.org/10.3389/fcimb.2022.903570

Rico-García E, Esteban Ortega-Torres A, Sabino Herrera-Matallana T (2023) Vermiliquer as a biostimulant and antioxidant in hydroponic lettuce (*Lactuca sativa*) production. https://doi.org/10.17221/5/2022-HORTSCI

Rodrigues de Queiroz A, Hines C, Brown J, Sahay S, Vijayan J, Stone JM et al (2023) The effects of exogenously applied antioxidants on plant growth and resilience. Phytochem Rev 22(2):407–447. https://doi.org/10.1007/s11101-023-09862-3

Rodríguez-Celma J, Vázquez-Reina S, Orduna J, Abadía A, Abadía J, Álvarez-Fernández A, López-Millán AF (2011) Characterization of flavins in roots of Fe-deficient strategy I plants, with a focus on Medicago truncatula. Plant Cell Physiol 52(12):2173–2189. https://doi.org/10.1093/pcp/pcr149

Rosado-Souza L, Fernie AR, Aarabi F (2020) Ascorbate and thiamin: metabolic modulators in plant acclimation responses. Plants 9(1):101. https://doi.org/10.3390/plants9010101

Ruchala J, Najdecka A, Wojdyla D, Liu W, Sibirny A (2025) Regulation of riboflavin biosynthesis in microorganisms and construction of the advanced overproducers of this vitamin. Int J Mol Sci 26(13):6243. https://doi.org/10.3390/ijms26136243

Schramm M, Wiegmann K, Schramm S, Gluschko A, Herb M, Utermöhlen O, Krönke M (2014) Riboflavin (vitamin B2) deficiency impairs NADPH oxidase 2 (Nox2) priming and defense against Listeria monocytogenes. Eur J Immunol 44(3):728–741. https://doi.org/10.1002/eji.201343940

Seregin IV, Kozhevnikova AD (2023) Phytochelatins: sulfur-containing metal (loid)-chelating ligands in plants. Int J Mol Sci 24(3):2430. https://doi.org/10.3390/ijms24032430

Singh GS (2023) Role of vitamins and minerals in treating certain health conditions: a review of current opinion. Pharmakeftiki 35(3). https://doi.org/10.60988/pj.v35i3.14

Song Z, Wang L (2003) Chemiluminescence inhibition assay for folic acid using flow injection analysis. Phytochem Anal 14(4):216–220. https://doi.org/10.1002/pca.704

Suzuki M, Wu S, Mimura M, Alseekh S, Fernie AR, Hanson AD, McCarty DR (2020) Construction and applications of a B vitamin genetic resource for investigation of vitamin-dependent metabolism in maize. Plant J 101(2):442–454. https://doi.org/10.1111/tpj.14535

Taheri P, Tarighi S (2010) Riboflavin induces resistance in rice against *Rhizoctonia solani* via jasmonate-mediated priming of phenylpropanoid pathway. J Plant Physiol 167(3):201–208. https://doi.org/10.1016/j.jplph.2009.08.003

Tripathi DK, Bhat JA, Antoniou C, Kandhol N, Singh VP, Fernie AR, Fotopoulos V (2024) Redox regulation by priming agents toward a sustainable agriculture. Plant Cell Physiol 65(7):1087–1102. https://doi.org/10.1093/pcp/pcae031

Umer S, Akram NA, Ashraf M, Abdel Latef AAH (2026) Pre-Sowing application of methionine and proline mitigates drought stress in two canola (Brassica napus L.) cultivars. Russ J Plant Physiol 73:48

Wang X, Zhai T, Zhang X, Tang C, Zhuang R, Zhao H et al (2021) Two stripe rust effectors impair wheat resistance by suppressing import of host Fe–S protein into chloroplasts. Plant Physiol 187(4):2530–2543. https://doi.org/10.1093/plphys/kiab434

Wang Z, Zhu W, Xing Y, Jia J, Tang Y (2022) B vitamins and prevention of cognitive decline and incident dementia: a systematic review and meta-analysis. Nutr Rev 80(4):931–949. https://doi.org/10.1093/nutrit/nuab057

Wu Y, Cheng S, Ding X, Lin X, Deng S, Peng L, Tian H (2024) Exogenous riboflavin application at different growth stages regulates photosynthetic accumulation and grain yield in fragrant rice. Agriculture 14(11):1979. https://doi.org/10.3390/agriculture14111979

Xiang N, Hu J, Wen T, Brennan MA, Brennan CS, Guo X (2020) Effects of temperature stress on the accumulation of ascorbic acid and folates in sweet corn (Zea mays L.) seedlings. J Sci Food Agric 100(4):1694–1701. https://doi.org/10.1002/jsfa.10184
Yang B, Sun Y, Fu S, Xia M, Su Y, Liu C et al (2021) Improving the production of riboflavin by introducing a mutant ribulose 5-phosphate 3-epimerase gene in *Bacillus subtilis*. Front Bioeng Biotechnol 9:704650. https://doi.org/10.3389/fbioe.2021.704650
Zhang S, Yang X, Sun M, Sun F, Deng S, Dong H (2009) Riboflavin-induced priming for pathogen defense in Arabidopsis thaliana. J Integr Plant Biol 51(2):167–174
Zhang JR, Ge YY, Liu PH, Wu DT, Liu HY, Li HB et al (2022) Biotechnological strategies of riboflavin biosynthesis in microbes. Engineering 12:115–127. https://doi.org/10.1016/j.eng.2021.03.018
Zhang X, Xu J, Wang M, Gan C, Ren Y, Zhao X, Yuan Z (2023) Riboflavin application regulates sugar and energy metabolism in strawberries during cold storage. Lwt 187:115301. https://doi.org/10.1016/j.lwt.2023.115301
Zhao S, Li M, Ren X, Wang C, Sun X, Sun M et al (2024) Enhancement of broad-spectrum disease resistance in wheat through key genes involved in systemic acquired resistance. Front Plant Sci 15:1355178. https://doi.org/10.3389/fpls.2024.1355178
Zhou YH, Zhang YL, Wang XM, Cui JX, Xia XJ, Shi K, Yu JQ (2011) Effects of nitrogen form on growth, CO_2 assimilation, chlorophyll fluorescence, and photosynthetic electron allocation in cucumber and rice plants. J Zhejiang Univ Sci B 12(2):126–134. https://doi.org/10.1631/jzus.B1000059
Zhou X, Jin E, Li S, Wang C, Qiao E, Wu G (2015) Effects of dietary supplementation of probiotics (*Bacillus subtilis*, *Bacillus licheniformis*, and *Bacillus natto*) on broiler muscle development and meat quality. Turk J Vet Anim Sci 39(2):203–210. https://doi.org/10.3906/vet-1406-67
Zhou W, Brockmöller T, Ling Z, Omdahl A, Baldwin IT, Xu S (2016) Evolution of herbivore-induced early defense signaling was shaped by genome-wide duplications in Nicotiana. elife 5:e19531. https://doi.org/10.7554/eLife.19531
Zhou G, Jiang W, Luo H, Li X, Wan Y (2023) Transcriptome and targeted metabolomic integrated analysis reveals mechanisms of B vitamin accumulation in Areca catechu nut development. Int J Biol Macromol 241:124570. https://doi.org/10.1016/j.ijbiomac.2023.124570
Zhou L, Ullah F, Zou J, Zeng X (2025) Molecular and physiological responses of plants that enhance cold tolerance. Int J Mol Sci 26(3):1157. https://doi.org/10.3390/ijms26031157
Zikankuba VL, Mteremko D, James AJEJON (2019) Staple crops biofortification linking agriculture, food and nutrition towards eliminating hidden hunger. Eur J Nutr Food Saf 9:112–121. https://doi.org/10.9734/EJNFS/2019/v9i230050
Zou M, Yu K, Huang Y, Sheng Q, Chen Y, Liu H, Dou D (2025) Activation of ROS signaling molecule by AgNPs: molecular mechanisms in roses under abiotic stress, exploration of stress memory, and impact on root microbiota. Plant Soil:1–22. https://doi.org/10.1007/s11104-025-07361-2

Chapter 3
Niacin: A Stimulant for Plant Stress Tolerance

Mona F. A. Dawood, Mamdouh Eissa, Samar G. Thabet, Murad Muhammad, Zainul Abideen, and Arafat Abdel Hamed Abdel Latef

Abstract Niacin (also known as nicotinic acid) and niacinamide (nicotinamide) are forms of vitamin B3. Niacin (nicotinic acid) is a vital pyridine derivative that plays a central role in plant metabolism and stress physiology. It serves as a precursor for the essential cofactors NAD^+ and $NADP^+$, which regulate cellular redox balance and energy transfer. In addition to its metabolic functions, Niacin acts as a signaling molecule influencing growth, development, and stress adaptation. Several studies have demonstrated that foliar or soil application of niacin enhances seed germination, seedling vigor, and overall biomass accumulation. Niacin also promotes photosynthetic efficiency by increasing chlorophyll content, pigment stability, and carbon assimilation. Under abiotic stresses such as salinity, drought, heavy metal toxicity, and temperature extremes, niacin improves plant tolerance by stimulating antioxidant defense systems and reducing oxidative damage caused by reactive

M. F. A. Dawood (✉)
Botany and Microbiology Department, Faculty of Science, Assiut University, Assiut, Egypt
e-mail: Mo_fa87@aun.edu.eg

M. Eissa
Department of Soils and Water, Faculty of Agriculture, Assiut University, Assiut, Egypt

S. G. Thabet
Department of Botany, Faculty of Science, Fayoum University, Fayoum, Egypt

M. Muhammad
State Key Laboratory of Desert and Oasis Ecology, Key Laboratory of Ecological Safety and Sustainable Development in Arid Lands, Xinjiang Institute of Ecology and Geography, Chinese Academy of Sciences, Beijing, China

University of Chinese Academy of Sciences, Beijing, China

Z. Abideen
Dr. Muhammad Ajmal Khan Institute of Sustainable Halophyte Utilization, University of Karachi, Karachi, Pakistan

Department of Applied Biology, University of Sharjah, Sharjah, United Arab Emirates

A. A. H. Abdel Latef
Department of Botany and Microbiology, Faculty of Science, Qena University, Qena, Egypt

A. A. H. Abdel Latef (ed.), *Harnessing Vitamins and Organic Acids for Plant Stress Resilience*, Sustainability Sciences in Asia and Africa,
https://doi.org/10.1007/978-981-95-8593-9_3

oxygen species. Furthermore, niacin applications have been shown to improve the accumulation of osmolytes, proteins, antioxidants, and secondary metabolites, thereby improving stress resilience and crop productivity. Its role as a biostimulant has been highlighted in various species such as cereals, legumes, and oilseed crops. Niacin also interacts with other growth regulators and natural extracts to synergistically enhance plant performance under stress conditions. Given its multifunctional benefits, niacin is emerging as a sustainable, eco-friendly approach to improving crop growth and yield across diverse environments. Future research should focus on unraveling the molecular pathways through which niacin mediates plant responses and on optimizing its application strategies for different crops and stress conditions.

Keywords Niacin · Nicotinic acid · Niacinamide · Vitamin B3 · Plant development · Stress mitigation

3.1 Introduction

In recent years, the frequency and intensity of abiotic stresses, such as drought, flooding, heat waves, and soil salinity, have markedly increased, becoming major constraints on crop productivity and quality (Abdelmoneim et al. 2025; Abdel Latef 2021, 2022, 2023, 2024, 2025, 2026a, b). These challenges pose serious threats to sustainable agriculture and global food security, especially as the rising world population drives a greater demand for food production under deteriorating environmental conditions. Therefore, enhancing the resilience of crops to heat, water shortages, and other extreme stresses is essential to minimize yield losses. Plants naturally cope with such stresses by reprogramming their transcriptome and metabolome, yet improving tolerance through human intervention has become a key research priority. Conventional strategies include germplasm screening, the development of genetically engineered or genome-edited crops, and, more recently, the application of chemical agents. The use of exogenous chemical compounds has emerged as a promising method, as these treatments can trigger molecular and physiological adjustments that fortify plants and help them withstand abiotic stress (Bashir et al. 2025). The best plant-active chemicals are those that are small in size, can easily move across cell membranes, and are effective even at low doses. They should interact specifically with their biological targets to avoid unwanted side effects (Serrano et al. 2015). In addition, they need to be safe for the environment, easy to produce, and inexpensive enough for large-scale use.

The screening process requires considerable effort, as it involves growing large numbers of plants or plant tissues under controlled stress or environmental conditions and then exposing them to many different small chemical compounds (Sako et al. 2021). Using this approach, researchers have successfully identified new compounds that reduce stress effects and promote plant growth (Sako et al. 2020).

Many naturally occurring organic compounds, such as plant growth protectants and vitamins, are widely applied to improve growth and increase productivity in various plant species. Among these vitamins, vitamin B3 (niacin), also known as nicotinic acid or nicotinamide, plays an important role in supporting vegetative growth and flowering. Niacin is essential for numerous metabolic processes because it forms part of the coenzymes NAD^+ and $NADP^+$, which are involved in redox reactions and energy metabolism. To avoid confusion with nicotine from tobacco, nicotinic acid is commonly referred to as niacin. It is a white, water-soluble compound that also dissolves in alkaline solutions and alcohol, but not in ether (Leonel et al. 2025). Chemically, niacin is a water-soluble monocarboxylic acid derived from pyridine and occurs in foods as nicotinic acid ($C_6H_5NO_2$) and nicotinamide ($C_6H_6N_2O$) (Ndaw et al. 2002). Both forms are biologically active because they can be interconverted within the organism and perform the same vitamin functions (Gehring 2004). Niacin gained attention since the discovery of NAD metabolism. Several reports have documented the role of niacin or derivatives in plant development and under stress. Thus, this chapter aims to summarize the physical and chemical properties of niacin, biosynthesis, and collect the main information on the role of niacin in plant growth and development, as well as its role in mitigating stress tolerance.

3.2 Vitamins

Since vitamins were first identified in connection with human nutritional deficiencies, early research mainly focused on their roles in humans and animals. As a result, their importance in plant biology was often overlooked (Smith et al. 2007). Plants, however, can synthesize a broad spectrum of vitamins that are crucial for the health of humans as well as the growth and metabolism of plants. These compounds function as cofactors, participate in redox reactions, and many possess strong antioxidant properties that protect cellular components (Rubio et al. 2008). More than 20 different vitamins have been identified uptill now, and they are generally categorized into two main groups: fat-soluble vitamins (A, D, E, and K) and vitamins soluble in water, which include nine B-complex vitamins (B1, B2, B3, B5, B6, B7, B9, and B12) along with vitamin C (Fitzpatrick et al. 2012; Li et al. 2021; Leonel et al. 2025). Vitamins' role as antioxidants in maintaining plant growth and development resistance to mitigate stresses has been well documented (Hanson et al. 2016). Disruptions in the synthesis of various B-vitamins (numbered from 1 to 9) and vitamin E induce serious problems for embryos or early-seedlings of various phenotypes, which improve when the corresponding vitamin is applied, thereby saving the phenotype (Hanson and Gregory 2011; Li et al. 2021). We will focus in the current chapter on vitamin B functions and distribution as the major group of niacin. Thiamine (vitamin B1) helps strengthen plant membranes, increases the accumulation of photosynthetic pigments, activates antioxidant enzymes, and enhances maize growth (Kaya et al. 2015). Several vitamins play important roles in plant growth and reproduction. Pantothenic acid (vitamin B5), ascorbic acid, and folate have been shown to enhance both vegetative growth and reproductive development (Rubio et al. 2008; Mehrshahi et al. 2010). Disruption of nicotinamide adenine dinucleotide (NAD; vitamin B3) biosynthesis leads to poor pollen fertility and reduced silique length (Hashida et al. 2009). In rice, niacin (vitamin B3) is involved in seed development, while its derivative trigonelline promotes increased seed width (Chen et al. 2016). Pyridoxine (vitamin B6) is required for normal post-embryonic root growth (Chen and Xiong 2005) and supports the expansion of aerial organs in Arabidopsis (Raschke et al. 2011). Ascorbate also stimulates early seedling growth, although it may delay flowering (Foyer et al. 2020).

Vitamins are also closely linked to plant defense mechanisms. Thiamine application enhances resistance in rice against bacterial leaf blight caused by *Xanthomonas oryzae* pv. *oryzae* and against sheath blight disease (Ahn et al. 2005; Wang et al. 2006; Bahuguna et al. 2012). Riboflavin has been reported to protect plants from several pathogens, including *Pseudomonas syringae* pv. *tomato* and *Peronospora parasitica* in Arabidopsis, as well as Tobacco mosaic virus and *Alternaria alternata* in tobacco (Dong and Beer 2000; Zhang et al. 2009; Abdel-Monaim 2011). Folate also contributes to plant resistance against bacterial and viral pathogens such as *Pseudomonas syringae*, cucumber mosaic virus, and *Xanthomonas axonopodis* (Wittek et al. 2015; Song et al. 2013).

Beyond their role in defense, vitamins act as essential cofactors or coenzymes in many plants' metabolic processes (Miret and Munne-Bosch 2014). Thiamine is converted into thiamine pyrophosphate, which is required for the activity of several enzymes involved in carbohydrate metabolism (Rosado-Souza et al. 2019). Riboflavin (vitamin B2) is the precursor of flavin mononucleotide (FMN) and flavin adenine dinucleotide (FAD), both of which are crucial for redox reactions in primary metabolism (Fischer and Bacher 2005). Niacin and nicotinamide are precursors of NAD^+ and $NADP^+$, which are essential for redox reactions and energy metabolism involving carbohydrates and lipids (Wahlberg et al. 2000).

Pantothenic acid is required for the synthesis of coenzyme A and acyl carrier protein, which are key cofactors in lipid biosynthesis and degradation (Coxon et al. 2005). Vitamin B6 participates in numerous enzymatic reactions, including transamination, decarboxylation, and amino acid metabolism (Drewke and Leistner 2001). Biotin (vitamin B7) functions as a covalently bound cofactor for enzymes involved in fatty acid and carbohydrate metabolism (Alban et al. 2000). Finally, tetrahydrofolate is essential for one-carbon transfer reactions and plays a central role in the biosynthesis of DNA, purines, methionine, amino acids, and other key metabolites (Mehrshahi et al. 2010).

Vitamin levels differ widely across the plant kingdom. In plant tissues, most B-group vitamins are present at concentrations of a few milligrams per gram, whereas biotin occurs at much lower levels, typically in the nanogram per gram fresh weight range. B-group vitamins tend to accumulate mainly in seeds, as reported for *Medicago sativa* (Asensi-Fabado and Munne-Bosch 2010). Within plant cells, vitamins are unevenly distributed among organelles; B vitamins are predominantly localized in the cytosol, plastids, and mitochondria (Roje 2007).

Also, the translocation of vitamin B between different tissues was reported. In this sense, Colinas and Fitzpatrick (2015) noted that the genes responsible for vitamin B1 biosynthesis are mainly expressed in green tissues. Ajjawi et al. (2007) stated that the synthesizing enzymes of vitamin B1 are mainly located in chloroplasts. Vitamin B1 is active in many growing parts of the plant, including meristematic tissues, developing organs, and non-photosynthetic structures (Martinis et al. 2016). This reveals that vitamin B1 can be transported from leaves to other plant tissues. Using a complementation assay like the approach used to identify PUP1, Martinis et al. (2016) identified PUT3 (an amino acid permease) as a transporter responsible for moving vitamin B1 from the leaves to the roots.

3.3 Niacin

B complex vitamins are being studied as biostimulants in plants, providing greater resistance to plant tissues and increasing growth and development of crops, even if applied in low amounts (Berglund et al. 2016; Martins et al. 2016; Vendruscolo et al. 2019; Thomé et al. 2023). Niacin (vitamin B3) refers to a group of related compounds, including nicotinic acid (pyridine-3-carboxylic acid), nicotinamide (pyridine-3-carboxamide), and other derivatives that share the biological activity of nicotinamide (Freese and Lysne 2023). In plants, vitamin B3 acts as a precursor of pyridine alkaloids, particularly trigonelline, which is formed by methylation of nicotinic acid. Trigonelline accumulates under stress conditions, where it functions as an osmoprotectant and plays a role in regulating the cell cycle (Matsui et al. 2007). Niacin is involved in controlling cell division, promoting plant growth and differentiation, and improving stress tolerance (Miret and Munne-Bosch 2014). Its importance is largely due to its role as a structural component of the coenzymes NAD and NADP, which are essential for many metabolic reactions and for the regulation of macro- and micronutrient metabolism (Paruch et al. 2022).

Nicotinic acid (NA) is a key water-soluble nutrient in plants and mainly occurs in two forms, e.g., niacin and nicotinamide (Chaves and Oliveira 2004). Moreover, nicotinamide also reduces endogenous abscisic acid (ABA) levels, helping plants maintain better growth under stress conditions. By modulating these hormonal balances, nicotinamide supports cell division, elongation, and overall metabolic activity, which enhances plant tolerance to environmental stresses. Several studies have shown that nicotinamide application can improve physiological performance, boost antioxidant defense systems, and regulate osmolyte accumulation, thereby safeguarding plants during drought, salinity, and other adverse conditions (Sen and Puthur 2020; Sabagh et al. 2019). In this way, NA functions not only as a metabolic precursor but also as a key signaling molecule that contributes to stress resilience and improved plant development (Farooq et al. 2022).

Nicotinamide serves as a precursor of the biologically active cofactors nicotinamide adenine dinucleotide (NAD) and nicotinamide adenine dinucleotide phosphate (NADP), which are essential for many cellular reactions, including energy production and respiratory metabolism (Kirkland and Meyer-Ficca 2018a, b). Moreover, it shares in energy transport in the plant cell, contributing to plant growth and development because of the photosynthetic process in the pentose phosphate oxidative route and mitochondrial metabolism (Dong et al. 2015). Nicotinamide acts in cell elongation, in the accumulation of reserves and consequently in vegetative growth, even in adverse conditions imposed, for example, by salinity (Abdelhamid et al. 2018). El-Bassiouny et al. (2014) stated that nicotinamide instigates the increase of indole acetic acid levels and growth rates as indole-3-acetic acid (IAA) improves cell division, contributing to the metabolic activities of the plant.

3.3.1 History and Physico-Chemical Properties

Niacin is an organic compound ($C_6H_5NO_2$) and an essential nutrient in the human diet (Sharma 2009). It is chemically related to pyridine-3-carboxylic acid, also known as nicotinic acid, and includes compounds that share the same biological activity. Therefore, the term niacin may refer specifically to nicotinic acid or collectively to both nicotinic acid and nicotinamide present in the diet (Food and Nutrition Board 1998). Niacin was first isolated from rice bran in 1911. It is readily metabolized and excreted by the human body, with a recommended daily intake of approximately 14–16 mg (Meyer-Ficca 2018). Later studies showed that niacin consists of two closely related forms: nicotinic acid and nicotinamide. Nicotinamide, the amide form of nicotinic acid, was first isolated from horse red blood cells in 1935 (Carlson 2004). Unlike nicotinic acid, which is known to reduce blood lipid levels, nicotinamide does not affect lipid metabolism.

In detail, niacin was first identified in 1867 by organic chemists, long before its role as an essential nutrient was recognized. Historical accounts of its discovery and the consequences of deficiency have been described extensively (Harris 1919; McCollum 1957; Darby et al. 1975; Hankes 1984; Loosli 1991). Funk (1911–1913) isolated niacin from yeast and rice polishings during his search for a water-soluble anti-beriberi vitamin. However, its therapeutic potential was initially overlooked because it failed to cure beriberi in pigeons, although it enhanced the effect of thiamin when given together. In 1935, Warburg and colleagues demonstrated its biochemical significance by isolating nicotinic acid as a component of the hydrogen transport system. Prior to this, nicotinic acid had been obtained as an oxidation product of nicotine, with Huber first preparing it in 1867 by oxidizing nicotine with potassium dichromate. Later, Suzuki (1894) and Funk (1912) independently isolated it from rice bran and yeast (van Eys 1991). A breakthrough came when Elvehjem et al. (1937) identified nicotinamide from liver extracts as the factor curing black tongue in dogs, paving the way for clinical reports of its success in treating pellagra. Further research by Krehl et al. (1945) established that tryptophan was as effective as niacin in preventing pellagra. Heidelberger subsequently demonstrated that tryptophan serves as a precursor for nicotinic acid synthesis, showing through tracer studies that L-[14C] tryptophan was converted to labeled niacin in rats. This discovery explained why protein-rich foods, such as milk, could both prevent and cure pellagra.

Chemically, niacin is among the simplest vitamins and occurs in two main forms: nicotinic acid (pyridine-3-carboxylic acid) and nicotinamide (nicotinic acid amide). Together with their coenzyme derivatives, these compounds are referred to as vitamin B3. Nicotinamide represents the biologically active form, serving as an essential component of the coenzymes NAD and NADP. In animal tissues, niacin primarily exists in its reduced coenzyme states, NADH and NADPH. Over time, the vitamin has been described by several synonyms, including PP factor (pellagra-preventive factor), anti-pellagra vitamin, 3-pyridinecarboxylic acid, nicobid, and nicotinex (Informatics 1974). Niacin readily forms salts with metals such as calcium, aluminum, copper, and sodium. In acidic media, it produces quaternary ammonium compounds like nicotinic acid hydrochloride, which are water-soluble, while in alkaline solutions, it forms carboxylate salts.

Nicotinic acid and nicotinamide are white, odorless crystalline compounds that dissolve easily in water and alcohol. They are stable under heat, light, air, and alkaline conditions, which helps maintain their presence and activity in food products. However, under strong alkaline conditions at high temperatures, niacin undergoes decarboxylation. Nicotinamide has a molecular weight of 122.14 g/mol and a melting point of 129 °C, with a boiling point of 150–160 °C. Nicotinic acid has a molecular weight of 123.12 g/mol, melts at 236 °C, and can sublime at elevated temperatures. Both forms are soluble in water—nicotinic acid to 16.7 g/L, and in alcohol; nicotinamide is also soluble in ether and glycerin. Neither form dissolves in lipid solvents (Chand and Savitri 2016).

3.3.2 Biosynthesis of Niacin in Plants

Niacin (nicotinic acid) is a metabolic product of NAD^+ and $NADP^+$. It serves as a precursor to pyridine alkaloids, such as nicotine, trigonelline, and ricinine, in some plants (Noctor et al. 2006). In animals, the pyridine ring of NAD^+ is synthesized from L-tryptophan, while in plants it originates from L-aspartate, and in bacteria from either L-tryptophan or L-aspartate (Rongvaux et al. 2003; Noctor et al. 2006). Debate persists on whether grasses can synthesize nicotinic acid from L-tryptophan, as genomic evidence remains inconclusive (Katoh et al. 2006; Noctor et al. 2006). Since nicotinic acid can be recycled to NAD^+ through the pyridine nucleotide cycle (salvage pathway), it functions as a vitamin in animals unable to synthesize L-tryptophan.

The nutritional role of vitamin B3 was first established when nicotinic acid supplementation cured "black tongue" disease in dogs fed a pellagra-inducing maize diet (Elvehjem et al. 2002). Similar treatments proved effective in human pellagra patients (Kohn 1938) and pigs (Chick et al. 1938). Later, L-tryptophan was identified as the precursor of nicotinic acid in animals, replacing dietary supplementation (Krehl 1949). This discovery went into parallelism with studies investigating whether plants also synthesize nicotinic acid from L-tryptophan (Leete 1965; Arditti 1967; Arditti and Tarr 1979; Tarr and Arditti 1982).

An alternative pathway for nicotinic acid biosynthesis was suggested when L-tryptophan was shown not to be a precursor in *Escherichia coli* and *Bacillus subtilis* (Yanofsky 1954). Instead, precursors appeared to include a four-carbon dicarboxylic acid and glycerol or its metabolites (Ortega and Brown 1960). Incorporation studies in tobacco demonstrated that glycerol and L-aspartic acid contribute to nicotine biosynthesis, indicating similarity between plant and bacterial pathways (Jackanicz and Byerrum 1966). Quinolinic acid, proposed earlier as an intermediate from L-tryptophan, was later confirmed as a key precursor of nicotinic acid in both plants and *E. coli* (Hadwiger et al. 1963).

Further evidence for an aspartate-derived pathway was provided by the detection of aspartate oxidase in cotton callus (Hosokawa et al. 1983). Genes encoding aspartate oxidase and quinolinate synthase, both localized in plastids, were cloned from Arabidopsis and shown to be essential for growth (Katoh et al. 2006). Although an earlier report suggested a tryptophan-derived pathway in grasses, the corresponding sequences were later removed during annotation (Noctor et al. 2006). Overall, current evidence strongly supports quinolinic acid biosynthesis from aspartate in plants.

The conversion of quinolinic acid to nicotinate mononucleotide (NaMN) is catalyzed by quinolinate phosphoribosyl transferase. This enzyme has been purified from castor bean (Mann and Byerrum 1974), and complementary cDNAs were later isolated from Nicotiana rustica and N. tabacum, encoding proteins with mitochondrial targeting signals (Sinclair et al. 2000).

The pyridine nucleotide cycle in plants includes the synthesis of NAD^+ from NaMN, degradation of NAD^+ to nicotinic acid, and recycling back to NAD^+. Despite its importance, this cycle remains poorly studied as NAD pyrophosphatase, which hydrolyzes NAD^+ to nicotinamide mononucleotide, has been purified from plants (Kornberg and Pricer Jr. 1950). Metabolic studies in *Ricinus communis* and *Hordeum vulgare* confirmed the recycling of nicotinamide and nicotinic acid into NAD^+ (Waller et al. 1966; Ryrie and Scott 1969), consistent with earlier mammalian work (Preiss and Handler 1957, 1958). Studies in Picea glauca and Catharanthus roseus have shown that nicotinic acid is converted into NAD^+, nicotinamide mononucleotide, and nicotinamide in plant tissues (Ashihara et al. 2005). These findings indicate that the pyridine nucleotide cycle in plants is regulated by a group of approximately 6–8 enzymes. Key enzymes involved in this pathway include nicotinate phosphoribosyl transferase, NaMN adenyl transferase, NAD synthetase, NAD pyrophosphatase, NMN nucleosidase, 5′-nucleotidase, nicotinamide riboside nucleosidase, and nicotinamide deamidase (Ashihara et al. 2005; Zheng et al. 2005; Noctor et al. 2006).

3.3.3 *Biological Functions of Niacin in Plants*

Plants require vitamins in small quantities (Bassouny et al. 2008), acting as coenzymes, regulating metabolism, and increasing plant activities without toxic or mutagenic effects (Bronzetti et al. 2001). The application of niacin, nicotinamide, or nicotinic acid can promote plant growth. The main processes induced by niacin were discussed as follows:

3.3.3.1 Morphological Attributes

Niacin supports plant growth mainly through its role in plant development. It enhances root growth, helps plants tolerate biotic and abiotic stresses, and activates enzymes involved in carbon metabolism during photosynthesis. In addition, niacin is a key component of NAD^+ and $NADP^+$, which act as electron carriers essential for many metabolic processes in plants (Kirkland and Meyer-Ficca 2018a, b). Vendruscolo et al. (2015) reported that preapplication of bean seeds with niacin, the acid form of nicotinamide, decreases the time of seedling emergence and average seedling emergence time, hence germinability. Vendruscolo et al. (2017) denoted that niacin increased root length and dry mass of melon trees. Also, Vendruscolo et al. (2023) recommended that vitamins can influence the morphological and physiological properties of lettuce. Abdallah et al. (2016) reported that applying vitamin B3 to quinoa significantly improved the growth of both shoots and roots. Similar positive effects were observed by Vendruscolo et al. (2017) and Vendruscolo et al. (2019), who found that niacin treatments enhanced vegetative growth in curly mustard and common bean plants. These enhancements were reflected in greater leaf numbers, increased plant height, and higher fresh and dry biomass of the aerial parts. Such improvements are likely linked to the role of niacin in strengthening the plant's energy status and nutrient reserves, ultimately supporting better growth and productivity (El-Bassiouny et al. 2014). Leonel et al. (2025) stated that niacin increased gas exchange and fresh mass of carrots in hydroponic production systems.

3.3.3.2 Yield Development

Niacin regulates the development of yield and its attributes, such as the morphology of seeds; Chen et al. (2016) reported that trigonelline, a niacin derivative, enhanced the rice grains' width. Abreu et al. (2020) displayed that nicotinamide enhanced the productive and biometric traits in bean, hence improving all yield traits. Khudair et al. (2019) denoted that nicotinamide at various levels improved the vegetative growth in terms of plant length and their lateral branches, number and area of leaves, as well as flower growth(e.g., date and duration of flowering, number of inflorescences and florets), as well as biochemical traits such as the level of chlorophyll, anthocyanin, nitrogen, phosphorus and potassium. Lima et al. (2024) documented stimulation of soybean productivity when nicotinamide was applied to water-stressed plants, improving vegetative growth via increasing the expansion of cells, reserve accumulation, and promoting attributes and yield of various crops under drought and salinity stresses.

3.3.3.3 Metabolism

The application of nicotinamide have positive effects on growth, yield, metabolism, and/or physiological traits of various plants like *Triticum aestivum* (El Bassiouny et al. 2005), *Zea mays* (da Silva Colla et al. 2021; Hassanein et al. 2009), *Vicia faba* (Dawood et al. 2019), *Pisum sativum* (Sadak 2016), *Helianthus annuus* (Mostafa 2020), *Oryza sativa* (Vendruscolo et al. 2019), *Beta vulgaris* var. *saccharifera* (Abd El-Lateef et al. 2020), and *Phaseolus vulgaris* (Abreu et al. 2020). When niacin is applied, it affects nitrate assimilation of roots, sulfate transport by leaves, and photosynthetic efficiency (Taiz et al. 2017). Niacin and thiamine were reported to improve photosynthetic pigments and efficiency under stressful conditions (El-Bassiouny et al. 2014; Vendruscolo et al. 2023) and enhanced the growth and yield of organs (da Silva Colla et al. 2021). Ferreira et al. (2023) denoted that vitamin B3 has strong antioxidant properties in plant tissues, improving plant characteristics under stress through improving metabolic activity. Abd El-Lateef et al. (2020) reported that sugar accumulation in vegetables, producing suitable conditions, is needed for tuberous roots development in various sugar beet cultivars. Based on Ali and Al-Bayaty (2023), applying nano-iron oxide and vitamin B3 to peppermint plants significantly improves the mineral levels in the leaves, such as nitrogen, phosphorus, and potassium. Nicotinamide is a vitamin that dissolves in water, shares in the composition of the coenzymes NADH and NADPH, which take part in oxidation/reduction enzymatic reactions in cells (Kirkland and Meyer-Ficca 2018a, b) that help in the metabolism of carbohydrates, fats, and proteins, essential for cellular respiration and photosynthetic reactions. Additionally, nicotinamide is associated with stress signaling, inducing or regulating secondary metabolite accumulation and plant defense metabolism action (El-Bassiouny et al. 2014; Zaki et al. 2022). Nicotinamide has been found to perform functions related to energy transport in the plant cell, contributing to plant growth and development (Dong et al. 2015). Nicotinamide is a precursor of NAD^+, which favors energy gains and influences cellular metabolism, respiration, and plant growth (Berglund et al. 2016). Nicotinamide induces physiological alterations in the biosynthetic pathways of enzymes, nucleic acids, and proteins (Dawood et al. 2019). In addition, nicotinamide stimulates the production of endogenous hormones like gibberellic acid and indoleacetic acid that promote organ and whole growth of plants as was documented by El-Bassiouny et al. (2014) who recorded increasing of IAA content by nicotinamide increased the content of indole acetic acid, hence the growth rates of the plants. However, Dong et al. (2015) indirectly contributes the role of nicotinamide in enhancing plant growth via energy transport in the plant cell, while Berglund et al. (2016) linked the role of nicotinamide with improving growth and yield characteristics.

3.3.4 *Niacin and Stressful Conditions*

The accumulation of vitamins in plant tissues is affected by stressful conditions. For instance, plants accumulate thiamine and thiamine pyrophosphate (TPP) under oxidative stress by affecting the expression of key biosynthetic enzymes (Rapala-Kozik et al. 2008; Tunc-Ozdemir et al. 2009; Rapala-Kozik et al. 2012). Also, Miret and Munne-Bosch (2014) stated that water-deficit, radiation stress, and high temperature induce the accumulation of nicotinamide. Berglund and Ohlsson (1995) noted that the endogenous nicotinamide increased when oxidative stress burst, thereby inducing defensive metabolism. Thus, the external application of niacin could play important roles under stressful and non-stressful conditions. The participation of niacin can be shown by its exogenous supplementation, allowing improvements in cell elongation and expansion, as well as enhancement of yield traits in response to water-deficit and salinization of soil, and accumulation of energy and nutritional reserves (Abdelhamid et al. 2018; El-Bassiouny et al. 2014; Vendruscolo and Seleguini 2020). In the following sections, we will review the main findings of the role of niacin and derivatives on plant regulations.

3.3.4.1 Drought Stress

The frequent occurrence of drought spells has potentially undermined the sustainability of modern high-input agriculture systems worldwide (Dawood et al. 2025; Sallam et al. 2024a, b; Younes et al. 2024). To ameliorate the damaging consequences of water stress on plants, biological and eco-friendly approaches have been developed, such as using vitamin B3. Khurshid et al. (2023) reported that nicotinic acid application enhanced morphological traits such as roots and shoots characteristics of wheat plants, yield attributes, grain and biological yields (spike length, grains spike^{-1}, spikelets spike^{-1} and weight of 1000 grains), physiological parameters along with biosynthesis of antioxidants. The application of nicotinic acid also improves barley height, relative water content, leaf water potential, leaf area index, biological yield, grain yield, harvest index, catalase, peroxidase, and superoxide dismutase when applied as a foliar spray or soil in response to drought stress (Farooq et al. 2022). Soaking rice grains in niacin improved the speed of seedling emergence of rice plants, plant vegetative development (the number of tillers and panicles), relative indices of chlorophyll, and grain yield grown under drought stress (Vendruscolo et al. 2019).

3.3.4.2 Salinity Stress

The use of water with a high concentration of salt has been increasingly frequent due to poor water management and the use of wastewater (Dawood et al. 2023; Abdelrhim et al. 2024; Abdelmoneim et al. 2025, 2026). This decreases all aspects of plants, raising the need for stimulants to mitigate these deleterious effects (Dawood et al. 2022; Ragaey et al. 2022; Abdel Latef 2026a; Abdel Latef et al. 2026; Azooz et al. 2026), such as using niacin. According to Campos et al. (2008), biostimulants can influence germination and seed dry matter biomass and promote plant growth in height. The joint action of vitamins thiamine and niacin enables maize to have greater tolerance to water stress, as pre-seeding treatment in vitamin solutions reduces the salt stress in the initial development of sweet maize (Vendruscolo and Seleguini 2020). Other studies demonstrate that the application of nicotinamide makes it possible to increase the development of vegetables at different times of the cycle. In this sense, the application of niacin via seeds provides increases in the speed of initial plant establishment, as well as greater leaf emission in beans (Vendruscolo et al. 2015). Also, when applied as a foliar treatment, nicotinamide improves the fertility of rice spikelets (Vendruscolo et al. 2019), as well as increasing vegetative growth and maize productivity (Thomé et al. 2023). Abdelhamid et al. (2018) denoted that nicotinamide or niacin supplementation to leaves alleviated the adverse effects of salinity stress through increased the photosynthetic pigments, sucrose, total soluble sugars, amino acids, proline, N-content of foliar system, the length of shoot, number of leaves, biomass of shoot and seed yield, while reducing lipid peroxidation products and the oxidative enzymes. Also, nicotinamide inhibited the uptake of Na^+ and improved the shoot contents of P, K^+, Mg^{2+}, Ca^{2+}, Fe^{2+}, Mn^{2+}, Zn^{2+}, and Cu^{2+}. Nicotinamide acts in cell elongation, in the accumulation of reserves, and, consequently, in vegetative growth under salinity stress (Abdelhamid et al. 2018). Garcia et al. (2023) reported that the application of B vitamins (thiamine or niacin) increased the net photosynthesis when compared to the NaCl treatment. However, only the application of thiamine resulted in a mitigating effect on the losses of plant dry mass accumulation. Also, Salih and Abdulrraziq (2024) reported that the foliar application of nicotinic acid effectively mitigated salinity-induced reductions in growth and photosynthetic pigments, particularly at lower seawater concentrations. de Lima et al. (2024) displayed that the foliar application of nicotinamide on the soybean resulted in increases in the vegetative traits of plant height and the number of branches per plant in the productive characteristics in terms of the number of pods, number of grains, mass of grains per plant, mass of 1000 grains, and yield, were also improved and can be considered as a biostimulant. Recently, Abu-tahon et al. (2025) reported that the foliar application of nicotinamide, especially in combination with moringa oleifera leaf extract, effectively alleviated salt stress in faba bean by enhancing growth, yield, photosynthetic pigments, metabolites (proteins and carbohydrates), minerals, and antioxidant capacity (total phenol, anthocyanin, ascorbic acid, flavonoids and proline contents). Also, Santos et al. (2025) reported that nicotinamide reduced the negative impacts of salt stress on nasturtium (*Tropaeolum majus*), promoting improvements in gas exchange, such as increased net photosynthesis, water-use efficiency, and stomatal conductance. Additionally, nicotinamide increased the dry weight of shoot and flowers in terms of the number of flowers and flower buds (Table 3.1).

Table 3.1 The effect of different application methods of niacin or other derivatives on salinity-stressed plants

Crop	Application method	Key responses	References
Maize	Priming	Enhanced germination, dry matter biomass, and plant height	Campos et al. (2008)
Beans	Seed treatment (niacin)	Faster establishment via regulation of leaf gas exchange parameters	Vendruscolo et al. (2015)
Beans	Seed treatment (niacin)	Faster establishment via regulation of leaf gas exchange parameters	Vendruscolo et al. (2015)
Broad beans	Foliar (nicotinamide or niacin)	Improved pigments, carbohydrates, N, growth, mineral uptake; reduced oxidative enzymes and lipid peroxidation	Abdelhamid et al. (2018)
Rice	Foliar (nicotinamide)	Improved spikelet fertility	Vendruscolo et al. (2019)
Sweet maize	Priming with niacin	Increased water/salt stress tolerance during early development. It improved the relative chlorophyll a, height of shoot, leaf number, diameter of stem, total dry matter of shoot and roots, as well as reducing dry matter losses	Vendruscolo and Seleguini (2020)
Maize	Foliar (nicotinamide)	Increased vegetative growth and overall productivity	Thomé et al. (2023)
Lettuce	Foliar (thiamine or niacin)	Improved dry matter, net photosynthetic rate, and upregulated stomatal conductance, transpiration, as well as the level of intra-cellular CO_2	Garcia et al. (2023)
Durum wheat	Foliar (nicotinic acid)	Mitigated salinity-induced loss of growth and pigments especially at low salinity	Salih and Abdulrraziq (2024)
Soybean	Foliar (nicotinamide)	Increased plant height, branch number, pods, grain mass, and yield	de Lima et al. (2024)
Faba bean	Foliar (nicotinamide)	Strong improvement in growth, yield, pigments, osmolytes, minerals, antioxidants under salt stress	Abu-tahon et al. (2025)
Nasturtium (*Tropaeolum majus*)	Nicotinamide	Reduced the negative impacts of salt stress by upregulating gas exchange parameters, enhancing net photosynthetic rate, the efficiency of water-use, and stomatal conductance causing higher vegetative flowering, and reproductive growth	Santos et al. (2025)

3.3.4.3 Temperature Stress

The biostimulant potential of the nicotinamide was documented in the cultivation of quinoa in sandy soils when applied at 100 mg L^{-1} (Abdallah et al. 2016), hydroponic soybean (Lima et al. 2024), and *Chlorella vulgaris* (Ferreira et al. 2023). Araújo et al. (2024) showed that nicotinamide priming, especially at 200 and

300 mg L^{-1} and temperatures, increases radical development under stressful temperatures and enhances gas exchange processes and initial growth, mainly in the aerial part.

Mahmood et al. (2012) reported that niacin significantly contributes to heat tolerance in maize by enhancing antioxidant capacity and reducing oxidative damage. Under high temperature stress, tolerant maize genotypes maintain higher or stable levels of niacin, along with other antioxidants such as ascorbic acid and riboflavin, which collectively improve redox balance and protect cellular structures and reduce H_2O_2 and lipid peroxidation in the shoots (Table 3.2).

3.3.4.4 Heavy Metals Stress

Several studies indicate that applying niacin or nicotinamide externally can reduce the harmful effects of heavy metals and improve plant tolerance, although the protective responses differ among plant species and metal types. For example, Ohlsson et al. (2008) showed that nicotinamide increased tolerance to zinc, copper, and cadmium in sensitive *Salix viminalis* clones by supporting plant growth, without affecting metal uptake or relying on glutathione production. Smolik and Sędzik-Wójcikowska (2024) reported that foliar spraying and soil irrigation of nicotinamide (100 μM) effectively alleviates lead-induced toxicity in barley by improving growth and enhancing stress tolerance mechanisms. The authors reported that exogenous niacin restored morphological performance by increasing the content of total chlorophyll and carotenoids and reducing the malondialdehyde (MDA) and hydrogen peroxide contents compared to plants grown with lead alone, while reducing catalase and peroxidase activities. Similarly, Bouhadi et al. (2024) observed that nicotinic acid boosted the antioxidant capacity, photosynthetic pigment, and mineral nutrition of Pb-stressed *Lens culinaris* seedlings. More recently, Li et al. (2024) demonstrated that nicotinamide repaired the damaging impact of cadmium stress on photosystem II of *Pistia stratiotes* L. through upregulating the expression of photosynthesis-related genes. Also, Yu et al. (2025) reported that applying niacin at 200 mg L^{-1} significantly lowered cadmium accumulation in leaves and improved plant biomass under Cd stress. Niacin helped protect photosystem I (PSI) by alleviating acceptor-side limitations and restoring electron flow by recovering plastocyanin and ferredoxin. It further enhanced the antioxidant defense system by stimulating the activities of key antioxidant enzymes and increasing the production of glutathione (GSH) and ascorbate (ASA). In addition, niacin upregulated genes related to $NAD(P)^+$ metabolism, including those encoding $NAD(P)^+$-nucleotidase and uridine nucleosidase, thereby boosting NAD^+ pool levels required for efficient electron transport and cellular redox balance. It also activated genes involved in riboflavin metabolism and flavin biosynthesis, providing further support for the electron transport machinery (Table 3.2).

Table 3.2 The effect of different application methods of niacin or other derivatives on plants exposed to various stresses

Plant	Stress applied	Method of application	Main plant responses	Reference
Chenopodium quinoa	High temperature/sandy soil	Foliar spray (100 mg L^{-1})	Improved growth, stress mitigation	Abdallah et al. (2016)
Soybean (hydroponic)	Temperature stress	Foliar nicotinamide	Enhanced vegetative attributes, improved initial growth	Lima et al. (2024)
Chlorella vulgaris	Temperature stress	Nicotinamide in medium	Increased biomass metabolism	Ferreira et al. (2023)
Maize	High temperature	Foliar	Higher niacin levels, antioxidant reinforcement, ↓ H_2O_2 & MDA	Mahmood et al. (2012)
Salix viminalis	Zn, Cu, Cd toxicity	Nicotinamide / nicotinic acid (combined with metal)	Growth improved, tolerance ↑, metal uptake unchanged	Ohlsson et al. (2008)
Barley	Pb stress	Foliar & soil irrigation (100 µM NIC)	Restored chlorophyll, ↓ MDA & H_2O_2, improved morphology	Smolik and Sędzik-Wójcikowska (2024)
Lens culinaris	Pb stress	Foliar nicotinic acid	↑ Antioxidant capacity, pigments, mineral nutrition	Bouhadi et al. (2024)
Pistia stratiotes	Cd stress	Foliar nicotinamide	Improved PSII repair, ↑ photosynthesis gene expression	Li et al. (2024)
Maize / various	Cd stress	Foliar niacin (200 mg L^{-1})	↓ Cd accumulation, ↑ biomass, activated antioxidant & $NAD(P)^+$ metabolism	Yu et al. (2025)
Multiple species	UV-B / AAPH / vanadyl sulfate	Stress induction (endogenous response)	↑ Nicotinamide & trigonelline as signaling molecules	Berglund et al. (1996)
Maize	Non-specific oxidative / reclaimed soil	Foliar nicotinamide	↑ Photosynthesis, pigments, metabolic activity	El-Bassiouny et al. (2014)
Urochloa brizantha	Growth improvement / nutrient stress	Niacin + N fertilization	↑ Biomass & dry matter acquisition	de Oliveira Silva et al. (2022)
Kiwi	Short-term temperature and drought	Niacin spray	↑ NADPH oxidase activity, ↓ ROS, ↑ tolerance	Zhu et al. (2018)
Spruce seedlings	Biotic—pine weevil	Foliar nicotinamide	↑ Yield traits under insect attack	Berglund et al. (2016)

3.3.4.5 Other Stresses

It has been reported that the exposure to oxidative stressors such as 2,2'-Azobis (2-amidinopropane) dihydrochloride (AAPH), vanadyl sulfate, or UV-B radiation has been shown to elevate nicotinamide and trigonelline levels in plant tissues, supporting the hypothesis that niacin-related metabolites act as signaling molecules in stress responses, likely mediated through poly(ADP-ribose) polymerase–dependent defense activation (Berglund et al. 1996). Nicotinamide boosted the net photosynthesis in maize plants, increased production capacity of photosynthetic pigments, promoted photosystem functionalities, and increased metabolic activities, which are relevant for growth by stimulating cell division and/or cell enlargement in newly reclaimed soils (El-Bassiouny et al. 2014). de Oliveira Silva et al. (2022) displayed that niacin, when applied in combination or after nitrogen fertilization, gives better conditions for developing and dry matter accumulation of *Urochloa brizantha*. Zhu et al. (2018) showed that niacin spraying enhanced the contents of NADPH and $NADP^+$ and reduced the reactive oxygen species (O_2^- and H_2O_2) in the leaves of Kiwi crops. In the kiwi fruit, the level of NADPH is enhanced during early fruit development, then attenuated later without a significant effect on $NADP^+$ levels throughout fruit development. Thus, niacin increased the activity of NADPH oxidases, thus increasing stress-tolerance in kiwifruit during the encounter of short-term stressful conditions of temperature and drought. Also, the role of niacin was reported under biotic stress. Berglund et al. (2016) reported that nicotinamide presents an efficient relation to the characteristics of yield production of spruce seedlings exposed to pine weevil attacks (Table 3.2).

3.4 Future Prospects

Future research should explore the molecular and signaling pathways underlying niacin's protective effects to better integrate it into stress management strategies. Optimizing application rates, methods, and timing across different crop species and environments will be critical for maximizing benefits. The synergistic use of niacin with other biostimulants, phytohormones, or organic extracts should be further investigated to develop comprehensive stress-resilience strategies, especially under multi-stress conditions. Advances in omics technologies can also help unravel how niacin modulates gene expression, metabolite profiles, and protein networks under stress. Finally, large-scale field trials are necessary to validate laboratory and pot experiment findings, ensuring that niacin-based applications contribute effectively to sustainable agriculture and food security under changing climatic conditions.

3.5 Conclusion

Niacin, also referred to as nicotinic acid, and its amide form niacinamide (nicotinamide), are vital forms of vitamin B3 that play integral roles in plant metabolism and stress regulation. Niacin (nicotinic acid) is a crucial metabolite serving dual functions: internally as a biochemical precursor for the synthesis of NAD^+ and $NADP^+$, central cofactors required for redox reactions, energy metabolism, and cellular homeostasis, and externally as a potent biostimulant that enhances plant growth, photosynthetic efficiency, and yield performance. Beyond its metabolic role, exogenous application of niacin has demonstrated significant stress-mitigating potential under a range of abiotic stresses such as drought, salinity, temperature extremes, and heavy metal toxicity. It achieves this by boosting antioxidant enzyme activities, promoting osmolyte synthesis (e.g., proline and soluble sugars), and stabilizing cellular metabolic functions. Additionally, niacin improves key physiological and developmental attributes, including chlorophyll concentration, membrane integrity, seed germination rate, and reproductive success, ultimately contributing to improved plant resilience and productivity. Given its multifunctional role in plant physiology and defense, niacin emerges as a sustainable, eco-friendly strategy for enhancing crop performance under both optimal and stress conditions. Its integration into modern agricultural practices can support climate-resilient farming systems and promote higher yield quality and stability, highlighting its promising potential in future crop improvement programs.

References

Abdallah MMS, El Habbasha SF, El Sebai T (2016) Comparison of yeast extract and Nicotinaminde foliar applications effect on quinoa plants grown under sandy soil condition. Int J PharmTech Res 9:24–32

Abdelhamid MA, Sadak MBS, Schmidhalter UBRBS, El-Saady ABKBM (2018) Interactive effects of salinity stress and nicotinamide on physiological and biochemical parameters of faba bean plant. Acta Biológica Colombiana 18:499–509

Abdel-Monaim MF (2011) Role of riboflavin and thiamine in induced resistance against charcoal rot disease of soybean. Afr J Biotechnol 10:10842–10855

Abdelmoneim MS, Dawood MF, Hafez EE, Hammad SF, Ghazy MA (2025) Unveiling the role of cadaverine in mitigating salinity and/or Bisphenol A toxicity in tomato plants and reduced Bisphenol A accumulation in tomato roots. Plant Physiol Biochem 224:109799

Abdelmoneim MS, Ghazy MA, Hafez EE, Hammad SF, Dawood MF (2026) Mitigating the impact of salinity and/or bisphenol A stress on the performance of the tomato crop by spraying with an algal extract of Ascophyllum nodosum L. J Hazard Mater 141787

Abdelrhim AS, Hemeda NF, Mwaheb MA, Omar MO, Dawood MF (2024) The role of Trichoderma koningii and Trichoderma harzianum in mitigating the combined stresses motivated by Sclerotinia sclerotiorum and salinity in common bean (Phaseolus vulgaris). Plant Stress 11(100370):10–1016

Abd El-Lateef EM, Abd El-Salam MS, Mekki BB, Yousef AR, Hussein HAA (2020) Response of sugar beet varieties to foliar treatments with bio stimulant growth substances under sandy soil conditions. GJER 14:29–36

Abdel Latef AAH (2021) Organic Solutes, Oxidative Stress, and Antioxidant Enzymes under Abiotic Stressors. CRC Press/Taylor & Francis Group. https://doi.org/10.1201/9781003022879
Abdel Latef AAH (2022) Sustainable remedies for abiotic stress in cereals. Springer Nature, Singapore Pte Ltd. https://doi.org/10.1007/978-981-19-5121-3
Abdel Latef AAH (2023) Medicinal plant responses to stressful conditions (a volume in the Exploring Medicinal Plants series). CRC Press/Taylor & Francis Group. https://doi.org/10.1201/9781003242963
Abdel Latef AAH (2024) Biochar in Mitigating Abiotic Stress in Plants. Academic Press. Elsevier Inc. https://doi.org/10.1016/C2023-0-01389-9
Abdel Latef AAH (2025) Oilseed crops under abiotic stress: Mitigation strategies and future perspectives. Springer Nature, Singapore Pte Ltd. https://doi.org/10.1007/978-981-96-8346-8
Abdel Latef AAH, Hameed M, Khan Z, Khan S, Imran A, Fatima N (2026) Foliar stimulants: A sustainable strategy for enhancing plant resilience to abiotic and biotic stresses. Physiologia Plantarum 178(1):e70741
Abdel Latef AAH (2026a) Using stimulants to improve plant health: from stress to strength. Elsevier. https://doi.org/10.1016/C2025-0-01048-7
Abdel Latef AAH (2026b) Chitosan nanoparticles for plant resilience under abiotic and biotic stressors. Springer Singapore. https://doi.org/10.1007/978-981-95-6422-4
Abreu MS, Lima SF, Neto FMO, Garcia DH, Taveira AC, Thomé SEM, Quirinto TS (2020) *Ascophyllum nodosum* e nicotinamida afetam produtividade do feijoeiro comum. Res Soc Develop 9:01–16
Abu-tahon M, Hanafy R, Akladious S, Shahhat I (2025) Stimulatory effects of nicotinamide and Moringa oleifera leaf extract on growth and biochemical responses of Vicia faba under salt-induced stress. Egypt J Bot 65(3):365–376
Ahn IP, Kim S, Lee YH (2005) Vitamin B-1 functions as an activator of plant disease resistance. Plant Physiol 138:1505–1515
Ajjawi I, Milla MAR, Cushman J, Shintani DK (2007) Thiamin pyrophosphokinase is required for thiamin cofactor activation in Arabidopsis. Plant Mol Biol 65:151–162
Alban C, Job D, Douce R (2000) Biotin metabolism in plants. Annu Rev Plant Phys 51:17–47
Ali WS, Al-Bayaty AJAR (2023) Enhancing-nutrient-elements and-phenolic compounds in mentha piperita L. through the application of nano iron oxide and vitamin b3 (niacin) spraying treatments. J Surv Fish Sci 10:5307–5317
Araújo TADND, Vendruscolo EP, Binotti FFDS, Costa E, Lima SFD, Sant'Ana GR, Bortolheiro FPDAP (2024) Nicotinamide increases the physiological performance and initial growth of maize plant. Int J Agron 2024(1):5567314
Arditti J (1967) Niacin biosynthesis in germinating Laeliocattleya orchid embryos and young seedlings. Am J Bot 54:291–298
Arditti J, Tarr JB (1979) Niacin biosynthesis in plants. Am J Bot 66:1105–1113
Asensi-Fabado MA, Munne-Bosch S (2010) Vitamins in plants: occurrence, biosynthesis and antioxidant function. Trends Plant Sci 15:582–592
Ashihara H, Stasolla C, Yin YL, Loukanina N, Thorpe TA (2005) De novo and salvage biosynthetic pathways of pyridine nucleotides and nicotinic acid conjugates in cultured plant cells. Plant Sci 169:107–114
Azooz MM, Abdel Latef AAH, Abdelmajeed WG et al. (2026) Biochar ameliorates salt stress in maize by modulating physiological responses and expression of SOS1 and NHX2 genes. J Soil Sci Plant Nutr. https://doi.org/10.1007/s42729-026-03123-w
Bahuguna RN, Joshi R, Shukla A, Pandey M, Kumar J (2012) Thiamine primed defense provides reliable alternative to systemic fungicide carbendazim against sheath blight disease in rice (Oryza sativa L.). Plant Physiol Bioch 57:159–167
Bashir K, Todaka D, Sako K, Ueda M, Aziz F, Seki M (2025) Chemical application improves stress resilience in plants. Plant Mol Biol. 115(2):47. https://doi.org/10.1007/s11103-025-01566-w. Erratum in: Plant Mol Biol. 2025 Jul 23;115(4):91. https://doi.org/10.1007/s11103-025-01621-6. PMID: 40105987; PMCID: PMC11922999

Bassouny FM, Hassanein RA, Baraka DM, Khalil RR (2008) Physiological effects of nicotinamide and ascorbic acid on Zea mays plant grown under salinity stress. II-changes in nitrogen constituents, protein profiles, protease enzymes and certain inorganic cations. Aust J Basic Appl Sci 2(3):350–359

Berglund T, Kalbin G, Strid Å, Rydström J, Ohlsson AB (1996) UV-B-and oxidative stress-induced increase in nicotinamide and trigonelline and inhibition of defensive metabolism induction by poly (ADP-ribose) polymerase inhibitor in plant tissue. FEBS Lett 380(1–2):188–193

Berglund T, Lindström A, Aghelpasand H, Stattin E, Ohlsson AB (2016) Protection of spruce seedlings against pine weevil attacks by treatment of seeds or seedlings with nicotinamide, nicotinic acid and jasmonic acid. Forestry 89(2):127–135. https://doi.org/10.1093/forestry/cpv040. 2-s2.0-85032069291

Berglund T, Ohlsson AB (1995) Defensive and secondary metabolism in plant tissue cultures, with special reference to nicotinamide, glutathione and oxidative stress. Plant Cell Tissue Organ Cult 43(2):137–145

Bouhadi M, El Kouali M H, Samir K, Elbouhmadi K, Talbi M, Fougrach H (2024) Exogenous application of thiamine and nicotinic acid improves tolerance and morpho-physiological parameters of Lens culinaris under lead (Pb) exposure. J Plant Growth Regul 43:4185–4198

Bronzetti G, Cini M, Andreoli E, Caltavuturo L, Panunzio M, Croce CD (2001) Protective effects of vitamins and selenium compounds in yeast. Mutat Res 496:105–115

Campos MF, Ono EO, Boaro CSF, Rodrigues JD (2008) Análise de crescimento em plantas de soja tratadas com substâncias reguladoras. Biotemas 21. https://doi.org/10.5007/2175-7925.2008v21n3p53

Carlson LA (2004) Niaspan, the prolonged release preparation of nicotinic acid (niacin), the broad-spectrum lipid drug. Int J Clin Pract 58:706–713. https://doi.org/10.1111/j.1368-5031.2004.00233.x

Chand T, Savitri B (2016) Vitamin B3, niacin. In: Industrial biotechnology of vitamins, biopigments, and antioxidants, pp 41–65

Chaves MM, Oliveira MM (2004) Mechanisms underlying plant resilience to water deficits: prospects for water-saving agriculture. J Exp Bot 55:2365–2384

Chen H, Xiong LM (2005) Pyridoxine is required for post-embryonic root development and tolerance to osmotic and oxidative stresses. Plant J 44:396–408

Chen W, Wang W, Peng M, Gong L, Gao Y, Wan J, Wang S, Shi L, Zhou B, Li Z, Peng X, Yang C, Qu L, Liu X, Luo J (2016) Comparative and parallel genome-wide association studies for metabolic and agronomic traits in cereals. Nat Commun 7:12767

Chick H, Macrae TF, Martin AJP, Martin CJ (1938) Curative action of nicotinic acid on pigs suffering from the effects of a diet consisting largely of maize. Biochem J 32(1):10

Colinas M, Fitzpatrick TB (2015) Natures balancing act: examining biosynthesis de novo, recycling and processing damaged vitamin B metabolites. Curr Opin Plant Biol 25:98–106

Coxon KM, Chakauya E, Ottenhof HH, Whitney HM, Blundell TL, Abell C, Smith AG (2005) Pantothenate biosynthesis in higher plants. Biochem Soc T 33:743–746

Darby WJ, McNutt KW, Todhunter EN (1975) Niacin. Nutr Rev 33:289–297

da Silva Colla RE, de Lima SF, Vendruscolo E, Secco VA, Piati GL, dos Santos OF (2021) Does foliar nicotinamide application affect second crop corn (Zea mays)?. Revista de la Facultad de Ciencias Agrarias UNCuyo 53(2):64–70

Dawood MG, Abdel-Baky YR, El-Awadi MES, Bakhoum GS (2019) Enhancement quality and quantity of faba bean plants grown under sandy soil conditions by nicotinamide and/or humic acid application. Bull Natl Res Cent 43(1):28

Dawood MF, Abu-Elsaoud AM, Sofy MR, Mohamed HI, Soliman MH (2022) Appraisal of kinetin spraying strategy to alleviate the harmful effects of UVC stress on tomato plants. Environ Sci Pollut Res 29(35):52378–52398

Dawood MF, Tahjib-Ul-Arif M, Shirazy BJ, Latef AAHA (2025) Unraveling the role of κ-carrageenan on the combined effect of drought and chromium stress in wheat (Triticum aestivium L.). Plant Physiol Biochem 223:109781

Dawood MF, Tahjib-Ul-Arif M, Sohag AAM, Abdel Latef AAH (2023) Role of acetic acid and nitric oxide against salinity and lithium stress in Canola (Brassica napus L.). Plants 13(1):51

de Lima SF, Vendruscolo EP, Alves VCD, Arguelho JC, Pião JDA, Seron CDC et al (2024) Nicotinamide as a biostimulant improves soybean growth and yield. Open Agric 9(1):20220259

de Oliveira Silva HJ, de Souza MI, Vendruscolo EP, Rochetti ACA, Sant'Ana GR (2022) Concentration and moment of niacin application on Urochloa brizantha in relation to nitrogen fertilization. Revista de Ciências Agroveterinárias 21(3):344–348

Dong H, Beer SV (2000) Riboflavin induces disease resistance in plants by activating a novel signal transduction pathway. Phytopathology 90:801–811

Dong W, Stockwell VO, Goyer A (2015) Enhancement of thiamin content in *Arabidopsis thaliana* by metabolic engineering. Plant Cell Physiol 56(12):2285–2296. https://doi.org/10.1093/pcp/pcv148. 2-s2.0-84952837928

El-Bassiony HM (2005) Physiological responses of wheat to salinity alleviation by nicotinamide and tryptophan. Int J Agric Biol 7:653–659

Drewke C, Leistner E (2001) Biosynthesis of vitamin B6 and structurally related derivatives. Vitam Horm 61:121–155

El-Bassiouny HSM, Bakry BA, Attia AAEM, Allah MMA (2014) Physiological role of humic acid and nicotinamide on improving plant growth, yield, and mineral nutrient of wheat (*Triticum durum*) grown under newly reclaimed sandy soil. Agric Sci 5:687–700

Elvehjem CA, Madden RJ, Strong FM, Wooley DW (1937) Relation of nicotinic acid and nicotinic acid amide to canine black tongue. J Am Chem Soc 59:1767–1768

Elvehjem CA, Madden RJ, Strong FM, Wolley DW (2002) The isolation and identification of the anti-black tongue factor. 1937. J Biol Chem 277:e22

Farooq TH, Bukhari MA, Irfan MS, Rafay M, Shakoor A, Rashid MHU, Lin Y, Saqib M, Malik Z, Khurshid N (2022) Effect of exogenous application of nicotinic acid on morpho-physiological characteristics of Hordeum vulgare L. under water stress. Plants 11. https://doi.org/10.3390/plants11182443

Ferreira ES, da Silva Binotti FF, Costa E, Vendruscolo EP, Binotti EDC, Salles JS, Salles JS (2023) Vitamin B3 with action on biological oxide/reduction reactions and growth biostimulant in Chlorella vulgaris cultivation. Algal Res 76:103306

Fischer M, Bacher A (2005) Biosynthesis of flavocoenzymes. Nat Prod Rep 22:324–350

Fitzpatrick TB, Basset GJ, Borel P, Carrari F, DellaPenna D, Fraser PD, Hellmann H, Osorio S, Rothan C, Valpuesta V, Caris-Veyrat C, Fernie AR (2012) Vitamin deficiencies in hu-mans: Can plant science help? Plant Cell 24:395–414

Food and Nutrition Board (1998) Dietary reference intakes: thiamin, riboflavin, niacin, vitamin B-6, vitamin B-12, pantothenic acid, biotin, and choline. National Academy Press, Washington, DC, pp 123–149

Foyer CH, Kyndt T, Hancock RD (2020) Vitamin C in plants: novel concepts, new perspectives, and outstanding issues. Antioxid Redox Sign 32:463–485

Freese R, Lysne V (2023) Niacin—a scoping review for Nordic nutrition recommendations 2023. Food Nutr Res. https://doi.org/10.29219/fnr.v67.10299

Garcia AA, Vendruscolo EP, Ferreira de Lima S, de Castro Seron C, Battistuzzi Martins M, Rodrigues Sant'Ana G (2023) Effect of B vitamins on lettuce plants subjected to saline stress. Agronomía Colombiana 41(1):1

Gehring W (2004) Nicotinic acid/niacinamide and the skin. J Cosmet Dermatol 3:88–93. https://doi.org/10.1111/j.1473-2130.2004.00115.x

Hadwiger LA, Badiei SE, Waller GR, Gholson RK (1963) Quinolinic acid as a precursor of nicotinic acid and its derivatives in plants. Biochem Biophys Res Commun 13:466–471

Hankes LV (1984) Nicotinic acid and nicotinamide. In: Machlin LJ (ed) Handbook of vitamins. Marcel Dekker, New York, pp 329–377

Hanson AD, Gregory JF (2011) Folate biosynthesis, turnover, and transport in plants. Annu Rev Plant Biol 62:105–125

Hanson AD, Beaudoin GA, McCarty DR, Gregory JF 3rd (2016) Does abiotic stress cause functional B vitamin deficiency in plants? Plant Physiol 172:2082–2097

Harris HF (1919) Pellagra. The Macmillan Co., New York

Hashida S-n, Takahashi H, Uchimiya H (2009) The role of NAD biosynthesis in plant development and stress responses. Ann Bot 103:819–824

Hassanein RA, Bassony FM, Barakat DM, Khalil RR (2009) Physiological Effects of Nicotinamide and Ascorbic Acid on Zea mays Plant Grown under Salinity Stress. 1-Changes in Growth, Some Relevant Metabolic Activities and Oxidative Defense Systems. Res J Agric Biol Sci 5:72–81

Hosokawa Y, Mitchell E, Gholson RK (1983) Higher plants contain L-aspartate oxidase, the first enzyme of the Escherichia coli quinolinate synthetase system. Biochem Biophys Res Commun 111:188–193

Informatics (1974) Monograph on Nniacin. (Submitted by FDA in response to an FOI request–2000, 416 pages).2

Jackanicz TM, Byerrum RU (1966) Incorporation of aspartate and malate into the pyridine ring of nicotine. J Biol Chem 241:1296–1299

Katoh A, Uenohara K, Akita M, Hashimoto T (2006) Early steps in the biosynthesis of NAD in Arabidopsis start with aspartate and occur in the plastid. Plant Physiol 141:851–857

Kaya C, Ashraf M, Sonmez O, Tuna AL, Polat T, Aydemir S (2015) Exogenous application of thiamin promotes growth and antioxidative defense system at initial phases of development in salt- stressed plants of two maize cultivars differing in salinity tolerance. Acta Physiol Plant 37: 1741

Khudair TY, Albbas FAA, Kreem KAA (2019) Effect of niacin (nicotinamide) and humic acid on growth and chemical traits of Pelargonium hortorum L. Indian J Ecol 46:173–178

Khurshid N, Bukhari MA, Ahmad T, Ahmad Z, Jatoi WN, Abbas SM et al (2023) Exogenously applied nicotinic acid alleviates drought stress by enhancing morpho-physiological traits and antioxidant defense mechanisms in wheat. Ecotoxicol Environ Saf 263:115350

Kirkland JB, Meyer-Ficca ML (2018a) Niacin. In: Advances in food and nutrition research, vol 83. Academic Press, Cambridge, MA, pp 83–149

Kirkland JB, Meyer-Ficca ML (2018b) Niacin. Adv Food Nutr Res 83:83–149

Kohn HI (1938) The concentration of coenzyme-like substance in blood following the administration of nicotinic acid to normal individuals and pellagrins. Biochem J 32:2075–2083

Kornberg A, Pricer WE Jr (1950) Nucleotide pyrophosphatase. J Biol Chem 186:763–778

Krehl WA, Teply LJ, Elvehjem CA (1945) Effect of corn grits on nicotinic acid requirements of the dog. Proc Soc Exp Biol Med 58(4):334–337

Krehl WA (1949) Niacin in amino acid metabolism. In Vitamins & Hormones (Vol. 7, pp. 111–146). Academic Press

Leete E (1965) Biosynthesis of alkaloids. Science 147:1000–1006

Leonel EAS, Vendruscolo EP, Araújo TO, Seron CDC, Sant'Ana GR, Bortolheiro FPDAP et al (2025) Niacin as a biostimulant in carrot production in a minimum soil tillage system in Brazil. Int J Horticul Sci Technol 12(2):465–478

Li Y, Yang C, Ahmad H, Maher M, Fang C, Luo J (2021) Benefiting others and self: production of vitamins in plants. J Integr Plant Biol 63(1):210–227

Li X, Hu N, Li YS, Tang HS, Huang XM, Yang T, Xu JC (2024) Integrated ultrastructural, physiological, transcriptomic, and metabolomic analysis uncovers the mechanisms by which nicotinamide alleviates cadmium toxicity in Pistia stratiotes L. J Hazard Mater 467:14

Lima SF, Vendruscolo EP, Alves VCD, Arguelho JC, Pião JDA, Seron CDC, Martins MB, Witt TW, Serafim GM, Contardi LM (2024) Nicotinamide as a biostimulant improves soybean growth and yield. Open Agric 9:20220259

Loosli JK (1991) In: Putnam PA (ed) Handbook of animal science. Academic Press, San Diego

Mahmood S, Wahid A, Rasheed R, Hussain I, Basra SM (2012) Possible antioxidative role of endogenous vitamins biosynthesis in heat stressed maize (Zea mays). Int J Agric Biol 14(5)

Mann DF, Byerrum RU (1974) Quinolinic acid phosphoribosyltransferase from castor bean endosperm. I. Purification and characterization. J Biol Chem 249:6817–6823

Martinis J, Gas-Pascual E, Szydlowski N, Crevecoeur M, Gisler A, Burkle L, Fitzpatrick TB (2016) Long-distance transport of thiamine (vitamin B-1) is concomitant with that of polyamines. Plant Physiol 171:542–553

Martins AG, Rampim L, Rosset JS, Prior M, Coppo JC, Coppo J (2016) Aplicação de bioestimulante em sementes de milho cultivado em solos de diferentes texturas. Scientia Agraria Paranaensis 15(4):440–445. https://doi.org/10.18188/1983-1471/sap.v15n4p440-445

Matsui A, Yin Y, Yamanaka K et al (2007) Metabolic fate of nicotinamide in higher plants. Physiol Plant 131:191–200. https://doi.org/10.1111/j.1399-3054.2007.00959.x

McCollum EV (1957) A history of nutrition. Houghton Mifflin, Boston

Mehrshahi P, Gonzalez-Jorge S, Akhtar TA, Ward JL, Santoyo-Castelazo A, Marcus SE, Lara-Nunez A, Ravanel S, Hawkins ND, Beale MH, Barrett DA, Knox JP, Gregory JF, Hanson AD, Bennett MJ, DellaPenna D (2010) Functional analysis of folate polyglutamylation and its essential role in plant metabolism and development. Plant J 64:267–279

Miret JA, Munné-Bosch S (2014) Plant amino acid-derived vitamins: biosynthesis and function. Amino Acids 46(4):809–824

Mostafa N (2020) Effect of foliar application of some vitamins and irrigation intervals on vegetative growth, flowering, and some biochemical constituents of Helianthus annuus L. plants. Sci J Agric Sci 2(2):1–8

Ndaw S, Bergaentzle M, Aoude-Werner D, Hasselmann C (2002) Enzymatic extraction procedure for the liquid chromatographic determination of niacin in foodstuffs. Food Chem 78:129–134. https://doi.org/10.1016/S0308-8146(02)00205-4

Noctor G, Queval G, Gakie're B (2006) NAD(P) synthesis and pyridine nucleotide cycling in plants and their potential importance in stress conditions. J Exp Bot 57:1603–1620

Ohlsson AB, Landberg T, Berglund T, Greger M (2008) Increased metal tolerance in Salix by nicotinamide and nicotinic acid. Plant Physiol Biochem 46:655–664

Ortega MV, Brown GM (1960) Precursors of nicotinic acid in Escherichia coli. 2939–2945

Paruch K, Biernasiuk A, Khylyuk D, Paduch R, Wujec M, Popiołek Ł (2022) Synthesis, biological activity and molecular docking studies of novel nicotinic acid derivatives. Int J Mol Sci 23:2823. https://doi.org/10.3390/ijms23052823

Preiss J, Handler P (1957) Enzymatic synthesis of nicotinamide mononucleotide. J Biol Chem 225: 759–770

Preiss J, Handler P (1958) Biosynthesis of diphosphopyridine nucleotide. I. Identification of intermediates. J Biol Chem 233:488–492

Ragaey MM, Sadak MS, Dawood MF, Mousa NH, Hanafy RS, Latef AAHA (2022) Role of signaling molecules sodium nitroprusside and arginine in alleviating salt-induced oxidative stress in wheat. Plants 11(14):1786

Rapala-Kozik M, Kowalska E, Ostrowska K (2008) Modulation of thiamine metabolism in Zea mays seedlings under conditions of abiotic stress. J Exp Bot 59:4133–4143

Rapala-Kozik M, Wolak N, Kujda M, Banas AK (2012) The upregulation of thiamine (vitamin B-1) biosynthesis in Arabidopsis thaliana seedlings under salt and osmotic stress conditions is mediated by abscisic acid at the early stages of this stress response. BMC Plant Biol 12:2

Raschke M, Boycheva S, Crevecoeur M, Nunes-Nesi A, Witt S, Fernie AR, Amrhein N, Fitzpatrick TB (2011) Enhanced levels of vitamin B-6 increase aerial organ size and positively affect stress tolerance in Arabidopsis. Plant J 66:414–432

Roje S (2007) Vitamin B biosynthesis in plants. Phytochemistry 68:1904–1921

Rongvaux A, Andris F, Van Gool F, Leo O (2003) Reconstructing eukaryotic NAD metabolism. BioEssays 25:683–690

Rosado-Souza L, Proost S, Moulin M, Bergmann S, Bocobza SE, Aharoni A, Fitzpatrick TB, Mutwil M, Fernie AR, Obata T (2019) Appropriate thiamin pyrophosphate levels are required for acclimation to changes in photoperiod. Plant Physiol 180:185–197

Rubio S, Whitehead L, Larson TR, Graham IA, Rodriguez PL (2008) The coenzyme A biosynthetic enzyme phosphopantetheine adenylyltransferase plays a crucial role in plant growth, salt/osmotic stress resistance, and seed lipid storage. Plant Physiol 148:546–556

Ryrie IJ, Scott KJ (1969) Nicotinate, quinolinate and nicotinamide as precursors in the biosynthesis of nicotinamide-adenine dinucleotide in barley. Biochem J 115:679–685

Sabagh AEL, Hossain A, Islam MS, Barutcular C, Hussain S, Hasanuzzaman M, Akram T, Mubeen M, Nasim W, Fahad S (2019) Drought and salinity stresses in barley: consequences and mitigation strategies. Aust J Crop Sci 13:810–820
Sadak MS (2016) Physiological role of yeast extract and nicotinamide on Pisum sativum L. plants under heat stress. Int J PharmTech Res 9(9):170-178
Sako K, Futamura Y, Shimizu T, Matsui A, Hirano H, Kondoh Y, Muroi M, Aono H, Tanaka M, Honda K, Shimizu K, Kawatani M, Nakano T, Osada H, Noguchi K, Seki M (2020) Inhibition of mitochondrial complex I by the novel compound FSL0260 enhances high salinity-stress tolerance in *Arabidopsis thaliana*. Sci Rep. https://doi.org/10.1038/s41598-020-65614-9
Sako K, Nguyen HM, Seki M (2021) Advances in chemical priming to enhance abiotic stress tolerance in plants. Plant Cell Physiol 61:1995–2003. https://doi.org/10.1093/pcp/pcaa119
Sallam A, El-Defrawy MM, Dawood MF, Hashem M (2024a) Screening wheat genotypes for specific genes linked to drought tolerance. Genes 15(9):1119
Sallam A, Dawood MF, Jarquín D, Mohamed EA, Hussein MY, Börner A, Ahmed AA (2024b) Genome-wide scanning to identify and validate single nucleotide polymorphism markers associated with drought tolerance in spring wheat seedlings. Plant Genome 17(2):e20444
Salih SM, Abdulrraziq AA (2024) Effect of exogenous application of nicotinic acid on genotypes of durum wheat (Triticum aestivum L.) under salt stress. Sci J Faculty Sci Sirte Univ 4(1):109–116
Santos TS, Correia MRS, Sena LS, Santana LPS, Silva GBG, Lima KS, Dutra EVS, Adas ME, Ribeiro MCBO, Ribeiro JES et al (2025) The combination of salicylic acid, nicotinamide, and Proline mitigates the damage caused by salt stress in nasturtium (*Tropaeolum majus*). Plants 14(8):1156. https://doi.org/10.3390/plants14081156
Sen A, Puthur JT (2020) Seed priming-induced physiochemical and molecular events in plants coupled to abiotic stress tolerance: an overview. In: Priming-Med Stress Cross-Stress Toler Crop Plants, pp 303–316
Serrano M, Kombrink E, Meesters C (2015) Considerations for designing chemical screening strategies in plant biology. Front Plant Sci 6:131. https://doi.org/10.3389/fpls.2015.00131
Sharma NN (2009) Hyper induction of nitrilase in *Nocardia globerula* NHB-2 and its application in biotransformation of N-substituted aromatic nitriles. PhD thesis. Himachal Pradesh University, Shimla
Sinclair SJ, Murphy KJ, Birch CD, Hamill JD (2000) Molecular characterization of quinolinate phosphoribosyltransferase (QPRtase) in Nicotiana. Plant Mol Biol 44:603–617
Smolik B, Sędzik-Wójcikowska M (2024) Examining nicotinamide application methods in alleviating lead-induced stress in spring barley. Agronomy 14:1314. https://doi.org/10.3390/agronomy14061314
Song GC, Choi HK, Ryu C-M (2013) The folate precursor para- aminobenzoic acid elicits induced resistance against cucumber mosaic virus and Xanthomonas axonopodis. Ann Bot 111:925–934
Smith AG, Croft MT, Moulin M, Webb ME (2007) Plants need their vitamins too. Curr Opin Plant Biol 10:266–275
Taiz L, Zeiger E, Moller IM, Murphy A (2017) Fisiologia e desenvolvimento vegetal, 6th edn. Artmed, Porto Alegre
Tarr JB, Arditti J (1982) Niacin biosynthesis in seedlings of Zea mays. Plant Physiol 69:553–556
Thomé SEM, Lima SF, Oliveira IC, Contardi LM, Vendruscolo EP, Andrade MGO, Cordeiro MAS, Arguelho JC, Oliveira JJ (2023) Biostimulants increase growth and yield of second-crop maize. Revista Brasileira de Engenharia Agrícola e Ambiental 27:550–558. https://doi.org/10.1590/1807-1929/agriambi.v27n7p550-558
Tunc-Ozdemir M, Miller G, Song LH, Kim J, Sodek A, Koussevitzky S, Misra AN, Mittler R, Shintani D (2009) Thiamin confers enhanced tolerance to oxidative stress in Arabidopsis. Plant Physiol 151:421–432
van Eys J (1991) Nicotinic acid. In: Machlin L (ed) Handbook of vitamins. Marcel Dekker, New York, pp 311–340

Vendruscolo EP, Seleguini A (2020) Effects of vitamin pre-sowing treatment on sweet maize seedlings irrigated with saline water. Acta Agronomica 69(1):20–25. https://doi.org/10.15446/acag.v69n1.67528

Vendruscolo EP, Rodrigues AHA, Martins APB, Campos LFC, Seleguini A (2015) Tratamento de sementes com niacina ou tiamina promove o desenvolvimento e a produtividade do feijoeiro. Revista de Ciências Agroveterinárias 17(1):83–90. https://doi.org/10.5965/223811711712018083. 2-s2.0-85046963109

Vendruscolo EP, Oliveira PR, Seleguini A (2017) Aplicação de niacina ou tiamina promovem incremento no desenvolvimento de mostarda. Cultura Agronômica 26:433–442. https://doi.org/10.32929/2446-8355.2017v26n3p433-442

Vendruscolo EP, Rodrigues AHA, Oliveira PR, Leitão RA, Campos LFC, Seleguini A, Lima SF (2019) Exogenous application of vitamins in upland rice. Revista de Agricultura Neotropical 6(2):1–6. https://doi.org/10.32404/rean.v6i2.3241

Vendruscolo EP, Seron CDC, Leonel EA, Lima SFD, Araujo SL, Martins MB, Oliveira JJD (2023) Do vitamins affect the morphophysiology of lettuce in a hydroponic system? Revista Brasileira de Engenharia Agrícola e Ambiental 27:698–703. https://doi.org/10.1590/1807-1929/agriambi.v27n9p698-703

Wahlberg G, Adamson U, Svensson J (2000) Pyridine nucleotides in glucose metabolism and diabetes: a review. Diabetes-Metab Res 16:33–42

Waller GR, Yang KS, Gholson RK, Hadwiger LA, Chaykin S (1966) The pyridine nucleotide cycle and its role in the biosynthesis of ricinine by Ricinus communis L. J Biol Chem 241:4411–4418

Wang GN, Ding XH, Yuan M, Qiu DY, Li XH, Xu CG, Wang SP (2006) Dual function of rice OsDR8 gene in disease resistance and thiamine accumulation. Plant Mol Biol 60:437–449

Wittek F, Kanawati B, Wenig M, Hoffmann T, Franz-Oberdorf K, Schwab W, Schmitt-Kopplin P, Vlot AC (2015) Folic acid induces salicylic acid-dependent immunity in Arabidopsis and enhances susceptibility to Alternaria brassicicola. Mol Plant Pathol 16:616–622

Younes NA, El-Sherbiny M, Alkharpotly AA, Sayed OA, Dawood AF, Hossain MA, Dawood MF (2024) Rice-husks synthesized-silica nanoparticles modulate silicon content, ionic homeostasis, and antioxidants defense under limited irrigation regime in eggplants. Plant Stress 11:100330

Yu Z, Li H, Dai Y, Huang X (2025) Exogenous niacin enhances Cd tolerance in pepper via NADP (H) pool expansion, photosynthetic electron transport recovery and antioxidant system activation. Sci Hortic 352:114438

Zaki FS, Khater MA, El-Awadi ME, Dawood MG, Shamoon MS, Shalaby MA, El-Din KGM (2022) The impact of α-tocopherol and nicotinamide on performance of lupine plant grown under sandy soil conditions. Egyptian Journal of Chemistry 65(132):1231–1240

Zhang SJ, Yang X, Sun MW, Sun F, Deng S, Dong HS (2009) Riboflavin-induced priming for pathogen defense in Arabidopsis thaliana. J Integr Plant Biol 51:167–174

Zheng XQ, Hayashibe E, Ashihara H (2005) Changes in trigonelline (N-methylnicotinic acid) content and nicotinic acid metabolism during germination of mungbean (Phaseolus aureus) seeds. J Exp Bot 56:1615–1623

Zhu X, Pan L, Xiao T, Ren X, Liu Z (2018) Exogenous niacin treatment increases NADPH oxidase in kiwifruit. Braz J Biol 78(4):686–690. https://doi.org/10.1590/1519-6984.173709. Epub 2018 Feb 1. PMID: 29412249

Chapter 4
Pyridoxine: Regulation Plant Defense Mechanisms

Mona F. A. Dawood, Mamdouh Eissa, Samar G. Thabet, Murad Muhammad, Zainul Abideen, Mohamed Imran Mir, and Arafat Abdel Hamed Abdel Latef

Abstract In plants, pyridoxine (Vitamin B6) plays a vital role in growth, development, and stress response. It acts as a cofactor for many enzymes involved in amino acid metabolism, hormone production, and other key biochemical pathways. Beyond its metabolic function, pyridoxine serves as a strong antioxidant, directly neutralizing reactive oxygen species (ROS) and protecting plant cells from oxidative damage caused by environmental stresses like high light intensity. Adequate levels of pyridoxine are associated with increased photosynthetic efficiency, better root structure, and overall plant health. On the other hand, a deficiency of this vitamin often leads to stunted growth, reduced photosynthetic activity, and increased

M. F. A. Dawood (✉)
Botany and Microbiology Department, Faculty of Science, Assiut University, Assiut, Egypt
e-mail: Mo_fa87@aun.edu.eg

M. Eissa
Department of Soils and Water, Faculty of Agriculture, Assiut University, Assiut, Egypt

S. G. Thabet
Department of Botany, Faculty of Science, Fayoum University, Fayoum, Egypt

M. Muhammad
State Key Laboratory of Desert and Oasis Ecology, Key Laboratory of Ecological Safety and Sustainable Development in Arid Lands, Xinjiang Institute of Ecology and Geography, Chinese Academy of Sciences, Urumqi, China

University of Chinese Academy of Sciences, Beijing, People's Republic of China

Z. Abideen
Dr. Muhammad Ajmal Khan Institute of Sustainable Halophyte Utilization, University of Karachi, Karachi, Pakistan

Department of Applied Biology, University of Sharjah, Sharjah, United Arab Emirates

M. I. Mir
Department of Plant Sciences, School of Life Sciences, University of Hyderabad, Hyderabad, Telangana, India

A. A. H. Abdel Latef
Department of Botany and Microbiology, Faculty of Science, Qena University, Qena, Egypt

A. A. H. Abdel Latef (ed.), *Harnessing Vitamins and Organic Acids for Plant Stress Resilience*, Sustainability Sciences in Asia and Africa,
https://doi.org/10.1007/978-981-95-8593-9_4

vulnerability to abiotic stresses. By combining its roles in enzyme function and antioxidant activity, pyridoxine supports both the fundamental metabolism and defense systems of plants, making it essential for maintaining plant productivity under normal and stressful conditions.

Keywords Pyridoxine · Vitamin B6 · Biosynthesis · Biotic stress · Abiotic stress · Genetic engineering

4.1 Introduction

Boosting crop production is essential for achieving food security, especially as the global population continues to grow and environmental conditions become more challenging (Hickey et al. 2019; Younes et al. 2021; Dawood 2025; Abdel Latef 2026). Chemical treatments applied to plants by various methods in the form of priming, foliar spray, or soil fertigation have emerged as a sustainable solution for enhancing crop production under changing environmental conditions (Rezvi et al. 2023; Abdel Latef et al. 2026). Vitamins are vital for the metabolism of both plants and animals, so applying different forms of vitamin B can be an effective way to strengthen plant tolerance to environmental stresses. Their importance comes from their redox properties, their function as cofactors in many biochemical reactions, and their strong antioxidant activity (El-Bassiouny et al. 2023). Pyridoxine (vitamin B6) in its active form represents an essential metabolite in all organisms. It can act as a coenzyme for numerous metabolic enzymes and is a potent antioxidant (Dolatabadian and Modarressanavy 2008). Moreover, pyridoxine is an essential cofactor for numerous metabolic pathways, such as amino acid metabolism and antibiotic production. Generally, pyridoxine plays a role as an antioxidant defense in plant tissues and studies mentioned the presence of vitamin B6 compounds in chloroplasts; thus, Havaux et al. (2009) stated that vitamin B6 functions as a photoprotector that limits active oxygen species accumulation and prevents superoxide anion-mediated oxidative damage in chloroplasts. However, studies on the application of pyridoxine to plants are not enough, and further studies are needed at this point to explain its vital role in plant development and its applicability to agricultural fields. Thus, more focus on recent studies gathering the role of pyridoxine and other related vitamin B6 vitamers in plants under biotic and abiotic stress is needed.

4.2 Vitamins and Pyridoxine

Vitamin B6 includes several related compounds, such as pyridoxine, pyridoxal, pyridoxamine, and their phosphorylated forms. All living organisms depend on this vitamin, mainly in its active form, pyridoxal 5′-phosphate (PLP), which functions as an essential metabolic cofactor (Percudani and Peracchi 2003). Numerous metabolic enzymes may utilize it as a coenzyme and it is also an effective antioxidant (Havaux et al. 2009). Plants and many microbes can produce vitamin B6 through a de novo biosynthesis pathway, and some plant species depend on this pathway for proper growth and development (Tambasco-Studart et al. 2005). However, not all plants can make vitamin B6 in their roots. In some species, the roots rely on vitamin B6 that is transported from the shoots (Bonner 1942). For this reason, foliar application can be an effective way to supply vitamin B6, especially for plants whose roots cannot synthesize it (Mozafar and Oertli 1992). Vitamin B6 acts as a cofactor for many metabolic enzymes, including those involved in amino acid metabolism and the synthesis of important biological compounds. Vitamin B6, as previously mentioned, is a group of six water-dissolved form, where PLP is the most crucial compound because of being cofactor for more than 140 enzymatic reactions in the cell (Hellmann and Mooney 2010). These reactions include processes related to amino acids, sugars, and fatty acids (Percudani and Peracchi 2003). Only the groups, e.g., fungi, bacteria, and plants biosynthesize vitamin B6 in their tissues, while animals and humans must take it externally from their diet. Therefore, vitamin B6 deficiency can occur when food intake is low, diets lack diversity, or chronic health conditions are present (di Salvo et al. 2012).

Since plants can synthesize and store vitamin B6 (Raschke et al. 2011; Chen and Xiong 2009), promoting its content in crops has strong potential to improve both human nutrition and plant performance (Bhullar and Gruissem 2013). Research over the past decade has identified the major genes that control vitamin B6 biosynthesis in *Arabidopsis*, highlighting the vitamin's importance in plant growth, photosynthesis, and stress tolerance (Mooney and Hellmann 2010). This knowledge opens opportunities to develop crop varieties with higher vitamin B6 content to enhance yield, stress resilience, and nutritional value. Studies have also shown that external applications of vitamin B6 can improve plant performance. Mahdi et al. (2021) reported an increase in pyridoxine concentration up to 1750 mg. Earlier work by Desouky (1995) and Hamada and Khulaef (2000) found that priming the seeds of faba bean in pyridoxine enhanced photosynthetic pigments and increased photosynthetic activity. Similarly, Khan et al. (2001) recorded that wheat plants treated with an optimal level of vitamin B6 showed the greatest improvement in growth traits.

4.3 History of Discovery

The chemical structure of vitamin B6 and its related compounds (hereafter referred to as vitB6) was first reported by Ohdake in 1932. During his research on isolating "Oryzanin" (vitamin B1) from rice polishings, vitamin B6 was unintentionally obtained as a secondary product (Ohdake 1932; Wiardy 1938). Ohdake first described the chemical formula of this compound, although he did not realize that he had discovered a vitamin or understand its biological importance. During this period, several researchers were actively studying and characterizing members of the vitamin B family (György 1934, 1938; Ichiba and Michi 1938; Keresztesy and Stevens 1938; Kuhn and Wendt 1938; Lepkovsky 1938; György and Eckardt

1939; Harris and Folkers 1939). Their goal was to identify the "rat pellagra prevention factor," a substance capable of curing acrodynia, a skin disorder in rats that resembles pellagra. These scientists discovered that this condition could be treated by adding a specific yeast extract to the diet. Paul György, a Hungarian-born researcher, was the first to identify vitamin B6 as the active compound responsible for this protective effect. (György 1934). In 1938, other researchers isolated the crystalline vitB6 from yeast (György 1938; Ichiba and Michi 1938; Keresztesy and Stevens 1938; Kuhn and Wendt 1938; Lepkovsky 1938). After the determination of the structure of vitB6 in 1939, György called the discovered vitamin pyridoxine because of its structural similarity to pyridine (György and Eckardt 1939). Then, the synthesis of vitB6 was accomplished by Harris and Folkers (1939).

Pyridoxine is an essential molecule for overall cellular metabolism (Hellmann and Mooney 2010). Since it was first identified in 1934 by György and colleagues, vitamin B6 has been recognized as a cofactor in more than 140 biochemical reactions inside the cell (Mooney et al. 2009). While many of these reactions involve the synthesis and breakdown of amino acids, vitamin B6 also plays important roles in fatty acid metabolism, the mobilization of storage compounds in both plants and animals, and the biosynthesis of plant hormones, neurotransmitters, and organelle-specific molecules such as chlorophyll (Adams et al. 2006; Choudhury et al. 2010; Cuesta-Seijo et al. 2017; Geng et al. 1995; Plecko and Stöckler 2009; Tsang et al. 2003; Ercan-Fang et al. 2005).

Vitamin B6 also has antioxidant properties, as it can neutralize reactive oxygen species (ROS) (Havaux et al. 2009). Because of its contribution to ROS scavenging and its role in chlorophyll formation, vitamin B6 supports photosynthesis and is considered a promising factor for improving plant tolerance to different abiotic and biotic stresses (Havaux et al. 2009). Vitamin B6 refers to a group of six closely related compounds that all share a pyridine ring structure. They differ in the chemical group attached to the 4′ position of the ring: an amino-methyl group in pyridoxamine (PM), a hydroxymethyl group in pyridoxine (PN), or an aldehyde group in pyridoxal (PL) (Parra et al. 2018).

4.4 Pyridoxine Biosynthesis

Vitamin B6, as PLP, is produced in plants, fungi, and eubacteria from glyceraldehyde 3-phosphate, ribose-5-phosphate, and glutamine through the multisubunit complex, 24-polypeptide, as revealed in studies on the *Bacillus subtilis*. This complex has 12 units of PDX1 synthase protein unit besides 12 PDX2 glutaminase protein units. In plants, a single gene encodes PDX2—its mutation being lethal—while three genes (PDX1.1–PDX1.3) encode PDX1 proteins. Among these, PDX1.1 and PDX1.3 function as enzymatically active proteins (Wagner et al. 2006). In contrast, PDX1.2 is not catalytically active; rather, it serves a regulatory role in vitamer biosynthesis by interacting with the multisubunit enzyme complex, particularly under stress conditions (Moccand et al. 2014; Dell'Aglio et al. 2017).

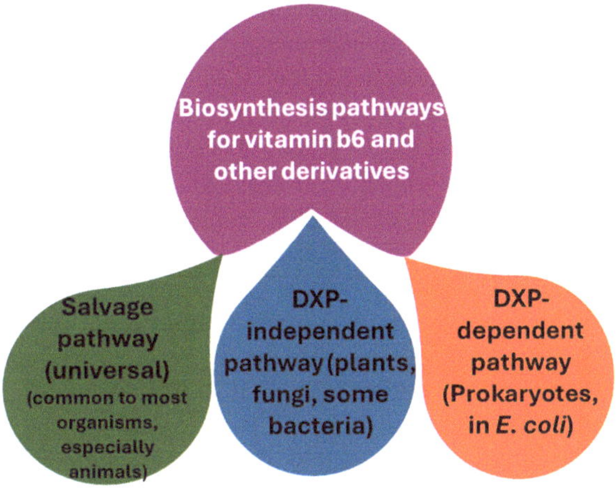

Fig. 4.1 The biosynthesis pathways of vitamin B6 and other derivatives

Environmental factors inducing stress in *Arabidopsis thaliana* plants mutated in one of the PDX1 subunit genes include salt and osmotic stress (Chen and Xiong 2005; Titiz et al. 2006), high light and photo-oxidative stress (Raschke et al. 2011; Moccand et al. 2011), high temperature (Havaux et al. 2009; Dell'Aglio et al. 2017), and ultraviolet-B radiation (Ristilä et al. 2011; Czégény et al. 2014).

Pyridoxine biosynthesis occurs through two de novo pathways, the bacterial DXP-dependent route and the more widespread DXP-independent (PDX1/PDX2) route, along with a universal salvage pathway that maintains PLP homeostasis (Fig. 4.1). While animals rely exclusively on the salvage pathway, plants and fungi depend on both systems, with salvage functions becoming especially critical under stress or for organelle PLP supply, as details have been provided by Parra et al. (2018). Pyridoxine occurs in several forms (pyridoxine, pyridoxal, pyridoxamine, and their phosphorylated derivatives), with PLP being the biologically active coenzyme form. In the DXP-dependent pathway (classical, in *E. coli*), it is long considered the only known pathway, utilizing deoxyxylulose 5′-phosphate (DXP) and 4-phosphohydroxy-L-threonine with key enzymes: PDX A (4-hydroxythreonine-4-phosphate dehydrogenase) and PDX J (pyridoxine 5′-phosphate synthase). In this pathway, pyridoxine 5′-phosphate (PNP) is produced, which is then converted to PLP by FMN-dependent PNP oxidase (PNPox) (Drewke et al. 1996; Notheis et al. 1995; Mittenhuber 2001).

The DXP-independent pathway (plants, fungi, some bacteria) was discovered later and considered the main de novo pathway in eukaryotes. This pathway included various precursors such as a 3-carbon sugar (glyceraldehyde-3-phosphate or dihydroxyacetone phosphate), a pentose phosphate (ribose-5-P or ribulose-5-P), and glutamine. This pathway is catalyzed by a large PLP synthase complex formed by PDX1 and PDX2 enzymes (Tambasco-Studart et al. 2007; Tambasco-Studart et al. 2005; Titiz et al. 2006; Wagner et al. 2006). It is essential in plants, as shown by *Arabidopsis* mutants of PDX1, which displayed stunted growth, delayed flowering, and altered seed composition (Titiz et al. 2006; Wagner et al. 2006; Boycheva et al. 2015; Raschke et al. 2011). However, the salvage pathway converts dietary or intracellular VitB6 vitamers into PLP. This pathway depends on enzymes like PDXK (kinase), phosphorylates pyridoxine (PN), pyridoxamine (PM), or pyridoxal (PL), DXH (oxidase) convert PNP and PMP to PLP, and PRL (pyridoxal reductase) interconvert vitamers (Gonzalez et al. 2007; Yang et al. 1996, 1998; Tang et al. 2005; Di Salvo et al. 2011; Biswal et al. 2006; Safo et al. 2005; Herrero et al. 2011). This pathway is critical in animals (which lack de novo pathways) and important in plants/fungi for organelle PLP supply and stress tolerance (Gonzalez et al. 2007; Sang et al. 2007; Shi et al. 2002; Colinas et al. 2016; Kirk and Leech 1972; Rueschhoff et al. 2013).

Both de novo and salvage pathways are tightly co-regulated. In *Arabidopsis*, loss of salvage enzymes (e.g., PDXH) increased stress sensitivity and altered nitrogen metabolism due to accumulation of PMP, which inhibits nitrate reductase (Gonzalez et al. 2007; Colinas et al. 2016). However, in animals, PLP binds serum albumin for systemic transport, requiring dephosphorylation before uptake (Schaeffer et al. 1989; Merrill et al. 1978; Narisawa et al. 2001; Whyte et al. 1985, 1988).

4.5 Transport of Pyridoxine

In plants, vitamin B6 transport has been reported despite their capacity for de novo biosynthesis. Only a few transporters have been recognized at the cell membrane and intracellular organs, and their functional characterization remains limited (Whittaker 2016). Current evidence indicates that, like yeast, plants possess specialized transport systems that facilitate the cellular uptake and intracellular distribution of vitamin B6 vitamers, ensuring sufficient availability for metabolic and antioxidant functions (Stolz and Vielreicher 2003; Stolz et al. 2005). Although vitamin B_6 is synthesized de novo in plants, the successful external application of PL, PM, and PN (Huang et al. 2011) indicates that plants can utilize the externally applied non-phosphorylated form of B_6 vitamers. This is further supported by the rescue of root and leaf developmental defects observed in *Arabidopsis pdx1.1* and *pdx1.3* mutants (Titiz et al. 2006; Wagner et al. 2006), as well as by the restoration of embryo development in the *Arabidopsis pdx2* mutant, whose embryos arrest at the globular stage, through direct pyridoxine application (Tambasco-Studart et al. 2007). The latter finding also suggests that vitamin B_6 is transferred between the embryo and surrounding maternal tissues. Furthermore, a purine permease (PUP1) was shown to contribute to vitamin B_6 recycling during guttation (Szydlowski et al. 2013), yet no additional B6 transporters have been identified in plants. Because de novo vitamin B_6 biosynthesis takes place in the cytosolic region (Tambasco-Studart et al. 2005), diffusion or active transport across organelle membranes must occur to sustain the activity of organelle-localized enzymes that utilize vitamin B6 as a cofactor (Gerdes et al. 2012). Given the high polarity of phosphorylated forms of B6, passive diffusion is unlikely (Mooney and Hellmann 2010), further implying the necessity of specific transporters.

4.6 Pyridoxine and Plant Development Under Non-Stress Conditions

The active B6 vitamer acts as a cofactor for more than 200 enzymatic reactions, many of which are stimulants for plant growth, development, and stress responses (Fitzpatrick 2024; Liu et al. 2022a, b). The significance of B6 vitamers for plants can be illustrated by studying mutant genotypes that suffer from vitamin B6 biosynthesis deficiency. Wagner et al. (2006) reported that a decrease in de novo production of *Arabidopsis* mutant, *rsr4-1*, changes the concentrations of metabolites like free amino acids and organic acids. Additionally, *Arabidopsis* mutants, i.e., *rsr4-1, pdx1.1, pdx1.3,* that exhibited lower concentrations of vitamin B6 are phenotypically different from wild-type plants, where damaged growth symptoms like deteriorated development of the seedling, decreased plant growth, and postponed flowering (Chen and Xiong 2005; Tambasco-Studart et al. 2005, 2007; Titiz et al. 2006; Wagner et al. 2006). While the damage of embryos was denoted in complete biosynthesis knockout mutants (i.e., *pdx2, pdx1.1, pdx1.3*) as reported in previous research (Tambasco-Studart et al. 2005; Titiz et al. 2006). Several studies have shown that certain forms of vitamin B6 act as strong antioxidants, thus this vitamin helps protect cells from oxidative damage (Denslow et al. 2005). This role is also supported by metabolite analyses of the *Arabidopsis* rsr4-1 mutant, which cannot make vitamin B6. These mutants display higher levels of lipid peroxidation, indicating increased oxidative stress (Lytovchenko et al. 2009). Pyridoxine and its related forms play two major roles in plants: normal plant growth and development (Raschke et al. 2011; Boycheva et al. 2015) as well as improving plant tolerance to stress (Raschke et al. 2011; Dell'Aglio et al. 2017). The developmental functions of vitamin B6 largely come from its active form, PLP, which acts as an essential cofactor for many enzymes implicated in producing key cellular components (Percudani and Peracchi 2003). However, the ability of B6 vitamers to neutralize reactive oxygen species is a major reason why they help plants survive stress (Czégény et al. 2014; Havaux et al. 2009). Pyridoxine is especially important for root growth after germination (Chen and Xiong 2005) and contributes to the expansion of shoot tissues in *Arabidopsis* (Raschke et al. 2011). Many studies have shown that pyridoxine improves overall plant growth and development (Table 4.1). PLP, the active form of pyridoxine, functions as a coenzyme for many metabolic reactions in all living organisms (Di Salvo et al. 2011). Recent work has also confirmed its strong antioxidant activity (Ristilä et al. 2011). In plants, pyridoxine acts as an effective antioxidant capable of quenching reactive oxygen species, including superoxide and singlet oxygen (Chen and Xiong 2005). Earlier research reported that pyridoxine promotes root system development (Shimasaki and Fukumoto 1998), improves nutrient uptake, and increases crop yield (Lone et al., 1999). Its importance for growth and differentiation has also been documented in several plant species (Proebsting et al. 1990). Hamada and Khulaef (2000) reported that exogenous application of pyridoxine as priming of faba bean seeds or spraying of faba bean seedlings enhanced the dry weights, photosynthetic pigments, and the

Table 4.1 The effect of pyridoxine on the development of plants under non-stressed conditions

Plant	Pyridoxine addition way	Effect of pyridoxine on plants	References
Bean (*Phaseolus vulgaris*)	Soaking seeds or spraying (100 ppm)	Stimulated biomass and enhanced pigments-related to photosynthesis with increment of net photosynthetic rate	Hamada and Khulaef (2000)
Mustard (*Brassica juncea*) and wheat (*Triticum aestivum*)	Pyridoxine (0.02%)	Improved growth and yield parameters	Khan et al. (2001)
Sunflower (*Helianthus annuus*) and rapeseed (*Brassica napus*)	Pyridoxine (100, 200, or 400 ppm)	Stimulated germination, attenuated the activity of catalase and peroxidation of lipids in the germinated seedlings	Dolatabadian and Modarressanavy (2008)
Corn (*Zea mays*)	Pyridoxine pre-sowing seed (200 ppm)	Improved the germinability of seeds and early growth traits of seedlings	Asli and Houshmandfar (2011)
Lupin (*Lupinus termis*)	Spraying	Improved growth, yield, seed quality, and anatomical features	Boghdady (2013)

rate of net photosynthesis. Khan et al. (2001) reported that application of 0.02% pyridoxine improved growth and yield traits for mustard and wheat plants. Also, Dolatabadian and Modarressanavy (2008) observed that pyridoxine stimulated germinability and retarded the accumulation of lipid peroxidation in germinated seeds of sunflower and rape seed. Similarly, Asli and Houshmandfar (2011) reported that pyridoxine priming prompted seed germination and growth characteristics of maize seedlings. Boghdady (2013) reported that foliar application of 300 ppm pyridoxine markedly enhanced the anatomical structure of *Lupinus termis* by increasing stem and leaf thickness. This enhancement was primarily associated with a greater development of epidermal, cortical, and vascular tissues, as well as an expansion of the parenchymatous area and pith diameter. Moreover, pyridoxine treatment increased the thickness of both the midvein and lamina, mainly through the enlargement of spongy tissues, while vascular bundles exhibited wider vessels, reflecting improved structural adaptation and potential physiological efficiency under growth conditions. As regards the photosynthetic pigments, Hendawy and Ezz El-Din (2010) explained that the B-vitamin complex functions as a group of co-enzymes that support many metabolic reactions in plants. These vitamins help enzymes break down carbohydrates, fats, and proteins, and they also play important roles in both photosynthesis and respiration. Foliar application of 200 ppm pyridoxine in *Sesamum indicum* cv. Shandaweel-3 markedly influenced the composition of seed oil fatty acids. The treatment caused a slight reduction in the proportion of unsaturated fatty acids, mainly due to a decrease in linoleic acid content, while oleic acid showed a slight increase compared to the control. On the other hand, saturated fatty acids increased, reflecting higher levels of palmitic, stearic, and arachidic acids in the oil of treated plants. These results indicate that pyridoxine application alters lipid metabolism, promoting a relative increase in saturated fatty acids and potentially enhancing oil stability and quality in sesame seeds (Nassar et al. 2017). Danon et al. (2005) stated that the addition of pyridoxine to the conditional *flu* mutant of *Arabidopsis* in the growth medium attenuated the production of singlet oxygen that causes cell death. This protective effect was also observed in plants engineered to accumulate higher levels of vitamin B6, where the *flu* mutant phenotype was noticeably weakened (Raschke et al. 2011). Likewise, adding pyridoxine was shown to recover the levels of the D1 protein in the *Arabidopsis pdx1.3* mutant, which otherwise shows reduced accumulation of this photosystem II component (Titiz et al. 2006). The breakdown of this protein is caused by singlet oxygen produced in the photosystem II reaction center; therefore, its stability indicates the oxidative status of the chloroplast. Treatment of leaf discs with vitamin B6 was shown to reduce singlet oxygen accumulation under high-light conditions in both wild-type plants and pdx1.3 mutants (Havaux et al. 2009).

4.7 The Role of Pyridoxine Under Stressful Conditions

4.7.1 Abiotic Stress

Arabidopsis plants increase their vitamin B6 content when exposed to UV-B radiation (Ristilä et al. 2011), and transgenic plants with higher vitamin B6 levels show greater tolerance to oxidative stress (Raschke et al. 2011). In contrast, mutants with reduced vitamin B6 content are more sensitive to stressful conditions, including high sucrose, salt, and mannose levels, as well as high light intensity, UV-B exposure, and oxidative stress (Chen and Xiong 2005; Titiz et al. 2006; Havaux et al. 2009; Ristilä et al. 2011), revealing its significance in plant stress responses besides previously mentioned developmental criteria. Vitamin B6 protects cells from oxidative damage due to the antioxidant activity of vitamin B6 that outperforms vitamins C and E (Ehrenshaft et al. 1998; Ehrenshaft et al. 1999a; Bilski et al. 2000, 2002; Graham et al. 2004). Mutations in genes required for the salvage pathway or the de novo production of vitamin B6 cause several plant phenotypes that are highly sensitive to salt stress and ROS (Ehrenshaft et al. 1999a, b; Gonzalez et al. 2007). Evidence from different organisms also displays that vitamin B6 is crucial for coping with oxidative and other abiotic stresses. For instance, loss of the *Arabidopsis thaliana* gene PDX1.3 results in strong sensitivity to Rose Bengal, a compound that generates ROS (Chen and Xiong 2005). *Arabidopsis pdx1.3* mutants are also more vulnerable to salt stress and UV-B exposure (Titiz et al. 2006). Interestingly, mutations in the vitamin B6 salvage pathway also cause abnormal stress responses, highlighting that proper vitamin B6 metabolism is necessary for abiotic stress tolerance. A well-known example is the SOS4 mutant of *Arabidopsis*, which carries a defect in the pyridoxine kinase PDXK and shows extreme sensitivity to salt stress (Shi et al. 2002). However, unlike the mutants defective in PDX1 genes, which consistently display reduced vitamin B6 levels, the sos4 mutant accumulates higher than normal amounts of vitamin B6 (Titiz et al. 2006; Wagner et al. 2006; Leuendorf et al. 2008; Gonzalez et al. 2007). Thus, pyridoxine plays a vital role in promoting plant growth and enhancing tolerance to abiotic stresses through its involvement in numerous metabolic and antioxidant processes. It functions as a coenzyme in amino acid, carbohydrate, and lipid metabolism, thereby supporting cell division and overall plant development. Under salinity stress, pyridoxine contributes to maintaining ion balance and protecting cellular structures by stimulating the antioxidant defense system. In *Vicia faba*, foliar application of pyridoxine at 50 ppm significantly enhanced seedling growth by approximately 35% compared to untreated controls and improved the accumulation of photosynthetic pigments, soluble sugars, and proteins under saline conditions. This effect is attributed to its strong antioxidant capacity, which aids in scavenging reactive oxygen species and stabilizing cellular metabolism under stress. Thus, pyridoxine acts not only as a growth stimulant but also as a protective agent, promoting plant resilience through improved redox homeostasis and metabolic regulation under salinity stress (Ahmed and Sattar 2024). Pereyra et al. (2012) reported that a change in vessel size caused by pyridoxine application has an important role in the adaptation to unfavorable environmental conditions.

Previous studies illustrated that oxidative stress strongly induced PDX1.2 expression, which is shared in the biosynthesis of vitamin B6 in *Arabidopsis* (Moccand et al. 2014). Bagri et al. (2018) stated that high expression of the *PDX-II* gene also increased the resistance of potato to salinity. In addition, vitamin B6 biosynthesis-deficient mutant plants showed high sensitivity to salinity stress (Gonzalez et al. 2007; Titiz et al. 2006). In maize, mutants that cannot properly produce vitamin B6 show severe developmental problems. For example, the small kernel2 (smk2) homozygous mutant has an embryonic lethal phenotype, meaning the embryos cannot develop normally. The SMK2 gene in maize performs a role similar to PDX2.1 in *Arabidopsis*, which is also essential for vitamin B6 biosynthesis (Yang et al. 2017). In *Arabidopsis thaliana*, vitamin B6 plays both direct and indirect roles in leaf acclimation to UV-B radiation. Studies using the vitamin B6-deficient mutant rsr4-1 and the wild type C24 revealed that after exposure to UV-B ($3.9\,kJ\,m^{-2}\,d^{-1}$ for 4 days), rsr4-1 exhibited a greater decline in photochemical yield compared to the control. In the absence of UV-B, rsr4-1 leaves contained significantly lower levels of pyridoxal-5′-phosphate (PLP), which increased to wild-type levels upon UV-B exposure. Furthermore, class-III ascorbate and glutathione peroxidase activities rose in C24 under UV-B but not in rsr4-1, while superoxide dismutase activity dropped by over 50% in the mutant. Although PLP is a strong antioxidant in vitro, these results indicate that vitamin B6 contributes to UV-B tolerance mainly indirectly, by supporting peroxidase-based antioxidant defense rather than directly scavenging reactive oxygen species. This protective effect is likely mediated through PLP-dependent biosynthesis of cysteine and heme, essential components for peroxidase function (Czégény et al. 2019).

4.7.2 Biotic Stress

The biosynthesis of vitamin B6 is tightly regulated, and its levels can be regulated in response to stress conditions like pathogen attacks (Zhang et al. 2014; Samsatly et al. 2020). The accumulation of vitamin B6 has defense signaling pathways of plant immunity. Knockout and knockdown mutants of PDX1 or SOS4 in various plants, such as *Arabidopsis* (Samsatly et al. 2020), tomato (Zhang et al. 2014), and rice (Liu et al. 2022a), where the content of VB6 was reduced, causing compromised basal immunity against *Pseudomonas syringae* pv. tomato (Pst) DC3000, *Botrytis cinerea*, *Xanthomonas oryzae* pv. *oryzicola* (Xoc), and *Rhizoctonia solani*. The Xoc effector protein AvrRxo1, an ATP-dependent protease, targets rice PDX1 for degradation, leading to decreased abscisic acid (ABA) levels and the reopening of closed stomata, which in turn inhibits stomatal immunity and enhances bacterial virulence (Liu et al. 2022b). The biocontrol agent *Bacillus subtilis* CBR05 enhances tomato defense against *Xanthomonas campestris* pv. *vesicatoria* by upregulating *SlPDX1.2* and *SlPDX1.3*, but not *SlSOS4*, thereby activating induced systemic resistance (ISR) (Chandrasekaran et al. 2019). In potatoes, higher vitamin B6 levels correlate with greater resistance to *Rhizoctonia solani* (Samsatly et al. 2020). In *Arabidopsis*, *pdx3* mutants show elevated salicylic acid (SA) levels and upregulation of defense genes, resulting in autoimmunity and developmental defects such as malformed, narrow leaves and reduced lamina area (Colinas and Fitzpatrick 2016; Colinas et al. 2016; Steensma et al. 2023). Disruption of *NPR1* signaling in *pdx3-3 npr1-2* double mutants worsens these defects, whereas *pdx3-3 sid2-1* and *pdx3-3 NahG* mutants, which reduce SA synthesis or enhance degradation, partially recover growth, indicating that SA contributes to but does not fully cause the abnormalities (Steensma et al. 2023). This suggests crosstalk between VB6 homeostasis and NPR1 in regulating immunity.

Pathogen-derived VB6 can act as a virulence factor. *Fusarium* spp. suppresses host resistance genes using VB6 to promote co-infection with *Phytophthora sojae* (Wang et al. 2023), while *R. solani* exploits VB6 to increase antioxidant capacity and evade host immunity (Samsatly et al. 2020). ROS sensitivity related to vitamin B6 was first discovered in the fungal pathogen *Cercospora nicotianae*. Mutant strains of this fungus were unable to tolerate cercosporin, a toxin they normally produce. Cercosporin acts as a photosensitizer and generates singlet oxygen and superoxide when exposed to light (Ehrenshaft et al. 1999a, b). When the defective genes from these sensitive mutants were cloned, researchers found that the mutations occurred in a gene corresponding to a PDX1 ortholog (Ehrenshaft et al. 1998; Ehrenshaft et al. 1999a, b). This discovery was significant because, before these studies, vitamin B6 had not been linked to protection against singlet oxygen (Bilski et al. 2000). Liu et al. (2025) link VB6 homeostasis, modulated by Δ^1-piperideine-6-carboxylic acid (P6C) from lysine catabolism, to systemic acquired resistance (SAR) in *Arabidopsis*. Pathogen infection induces P6C accumulation via pipecolate oxidase (PipOX), forming a Pip–P6C branch that suppresses SAR by sequestering VB6 vitamers such as PLP, PL, PN, and PM. Overexpression of *PipOX* leads to ~15-fold higher

P6C levels and impaired SAR (Liu et al. 2025). This depletion of PLP limits its role as a cofactor for key enzymes in SAR, SA, and ROS signaling. PLP-dependent *ALDH7B4* activity is also reduced, causing accumulation of α-aminoadipic semialdehyde (AASA) and P6C, further weakening immune responses (Liu et al. 2025). Application of VB6 vitamers (10–100 μM) restores SAR in *35S–PipOX* plants but not in *ald1*, *sard4*, or *fmo1* mutants, confirming the essential role of P6C-sequestered VB6 in SAR signaling (Liu et al. 2025). Mutants *pdx1.3* and *sos4*, which have reduced PLP levels, show compromised SAR, but treatment with PLP restores the phenotype, highlighting the key role of PLP and the salvage pathway (Liu et al. 2025).

Proper VB6 balance is required for SAR, as high concentrations (1 mM) of VB6 inhibit SAR but enhance local resistance (Liu et al. 2025). The Pip–P6C and Pip–NHP branches act independently to regulate SAR, with Pip and NHP promoting SAR and P6C inhibiting it by depleting VB6 (Liu et al. 2025). VB6 also serves as a cofactor for hundreds of enzymes (Colinas and Fitzpatrick 2016; Colinas et al. 2016) and for aldehyde oxidase, which catalyzes the final step of ABA biosynthesis and contributes to stomatal immunity in rice (Liu et al. 2022a, b). These findings reveal that VB6 is a key regulator of plant immunity, acting through SA, ABA, and ROS pathways to balance growth and defense. Future studies should explore how VB6 vitamers, especially PLP, influence hormone signaling and immune responses, and how pathogens exploit VB6 to enhance virulence (Rodrigues et al. 2006; Yanaka et al. 2005; Liu et al. 2025).

4.8 Genetic Engineering Strategies to Increase Vitamin B6 in Crops

From the viewpoint of genetic engineering, plants synthesize vitamin B6 through a de novo pathway that requires only two enzymes, PDX1 and PDX2 (Tambasco-Studart et al. 2005). This biosynthetic route, known as the DXP-independent pathway, is distinct from the DXP-dependent pathway found in *Escherichia coli*, which involves seven enzymes and is limited to certain bacterial species (Mittenhuber 2001; Fitzpatrick et al. 2007). In *Arabidopsis thaliana*, three PDX1 genes have been identified; PDX1.1 and PDX1.3 actively participate in vitamin B6 production, whereas the role of PDX1.2 is still unclear. Only a single PDX2 gene has been reported in this species (Tambasco-Studart et al. 2005). Comparable gene patterns have also been identified in rice and cassava (Ouyang et al. 2007; Prochnik et al. 2012). Efforts to enhance vitamin B6 accumulation in plants have mainly targeted genes of the DXP-independent pathway (Table 4.2). Heterologous expressions of *Cercospora nicotianae* PDX1 and PDX2 in tobacco increased total vitamin B6 content by about 1.2-fold but caused delayed seed germination and growth, suggesting tight regulation of vitamin B6 homeostasis (Herrero et al. 2011). Overexpression of endogenous *Arabidopsis* PDX1.3 and/or PDX2 under the

Table 4.2 Genetic engineering strategies to enhance pyridoxine biosynthesis in plants

Plant species	Gene(s) involved/ pathway	Efforts to enhance vitamin B6 accumulation	References
Arabidopsis thaliana	PDX1 and PDX2 (de novo, DXP-independent pathway)	Vitamin B_6 synthesis controlled by two enzymes, PDX1 and PDX2; three PDX1 homologs (PDX1.1, PDX1.2, PDX1.3) and one PDX2 homolog identified; PDX1.2 function unknown	Tambasco-Studart et al. (2005)
Oryza sativa (rice) *and Manihot esculenta* (cassava)	PDX1 and PDX2	A similar number of PDX homologs as in *Arabidopsis*, indicating conservation of the pathway	Ouyang et al. (2007) and Prochnik et al. (2012)
Nicotiana tabacum (tobacco, transgenic)	Cercospora nicotianae PDX1 & PDX2 (heterologous expression)	1.2-fold increase in total vitamin B_6; altered endogenous gene expression; delayed germination and growth; suggests tight regulation of vitamin B_6 homeostasis	Herrero et al. (2011)
Arabidopsis thaliana	PDX1.3 and/or PDX2 under CaMV 35S promoter	Increase in total vitamin B_6 mainly in seeds; PDX2 overexpression had a greater effect than PDX1; no growth inhibition observed	Chen and Xiong (2009)
Arabidopsis thaliana	PDX1.3 and PDX2 under seed-specific promoter	Up to threefold increase in vitamin B_6 content in seeds; no adverse growth effects	Chen and Xiong (2009)
Arabidopsis thaliana	PDX1.1 under CaMV 35S promoter	Fourfold increase in vitamin B_6 in shoots and seeds; enhanced antioxidant capacity and oxidative stress tolerance; associated with higher PDX2 expression	Raschke et al. (2011)
Arabidopsis thaliana	PDX1.1 + PDX2 co-overexpression	Highest vitamin B_6 accumulation; increased organ and seed size; potential for crop biofortification	Raschke et al. (2011)
Arabidopsis thaliana	PDX1.3 overexpression	No increase in protein level despite higher transcript; PDX1.3 targeted for ubiquitination and degradation	Manzano et al. (2008)
Arabidopsis thaliana	Salvage pathway (PDX3 oxidase, SOS4 kinase, PL reductase)	Converts different vitamers (PN, PM, PLP); cross-talk between de novo and salvage pathways regulates homeostasis	Tanaka et al. (2005), Sang et al. (2011), and Herrero et al. (2011)

(continued)

Table 4.2 (continued)

Plant species	Gene(s) involved/ pathway	Efforts to enhance vitamin B6 accumulation	References
Arabidopsis thaliana	Vitamin B_6 bioavailability	Enhanced vitamin B_6 in transgenics is sufficient to meet RDA; need to assess free, phosphorylated, and glycosylated vitamers for optimal biofortification	Fitzpatrick et al. (2012)

constitutive CaMV 35S promoter similarly enhanced total vitamin B6 levels, mainly in seeds (Chen and Xiong 2009). When both genes were expressed under a seed-specific promoter, vitamin B6 content increased up to threefold without negative effects on growth. In contrast, overexpression of *Arabidopsis* PDX1.1 resulted in a fourfold elevation of vitamin B6 in shoots and seeds, accompanied by increased PDX2 expression and improved oxidative stress tolerance (Raschke et al. 2011). Interestingly, while elevated PDX1.1 transcript levels correlated with higher protein accumulation, overexpression of PDX1.3 did not, likely due to post-translational regulation and proteasomal degradation (Manzano et al. 2008).

Overexpression of both PDX1.1 and PDX2 led to the greatest enrichment of vitamin B6, also enhancing organ size and seed weight (Raschke et al. 2011). Enzymatic activities of PDX1 and PDX2 directly produce pyridoxal 5′-phosphate (PLP), though increases in pyridoxine (PN), pyridoxamine (PM), and PLP were all observed in transgenic lines (Tambasco-Studart et al. 2005; Raschke et al. 2011). Beyond de novo synthesis, plants also employ a "salvage pathway" that interconverts B6 vitamers via PDX3 oxidase, SOS4 kinase, PL reductase, and possible non-specific phosphatases (Tanaka et al. 2005; Herrero et al. 2011; Sang et al. 2011; Leasure et al. 2011). Crosstalk between these two pathways likely contributes to vitamin homeostasis and should be considered in future biofortification strategies. The increase in vitamin B6 levels reported by Raschke et al. (2011) could potentially meet human recommended dietary allowances (Fitzpatrick et al. 2012). Future studies should also explore the regulation of free, phosphorylated, and glycosylated vitamers, given their influence on both bioavailability and agronomic performance.

4.9 Conclusion

Pyridoxine, particularly in its active form, pyridoxal-5′-phosphate (PLP), is a vital coenzyme that participates in numerous metabolic, developmental, and stress-related processes in plants. Beyond its classical role in amino acid and carbohydrate metabolism, pyridoxine has emerged as a multifunctional molecule with strong antioxidant capacity, contributing to redox homeostasis and protection against oxidative damage induced by abiotic and biotic stresses. Exogenous application of pyridoxine enhances seed germination, growth, photosynthetic efficiency, and stress tolerance by improving antioxidant defense and osmotic adjustment. From a biotechnological perspective, manipulating genes involved in vitamin B_6 biosynthesis (such as PDX1 and PDX2) offers promising avenues for metabolic engineering to increase vitamin content, improve crop resilience, and enrich nutritional value. However, the complex regulation between de novo and salvage pathways and the balance among different B_6 vitamers require further investigation. Overall, pyridoxine represents an eco-friendly, natural growth regulator with great potential for sustainable agriculture and biofortification strategies aimed at improving both plant performance and human nutrition.

References

Abdel Latef AAH (2026) Using stimulants to improve plant health: From stress to strength. Elsevier. https://doi.org/10.1016/C2025-0-01048-7

Abdel Latef AAH, Hameed M, Khan Z, Khan S, Imran A, Fatima N (2026) Foliar stimulants: A sustainable strategy for enhancing plant resilience to abiotic and biotic stresses. Physiologia Plantarum 178(1):e70741

Adams JB, George F, Audhya T (2006) Abnormally high plasma levels of vitamin B6 in children with autism not taking supplements compared to controls not taking supplements. J. Altern. Complement. Med. (N. Y. NY) 12:59–63. https://doi.org/10.1089/acm.2006.12.59

Ahmed EZ, Sattar AMAE (2024) Improvement of Vicia faba plant tolerance under salinity stress by the application of thiamine and pyridoxine vitamins. Sci Rep 14(1):22367

Asli DE, Alireza Houshmandfar A (2011) Seed Germination and Early Seedling Growth of Corn (Zea Mays L.) As Affected by Different Seed Pyridoxine-priming Duration. https://www.thefreelibrary.com/Seedgerminationandearlyseedlinggrowthofcorn(ZeaMaysL.)as...-a0256863987. Accessed 13 Sep 2024

Bagri DS, Upadhyaya DC, Kumar A, Upadhyaya CP (2018) Overexpression of PDX-II gene in potato (Solanum tuberosum L.) leads to the enhanced accumulation of vitamin B6 in tuber tissues and tolerance to abiotic stresses. Plant Sci 272:267–275. https://doi.org/10.1016/j.plantsci.2018.04.024

Boghdady MS (2013) Efficiency of pyridoxine on the growth, yield, seed quality and anatomy of Egyptian lupine (Lupinus termis Forssk.)

Bilski P, Li MY, Ehrenshaft M, Daub ME, Chignell CF (2000) Vitamin B6 (pyridoxine) and its derivatives are efficient singlet oxygen quenchers and potential fungal antioxidants. Photochem Photobiol 71:129–134.

Bilski P, Daub ME, Chignell CF (2002) Direct detection of singlet oxygen via its phosphorescence from cellular and fungal cultures. Meth Enzymol 352:41–52. https://doi.org/10.1016/s0076-6879(02)52005-x. [DOI] [PubMed] [Google Scholar]

Biswal BK, Au K, Cherney MM, Garen C, James MNG (2006) The molecular structure of Rv2074, a probable pyridoxine 5′-phosphate oxidase from mycobacterium tuberculosis, at 1.6 angstroms resolution. Acta Crystallograph Sect F Struct Biol Cryst Commun 62:735–742. https://doi.org/10.1107/S1744309106025012

Bonner J (1942) Transport of thiamin in the tomato plant. Am J Bot 29:136–142

Boycheva S, Dominguez A, Rolcik J, Boller T, Fitzpatrick TB (2015) Consequences of a deficit in vitamin B6 biosynthesis de novo for hormone homeostasis and root development in Arabidopsis. Plant Physiol 167:102–117. https://doi.org/10.1104/pp.114.247767

Bhullar NK, Gruissem W (2013) Nutritional enhancement of rice for human health: the contribution of biotechnology. Biotechnol Adv 31(1):50–57

Chandrasekaran M, Paramasivan M, Chun SC (2019) Bacillus subtilis CBR05 induces Vitamin B6 biosynthesis in tomato through the de novo pathway in contributing disease resistance against Xanthomonas campestris pv. vesicatoria. Sci Rep 9:6495. https://doi.org/10.1038/s41598-019-41888-6

Chen H, Xiong L (2005) Pyridoxine is required for post-embryonic root development and tolerance to osmotic and oxidative stresses. Plant J 44:396–408. https://doi.org/10.1111/j.1365-313X.2005.02538.x

Chen H, Xiong LM (2009) Enhancement of vitamin B6 levels in seeds through metabolic engineering. Plant Biotechnol J 7:673–681

Choudhury SR, Singh SK, Roy S, Sengupta DN (2010) An insight into the sequential, structural and phylogenetic properties of banana 1-aminocyclopropane-1-carboxylate synthase 1 and study of its interaction with pyridoxal-5′-phosphate and aminoethoxyvinylglycine. J Biosci 35:281–294

Colinas M, Fitzpatrick TB (2016) Interaction between vitamin B6 metabolism, nitrogen metabolism and autoimmunity. Plant Signal Behav 11(4):e1161876. https://doi.org/10.1080/15592324.2016.1161876

Colinas M, Eisenhut M, Tohge T, Pesquera M, Fernie AR, Weber APM, Fitzpatrick TB (2016) Balancing of B6 Vitamers is essential for plant development and metabolism in Arabidopsis. Plant Cell 28:439–453. https://doi.org/10.1105/tpc.15.01033

Chen H, Xiong L (2005) Pyridoxine is required for post-embryonic root development and tolerance to osmotic and oxidative stresses. The plant journal 44(3):396–408

Cuesta-Seijo JA, Ruzanski C, Krucewicz K, Meier S, Hägglund P, Svensson B, Palcic MM (2017) Functional and structural characterization of plastidic starch phosphorylase during barley endosperm development. PLoS One 12:e0175488

Czégény G et al (2014) Hydrogen peroxide contributes to the ultraviolet-B (280-315 nm) induced oxidative stress of plant leaves through multiple pathways. FEBS Lett 588:2255–2261

Czégény G, Kőrösi L, Strid Å et al (2019) Multiple roles for Vitamin B_6 in plant acclimation to UV-B. Sci Rep 9:1259. https://doi.org/10.1038/s41598-018-38053-w

Danon A, Miersch O, Felix G, Camp RG, Apel K (2005) Concurrent activation of cell death-regulating signaling pathways by singlet oxygen in *Arabidopsis thaliana*. Plant J 41:68–80

Dawood MF (2025) A new frontier for plant-microbe interaction under changing climate conditions. Glob Chang Biol 31(7):e70393

Dell'Aglio E, Boycheva S, Fitzpatrick TB (2017) The pseudoenzyme PDX1.2 sustains vitamin B6 biosynthesis as a function of heat stress. Plant Physiol 174:2098–2112

Denslow SA, Walls AA, Daub ME (2005) Regulation of biosynthetic genes and antioxidant properties of vitamin B6 vitamers during plant defense responses. Physiol Mol Plant Pathol 66:244–255

Desouky SA (1995) Effect of some organic additives on salinized Chlorella vulgaris (Doctoral dissertation, Ph. D. Thesis, Faculty of Science, Assiut University, Egypt)

Di Salvo ML, Contestabile R, Safo MK (2011) Vitamin B(6) salvage enzymes: mechanism, structure and regulation. Biochim Biophys Acta 1814:1597–1608. https://doi.org/10.1016/j.bbapap.2010.12.006

di Salvo ML, Safo MK, Contestabile R (2012) Biomedical aspects of pyridoxal 5′-phosphate availability. Front Biosci (Elite Ed) 4(3):897–913

Drewke C, Klein M, Clade D, Arenz A, Müller R, Leistner E (1996) 4-O-phosphoryl-L-threonine, a substrate of the pdxC(serC) gene product involved in vitamin B6 biosynthesis. FEBS Lett 390:179–182. https://doi.org/10.1016/0014-5793(96)00652-7

Dolatabadian A, Modarressanavy SAM (2008) Effect of ascorbic acid, pyridoxine and hydrogen peroxide treatments on germination, catalase activity, protein and malondialdehyde content of three oil seeds. Notulae Botanicae Horti Agrobotanici Cluj-Napoca (online) 36:61–66

Ehrenshaft M, Jenns AE, Chung KR, Daub ME (1998) SOR1, a gene required for photosensitizer and singlet oxygen resistance in Cercospora fungi, is highly conserved in divergent organisms. Mol Cell 1:603–609. https://doi.org/10.1016/S1097-2765(00)80060-X

Ehrenshaft M, Bilski P, Li MY, Chignell CF, Daub ME (1999a) A highly conserved sequence is a novel gene involved in de novo vitamin B6 biosynthesis. Proc Natl Acad Sci USA 96:9374–9378. https://doi.org/10.1073/pnas.96.16.9374

Ehrenshaft M, Chung KR, Jenns AE, Daub ME (1999b) Functional characterization of SOR1, a gene required for resistance to photosensitizing toxins in the fungus Cercospora nicotianae. Curr Genet 34:478–485. https://doi.org/10.1007/s002940050423

El-Bassiouny H, Sadak MS, Mahfouz S, El-Enany M, Elewa T (2023) Use of thiamine, pyridoxine and bio stimulant for better yield of wheat plants under water stress: growth, osmoregulations, antioxidantive defence and protein pattern. Egypt J Chem 66(4):407–424

Ercan-Fang N, Taylor MR, Treadway JL, Levy CB, Genereux PE, Gibbs EM, Rath VL, Kwon Y, Gannon MC, Nuttall FQ (2005) Endogenous effectors of human liver glycogen phosphorylase modulate effects of indole-site inhibitors. Am J Physiol Endocrinol Metab 289:E366–E372

Fitzpatrick TB (2024) B vitamins: an update on their importance for plant homeostasis. Annu Rev Plant Biol 75:67–93. https://doi.org/10.1146/annurev-arplant-060223-025336

Fitzpatrick TB, Amrhein N, Kappes B, Macheroux P, Tews I, Raschle T (2007) Two independent routes of de novo vitamin B6 biosynthesis: not that different after all. Biochem J 407:1–13

Fitzpatrick TB, Basset GJ, Borel P, Carrari F, Dellapenna D, Fraser PD et al (2012) Vitamin deficiencies in humans: can plant science help? Plant Cell 24:395–414

Geng MY, Saito H, Katsuki H (1995) Effects of vitamin B6 and its related compounds on survival of cultured brain neurons. Neurosci Res 24:61–65. https://doi.org/10.1016/0168-0102(96)81279-X

Gerdes S, Lerma-Ortiz C, Frelin O, Seaver SM, Henry CS, De Crecy-Lagard V et al (2012) Plant B vitamin pathways and their compartmentation: a guide for the perplexed. J Exp Bot 63:5379–5395

Gonzalez E, Danehower D, Daub ME (2007) Vitamer levels, stress response, enzyme activity, and gene regulation of Arabidopsis lines mutant in the pyridoxine/pyridoxamine 5′-phosphate oxidase (PDX3) and the pyridoxal kinase (SOS4) genes involved in the vitamin B6 salvage pathway. Plant Physiol 145:985–996. https://doi.org/10.1104/pp.107.105189

Graham CM, Ehrenshaft M, Hausner G, Reid DM (2004) A highly conserved gene for vitamin B biosynthesis may have consequences for stress and hormone responses in plants. Physiol Plant 121:8–14. https://doi.org/10.1111/j.0031-9317.2004.00283.x

György P (1934) Vitamin B2 and the pellagra-like dermatitis in rats. Nature 133:498–499. https://doi.org/10.1038/133498a0

György P (1938) Crystalline vitamin B6. J Am Chem Soc 60:983–984. https://doi.org/10.1021/ja01271a505

Hamada AM, Khulaef EM (2000) Simulative effects of ascorbic acid, thiamin or pyridoxine on Viciafaba growth and some related metabolic activities. Pakistan J. Biol. Sci.3(8), 1330–1332. https://doi.org/10.3923/pjbs.2000.1330.1332

György P, Eckardt RE (1939) Vitamin B6 and skin lesions in rats. Nature 144:512–512

Harris SA, Folkers K (1939) Synthetic vitamin B6. Science 89:347. https://doi.org/10.1126/science.89.2311.347

Havaux M, Ksas B, Szewczyk A, Rumeau D, Franck F, Caffarri S, Triantaphylidès C (2009) Vitamin B6 deficient plants display increased sensitivity to high light and photo-oxidative stress. BMC Plant Biol 9:130. https://doi.org/10.1186/1471-2229-9-130

Hellmann H, Mooney S (2010) Vitamin B6: a molecule for human health? Mol Basel Switz 15:442–459. https://doi.org/10.3390/molecules15010442

Hendawy SF, El-Din AE (2010) Growth and yield of Foeniculum vulgare var. azoricum as influenced by some vitamins and amino acids. Ozean. J. Appl. Sci 3(1):113–122

Herrero S, González E, Gillikin JW, Vélëz H, Daub ME (2011) Identification and characterization of a pyridoxal reductase involved in the vitamin B6 salvage pathway in Arabidopsis. Plant Mol Biol 76:157–169. https://doi.org/10.1007/s11103-011-9777-x

Hickey LT, Hafeez AN, Robinson H, Jackson SA, Leal-Bertioli SC, Tester M et al (2019) Breeding crops to feed 10 billion. Nat Biotechnol 37(7):744–754

Huang S, Zeng H, Zhang J, Wei S, Huang L (2011) Interconversion of different forms of vitamin B6 in tobacco plants. Phytochemistry 72:2124–2129
Ichiba A, Michi K (1938) Cristalline vitamin B6. Sci. Pap. Inst. Phys. Chem. Res. 34:623–626. (Tokyo)
Keresztesy JC, Stevens JR (1938) Vitamin B-6. Proc Soc Exp Biol Med 38:64–65. https://doi.org/10.3181/00379727-38-9740P
Khan MA, Gul B, Weber DJ (2001) Seed germination characteristics of Halogetonglomeratus. Can J Bot 79(10):1189–1194
Kirk PR, Leech RM (1972) Amino acid biosynthesis by isolated chloroplasts during photosynthesis. Plant Physiol 50:228–234. https://doi.org/10.1104/pp.50.2.228
Kuhn R, Wendt G (1938) Über das antidermatitische Vitamin der Hefe. Ber Chem Ges 71:780–782
Lepkovsky S (1938) Crystalline factor I. Science 87:169–170. https://doi.org/10.1126/science.87.2251.169
Leuendorf JE, Genau A, Szewczyk A, Mooney S, Drewke C, Leistner E, Hellmann H (2008) The Pdx1 family is structurally and functionally conserved between Arabidopsis thaliana and Ginkgo biloba. FEBS J 275:960–969. https://doi.org/10.1111/j.1742-4658.2008.06275.x
Liu H, Iyer LM, Norris P (2025) Piperideine-6-Carboxylic Acid Regulates Vitamin B6 Homeostasis and Modulates Systemic Immunity in Plants. Nature Plants 11, no. 2: 263–278. https://doi.org/10.1038/s41477-025-01906-0
Liu H, Lu C, Li Y et al (2022a) The bacterial effector AvrRxo1 inhibits vitamin B6 biosynthesis to promote infection in Rice. Plant Commun 3(3):100324. https://doi.org/10.1016/j.xplc.2022.100324
Liu Z, Farkas P, Wang K, Kohli MO, Fitzpatrick TB (2022b) B vitamin supply in plants and humans: the importance of vitamer homeostasis. Plant J 111(3):662–682. https://doi.org/10.1111/tpj.15859
Lone NA, Khan NA, Hayat S, Azam ZM, Samiullah N (1999) Evaluation of effect of some B-Vitamins on root development of mustard. Ann Appl Biol 134 (Supplement):30–37
Lytovchenko A, Beleggia R, Schauer N, Isaacson T, Leuendorf JE, Hellmann H et al (2009) Application of GC-MS for the detection of lipophilic compounds in diverse plant tissues. Plant Methods 5:4–14
Mahdi W, Ali M, Albadry S, Ibrahim I (2021) Effect of pyridoxine (vitamin B6) and saline stress on the growth and antioxidant enzymes of wheat Triticumaestivum L. Ann. Rom. Soc. Cell Biol 25:4470–4476
Manzano C, Abraham Z, LópezTorrejón G, Del Pozo J (2008) Identification of ubiquitinated proteins in *Arabidopsis*. Plant Mol Biol 68:145–158
Merrill AH, Horiike K, McCormick DB (1978) Evidence for the regulation of pyridoxal 5-′-phosphate formation in liver by pyridoxamine (pyridoxine) 5′-phosphate oxidase. Biochem Biophys Res Commun 83:984–990. https://doi.org/10.1016/0006-291X(78)91492-4
Mittenhuber G (2001) Phylogenetic analyses and comparative genomics of vitamin B6 (pyridoxine) and pyridoxal phosphate biosynthesis pathways. J Mol Microbiol Biotechnol 3:1–20
Moccand C, Kaufmann M, Fitzpatrick TB (2011) It takes two to tango: defining an essential second active site in pyridoxal 5′-phosphate synthase. PLoS One 6:e16042
Moccand C et al (2014) The pseudoenzyme PDX1.2 boosts vitamin B6 biosynthesis under heat and oxidative stress in Arabidopsis. J Biol Chem 289:8203–8216
Mooney S, Hellmann H (2010) Vitamin B6: killing two birds with one stone? Phytochemistry 71: 495–501
Mooney S, Leuendorf J-E, Hendrickson C, Hellmann H (2009) Vitamin B6: a long known compound of surprising complexity. Molecules 14:329–351. https://doi.org/10.3390/molecules14010329
Mozafar A, Oertli JJ (1992) Uptake and transport of thiamin (vitamin B 1) by barley and soybean. J Plant Physiol 139(4):436–442

Narisawa S, Wennberg C, Millán JL (2001) Abnormal vitamin B6 metabolism in alkaline phosphatase knock-out mice causes multiple abnormalities, but not the impaired bone mineralization. J Pathol 193:125–133
Nassar R, Arafa S, Farouk S (2017) Effect of foliar spray with pyridoxine on growth, anatomy, photosynthetic pigments, yield characters and biochemical constituents of seed oil of sesame plant (Sesamum indicum L.). Middle East J Appl Sci 7:80–91
Notheis C, Drewke C, Leistner E (1995) Purification and characterization of the pyridoxol-5′-phosphate:oxygen oxidoreductase (deaminating) from Escherichia coli. Biochim Biophys Acta 1247:265–271. https://doi.org/10.1016/0167-4838(94)00235-9
Ohdake S (1932) Isolation of "Oryzanin" (Antineuritic vitamin) from Rice-polishings. Bull Agri Chem Soc Japan 8:11–46. https://doi.org/10.1271/bbb1924.8.11
Ouyang S, Zhu W, Hamilton J, Lin H, Campbell M, Childs K et al (2007) The TIGR rice genome annotation resource: improvements and new features. Nucleic Acids Res 35:883–887
Parra M, Stahl S, Hellmann H (2018) Vitamin B_6 and its role in cell metabolism and physiology. Cells 7(7):84. https://doi.org/10.3390/cells7070084
Percudani R, Peracchi A (2003) A genomic overview of pyridoxal-phosphate-dependent enzymes. EMBO Rep 4:850–854
Pereyra, MA, Garcia P, Colabelli MN, Barassi CA, CREUS CM (2012) A better water status in wheat seedlings induced by Azospirillum under osmotic stress is related to morphological changes in xylem vessels of thebr coleoptile. Applied Soil Ecology 53:94–97
Plecko B, Stöckler S (2009) Vitamin B6 dependent seizures. Can J Neurol Sci 36(Suppl. 2):S73–S77
Prochnik S, Marri PR, Desany B, Rabinowicz PD, Kodira C, Mohiuddin M et al (2012) The *cassava* genome: current progress, future directions. Trop Plant Biol 5:88–94
Proebsting WM, Maggard SP, Guo WW (1990) The relationship of thiamin to the alt locus of Pisum sativum L. J. Plant Physiol 136 231–235
Raschke M, Boycheva S, Crèvecoeur M, Nunes-Nesi A, Witt S, Fernie AR, Amrhein N, Fitzpatrick TB (2011) Enhanced levels of vitamin B(6) increase aerial organ size and positively affect stress tolerance in Arabidopsis. Plant J 66:414–432. https://doi.org/10.1111/j.1365-313X.2011.04499.x
Rezvi HUA, Tahjib-Ul-Arif M, Azim MA, Tumpa TA, Tipu MMH, Najnine F et al (2023) Rice and food security: climate change implications and the future prospects for nutritional security. Food Energy Secur 12(1):e430
Ristilä M, Strid H, Eriksson LA, Strid Å, Sävenstrand H (2011) The role of the pyridoxine (vitamin B6) biosynthesis enzyme PDX1 in ultraviolet-B radiation responses in plants. Plant Physiol Biochem 49:284–292
Rodrigues SM, Andrade MO, Gomes APS, DaMatta FM, Baracat-Pereira MC, Fontes EP (2006) Arabidopsis and tobacco plants ectopically expressing the soybean Antiquitin-like ALDH7 gene display enhanced tolerance to drought, salinity, and oxidative stress. J Exp Bot 57(9): 1909–1918. https://doi.org/10.1093/jxb/erj132
Rueschhoff EE, Gillikin JW, Sederoff HW, Daub ME (2013) The SOS4 pyridoxal kinase is required for maintenance of vitamin B6-mediated processes in chloroplasts. Plant Physiol Biochem 63:281–291. https://doi.org/10.1016/j.plaphy.2012.12.003
Safo MK, Musayev FN, Schirch V (2005) Structure of Escherichia coli pyridoxine 5′-phosphate oxidase in a tetragonal crystal form: insights into the mechanistic pathway of the enzyme. Acta Crystallogr D Biol Crystallogr 61:599–604. https://doi.org/10.1107/S0907444905005512
Samsatly J, Bayen S, Jabaji SH (2020) Vitamin B6 is under a tight balance during disease development by *Rhizoctonia Solani* on different cultivars of potato and on *Arabidopsis thaliana* mutants. Front Plant Sci 11:875. https://doi.org/10.3389/fpls.2020.00875
Sang Y, Barbosa JM, Wu H, Locy RD, Singh NK (2007) Identification of a pyridoxine (pyridoxamine) 5′-phosphate oxidase from Arabidopsis thaliana. FEBS Lett 581:344–348. https://doi.org/10.1016/j.febslet.2006.12.028

Sang Y, Locy RD, Goertzen LR, Rashotte AM, Si Y, Kang K et al (2011) Expression, in vivo localization and phylogenetic analysis of a pyridoxine 5′-phosphate oxidase in *Arabidopsis thaliana*. Plant Physiol Biochem 49:88–95

Schaeffer MC, Sampson DA, Skala JH, Gietzen DW, Grier RE (1989) Evaluation of vitamin B-6 status and function of rats fed excess pyridoxine. J Nutr 119:1392–1398. https://doi.org/10.1093/jn/119.10.1392

Shi H, Xiong L, Stevenson B, Lu T, Zhu J-K (2002) The Arabidopsis salt overly sensitive 4 mutants uncover a critical role for vitamin B6 in plant salt tolerance. Plant Cell 14:575–588. https://doi.org/10.1105/tpc.010417

Shimasaki K, Fukumoto Y (1998) Effects of B vitamins and bezyl aminopurine on adventitious shoot formation from hypocotyls segments of snapdragon (Antirrhinum majus L.). Plant Biotech 15(4):239–240

Steensma P, Eisenhut M, Colinas M et al (2023) Pyridox (AM) INE 5′-phosphate oxidase3 of *Arabidopsis thaliana* maintains carbon/ nitrogen balance in distinct environmental conditions. Plant Physiol 193(2):1433–1455

Stolz J, Vielreicher M (2003) Tpn1p, the plasma membrane vitamin B6 transporter of Saccharomyces cerevisiae. J Biol Chem 278:18990–18996. https://doi.org/10.1074/jbc.M300949200

Stolz J, Wöhrmann HJP, Vogl C (2005) Amiloride uptake and toxicity in fission yeast are caused by the pyridoxine transporter encoded by bsu1+ (car1+) Eukaryot. Cell 4:319–326. https://doi.org/10.1128/EC.4.2.319-326.2005

Szydlowski N, Bürkle L, Pourcel L, Moulin M, Stolz J, Fitzpatrick TB (2013) Recycling of pyridoxine (vitamin B6) by PUP1 in *Arabidopsis*. Plant J 75(1):40–52. https://doi.org/10.1111/tpj.12195

Tambasco-Studart M, Titiz O, Raschle T, Forster G, Amrhein N, Fitzpatrick TB (2005) Vitamin B6 biosynthesis in higher plants. Proc Natl Acad Sci USA 102:13687–13692. https://doi.org/10.1073/pnas.0506228102

Tambasco-Studart M, Tews I, Amrhein N, Fitzpatrick TB (2007) Functional analysis of PDX2 from Arabidopsis, a glutaminase involved in vitamin B6 biosynthesis. Plant Physiol 144:915–925. https://doi.org/10.1104/pp.107.096784

Tanaka T, Tateno Y, Gojobori T (2005) Evolution of vitamin B6 (pyridoxine) metabolism by gain and loss of genes. Mol Biol Evol 22:243–250

Tang L, Li M-H, Cao P, Wang F, Chang W-R, Bach S, Reinhardt J, Ferandin Y, Galons H, Wan Y et al (2005) Crystal structure of pyridoxal kinase in complex with roscovitine and derivatives. J Biol Chem 280:31220–31229. https://doi.org/10.1074/jbc.M500805200

Titiz O, Tambasco-Studart M, Warzych E, Apel K, Amrhein N, Laloi C, Fitzpatrick TB (2006) PDX1 is essential for vitamin B6 biosynthesis, development and stress tolerance in Arabidopsis. Plant J Cell Mol Biol 48:933–946. https://doi.org/10.1111/j.1365-313X.2006.02928.x

Tsang EWT, Hu Z, Chang Q, McGregor DI, Keller WA (2003) Expression of a Brassic napus glutamate 1-semialdehyde aminotransferase in Escherichia coli and characterization of the recombinant protein. Protein Expr Purif 29:193–201

Wagner S, Bernhardt A, Leuendorf JE, Drewke C, Lytovchenko A, Mujahed N, Gurgui C, Frommer WB, Leistner E, Fernie AR et al (2006) Analysis of the Arabidopsis rsr4-1/pdx1-3 mutant reveals the critical function of the PDX1 protein family in metabolism, development, and vitamin B6 biosynthesis. Plant Cell 18:1722–1735. https://doi.org/10.1105/tpc.105.036269

Wang S, Zhang X, Zhang Z et al (2023) *Fusarium*-produced vitamin B6 promotes the evasion of soybean resistance by *Phytophthora sojae*. J Integr Plant Biol 65(9):2204–2217. https://doi.org/10.1111/jipb.13505

Whittaker JW (2016) Intracellular trafficking of the pyridoxal cofactor. Implications for health and metabolic disease. Arch Biochem Biophys 592:20–26. https://doi.org/10.1016/j.abb.2015.11.031

Whyte MP, Mahuren JD, Vrabel LA, Coburn SP (1985) Markedly increased circulating pyridoxal-5′-phosphate levels in hypophosphatasia. Alkaline phosphatase acts in vitamin B6 metabolism. J Clin Invest 76:752–756. https://doi.org/10.1172/JCI112031

Whyte MP, Mahuren JD, Fedde K, Cole F, McCabe ERB, Coburn SP (1988) Perinatal hypophosphatasia: tissue levels of vitamin B6 are unremarkable despite markedly increased circulating concentrations of pyridoxal-5′-phosphate. Evidence for an ectoenzyme role for tissue-nonspecific alkaline phosphatase. J Clin Invest 81:1234–1239. https://doi.org/10.1172/JCI113440

Wiardy PW (1938) Crystalline Vitamin B6 (Adermin). Nature 142:1158–1158. https://doi.org/10.1038/1421158b0

Yanaka N, Koyama TA, Komatsu SI, Nakamura E, Kanda M, Kato N (2005) Vitamin B6 suppresses NF-κB activation in LPSStimulated mouse macrophages. Int J Mol Med 16(6): 1071–1075. https://doi.org/10.3892/ijmm.16.6.1071

Yang Y, Zhao G, Winkler ME (1996) Identification of the pdxK gene that encodes pyridoxine (vitamin B6) kinase in Escherichia coli K-12. FEMS Microbiol Lett 141:89–95. https://doi.org/10.1111/j.1574-6968.1996.tb08368.x

Yang Y, Tsui HC, Man TK, Winkler ME (1998) Identification and function of the pdxY gene, which encodes a novel pyridoxal kinase involved in the salvage pathway of pyridoxal 5-′-phosphate biosynthesis in Escherichia coli K-12. J Bacteriol 180:1814–1821. https://doi.org/10.1128/jb.180.7.1814-1821.1998

Yang YZ, Ding S, Wang Y, Li CL, Shen Y, Meeley R, McCarty DR, Tan BC (2017) Small kernel2 encodes a Glutaminase in vitamin B6 biosynthesis essential for maize seed development. Plant Physiol 174:1127–1138. https://doi.org/10.1104/pp.16.01295

Younes NA, Rahman MM, Wardany AA, Dawood MF, Mostofa MG, Keya SS et al (2021) Antioxidants and bioactive compounds in licorice root extract potentially contribute to improving growth, bulb quality and yield of onion (Allium cepa). Molecules 26(9):2633

Zhang Y, Liu B, Li X et al (2014) The *de novo* biosynthesis of vitamin B6 is required for disease resistance against *Botrytis cinerea* in tomato. Mol Plant-Microbe Interact 27(7):688–699. https://doi.org/10.1094/MPMI-01-14-0020-R

Chapter 5
Folic Acid: Boosting Stress Adaptation in Plants

Boujemaa Fassih, Redouane Ouhaddou, Aicha Nait Douch, Mohamed Ait-El-Mokhtar, Raja Ben-Laouane, Abdelilah Meddich, and Marouane Baslam

Abstract Folic acid (FA), a member of the B-group vitamins, plays a critical role in plant development and environmental stress resilience. Under abiotic stresses—such as drought, salinity, heavy metals, and temperature extremes—and biotic challenges including pathogen infections and pest attacks, FA has emerged as a bioactive molecule regulating stress perception and defense activation. FA contributes to DNA synthesis and repair, redox regulation, amino acid metabolism, and methylation pathways. Recent findings highlight its role in enhancing antioxidant enzyme activities, modulating the expression of stress-related genes, stabilizing cellular

B. Fassih · R. Ouhaddou · A. N. Douch
Centre d'Agrobiotechnologie et Bioingénierie, Unité de Recherche Labellisée CNRST (Centre AgroBiotech-URL-CNRST-05), Cadi Ayyad University, Marrakesh, Morocco

M. Ait-El-Mokhtar
Centre d'Agrobiotechnologie et Bioingénierie, Unité de Recherche Labellisée CNRST (Centre AgroBiotech-URL-CNRST-05), Cadi Ayyad University, Marrakesh, Morocco

Laboratory of Biotechnology, Agri-food, Materials, and Environment (LBAME), Department of Biology, Faculty of Science and Techniques Mohammedia, Hassan II University of Casablanca, Mohammedia, Morocco

R. Ben-Laouane
Laboratory of Environment and Health, Department of Biology, Faculty of Science and Techniques, Moulay Ismail University, Errachidia, Morocco

A. Meddich
Centre d'Agrobiotechnologie et Bioingénierie, Unité de Recherche Labellisée CNRST (Centre AgroBiotech-URL-CNRST-05), Cadi Ayyad University, Marrakesh, Morocco

African Sustainable Agriculture Research Institute (ASARI), University Mohamed VI Polytechnic (UM6P), Laayoune, Morocco

M. Baslam (✉)
Centre d'Agrobiotechnologie et Bioingénierie, Unité de Recherche Labellisée CNRST (Centre AgroBiotech-URL-CNRST-05), Cadi Ayyad University, Marrakesh, Morocco

GrowSmart, Amsterdam, The Netherlands

A. A. H. Abdel Latef (ed.), *Harnessing Vitamins and Organic Acids for Plant Stress Resilience*, Sustainability Sciences in Asia and Africa,
https://doi.org/10.1007/978-981-95-8593-9_5

structures, and supporting systemic resistance. Furthermore, FA appears to interact with phytohormonal signaling and nutrient transporters under stress.

This chapter presents a comprehensive examination of FA's molecular, physiological, and agronomic impacts under complex environmental conditions. It also explores the potential of exogenous FA applications as a low-cost, scalable solution to support crop productivity and sustainable agriculture.

Keywords Folic acid · Abiotic stress · Biotic stress · Gene regulation · Antioxidant enzymes · Methylation · Systemic resistance · Nutrient transport · Stress-responsive signaling

5.1 Introduction

Climate change represents one of the most formidable challenges to global agriculture, exerting profound effects on plant productivity, and ecosystem stability (Tramblay et al. 2020). Rising temperatures, recurrent droughts, progressive soil salinization, and erratic precipitation patterns collectively intensify the frequency and severity of abiotic stresses such as drought, salinity, extreme heat, cold, and metal toxicity (Arif et al. 2020; Fassih et al. 2024; Shin et al. 2021; Badr Elsayed et al. 2025; Zhang et al. 2015; Anli et al. 2026). These environmental constraints disrupt fundamental physiological and metabolic processes—impairing nutrient uptake, enhancing the overproduction of reactive oxygen species (ROS), destabilizing membranes, and diminishing photosynthetic efficiency—all of which lead to declines in plant growth, yield, and quality (Sharma et al. 2012; Huchzermeyer et al. 2022; Abdel Latef 2026). Compounding these abiotic threats are biotic stresses—pathogens, pests, and their interactions—that are becoming more frequent and virulent under changing climatic regimes, further undermining the resilience of agricultural systems (Fassih et al. 2025; Iqbal et al. 2021; Lal et al. 2023).

Given this growing complexity, conventional management strategies centered on genetic improvement or chemical protection are no longer sufficient to safeguard productivity. There is thus a critical need for sustainable, adaptive, and biologically based interventions capable of enhancing plant resilience to multifactorial stress environments. Among emerging approaches, the use of natural physiological regulators—particularly low-molecular-weight biomolecules—offers a promising direction.

Folic acid (FA), a water-soluble B-group vitamin (vitamin B9), has recently gained attention as a natural, non-toxic, and biodegradable biostimulant with remarkable potential for sustainable agriculture (Alsamadany et al. 2022). Its ease of exogenous application—via foliar sprays or soil incorporation—combined with its low cost and compatibility across cropping systems makes it an attractive candidate for both large-scale and precision farming (Badr Elsayed et al. 2025). Unlike several biostimulants whose activity is often species- or context-specific, FA has demonstrated broad-spectrum efficacy across diverse crop species and environmental conditions (Alaa El-Den et al. 2022; Badr Elsayed et al. 2025; Kilic and Aca 2016).

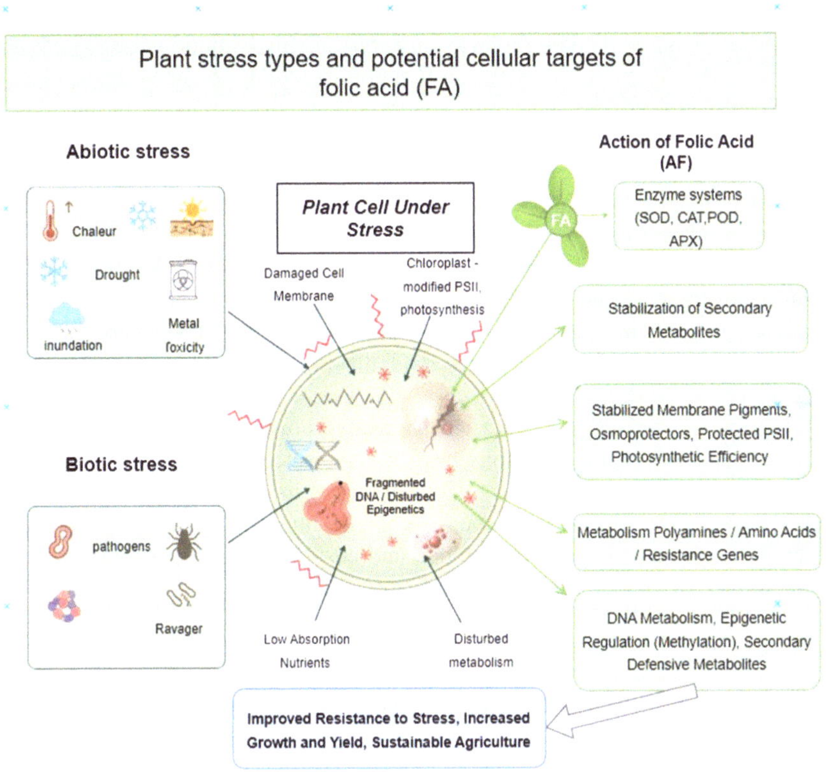

Fig. 5.1 Plant stress types and potential cellular targets of folic acid (FA). Folic acid (FA) functions as an essential cellular protector under both abiotic (heat, drought, salinity, flooding, and metal toxicity) and biotic (pathogen and pest) stresses that disrupt cellular homeostasis. Stress conditions damage membranes, chloroplasts, photosystem II (PSII), DNA, and nutrient assimilation pathways. Exogenous or endogenous FA mitigates these effects through multiple mechanisms: activation of enzymatic antioxidant systems (SOD, CAT, POD, APX); stabilization of secondary metabolites and photosynthetic pigments; enhancement of osmoprotectant and polyamine metabolism; and regulation of gene expression through DNA methylation. From a holistic perspective, these actions sustain photosynthetic efficiency, reinforce membrane integrity, and strengthen stress-responsive signaling, resulting in improved plant growth, yield, and resilience—key drivers of sustainable agriculture. Abbreviations: *SOD* superoxide dismutase, *CAT* catalase, *POD* peroxidase, *APX* ascorbate peroxidase

Under abiotic stress, FA fortifies plant tolerance through the activation of antioxidant defense systems—notably superoxide dismutase (SOD), catalase (CAT), peroxidase (POD), and ascorbate peroxidase (APX)—which scavenge excess ROS and preserve cellular redox homeostasis (Fig. 5.1) (Aljuaid et al. 2022; Tanveer et al. 2024). In parallel, FA promotes the accumulation of osmoprotectants such as proline and soluble sugars, stabilizes chlorophyll pigments and photosystem II, and thereby sustains photosynthetic efficiency and biomass production under adverse conditions (Al-Elwany et al. 2022).

At the metabolic level, FA participates in the regulation of amino acid and polyamine biosynthesis, key pathways for maintaining osmotic balance and stress signaling (Mehrshahi et al. 2010). Its function as a donor of one-carbon units via tetrahydrofolate-mediated reactions connects it to the synthesis of methionine, S-adenosylmethionine (SAM), and nucleotides—molecules essential for methylation, DNA repair, and epigenetic regulation.

Beyond abiotic adaptation, FA modulates defense signaling and secondary metabolism in response to pathogens and pests (Mehrshahi et al. 2010; Meng et al. 2025). It enhances the biosynthesis of phenylpropanoids, flavonoids, and alkaloids, reinforcing structural and chemical barriers against infection. Moreover, by influencing DNA methylation patterns and chromatin remodeling, FA contributes to the activation of resistance-related genes and the maintenance of genomic stability during stress exposure (Fig. 5.1).

This chapter provides a comprehensive analysis of folic acid's molecular, physiological, and agronomic functions in plants subjected to complex environmental constraints. It synthesizes current understanding of its mechanisms of action, highlights recent advances in exogenous FA application strategies, and evaluates its potential as an eco-compatible, scalable, and cost-effective solution for improving crop productivity under climate-induced stress conditions.

5.2 Biochemical Role and Biosynthesis of Folic Acid in Plants

5.2.1 Structure, Synthesis, and Folate Cycle

In plants, FA encompasses a family of compounds collectively known as folates, characterized by a pteridine ring, a p-aminobenzoic acid (pABA) moiety, and a polyglutamate tail that enhances cellular retention and enzymatic affinity (Fig. 5.2) (Wusigale and Liang 2020). Folate biosynthesis is a highly compartmentalized process involving coordinated reactions in plastids, mitochondria, and the cytosol (Hanson and Gregory 2011).

The pathway initiates in plastids, where guanosine triphosphate (GTP) is converted into pteridine by GTP cyclohydrolase I. Concurrently, pABA is synthesized from chorismate through aminodeoxychorismate synthase and lyase. These two precursor molecules are subsequently transported into mitochondria, where dihydropteroate synthase catalyzes their condensation to form dihydropteroate. This intermediate is transformed into dihydrofolate, then reduced to tetrahydrofolate (THF) by dihydrofolate reductase, and finally polyglutamylated by folylpolyglutamate synthase to generate metabolically active folates.

Within the folate cycle, THF and its derivatives—5,10-methylene-THF, 5-methyl-THF, and 10-formyl-THF—serve as one-carbon carriers essential for nucleotide biosynthesis (purines and thymidylate), amino acid metabolism (serine, glycine, methionine), and methylation reactions that regulate DNA and histone

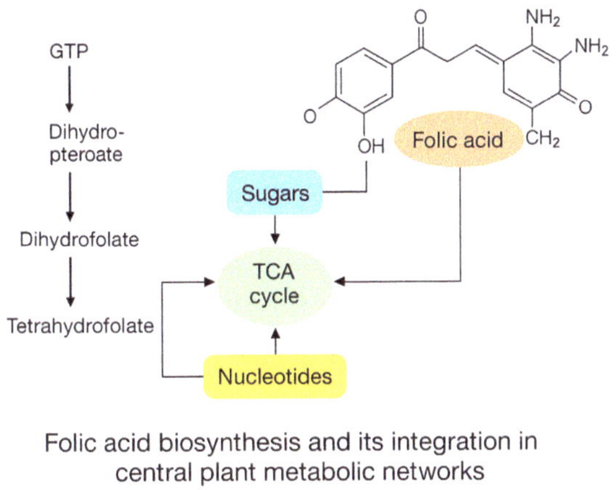

Fig. 5.2 Folic acid biosynthesis and its integration into central plant metabolic networks. Folic acid (vitamin B_9) originates from the nucleotide guanosine triphosphate (GTP), which is sequentially converted into dihydropteroate, dihydrofolate, and finally reduced to its active form, tetrahydrofolate (THF). Active THF participates in one-carbon metabolism, providing essential methyl and formyl groups for the synthesis of nucleotides (DNA and RNA precursors) and for amino acid metabolism. Through its interconnection with the tricarboxylic acid (TCA) cycle—fueled by sugars—folate metabolism sustains carbon flux, redox balance, and cellular proliferation, thereby linking energy production with biosynthetic and regulatory functions in plant cells.

modification (Hanson and Gregory 2011; Meng et al. 2025). Through these interconnected roles, folates sustain primary metabolism, drive growth and development, and underpin plant adaptation to environmental and biotic stresses.

5.2.2 Role in One-Carbon Metabolism, Methylation, and DNA Repair

Folic acid is a central mediator of one-carbon metabolism, a network of reactions that transfers single-carbon units required for the synthesis of nucleotides, amino acids, and methyl groups (Ulrich et al. 2008). In plants, folate derivatives such as THF function as carbon-unit carriers, enabling the methylation, methylene, and formyl transfer reactions that form the backbone of purine and thymidylate biosynthesis—key processes in DNA replication and cell division (Gorelova et al. 2019).

A pivotal downstream consequence of folate metabolism is the generation of S-adenosylmethionine (SAM), the universal methyl donor. SAM is produced via folate-dependent methylation of homocysteine to methionine and fuels a spectrum of epigenetic and biochemical methylation reactions, including DNA and histone modification. These methylation events are vital for regulating gene expression, chromatin dynamics, and stress–responsive transcriptional reprogramming.

Folate-driven one-carbon metabolism also underpins DNA repair and genome stability. During environmental stress, when ROS accumulation leads to DNA strand breaks and base damage, folate-dependent pathways ensure adequate supplies of methyl groups and nucleotides for both repair and replication (Ulrich et al. 2008; Krushkal et al. 2016). Thus, sufficient folate availability sustains genome integrity, epigenetic plasticity, and metabolic resilience—core determinants of stress tolerance in plants.

5.2.3 Impact on Primary Metabolism During Stress Conditions

Folic acid exerts a metabolic buffering effect during abiotic and biotic stress by stabilizing primary metabolism, particularly at the intersection of carbon and nitrogen assimilation. Acting through one-carbon metabolism, FA supports the synthesis of amino acids, nucleotides, and organic acids that maintain cellular energy and redox balance (Ravanel et al. 2011). Under drought, salinity, or temperature extremes, plants often experience metabolic disequilibrium manifested as altered sugar utilization, reduced ATP generation, and disrupted redox homeostasis (Alaa El-Den et al. 2022).

Through its integrative biochemical role, FA contributes to restoring this balance by:

- Enhancing amino acid biosynthesis, including stress-related metabolites such as proline that function as osmoprotectants.
- Regulating carbohydrate metabolism and promoting the synthesis of TCA-linked organic acids for sustained ATP production.
- Supporting antioxidant defenses via the generation of glutathione precursors, thus minimizing H_2O_2 accumulation and oxidative injury (Meng et al. 2025).

These processes enable folic acid to act as a metabolic stabilizer under stress, reinforcing both energy metabolism and redox homeostasis. By linking primary metabolism with stress signaling, folic acid enhances plant resilience and maintains productivity in challenging environments.

5.3 Folic Acid-Mediated Tolerance to Abiotic Stresses

5.3.1 Drought and Salinity

Folic acid plays a vital role in mitigating drought- and salinity-induced stress by sustaining osmotic homeostasis, stabilizing metabolic fluxes, and reinforcing redox regulation (Fig. 5.3) (Aljuaid et al. 2022). Under conditions of water deficit or high salinity, plants experience osmotic imbalance, ionic toxicity, and impaired

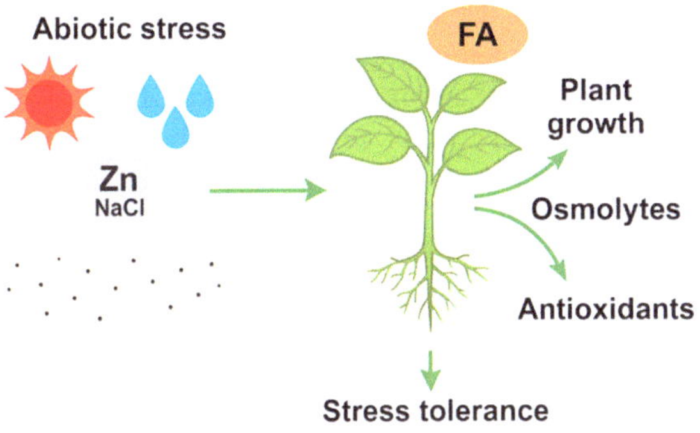

Fig. 5.3 Effects of FA on plant growth and stress response under abiotic stress. Folic acid acts as an essential regulator for plant growth and stress tolerance. When faced with abiotic stress (heat, drought, metal toxicity, salinity), FA improves plant growth by stimulating the synthesis of osmolytes (or osmoprotectants) and antioxidants. At the cellular level, this action is mediated by FA, which activates enzymatic defense systems, repairs fragmented DNA, stabilizes membrane pigments, and protects Photosystem PSII, resulting in increased stress tolerance and improved yields for sustainable agriculture

photosynthesis, ultimately restricting growth and yield (Ait-El-Mokhtar et al. 2022; Anli et al. 2022; Raja et al. 2022; Ismael and Ibrahim 2025). FA alleviates these constraints by promoting the accumulation of compatible osmolytes—including proline, soluble sugars, and glycine betaine—which preserve cellular turgor and protect macromolecules.

FA also contributes to chlorophyll stabilization and CO_2 assimilation by maintaining the activity of ribulose-1,5-bisphosphate carboxylase/oxygenase (RuBisCO), thereby supporting photosynthetic efficiency. At the metabolic level, FA enhances nitrogen assimilation and sustains amino acid and TCA-linked organic acid synthesis, ensuring a continuous energy supply under stress (Badr Elsayed et al. 2025). In parallel, FA strengthens the antioxidant defense system by facilitating glutathione biosynthesis, mitigating H_2O_2 accumulation, and preventing ROS-induced oxidative injury (Abdiazar et al. 2024; Alsamadany et al. 2022). Together, these multifaceted responses position FA as a central regulator of metabolic stability and oxidative protection under osmotic stress.

5.3.2 *Temperature Extremes*

Exposure to extreme temperatures disrupts membrane integrity, enzymatic activity, and redox equilibrium. Folic acid mitigates these effects through its dual capacity to stabilize macromolecular structures and sustain cellular energy metabolism. Under high-temperature stress, FA enhances the synthesis of heat shock proteins and preserves photosystem integrity, thereby reducing thermally induced photoinhibition. Under chilling stress, FA promotes the accumulation of cryoprotective metabolites—including amino acids and soluble sugars—that maintain membrane fluidity and prevent intracellular ice formation (Xiang et al. 2020).

By reinforcing the TCA cycle and maintaining ATP synthesis, FA ensures metabolic continuity during temperature fluctuations. Its capacity to restore redox balance and activate enzymatic antioxidants such as SOD, CAT, and POD provides a critical defense against temperature-induced oxidative perturbations (Gouda et al. 2020). Consequently, FA enhances both heat and cold tolerance by integrating metabolic resilience with cellular protection.

5.3.3 *Heavy Metal Toxicity*

Heavy metals such as cadmium, lead, and arsenic impose severe oxidative and metabolic stress, disrupting nutrient uptake, enzyme function, and photosynthesis (El-Sappah et al. 2024). Folic acid counteracts these effects by promoting metal chelation and detoxification mechanisms. It contributes to the synthesis of phytochelatins and glutathione, key thiol-rich peptides that sequester toxic ions into vacuoles, reducing their cytosolic availability (Sahito et al. 2024).

Moreover, FA supports nitrogen assimilation and amino acid metabolism, processes frequently impaired under metal exposure, ensuring sustained protein synthesis and metabolic activity. By maintaining chlorophyll stability, protecting membrane integrity, and modulating the expression of metal transporter and detoxification genes, FA acts as a potent metabolic regulator that limits the physiological and biochemical disruptions caused by heavy metal toxicity (Sahito et al. 2024).

5.3.4 *Oxidative Stress*

Oxidative stress arises from excessive accumulation of ROS, including hydrogen peroxide (H_2O_2), superoxide radicals, and hydroxyl ions, which damage cellular lipids, proteins, and nucleic acids (Ouhaddou et al. 2022; Sharma et al. 2020). Folic acid is a key orchestrator of antioxidant defense and redox homeostasis. It reinforces the ascorbate–glutathione (AsA–GSH) cycle by promoting glutathione synthesis and supporting NADPH regeneration, both of which are crucial for ROS detoxification (Alsamadany et al. 2022).

FA also upregulates the activities of SOD, CAT, and POD, reducing oxidative damage at the cellular level (Gorelova et al. 2019; Hanson and Gregory 2011). Furthermore, by stabilizing the photosynthetic and mitochondrial electron transport chains, FA prevents excess ROS formation at their source (Wusigale and Liang 2020). These mechanisms highlight FA's central role in oxidative stress mitigation and maintenance of cellular redox integrity.

5.4 Folic Acid-Mediated Tolerance to Biotic Stresses

Folic acid and its derivatives are indispensable cofactors in one-carbon (C_1) metabolism, driving the synthesis of nucleic acids, amino acids (methionine, serine, glycine), and the universal methyl donor S-adenosylmethionine (SAM) (Aslam et al. 2021; Pieterse et al. 2014). These functions not only sustain basic metabolism but also underpin immune and repair processes, giving FA an indirect yet crucial role in enhancing tolerance to pathogen- and pest-induced oxidative stress (El-Den et al. 2022) (Fig. 5.4 and Table 5.1).

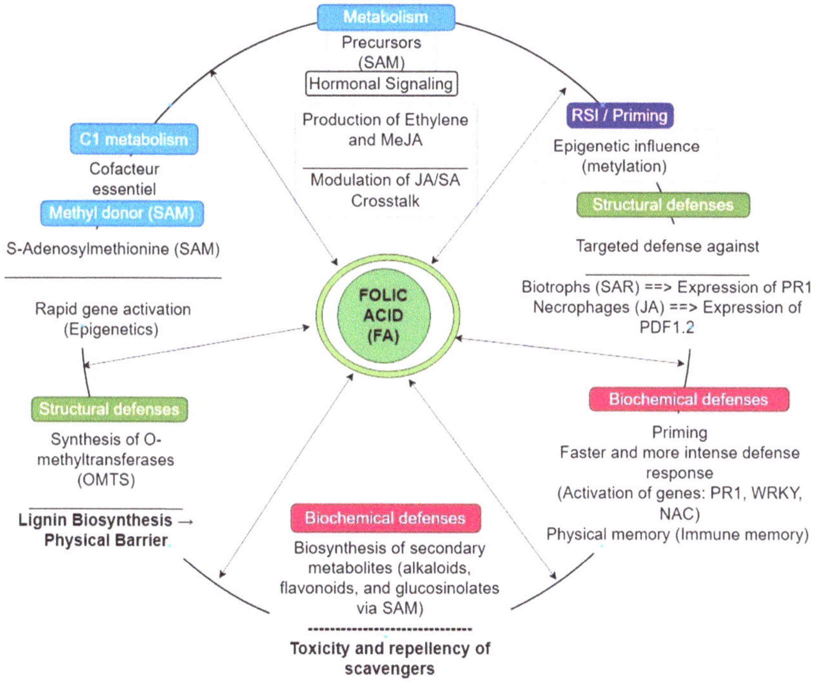

Fig. 5.4 Molecular map of folic acid-induced resistance under biotic stress conditions. Folic acid plays a central and multifaceted role in the activation and regulation of plant defense mechanisms. It positions folic acid as a metabolic (*via* the SAM cycle) and hormonal pivot, directly or indirectly influencing structural defenses (biosynthesis of OMTS, lignin), biochemical defenses (secondary metabolites), and the activation of targeted immune responses (SAR/PR1 against biotrophs, JA/PDF1.2 against necrotrophs) as well as the phenomenon of priming (preparation for a faster and more intense defense response). Abbreviations: *FA* Folic Acid, *SAM* S-Adenosylmethionine, *MeJA* Methyl Jasmonate, *JA* Jasmonic Acid, *SA* Salicylic Acid, *RSI* Systemic Induced Resistance, *SAR* Systemic Acquired Resistance, *PR1* Pathogenesis-Related protein 1, *PDF1.2* Plant Defensin 1.2, *OMTS* O-methyltransferases, *PR/WRKY/NAC* Pathogenesis-Related genes

Table 5.1 Summary of FA-mediated biotic stress responses in different crops

Crop	Biotic stress agent	Effect of folic acid or folate	Mechanism or suggested role	Reference
Arabidopsis (*Arabidopsis thaliana)*	Biotrophic bacterium: *Pseudomonas syringae*	Increase in systemic, local resistance	Induction of acquired systemic immunity (SAR), mediated by the salicylic acid (SA) signaling pathway	González and Vera (2019)
Arabidopsis (*A. thaliana)*	Necrotrophic fungus: *Alternaria brassicicola*	Increase in local susceptibility	The activation of the SA pathway by FA antagonizes the defense dependent on jasmonic acid (JA), essential against necrotrophs	Wittek et al. (2015)
Tomato (*Solanum lycopersicum* L.)	Necrotrophic fungus: *Alternaria solani* (Early burning)	Biotic stress mitigation	Increase in defense enzymes (POD, SOD, CAT) and pathogenesis-related proteins (PR-proteins)	El-Den et al. (2022)
Microalga (*Haematococcus pluvialis*)	Oomycete/fungus: *Paraphysoderma sedebokerense*	Inhibition of infection (through the application of antifolates)	The C1 metabolism of folates is identified as an essential antifungal target for the pathogenicity of the microbe. (Indirect effect on the plant, targeting the pathogen)	Yan et al. (2023)

5.4.1 Defense Against Pathogens

Folic acid and its derivatives (folates) play an essential role in defending plants against pathogens by acting as a chemical elicitor and providing crucial metabolic support (Fig. 5.4 and Table 5.1). As an elicitor, the application of FA or its precursor, 7,8-dihydroneopterin (DHN), induces systemic acquired resistance (SAR) against biotrophic pathogens (e.g., *Pseudomonas syringae*), a mechanism dependent on the salicylic acid (SA) pathway and confirmed by increased expression of the marker gene *PATHOGENESIS-RELATED1* (*PR1*) (Rahman et al. 2015). Furthermore, FA modulates the classic antagonistic crosstalk between the SA pathway and the jasmonic acid (JA) pathway, as evidenced by increased susceptibility to necrotrophs (e.g., *Alternaria brassicicola*) following treatment (Rahman et al. 2015). Metabolically, folates are central to the metabolism of monocarbon units (C1), ensuring the synthesis of SAM, the universal methyl donor (Aslam et al. 2021; Pieterse et al. 2014). This methyl supply is vital for the biosynthesis of lignin, a phenolic polymer that stiffens the cell wall and provides an essential physical barrier against pathogen intrusion. Under stress, ensuring adequate levels of SAM through FA removes the metabolic bottleneck for O-methyltransferases (OMTs), enabling the rapid and localized lignification necessary for tissue reinforcement (Haghpanah et al. 2024; Kim et al. 2022; Pieterse et al. 2014).

5.4.2 Insect and Pest Resistance

Folic acid exerts pivotal metabolic control over plant defense against herbivores and necrophages by regulating the pool of SAM, the universal methyl donor, whose synthesis depends on FA via the C_1 cycle (Jenkins et al. 2022). S-adenosylmethionine is the immediate precursor of ethylene and acts synergistically with JA, the key hormone in pest resistance (Rabari et al. 2023). Adequate availability of SAM, guaranteed by FA, ensures efficient JA/ethylene production and signaling, allowing the plant to fine-tune its defense strategy against aggressors (Aljuaid et al. 2022). Furthermore, SAM is vital for the final methylation steps of the biosynthesis of secondary metabolites such as alkaloids, terpenes, and phenolic compounds (Adinath Divekar et al. 2022). These methylation reactions are often required for these compounds to reach full biological activity, including toxicity or repellence (Adinath Divekar et al. 2022). In addition, FA supports the methylation of lignin precursors, thus strengthening cell wall rigidity for increased structural defense against tissue penetration and mastication (Alsamadany et al. 2022). This metabolite, therefore, influences pest resistance by ensuring that hormonal pathways are fully operational and that chemical defenses are properly synthesized and activated, providing an integrated biochemical shield against herbivory (Aljuaid et al. 2022).

5.4.3 Induced Systemic Resistance (ISR)

Folic acid functions as an indirect inducer and powerful amplifier of induced systemic resistance (ISR), a defense mechanism activated by beneficial microorganisms such as plant growth-promoting rhizobacteria (PGPR) and based on priming (Zhu et al. 2022). The role of FA in ISR is to establish immune memory through epigenomic regulation. By modulating SAM metabolism, FA influences histone methyltransferase (HMT) activity, notably promoting the deposition of the activation mark H3K4me3 in the promoter regions of defense-related genes such as WRKY and NAC (Liu et al. 2025). This chromatin remodeling maintains these genes in a primed state, enabling faster and stronger activation during pathogen or microbial challenge (Escaray et al. 2022). FA also enhances general cellular resilience by supporting antioxidant systems and ionic homeostasis (Aljuaid et al. 2022; Alsamadany et al. 2022), allowing the plant to invest resources efficiently in defense signaling without compromising growth. By maintaining an optimal metabolic status for SAM and reinforcing JA/ethylene signaling networks, FA acts as a crucial metabolic cofactor that bridges energy metabolism, hormonal crosstalk, and epigenetic priming to orchestrate a robust and sustainable systemic immune response (Rabari et al. 2023).

5.5 Gene Regulation and Signal Transduction

5.5.1 Stress-Responsive Gene Expression

One-carbon metabolism and methylation, the most fundamental mechanism by which FA influences gene expression, rely on its participation in the methionine cycle and the production of S-adenosylmethionine (SAM), the universal donor of methyl groups (CH_3) in plant cells (Fig. 5.5). SAM serves as the substrate for DNA

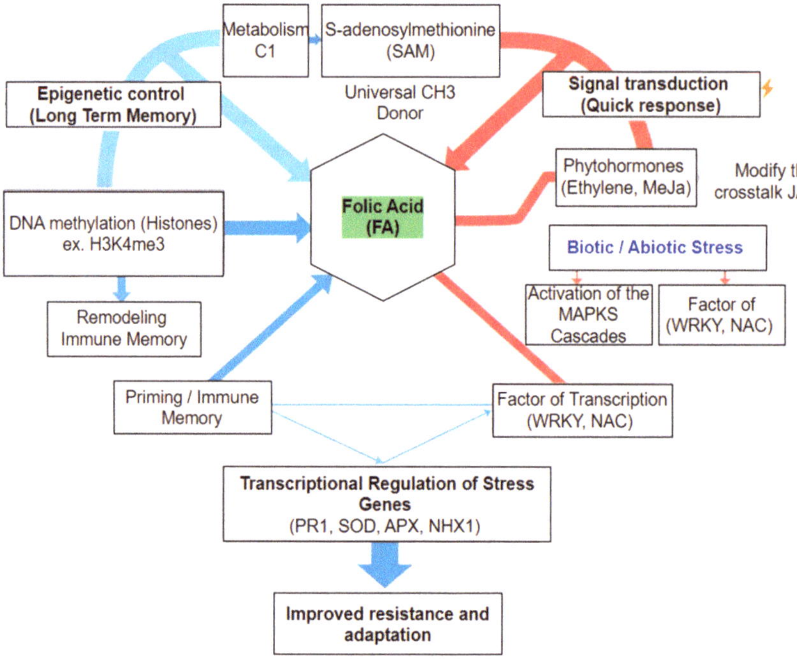

Fig. 5.5 Overview of FA-mediated transcriptional and epigenetic regulation under stress. Folic acid (FA) maintains plant resistance and adaptation to biotic and abiotic stresses. Folic acid acts as a regulatory hub via two main mechanisms (blue and orange arrows): on the one hand, it is a universal CH_3 donor via SAM, supporting C1 metabolism and epigenetics (long-term memory, e.g., DNA/histone methylation such as H3K4me3) to remodel immune memory and priming. On the other hand, FA is involved in signal transduction (rapid response) by modifying the balance of phytohormones (ethylene, MeJA) and JA/SA crosstalk, activating MAPK cascades and key transcription factors (WRKY, NAC). The integration of these pathways leads to the transcriptional regulation of stress genes (*PR1*, *SOD*, *APX*, *NHX1*), ultimately resulting in improved plant resistance and adaptation. Abbreviations: *FA* Folic Acid, *SAM* S-adenosylmethionine, *CH_3* Methyl group, *MeJA* Methyl Jasmonate, *JA* Jasmonic Acid, *SA* Salicylic Acid, *MAPKs* Mitogen-Activated Protein Kinases, *WRKY* WRKY, *NAC* NAC, *PR1* Pathogenesis-Related protein 1, *SOD* Superoxide Dismutase, *APX* Ascorbate Peroxidase, *NHX1* Na^+/H^+ antiporter 1, *H3K4me3* Histone H3 Lysine 4 trimethylation, *C1* C1 metabolism

methyltransferases (DNMTs) and histone methyltransferases (HMTs), enzymes that transfer methyl groups to DNA and histones, thereby shaping chromatin structure and gene accessibility (Escaray et al. 2022). DNA methylation in promoter regions is generally associated with transcriptional repression, whereas histone methylation—such as *H3K4* methylation—can promote gene activation or silencing, depending on context. Evidence from Arabidopsis demonstrates that folate homeostasis exerts a methylation-dependent control over immunity: pharmacological inhibition of the folate pathway (via antifolates) perturbs SAM levels, leading to altered DNA methylation patterns and the transcriptional activation of defense genes, which enhances resistance to *Pseudomonas syringae* (Escaray et al. 2022). This mechanism underlies epigenetic priming, in which FA modulates the methylation state of key defense loci (e.g., *PR* genes), preparing the genome for accelerated transcription upon secondary challenge. The resulting "immune memory" enables plants to mount faster responses without the continuous metabolic burden of constitutive defense activation (Escaray et al. 2022).

Impact on Antioxidant Response and Homeostasis Genes
The regulatory influence of FA extends to the modulation of genes encoding antioxidant and homeostatic effectors. Folates possess intrinsic antioxidant properties, and their application enhances tolerance to abiotic stresses such as salinity and metal toxicity by limiting ROS accumulation (Aljuaid et al. 2022; Alsamadany et al. 2022). This effect involves the transcriptional up-regulation of antioxidant enzymes. In green beans exposed to salt stress, FA treatment significantly increases the expression of SOD, CAT, and APX genes (Alsamadany et al. 2022). Likewise, under ionic and osmotic stress, FA modulates the transcription of homeostasis genes: in maize subjected to sodium-alkali stress, FA elevates the expression of the potassium/sodium exchanger ZmHKT1 and the photosystem II component D2 protein (Aljuaid et al. 2022). In green beans, the Na^+/H^+ antiporter (NHX1), Salt Overly Sensitive 1 (SOS1), and osmotin genes are similarly overexpressed (Alsamadany et al. 2022). This sophisticated gene regulation not only safeguards cellular survival but also sustains the energy state required for the activation of broader defense pathways.

Interaction with Hormone Signaling Pathways (Crosstalk)
The ability of FA to modulate phytohormonal signaling (as detailed in Sect. 5.4.2) operates largely through transcriptional regulation. Application of FA or its precursor 7,8-dihydroneopterin (DHN) induces systemic resistance mediated by increased expression of the SA marker gene PATHOGENESIS-RELATED 1 (PR1) (Zhu et al. 2022). This activation of PR1 constitutes direct molecular evidence of FA's impact on SAR-type signal transduction cascades. Furthermore, as a metabolic regulator influencing the biosynthesis of JA, ethylene, and SA, FA determines the activation or repression of major transcription-factor (TF) families—including WRKY, MYB, and AP2/ERF—which orchestrate the expression of hundreds of defense-related and secondary-metabolite genes (Adinath Divekar et al. 2022). In synthesis, FA acts as a metabolic–signaling co-regulator that uses the SAM-dependent C_1 cycle to shape the epigenetic landscape of the genome and amplify the transcriptional responses of key

gene networks governing antioxidant defense, ionic homeostasis, and pathogen resistance under both biotic and abiotic stress, ensuring an integrated and energetically efficient protective response.

5.5.2 *Modulation of Signaling Cascades*

Folic acid and its derivatives transcend their classical function as metabolic cofactors by acting as modulators of signaling cascades and epigenetic regulators—two central mechanisms underlying plant resilience. The role of FA in signal transduction is primarily indirect, as it initiates the metabolism of one-carbon (C_1) units leading to SAM synthesis. SAM serves as the biochemical precursor for the phytohormones ethylene and methyl jasmonate (MeJA), which together fine-tune the JA/SA signaling balance. Fluctuations in SAM availability consequently modulate the activity of Mitogen-Activated Protein Kinases (MAPKs), which determine the amplitude and duration of downstream defense responses (González and Vera 2019). The activation of the SA pathway by FA or its precursor reflects this upstream modulation of kinase inputs, while downstream effects manifest in the altered activity of transcription factors—the ultimate targets of MAPK signaling.

Induction of SA-dependent resistance by FA necessarily involves the activation of WRKY-type TFs, which govern systemic acquired resistance (SAR) and the expression of the hallmark gene PR1 (Aljuaid et al. 2022; Zhou et al. 2024). Concurrently, FA-mediated tolerance to salinity and osmotic stress correlates with the positive regulation of NAC-family TFs, key coordinators of ionic homeostasis and cell-protection genes (Aljuaid et al. 2022). Recent transcriptomic analyses indicate that WRKY and NAC loci are among the most enriched in accessible chromatin regions during stress exposure, suggesting that FA-mediated chromatin remodeling specifically enhances the binding capacity of these TF families (Luo et al. 2025).

Epigenetic regulation represents the most profound layer of control, essential for long-term adaptation and immune memory. FA's central role lies in regulating SAM supply, the universal methyl donor for DNMTs (Aarabi et al. 2015). Foundational work has demonstrated that folate metabolism maintains a globally methylated DNA state that exerts negative control over basal immunity (Escaray et al. 2022; González and Vera 2019). Disruption of the FA/SAM cycle leads to localized DNA hypomethylation or hypermethylation and subsequent de-repression of silenced defense genes, thereby enhancing resistance to pathogens (Escaray et al. 2022). Beyond DNA, SAM-dependent regulation extends to chromatin dynamics via histone methylation catalyzed by HMTs, which fine-tune gene activation or repression. The trimethylation of histone H3 at lysine 4 (H3K4me3) is particularly associated with transcriptionally active chromatin in immune gene promoters (Liu et al. 2025). By stabilizing favorable chromatin states around defense loci, FA not only triggers immediate transcriptional responses but also installs a genomic memory that enables faster and more resource-efficient activation during subsequent stress events.

Through these coordinated mechanisms, FA establishes a mechanistic bridge between cellular metabolism and the epigenetic plasticity that underpins durable stress resilience.

5.5.3 Epigenetic Effects

Epigenetic mechanisms constitute the deepest and most durable form of gene regulation orchestrated by FA, providing the molecular foundation for long-term stress adaptation and the establishment of immune memory (priming) (González and Vera 2019).

DNA Methylation and Negative Control of Immunity
The epigenetic function of FA derives directly from its central role in C_1 metabolism, as it governs SAM availability. Pioneering studies have shown that the maintenance of global DNA methylation through folate metabolism exerts negative regulation over plant immunity (González and Vera 2019). Disruption of the FA/SAM cycle—via antifolate treatment, for example—induces widespread alterations in DNA methylation, especially within non-CG contexts, resulting in the desilencing and transcriptional activation of previously repressed defense genes. This shift enhances resistance to *Pseudomonas syringae* (González and Vera 2019), positioning FA as a pivotal metabolic determinant of the basal immune state.

Chromatin Remodeling and Transgenerational Memory
SAM-dependent control extends beyond DNA methylation to dynamic chromatin remodeling through histone modifications. SAM provides the methyl groups required by HMTs to establish histone methylation marks, such as H3K4me3 on histone H3 lysine 4, which are closely associated with transcriptional activation (Liu et al. 2025). Emerging evidence from model systems demonstrates that exogenous SAM or FA can induce these histone modifications and that they may be transmitted across generations (F_1 and F_2), suggesting a mechanism for transgenerational priming (Liu et al. 2025). Although most of these data originate from non-plant models, the conserved role of SAM and H3K4me3 in plant immunity supports the hypothesis that FA contributes to heritable chromatin plasticity. By regulating SAM pools, FA modulates chromatin organization at stress-responsive promoters, thereby installing a heritable genomic memory that enables faster, more efficient reactivation of defense pathways upon re-exposure. Folates thus function as a molecular nexus linking central carbon metabolism to the epigenetic flexibility essential for plant survival and evolutionary adaptation (Liu et al. 2025).

5.6 Interaction with Hormonal and Nutrient Pathways

Folates serve as essential cofactors in one-carbon transfer reactions and nucleotide biosynthesis, forming the biochemical foundation for numerous cellular and developmental processes in plants (Fig. 5.6). They play pivotal roles in the biosynthesis of amino acids, phytohormones, photosynthetic pigments, and the lipid-soluble antioxidant vitamin E (Aljuaid et al. 2022). Through their involvement in amino acid metabolism and the provision of methyl groups, folates are indispensable to

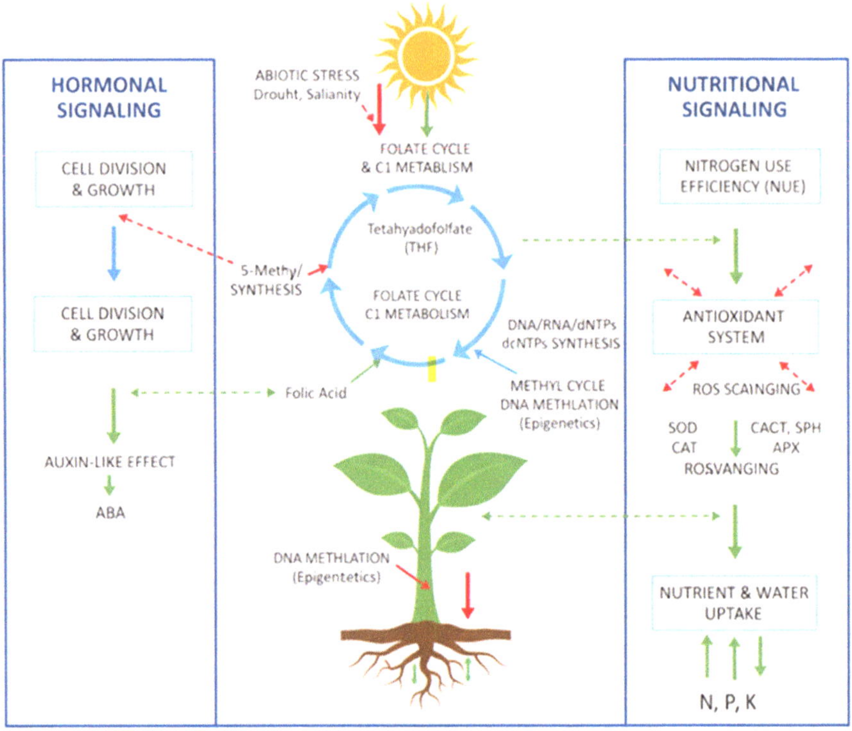

Fig. 5.6 Integration of FA with hormonal and nutritional signaling under stress. Folic acid (FA), or folates, is an essential molecule in the metabolism and stress tolerance of plants, whose biosynthesis begins with GTP and passes through Dihydrofolate to produce Tetrahydrofolate (THF). THF is crucial because it is directly involved in the synthesis of Nucleotides (DNA/RNA) and in C1 Metabolism (methylation), thus connecting the production of carbonaceous matter (Sugars, TCA Cycle) to cell division, growth and epigenetic regulation. In the face of abiotic (heat, drought, salinity) and biotic stresses, FA improves resistance by stabilizing membrane pigments, protecting the PSII (Photosystem II), repairing fragmented DNA, and activating the antioxidant system (*via* SOD, CAT, and APX), leading to improved resistance to stress. Abbreviations: *FA* Folic Acid, *THF* Tetrahydrofolate, *5-Methyl-THF* 5-Methyltetrahydrofolate, *C1* One-Carbon, *DNA* Deoxyribonucleic Acid, *RNA* Ribonucleic Acid, *N* Nitrogen, *P* Phosphorus, *K* Potassium, *NUE* Nitrogen Use Efficiency, *ROS* Reactive Oxygen Species, *SOD* Superoxide Dismutase, *CAT* Catalase, *APX* Ascorbate Peroxidase, *ABA* Abscisic Acid

methylation reactions that regulate gene expression and govern the synthesis of proteins, lipids, chlorophyll, and lignin (Kołton et al. 2022). Moreover, FA interacts with several phytohormonal signaling pathways—particularly abscisic acid (ABA) and salicylic acid (SA)—and exerts a broad regulatory influence over growth, defense, and metabolic coordination. Experimental studies demonstrate that FA can stimulate plant growth and development by enhancing protein and nucleic acid biosynthesis while simultaneously modulating natural hormone activity, including auxins (IAA), cytokinins, and gibberellins (Al-Elwany et al. 2022; Ravanel et al. 2011). In *Arabidopsis thaliana*, the exogenous application of FA or its biosynthetic precursor 7,8-dihydroneopterin (DHN) enhances resistance to *Pseudomonas syringae*, concomitant with increased transcript accumulation of the SA-responsive marker gene PATHOGENESIS-RELATED 1 (PR1) (Wittek et al. 2015). Conversely, González and Vera (2019) showed that the folate metabolic pathway can act as an SA-independent negative regulator of plant defense. Chemical or genetic disruption of folate biosynthesis enhances resistance to *Pseudomonas syringae* pv. tomato DC3000 by establishing a primed immune state, in which plants exhibit heightened defensive readiness and stronger inducible responses upon pathogen challenge.

Para-aminobenzoic acid (PABA), a precursor of folate, participates in critical hormonal crosstalk with the auxin pathway, modulating auxin–ethylene interactions that influence root gravitropism and developmental plasticity. PABA also fine-tunes auxin sensitivity by mediating sucrose signaling, which in turn affects auxin distribution and contributes to adaptive developmental processes (Mishra et al. 2022). Elevated PABA levels alter root system architecture by modifying the auxin–cytokinin balance; increased PABA concentrations lead to reduced meristem size and premature cell differentiation, reflecting a hormonal interplay that shapes root growth (Lasok et al. 2023). Likewise, FA itself modulates root architecture by influencing auxin distribution and regulating the expression of auxin transporter genes—including PIN1, PIN3, and PIN7—which determine auxin flow and lateral root patterning (Ayala-Rodríguez et al. 2017; Li et al. 2023).

Beyond hormonal regulation, folates are deeply integrated into nutrient assimilation and metabolic homeostasis, serving as essential cofactors in carbohydrate and nitrogen metabolism. FA and its derivatives participate in amino acid and plant growth regulator biosynthesis, indirectly improving nutrient uptake by promoting cell division, genome stability, and root development (Alsamadany et al. 2022; Sahito et al. 2024). In Arabidopsis, mutants with perturbed folate metabolism exhibit differential expression of genes linked to nitrate uptake and root system architecture, revealing that folate homeostasis influences nitrate transport and nitrogen assimilation pathways (Jiang et al. 2013). Altered folate balance affects the expression of nitrate reductase and glutamine synthetase, suggesting that FA status may act as a metabolic signal adjusting nutrient uptake and assimilation in response to environmental constraints.

Under salinity stress, FA supplementation enhances the transcription of ion-transport and homeostasis genes. For instance, in *Phaseolus vulgaris*, FA upregulates the expression of the Na^+/H^+ antiporter (NHX1), the SOS1 regulator, and the osmoprotective protein Osmotin, collectively improving ion homeostasis and nutrient uptake under saline conditions (Alsamadany et al. 2022). FA also promotes the accumulation of osmoprotectants such as proline, soluble sugars, and free amino acids, which stabilize osmotic potential, enhance water retention, and protect cytosolic enzymes from ionic toxicity (Alsamadany et al. 2022). Similarly, FA enhances K^+ and Ca^{2+} uptake while reducing Na^+ accumulation, improving the K^+/Na^+ ratio and maintaining enzymatic integrity under sodic–alkaline stress (Aljuaid et al. 2022). This ionic optimization is achieved through the up-regulation of ZmHKT1, D2-Protein, and PEP-case, coupled with the down-regulation of ZmSOS1 and ZmNHX1 in maize exposed to alkaline stress. These findings underscore the multifaceted role of FA as a metabolic integrator and signaling modulator that bridges hormonal regulation, nutrient homeostasis, and stress adaptation. By coordinating carbon, nitrogen, and hormone pathways, folates ensure balanced growth, sustained photosynthetic efficiency, and enhanced resilience against environmental fluctuations.

5.7 Exogenous Application Strategies and Agricultural Implications

Exogenous application of FA is emerging as a promising strategy to enhance crop growth, stress tolerance, and yield (Khan et al. 2022). Among the various approaches, foliar spraying remains the most widely employed method, typically applied at concentrations ranging from 25 to 200 μM for vegetables and up to 150 mg L^{-1} for leafy greens. Application timing and frequency are optimized according to crop species and prevailing environmental conditions, often coinciding with key developmental or stress-sensitive stages (Alsamadany et al. 2022; Khan et al. 2022; Kilic and Aca 2016). For instance, foliar treatment with 50 mM FA in *Coriandrum sativum* under drought stress markedly improved plant height, biomass, physiological performance, gas exchange parameters, and yield (Khan et al. 2022). In cabbage, applications of 100–200 μM FA alleviated drought-induced growth reduction (Kul 2025), whereas spinach treated with 150 mg L^{-1} FA displayed greater plant height, expanded leaf area, increased yield, and reduced oxalate and nitrate contents (Al-Mharib et al. 2022). Similarly, foliar application of 250–500 mg L^{-1} FA in wheat enhanced plant height, tiller number, dry weight, and chlorophyll content (Baqir and Zeboon 2024). In maize seedlings exposed to sodic–alkaline stress, FA foliar treatments resulted in pronounced improvements in biomass and physiological vigor, effects attributed to FA's role in promoting cell division, maintaining genome stability, and regulating stress-responsive gene expression (Aljuaid et al. 2022).

In fruit crops, FA also exerts a pronounced impact on yield and quality. In strawberry, foliar spraying with 40 mM FA at 30 days after planting significantly enhanced yield components—including the weight of primary and secondary fruits and the number of achenes—and improved fruit quality parameters such as total soluble solids, reducing sugars, titratable acidity, anthocyanin, phenolic compounds, and vitamin C concentration (Raeisi et al. 2017). Beyond foliar use, seed priming with FA has demonstrated similar benefits: supplementation reduced cadmium (Cd) toxicity and improved germination and biomass accumulation in *Solanum nigrum* by enhancing antioxidant activity, lowering oxidative injury, and increasing Cd hyperaccumulation capacity (Sahito et al. 2024). Pre-treatment of barley seeds with 50 μM FA effectively alleviated salinity-induced germination inhibition and improved seedling vigor (Kilic and Aca 2016). Moreover, post-harvest application of FA (5 mg L^{-1}) has been shown to prolong shelf life and preserve the nutritional integrity of broccoli during storage (Zhao et al. 2024). These findings underscore FA's versatility as a bioactive molecule capable of enhancing growth, physiological resilience, and product quality across diverse species and environments.

The mechanisms underlying these benefits are largely associated with FA's biochemical functions—enhancing antioxidant defense, osmoprotectant accumulation, and membrane stability—thereby mitigating oxidative and osmotic stress (Alsamadany et al. 2022). These physiological improvements translate directly into enhanced yield and stress resilience in leafy greens, legumes, cereals, and fruiting vegetables. The integration of FA into conventional crop management systems therefore represents a promising and sustainable biostimulant approach, particularly relevant to regions facing increasing environmental constraints (Al-Elwany et al. 2022; Alsamadany et al. 2022; Khan et al. 2022; Kul 2025).

Recent research further demonstrates that B-group vitamins—including FA—can be combined synergistically with other biostimulants such as amino acids, seaweed extracts, humic and fulvic acids, and chitosan without antagonistic effects (Malik et al. 2020; Mosa et al. 2022; Soppelsa et al. 2019). These combined formulations often yield additive or even synergistic effects on plant growth, biomass accumulation, yield, and quality, particularly under nutrient deficiency or abiotic stress conditions. For instance, foliar treatments with B-complex vitamins alongside natural biostimulants have enhanced root and fruit biomass, leaf expansion, and fruit quality in strawberry, apple, and grape crops compared with single-component applications (Mosa et al. 2021, 2022; Soppelsa et al. 2019). Similarly, in turnip, the application of vitamin B complexes (including FA) outperformed chemical fertilizers in improving physiological and morphological traits, while the addition of amino acids further enhanced photosynthetic efficiency and yield (Raza et al. 2023). These studies highlight FA's potential as a bio-based, eco-compatible growth enhancer, supporting an integrative strategy for climate-resilient and nutritionally optimized agriculture.

5.8 Conclusions and Future Perspectives

In summary, folic acid and its active derivatives—tetrahydrofolate, 5-methyltetrahydrofolate, and folinic acid—occupy a central and integrative position in plant stress physiology. They reinforce antioxidant defense systems by upregulating key enzymes such as SOD, CAT, APX, and POD, thereby mitigating oxidative damage under abiotic stresses, including drought, salinity, and temperature extremes. FA further regulates phytohormonal networks, modulating ABA-dependent responses for abiotic stress adaptation, SA-mediated immune signaling against pathogens, and auxin-driven root development and nutrient uptake. It also enhances nutrient acquisition by influencing the expression of ion transporters and homeostasis genes, maintaining K^+/Na^+ balance and cellular stability under stress.

These interlinked biochemical and molecular pathways converge to support robust plant growth, developmental plasticity, and adaptive resilience in challenging environments. As global climate change amplifies environmental stress pressures, integrating FA into modern crop management offers a sustainable, low-cost, and scalable strategy to strengthen food security while minimizing chemical inputs. Future research should prioritize a deeper elucidation of FA's molecular mechanisms—particularly its interactions with the one-carbon network, hormonal crosstalk, and epigenetic regulation—across diverse species and environmental contexts. Such efforts will enable the optimization of FA-based biostimulant applications as a cornerstone of climate-smart, resource-efficient, and sustainable agriculture.

References

Aarabi M, San Gabrie MC, Chan D, Behan NA, Caron M, Pastinen T, Bourque G, MacFarlane AJ, Zini A, Trasler J (2015) High-dose folic acid supplementation alters the human sperm methylome and is influenced by the MTHFR C677T polymorphism. Hum Mol Genet 24(22): 6301–6313. https://doi.org/10.1093/hmg/ddv338

Abdiazar N, Zahedi H, Sharghi Y, Modarres-Sanavy SAM, Alipour A (2024) Comparing effects of folic acid, epibrassinolide, chitosan and glutathione foliar treatments on safflower's physiology and yield during water stress. Russ J Plant Physiol 71(1):35. https://doi.org/10.1134/S1021443723603312

Abdel Latef AAH (2026) Using stimulants to improve plant health: From stress to strength. Elsevier. https://doi.org/10.1016/C2025-0-01048-7

Adinath Divekar P, Narayana S, Adinath Divekar B, Kumar R, Gadratagi Gowda B, Ray A, Kumar Singh A, Rani V, Singh V, Singh Kumar A, Kumar A, Pratap Singh R, Shyam Meena R, Kanti Behera T (2022) Plant secondary metabolites as defense tools against herbivores for sustainable crop protection. Int J Mol Sci 23(5):2690. https://doi.org/10.3390/ijms23052690

Ait-El-Mokhtar M et al. (2022) Cereals and phytohormones under drought stress. In: Abdel Latef, A.A.H. (eds) Sustainable Remedies for Abiotic Stress in Cereals. Springer, Singapore. https://doi.org/10.1007/978-981-19-5121-3_13

Alaa El-Den HR, M.Abd El-Wahab G, Masoud SA (2022) Using salicylic acid, folic acid and/or mancozeb in controlling tomato early blight biotic stress and their effects on growth, yield, fruit quality, and stress-related enzymes. Iraqi J Agric Sci 53(6):1548–1559. https://doi.org/10.36103/ijas.v53i6.1670

Al-Elwany OAAI, Hemida KA, Abdel-Razek MA, El-Mageed TAA, El-Saadony MT, AbuQamar SF, El-Tarabily KA, Taha RS (2022) Impact of folic acid in modulating antioxidant activity, Osmoprotectants, anatomical responses, and photosynthetic efficiency of Plectranthus

amboinicus under salinity conditions. Front Plant Sci 13(July):1–15. https://doi.org/10.3389/fpls.2022.887091
Aljuaid BS, Mukherjee S, Sayed AN, El-Gabry YAEG, Omar MMA, Mahmoud SF, Alsubeie MS, Darwish DBE, Al-Qahtani SM, Al-Harbi NA, Alzuaibr FM, Basahi MA, Hamada MMA (2022) Folic acid reinforces maize tolerance to sodic-alkaline stress through modulation of growth, biochemical and molecular mechanisms. Life 12(9):1327. https://doi.org/10.3390/life12091327
Al-Mharib MZK, Al-Ubaidy RM, Mohammed MM (2022) Effect of spraying with folic acid and yeast extract on the growth, yield and calcium oxalate concentration of spinach (Spinacia oleracea L.). Appl Ecol Environ Res 20(3):2763–2768. https://doi.org/10.15666/aeer/2003_27632768
Alsamadany H, Mansour H, Elkelish A, Ibrahim MFM (2022) Folic acid confers tolerance against salt stress-induced oxidative damages in snap beans through regulation growth, metabolites, antioxidant machinery and gene expression. Plants 11(11):1459. https://doi.org/10.3390/plants11111459
Anli M et al. (2022) Use of biostimulants to improve drought tolerance in cereals. In: Abdel Latef, A.A.H. (eds) Sustainable Remedies for Abiotic Stress in Cereals. Springer, Singapore. https://doi.org/10.1007/978-981-19-5121-3_20
Anli M, Boutasknit A, Ait-El-Mokhtar M, Ben-Laouane R, Abdouroihamane H, Meddich A, Baslam M (2026) Plant extracts for boosting plant stress tolerance. In: Abdel Latef AAH (eds) Using Stimulants to Improve Plant Health: From Stress to Strength. Elsevier. https://doi.org/10.1016/B978-0-443-45576-6.00009-1
Arif Y, Singh P, Siddiqui H, Bajguz A, Hayat S (2020) Plant physiology and biochemistry salinity induced physiological and biochemical changes in plants: an omic approach towards salt stress tolerance. Plant Physiol Biochem 156(August):64–77. https://doi.org/10.1016/j.plaphy.2020.08.042
Aslam S, Gul N, Mir MA, Asgher M, Al-Sulami N, Abulfaraj AA, Qari S (2021) Role of Jasmonates, calcium, and glutathione in plants to combat abiotic stresses through precise signaling cascade. Front Plant Sci 12(July):1–29. https://doi.org/10.3389/fpls.2021.668029
Ayala-Rodríguez JÁ, Barrera-Ortiz S, Ruiz-Herrera LF, López-Bucio J (2017) Folic acid orchestrates root development linking cell elongation with auxin response and acts independently of the TARGET OF RAPAMYCIN signaling in Arabidopsis thaliana. Plant Sci 264 (September):168–178. https://doi.org/10.1016/j.plantsci.2017.09.011
Badr Elsayed EAE, Bakhom GS, Sadak MS, Amin GA (2025) Exogenous folic acid alleviates drought stress effect on growth and productivity via enhancing physiological aspects in barley plant. Egypt J Agron 47(1):157–165. https://doi.org/10.21608/agro.2025.310178.1485
Baqir HA, Zeboon NH (2024) Response of some characteristics of wheat growth to spraying with vitamin B9 and E. Iraqi J. Agric. Sci 55:803–812
El-Den A, Abd El-Wahab GM, Masoud SA (2022) Using salicylic acid, folic acid and/or Mancozeb in controlling tomato early blight biotic stress and their effects on growth, yield, fruit quality, and stress-related enzymes. Iraqi J. Agric. Sci 2022(6):1548–1559
El-Sappah AH, Zhu Y, Huang Q, Chen B, Soaud SA, Abd Elhamid MA, Yan K, Li J, El-Tarabily KA (2024) Plants' molecular behavior to heavy metals: from criticality to toxicity. Front Plant Sci 15:1423625. https://doi.org/10.3389/fpls.2024.1423625
Escaray F, Felipo-Benavent A, Vera P (2022) Linking plant metabolism and immunity through methionine biosynthesis. Mol Plant 15(1):6–8. https://doi.org/10.1016/j.molp.2021.12.007
Fassih B, Ait-el-mokhtar M, Douch AN, Boutasknit A, Ben-Laouane R, Aganchich B, Wahbi S (2024) Combined effect of subsurface water retention technology and arbuscular mycorrhizal fungi on growth, physiology and biochemistry of Argan seedlings under field conditions. Plant 13(15):2098. https://doi.org/10.3390/plants13152098
Fassih B, Ait-El-Mokhtar M, Douch AN, Boutasknit A, Ben-Laouane R, Aganchich B, Wahbi S (2025) The impact of browsing intensity on argan trees in the Essaouira region of Morocco. J Arid Environ 231(April):105472. https://doi.org/10.1016/j.jaridenv.2025.105472
González B, Vera P (2019) Folate metabolism interferes with plant immunity through 1C methionine synthase-directed genome-wide DNA methylation enhancement. Mol Plant 12(9): 1227–1242. https://doi.org/10.1016/j.molp.2019.04.013

Gorelova V, Bastien O, De Clerck O, Lespinats S, Rébeillé F, Van Der Straeten D (2019) Evolution of folate biosynthesis and metabolism across algae and land plant lineages. Sci Rep 9(1):5731. https://doi.org/10.1038/s41598-019-42146-5

Gouda A, Amer SA, Gabr S, Tolba SA (2020) Effect of dietary supplemental ascorbic acid and folic acid on the growth performance, redox status, and immune status of broiler chickens under heat stress. Trop Anim Health Prod 52(6):2987–2996. https://doi.org/10.1007/s11250-020-02316-4

Haghpanah M, Hashemipetroudi S, Arzani A, Araniti F (2024) Drought tolerance in plants: physiological and molecular responses. Plants 13(21):1–30. https://doi.org/10.3390/plants13212962

Hanson AD, Gregory JF (2011) Folate biosynthesis, turnover, and transport in plants. Annu Rev Plant Biol 62(1):105–125. https://doi.org/10.1146/annurev-arplant-042110-103819

Huchzermeyer B, Menghani E, Khardia P, Shilu A (2022) Metabolic pathway of natural antioxidants, antioxidant enzymes and ROS providence. Antioxidants 11(4):761. https://doi.org/10.3390/antiox11040761

Iqbal Z, Iqbal MS, Hashem A, Abd Allah EF, Ansari MI (2021) Plant defense responses to biotic stress and its interplay with fluctuating dark/light conditions. Front Plant Sci 12(March):1–22. https://doi.org/10.3389/fpls.2021.631810

Ismael MB, Ibrahim DA (2025) The effect of folic and salicylic acid on growth of cotton (Gossypium hirsutum L.) under drought conditions. Sci J Univ Zakho 13(2):114–124. https://doi.org/10.25271/sjuoz.2025.13.2.1465

Jenkins T, Aston K, Carrell D, DeVilbiss E, Sjaarda L, Perkins N, Mills JL, Chen Z, Sparks A, Clemons T, Chaney K, Peterson CM, Emery B, Hotaling J, Johnstone E, Schisterman E, Mumford SL (2022) The impact of zinc and folic acid supplementation on sperm DNA methylation: results from the folic acid and zinc supplementation randomized clinical trial (FAZST). Fertil Steril 117(1):75–85. https://doi.org/10.1016/j.fertnstert.2021.09.009

Jiang L, Liu Y, Sun H, Han Y, Li J, Li C, Guo W, Meng H, Li S, Fan Y, Zhang C (2013) The mitochondrial folylpolyglutamate synthetase gene is required for nitrogen utilization during early seedling development in arabidopsis. Plant Physiol 161(2):971–989. https://doi.org/10.1104/pp.112.203430

Khan MT, Ahmed S, Shah AA (2022) Regulatory role of folic acid in biomass production and physiological activities of Coriandrum sativum L. under irrigation regimes. Int J Phytoremediation 24(10):1025–1038. https://doi.org/10.1080/15226514.2021.1993785

Kilic S, Aca HT (2016) Role of exogenous folic acid in alleviation of morphological and anatomical inhibition on salinity-induced stress in barley. Ital J Agron 11(4):777. https://doi.org/10.4081/ija.2016.777

Kim D, Jeon SJ, Yanders S, Park SC, Kim HS, Kim S (2022) MYB3 plays an important role in lignin and anthocyanin biosynthesis under salt stress condition in Arabidopsis. Plant Cell Rep 41(7):1549–1560. https://doi.org/10.1007/s00299-022-02878-7

Kołton A, Długosz-Grochowska O, Wojciechowska R, Czaja M (2022) Biosynthesis regulation of folates and phenols in plants. Sci Hortic 291(September 2021):110561. https://doi.org/10.1016/j.scienta.2021.110561

Krushkal J, Zhao Y, Hose C, Monks A, Doroshow JH, Simon R (2016) Concerted changes in transcriptional regulation of genes involved in DNA methylation, demethylation, and folate-mediated one-carbon metabolism pathways in the NCI-60 cancer cell line panel in response to cancer drug treatment. Clin Epigenetics 8(1):73. https://doi.org/10.1186/s13148-016-0240-3

Kul R (2025) Foliar application of folic acid on cabbage seedlings grown under restricted irrigation conditions can alleviate the negative effects. Turk J Agric Food Sci Technol 13(3):760–765. https://doi.org/10.24925/turjaf.v13i3.760-765.7453

Lal MK, Tiwari RK, Altaf MA, Kumar A, Kumar R (2023) Editorial: abiotic and biotic stress in horticultural crops: insight into recent advances in the underlying tolerance mechanism. Front Plant Sci 14:1212982. https://doi.org/10.3389/fpls.2023.1212982

Lasok H, Nziengui H, Kochersperger P, Ditengou FA (2023) Arabidopsis root development regulation by the endogenous folate precursor, Para-Aminobenzoic acid, via modulation of the root cell cycle. Plants 12(24):4076. https://doi.org/10.3390/plants12244076

Li Y, Luo J, Chen R, Zhou Y, Yu H, Chu Z, Lu Y, Gu X, Wu S, Wang P, Kuang H, Ouyang B (2023) Folate shapes plant root architecture by affecting auxin distribution. Plant J 113(5): 969–985. https://doi.org/10.1111/tpj.16093

Liu F, Wang Q, Xiong J, Wang M, Zhou H, Xiao Y (2025) Parental S-adenosylmethionine diet defines offspring immune response via histone H3K4me3 complex and endoplasmic reticulum UPR. Cell Commun Signal 23(1):397. https://doi.org/10.1186/s12964-025-02386-7

Luo L, Qu Q, Cao M, Zhang Y, Sun Y, Mao F, Chen J, Zhu Y, Yang Y, Liu H, Li C, Lin D, Lu G, Lin Z, Zhu F, Xiao J (2025) Epigenetic maps of pearl millet reveal a prominent role for CHH methylation in regulating tissue-specific gene expression. aBIOTECH 6:394–410. https://doi.org/10.1007/s42994-025-00243-2

Malik A, Mor VS, Tokas J, Punia H, Malik S, Malik K, Sangwan S, Tomar S, Singh P, Singh N, Himangini, Vikram, Nidhi, Singh G, Vikram, Kumar V, Sandhya, Karwasra A (2020) Biostimulant-treated seedlings under sustainable agriculture: a global perspective facing climate change. Agronomy 11(1):14. https://doi.org/10.3390/agronomy11010014

Mehrshahi P, Gonzalez-Jorge S, Akhtar TA, Ward JL, Santoyo-Castelazo A, Marcus SE, Lara-Núñez A, Ravanel S, Hawkins ND, Beale MH, Barrett DA, Paul Knox J, Gregory JF, Hanson AD, Bennett MJ, DellaPenna D (2010) Functional analysis of folate polyglutamylation and its essential role in plant metabolism and development. Plant J 64(2):267–279. https://doi.org/10.1111/j.1365-313X.2010.04336.x

Meng H, Huang B, Guo C, Zheng M, Ye T, He E, Guo Y, Wang W (2025) Folic acid coordinates metabolomic and transcriptomic reprogramming to enhance growth plasticity, photosynthetic efficiency, flavonoid biosynthesis, and glycosylation in dendrobium officinale. Plant Stress 18: 101066. https://doi.org/10.1016/j.stress.2025.101066

Mishra BS, Sharma M, Laxmi A (2022) Role of sugar and auxin crosstalk in plant growth and development. Physiol Plant 174(1):1–21. https://doi.org/10.1111/ppl.13546

Mosa WFA, Salem MZM, Al-huqail AA, Hayssam MA (2021) Application of glycine, folic acid, and moringa extract as bio-stimulants for enhancing the production of 'Flame Seedless' grape cultivar. Bioresources 16(2):3391–3410

Mosa WFA, Sas-Paszt L, Głuszek S, Górnik K, Anjum MA, Saleh AA, Abada HS, Awad RM (2022) Effect of some biostimulants on the vegetative growth, yield, fruit quality attributes and nutritional status of apple. Horticulturae 9(1):32. https://doi.org/10.3390/horticulturae9010032

Ouhaddou R, Ben-Laouane R, Lahlali R, Anli M, Ikan C, Boutasknit A, Slimani A, Oufdou K, Baslam M, Ait Barka E, Meddich A (2022) Application of indigenous rhizospheric microorganisms and local compost as enhancers of lettuce growth, development, and salt stress tolerance. Microorganisms 10(8):1625. https://doi.org/10.3390/microorganisms10081625

Pieterse MJ, Zamioudis C, Berendsen RL, Weller DM, Van Wees SCM, Bakker PAHM (2014) Induced systemic resistance by beneficial microbes. Annu Rev Phytopathol 52:347–375. https://doi.org/10.1146/annurev-phyto-082712-102340

Rabari A, Ruparelia J, Jha CK, Sayyed RZ, Mitra D, Priyadarshini A, Senapati A, Panneerselvam P, Das Mohapatra PK (2023) Articulating beneficial rhizobacteria-mediated plant defenses through induced systemic resistance: a review. Pedosphere 33(4):556–566. https://doi.org/10.1016/j.pedsph.2022.10.003

Raeisi J, Pakkish Z, Saffari VR (2017) Efficiency of folic acid in improving yield and fruit quality of strawberry. J. Plant Physiol. Breed. 7(1):15–25

Rahman A, Uddin W, Wenner NG (2015) Induced systemic resistance responses in perennial ryegrass against Magnaporthe oryzae elicited by semi-purified surfactin lipopeptides and live cells of Bacillus amyloliquefaciens. Mol Plant Pathol 16(6):546–558. https://doi.org/10.1111/mpp.12209

Raja BL et al. (2022) Use of biostimulants to improve salinity tolerance in cereals. In: Abdel Latef, A.A.H. (eds) Sustainable Remedies for Abiotic Stress in Cereals. Springer, Singapore. https://doi.org/10.1007/978-981-19-5121-3_19

Ravanel S, Douce R, Rébeillé F (2011) Metabolism of folates in plants. Adv Bot Res 59:67–106. https://doi.org/10.1016/B978-0-12-385853-5.00004-0

Raza Q, Bashir MA, Rehim A, Geng Y, Raza HMA, Hussain S, Ahmad I, Wasif M (2023) Identifying the role of biostimulants in turnip (Brassica rapa L.) production compared with chemical fertilization. Sustainability 15(15):11851. https://doi.org/10.3390/su151511851

Sahito ZA, Zehra A, Yu S, Chen S, Arif MAR, Raza ST, Lahori AH, Mwaheb MA, He Z, Yang X (2024) Folic acid supplementation improves seed germination, seedling growth and cadmium uptake in a mining ecotype of Solanum nigrum L. Environ Technol Innov 34(March):103600. https://doi.org/10.1016/j.eti.2024.103600

Sharma P, Jha AB, Dubey RS, Pessarakli M (2012) Reactive oxygen species, oxidative damage, and antioxidative defense mechanism in plants under stressful conditions. J Bot 2012:1–26. https://doi.org/10.1155/2012/217037

Sharma A, Soares C, Sousa B, Martins M, Kumar V, Shahzad B, Sidhu GPS, Bali AS, Asgher M, Bhardwaj R, Thukral AK, Fidalgo F, Zheng B (2020) Nitric oxide-mediated regulation of oxidative stress in plants under metal stress: a review on molecular and biochemical aspects. Physiol Plant 168(2):318–344. https://doi.org/10.1111/ppl.13004

Shin YK, Bhandari SR, Jo JS, Song JW, Lee JG (2021) Effect of drought stress on chlorophyll fluorescence parameters, phytochemical contents, and antioxidant activities in lettuce. Horticulturae 7(8):238. https://doi.org/10.3390/horticulturae7080238

Soppelsa S, Kelderer M, Casera C, Bassi M, Robatscher P, Matteazzi A, Andreotti C (2019) Foliar applications of biostimulants promote growth, yield and fruit quality of strawberry plants grown under nutrient limitation. Agronomy 9(9):483. https://doi.org/10.3390/agronomy9090483

Tanveer Y, Yasmin H, Nosheen A, Farah MA, Altaf MA (2024) Synergizing Bacillus halotolerans, Pseudomonas sihuiensis and Bacillus atrophaeus with folic acid for enhanced drought resistance in wheat by metabolites and antioxidants. BMC Plant Biol 24(1):1003. https://doi.org/10.1186/s12870-024-05609-0

Tramblay Y, Koutroulis A, Samaniego L, Vicente-Serrano SM, Volaire F, Boone A, Le Page M, Llasat MC, Albergel C, Burak S, Cailleret M, Kalin KC, Davi H, Dupuy JL, Greve P, Grillakis M, Hanich L, Jarlan L, Martin-StPaul N et al (2020) Challenges for drought assessment in the Mediterranean region under future climate scenarios. Earth Sci Rev 210(August):103348. https://doi.org/10.1016/j.earscirev.2020.103348

Ulrich CM, Reed MC, Nijhout HF (2008) Modeling folate, one-carbon metabolism, and DNA methylation. Nutr Rev 66:S27–S30. https://doi.org/10.1111/j.1753-4887.2008.00062.x

Wittek F, Kanawati B, Wenig M, Hoffmann T, Franz-Oberdorf K, Schwab W, Schmitt-Kopplin P, Vlot AC (2015) Folic acid induces salicylic acid-dependent immunity in Arabidopsis and enhances susceptibility to Alternaria brassicicola. Mol Plant Pathol 16(6):616–622. https://doi.org/10.1111/mpp.12216

Wusigale, Liang L (2020) Folates: stability and interaction with biological molecules. J Agric Food Res 2:100039. https://doi.org/10.1016/j.jafr.2020.100039

Xiang N, Hu J, Wen T, Brennan MA, Brennan CS, Guo X (2020) Effects of temperature stress on the accumulation of ascorbic acid and folates in sweet corn (Zea mays L.) seedlings. J Sci Food Agric 100(4):1694–1701. https://doi.org/10.1002/jsfa.10184

Yan H, Ding M, Lin J, Zhao L, Han D, Hu Q (2023) Folate-mediated one-carbon metabolism as a potential antifungal target for the sustainable cultivation of microalga Haematococcus pluvialis. Biotechnol Biofuels Bioprod 16(1):1–14. https://doi.org/10.1186/s13068-023-02353-9

Zhang H, Voroney RP, Price GW (2015) Effects of temperature and processing conditions on biochar chemical properties and their influence on soil C and N transformations. Soil Biol Biochem 83:19–28. https://doi.org/10.1016/j.soilbio.2015.01.006

Zhao Y, Shi J, Feng B, Yuan S, Yue X, Shi W, Yan Z, Xu D, Zuo J, Wang Q (2024) Multi-omic analysis of the extension of broccoli quality during storage by folic acid. J Adv Res 59:65–78. https://doi.org/10.1016/j.jare.2023.07.001

Zhou X, Lei Z, An P (2024) Post-translational modification of WRKY transcription factors. Plants 13(15):1–14. https://doi.org/10.3390/plants13152040

Zhu L, Huang J, Lu X, Zhou C (2022) Development of plant systemic resistance by beneficial rhizobacteria: recognition, initiation, elicitation and regulation. Front Plant Sci 13 (August):1–16. https://doi.org/10.3389/fpls.2022.952397

Chapter 6
Ascorbic Acid: An Antioxidant for Plant Stress Mitigation

Biswajit Pramanick, Srilekha Das, Shyam Sundar Das, Sumanta Daw, and Rajeev Kumar Yadav

Abstract Plants are continuously exposed to a range of abiotic and biotic stresses that compromise their growth, development, and productivity. These stresses, including drought, salinity, extreme temperatures, heavy metals, and pathogen attacks, commonly lead to an excessive accumulation of reactive oxygen species (ROS). While ROS act as signaling molecules at low concentrations, their uncontrolled build-up causes oxidative damage to proteins, lipids, and nucleic acids, severely affecting plant metabolism. To counteract oxidative stress, plants have evolved a complex antioxidant defense system comprising enzymatic and non-enzymatic antioxidants. Among these, ascorbic acid (AsA), or vitamin C, is one of the most abundant, versatile, and effective molecules in plants. Ascorbic acid plays multiple roles: it directly scavenges ROS, regenerates other antioxidants, regulates redox signaling, and participates in hormone biosynthesis and stress signaling cascades. Its biosynthesis primarily occurs through the L-galactose (Smirnoff–Wheeler) pathway, though alternative pathways also contribute under specific conditions. The ascorbate–glutathione cycle further enhances the efficiency of AsA by recycling it in reduced form. Through these mechanisms, AsA protects photosynthetic machinery, stabilizes membranes, and supports cellular homeostasis under diverse stress conditions. Beyond its endogenous functions, exogenous application of AsA through foliar sprays or seed priming has been demonstrated to alleviate stress-induced damage and improve crop performance. Advances in biotechnology, including transgenic approaches and CRISPR/Cas-mediated genome editing, offer novel opportunities to enhance AsA levels in crops for improved

B. Pramanick (✉) · S. Das · S. S. Das
Department of Agronomy, Dr Rajendra Prasad Central Agricultural University, Pusa, Samastipur, Bihar, India

S. Daw
Department of Microbiology, Maulana Abul Kalam Azad University of Technology, Kolkata, West Bengal, India

R. K. Yadav
Department of Horticulture, Dr Rajendra Prasad Central Agricultural University, Pusa, Samastipur, Bihar, India

A. A. H. Abdel Latef (ed.), *Harnessing Vitamins and Organic Acids for Plant Stress Resilience*, Sustainability Sciences in Asia and Africa,
https://doi.org/10.1007/978-981-95-8593-9_6

resilience against climate change-induced stressors. This chapter provides a comprehensive overview of the biochemical, physiological, and molecular roles of ascorbic acid in stress mitigation, with emphasis on its biosynthesis, metabolism, interaction with other antioxidants and hormones, and its potential application in sustainable crop improvement strategies.

Keywords Ascorbic acid · Antioxidants · ROS · Abiotic stress · Biotic stress · Ascorbate–glutathione cycle · Stress tolerance

6.1 Introduction

The environment has a significant influence on plant survival and productivity. Unlike animals, plants are immobile and thus cannot escape unfavorable conditions. Instead, they must adapt physiologically, biochemically, and genetically to withstand stress factors. Abiotic stresses such as drought, salinity, heat, chilling, and heavy metal toxicity, alongside biotic stresses like pathogen and insect attacks, are among the primary causes of yield reduction in agriculture. It is estimated that global crop productivity suffers losses of more than 50% annually due to combined stresses, posing a significant challenge to food and nutritional security (Singh et al. 2017; Abdel Latef 2026). A common consequence of stress exposure is the excessive generation of reactive oxygen species (ROS), including singlet oxygen ($^1O_2$), superoxide anion (O_2^-), hydrogen peroxide (H_2O_2), and hydroxyl radicals (•OH). While ROS at basal levels is integral to signaling and defense mechanisms, its overaccumulation under stress conditions leads to oxidative damage, disrupting membranes, proteins, and nucleic acids. Thus, the ability of plants to efficiently regulate ROS levels is a decisive factor for stress tolerance (Jomova et al. 2023; Tuder et al. 2003).

To combat ROS toxicity, plants have developed an intricate antioxidant defense system, broadly divided into enzymatic and non-enzymatic components. Enzymatic antioxidants include superoxide dismutase (SOD), catalase (CAT), peroxidase (POD), and glutathione reductase (GR), which actively detoxify ROS. Non-enzymatic antioxidants, on the other hand, encompass ascorbic acid, glutathione, tocopherols, carotenoids, and flavonoids. Together, these antioxidants work synergistically to maintain redox homeostasis and prevent oxidative damage. Among these molecules, ascorbic acid (AsA) is considered the most versatile and multifunctional antioxidant. It is present in virtually all cellular compartments, including the chloroplasts, mitochondria, peroxisomes, vacuoles, and the apoplast, which highlights its ubiquity and indispensability. Ascorbic acid, commonly known as vitamin C, is a six-carbon lactone with potent reducing properties. In plants, it constitutes up to 10 mM in cellular compartments, making it one of the most abundant soluble antioxidants. Unlike many other antioxidants, AsA not only scavenges ROS directly but also functions in multiple physiological and developmental processes. It serves as a cofactor for several enzymes, regulates photosynthesis, contributes to cell wall synthesis, and influences gene expression. The ascorbate–glutathione (AsA–GSH) cycle, a highly efficient detoxification pathway, enables AsA to be recycled repeatedly, ensuring continuous ROS scavenging capacity. This regenerative ability distinguishes AsA as a central hub in the antioxidant network of plants (Jomova et al. 2024; Shen et al. 2021).

The discovery of vitamin C in the early twentieth century initially focused on its role in human health, particularly in preventing scurvy. Its presence and importance in plants, however, gained recognition later. By the mid-twentieth century, researchers identified AsA's abundance in photosynthetic tissues, linking it to light-driven processes. In the 1990s, the elucidation of the Smirnoff–Wheeler pathway marked a turning point, as it provided a clear understanding of AsA biosynthesis in higher plants. Since then, significant progress has been made in unraveling the molecular genetics, biochemical regulation, and physiological roles of AsA, especially in relation to stress tolerance (Carpenter 2012; Alberts et al. 2025; Pisalsitsakul et al. 2022).

Under stress conditions, plants frequently exhibit altered levels of AsA. For instance, drought or salinity often decreases the AsA pool due to higher consumption in ROS detoxification. Conversely, certain stresses may induce enhanced biosynthesis as part of an adaptive mechanism. These dynamic changes in AsA metabolism highlight its central role in plant stress responses. Studies across diverse crops provide compelling evidence of the importance of AsA. In wheat, higher leaf AsA content correlates with drought tolerance. In rice, salt-tolerant cultivars often exhibit enhanced AsA–GSH cycle activity compared to sensitive ones. Similarly, in horticultural crops, AsA not only mitigates stress but also improves post-harvest fruit quality by reducing oxidative spoilage.

In the context of global climate change and increasing environmental stress, enhancing plant resilience is a major goal of modern agriculture. The role of antioxidants, particularly AsA, offers promising avenues for sustainable crop production. Unlike synthetic agrochemicals, which often have environmental side effects, AsA is naturally occurring, biodegradable, and non-toxic, making it suitable for eco-friendly crop management strategies. The application of AsA through foliar sprays, seed priming, or fertigation has been shown to alleviate stress symptoms in cereals, legumes, fruits, and vegetables. At the same time, advances in molecular breeding and biotechnology allow researchers to develop high-AsA cultivars through genetic modification or targeted gene editing. Together, these approaches provide a dual strategy: exogenous supplementation for short-term stress relief and genetic enhancement for long-term resilience (El-Hawary et al., 2023).

This chapter provides a comprehensive exploration of the multifaceted roles of ascorbic acid in plants. The following sections will discuss the chemistry, distribution, and biosynthesis of AsA in plants, its functional role in mitigating abiotic and biotic stresses, interactions with other antioxidants and phytohormones, the potential of exogenous application in crop management, and future perspectives involving biotechnology and precision agriculture. By integrating physiological, biochemical, and molecular insights, this chapter aims to highlight the potential of AsA as a key molecule for stress tolerance, crop productivity, and sustainable agriculture.

6.2 Ascorbic Acid: Chemistry and Cellular Distribution

6.2.1 Chemical Structure and Properties

Ascorbic acid, commonly known as vitamin C, is a water-soluble, six-carbon lactone derived from glucose metabolism. It exists primarily in its reduced form (ascorbate), which is the biologically active antioxidant (Malik et al. 2022). Structurally, it possesses an enediol group in its lactone ring, which is responsible for its strong reducing capacity. This enediol group readily donates electrons to neutralize ROS. The oxidized form of ascorbate, known as monodehydroascorbate (MDHA), is a short-lived radical that can either be reduced back to ascorbate by monodehydroascorbate reductase (MDHAR) or undergo spontaneous disproportionation to yield dehydroascorbate (DHA). DHA, in turn, can be reduced to ascorbate by dehydroascorbate reductase (DHAR) in a glutathione-dependent reaction. This reversible interconversion between reduced and oxidized states ensures that ascorbate maintains its antioxidant capacity over time, making it a highly efficient molecule in cellular redox regulation. Physicochemically, ascorbic acid is stable at slightly acidic pH but unstable under alkaline conditions, where it undergoes rapid oxidation. It is also sensitive to light and heat, which has implications for its stability during stress conditions in planta as well as in post-harvest storage of crops.

6.2.2 Subcellular Localization

The effectiveness of AsA in stress mitigation is not only dependent on its concentration but also on its precise distribution within the plant cell. It is ubiquitously distributed in almost all compartments, which reflects its multifaceted roles:

Chloroplasts AsA is highly concentrated in chloroplasts, where it protects the photosynthetic machinery from photooxidative stress. During photosynthesis, the generation of ROS in the thylakoid membranes is inevitable, especially under high light intensity. AsA directly scavenges singlet oxygen ($^1O_2$) and reduces H_2O_2 via the ascorbate peroxidase (APX) reaction. It also plays a key role in the xanthophyll cycle, which dissipates excess energy as heat.

Mitochondria In mitochondria, AsA regulates oxidative phosphorylation by detoxifying ROS generated during electron transport. It ensures membrane integrity and supports ATP production under stress conditions such as drought and salinity.

Peroxisomes AsA detoxifies H_2O_2 produced during photorespiration and fatty acid β-oxidation. The presence of APX in peroxisomes highlights the essential role of AsA in this organelle.

Cytosol The cytosolic pool of AsA acts as a buffer, regulating redox signaling and maintaining cellular homeostasis. It provides a link between organelle-specific AsA pools through transport and recycling.

Apoplast AsA is found in the apoplast (the cell wall region and intercellular spaces), where it participates in defense against pathogens and regulation of cell wall growth. The apoplastic AsA/DHA ratio is critical for redox signaling during stress and for modulation of cell expansion.

Vacuoles Vacuoles contain AsA primarily as a storage site. During stress, vacuolar AsA can be mobilized to other compartments to support antioxidant defense.

6.2.3 Tissue and Organ-Level Distribution

At the whole-plant level, AsA distribution varies across tissues and developmental stages. Generally, it is most abundant in photosynthetic tissues such as leaves, which are highly exposed to light stress and thus require robust antioxidant protection. Fruits, particularly citrus, guava, kiwi, and peppers, are also rich in AsA, where it functions not only in stress tolerance but also in influencing nutritional quality and taste. Roots typically contain lower levels of AsA, though under stress (e.g., salinity), AsA levels may rise to protect root tissues from oxidative damage. Young, actively growing tissues tend to have higher AsA concentrations compared to mature tissues, consistent with its role in cell division and growth regulation. In reproductive organs such as flowers and developing seeds, AsA influences pollen viability, fertilization, and embryogenesis (Malik et al. 2022).

6.2.4 Transport and Compartmentalization

The dynamic distribution of AsA between cellular compartments suggests the existence of specific transport mechanisms. Although the precise transporters are not fully characterized in plants, evidence suggests that membrane-bound carriers regulate AsA and DHA transport across chloroplast, mitochondrial, and plasma membranes. DHA is more readily transported than reduced AsA, and once inside the cell, DHA is reduced back to AsA, effectively shuttling antioxidant potential between compartments (Yang et al. 2021). This transport and recycling mechanism ensures that AsA levels are finely tuned in response to developmental cues and environmental stresses. For example, under high light or drought stress, more AsA is directed toward chloroplasts, where the demand for ROS scavenging is greatest.

6.2.5 Significance of Distribution in Stress Mitigation

The ubiquitous and compartment-specific distribution of AsA underscores its importance in stress mitigation. By being present in nearly every cellular location where ROS are generated, AsA provides a first line of defense. Moreover, its mobility and recycling ensure that no organelle is left unprotected during stress exposure. The apoplast–cytosol–chloroplast continuum of AsA also functions as a signaling axis, where changes in redox status are communicated across tissues, activating stress-responsive genes. Thus, the spatial distribution of AsA complements its chemical versatility, enabling it to serve as both a scavenger of ROS and a regulator of stress signaling pathways.

6.3 Significance in Stress Mitigation

The multiplicity of biosynthetic and recycling pathways ensures that plants maintain adequate AsA levels under fluctuating environmental conditions. The integration of AsA synthesis with photosynthesis, respiration, and cell wall metabolism provides a metabolic safeguard against stress-induced oxidative damage (Smirnoff and Wheeler 2024). Furthermore, the ability to regulate AsA synthesis in a tissue- and stress-specific manner highlights its role not only as an antioxidant but also as a redox signal that modulates stress-responsive gene expression and developmental programs. Figure 6.1 represents the plant stress mitigation role of AsA.

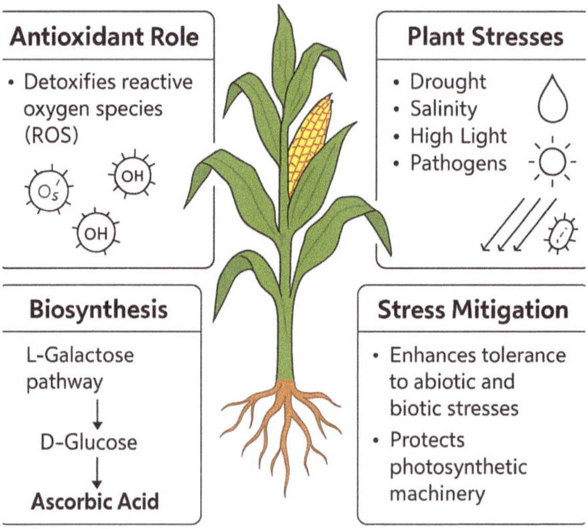

Fig. 6.1 Role of ascorbic acid in mitigating plant stress

6.4 Functional Role of Ascorbic Acid in Stress Mitigation

Plants are constantly challenged by environmental stresses, which perturb cellular metabolism and promote the overproduction of ROS. Uncontrolled ROS accumulation disrupts membranes, proteins, nucleic acids, and photosynthetic machinery, leading to growth inhibition and yield loss. Ascorbic acid (AsA) functions as a frontline antioxidant, directly detoxifying ROS and participating in signaling cascades that coordinate plant responses to stress. Below, we examine the specific roles of AsA in different stress contexts (Hasanuzzaman et al. 2020). Table 6.1 represents a brief description of the role of AsA on stress mitigation.

Table 6.1 Role of ascorbic acid (AsA) in mitigating stresses in plants

Attributes	Measured parameters	Observed AsA effects	References
Growth	Plant height, shoot/root dry weight, leaf area, biomass, fruit set/yield	Maintained/increased plant height and dry matter; improved leaf area, total biomass, and fruit set under drought	Athar et al. (2008), Noman et al. (2015) and Khan et al. (2011)
Physiology	Relative water content (RWC), stomatal conductance, chlorophyll fluorescence (Fv/Fm), chlorophyll content, photosynthetic rate	Higher RWC and stomatal conductance; preserved Fv/Fm and chlorophyll content; sustained photosynthetic CO_2 assimilation	Farooq et al. (2013)
Pigments	Chlorophyll a, chlorophyll b, total chlorophyll, carotenoids	Reduced pigment degradation; higher chlorophyll a/b and carotenoid levels under stress	Ali et al. (2015) and Gaafar et al. (2020)
Osmolytes	Proline, glycine betaine, soluble sugars, soluble proteins	Elevated proline, soluble sugars, and other osmolytes contribute to osmotic adjustment	Seminario et al. (2017)
Oxidative stress markers	Malondialdehyde (MDA), electrolyte leakage (EL)	Lower MDA and EL indicate reduced lipid peroxidation and membrane damage	Seminario et al. (2017)
Antioxidant enzymes	Superoxide dismutase (SOD), catalase (CAT), peroxidase (POD), and ascorbate peroxidase (APX)	Increased activities of SOD, CAT, POD, APX; enhanced AsA–GSH cycle functioning	Gallie (2013)
Molecular responses	Expression of antioxidant genes (SOD, APX), ABA-related signaling genes	Upregulation of antioxidant-related genes; modulation of ABA-linked stress signaling contributing to stomatal response and stress adaptation	Noman et al. (2015) and Gallie (2013)

6.5 Ascorbic Acid and Salinity Stress

Salinity stress is one of the most severe environmental challenges limiting crop productivity worldwide. High concentrations of soluble salts in the soil, particularly sodium chloride, impair water uptake, disrupt ionic balance, and lead to both osmotic and oxidative stress in plants. This ultimately suppresses photosynthesis, reduces growth, and lowers yields. The buildup of ROS such as superoxide radicals, hydrogen peroxide, and hydroxyl radicals under saline conditions is a major cause of cellular damage, resulting in lipid peroxidation, protein oxidation, and DNA fragmentation. To counter these harmful effects, plants rely on their antioxidant defense system, among which ascorbic acid, plays a pivotal role. AsA is not only a direct scavenger of ROS but also a cofactor in enzymatic reactions, a modulator of cell signaling, and an essential component in maintaining cellular redox homeostasis. Because of its central role in antioxidant defense and stress mitigation, AsA has been extensively studied in relation to plant tolerance under salinity stress.

When plants are exposed to salinity, one of the earliest consequences is osmotic stress, where high salt concentration in the rhizosphere lowers soil water potential, making it difficult for roots to absorb water. This reduces cell turgor, causes stomatal closure, and limits gas exchange. In addition, ionic toxicity arises when excessive sodium and chloride ions accumulate in the cytoplasm, displacing vital ions like potassium and calcium, which are essential for enzyme activity and membrane stability. The oxidative stress component, however, is often the most damaging, as ROS accumulation triggers a cascade of physiological dysfunctions. AsA, being one of the most abundant antioxidants in plant tissues, plays a crucial role in counteracting this oxidative burst. It directly neutralizes ROS and, through its involvement in the ascorbate–glutathione cycle, helps in detoxifying hydrogen peroxide with the aid of the enzyme ascorbate peroxidase. In this way, AsA prevents oxidative damage to membranes, chloroplasts, and other organelles, ensuring better survival of plants under saline conditions.

Several experimental studies provide strong evidence of AsA's protective role under salt stress. For example, Hassan et al. (2021) demonstrated that foliar application of AsA in barley significantly enhanced salinity tolerance by improving morpho-physiological traits, modulating ion uptake, and boosting antioxidant enzyme activity. Similarly, Kanwal et al. (2024) reported that pea plants treated with exogenous AsA exhibited higher activities of superoxide dismutase, catalase, and peroxidases under 100 mM NaCl stress. This was accompanied by greater accumulation of compatible solutes such as soluble sugars, proteins, and flavonoids, which helped maintain osmotic balance. Moreover, AsA application reduced sodium accumulation in shoots and roots while promoting potassium and calcium uptake, thereby maintaining ion homeostasis. These improvements translated into better yield attributes compared to untreated, salt-stressed controls. In quinoa seedlings, seed priming with low concentrations of AsA (0.1–0.5 μM) enhanced germination, photosynthetic pigment stability, and biomass accumulation under 200 mM NaCl, confirming that AsA application at the seedling stage could be particularly effective

(Shah et al. 2025). Tomato fruits under salt stress also accumulate AsA endogenously as part of their defense mechanism. Hasegawa et al. (2025) found that salt stress increased the expression of genes involved in the D-mannose/L-galactose pathway of AsA biosynthesis, leading to higher AsA accumulation in fruit tissues, although this sometimes-diverted carbon flux away from cell wall polysaccharide synthesis, indicating potential trade-offs. In wheat, AsA applied in combination with silver nanoparticles and leaf extracts improved salinity tolerance by enhancing chlorophyll content, osmolyte accumulation, and antioxidant enzyme activities, suggesting that AsA can also work synergistically with other stress-ameliorating agents (Pirzada et al. 2024).

The benefits of AsA under salinity stress extend beyond simple ROS scavenging. It protects photosynthetic machinery by preserving chlorophyll content and maintaining the functionality of photosystem II. AsA also contributes to the regeneration of tocopherol (vitamin E), another important antioxidant, thereby supporting the overall redox network. By promoting proline and glycine betaine accumulation, AsA helps plants to adjust osmotically, which ensures water retention and cellular hydration under saline conditions. At the ionic level, AsA treatment reduces the toxic buildup of sodium and chloride ions in tissues while sustaining potassium and calcium levels, which are critical for enzymatic reactions and cytoplasmic stability. The modulation of ion transporters and stress-responsive genes by AsA further enhances plant resilience. These integrated physiological and biochemical effects allow plants to maintain growth and productivity under otherwise damaging saline environments.

The effectiveness of AsA, however, depends on the concentration applied, the mode of application, the timing of treatment, and the plant species involved. Foliar sprays of 5–10 mM AsA are commonly reported to be beneficial in legumes, cereals, and vegetables, while seed priming with much lower concentrations (μM range) has proven effective during germination and early seedling stages. The timing of application also matters, with treatments applied before or during the onset of salt stress showing greater efficacy. Different species and even varieties within the same species respond differently, as highlighted in peas, where some genotypes benefited more than others (Kanwal et al. 2024). Likewise, the severity of stress influences the outcomes, with moderate salt stress being more amenable to mitigation by AsA than extreme stress levels. While exogenous AsA supplementation helps, it is also important to recognize that plants can upregulate their own AsA biosynthesis under stress, as seen in tomatoes. Thus, genetic approaches to enhance endogenous AsA synthesis may also be a viable long-term strategy.

Despite its promise, there are limitations and challenges in the practical application of AsA for salinity tolerance. Large-scale foliar spraying may not always be economically feasible for farmers, and optimal dosages vary widely among crops, making blanket recommendations difficult. Many studies demonstrating positive results are conducted in controlled environments, and more field-based validations are needed to assess performance under variable climatic and soil conditions. Additionally, plants often face multiple stresses simultaneously, such as drought combined with salinity, and the role of AsA under such combined stress scenarios remains less understood. There may also be metabolic trade-offs, as observed in tomato, where increased AsA synthesis diverted precursors away from other vital pathways. Nonetheless, given its central role in redox regulation, osmotic adjustment, and ion homeostasis, AsA remains a highly promising candidate for improving salinity tolerance in crops.

6.6 Ascorbic Acid and Drought Stress

Drought is one of the most pervasive abiotic stresses limiting agricultural productivity across the globe, particularly in arid and semi-arid regions where water scarcity coincides with rising temperatures and climate variability. Drought affects plants at multiple levels, beginning with reduced soil water potential, which restricts root water uptake and ultimately lowers plant tissue water content. This triggers osmotic stress, stomatal closure, inhibition of photosynthesis, reduction in leaf expansion, and premature senescence. One of the most damaging consequences of drought stress is the overproduction of ROS, which disrupt membranes, denature proteins, and impair DNA stability. The imbalance between ROS production and the antioxidant defense capacity leads to oxidative stress, a hallmark of drought-induced damage. To cope with these conditions, plants rely heavily on enzymatic and non-enzymatic antioxidant systems, among which AsA is a key non-enzymatic component. AsA is one of the most abundant, water-soluble antioxidants in plants, playing multiple roles in scavenging ROS, regenerating other antioxidants, regulating enzymatic activity, maintaining photosynthetic efficiency, and modulating stress signaling pathways. Because of its central role in plant metabolism and redox homeostasis, AsA is widely recognized as a crucial player in mitigating drought stress (Younis et al. 2023).

Under drought conditions, one of the earliest responses of plants is the closure of stomata to reduce water loss through transpiration. While this conserves water, it also limits CO_2 entry into leaves, leading to decreased photosynthetic carbon assimilation. Consequently, plants experience an imbalance between light energy absorbed and its utilization in photosynthesis, which enhances the generation of ROS in chloroplasts. Excessive ROS, if not neutralized, results in oxidative damage to the thylakoid membranes, degradation of chlorophyll pigments, and impairment of photosystem II activity. AsA, by acting as an efficient ROS scavenger, directly neutralizes superoxide and hydroxyl radicals and serves as a substrate for ascorbate peroxidase (APX) in the ascorbate–glutathione cycle, where hydrogen peroxide is detoxified into water. This enzymatic cycle not only maintains low levels of ROS but also sustains the redox balance essential for photosynthetic and metabolic functions. In addition, AsA contributes to the regeneration of other antioxidants such as α-tocopherol (vitamin E), thereby strengthening the antioxidant network and enhancing drought resilience.

Beyond its antioxidant role, AsA also contributes to osmotic adjustment under drought stress. Plants often accumulate compatible solutes such as proline, glycine betaine, and soluble sugars to maintain cell turgor and water balance when soil moisture is limited. Several studies have shown that exogenous application of AsA enhances the accumulation of these osmolytes, thereby supporting osmotic regulation. For example, Athar et al. (2008) reported that foliar application of AsA in wheat improved leaf relative water content, proline concentration, and photosynthetic efficiency under drought stress, which translated into higher biomass accumulation compared to untreated drought-stressed plants. Similarly, Noman et al. (2015) demonstrated that AsA treatment in maize improved osmotic adjustment, stabilized membranes, and enhanced photosynthetic performance under water deficit conditions. These findings suggest that AsA not only protects cellular structures from oxidative damage but also indirectly improves water relations and growth.

Another crucial effect of drought is the disruption of nutrient uptake and transport due to reduced soil water availability. Essential nutrients such as potassium, calcium, and nitrogen become less accessible under water-limited conditions. AsA plays a role in alleviating nutrient imbalance by modulating ion transport and protecting root cell membranes from oxidative damage, thereby improving nutrient acquisition. Moreover, AsA influences the expression of drought-responsive genes, including those encoding antioxidant enzymes such as SOD, CAT, and POD. Enhanced expression of these enzymes, together with higher AsA concentrations, establishes a more efficient antioxidant system capable of maintaining cellular homeostasis during drought. For instance, Khan et al. (2011) observed that AsA application in tomato under water deficit improved SOD and CAT activities, stabilized chlorophyll pigments, and enhanced fruit yield, demonstrating its role in stress signaling and metabolic adjustments.

Seed priming and foliar sprays with AsA have emerged as practical strategies to improve drought tolerance in diverse crops. Seed priming with AsA before germination can prepare seedlings for water-limited conditions by boosting antioxidant enzyme activity, promoting early growth, and ensuring higher vigor. Foliar sprays, on the other hand, directly deliver AsA to leaves where drought-induced ROS production is most intense. The effectiveness of these approaches, however, depends on the concentration of AsA applied, the timing of treatment, and the species or genotype under study. Optimal concentrations vary, but moderate doses are typically beneficial, while excessively high concentrations may not enhance tolerance further and sometimes even impose metabolic imbalances. Nonetheless, across species such as wheat, maize, tomato, sunflower, and soybean, AsA has consistently demonstrated positive effects on physiological parameters, biomass, and yield under drought stress (Smirnoff 2018).

Recent studies have also highlighted the role of AsA in photosystem protection and chloroplast stability under drought. AsA participates in the Mehler peroxidase reaction, which helps detoxify ROS generated during photosynthetic electron transport, thus preventing photo-inhibition. It also contributes to maintaining the ascorbate pool in chloroplasts, which is essential for the operation of the water–water cycle, another mechanism to alleviate excess excitation energy under drought. AsA accumulation in chloroplasts correlates strongly with drought tolerance in many species, indicating its pivotal role in sustaining photosynthesis under stress (Gallie

2013). Moreover, AsA modulates the biosynthesis of plant hormones such as abscisic acid (ABA), which regulates stomatal closure during drought. By interacting with hormonal signaling pathways, AsA not only mitigates oxidative damage but also contributes to stress perception and adaptive responses.

Despite these benefits, there are challenges in the large-scale use of AsA for improving drought tolerance. One concern is the economic feasibility of repeated foliar sprays under field conditions, particularly for resource-limited farmers. Another limitation is the variability in response among different crops and genotypes, which complicates the standardization of application protocols. Furthermore, most research on AsA under drought has been conducted under controlled greenhouse or growth chamber conditions, with relatively fewer field-based studies. Thus, while experimental results are promising, the scalability of AsA application under variable field environments remains to be fully tested. There is also the possibility of metabolic trade-offs, as increasing AsA synthesis or exogenous supplementation might redirect resources from other essential processes. Nonetheless, these limitations do not undermine the potential of AsA but rather emphasize the need for integrated research combining physiology, molecular biology, and agronomy.

Future research should focus on exploring genetic approaches to enhance endogenous AsA biosynthesis in crops, such as overexpression of key enzymes in the AsA biosynthetic pathway, which may provide sustainable drought tolerance without the need for external supplementation. Additionally, studies examining the interaction of AsA with other protective compounds, including polyamines, nitric oxide, and silicon, may reveal synergistic effects that could further enhance plant drought resistance. Systems biology approaches, integrating transcriptomic, proteomic, and metabolomic data, can provide deeper insights into how AsA regulates plant responses at multiple levels under drought stress. Moreover, long-term field trials under diverse agro-ecological conditions are essential to validate the practical utility of AsA in crop production.

6.7 Role of Ascorbic Acid in Waterlogging/Flooding Stress

Waterlogging and flooding are serious abiotic stresses that restrict crop productivity in many parts of the world, particularly in low-lying areas with heavy rainfall, poor drainage, or riverine ecosystems. Unlike drought or salinity, where plants suffer from a lack of water or excess salts, flooding causes an excess of water around the root zone, drastically reducing oxygen availability. This hypoxic or even anoxic condition impairs aerobic respiration, disrupts root metabolism, and leads to severe energy deficits in plants. Under waterlogged soils, nutrient imbalances also occur because oxygen deficiency alters redox potential, solubilizing some ions while restricting the uptake of others. The metabolic imbalance caused by hypoxia enhances the production of ROS, particularly upon re-oxygenation when the floodwaters recede. This oxidative burst is often more damaging than the flooding itself, leading to lipid peroxidation, protein degradation, chlorophyll breakdown, and DNA injury. To withstand such stress, plants depend on antioxidant systems, and among them, AsA plays a crucial role as a central component of redox regulation and ROS detoxification.

AsA is one of the most abundant water-soluble antioxidants in plant cells, found in chloroplasts, cytosol, mitochondria, and apoplast. Under waterlogging or flooding stress, AsA functions at multiple levels. First, it acts directly as a scavenger of ROS, neutralizing hydrogen peroxide and hydroxyl radicals that accumulate due to impaired mitochondrial and chloroplast activity under oxygen-deprived conditions. Second, it operates within the ascorbate–glutathione cycle, in which APX uses AsA to detoxify hydrogen peroxide into water, simultaneously generating monodehydroascorbate, which can be recycled back to AsA. This cycle is especially critical when plants undergo re-oxygenation after flooding, as the sudden restoration of oxygen triggers a sharp rise in ROS levels. By maintaining a dynamic redox pool, AsA prevents irreversible oxidative damage and supports recovery. Third, AsA is involved in maintaining photosynthetic performance during and after flooding. Although photosynthesis is reduced due to stomatal closure and chlorophyll loss during submergence, AsA helps preserve chloroplast integrity, stabilizes photosystem II, and contributes to non-photochemical quenching mechanisms that dissipate excess energy safely.

Experimental evidence strongly supports the protective role of AsA under flooding stress. For instance, Ullah et al. (2017) demonstrated that exogenous application of AsA mitigated flood-induced damage in *Vigna angularis* by enhancing antioxidant enzyme activities, preserving chlorophyll pigments, and reducing oxidative injury, which ultimately improved plant growth and biomass. This study clearly highlights the capacity of AsA to strengthen the antioxidant defense system and maintain physiological stability under waterlogged conditions. Another important function of AsA under waterlogging is its role in hormone signaling and stress adaptation. Flooding induces ethylene accumulation, which triggers adaptive responses such as adventitious root formation and aerenchyma development. AsA, being a cofactor for enzymes involved in hormone biosynthesis, it can influence ethylene metabolism and thus indirectly regulate morphological adaptations to flooding. Moreover, AsA interacts with ABA and gibberellins, which control processes such as stomatal regulation and stem elongation, further contributing to plant survival. For example, under complete submergence, rice varieties with higher endogenous AsA levels exhibit better elongation responses that allow shoots to reach the water surface, ensuring gas exchange.

At the cellular level, AsA also helps stabilize membranes and proteins under hypoxic stress. By preventing ROS-induced lipid peroxidation, it maintains plasma membrane integrity, which is essential for ion homeostasis and nutrient uptake. Additionally, AsA supports mitochondrial function by protecting respiratory enzymes from oxidative inactivation, thereby facilitating energy metabolism even under oxygen-deficient conditions. Its presence in apoplastic space helps regulate redox status and signaling, which is crucial for activating downstream stress responses.

The mode of AsA application influences its effectiveness against flooding. Seed priming with AsA before planting has been reported to enhance seedling vigor and improve tolerance to waterlogging by preconditioning antioxidant systems. Foliar sprays during vegetative stages have also been effective in boosting tolerance by directly increasing leaf antioxidant pools. Moreover, breeding approaches that enhance endogenous AsA biosynthesis may provide long-term resilience. For instance, transgenic approaches overexpressing GDP-mannose pyrophosphorylase, a key enzyme in the AsA biosynthetic pathway, have been associated with improved oxidative stress tolerance, suggesting that boosting internal AsA production can be a sustainable strategy for flood-prone regions.

Nevertheless, the role of AsA under flooding is not without limitations. The effectiveness of exogenous AsA varies with plant species, developmental stage, and severity of flooding. Excessively prolonged submergence may overwhelm even strengthened antioxidant systems, leading to plant death regardless of AsA supplementation. Additionally, the cost and practicality of large-scale AsA application in field conditions may pose challenges. More field-based research is required to validate laboratory and greenhouse findings and to optimize application methods. Moreover, interactions with other metabolites and signaling molecules under complex field environments remain to be elucidated.

6.8 Ascorbic Acid and Extreme Temperature Stress

Extreme temperatures, both high and low, are among the most critical abiotic stresses influencing plant survival, growth, and productivity. Global climate change has intensified the occurrence of heat waves and cold spells, thereby threatening agricultural sustainability. High temperature stress accelerates respiration, disrupts photosynthesis, denatures proteins, and damages membranes, while low temperature or chilling/freezing stress alters membrane fluidity, reduces enzyme activity, and impairs photosynthetic performance. Both extremes lead to overproduction of reactive oxygen species (ROS) which trigger oxidative stress and cellular injury. To mitigate these effects, plants rely on antioxidant defense systems, and ascorbic acid plays a central role as a non-enzymatic antioxidant, cofactor, and regulator of stress signaling pathways.

AsA is one of the most abundant antioxidants in plants, distributed across chloroplasts, mitochondria, cytosol, vacuoles, and the apoplast. Under extreme temperatures, AsA functions as a frontline ROS scavenger, preventing oxidative damage to membranes, proteins, and nucleic acids. Through its involvement in the ascorbate–glutathione (AsA–GSH) cycle, AsA acts as a substrate for APX, which detoxifies hydrogen peroxide into water. This cycle becomes particularly crucial under high-temperature stress, where enhanced photorespiration and disrupted photosynthetic electron transport generate excess ROS. For example, Gallie (2013) reported that AsA protects photosystem II and sustains electron transport under heat stress by maintaining a favorable redox balance. Similarly, Smirnoff (2018) highlighted the role of AsA in stabilizing thylakoid membranes and protecting chlorophyll from thermal degradation.

Under chilling and freezing stress, AsA contributes to the maintenance of membrane integrity by limiting lipid peroxidation and stabilizing antioxidant enzymes. Low temperatures often inhibit enzyme activity, but AsA ensures that ROS accumulation is kept under check, allowing for recovery when favorable conditions return. Exogenous application of AsA has been shown to enhance chilling tolerance in several crops. For instance, Anjum et al. (2017) observed that AsA treatment in maize improved photosynthetic efficiency, increased proline accumulation, and reduced malondialdehyde content under low-temperature stress, suggesting reduced membrane lipid damage. In tomato, foliar AsA application enhanced antioxidant enzyme activities, preserved chlorophyll content, and improved fruit set under heat stress (Khan et al. 2011). These findings demonstrate that AsA not only scavenges ROS but also enhances osmolyte accumulation and enzymatic defense, thereby providing multiple layers of protection.

In addition to its antioxidant function, AsA interacts with plant hormones that mediate stress responses. High temperature stress often induces ethylene biosynthesis, which can accelerate senescence. AsA acts as a cofactor for 1-aminocyclopropane-1-carboxylic acid oxidase, a key enzyme in ethylene biosynthesis, thereby influencing ethylene-mediated responses to heat. Under chilling stress, AsA has been linked with abscisic acid signaling, which regulates stomatal closure and water conservation. Thus, AsA operates not only as an antioxidant but also as a regulator of hormonal and metabolic adjustments under temperature extremes.

The mode and timing of AsA application determine its effectiveness. Foliar sprays, seed priming, or inclusion in nutrient solutions have all been shown to improve thermotolerance. However, responses vary depending on species, genotype, and stress severity. While exogenous AsA supplementation can be beneficial, genetic strategies to enhance endogenous AsA biosynthesis are increasingly explored. Overexpression of genes encoding enzymes in the AsA biosynthetic pathway, such as GDP-mannose pyrophosphorylase, has been associated with improved thermotolerance in transgenic plants, highlighting the potential of metabolic engineering approaches.

6.9 AsA and Heavy Metals Stress

Heavy metal contamination of soils has become a serious environmental issue due to rapid industrialization, mining, sewage irrigation, and excessive use of agrochemicals. Toxic heavy metals such as cadmium (Cd), lead (Pb), mercury (Hg), chromium (Cr), nickel (Ni), and arsenic (As) accumulate in the rhizosphere and are absorbed by plants, leading to profound physiological and biochemical disturbances. While certain metals, such as copper (Cu), iron (Fe), and zinc (Zn), are essential micronutrients in trace amounts, their excess, as well as the presence of non-essential metals like Cd and Pb, exert toxic effects. The primary manifestations of heavy metal stress include chlorosis, stunted root and shoot growth, nutrient imbalance, inhibition of photosynthesis, premature senescence, and ultimately yield loss.

At the cellular level, heavy metal toxicity disrupts homeostasis through two major mechanisms: (i) displacement or substitution of essential nutrients from enzyme active sites and membranes, and (ii) overproduction of ROS. The uncontrolled accumulation of ROS causes oxidative stress, which damages lipids, proteins, nucleic acids, and cellular membranes. To counteract this, plants deploy a sophisticated antioxidant defense system, which includes enzymatic antioxidants like SOD, CAT, POD, and APX, as well as non-enzymatic antioxidants such as AsA, GSH, carotenoids, and flavonoids. Among these, AsA occupies a central position due to its abundance, versatile roles in redox homeostasis, and function as both a direct antioxidant and a cofactor in enzymatic reactions.

AsA is ubiquitously distributed in plant tissues, with high concentrations in chloroplasts, cytosol, vacuoles, mitochondria, and apoplast. Its role under heavy metal stress is multifaceted. First, AsA directly scavenges ROS, thereby preventing lipid peroxidation and maintaining membrane integrity. Second, AsA is a key component of the AsA–GSH cycle, where it serves as a substrate for APX to reduce hydrogen peroxide into water. The oxidized form of AsA, MDHA, is recycled back to AsA by MDHA reductase or DHAR using reducing equivalents from NADPH and GSH. This recycling ensures a continuous supply of reduced AsA for ROS detoxification. Third, AsA participates in maintaining photosynthetic performance. Heavy metals such as Cd and Pb inhibit chlorophyll biosynthesis, damage thylakoid membranes, and disrupt photosystem II activity. AsA protects chlorophyll, supports the xanthophyll cycle, and safeguards electron transport, thereby sustaining photosynthetic efficiency even under stress.

Experimental studies provide compelling evidence that AsA mitigates heavy metals stress across different crops. In maize, exogenous AsA reduced Cd uptake, enhanced antioxidant activity, and improved growth, highlighting its role in detoxification (Zhang et al. 2019). Similarly, in wheat seedlings, AsA conferred Cd tolerance via nitric oxide-mediated signaling pathways, lowering oxidative damage and maintaining membrane integrity (Wang et al. 2017). In barley, foliar AsA application combined with biochar improved physio-biochemical responses under Cd-contaminated soil, strengthening antioxidant defenses (Yaseen et al. 2021), while another study showed endogenous AsA levels correlated with Cd tolerance between contrasting barley genotypes (Wu et al. 2004). Beyond cadmium, AsA also alleviates Pb toxicity in wheat by enhancing photosynthetic efficiency and boosting enzymatic and non-enzymatic antioxidants (Alamri et al. 2018). Together, these findings demonstrate that AsA not only detoxifies heavy metals but also fortifies redox homeostasis, photosynthesis, and stress signaling, thereby improving plant resilience.

Beyond antioxidant functions, AsA also contributes to heavy metal detoxification. It indirectly promotes the synthesis of phytochelatins and metallothioneins-thiol-rich peptides and proteins that chelate heavy metals and sequester them into vacuoles, thereby reducing their cytoplasmic toxicity. AsA also enhances cell wall binding of metals by maintaining apoplastic redox balance, limiting their translocation to sensitive tissues. Furthermore, AsA modulates nutrient uptake under heavy metal stress. For example, Cd often competes with Ca^{2+}, Mg^{2+}, and Zn^{2+}, disrupting nutrient balance. By protecting membrane integrity and transporter activity, AsA helps maintain ion homeostasis.

At the molecular level, AsA influences gene expression under metal toxicity. Bulley et al. (2021) demonstrated that *Arabidopsis* plants overexpressing GDP-L-galactose phosphorylase (GGP), a key enzyme in the AsA biosynthetic pathway, showed higher AsA content and improved Cd tolerance through enhanced regulation of ROS detoxification and metal transporter genes. Such findings underscore the importance of both endogenous biosynthesis and exogenous supplementation of AsA in conferring tolerance.

Practical strategies for exploiting AsA include seed priming, foliar spraying, and nutrient solution supplementation. Seed priming with AsA enhances germination and seedling vigor under Cd or Pb stress by preconditioning the antioxidant machinery. Foliar sprays provide direct reinforcement of leaf antioxidant pools, which are most affected under metal-induced oxidative stress. However, the success of these strategies depends on optimal dosage, timing, and plant species. Moderate concentrations are usually beneficial, whereas excessively high levels may disrupt ROS signaling, which is necessary for activating stress defense pathways.

Despite the promising results, there are limitations. Many studies have been conducted under controlled environments, and field validations are limited. Variability among plant genotypes and environmental conditions complicates the development of universal protocols. In addition, while exogenous AsA is effective, large-scale applications may not always be cost-effective. Therefore, genetic engineering and breeding approaches to enhance endogenous AsA biosynthesis represent sustainable alternatives. Transgenic crops with elevated AsA levels have shown improved tolerance to multiple abiotic stresses, including heavy metals, highlighting the potential of this strategy.

6.10 Conclusion

Ascorbic acid is a central metabolite in plants, serving as a versatile antioxidant, a redox regulator, and a key signaling molecule that coordinates growth, development, and stress responses. Its multifaceted role in ROS, protecting cellular structures, and modulating stress-responsive pathways underscores its indispensable function in plant physiology. The biosynthesis of AsA, predominantly via the L-galactose pathway, is complemented by alternative routes such as the myo-inositol, D-galacturonic acid, and L-glucose pathways, providing metabolic flexibility to maintain adequate AsA pools under diverse developmental and environmental conditions. The tight regulation of AsA biosynthesis at transcriptional, feedback, environmental, and hormonal levels ensures that plants can dynamically adjust their antioxidant capacity to meet changing demands.

Through its participation in the ascorbate–glutathione cycle and interaction with other antioxidants and phytohormones, AsA not only mitigates oxidative damage but also functions as a signaling hub, integrating stress perception with adaptive responses. Its protective roles span multiple abiotic stresses, including drought, salinity, extreme temperatures, and heavy metal toxicity, as well as biotic stresses caused by pathogens and herbivores. Empirical evidence from genetic, biochemical, and physiological studies demonstrates that both endogenous AsA accumulation and exogenous application can enhance plant tolerance to environmental challenges, improve photosynthetic efficiency, stabilize membranes, and maintain cellular redox homeostasis.

In summary, ascorbic acid represents a critical determinant of plant resilience and productivity. Understanding its biosynthesis, regulation, and functional roles provides valuable insights into plant stress physiology and offers practical opportunities for crop improvement through metabolic engineering, exogenous supplementation, or breeding strategies aimed at enhancing AsA content. By harnessing the protective and regulatory potential of AsA, it is possible to develop crops better equipped to withstand the increasing frequency and intensity of environmental stresses, thereby supporting sustainable agricultural productivity and food security.

References

Abdel Latef AAH (2026) Chitosan nanoparticles for plant resilience under abiotic and biotic stressors. Smart Nanomaterials Technology. Springer, Singapore, pp 1–353. https://doi.org/10.1007/978-981-95-6422-4

Alamri SA, Siddiqui MH, Al-Khaishany MY, Khan MN, Ali HM, Alaraidh IA, Alsahli AA, Al-Rabiah H, Mateen M (2018) Ascorbic acid improves the tolerance of wheat plants to lead toxicity. J Plant Interact 13(1):409–419. https://doi.org/10.1080/17429145.2018.1491067

Alberts A, Moldoveanu ET, Niculescu AG, Grumezescu AM (2025) Vitamin C: a comprehensive review of its role in health, disease prevention, and therapeutic potential. Molecules 30(3):748. https://doi.org/10.3390/molecules30030748

Ali N, Raza MH, Naheed F, Ali Q, Farid M, Rizwan M, Irshad MK (2015) Foliar application of ascorbate enhances the physiological and biochemical attributes of maize (Zea mays L.) cultivars under drought stress. Arch Agron Soil Sci 61(12):1659–1672. https://doi.org/10.1080/03650340.2015.1028379

Anjum SA, Ashraf U, Tanveer M, Khan I, Hussain S, Shahzad B, Zohaib A, Abbas F, Saleem MF, Ali I, Wang LC (2017) Drought induced changes in growth, osmolyte accumulation and antioxidant metabolism of three maize hybrids. Front Plant Sci 8:69. https://doi.org/10.3389/fpls.2017.00069

Athar HR, Khan A, Ashraf M (2008) Exogenously applied ascorbic acid alleviates salt-induced oxidative stress in wheat. Environ Exp Bot 63(1–3):224–231. https://doi.org/10.1016/j.envexpbot.2007.10.018

Bulley SM, Cooney JM, Laing W (2021) Elevating ascorbate in Arabidopsis stimulates the production of abscisic acid, phaseic acid, and to a lesser extent auxin (IAA) and jasmonates, resulting in increased expression of DHAR1 and multiple transcription factors associated with abiotic stress tolerance. Int J Mol Sci 22(13):6743. https://doi.org/10.3390/ijms22136743

Carpenter KJ (2012) The discovery of vitamin C. Ann Nutr Metab 61(3):259–264. https://doi.org/10.1159/000343121

El-Hawary MM, Hashem OSM, Hasanuzzaman M (2023) Seed priming and foliar application with ascorbic acid and salicylic acid mitigate salt stress in wheat. Agron 13(2): 493. https://doi.org/10.3390/agronomy13020493

Farooq M, Irfan MS, Aziz T, Ahmad IN, Cheema SA (2013) Seed priming with ascorbic acid improves drought resistance of wheat. J Agron Crop Sci 199(1):12–22. https://doi.org/10.1111/j.1439-037X.2012.00521.x

Gaafar AA, Ali SI, El-Shawadfy MA, Salama ZA, Sękara A, Ulrichs C, Abdelhamid MT (2020) Ascorbic acid induces the increase of secondary metabolites, antioxidant activity, growth, and productivity of the common bean under water stress conditions. Plants 9(5):627. https://doi.org/10.3390/plants9050627

Gallie DR (2013) The role of l-ascorbic acid recycling in responding to environmental stress and in promoting plant growth. J Exp Bot 64(2):433–443. https://doi.org/10.1093/jxb/ers330

Hasanuzzaman M, Bhuyan MHMB, Parvin K, Bhuiyan TF, Anee TI, Nahar K, Hossen MS, Zulfiqar F, Alam MM, Fujita M (2020) Regulation of ROS metabolism in plants under environmental stress: a review of recent experimental evidence. Int J Mol Sci 21(22):8695. https://doi.org/10.3390/ijms21228695

Hasegawa C, Yamamoto K, Ito H (2025) Salt stress-induced ascorbic acid accumulation and its regulation by mannose metabolism in tomato plants. Horticulturae 11(4):400. https://doi.org/10.3390/horticulturae11040400

Hassan A, Amjad SF, Saleem MH et al (2021) Foliar application of ascorbic acid enhances salinity stress tolerance in barley (Hordeum vulgare L.) through modulation of morpho-physio-biochemical attributes, ion uptake, osmo-protectants and stress response genes expression. Saudi J Biol Sci 28(8):4276–4290

Jomova K, Raptova R, Alomar SY, Alwasel SH, Nepovimova E, Kuca K, Valko M (2023) Reactive oxygen species, toxicity, oxidative stress, and antioxidants: chronic diseases and aging. Arch Toxicol 97(10):2499–2574. https://doi.org/10.1007/s00204-023-03562-9

Jomova K, Alomar SY, Alwasel SH et al (2024) Several lines of antioxidant defense against oxidative stress: antioxidant enzymes, nanomaterials with multiple enzyme-mimicking activities, and low-molecular-weight antioxidants. Arch Toxicol 98:1323–1367. https://doi.org/10.1007/s00204-024-03696-4

Kanwal R, Maqsood MF, Shahbaz M et al (2024) Exogenous ascorbic acid as a potent regulator of antioxidants, osmo-protectants, and lipid peroxidation in pea under salt stress. BMC Plant Biol 24:247. https://doi.org/10.1186/s12870-024-04947-3

Khan TA, Mazid M, Mohammad F (2011) A review of ascorbic acid potentialities against oxidative stress induced in plants. J Agrobiol 28(2):97–111

Malik M, Narwal V, Pundir CS (2022) Ascorbic acid biosensing methods: a review. Process Biochem 118:11–23

Noman A, Ali S, Naheed F, Ali Q, Farid M, Rizwan M, Irshad MK (2015) Foliar application of ascorbate enhances the physiological and biochemical attributes of maize (*Zea mays* L.) cultivars under drought stress. Arch Agron Soil Sci 61(12):1659–1672. https://doi.org/10.1080/03650340.2015.1028379

Pirzada AM, Anwar T, Qureshi WA et al (2024) Salinity stress mitigation in wheat through synergistic application of ascorbic acid, silver nanoparticles, and Salvadora oleoides aqueous leaf extract: physiological and biochemical responses. Sci Rep 14:30687. https://doi.org/10.1038/s41598-024-76194-3

Pisalsitsakul N, Pinnoi C, Sutanthavibul N, Kamolratanakul P (2022) Taking 200 mg vitamin C three times per day improved extraction socket wound healing parameters: a randomized clinical trial. Int J Dent 2022:6437200. https://doi.org/10.1155/2022/6437200

Seminario A, Song L, Zulet A, Nguyen HT, González EM, Larrainzar E (2017) Drought stress causes a reduction in the biosynthesis of ascorbic acid in soybean plants. Front Plant Sci 8:1042. https://doi.org/10.3389/fpls.2017.01042

Shah S, Khan Y, Cheng Z, Bouskout M, Zhang T, Yan H, Wang M (2025) Priming effect of ascorbic acid on the growth and biomass of quinoa under saline conditions. Front Plant Sci 16: 1600423. https://doi.org/10.3389/fpls.2025.1600423

Shen J, Griffiths PT, Campbell SJ et al (2021) Ascorbate oxidation by iron, copper and reactive oxygen species: review, model development, and derivation of key rate constants. Sci Rep 11: 7417. https://doi.org/10.1038/s41598-021-86477-8

Singh R, Pramanick B, Singh AP, Neelam, Kumar S, Kumar A, Singh G (2017) Bio-efficacy of fenoxaprop-p-ethyl for grassy weed control in onion and its residual effect on succeeding maize crop. Indian J Weed Sci 49(1):63–66. https://doi.org/10.5958/0974-8164.2017.00015.6

Smirnoff N (2018) Ascorbic acid metabolism and functions: a comparison of plants and mammals. Free Radic Biol Med 122:116–129. https://doi.org/10.1016/j.freeradbiomed.2018.03.033

Smirnoff N, Wheeler GL (2024) The ascorbate biosynthesis pathway in plants is known, but there is a way to go with understanding control and functions. J Exp Bot 75(9):2604–2630. https://doi.org/10.1093/jxb/erad505

Tuder RM, Zhen L, Cho CY, Taraseviciene-Stewart L, Kasahara Y, Salvemini D, Voelkel NF, Flores SC (2003) Oxidative stress and apoptosis interact and cause emphysema due to vascular endothelial growth factor receptor blockade. Am J Respir Cell Mol Biol 29(1):88–97. https://doi.org/10.1165/rcmb.2002-0228OC

Ullah I, Waqas M, Khan MA, Hamayun M, Hussain A, Iqbal A, Jan R, Lee IJ, Khan A (2017) Exogenous ascorbic acid mitigates flood stress damages of *Vigna angularis*. Appl Biol Chem 60(6):603–614. https://doi.org/10.1007/s13765-017-0316-6

Wang Z, Li W, Wu W, Guo J, Yang Y (2017) Cadmium stress tolerance in wheat seedlings induced by ascorbic acid was mediated by NO signaling pathways. Ecotoxicol Environ Saf 135:75–81. https://doi.org/10.1016/j.ecoenv.2016.09.013

Wu F-B, Chen F, Wei K, Zhang G-P (2004) Effect of cadmium on free amino acid, glutathione and ascorbic acid concentrations in two barley genotypes (*Hordeum vulgare* L.) differing in cadmium tolerance. Chemosphere 57(6):447–454. https://doi.org/10.1016/j.chemosphere.2004.06.042

Yang X, Lu M, Wang Y, Wang Y, Liu Z, Chen S (2021) Response mechanism of plants to drought stress. Horticulturae 7(3):50. https://doi.org/10.3390/horticulturae7030050

Yaseen S, Amjad SF, Mansoora N, Kausar S, Shahid H, Alamri SAM, Alrumman SA, Eid EM, Ansari MJ, Danish S, Datta R (2021) Supplemental effects of biochar and foliar application of ascorbic acid on physio-biochemical attributes of barley (*Hordeum vulgare* L.) under cadmium-contaminated soil. *Sustainability 13*(16):9128. https://doi.org/10.3390/su13169128

Younis M, Akram NA, Abdel Latef AAH, Ashraf M (2023) Appraisal of improvement in physiological and metabolic processes by exogenously applied natural and synthetic ascorbic acid in okra (Abelmoschus esculentus l.) fruit subjected to water deficit stress. Phyton-International Journal of Experimental Botany 92(10):2761–2784

Zhang K, Wang G, Bao M, Wang L, Xie X (2019) Exogenous application of ascorbic acid mitigates cadmium toxicity and uptake in maize (Zea mays L.). Environ Sci Pollut Res 26(19): 19261–19271. https://doi.org/10.1007/s11356-019-05265-0

Chapter 7
Tocopherol: Enhancing Plant Tolerance to Stress

Murat Dikilitas, Behzat Baran, Berrin Arslan, and Arafat Abdel Hamed Abdel Latef

Abstract Global climate change has intensified its impact via drought, temperature fluctuations, high-light intensity, salinity, flooding, freezing, and pollution in nearly all areas. On the other hand, biotic stress factors would still carry out their pathogenicity under the changing environmental conditions. The majority of the pathogenic microorganisms can adapt to abiotic stress conditions when compared to most crop plants. The adaptation of pathogens to adverse conditions makes them a potential danger in every situation, and they can be involved in simultaneous or consecutive stresses with those of abiotic stress factors. Due to global climate change, the combined effects of abiotic and biotic stress factors may replace the single stress factors. Developing multiple-stress-tolerant crop plants would still require a bit more time, possibly after revealing the molecular and biochemical pathways of crop plants under multiple stress conditions. Removing stressors from the vicinity of crop plants is almost impossible, and generating stress-resilient plants would take time; however, we could improve the conditions of plants by evaluating and applying beneficial chemicals. Several approaches have been proposed to enhance the conditions of crop plants. One of the key compounds that has been recently proposed to enhance plant tolerance is tocopherol. In this chapter, we evaluated the role of tocopherol and its potential use to enhance plant resistance or tolerance and the

M. Dikilitas (✉)
Faculty of Agriculture, Department of Plant Protection, Harran University, S. Urfa, Türkiye

B. Baran
Plant Protection Research Institute, Sur, Diyarbakır, Türkiye

B. Arslan
Faculty of Agriculture, Department of Plant Protection, Siirt University, Siirt, Türkiye

A. A. H. Abdel Latef
Department of Botany and Microbiology, Faculty of Science, Qena University, Qena, Egypt

A. A. H. Abdel Latef (ed.), *Harnessing Vitamins and Organic Acids for Plant Stress Resilience*, Sustainability Sciences in Asia and Africa,
https://doi.org/10.1007/978-981-95-8593-9_7

possible combination with other compounds for sustainable agriculture under stress conditions.

Keywords Tocopherol · Combined stress · Drought · Pathogen · Quality

7.1 Introduction

Plants are sessile organisms that are constantly exposed to a variety of environmental stresses, including abiotic factors such as drought, salinity, temperature extremes, and UV radiation, as well as biotic factors such as pathogens and herbivores, or combinations of these. These stress factors induce reactive oxygen species (ROS) that can lead to lipid peroxidation, protein and carbohydrate denaturation, DNA damage, etc. The overproduction of ROS can lead to oxidative damage affecting the cellular machinery (Baysal et al. 2021; Mishra et al. 2023; Rady et al. 2025; Abdel Latef et al. 2026). This may eventually disrupt the whole metabolism and result in significant crop reduction. Major stress factors, followed by global climate changes, have, therefore, started threatening our food security. It has been clearly revealed that plants have developed a defense mechanism consisting of enzymatic and non-enzymatic antioxidants (Dikilitas et al. 2021; Vicidomini et al. 2024). Non-enzymatic antioxidants consist of carotenoids, ascorbate, tocopherol, and glutathione (Zha et al. 2019). Although major pathways have been known, new molecular pathways or genes responsible for the tolerance or resistance of the crop plants should also be revealed in detail. Great progress has been made so far in agriculture; however, this may not be enough to feed the increasing population under severe stress conditions. Even if we increase crop production, we should also keep the quality of crops at certain levels.

To sustain a quality of life, we always need to keep vitamin levels as high as possible. To extend and increase resistance or tolerance levels of crop plants, vitamins have been recognized as valuable chemicals. One of the most appropriate ways of increasing vitamin levels in crop plants is to make an exogenous application. Various antioxidants have been suggested to increase resistance in crop plants. One of the most promising antioxidants belongs to tocopherol group (Fig. 7.1). Tocopherols, commonly known as vitamin E, are lipid-soluble antioxidants synthesized only in photosynthetic organisms, however, they are also synthesized in various plant organs, including leaves, seeds, and roots in changing concentrations

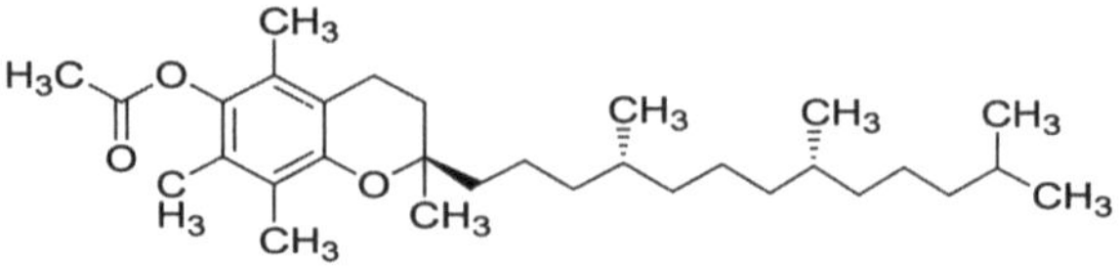

Fig. 7.1 Chemical structure of tocopherol. (https://pubchem.ncbi.nlm.nih.gov/compound/alpha-Tocopherol) and (https://www.kegg.jp/entry/C13202). Molecular formula ($C_{29}H_{50}O_2$)

depending on the plant species, developmental stage, and environmental conditions (Maeda et al. 2006). Tocopherols play significant roles in maintaining redox homeostasis, particularly within cellular membranes, which are vital for plant stress tolerance. Of all the tocopherol isoforms, α-tocopherol exhibits the greatest vitamin E activity and is the dominant form in photosynthetic tissues (Munné-Bosch and Alegre 2002). However, other tocopherol forms, such as δ-, γ-, and β-tocopherol, also contribute to the overall antioxidant capacity of plants (Delgado et al. 2020). Apart from tocopherols, tocotrienols have also gained interest because of their health benefits (Montagani Marelli et al. 2019). It is important to note that the human body preferentially uses α-tocopherol, and only α-tocopherol is defined as vitamin E and supplementation of α-tocopherol is considered to reverse vitamin E deficiency. Plant seeds and nuts, especially sunflower seeds, hazelnuts and, almonds, are rich sources of tocopherol. Tomatoes, avocados, spinach, asparagus, and broccoli are also rich in tocopherol. Similarly, although plant stress tolerance is increased by tocopherols, decreased levels make oxidative damage easier.

Tocopherols, among the crucial components of the antioxidant defense machinery in plants, stand out due to their lipophilic nature and ability to protect membrane lipids from peroxidation. Tocopherols function predominantly by scavenging lipid peroxyl radicals (LOO*) through a hydrogen atom donation mechanism, forming tocopheroxyl radicals recycled by ascorbate or glutathione cycles. Tocopherols consist of a chromanol head group and a lipophilic prenyl tail (Voll and Abbasi 2007). They stated that leaves were dominated by α-tocopherol, while seeds were dominated by γ-tocopherol in seeds. Similar issues were made by Velasco et al. (2013) who stated that α-tocopherol was more pronounced in leaves and flowers, while γ-tocopherol was accumulated in seeds. Tocopherols can regulate stress-responsive gene expressions via redox-sensitive transcription factors and modulate non-enzymatic antioxidant capacity through synergistic interactions with ascorbate, carotenoids, and glutathione (Munoz and Munne-Bosh 2019). Tocopherols are synthesized via the shikimate and methylerythritol phosphate (MEP) pathways. The terminal methylation step by γ-tocopherol methyltransferase (VTE4) converts γ- to α-tocopherol, the most active form in terms of antioxidant capacity (Vinutha et al. 2017).

Tocopherols are localized in chloroplast envelopes, thylakoid and plastid membranes, act as safeguards during photosynthesis, regulate growth and development, and take significant roles under stress conditions. As stated above, tocopherols confer protection against various abiotic or biotic stresses via maintaining membrane integrity and reducing oxidative damage. For example, under drought stress, tocopherol accumulation limits lipid peroxidation, which is the formation of malondialdehyde (MDA) in thylakoid membranes, thus maintaining photosynthetic efficiency (Faizan et al. 2023). In salinity stress, tocopherols stabilize ion homeostasis and mitigate Na^+-induced oxidative stress (Guo et al. 2022). Tasaka et al. (2023) found that during high light and UV-B exposure, α-tocopherol effectively quenches singlet oxygen and prevents photoinhibition, thereby maintaining the functionality and fluidity of photosynthetic membranes. Since cold and heat stresses similarly induce ROS accumulation, tocopherols improve the conditions of plants by reducing peroxyl radicals and preventing chain lipid oxidation (Rajput et al. 2021).

During pathogen infection, plants initiate an oxidative burst via rapid ROS generation, as occurs in abiotic stress conditions. So, tocopherols can play significant roles in inducing disease resistance.

Tocopherols can modulate the redox status to avoid excessive damage while triggering defense signaling (Hidalgo et al. 2020). Tocopherols play a significant role in maintaining membrane stability, as well as in salicylic acid (SA), jasmonic acid (JA) and abscisic acid (ABA) pathways (Stahl et al. 2019). It has been concluded that tocopherol-deficient mutants exhibited increased susceptibility to pathogens due to impaired ROS homeostasis and disruption in signaling cascades (Mesa et al. 2024). Genetic manipulation of tocopherol biosynthesis has yielded plants with enhanced tolerance to multiple stresses. Overexpression of *VTE2*, *VTE1*, or *VTE4* genes in *Arabidopsis*, tobacco, and maize resulted in elevated α-tocopherol levels and improved resistance to drought, salinity, and high light (Albert et al. 2022). Similarly, Traber (2024) reported that α-tocopherol deficiency in zebrafish embryos increased lipid peroxidation and induced deficiencies in secondary metabolites. The author stated that α-tocopherol had a significant role in limiting lipid peroxidation chain reactions and controlling embryonic development. The neurodevelopmental process was highly dependent on α-tocopherol accumulation.

Enhancing endogenous tocopherol levels can augment plant health under stress and extend shelf life by preventing oxidative deterioration during growth and storage periods. It also helps human metabolism. Future research should elucidate the precise signaling networks interfacing tocopherol function with hormonal and ROS signaling under combined stress scenarios.

This chapter aims to elucidate the multifaceted role of tocopherols in enhancing plant resilience to a wide array of environmental challenges. These challenges include abiotic stresses, such as heavy metal toxicity, salinity, drought, and temperature extremes, as well as biotic stresses arising from pathogen attacks and other biotic factors. This chapter also highlighted the potential use of tocopherols in biotechnology and in combination with other methods of application.

7.2 Biosynthesis and Transport of Tocopherols in Plants and Their Main Functions in Organisms

Plants have evolved complex antioxidant defense systems to mitigate the damaging effects of ROS, including enzymatic and non-enzymatic antioxidants. Antioxidant enzymes such as peroxidase (POD), catalase (CAT), glutathione reductase (GR), and superoxide dismutase (SOD), as well as non-enzymatic antioxidants including carotenoids, flavonoids, and ascorbic acid, play a vital role in helping plants tolerate abiotic and biotic stress factors (Abdel Latef 2021; Bidabadi et al. 2021; Dawood et al. 2021; Bhat et al. 2022). These antioxidants work together to maintain cellular redox homeostasis and scavenge ROS. Vitamin E is a phenolic antioxidant that neutralizes free radicals by converting the hydrogen atom in its ring structure's hydroxyl (-OH) group. It is primarily found in the phospholipid bilayer of cell membranes. Tocopherols, like lipid-soluble antioxidants, play a critical role in protecting cellular membranes from lipid peroxidation, a major consequence of oxidative stress, which is associated with numerous diseases and clinical conditions. Tocopherols act as chain-breaking antioxidants, scavenging lipid peroxyl radicals and preventing the propagation of lipid peroxidation in cellular membranes (Rinnerthaler et al. 2015). In addition to scavenging free radicals directly, tocopherols can also indirectly regulate the activity of other antioxidant enzymes and signaling molecules (Zhao et al. 2025). Tocopherol may also regulate the expression of genes that respond to stress, including those that encode antioxidant enzymes, heat shock proteins, and osmoprotectants. This further enhances the plant's tolerance to environmental challenges. Although most of what is known about the biological functions of tocopherol comes from the studies of mammalian cells, recent studies of tocopherols have shown increasing trends in plant science. Tocopherol biosynthesis is simply illustrated in Fig. 7.2. Its production during stress has been linked to adaptation to changing environmental conditions. Understanding how plants

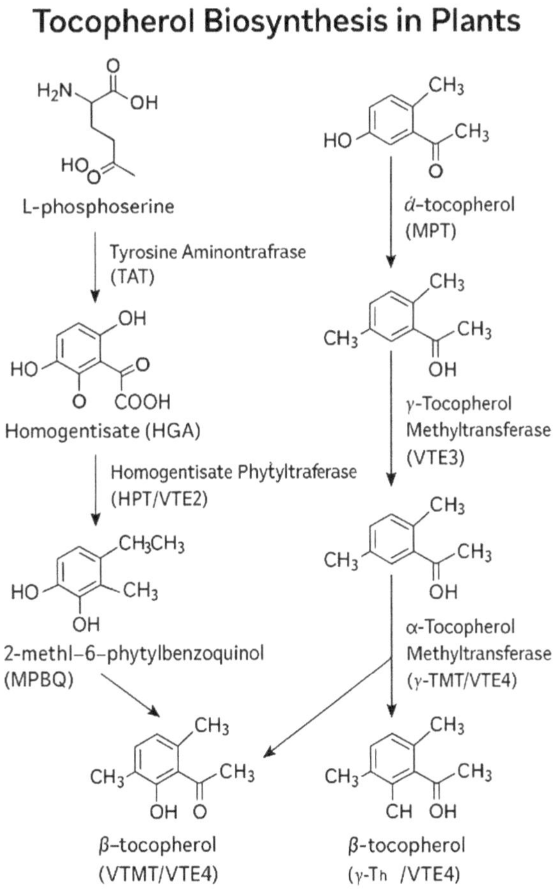

Fig. 7.2 Overview of tocopherol biosynthesis in plants. The illustration for tocopherol biosynthesis is simplified from the source of KEGG Pathway (https://www.kegg.jp). The synthesis of the tocopherol pathway begins with tyrosine and homogentisate (HGA) and passes through 2-methyl-6-phytylbenzquinol (MPBQ). Depending on enzyme activity, the pathway subsequently produces α-, β-, δ-, and γ-tocopherols. These consist of a hydrophobic membrane tail located in the membrane core and a polar chromanol head group located on the membrane surface

recognize stress signals and convey them to trigger adaptation or stress responses is a crucial step towards improvement in crop production in terms of both quantity and quality. Tocopherol accumulation in plants has strong connections with abiotic stress tolerance. Conversely, a deficiency in tocopherol increases oxidative stress, causing damage to proteins, DNA, and lipids, and eventually leading to cell death. The function of tocopherols is related to their ability to scavenge free radicals and quench singlet oxygen. They donate hydrogen atoms from their hydroxyl group to lipid peroxyl radicals, converting them into stable hydroperoxides and terminating the chain reaction of lipid peroxidation (Mishra et al. 2023). Tocopherols can protect cells from DNA damage and reduce the risk of mutation and cancer. They can act as signaling molecules, modulating the activity of protein kinases and other signaling components involved in stress responses (Fanta Yadang et al. 2019). Functions of tocopherols in plants and higher organisms are summarized in Table 7.1.

The regulation of tocopherol biosynthesis is influenced by light, temperature, nutrient availability, and stress conditions. Tocopherol content in plant tissues can be manipulated through genetic engineering, providing a promising strategy for enhancing the nutritional value of crops. Its biosynthesis is coordinated with other metabolic pathways, such as isoprenoid and phenylpropanoid biosynthesis, to ensure optimal resource allocation and metabolic balance. The regulation of tocopherol biosynthesis is crucial for maintaining the correct levels of this compound in plant tissues. This is essential for stress tolerance, as well as for plant growth and development. Recently, it has been reported that tocopherols, particularly α-tocopherol, contribute to the fine-tuning of signal transmission outside chloroplasts (Muñoz and Munné-Bosch 2019).

7.3 Role of Tocopherol in Abiotic/Biotic Stress Tolerance/Resistance

Plants cannot move; hence, they must endure abiotic stressors, including drought, salinity, and extreme temperatures (Zhang et al. 2020; Khan 2023; Zhang et al. 2023; Barman et al. 2026). Plants respond to these stresses through a combination of morphological, physiological, and biochemical adaptations.

The mechanisms for tolerance involve perceiving stress signals, activating defense responses, and restoring cellular homeostasis. Epigenetic modifications, such as DNA methylation, chromatin remodeling, and histone modification, enable plants to alter their gene expression patterns in order to adapt to changing environmental conditions (Kumar and Mohapatra 2024). These responses could be categorized into two steps. The first one involves primary metabolites such as amino acids, carbohydrates, lipids, polyamines, and glycine betaine, which act as osmoprotectants for plants facing environmental stressors (Salam et al. 2023). The second one involves the secondary metabolites, which include phenolics, terpenoids, and nitrogen-containing compounds that protect plants from biotic stressors like

Table 7.1 Functions of tocopherol in human and plant metabolisms

Biological function in humans	Deficiency symptoms in humans	Biological functions in plants	Deficiency symptoms in plants	Etiology of vitamin deficiency in humans/plants	References
Lipid-soluble antioxidant; protects cell membranes	Neuromuscular problems, ataxia, anemia, impaired vision	Protect chloroplast membranes from oxidative stress	Chlorosis, necrosis, and reduced photosynthetic efficiency	Humans: low-fat diets, premature infants Plants: environmental and biotic stresses	El-Beltagi et al. (2013)
Protects the immune system	Increased susceptibility to infections and decreased immunity	Stabilizes thylakoid membranes and photosystem II	Photoinhibition, reduced chlorophyll content	Humans: genetic defects Plants: impaired tocopherol biosynthetic genes	Lewis et al. (2019)
Essential for neurological function, it prevents oxidative damage in neurons	Peripheral neuropathy, muscle weakness	Scavenges ROS during abiotic and biotic stresses	Accumulation of ROS, premature senescence	Humans: cystic fibrosis Plants: Oxidative stress	Ali et al. (2020a, b)
Reduces the risk of cardiovascular diseases	Increased risk of atherosclerosis and cardiovascular dysfunction	Modulates plant hormone signaling (JA, ABA, SA) and seed development	Defective seed germination, impaired embryo development	Humans: low vitamin E intake in the diet Plants: nutrient imbalance, stress impairing the tocopherol pathway	Poudel et al. (2023)
Maintains skin health, slows ageing, and aids wound healing	Skin dryness, impaired wound healing	Prevents lipid peroxidation in seeds	Reduced seed viability and storability	Humans: malnutrition, low-fat diets Plants: genetic mutations in tocopherol biosynthesis	Styrczewska et al. (2019)

herbivores, insects, and pathogens (Salam et al. 2023). Under stress conditions, plants experience an imbalance between the production and scavenging of ROS, leading to oxidative stress, which can eventually cause cellular damage and disrupt normal physiological processes (Hernández-Carranza et al. 2023; Pereira et al. 2025). The importance of tocopherols in plants has been increasing as the stress factors get more complex and are combined with each other. The revealing of detailed pathways exhibits the importance of secondary metabolites under stress conditions. Its antioxidant activity is in good correlation with quinone structure.

Biosynthesis and functions of tocopherols have been considered as the major host plant responses to cope with stress. Tocopherols help maintain photosynthetic efficiency and enable plants to continue producing energy even under stress conditions (Bai et al. 2020). Chishti et al. (2025) also reported that α-tocopherol not only protected radish plants via enhancing antioxidant enzymes under drought stress but also enhanced total soluble proteins to maintain cell internal osmotic potential. The α-tocopherol maintained stomatal conductance and transpiration rate, and regulated the beneficial inorganic ions (K^+, Ca^{2+}, PO_4^{3-}) in the shoots.

Understanding the intricate interplay between tocopherols and plant stress responses is crucial for developing strategies to enhance crop resilience and improve food security in the face of increasingly challenging environmental conditions.

7.3.1 Drought Stress

Drought stress triggers a series of biochemical, physiological, and molecular responses in plants. These include osmotic adjustment, the synthesis of protective proteins, and the activation of antioxidant systems (Nazari et al. 2023; Haghpanah et al. 2024; Dawood et al. 2025; Jameel et al. 2026). Drought is perhaps the most pervasive and economically damaging abiotic stress faced by agricultural systems. It occurs when there is an insufficient supply of water to meet the demands of crops. It leads to impaired growth, reduction in yields, and quality. Drought stress disrupts plant water balance, inhibits photosynthesis, and compromises cellular integrity, resulting in wilting, leaf rolling, and premature senescence. Tocopherols can alleviate these effects by scavenging free radicals, stabilizing cell membranes, and protecting photosynthetic machinery (Dash et al. 2025).

Loyola et al. (2012) conducted physiological and molecular analyses to investigate the physiological adaptations and increased levels of α-tocopherol that play a role in the adaptive responses of *Solanum chilense* to drought. The study demonstrates that *S. chilense* exhibits increased tolerance and survival rates in response to drought stress, accompanied by elevated levels of α-tocopherol. The study suggests that increasing tocopherol concentrations could form part of a strategy for cultivating more drought-resistant crops.

Sarafraz-Ardakani et al. (2014), meanwhile, examined two wheat varieties with different levels of drought tolerance and found that α-tocopherol levels were significantly higher in the drought-tolerant variety. Field experiments were conducted to evaluate physiological responses, including antioxidant regulation, under conditions of water deficiency. An increase in α-tocopherol was found to be associated with a decrease in lipid peroxidation and oxidative stress markers. This indicates that tocopherol plays a significant role in protecting plants from drought-induced damage and increasing their tolerance to drought. Kalinova et al. (2006) evaluated the effects of climate factors on tocopherol abundance in buckwheat plants and found that α-tocopherol levels increased with sunlight and temperature, but decreased with rainfall. The authors also concluded that targeted cultivation practices in specific weather conditions are required to maximize tocopherol production and enhance drought tolerance. Akram et al. (2023) demonstrated the effectiveness of applying α-tocopherol externally to eggplants under drought stress, showing significant improvements in growth and metabolism, increased proline accumulation, and enhanced antioxidant activity. The authors stated that α-tocopherol application facilitates moisture retention and improves overall plant health by reducing the effects of stress. They also suggested that managing tocopherol levels could be essential for enhancing crop productivity in drought conditions.

In summary, tocopherols, particularly α-tocopherol, play a vital role in enhancing plant resistance to drought stress. They can mitigate the negative effects of water deficit through various physiological mechanisms, such as antioxidant activity and membrane stabilization, as demonstrated by numerous studies. Effectively managing and increasing tocopherol content in plants could be a viable strategy for enhancing drought tolerance and improving overall plant health in response to changing environmental conditions.

7.3.2 *Salinity Stress*

Salinity stress arises from the accumulation of soluble salts, primarily sodium chloride, in the soil and irrigation water. High soil salinity levels disrupt osmotic balance, impair nutrient uptake, and induce ion toxicity in plants. Symptoms of salinity stress include leaf burn, stunted growth, and chlorosis, ultimately diminishing crop productivity and quality. Tocopherol enhances the capacity of plants to withstand salt-induced oxidative damage Gruda et al. 2024). It maintains the integrity of cell membranes, which is essential for controlling ion movement and preventing the harmful entry of sodium ions into cells. Plants adapt to saline conditions through various mechanisms, including the regulation of ion transport and the synthesis of compatible solutes. In response to salt stress, plants utilize phytohormone-regulated salt-signaling pathways, which improve defense capabilities. When exposed to saline-sodic conditions, plants may develop micronutrient deficiencies such as zinc and manganese. Zinc forms complexes with hydroxyl ions, which reduces its bioavailability (Raza et al. 2023a, b). Plants must quickly detect and react to salt stress to survive by triggering adaptive molecular processes. These responses include the activation of signaling cascades and the synthesis of protective proteins and metabolites (Yu et al. 2022).

Recent studies have emphasized the important function of tocopherols, especially α-tocopherol, in improving plant resistance to salt stress. The primary function of tocopherols involves preventing oxidative stress by scavenging ROS, which are formed during salt stress (Hassan et al. 2021). The importance of tocopherols lies in their distinct protective capacities against the increased ionic strength and osmotic pressure caused by high salinity. Naqve et al. (2021) reported that tocopherols improved growth conditions and appearance characteristics under salt stress. They also found that tocopherols increased the degradation of antioxidant enzymes, reduced oxidative damage, and improved chlorophyll recovery.

In summary, tocopherols, particularly α-tocopherol, play a pivotal role in enhancing plant growth in response to salt stress and provide a promising approach to improving salt management strategies.

7.3.3 Temperature Stress

Temperature extremes, encompassing both heat and cold stress, exert profound effects on crop growth and development. Heat stress occurs when temperatures exceed the optimal range for plant metabolism, leading to cellular damage, protein denaturation, and oxidative stress. Symptoms include leaf wilting, leaf scorching, and flower abortion, ultimately resulting in yield losses (Bidabadi et al. 2022; Shikari et al. 2022; Zahra et al. 2023). Conversely, cold stress occurs when temperatures fall below the threshold for normal growth, causing chilling injury, frost damage, and impaired photosynthesis. Cold stress, like heat stress, causes leaf necrosis, tissue discoloration, and reduces fruit set, and poses a significant reduction in crop production in temperate and high-altitude regions. When compared to both extreme temperatures, from hot to freezing temperatures, frost is extremely important due to its causes of significant economic losses. In fact, plants have two mechanisms for freezing: avoidance and tolerance. In avoidance mechanisms, plants try to prevent ice formation, while in the tolerance mechanism, a series of tools, such as metabolites and proteins (cryoprotectants), have been used to survive with apoplastic ice formation to maintain cell membrane and macromolecule stability.

Tocopherols are essential for maintaining optimal membrane function and cellular homeostasis in plants, thereby promoting overall plant health and productivity. Tocopherols effectively maintain the stability and integrity of cell membranes. On the other hand, tocopherols protect plants from damage caused by high light intensity by quenching singlet oxygen and scavenging free radicals (Yeshi et al. 2022). As stated above, membrane stabilization via tocopherol application could play a crucial role in protecting plants from the negative effects of stress and enhancing the growth and survival of plants in challenging environments. Similarly, Pérez-Llorca et al. (2019) reported that carotenoids and α-tocopherol played a significant role in protecting photosynthetic organelles under severe stress conditions. The authors concluded that adjustment of α-tocopherol concentration as early as 3 h before maximum light intensity significantly protected the photosynthetic apparatus. Spicher et al. (2017) reported that α-tocopherol and plastoquinone/plastoquinol were significantly upregulated in tomato plants under temperature and high light stress.

7.3.4 Heavy Metals Stress

The relationship between tocopherols and heavy metals has received considerable attention in recent literature, particularly in view of the role of tocopherols as antioxidants in mitigating the oxidative stress caused by heavy metals. Alsherif et al. (2023) investigated the impact of heavy metal exposure on the production of alpha and beta tocopherols in *Sesuvium portulacastrum*. They reported persistent increases in tocopherol content in both the shoot and root tissues. Tocopherols are important non-toxic antioxidants that combat ROS produced by metal-induced oxidative stress. The study concluded that increasing tocopherol biosynthesis could be a viable strategy for conferring heavy metal tolerance to plants. Alsherif et al. (2023) propose further research into the specific mechanisms by which tocopherols exhibit this tolerance and report that understanding these pathways could facilitate the cultivation of heavy metal-tolerant plant species in contaminated soils.

Highlighting the significance of tocopherols in counteracting heavy metal-induced oxidative stress (Jin and Daniell 2014; Cetin et al. 2014), the researchers demonstrated that the expression of γ-tocopherol methyltransferase (TMT) in chloroplasts results in elevated tocopherol levels. Consequently, plants such as *Arabidopsis thaliana* become less susceptible to metals stress. Researchers have reported that plants with engineered tocopherol profiles exhibit reduced lipid peroxidation and improved membrane integrity under heavy metal stress.

Ali et al. (2020a, b) investigated the role of native tocopherols in enhancing tolerance to cadmium stress in rapeseed. They emphasized that tocopherols are critical components of the antioxidant defense system and effectively detoxify ROS accumulation due to cadmium exposure. The beneficial effects of tocopherol have also been observed in conditions of Cd stress. Ali et al. (2020a, b) reported that rapeseed genotypes with a high seed oil content were more resistant to Cd stress than those with a low seed oil content. They also noted a positive correlation between high seed oil content and tocopherol content, the latter of which limits Cd uptake by roots and transport to shoots.

The results show that plants with higher levels of tocopherol exhibit lower oxidative stress and better overall health when exposed to cadmium. This confirms the essential role of tocopherols in the defense mechanisms of plants. These studies collectively demonstrate the protective role of tocopherols against heavy metal-induced oxidative stress in various biological systems. The consistent finding that higher tocopherol levels are associated with reduced oxidative damage reinforces the importance of these compounds as nutrients in agriculture. Future research recommendations include investigating ways to increase tocopherol biosynthesis in crops through genetic manipulation, as well as studying the effects of tocopherol status on plant health in populations exposed to heavy metals.

7.3.5 Pathogen Stress

Tocopherol-related studies have also taken place in biotic stress conditions. Tocopherols are renowned for their antioxidant properties and are crucial in plant defense mechanisms against biotic stresses, including attacks by pathogens (Savignac et al. 2022; Khan et al. 2025). The response of plants to pathogens typically involves an increased production of ROS, which leads to oxidative stress (Georgiadou et al. 2020). Waśkiewicz et al. (2022) reported significant differences in tocochromanol profiles following inoculation of sweet corn cultivars infected with *Fusarium verticillioides* and *F. proliferatum*, indicating that tocopherol levels are important in protecting against fungal pathogens. According to Cela et al. (2018), changes in the composition of tocopherols in chloroplasts may negatively impact the way *Arabidopsis thaliana* responds to biotic stress. This occurs through changes in the composition of membrane fatty acids, increased lipid peroxidation, and delayed defense activation when the plant is exposed to *Botrytis cinerea*. Chhabra et al. (2022) reported that biochemical parameters play a role in determining resistance to root rot caused by *Fusarium fujikuroi*, which results in significant yield losses in rice. They concluded that future studies should directly measure tocochromanol levels to confirm this relationship.

In summary, the cumulative data demonstrate a strong correlation between pathogen stress and tocopherol levels in plants. Not only do tocopherols neutralize increased ROS during pathogen attacks, but they also enhance plant health and resistance to biotic stresses. Future studies should, therefore, focus on genetic modifications to increase tocopherol synthesis in crops, providing a sustainable method of enhancing disease resistance that does not rely on pesticides.

7.4 Molecular Mechanisms and Signaling

Tocopherols facilitate seed germination, seedling growth, root development, and leaf expansion by modulating oxidative stress (Ghosh et al. 2022). Tocopherols impact hormone signaling pathways, influencing various developmental processes, including flowering time, fruit ripening, and senescence. Its deficiency can result in developmental abnormalities such as stunted growth, chlorosis, and premature senescence (Almeida et al. 2015). The molecular and signaling mechanisms of tocopherol are briefly summarized in Fig. 7.3.

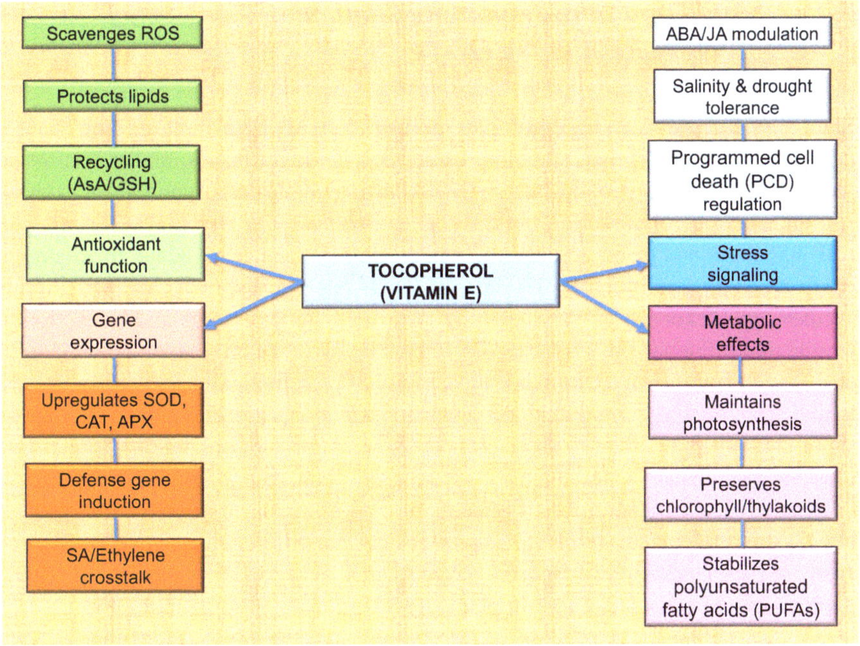

Fig. 7.3 Tocopherol mechanisms in plants. This figure was prepared with the help of multiple references (Fritsche et al. 2017; Ali et al. 2022; Ghosh et al. 2022; Meena et al. 2025)

7.4.1 Membrane Stabilization and ROS Detoxification

Tocopherols prevent lipid peroxidation by reacting with lipid radicals, which stops the chain reaction of lipid damage and preserves the integrity of membrane structures. They can deactivate photoproduced ROS and prevent lipids from oxidation (Ma et al. 2020). They physically stabilize membrane structures and lessen the harmful effects of stress. Tocopherols react with phospholipid bilayers and interact with unsaturated fatty acids and other membrane components. Under stress conditions, tocopherols can help prevent membrane rigidification and maintain membrane fluidity, which is essential for maintaining proper cellular function.

Membrane stability is crucial for the optimal operation of membrane-bound enzymes and transport proteins, which are involved in a wide array of metabolic activities (Delgado et al. 2020). The phenolic structure of tocopherols allows them to donate hydrogen atoms to free radicals, neutralizing their reactivity and preventing chain reactions that cause cellular damage (Es-Sai et al. 2025). It has been reported that tocopherols provide a comprehensive antioxidant defense system by working synergistically with other antioxidant enzymes, such as glutathione peroxidase, catalase, and superoxide dismutase (Al-Harthi et al. 2021).

7.4.2 *Signaling Pathways and Regulation of Stress-Responsive Genes*

Plant signaling pathways are complex networks involving a multitude of components, including receptors, signaling molecules, protein kinases, and transcription factors, all intricately coordinated to orchestrate cellular responses to external stimuli. Tocopherols function not only as antioxidants but also play great roles in cell signaling and gene regulation, which enables plants to adapt to changing environmental conditions (Ali et al. 2022). It has been shown that tocopherols regulate gene expression by interacting with regulatory proteins and transcription factors that influence the synthesis of proteins involved in stress tolerance. Tocopherols have been shown to influence the activity of protein kinases and phosphatases, which are essential components of signal transduction cascades involved in stress signaling (Liu et al. 2024). Tocopherols exert regulatory control over the expression of stress-responsive genes, thereby modulating plant responses to environmental challenges. The concentration of phenols in plant tissue is an effective indicator of abiotic stress tolerance, which varies greatly between different plant species (Sharma et al. 2019). Transcription factors—proteins that bind to DNA and regulate gene expression—play a crucial role in plant stress responses (Yao et al. 2018).

7.5 Application of Tocopherol in Crop Management

The application of exogenous tocopherol to plants can provide direct antioxidant protection and enhance their tolerance to oxidative stress caused by various stress factors. Tocopherol supplementation can mitigate the adverse effects of stress, improve plant growth and productivity under adverse conditions (Table 7.2). Plants adapt to adverse environments through the biosynthesis of compounds such as phytohormones and metabolites. Understanding the regulatory mechanisms of tocopherol biosynthesis is crucial for developing strategies to enhance tocopherol content in crops and improve their nutritional value and stress tolerance. Furthermore, exploring the interaction between tocopherol biosynthesis and other metabolic pathways could provide insights into optimizing resource allocation and metabolic balance in plants under stress conditions. Improving plant stress tolerance by modulating transcription factors is now a frequent strategy due to their early stress response and influence on downstream target genes (Ali et al. 2022). Tocopherols offer considerable advantages in improving plant resilience to several environmental stresses. They serve as a key component of the antioxidant defense system of plants, counteracting the detrimental effects of ROS brought on by environmental challenges (Parvin et al. 2019). Tocopherols, particularly α-tocopherol, are essential for maintaining membrane integrity, preventing lipid peroxidation, and scavenging free radicals during stress conditions. Advancements in genetic engineering, metabolic engineering, and agronomic practices offer opportunities to optimize tocopherol

Table 7.2 Application of tocopherol to plants

Tocopherol application	Plant species	Stress type	Mode of action	References
Foliar α-tocopherol spray	*Zea mays* (maize)	Drought stress	Improves photosynthetic pigments, water relations, antioxidant defense, and nutrient uptake	Ali et al. (2020a, b)
Exogenous α-tocopherol application	*Lens culinaris* (lentil)	Drought stress	Enhances antioxidant enzymes (SOD, CAT, APX), reduce lipid peroxidation and oxidative damage	Shah et al. (2021)
Foliar α-tocopherol application	*Triticum aestivum* (wheat)	Drought stress	Enhance biomass, yield, water relations, antioxidants, and nutritional quality	Ali et al. (2019)
α-Tocopherol fertigation	*Raphanus sativus* (radish)	Drought stress	Improves membrane protection, photosynthetic pigments, water use efficiency, and antioxidant enzymes	Chishti et al. (2025)
Foliar α-tocopherol (with salicylic acid)	*Triticum aestivum* (wheat)	Sowing stress	Enhances germination, chlorophyll, and antioxidant enzyme activity under deep-sowing stress	Saeed et al. (2024)
Foliar α-tocopherol (with ascorbate, GSH)	*Gossypium* spp. (cotton)	Drought stress	Enhance pigments, sugar, amino acids, and antioxidant enzyme activities under water stress	Mekki et al. (2015)
Foliar α-tocopherol spray	Chickpea (*Cicer arietinum)*	Drought stress	Enhances antioxidant enzymes, reduces MDA and H_2O_2, and maintains protein content	El-Beltagi et al. (2022)
Foliar α-tocopherol in salinity	Various crops	Salinity	Improves photosystem II, ROS scavenging, and growth under saline conditions	Naqve et al. (2021)

SOD superoxide dismutase, *CAT* catalase, *APX* ascorbate peroxidase, *MDA* malondialdehyde, *GSH* glutathione

levels in plants and harness their protective effects against environmental stresses (Verma et al. 2021). Enhanced tocopherol biosynthesis through genetic manipulation has been shown to improve plant tolerance to various stresses, including drought, salinity, and oxidative stress. Furthermore, agronomic practices such as optimizing nutrient availability and irrigation regimes can also enhance tocopherol accumulation in plant tissues, thereby bolstering their stress resilience. Breeding strategies can be employed to select for high-tocopherol varieties, further enhancing plant tolerance to stress. Enhancing tocopherol content in plants can protect photosynthetic machinery under stress conditions, which in turn contributes to improved plant performance and productivity. Tocopherol overproduction can be achieved by genetic manipulation (Vinocur and Altman 2005). Genetic engineering, conventional breeding, and agronomic practices are viable strategies to enhance tocopherol production in plants. Overexpression of genes involved in tocopherol biosynthesis increases tocopherol content and enhances plant stress tolerance.

The application of methyl jasmonate and plant growth regulators, such as salicylic acid, strengthens a plant's resistance to environmental stress and increases its levels of tocopherol. Nutrient management, including optimizing nitrogen, phosphorus, and potassium supply, positively influences tocopherol biosynthesis and enhances plant growth under stress conditions. Genetic engineering approaches, including transcription factors, have emerged to improve plant tolerance. The complex cascades of gene expression during osmotic stress response have been analyzed. Enhancing plant tolerance by manipulating the expression of transcription factors has become a popular approach (Lan Thi Hoang et al. 2017; Lata and Shivhare 2021). Biotechnological tools, such as RNA-seq and CRISPR-Cas9, are useful in identifying target genes and manipulating them to achieve tolerance (Imran et al. 2021). Molecular markers accelerate genetic introgression using molecular markers or site-directed mutagenesis (Villalobos-López et al. 2022). The integration of omics technologies with genetic engineering strategies offers great potential for developing crops with broad-spectrum stress tolerance (Cappetta et al. 2020). Crops from now on rarely face a single type of stress; in fact, multiple stresses would replace a single stress. Climate-smart crops can be created by increasing tolerance to abiotic and biotic stresses and high-yielding biofortified crops (Kaur et al. 2025). Genome engineering has been reported to enable plants to develop resistance to stresses and increase nutrient uptake by improving nutrient utilization efficiency. In short, this technology can help to overcome challenges related to food and nutrition security (Husaini 2022). Development of climate-resilient crops is essential to ensure food security under changing climatic conditions. In some cases, increases in vitamin E contents were evident with other chemicals in plants under stress conditions. For example, Mesa et al. (2024) stated that foliar application of methyl jasmonate at the flowering stage in tomato under the combined heat and salt stress improved vitamin E accumulation although it did not improve fruit production. Similarly, Hasanuzzaman et al. (2023) used ascorbic acid (1 mmol L^{-1}) and α-tocopherol (0.5 mmol L^{-1}) to reduce the negative impact of salt stress on rapeseed plants (*Brassica campestris* cv. BARI Sarisha-17). They stated that applications of α-tocopherol and ascorbic acid enhanced the levels of endogenous ascorbate and glutathione, as well as the antioxidant enzymes, thereby enhancing the antioxidant defense system, as indicated by reduced hydrogen peroxide generation, membrane lipid peroxidation, and electrolyte leakage. To increase the efficacy of tocopherols, ascorbic acid and α-tocopherol combination were also applied to chickpea for sustainable growth under drought stress (El-Beltagi et al. 2022). Similarly, Saeed et al. (2023) reported that salicylic acid and α-tocopherol combination enhanced the salinity tolerance of wheat cultivars by modulating their physiological and biochemical mechanisms when compared to those individual treatments. Abdullah et al. (2022) reported that foliar spraying of α-tocopherol improved linseed (*Linum usitatissimum* L.) growth under salinity stress. Zaman and Ozdemir (2022) reported that thermopriming *Triticum aestivum* L. seeds by soaking them and spraying them with α-tocopherol (150 mmol L^{-1}) at low (4 °C) and high (80 °C) temperatures increased seedling establishment and crop production under drought stress conditions. Vitamin E has been suggested as a supplement for helping quite a few health conditions in humans due to its antioxidant activity and enabling integrity in cell membranes. Kim et al. (2021) reported that low molecular weight antioxidants are not only important for plant protection but also have nutritional values for humans as antiaging and anti-disease molecules. Similar issues were made by Taie and Rady (2024) who stated that reduction in lipid peroxidation and hydrogen peroxide followed by the application 200 ppm of α-tocopherol improved the growth of soybean under saline conditions. Zhang et al. (2020) also reported that the application of α-tocopherol significantly inhibited the synthesis of ethylene and delayed seedling senescence caused by microelement deficiency in *Arabidopsis*.

In some studies, tocopherol accumulation was indirectly induced rather than applying tocopherol directly onto plants. Kusajima et al. (2021) reported that heat shock treatment in *Arabidopsis* activated salicylic acid biosynthesis to enhance resistance to biotic stress. The authors stated that activation of shikimate pathway towards synthesis of salicylic acid upregulated other related metabolic pathways that regulate tocopherol biosynthesis. Tocopherol biosynthesis genes, *HPPD, VTE1*, *VTE2*, *VTE3*, and *VTE4* were upregulated upon heat stress treatment.

Tocopherols were also used for seed priming. Hameed et al. (2021) reported that 100 ppm α-tocopherol resulted in improved plant development, increased levels of osmoprotectants, and fortified cellular defense system in carrot plants exposed to abiotic stress. The authors stated that increased uptake of essential nutrients, upregulation of the defense related-genes, water status, stomatal behavior, and increased efficacy of the photosynthetic apparatus played significant roles in improving carrot plants. Kim et al. (2020) reported that new crop cultivars that can overcome stress are needed for global food security and energy sources. They stated that sweet potato (*Ipomoea batatas* L.), which contains rich low-molecular-weight antioxidants (carotenoids, ascorbate, tocopherols, polyphenols), could be preserved, and stress tolerance was enabled through engineering in biosynthetic and metabolic genes related to low-molecular-weight antioxidants.

Vitamin E can interact with some medications and other supplements. It is mostly used to support skin health, skin ageing, immune function, cardiovascular diseases, Alzheimer's disease, cancer, and other immune-related disorders. Quite a few studies have shown that people who consume foods with more vitamin E, or who take dietary supplements with vitamin E, have a lower incidence of immune-related disorders. However, there are controversial results about the supplementary intakes of vitamin E, possibly due to the placebo effect. But there is a certain finding that a deficiency of vitamin E can cause neurological and immune-related disorders. Péter et al. (2015) reported that the majority of the world population cannot adequately intake vitamin E. This is also valid for newborns.

7.6 Conclusion and Future Prospects

Plants have been exposed to various types of abiotic and biotic stress factors throughout their life cycle. Since they cannot escape from stress like animals and humans, they must combat and modify their defense systems. Biochemical, physiological, and molecular mechanisms have been developed in plants to combat new and oncoming stresses. Under these circumstances, new pathways and chemicals could be found to improve the conditions of plants. The external application of antioxidants may benefit plants by helping them to combat stress factors. This simple approach, which is both culturally and economically viable, may be preferable to complex techniques such as genetic engineering, which have not yielded satisfactory results for decades. Advancements in analytical techniques such as metabolomics and proteomics offer opportunities to comprehensively characterize tocopherol profiles in plants under different stress conditions. These innovative approaches can identify novel tocopherol derivatives and their roles in stress response pathways. This knowledge can be harnessed to develop targeted strategies for enhancing tocopherol biosynthesis and improving crop resilience. The study of tocopherols in plant stress tolerance holds great promise for developing crops that can thrive in adverse environments. Further research is needed to fully elucidate the intricate mechanisms by which tocopherols contribute to stress resilience and to explore novel strategies for enhancing their biosynthesis and utilization in plants (Shahinnia et al. 2021). Biostimulants are now being frequently employed in agricultural practices to improve crop productivity and quality. One of the handicaps in the use of biostimulants is their expense and application to large areas. Therefore, major studies should focus on identifying novel genes involved in tocopherol biosynthesis and new target genes with metabolomic functions. Understanding the signaling pathways will enable the development of tolerant plants.

Investigating the interactions between tocopherol and other antioxidant compounds in plant defense mechanisms should be well evaluated so that we can produce solutions for multiple stress factors. Because abiotic stresses, including salinity, drought, cold, and heat, can devastate plant growth and productivity, with combined stresses having particularly severe effects (Wang et al. 2016).

We could classify our targets into various sections: genetic engineering approaches in which the overexpression of genes involved in tocopherol biosynthesis, genome editing, and trans-genesis (Husaini 2022), a combination of multiple stress-tolerance genes, including those involved in tocopherol biosynthesis, can be suggested (Lata and Shivhare 2021). Agronomic practices can have a major effect on crop yields, and they can be optimized to enhance tocopherol accumulation in plant tissues. Optimizing nutrient availability, particularly microelements such as copper and manganese, which are essential for tocopherol biosynthesis, can enhance tocopherol accumulation. Furthermore, proper irrigation management can alleviate water stress and promote tocopherol production in plants. By carefully managing soil fertility, water availability, and other environmental factors, growers can create conditions that favor tocopherol biosynthesis and enhance plant tolerance to stress. Breeding strategies can be employed to select for high-tocopherol varieties, further enhancing plant tolerance to stress. Traditional breeding approaches, such as marker-assisted selection, can be used to identify and select plants with superior tocopherol biosynthesis capacity. These approaches can be combined with advanced genomic tools to accelerate the breeding process and develop tolerant varieties with enhanced stress tolerance. Genome editing technologies, such as CRISPR-Cas9, provide effective and precise methods of modifying genes involved in stress response pathways and tocopherol biosynthesis.

These technologies enable targeted manipulation of plant genomes, allowing for the creation of stress-tolerant crops with improved tocopherol content and enhanced resilience to environmental challenges (Cappetta et al. 2020; Nascimento et al. 2023). Exploiting microorganisms enhancing tocopherol concentrations could also improve plant resistance and overcome stress via accumulating secondary producing antioxidant enzymes and metabolites (Lan Thi Hoang et al. 2017; Koza et al. 2022).

Further research is required to elucidate the complex regulatory networks that govern the biosynthesis of tocopherols and the stress response pathways in plants. This will allow scientists to have a better understanding of the subject. Furthermore, investigating the interactions between tocopherol and other antioxidant compounds in plant defense mechanisms can provide insights into synergistic strategies for enhancing stress tolerance.

References

Abdel Latef AAH (2021). Organic solutes, oxidative stress, and antioxidant enzymes under abiotic stressors. CRC Press/Taylor & Francis Group. pp. 1–414. https://doi.org/10.1201/9781003022879

Abdel Latef AAH, Hameed M, Khan Z, Khan S, Imran A, Fatima N (2026) Foliar stimulants: A sustainable strategy for enhancing plant resilience to abiotic and biotic stresses. Physiologia Plantarum 178 no. 1:e70741

Abdullah, Mahmood A, Bibi S, Naqve M, Javaid MM, Zia MA et al (2022) Physiological, biochemical, and yield responses of linseed (*Linum usitatissimum* L.) in α-tocopherol-mediated alleviation of salinity stress. Front Plant Sci 13:867172. https://doi.org/10.3389/fpls.2022.867172

Akram NA, Bashir R, Ashraf G, Bashir S, Ashraf M, Alyemeni MN et al (2023) Exogenous α-tocopherol regulates the growth and metabolism of eggplant (*Solanum melongena* L.) under drought stress. Plants 12(2):237. https://doi.org/10.3390/plants12020237

Albert E, Kim S, Magallanes-Lundback M, Bao Y, Deason N, Danilo B et al (2022) Genome-wide association identifies a missing hydrolase for tocopherol synthesis in plants. Proc Natl Acad Sci 119(23):e2113488119. https://doi.org/10.1073/pnas.2113488119

Al-Harthi MM, Bafeel SO, El-Zohri M (2021) Gibberellic acid and jasmonic acid improve salt tolerance in summer squash by modulating some physiological parameters symptomatic for oxidative stress and mineral nutrition. Plants 10(12):2768. https://doi.org/10.3390/plants10122768

Ali Q, Ali S, Iqbal N, Javed MT, Rizwan M, Khaliq R et al (2019) Alpha-tocopherol fertigation confers growth physio-biochemical and qualitative yield enhancement in field grown water deficit wheat (*Triticum aestivum* L.). Sci Rep 9(1):12924. https://doi.org/10.1038/s41598-019-49481-7

Ali E, Hassan Z, Irfan M, Hussain S, Rehman HU, Shah JM et al (2020a) Indigenous tocopherol improves tolerance of oilseed rape to cadmium stress. Front Plant Sci 11:547133. https://doi.org/10.3389/fpls.2020.547133

Ali Q, Tariq Javed M, Haider MZ, Habib N, Rizwan M, Perveen R et al (2020b) α-Tocopherol foliar spray and translocation mediates growth, photosynthetic pigments, nutrient uptake, and oxidative defense in maize (*Zea mays* L.) under drought stress. Agronomy 10(9):1235. https://doi.org/10.3390/agronomy10091235

Ali E, Hussain S, Hussain N, Kakar KU, Shah JM, Zaidi SHR et al (2022) Tocopherol as plant protector: An overview of tocopherol biosynthesis enzymes and their role as antioxidant and signaling molecules. Acta Physiol Plant 44(2):20. https://doi.org/10.1007/s11738-021-03350-x

Almeida J, Asís R, Molineri VN, Sestari I, Lira BS, Carrari F et al (2015) Fruits from ripening impaired, chlorophyll degraded and jasmonate insensitive tomato mutants have altered tocopherol content and composition. Phytochemistry 111:72–83. https://doi.org/10.1016/j.phytochem.2014.11.007

Alsherif EA, Yaghoubi Khanghahi M, Crecchio C, Korany SM, Sobrinho RL, AbdElgawad H (2023) Understanding the active mechanisms of plant (*Sesuvium portulacastrum* L.) against heavy metal toxicity. Plants 12(3):676. https://doi.org/10.3390/plants12030676

Bai F, Gusbeth C, Frey W, Nick P (2020) Nanosecond pulsed electric fields modulate the expression of the astaxanthin biosynthesis genes psy, crtR-b and bkt 1 in *Haematococcus pluvialis*. Sci Rep 10(1):15508. https://doi.org/10.1038/s41598-020-72479-5

Barman M, Das S, Kundu R, Mukherjee S, Singh SK, Abdel Latef AAH (2026) Overview of abiotic stress and its agricultural impact. In: Abdel Latef, A.A.H. (eds) Chitosan Nanoparticles for Plant Resilience Under Abiotic and Biotic Stressors. Smart Nanomaterials Technology. Springer, Singapore. https://doi.org/10.1007/978-981-95-6422-4_2

Baysal İ, Ekizoğlu M, Ertaş A, Temiz B, Ağalar HG, Yabanoğlu-Çiftçi S, Temel H, Uçar G, Türkmenoğlu FP (2021) Identification of phenolic compounds by LC-MS/MS and evaluation of bioactive properties of two edible halophytes: *Limonium effusum* and *L. sinuatum*. Molecules 26(13):4040. https://doi.org/10.3390/molecules26134040

Bhat MY, Gul MZ, Dar JS (2022) Gene expression and role of antioxidant enzymes in crop plants under stress. In: Antioxidant defense in plants: molecular basis of regulation. Springer Nature Singapore, Singapore, pp 31–56. https://doi.org/10.1007/978-981-16-7981-0_3

Bidabadi SS, Sharifi P, Abdel Latef AAH (2021) Antioxidant enzymes and salt stress in plants. In: Abdel Latef AAH (ed.) Organic Solutes, Oxidative Stress, and Antioxidant Enzymes under Abiotic Stressors. CRC Press/Taylor & Francis Group. https://doi.org/10.1201/9781003022879-11

Bidabadi SS, Abdel Latef AAH (2022) Cereals and phytohormones under temperature stress. In: Abdel Latef, A.A.H. (eds) Sustainable Remedies for Abiotic Stress in Cereals. Springer, Singapore. https://doi.org/10.1007/978-981-19-5121-3_14

Cappetta E, Andolfo G, Di Matteo A, Ercolano MR (2020) Empowering crop resilience to environmental multiple stress through the modulation of key response components. J Plant Physiol 246:153134. https://doi.org/10.1016/j.jplph.2020.153134

Cela J, Tweed JK, Sivakumaran A, Lee MR, Mur LA, Munné-Bosch S (2018) An altered tocopherol composition in chloroplasts reduces plant resistance to Botrytis cinerea. Plant Physiol Biochem 127:200–210. https://doi.org/10.1016/j.plaphy.2018.03.033

Cetin ES, Babalik Z, Hallac-Turk F, Gokturk-Baydar N (2014) The effects of cadmium chloride on secondary metabolite production in Vitis vinifera cv. cell suspension cultures. Biol Res 47(1): 47. https://doi.org/10.1186/0717-6287-47-47

Chhabra R, Kaur N, Bala A (2022) Biochemical and anatomical characteristics of basmati and non-basmati rice (Oryza sativa L.) for resistance to foot rot. Bangladesh J Bot 51(1):29–36. https://doi.org/10.3329/bjb.v51i1.58817

Chishti MS, Shahbaz M, Kaleem M, Shafi S, Mehmood A, Qingzhu Z et al (2025) Fertigation with alpha-tocopherol enhances morphological, physiological, and antioxidant responses in radish (Raphanus sativus L.) under drought stress. BMC Plant Biol 25(1):30. https://doi.org/10.1186/s12870-025-06052-5

Dash L, Singh A, Roychoudhury A (2025) Non-enzymatic antioxidant defense systems in plants. In: Role of antioxidants in mitigating plant stress. Academic Press, pp 51–61. https://doi.org/10.1016/B978-0-443-26799-4.00013-X

Dawood et al. (2021) Insights into the enzymatic antioxidants and their genetic expressions responses of plants to heavy metals. In: Abdel Latef AAH (ed.) Organic Solutes, Oxidative Stress, and Antioxidant Enzymes under Abiotic Stressors. CRC Press/Taylor & Francis Group. https://doi.org/10.1201/9781003022879-14

Dawood MFA, Tahjib-Ul-Arif MD, Shirazy BJ, Abdel Latef AAH (2025) Unraveling the role of κ-carrageenan on the combined effect of drought and chromium stress in wheat (Triticum aestivium L.). Plant Physiology and Biochemistry 223:J109781
Delgado A, Al-Hamimi S, Ramadan MF, Wit MD, Durazzo A, Nyam KL, Issaoui M (2020) Contribution of tocols to food sensorial properties, stability, and overall quality. J Food Qual 2020(1):8885865. https://doi.org/10.1155/2020/8885865
Dikilitas M, Simsek E, Karakas S, Abdel Latef AAH (2021) Abiotic stresses and their interactions with each other on plant growth, development and defense mechanisms. In: Abdel Latef AAH (ed.) Organic Solutes, Oxidative Stress, and Antioxidant Enzymes under Abiotic Stressors. CRC Press/Taylor & Francis Group. https://doi.org/10.1201/9781003022879-1
El-Beltagi HS, Mohamed HI (2013) Reactive oxygen species, lipid peroxidation and antioxidative defense mechanism. Notulae Botanicae Horti Agrobotanici Cluj-Napoca 41(1):44–57. https://doi.org/10.15835/nbha4118929
El-Beltagi HS, Shah S, Ullah S, Sulaiman, Mansour AT, Shalaby TA (2022) Impacts of ascorbic acid and alpha-tocopherol on chickpea (*Cicer arietinum* L.) grown in water deficit regimes for sustainable production. Sustainability 14(14):8861. https://doi.org/10.3390/su14148861
Es-Sai B, Wahnou H, Benayad S, Rabbaa S, Laaziouez Y, El Kebbaj R et al (2025) Gamma-tocopherol: a comprehensive review of its antioxidant, anti-inflammatory, and anticancer properties. Molecules 30(3):653. https://doi.org/10.3390/molecules30030653
Faizan M, Alam P, Rajput VD, Shareen KK, Faraz A et al (2023) Potential role of tocopherol in protecting crop plants against abiotic stresses. Physiol Mol Biol Plants 29(10):1563–1575. https://doi.org/10.1007/s12298-023-01354-0
Fanta Yadang SA, Taiwe Sotoing G, Ngatcha Zouakeu KS, Khan MA, Agbor GA, Ur-Rahman N, Ngo Bum E (2019) Quantification of bioactive compounds and evaluation of the antioxidant activity of *Carissa edulis* Valh (Apocynaceae) leaves. Sci World J 2019(1):7549620. https://doi.org/10.1155/2019/7549620
Fritsche S, Wang X, Jung C (2017) Recent advances in our understanding of tocopherol biosynthesis in plants: an overview of key genes, functions, and breeding of vitamin E improved crops. Antioxidants 6(4):99. https://doi.org/10.3390/antiox6040099
Georgiadou EC, Manganaris GA, Fotopoulos V (2020) The diverse roles of vitamin E, its occurrence and regulation in different plant tissues. Food Sci Technol Qual 27(4):113–126. https://hdl.handle.net/20.500.14279/23137
Ghosh UK, Hossain MS, Islam MN, Khan MAR (2022) Role of tocopherol in conferring abiotic stress tolerance in plants. In: Antioxidant defense in plants: molecular basis of regulation. Springer Nature Singapore, Singapore, pp 215–233. https://doi.org/10.1007/978-981-16-7981-0_10
Gruda NS, Dong J, Li X (2024) From salinity to nutrient-rich vegetables: strategies for quality enhancement in protected cultivation. Crit Rev Plant Sci 43(5):327–347. https://doi.org/10.1080/07352689.2024.2351678
Guo Y, Li D, Liu T, Liao M, Li Y, Zhang W et al (2022) Effect of overexpression of γ-tocopherol methyltransferase on α-tocopherol and fatty acid accumulation and tolerance to salt stress during seed germination in *Brassica napus* L. Int J Mol Sci 23(24):15933. https://doi.org/10.3390/ijms232415933
Haghpanah M, Hashemipetroudi S, Arzani A, Araniti F (2024) Drought tolerance in plants: physiological and molecular responses. Plants 13(21):2962. https://doi.org/10.3390/plants13212962
Hameed A, Akram NA, Saleem MH, Ashraf M, Ahmed S, Ali S et al (2021) Seed treatment with α-tocopherol regulates growth and key physio-biochemical attributes in carrot (*Daucus carota* L.) plants under water limited regimes. Agronomy 11(3):469. https://doi.org/10.3390/agronomy11030469
Hasanuzzaman M, Raihan MRH, Alharby HF, Al-Zahrani HS, Alsamadany H, Alghamdi KM et al (2023) Foliar application of ascorbic acid and tocopherol in conferring salt tolerance in rapeseed by enhancing K+/Na+ homeostasis, osmoregulation, antioxidant defense, and glyoxalase system. Agronomy 13(2):361. https://doi.org/10.3390/agronomy13020361

Hassan A, Amjad S, Saleem M, Yasmin H, Imran M, Riaz M et al (2021) Foliar application of ascorbic acid enhances salinity stress tolerance in barley (*Hordeum vulgare* L.) through modulation of morpho-physio-biochemical attributes, ion's uptake, osmo-protectants and stress response genes expression. Saudi J Biol Sci 28(8):4276–4290. https://doi.org/10.1016/j.sjbs.2021.03.045

Hernández-Carranza P, Avila-Sosa R, Vera-López O, Navarro-Cruz AR, Ruíz-Espinosa H, Ruiz-López II, Ochoa-Velasco CE (2023) Uncovering the role of hormones in enhancing antioxidant defense systems in stressed tomato (*Solanum lycopersicum*) plants. Plants 12(20):3648. https://doi.org/10.3390/plants12203648

Hidalgo M, Rodríguez V, Kreindl C, Porras O (2020) Biological redox impact of tocopherol isomers is mediated by fast cytosolic calcium increases in living caco-2 cells. Antioxidants 9(2):155. https://doi.org/10.3390/antiox9020155

Husaini AM (2022) High-value pleiotropic genes for developing multiple stress-tolerant biofortified crops for 21st-century challenges. Heredity 128(6):460–472. https://doi.org/10.1038/s41437-022-00500-w

Imran QM, Falak N, Hussain A, Mun BG, Yun BW (2021) Abiotic stress in plants; stress perception to molecular response and role of biotechnological tools in stress resistance. Agronomy 11(8):1579. https://doi.org/10.3390/agronomy11081579

Jameel A, Akram NA, Bashir R, Ashraf M, Abdel Latef AAH (2026) Salicylic acid enhances yield and alleviates drought-induced oxidative damage in two carrot cultivars with contrasting tolerance. Russian Journal of Plant Physiology 73:23

Jin S, Daniell H (2014) Expression of γ-tocopherol methyltransferase in chloroplasts results in massive proliferation of the inner envelope membrane and decreases susceptibility to salt and metal-induced oxidative stresses by reducing reactive oxygen species. Plant Biotechnol J 12(9): 1274–1285. https://doi.org/10.1111/pbi.12224

Kalinova J, Triska J, Vrchotova N (2006) Distribution of vitamin E, squalene, epicatechin, and rutin in common buckwheat plants (*Fagopyrum esculentum* Moench). J Agric Food Chem 54(15): 5330–5335. https://doi.org/10.1021/jf060521r

Kaur N, Qadir M, Francis DV, Alok A, Tiwari S, Ahmed ZF (2025) CRISPR/Cas9: a sustainable technology to enhance climate resilience in major staple crops. Front Genome Editing 7: 1533197. https://doi.org/10.3389/fgeed.2025.1533197

Khan AL (2023) The phytomicrobiome: solving plant stress tolerance under climate change. Front Plant Sci 14:1219366. https://doi.org/10.3389/fpls.2023.1219366

Khan A, Kanwal F, Ullah S, Fahad M, Tariq L, Altaf MT et al (2025) Plant secondary metabolites—central regulators against abiotic and biotic stresses. Meta 15(4):276. https://doi.org/10.3390/metabo15040276

Kim HS, Wang W, Kang L, Kim SE, Lee CJ, Park SC et al (2020) Metabolic engineering of low-molecular-weight antioxidants in sweetpotato. Plant Biotechnol Rep 14(2):193–205. https://doi.org/10.1007/s11816-020-00621-w

Kim H, Uzilday RÖ, Uzilday B, Kim S, Lee C, Park S et al (2021) Manipulation of lipophilic antioxidants to enhance oxidative stress tolerance andnutritional quality in transgenic sweetpotato. Turk J Bot 45(7):613–622. https://doi.org/10.3906/bot-2109-50

Koza NA, Adedayo AA, Babalola OO, Kappo AP (2022) Microorganisms in plant growth and development: roles in abiotic stress tolerance and secondary metabolites secretion. Microorganisms 10(8):1528. https://doi.org/10.3390/microorganisms10081528

Kumar S, Mohapatra T (2024) Epigenetic modifications in genome help remembering the stress tolerance strategy adopted by the plant. Front Biosci Landmark 29(3):126. https://doi.org/10.31083/j.fbl2903126

Kusajima M, Fujita M, Nishiuchi T, Nakashita H, Asami T (2021) Induction of tocopherol biosynthesis through heat shock treatment in Arabidopsis. Biosci Biotechnol Biochem 85(3): 502–509. https://doi.org/10.1093/bbb/zbaa053

Lan Thi Hoang X, Ngoc Hai Nhi D, Binh Anh Thu N, Phuong Thao N, Phan Tran LS (2017) Transcription factors and their roles in signal transduction in plants under abiotic stresses. Curr Genomics 18(6):483–497. https://doi.org/10.2174/1389202918666170227150057

Lata C, Shivhare R (2021) Engineering cereal crops for enhanced abiotic stress tolerance. Proc Indian Natl Sci Acad 87(1):63–83. https://doi.org/10.1007/s43538-021-00006-9

Lewis ED, Meydani SN, Wu D (2019) Regulatory role of vitamin E in the immune system and inflammation. IUBMB Life 71(4):487–494. https://doi.org/10.1002/iub.1976

Liu EH, Liao WZ, Chen HK, Huang XY, Li RX, Liang HW, Guo XG (2024) Association between serum vitamin E and bacterial vaginitis in women: a cross-sectional study. BMC Womens Health 24(1):316. https://doi.org/10.1186/s12905-024-03065-4

Loyola J, Verdugo I, González E, Casaretto JA, Ruiz-Lara S (2012) Plastidic isoprenoid biosynthesis in tomato: physiological and molecular analysis in genotypes resistant and sensitive to drought stress. Plant Biol 14(1):149–156. https://doi.org/10.1111/j.1438-8677.2011.00465.x

Ma J, Qiu D, Gao H, Wen H, Wu Y, Pang Y et al (2020) Over-expression of a γ-tocopherol methyltransferase gene in vitamin E pathway confers PEG-simulated drought tolerance in alfalfa. BMC Plant Biol 20(1):226. https://doi.org/10.1186/s12870-020-02424-1

Maeda H, Song W, Sage TL, DellaPenna D (2006) Tocopherols play a crucial role in low-temperature adaptation and phloem loading in Arabidopsis. Plant Cell 18(10):2710–2732. https://doi.org/10.1105/tpc.105.039404

Meena VK, Chand S, Shekhawat HVS, Choudhary K, Sharma JK, Sheera A et al (2025) Advances in plant tocopherol biosynthesis: from pathway elucidation to crop biofortification strategies. Discover Plants 2(1):9. https://doi.org/10.1007/s44372-025-00093-8

Mekki BED, Hussien HA, Salem H (2015) Role of glutathione, ascorbic acid and α-tocopherol in alleviation of drought stress in cotton plants. Int J ChemTech Res 8(4):1573–1581

Mesa T, Romero A, Munné-Bosch S (2024) Differential response of roots and leaves to combined heat and salinity stresses in tomato plants. Environ Exp Bot 226:105890. https://doi.org/10.1016/j.envexpbot.2024.105890

Mishra N, Jiang C, Chen L, Paul A, Chatterjee A, Shen G (2023) Achieving abiotic stress tolerance in plants through antioxidative defense mechanisms. Front Plant Sci 14:1110622. https://doi.org/10.3389/fpls.2023.1110622

Montagnani Marelli M, Marzagalli M, Fontana F, Raimondi M, Moretti RM, Limonta P (2019) Anticancer properties of tocotrienols: a review of cellular mechanisms and molecular targets. J Cell Physiol 234(2):1147–1164. https://doi.org/10.1002/jcp.27075

Munné-Bosch S, Alegre L (2002) The function of tocopherols and tocotrienols in plants. Crit Rev Plant Sci 21(1):31–57. https://doi.org/10.1080/0735-260291044179

Muñoz P, Munne-Bosch S (2019) Vitamin E in plants: biosynthesis, transport, and function. Trends Plant Sci 24(11):1040–1051. https://doi.org/10.1016/j.tplants.2019.08.006

Munoz P, Munne-Bosh S (2019) Vitamin E in plants: biosynthesis, transport and function. Trends Plant Sci 24(11):1040–1051. https://doi.org/10.1016/j.tplants.2019.08.006

Naqve M, Wang X, Shahbaz M, Mahmood A, Bibi S, Fiaz S (2021) Alpha tocopherol-induced modulations in the morphophysiological attributes of okra under saline conditions. Front Plant Sci 12:800251. https://doi.org/10.3389/fpls.2021.800251

Nascimento FDS, Rocha ADJ, Soares JMDS, Mascarenhas MS, Ferreira MDS, Morais Lino LS et al (2023) Gene editing for plant resistance to abiotic factors: a systematic review. Plants 12(2):305. https://doi.org/10.3390/plants12020305

Nazari M, Ghasemi-Soloklui AA, Kordrostami M, Abdel Latef AAH (2023) Deciphering the response of medicinal plants to abiotic stressors: A focus on drought and salinity. Plant Stress 10:100255

Parvin K, Hasanuzzaman M, Bhuyan MB, Nahar K, Mohsin SM, Fujita M (2019) Comparative physiological and biochemical changes in tomato (*Solanum lycopersicum* L.) under salt stress and recovery: role of antioxidant defense and glyoxalase systems. Antioxidants 8(9):350. https://doi.org/10.3390/antiox8090350

Pereira AG, Echave J, Jorge AO, Nogueira-Marques R, Nur Yuksek E, Barciela P et al (2025) Therapeutic and preventive potential of plant-derived antioxidant nutraceuticals. Foods 14(10):1749. https://doi.org/10.3390/foods14101749

Pérez-Llorca M, Casadesús A, Munné-Bosch S, Müller M (2019) Contrasting patterns of hormonal and photoprotective isoprenoids in response to stress in Cistus albidus during a Mediterranean winter. Planta 250(5):1409–1422. https://doi.org/10.1007/s00425-019-03234-y

Péter S, Friedel A, Roos FF, Wyss A, Eggersdorfer M, Hoffmann K, Weber P (2015) A systematic review of global alpha-tocopherol status as assessed by nutritional intake levels and blood serum concentrations. Int J Vitam Nutr Res 85(5–6):261–281. https://doi.org/10.1024/0300-9831/a000281

Poudel P, Petropoulos SA, Di Gioia F (2023) Plant tocopherols and phytosterols and their bioactive properties. In: Natural secondary metabolites: from nature, through science, to industry. Springer International Publishing, Cham, pp 285–319. https://doi.org/10.1007/978-3-031-18587-8_8

Rady MM, Abdelfattah MA, Desoky EM, Azzam CR, Abdel Latef AAH, Mahmoud AEM, Mohamed IAA (2025) A novel group of bio-stimulators: their attenuation of abiotic stress impacts in crop plants. Journal of Plant Nutrition 1–16

Rajput VD, Harish, Singh RK, Verma KK, Sharma L, Quiroz-Figueroa FR et al (2021) Recent developments in enzymatic antioxidant defence mechanism in plants with special reference to abiotic stress. Biology 10(4):267. https://doi.org/10.3390/biology10040267

Raza A, Tabassum J, Fakhar AZ, Sharif R, Chen H, Zhang C et al (2023a) Smart reprograming of plants against salinity stress using modern biotechnological tools. Crit Rev Biotechnol 43(7): 1035–1062. https://doi.org/10.1080/07388551.2022.2093695

Raza S, Zia-ur-Rehman M, Alghamdi SA, Alghanem SMS, Usman M, Ahmed R et al (2023b) Effects of zinc-enriched amino acids on rice plants (*Oryza sativa* L.) for adaptation in saline-sodic soil conditions: growth, nutrient uptake and biofortification of zinc. S Afr J Bot 162:370–380. https://doi.org/10.1016/j.sajb.2023.09.011

Rinnerthaler M, Bischof J, Streubel MK, Trost A, Richter K (2015) Oxidative stress in aging human skin. Biomolecules 5(2):545–589. https://doi.org/10.3390/biom5020545

Saeed S, Ullah A, Ullah S, Elshikh MS, Noor J, Eldin SM et al (2023) Salicylic acid and α-tocopherol ameliorate salinity impact on wheat. ACS Omega 8(29):26122–26135. https://doi.org/10.1021/acsomega.3c02166

Saeed S, Ullah S, Amin F, Al-Hawadi JS, Okla MK, Alaraidh IA et al (2024) Salicylic acid and tocopherol improve wheat (*Triticum aestivum* L.) physio-biochemical and agronomic features grown in deep sowing stress: a way forward towards sustainable production. BMC Plant Biol 24(1):477. https://doi.org/10.1186/s12870-024-05180-8

Salam U, Ullah S, Tang ZH, Elateeq AA, Khan Y, Khan J et al (2023) Plant metabolomics: an overview of the role of primary and secondary metabolites against different environmental stress factors. Life 13(3):706. https://doi.org/10.3390/life13030706

Sarafraz-Ardakani MR, Khavari-Nejad RA, Moradi F, Najafi F (2014) Abscisic acid and cytokinin-induced osmotic and antioxidant regulation in two drought-tolerant and drought-sensitive cultivars of wheat during grain filling under water deficit in field conditions. Notulae Scientia Biologicae 6(3):354–362. https://doi.org/10.15835/nsb639301

Savignac JM, Atanasova V, Chéreau S, Ortéga V, Richard-Forget F (2022) Role of tocochromanols in tolerance of cereals to biotic stresses: specific focus on pathogenic and toxigenic fungal species. Int J Mol Sci 23(16):9303. https://doi.org/10.3390/ijms23169303

Shah W, Ullah S, Ali S, Idrees M, Khan MN, Ali K et al (2021) Effect of exogenous alpha-tocopherol on physio-biochemical attributes and agronomic performance of lentil (*Lens culinaris* Medik.) under drought stress. PLoS One 16(8):e0248200. https://doi.org/10.1371/journal.pone.0248200

Shahinnia F, Carrillo N, Hajirezaei MR (2021) Engineering climate-change-resilient crops: new tools and approaches. Int J Mol Sci 22(15):7877. https://doi.org/10.3390/ijms22157877

Sharma A, Shahzad B, Rehman A, Bhardwaj R, Landi M, Zheng B (2019) Response of phenylpropanoid pathway and the role of polyphenols in plants under abiotic stress. Molecules 24(13):2452. https://doi.org/10.3390/molecules24132452

Shikari AB et al (2022) Cereals under abiotic stress: An Overview. In: Abdel Latef, A.A.H. (eds) Sustainable Remedies for Abiotic Stress in Cereals. Springer, Singapore. https://doi.org/10.1007/978-981-19-5121-3_1

Spicher L, Almeida J, Gutbrod K, Pipitone R, Dörmann P, Glauser G et al (2017) Essential role for phytol kinase and tocopherol in tolerance to combined light and temperature stress in tomato. J Exp Bot 68(21–22):5845–5856. https://doi.org/10.1093/jxb/erx356

Stahl E, Hartmann M, Scholten N, Zeier J (2019) A role for tocopherol biosynthesis in Arabidopsis basal immunity to bacterial infection. Plant Physiol 181(3):1008–1028. https://doi.org/10.1104/pp.19.00618

Styrczewska M, Zuk M, Boba A, Zalewski I, Kulma A (2019) Use of natural components derived from oil seed plants for treatment of inflammatory skin diseases. Curr Pharm Des 25(20):2241–2263. https://doi.org/10.2174/1381612825666190716111700

Taie HA, Rady MM (2024) α-Tocopherol mediates alleviation of salt stress effects in Glycine max through up-regulation of the antioxidant defense system and secondary metabolites. Acta Physiol Plant 46(4):39. https://doi.org/10.1007/s11738-024-03654-8

Tasaka T, Matsumoto T, Nagashima U, Nagaoka SI (2023) Potential energy curve for singlet-oxygen quenching reaction by vitamin E. J Photochem Photobiol A Chem 442:114749. https://doi.org/10.1016/j.jphotochem.2023.114749

Traber MG (2024) Deciphering the enigma of the function of alpha-tocopherol as a vitamin. Free Radic Biol Med 221:64–74. https://doi.org/10.1016/j.freeradbiomed.2024.05.028

Velasco L, García-Navarro E, Pérez-Vich B, Fernández-Martínez JM (2013) Selection for contrasting tocopherol content and profile in E thiopian mustard. Plant Breed 132(6):694–700. https://doi.org/10.1111/pbr.12084

Verma H, Kumar D, Kumar V, Kumari M, Singh SK, Sharma VK et al (2021) The potential application of endophytes in management of stress from drought and salinity in crop plants. Microorganisms 9(8):1729. https://doi.org/10.3390/microorganisms9081729

Vicidomini C, Palumbo R, Moccia M, Roviello GN (2024) Oxidative processes and xenobiotic metabolism in plants: mechanisms of defense and potential therapeutic implications. J Xenobiotics 14(4):1541–1569. https://doi.org/10.3390/jox14040084

Villalobos-López MA, Arroyo-Becerra A, Quintero-Jiménez A, Iturriaga G (2022) Biotechnological advances to improve abiotic stress tolerance in crops. Int J Mol Sci 23(19):12053. https://doi.org/10.3390/ijms231912053

Vinocur B, Altman A (2005) Recent advances in engineering plant tolerance to abiotic stress: achievements and limitations. Curr Opin Biotechnol 16(2):123–132. https://doi.org/10.1016/j.copbio.2005.02.001

Vinutha T, Bansal N, Kumari K, Prashat GR, Sreevathsa R, Krishnan V et al (2017) Comparative analysis of tocopherol biosynthesis genes and its transcriptional regulation in soybean seeds. J Agric Food Chem 65(50):11054–11064. https://doi.org/10.1021/acs.jafc.7b03448

Voll LM, Abbasi AR (2007) Are there specific in vivo roles for α-and γ-tocopherol in plants? Plant Signal Behav 2(6):486–488. https://doi.org/10.4161/psb.2.6.4499

Wang H, Wang H, Shao H, Tang X (2016) Recent advances in utilizing transcription factors to improve plant abiotic stress tolerance by transgenic technology. Front Plant Sci 7:67. https://doi.org/10.3389/fpls.2016.00067

Waśkiewicz A, Muzolf-Panek M, Stępień Ł, Czembor E, Uwineza PA, Górnaś P, Bryła M (2022) Variation in tocochromanols level and mycotoxins content in sweet maize cultivars after inoculation with *Fusarium verticillioides* and *F.* proliferatum. Foods 11(18):2781. https://doi.org/10.3390/foods11182781

Yao W, Zhao K, Cheng Z, Li X, Zhou B, Jiang T (2018) Transcriptome analysis of poplar under salt stress and over-expression of transcription factor NAC57 gene confers salt tolerance in transgenic Arabidopsis. Front Plant Sci 9:1121. https://doi.org/10.3389/fpls.2018.01121

Yeshi K, Crayn D, Ritmejerytė E, Wangchuk P (2022) Plant secondary metabolites produced in response to abiotic stresses has potential application in pharmaceutical product development. Molecules 27(1):313. https://doi.org/10.3390/molecules27010313

Yu Y, Gui Y, Li Z, Jiang C, Guo J, Niu D (2022) Induced systemic resistance for improving plant immunity by beneficial microbes. Plants 11(3):386. https://doi.org/10.3390/plants11030386

Zahra N, Hafeez MB, Ghaffar A, Kausar A, Al Zeidi M, Siddique KH, Farooq M (2023) Plant photosynthesis under heat stress: effects and management. Environ Exp Bot 206:105178. https://doi.org/10.1016/j.envexpbot.2022.105178

Zaman SAR, Ozdemir FA (2022) Effect of Thermopriming and alpha-tocopherol spray in *Triticum aestivum* L. under induced drought stress: a future perspective of climate change in the region. Sains Malaysiana 51(4):977–991. https://doi.org/10.17576/jsm-2022-5104-03

Zha L, Liu W, Zhang Y, Zhou C, Shao M (2019) Morphological and physiological stress responses of lettuce to different intensities of continuous light. Front Plant Sci 10:1440. https://doi.org/10.3389/fpls.2019.01440

Zhang ZW, Yang XY, Zheng XJ, Fu YF, Lan T, Tang XY et al (2020) Vitamin E is superior to vitamin C in delaying seedling senescence and improving resistance in Arabidopsis deficient in macro-elements. Int J Mol Sci 21(19):7429. https://doi.org/10.3390/ijms21197429

Zhang Y, Xu J, Li R, Ge Y, Li Y, Li R (2023) Plants' response to abiotic stress: mechanisms and strategies. Int J Mol Sci 24(13):10915. https://doi.org/10.3390/ijms241310915

Zhao Z, Jia M, An Y, Bao Y, Zhao Y (2025) The protective effect of *Boschnikia rossica* extract on free radical-induced oxidative damage of biomolecules and establishment of a method for determining the content of Oleanolic acid. Foods 14(10):1658. https://doi.org/10.3390/foods14101658

Chapter 8
Acetic Acid: Modulating Stress Responses in Plants

Iman Janah, Aouabe Abdelhamid, Redouane Ouhaddou, Raja Ben-Laouane, Mohamed Ait-El-Mokhtar, Abdelilah Meddich, and Marouane Baslam

Abstract In the face of escalating environmental challenges, both abiotic (e.g., drought, salinity, extreme temperatures, heavy metal toxicity) and biotic (e.g., pathogenic bacteria, fungi, viruses, insect herbivores) stressors significantly impede global agricultural productivity. These stressors disrupt plant metabolic pathways, leading to oxidative damage, compromised cellular functions, and diminished growth and yield. Acetic acid (AA), a simple carboxylic acid, has emerged as a pivotal metabolite in enhancing plant resilience against such adversities. This

I. Janah · A. Abdelhamid · R. Ouhaddou
Centre d'Agrobiotechnologie et Bioingénierie, Unité de Recherche Labellisée CNRST (Centre AgroBiotech-URL-CNRST-05), Cadi Ayyad University, Marrakesh, Morocco

R. Ben-Laouane
Centre d'Agrobiotechnologie et Bioingénierie, Unité de Recherche Labellisée CNRST (Centre AgroBiotech-URL-CNRST-05), Cadi Ayyad University, Marrakesh, Morocco

FSTE-FSM Joint Laboratory: NRHE-UMI, Bioresources, Environment and Health Research Team, Moulay Ismail University, Meknes, Morocco

M. Ait-El-Mokhtar
Centre d'Agrobiotechnologie et Bioingénierie, Unité de Recherche Labellisée CNRST (Centre AgroBiotech-URL-CNRST-05), Cadi Ayyad University, Marrakesh, Morocco

Laboratory of Biotechnology, Agri-food, Materials, and Environment (LBAME), Department of Biology, Faculty of Science and Techniques Mohammedia, Hassan II University of Casablanca, Mohammedia, Morocco

A. Meddich
Centre d'Agrobiotechnologie et Bioingénierie, Unité de Recherche Labellisée CNRST (Centre AgroBiotech-URL-CNRST-05), Cadi Ayyad University, Marrakesh, Morocco

African Sustainable Agriculture Research Institute (ASARI), University Mohamed VI Polytechnic (UM6P), Laayoune, Morocco

M. Baslam (✉)
Centre d'Agrobiotechnologie et Bioingénierie, Unité de Recherche Labellisée CNRST (Centre AgroBiotech-URL-CNRST-05), Cadi Ayyad University, Marrakesh, Morocco

GrowSmart, Amsterdam, Netherlands

A. A. H. Abdel Latef (ed.), *Harnessing Vitamins and Organic Acids for Plant Stress Resilience*, Sustainability Sciences in Asia and Africa,
https://doi.org/10.1007/978-981-95-8593-9_8

chapter delves into the multifaceted roles of AA in modulating plant responses to both abiotic and biotic stresses, emphasizing its involvement in osmoregulation, antioxidant defense activation, gene expression modulation, and interactions with phytohormones and microbial communities. We explore the biosynthesis and metabolic pathways of AA, its influence on stress-responsive signaling networks, and its practical applications in agriculture. By elucidating the mechanisms through which AA confers stress tolerance, this chapter aims to provide insights into sustainable strategies for improving crop resilience and productivity amidst environmental challenges.

Keywords Acetic acid · Plant stress responses · A(biotic) stress · Antioxidant defense · Gene expression regulation · Phytohormone interaction · Plant-microbe interaction

8.1 Introduction

As immobile organisms, plants are constantly exposed to multiple environmental constraints that negatively influence their growth, development, and reproductive success (El-Sappah and Rather, 2021; Ihtisham et al., 2023; Janah et al., 2025). These stressors are broadly classified into two categories: abiotic and biotic stresses, both of which can significantly alter plant physiology, reduce crop yields, and compromise food security, particularly in the context of global climate change and increasing biotic pressures (El-Sappah and Rather, 2021; Adedayo et al., 2025; Anli et al. 2026). To withstand these adverse conditions, plants have evolved sophisticated defense systems that rely on complex signaling networks and metabolic reprogramming.

Central to these adaptive responses are diverse metabolic regulators, small molecules that act as mediators between the perception of stress signals and the activation of downstream defense mechanisms. These regulators modulate gene expression, adjust cellular redox status, maintain osmotic balance, and coordinate hormonal crosstalk, thereby enabling plants to mount efficient and timely responses to environmental challenges (Fichman and Mittler, 2020; Theerawitaya et al., 2020; Nawaz et al., 2021; Sinha et al., 2021; Zandalinas et al., 2021). Metabolic regulators thus play a pivotal role in sensing environmental cues and modulating downstream pathways to maintain cellular homeostasis (Panchal et al., 2021). These small molecules, including organic acids, often function as signaling intermediates or metabolic switches that activate protective mechanisms (Panchal et al., 2021). They reprogram gene expressions, stabilize cellular structures, detoxify reactive oxygen species (ROS), and facilitate the accumulation of osmoprotectants. By fine-tuning energy allocation and stress-responsive pathways, metabolic regulators enhance plant tolerance to a wide range of abiotic and biotic stressors (Panchal et al., 2021).

Among organic acids, acetic acid (AA) has recently attracted considerable attention for its versatile role in enhancing plant tolerance to environmental stresses. Once regarded merely as a short-chain carboxylic acid involved in primary metabolism, AA is now recognized for its multifunctional role in plant stress responses (Kim et al., 2017; El-Rahman et al., 2019; Utsumi et al., 2019; Hossain et al., 2020, 2022). It has been shown to improve resilience to a range of abiotic challenges as well as biotic pressures. For instance, exogenous AA reduced cadmium-induced damage in sunflowers by enhancing antioxidant defenses (Hawrylak-Nowak et al., 2015), and at higher concentrations, it has been applied as a natural weed suppressant in potato cultivation (Ivany, 2010).

In addition to its role as a metabolic intermediate, AA functions as a signaling molecule capable of initiating cascades of defense responses. These include modulation of stomatal conductance (Dong et al., 2018), stimulation of antioxidant enzyme activities (Kappachery et al., 2024), and upregulation of stress-responsive genes (Li et al., 2021b). In *Arabidopsis*, AA has been shown to promote de novo biosynthesis of jasmonic acid (JA) and induce histone H4 acetylation, thereby contributing to drought tolerance through epigenetic regulation (Kim et al., 2017). Considering the central role of JA in defense against insect herbivores (Howe et al., 2018), AA is increasingly regarded as a promising priming agent for integrated stress management.

Moreover, AA's influence on phytohormonal dynamics is highly context-dependent, varying across plant species and environmental conditions. Under salinity stress, for example, AA elevated salicylic acid (SA) levels in grapevine, while in wheat it increased indole-3-acetic acid (IAA) and gibberellic acid (GA) but lowered abscisic acid (ABA) concentrations (Lv et al., 2021; Khan et al., 2023). These findings highlight the complexity of AA-mediated hormonal signaling and emphasize the need to account for species-specific and environmental interactions when evaluating its effects. Preliminary evidence also suggests that AA may influence rhizosphere composition and soil microbial communities (Kong et al., 2022), although this area remains underexplored and requires further investigation across diverse soil types and stress scenarios.

This chapter aims to provide a comprehensive overview of the role of acetic acid in plant stress physiology. We begin by examining its biosynthesis and metabolic integration within plant systems, followed by an in-depth discussion of its roles in abiotic and biotic stress responses. Furthermore, we analyze the molecular mechanisms underlying AA-induced tolerance, including gene regulation, signaling networks, and crosstalk with other metabolic and hormonal pathways. Finally, we evaluate the potential of acetic acid-based applications in agriculture, particularly in the context of sustainable crop management under adverse environmental conditions. By consolidating current knowledge and highlighting emerging research, this chapter seeks to establish acetic acid as a promising agent for improving plant performance.

8.2 Biosynthesis and Metabolic Pathways of Acetic Acid in Plants

Acetic acid, a small carboxylic acid molecule, is endogenously synthesized in plant tissues through distinct metabolic routes (Fig. 8.1). Its biosynthesis typically begins with the enzymatic decarboxylation of pyruvate, producing acetaldehyde. This intermediate is subsequently oxidized to acetate by aldehyde dehydrogenase (ALDH) (Rasheed et al., 2018). The resulting acetate is then converted into acetyl-CoA via acetyl-CoA synthetase, thereby establishing a direct link between acetate metabolism and core cellular functions (Rasheed et al., 2018).

Once formed, acetyl-CoA functions as a central metabolic hub. It enters the citric acid cycle to support respiration and ATP production, while also serving as a precursor for the biosynthesis of essential biomolecules such as fatty acids, amino acids, and secondary metabolites that underpin plant growth and stress tolerance (Tadege et al., 1999). Additionally, carbon derived from acetate can be redirected into carbohydrate biosynthesis through the glyoxylate cycle, particularly under stress or carbon-limited conditions, highlighting the metabolic flexibility of this pathway in meeting cellular demands (Tadege et al., 1999).

In addition to its metabolic contributions, acetyl-CoA also plays a key role in epigenetic regulation by donating acetyl groups for histone acetylation. This process

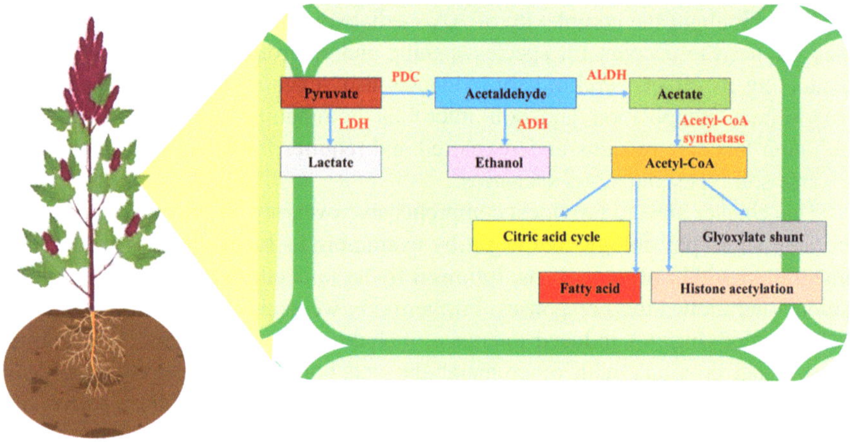

Fig. 8.1 Metabolic pathways of acetic acid in plants. Under anaerobic conditions, pyruvate derived from glycolysis can be converted to lactate by lactate dehydrogenase (LDH). Alternatively, it enters acetate biosynthesis via a two-step enzymatic process: pyruvate decarboxylase (PDC) converts pyruvate into acetaldehyde, which is subsequently oxidized to acetate by aldehyde dehydrogenase (ALDH). Acetaldehyde may also be reduced to ethanol through the action of alcohol dehydrogenase (ADH). The resulting acetate is converted to acetyl-CoA by acetyl-CoA synthetase. Acetyl-CoA then functions as a central metabolic hub—fueling the citric acid cycle, supporting fatty acid biosynthesis, contributing to carbohydrate production via the glyoxylate cycle, and providing acetyl groups for histone acetylation that regulate gene expression

influences chromatin architecture and gene expression, particularly under environmental stress, thereby linking acetate metabolism to plant adaptive responses (Kim et al., 2017).

8.3 Acetic Acid in Modulating Plant Responses to Abiotic Stressors

8.3.1 Drought Stress

Acetic acid plays a central role in plant metabolism as a precursor of acetyl-CoA, contributing to the citric acid cycle, fatty acid biosynthesis, and the production of secondary metabolites. Beyond its metabolic functions, AA contributes to stress responses by supporting key physiological and biochemical processes involved in adaptation and tolerance, including growth, root development, germination, photosynthesis, antioxidant defense, and osmoregulation (Ogawa et al., 2021; Kudo et al., 2023). The exogenous application of AA has emerged as a promising strategy for enhancing drought tolerance in a wide range of plant species, including both herbaceous and woody plants (Table 8.1). Recent studies have shown that AA plays a multifunctional role in mitigating drought stress by modulating physiological processes, activating hormonal signaling pathways, and priming plants for enhanced resilience under water-deficient conditions.

Allen and Allen (2021) reported that low concentrations of AA (10–40 mM) significantly delayed drought-induced mortality in ornamental Begonia × hybrida, while reducing water consumption without compromising flowering. This suggests that AA can function as an effective antitranspirant agent in horticultural production. In cotton, Li et al. (2021b) demonstrated that AA treatment reduced stomatal conductance and transpiration, ultimately maintaining higher soil water content and enhancing plant survival under drought conditions. Similarly, Sun et al. (2022) revealed through transcriptomic profiling of AA-treated apples under drought stress that AA activated hormone signal transduction pathways, including ABA- and JA-mediated mitogen-activated protein kinase (MAPK) cascades. This activation facilitated both metabolic adjustment and transcriptional reprogramming to alleviate drought-induced growth inhibition. Taken together, these findings suggest that AA induces a system-wide priming effect, enabling plants to respond more rapidly and effectively when drought occurs.

In tropical crops such as cassava, AA has also been shown to promote drought avoidance. Utsumi et al. (2019) found that AA-treated cassava plants maintained higher relative leaf water content, chlorophyll, and carotenoid levels during progressive drought, while exhibiting elevated leaf temperatures—an indicator of reduced transpiration due to stomatal closure. These physiological adjustments allowed cassava to sustain water status and delay the onset of severe stress without compromising growth. In rice, Ogawa et al. (2021) provided further mechanistic

Table 8.1 Summary of acetic acid-mediated responses to various abiotic stressors

Plant species	Application method	Stress	Key outcomes/benefits	Reference
Begonia	Foliar application (10–40 mM)	Drought	Delayed mortality, reduced water use, and maintained flowering	Allen and Allen (2021)
Cotton	Soil application (2, 4, 6, and 8 mM)	Drought	Increased ABA, JA, and gene upregulation, reduced transpiration	Li et al. (2021b)
Apple	Soil application (10 mM)	Drought	Activated hormone signaling (ABA, JA), improved metabolic response	Sun et al. (2022)
Cassava	Soil treatment (1 and 10 mM)	Drought	Maintained water content, increased ABA, induced stress genes	Utsumi et al. (2019)
Rice	Root application (30 mM)	Drought	Activated JA signaling in roots, reduced transpiration	Ogawa et al. (2021)
Soybean	Foliar spray (20 mM)	Drought	Improved photosynthesis, water efficiency, mineral balance	Rahman et al. (2021)
Maize	Seed priming (500 μM) Root treatment (25 mM)	Drought	Improved growth and proline biosynthesis	Mahmud et al. (2023)
Blueberry	Soil treatment (100 mM)	Drought	Enhanced rhizosphere microbiome, improved nitrogen uptake	Kong et al. (2022)
Mung bean	Foliar application (20 mM)	Salinity	Improved mineral uptake, antioxidant defense, and osmoprotection	Rahman et al. (2019)
Ryegrass	Pre-treatment of soil (20 mM)	Salinity	Improved K^+/Na^+ ratio, ROS detoxification	Zhang et al. (2021)
Wheat	Foliar spray (10 and 20 mM)	Salinity	Enhanced antioxidant enzymes and ionic balance	Khan et al. (2023)
Canola	Pre-treatment of seeds (16 mM)	Salinity + lithium	Improved growth, and detoxification via phytochelatins	Dawood et al. (2023)
Stiff sedge	Pretreatment to soil (30 mM)	Salinity	Improved membrane stability and lipid metabolism regulation	Hu et al. (2021)
Lentil	Sodium acetate (5–10 mM)	Salinity	Improved ion balance, and chlorophyll, and antioxidant defense	Hossain et al. (2019)
Strawberry	Foliar application (1 and 2 mM)	Salinity	Improved biomass, antioxidant enzymes, and hormone modulation	Mirfattahi and Eshghi (2023)
Mustard	Seed priming (250 μM)	Cadmium	Improved growth and antioxidant defense	Chowardhara et al. (2022)

(continued)

Table 8.1 (continued)

Plant species	Application method	Stress	Key outcomes/benefits	Reference
Sunflower (hydroponic)	AA was add to the nutrient solution (250–500 μM)	Cadmium	Reduced Cd toxicity, improved pigments, and root activity	Hawrylak-Nowak et al. (2015)
Sunflower (soil)	Soil application (1, 2, 3, 4, 5, 6 mmol/kg)	Cadmium	Improved phytoremediation, biomass, and enzyme activity	Wang et al. (2022b)
Tomato	Soil (100 mg/kg)	Cadmium	Reduced Cd uptake via JA signaling Improved Fe availability	Zhu et al. (2021)
Pteris vittata	Soil application (10 and 20 mM)	Drought + arsenic	Improved growth, As uptake, and soil pH	Wang et al. (2023)
Blueberry	Irrigating water (100 mM)	High temperature	Improved carbon metabolism and sex-specific microbial shifts	He et al. (2024)
Mexican lime	15 mM AA	Cold stress	Improved photosynthesis, antioxidant capacity, and pigment content	Sanie Khatam et al. (2023)

ABA abscisic acid, *JA* jasmonic acid, *ROS* reactive oxygen species, Na^+ sodium ion, K^+ potassium ion, Cl^- chloride ion, *Cd* cadmium, *As* arsenic, *AA* acetic acid, *Fe* iron, *μM* micromolar, *mM* millimolar

insight by demonstrating that AA activates JA signaling specifically in roots even before drought occurs. This root-localized activation induces drought-like responses such as reduced xylem sap flow and lower transpiration, thereby preconditioning shoots for subsequent stress. Interestingly, AA itself was metabolized into glutamine in the xylem sap, suggesting that it may not act directly as the long-distance signal but instead initiates cascades that lead to systemic JA signaling. This AA-induced root-to-shoot communication confers improved survival and drought readiness in rice, even in the absence of immediate water deficit (Ogawa et al., 2021).

In soybeans, foliar application of AA improved drought tolerance by enhancing root biomass, leaf area, photosynthesis, and water use efficiency (Rahman et al., 2021). Treated plants exhibited reduced oxidative stress through enhanced antioxidant enzyme activity and maintained osmotic balance via increased soluble sugars and amino acids. Furthermore, AA optimized mineral partitioning among organs, improving leaf/stem, stem/root, and leaf/root mineral ratios (Rahman et al., 2021). In maize, Mahmud et al. (2023) investigated AA through seed priming and root treatment, validating the results using *Arabidopsis* mutants. They showed that AA improved growth performance, leaf water content, and photosynthetic rate under drought conditions. These improvements were closely associated with elevated proline accumulation, and validation in *Arabidopsis* confirmed that AA's protective effects are tightly linked to proline metabolism—highlighting a regulatory crosstalk between AA signaling and osmotic adjustment pathways.

Beyond direct impacts on plant physiology and hormone signaling, AA also influences the rhizosphere microbiome and soil metabolic composition. In *Salix myrtillacea*, Kong et al. (2022) demonstrated that exogenous AA enhanced drought tolerance not only by altering plant traits but also by enriching beneficial soil microorganisms such as Azotobacter and Pseudomonas. These microbial shifts correlated with changes in soil metabolites, including increases in amino acids (e.g., glycyl-L-tyrosine, L-glutamine, seryl-tryptophan) and decreases in phenylpropanoids (e.g., fraxetin, sinapyl aldehyde). Together, these alterations facilitated improved nitrogen uptake and contributed to overall drought resilience.

8.3.2 Salinity Stress

Acetic acid has also shown strong potential in enhancing plant tolerance to salinity stress (Table 8.1). In mung beans exposed to seawater-induced salinity, AA-treated plants displayed significantly improved growth and physiological performance compared with untreated controls (Rahman et al., 2019). This protective effect was attributed to reduced sodium accumulation, improved uptake of essential minerals such as potassium, calcium, and magnesium, and enhanced water-use efficiency. AA also alleviated oxidative stress by boosting catalase activity and promoting the accumulation of osmoprotectants such as proline, free amino acids, and soluble sugars. Elevated levels of phenolics and flavonoids further strengthened the antioxidant defense system, contributing to overall stress mitigation (Rahman et al., 2019).

Similar effects were observed in perennial ryegrass, where AA pre-treatment significantly mitigated salt-induced damage by maintaining a higher K^+/Na^+ ratio and enhancing reactive oxygen species (ROS) detoxification through increased superoxide dismutase (SOD) and catalase (CAT) activities (Zhang et al., 2021). Khan et al. (2023) further demonstrated that AA mitigates salt stress by activating multiple defense mechanisms. Specifically, AA reduced the accumulation of harmful ROS and lipid peroxidation products, while enhancing the activity of antioxidant enzymes such as SOD, CAT, peroxidase (POD), and ascorbate peroxidase (APX). By strengthening antioxidant defenses, AA effectively prevented oxidative damage under salt stress. At the same time, AA improved ionic regulation by limiting the buildup of toxic Na^+ and Cl^- ions and promoting the uptake of essential nutrients (K^+, Ca^{2+}, and Mg^{2+}), thereby maintaining ionic homeostasis—a key determinant of salt tolerance (Khan et al., 2023).

The pivotal role of AA in salt tolerance has also been confirmed in canola. Dawood et al. (2023) reported that AA pre-treatment alleviated the combined effects of salinity and lithium stress by restoring growth, maintaining Na^+/K^+ balance, and enhancing antioxidant defenses. AA also increased osmolyte accumulation, stabilized pigments and secondary metabolites such as flavonoids and anthocyanins, and modulated key signaling molecules such as SA. Notably, AA promoted lithium detoxification through elevated levels of phytochelatins and metallothioneins (Dawood et al., 2023).

Additional evidence highlights AA's role in lipid remodeling under salinity stress. In *Carex rigescens*, AA priming under 300 mM NaCl reduced malondialdehyde (MDA) levels and electrolyte leakage, reflecting improved membrane stability. These effects were associated with increased levels of key phospholipids and glycolipids, decreased accumulation of toxic free fatty acids, and altered expression of lipid metabolism-related genes (CK, PLDα), underscoring lipid remodeling as a central mechanism of AA-mediated salt stress tolerance (Hu et al., 2021).

In lentil seedlings, exogenous application of AA in the form of sodium acetate (5–10 mM) alleviated the effects of 100 mM salinity stress by improving water balance, preserving chlorophyll content, and sustaining ion homeostasis. Long-term glasshouse trials further showed that treated plants exhibited greater biomass, reduced membrane injury, and lower Na^+ accumulation in shoots, confirming acetate's efficacy in promoting salt tolerance (Hossain et al., 2019).

Recent studies have also begun to explore AA's effects on fruit crops. In strawberries, AA improved plant biomass and fruit yield under salinity stress by enhancing antioxidant enzyme activities (SOD and CAT) and modulating hormonal balance. Specifically, AA reduced stress-induced abscisic acid (ABA) accumulation while elevating growth-promoting hormones such as gibberellins (GAs), salicylic acid, and cytokinins. This hormonal adjustment limited growth inhibition and strengthened overall stress resilience (Mirfattahi and Eshghi, 2023).

8.3.3 Heavy Metal Toxicity

Heavy metal accumulation severely disrupts plant metabolism, causing growth inhibition, oxidative stress, and cellular damage. In this context, acetic acid has emerged as a promising agent for enhancing plant tolerance and supporting phytoremediation (Table 8.1).

In *Brassica juncea*, exogenous application of AA (250 μM) significantly alleviated cadmium (Cd)-induced toxicity by improving biomass and chlorophyll content while reducing MDA and hydrogen peroxide levels. This protective effect was closely associated with enhanced antioxidant defense, particularly through activation of the ascorbate–glutathione (AsA–GSH) cycle and increased activities of key enzymes, including SOD, CAT, and glutathione reductase (GR), underscoring AA's ability to mitigate Cd stress (Chowardhara et al., 2022).

In a comparative hydroponic study assessing the effects of malic acid and AA on Cd toxicity in sunflower, AA treatment (250–500 μM) reduced oxidative stress, enhanced photosynthetic pigment levels, and improved root activity. Although AA was less effective than malic acid, it nonetheless limited Cd translocation to the shoots and promoted the accumulation of endogenous low molecular weight organic acids in roots—compounds known to facilitate detoxification and stress tolerance (Hawrylak-Nowak et al., 2015).

Further evidence highlights AA's role in enhancing phytoremediation efficiency in Cd-contaminated soils. In oilseed sunflower, AA application improved biomass accumulation and Cd removal rates by modulating soil enzyme activities, decreasing toxic Cd fractions, and increasing less bioavailable carbonate-bound Cd forms. These effects were linked to enhanced CAT and sucrase activity, as well as a shift in Cd speciation that supported better plant growth under contamination (Wang et al., 2022b).

In tomato, AA was shown to regulate Cd accumulation via modulation of JA-dependent signaling. Both exogenous AA application and AA produced by the endophytic bacterium *Lysinibacillus fusiformis* Cr33 significantly increased JA levels in roots, leading to downregulation of Fe uptake-related genes and reduced Cd absorption. This response was accompanied by decreased nitric oxide production and improved Fe availability in the rhizosphere and cell wall, thereby enhancing plant tolerance to Cd toxicity (Zhu et al., 2021).

Beyond cadmium, AA has also been implicated in arsenic (As) detoxification. In *Pteris vittata*, an established arsenic hyperaccumulator, application of AA—particularly in the form of calcium acetate—significantly enhanced both drought tolerance and arsenic remediation efficiency. Treated plants exhibited improved growth under water-deficient conditions and greater As accumulation in shoots. These benefits were associated with increased soil pH and improved As availability in the rhizosphere, demonstrating AA's capacity to support phytoremediation under combined abiotic stress conditions (Wang et al., 2023).

8.3.4 Temperature Extremes

Although relatively few studies have examined the role of AA in plant tolerance to extreme temperatures, existing evidence suggests that it may significantly enhance resilience to both heat and cold stress (Table 8.1).

In *Salix myrtillacea*, AA application produced distinct sex-specific benefits, particularly in females, by modulating the rhizosphere microbial community. Treated female plants exhibited an increased abundance of beneficial bacterial phyla and improved carbon metabolism, which enhanced carbon uptake and storage. These changes collectively contributed to improved tolerance to high-temperature stress (He et al., 2024).

Similarly, in *Citrus aurantifolia*, exogenous AA application (15 mM) enhanced tolerance to cold and freezing stress by reducing cellular damage markers such as electrolyte leakage and MDA, while also improving photosynthetic efficiency and antioxidant capacity (Sanie Khatam et al., 2023). At 0 °C, AA treatment enhanced both enzymatic antioxidant defenses—POD, CAT, and SOD—and non-enzymatic antioxidant systems. At −6 °C, the non-enzymatic system played a more dominant role. In addition, AA-treated plants showed elevated levels of pigments, total phenolics, and soluble carbohydrates, indicating that AA confers cold tolerance through multiple, complementary physiological and biochemical pathways (Sanie Khatam et al., 2023).

Despite these promising findings, further research is required to elucidate the precise molecular and physiological mechanisms by which AA enhances plant tolerance to temperature extremes. Critical areas for investigation include its involvement in signaling pathways, regulation of gene expression, hormonal crosstalk, and modulation of rhizosphere microbial communities. Moreover, comprehensive studies across diverse species and environmental contexts are essential to validate its efficacy and evaluate its potential for broader agricultural applications.

8.4 Acetic Acid in Modulating Plant Responses to Biotic Stressors

Acetic acid is increasingly recognized as an unconventional regulator of plant immune responses. While most research has historically focused on its role in tolerance to abiotic stresses, particularly drought, its contribution to biotic stress responses remains comparatively underexplored. This represents a promising but still emerging field in plant biology.

One key mechanism involves AA-mediated stimulation of the JA pathway, which plays a central role in defense against herbivorous insects and necrotrophic pathogens. This hormonal activation is often accompanied by the increased production of defense-related compounds, including phenolics, ROS, and pathogenesis-related proteins, thereby contributing to systemic acquired resistance (Ogawa et al., 2021).

Exogenous application of AA has also been shown to elevate endogenous ABA levels and activate ABA-responsive genes such as *Open Stomata 1 (OST1)* and *Protein Phosphatase 2Cs (PP2Cs)*, key components of ABA signaling. This enhances plant tolerance to both abiotic and biotic stresses by triggering adaptive physiological responses. A major effect of AA-induced ABA signaling is stomatal closure, which reduces water loss while simultaneously limiting pathogen entry—an early immune defense mechanism. Moreover, ABA promotes the accumulation of polyamines, proline, and sphingolipids, all of which support cell protection and resilience under stress. Collectively, these findings suggest that AA reinforces ABA-mediated immunity and stress adaptation (Bharath et al., 2021; Sun et al., 2022).

Beyond hormonal signaling, AA has demonstrated direct antimicrobial activity. In contaminated lettuce, AA reduced the growth of *Shigella* spp. by up to 2 log units. Although less effective than lactic acid, AA caused the highest proportion of sublethally injured bacterial cells and produced strong inhibition zones in diffusion assays, suggesting that AA can weaken pathogens and contribute to surface-level microbial control strategies (In et al., 2013).

With respect to insect control, Medrano et al. (2024) reported that AA alone exhibited no direct insecticidal activity against Aedes aegypti, as no mortality was observed in pure AA treatments at either 1 mL or 3 mL application rates. However, when combined with avocado seed oil extract, AA significantly enhanced the

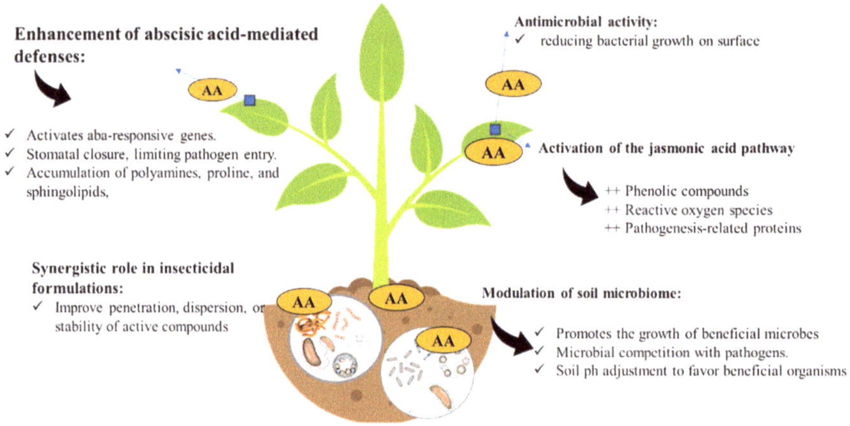

Fig. 8.2 Role of acetic acid (AA) in plant defense against biotic stressors. AA contributes to plant immunity through multiple, interconnected pathways. In aerial tissues, AA stimulates jasmonic acid (JA) and abscisic acid (ABA) signaling, leading to stomatal closure, activation of ABA-responsive genes (*OST1*, *PP2Cs*), and accumulation of protective metabolites (e.g., proline, polyamines, sphingolipids). These processes reduce water loss and limit pathogen entry while reinforcing systemic resistance. In parallel, AA enhances the production of defense-related compounds, including phenolics, reactive oxygen species (ROS), and pathogenesis-related proteins, which strengthen innate immune responses. At the soil interface, AA modulates rhizosphere microbiota by promoting beneficial plant growth–promoting rhizobacteria (PGPR) and biocontrol fungi, while suppressing pathogens through microbial competition and antimicrobial compound production. In addition, AA shows direct antimicrobial activity against bacterial pathogens and acts synergistically with natural extracts in botanical insecticides, amplifying their efficacy against insect pests. Collectively, these functions position AA as a versatile modulator of plant defense under biotic stress

insecticidal efficacy of the formulation. This indicates that AA may act as a synergistic agent, potentially improving the penetration, dispersion, or stability of bioactive compounds within botanical insecticides. While not effective as a standalone insecticide, AA can amplify the potency of natural plant extracts, supporting its potential use in eco-friendly pest management strategies (Fig. 8.2).

In addition, AA has been shown to modulate the soil microbiome by promoting the growth and activity of beneficial microorganisms such as plant growth-promoting rhizobacteria (PGPR) and certain biocontrol fungi (Kong et al., 2022). These microbes enhance plant immunity by competing with pathogens for space and nutrients, producing antimicrobial compounds, and strengthening host defenses. By altering soil pH and organic acid availability, AA creates a more favorable environment for beneficial microbes, thereby indirectly suppressing soil-borne pathogens (Kong et al., 2022; Al-Baldawy et al., 2023).

8.5 Interaction of Acetic Acid with Plant Hormones and Rhizosphere Microbes

Plant hormones, or phytohormones, play a central role in regulating growth, development, and plant responses to diverse environmental signals (El Sabagh et al., 2022). At the same time, the rhizosphere—the soil zone directly influenced by plant roots—harbors a complex and dynamic microbial community essential for plant nutrition, health, and stress tolerance (Vetterlein et al., 2020).

The interaction between AA and phytohormones under abiotic stress appears to be complex and highly species-dependent. For example, in grapevine, AA treatment increased salicylic acid levels, whereas in wheat, it elevated IAA and GA while reducing ABA concentrations (Lv et al., 2021; Khan et al., 2023). AA, a simple organic acid commonly found in the rhizosphere, can originate from both root exudation and microbial metabolism (Kong et al., 2022). Increasing evidence suggests that AA functions as a key modulator of interactions within the plant–soil–microbe continuum, influencing both hormonal crosstalk and microbial dynamics (Santoyo et al., 2017).

8.5.1 Crosstalk with Phytohormones

8.5.1.1 Auxins

As a fundamental metabolic product, AA may indirectly influence auxin production—particularly IAA—by altering precursor availability or modulating the activity of key enzymes in the biosynthetic pathway (Rahman et al., 2024). IAA biosynthesis requires specific precursors such as tryptophan, and AA could affect the balance of carbon and nitrogen metabolism, thereby influencing the pool of precursors available for IAA synthesis (Lang et al., 2019).

In addition, the role of AA in regulating the polar transport of auxin, a process crucial for establishing developmental gradients in plants, remains to be elucidated. Similarly, the effect of AA on the expression of auxin-responsive genes—central regulators of diverse growth and developmental processes—warrants further investigation.

Experimental evidence suggests a strong link between AA and auxin signaling. In perennial ryegrass, AA treatment significantly increased endogenous IAA levels and upregulated the expression of genes associated with auxin biosynthesis and signaling pathways (Zhang et al., 2021) (Fig. 8.3). These findings indicate that AA may enhance auxin accumulation, thereby supporting growth and developmental processes under stress conditions. Furthermore, AA may influence auxin signaling indirectly through interactions with other hormones such as JA and ABA (Kim et al., 2017; Sun et al., 2022).

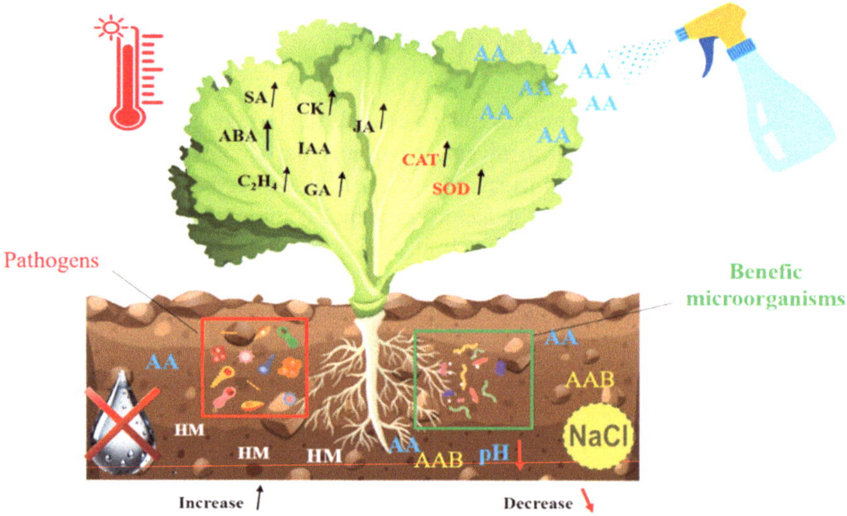

Fig. 8.3 Acetic acid (AA) as a central regulator of plant stress responses and rhizosphere interactions. Exogenous and endogenously produced AA acts as a hub connecting plant hormone signaling, antioxidant defense, and soil microbial dynamics. In shoots, AA modulates phytohormones including salicylic acid (SA), abscisic acid (ABA), jasmonic acid (JA), cytokinins (CK), gibberellins (GA), auxins (IAA), and ethylene (C_2H_4), thereby influencing growth, stomatal behavior, and stress adaptation. AA also enhances antioxidant capacity through activation of catalase (CAT) and superoxide dismutase (SOD), supporting tolerance to abiotic stressors such as salinity (NaCl), drought, heavy metals (HM), and temperature extremes. In roots and the rhizosphere, AA is produced by acetic acid bacteria (AAB) such as Acetobacter and Gluconobacter and influences soil pH and nutrient cycling. At low concentrations, AA promotes beneficial plant growth–promoting rhizobacteria (PGPR) and other microbes, while at higher doses it suppresses pathogens and harmful fungi. Collectively, these interactions position AA as a key integrator of plant physiology and soil microbial ecology, enhancing resilience against biotic and abiotic stressors. *Abbreviations*: AA, acetic acid; SA, salicylic acid; ABA, abscisic acid; JA, jasmonic acid; CK, cytokinins; GA, gibberellins; IAA, indole-3-acetic acid; C_2H_4, ethylene; CAT, catalase; SOD, superoxide dismutase; HM, heavy metals; AAB, acetic acid bacteria; PGPR, plant growth–promoting rhizobacteria

8.5.1.2 Cytokinins

Cytokinins are essential phytohormones that regulate cell division, differentiation, and overall plant development (Dello Ioio et al., 2008). The potential influence of AA on cytokinin biosynthesis or degradation could therefore have significant consequences for plant growth and stress adaptation. The most plausible mechanism involves AA-mediated modulation of gene expression and hormone signaling pathways (Rahman et al., 2024).

Evidence from perennial ryegrass supports this interaction. Under salt stress, AA treatment upregulated the *Pyruvate Decarboxylase 1 (LpPDC1)* gene, which is primarily involved in AA biosynthesis but has also been associated with increased cytokinin levels (Zhang et al., 2021). This suggests that the AA application may influence cytokinin accumulation through transcriptional regulation. Such regulation could involve transcription factors or signaling cascades activated by AA (Sun et al., 2022). Furthermore, AA's role in histone acetylation could facilitate expression of genes linked to cytokinin biosynthesis, degradation (e.g., *Cytokinin Oxidase/Dehydrogenase* [*CKX*]), and signaling (e.g., *Arabidopsis Histidine Kinase* [*AHK*], *Arabidopsis Response Regulator* [*ARR*]) (Tang et al., 2023).

Beyond direct modulation, AA may influence cytokinin signaling indirectly through its effects on auxin. Crosstalk between auxin and cytokinin is well established, particularly within the two-component signaling system that governs their integration. For example, auxin can rapidly regulate cytokinin biosynthesis and alter the size of the cytokinin pool (Jones et al., 2010), while cytokinins in turn influence auxin distribution and activity (Bajguz and Piotrowska, 2009). This interplay is especially relevant in plant–pathogen interactions, where pathogens manipulate hormone levels to facilitate infection. Cytokinins often act synergistically with SA in defense against biotrophic pathogens (Singh et al., 2022), while auxin–cytokinin dynamics are known to affect susceptibility or resistance outcomes (Akhtar et al., 2020).

Although direct evidence of AA's role in regulating cytokinins during pathogen infection remains limited, its established function in modulating global hormone signaling strongly suggests potential indirect effects. By altering auxin levels and signaling pathways, AA may influence cytokinin dynamics, thereby shaping plant responses under both stress and pathogen challenge (Li et al., 2021a; Sosnowski et al., 2023).

8.5.1.3 Gibberellins

Gibberellins constitute a large family of phytohormones that are fundamental to stem elongation, seed germination, and flowering (Castro-Camba et al., 2022). Direct evidence of AA influencing GA levels or activity in plants under pathogen or insect attack remains limited. Nonetheless, some studies suggest potential points of interaction.

Under biotic stress, it is plausible that AA could affect GA production or signaling. For instance, infestation of rice by the brown planthopper (*Nilaparvata lugens*) increases levels of inactive GAs, suggesting that GA metabolism is part of the plant's defensive response (Jin et al., 2023). Similarly, application of GA to tomato inhibited feeding and disrupted the biological performance of Spodoptera littoralis larvae, implying a role for GAs in plant–insect interactions (Anwaar et al., 2024). SA antagonistically regulates GA signaling during pathogen attack (Lazebnik et al., 2014), and since AA can modulate SA levels, it may indirectly affect GA-mediated defense responses. In addition, some endophytic fungi produce both GAs and indole-IAA, which promote plant growth under abiotic stress (Waqas et al., 2012). Whether AA influences these GA-mediated fungal effects remains an open question (Waqas et al., 2015).

Evidence of AA–GA interaction is clearer under abiotic stresses. In wheat subjected to salinity, AA increased levels of IAA and GAs while reducing ABA concentrations, thereby contributing to salt tolerance (Khan et al., 2022) (Fig. 8.3). Under drought, AA altered ABA levels (Rahman et al., 2024), potentially affecting the ABA/GA balance—a well-established determinant of water stress tolerance (Waqas et al., 2012; Shu et al., 2018). AA has also been shown to stimulate proline biosynthesis (Mahmud et al., 2023), which may involve GA regulation. Moreover, AA's capacity to modulate multiple hormones, including JA, IAA, cytokinins, and ABA (Zhang et al., 2021), suggests that it can indirectly influence GA signaling within broader hormonal networks.

With respect to extreme temperature stress, data remain scarce, but GAs are known to contribute to heat tolerance and growth regulation (Nagar et al., 2021). AA may interact with this pathway, either independently or in combination with GA treatments. For example, combined AA and GA_3 applications improved germination and growth under cold conditions (Cui et al., 2024), while also enhancing growth under heat stress (Khan et al., 2020). These findings suggest that AA could modulate GA signaling across a wide range of environmental contexts, though more targeted studies are required to clarify the underlying mechanisms.

8.5.1.4 Abscisic Acid

Abscisic acid is a central phytohormone that regulates numerous physiological and biochemical processes during normal growth as well as under diverse abiotic and biotic stresses (Bharath et al., 2021). In addition to its established role in stress responses, ABA is involved in developmental processes such as seed germination, dormancy, and regulation of stomatal closure (Rehman et al., 2021). ABA biosynthesis is strongly upregulated in plants encountering stressful environmental conditions (Singh and Roychoudhury, 2023).

Interactions between AA and ABA are emerging as an important area of study. In cotton, Li et al. (2021b) demonstrated that water stress combined with AA treatment increased ABA levels and induced expression of ABA biosynthetic genes, including *9-cis-epoxycarotenoid dioxygenase* (*NCED2*, *NCED3*, and *NCED9*), 3 days after treatment. In contrast, research in perennial ryegrass under salt stress showed that while stress itself and exogenous AA both elevated endogenous AA content, AA treatment resulted in reduced ABA content in leaves, suggesting a potentially antagonistic relationship between AA and ABA under salinity (Zhang et al., 2021) (Fig. 8.3). These contrasting outcomes indicate that the AA–ABA relationship may vary across plant species and stress contexts.

In *cassava* exposed to drought, AA was reported to regulate ABA levels and modulate stomatal movement, thereby contributing to drought avoidance (Orek, 2023). This highlights AA's potential role in fine-tuning ABA-mediated water conservation strategies. Reactive oxygen species and MAPK cascades are key components of ABA signaling in stress responses. ABA induces ROS generation, which acts as a signaling cue for downstream stress adaptation (Li et al., 2022). By contrast, AA has been shown to modulate ROS accumulation by enhancing antioxidant enzyme activities, particularly SOD and CAT, thereby reducing oxidative damage (Zhang et al., 2021). Moreover, MAPK cascades serve as a critical node of crosstalk between ABA and JA, regulating the expression of stress-responsive genes (Adhikari et al., 2023).

The interaction between AA and ABA also has important implications for biotic stress responses. ABA can influence plant immunity by modulating the expression of defense-related genes and interacting with other phytohormones, particularly JA and SA (Cao et al., 2011; Mine, 2019). Through these interconnected pathways, AA's regulation of ABA may indirectly affect resistance to pathogens and pests, adding a further layer of complexity to its role in plant stress physiology (Nassar and Adss, 2016).

8.5.1.5 Ethylene

Ethylene is a gaseous phytohormone that regulates fruit ripening, senescence, and plant responses to both biotic and abiotic stresses (Riyazuddin et al., 2020). Its biosynthesis follows a well-defined pathway: the amino acid methionine is converted into S-adenosylmethionine (SAM) by SAM synthetase, which is then converted into 1-aminocyclopropane-1-carboxylic acid (ACC) by ACC synthase (ACS)—the rate-limiting step of ethylene production. Finally, ACC is oxidized into ethylene by ACC oxidase (ACO) (Kim et al., 2021).

Ethylene perception and signaling occur via a family of receptors located in the endoplasmic reticulum membrane. Upon activation, these receptors regulate downstream signaling cascades involving the *Constitutive Triple Response 1* (*CTR1*) *kinase and Ethylene Insensitive 2* (*EIN2*). This signaling ultimately activates transcription factors such as *EIN3/Ethylene Insensitive-Like 1* (*EIL1*) and *Ethylene Response Factors* (*ERFs*), which control the expression of ethylene-responsive genes. Ethylene biosynthesis is often rapidly induced in plants under environmental stress, acting as an early signal that initiates adaptive defense responses (Chen et al., 2021).

The interaction between AA and ethylene is not yet fully understood. Research in perennial ryegrass under salt stress revealed that AA did not significantly affect ACC content, despite salinity itself increasing ACC levels (Kudo et al., 2023). This suggests that AA-mediated mitigation of salt stress may not directly involve ethylene biosynthesis. Nonetheless, interactions at the signaling level cannot be excluded. AA could potentially influence the expression or activity of ACS and ACO enzymes through epigenetic mechanisms such as histone acetylation (Kudo et al., 2023) or modulate downstream ethylene signaling components, including receptors, CTR1, EIN2, EIN3/EIL1, and ERFs (Husain et al., 2020).

Furthermore, given AA's established effects on auxin, it may indirectly regulate ethylene production and signaling through hormonal cross-regulation. This interaction could be particularly important in plant immunity, as injury-induced ethylene production is known to trigger defense mechanisms (Jasrotia and Jasrotia, 2022). By modulating this process, AA may fine-tune ethylene-mediated defense responses, further reinforcing its role as a versatile regulator of plant stress adaptation.

8.5.1.6 Jasmonic Acid

Jasmonic acid and its derivatives, collectively termed jasmonates, are lipid-derived phytohormones that regulate a wide range of plant processes, from growth and development to responses against both abiotic and biotic stresses (Wang et al., 2021). The JA signaling pathway is tightly regulated and involves several key components, including the *Coronatine Insensitive 1 (COI1) receptor*, a family of Jasmonate Zim-Domain (JAZ) repressor proteins, and transcription factors such as *MYC2* (Yang et al., 2019). Upon stress perception, JA biosynthesis is rapidly induced, activating downstream defense and stress response mechanisms (Wang et al., 2020).

A growing body of evidence highlights the strong influence of acetic acid on JA pathways in plants under stress. Exogenous AA application has consistently been shown to enhance JA biosynthesis and activate JA signaling across multiple plant species, including rice (Ogawa et al., 2021) and tomato (Zhu et al., 2021) (Fig. 8.3). In salt-stressed plants, AA treatment increased the endogenous levels of jasmonates such as JA, JA-isoleucine (JA-Ile), and cis-12-oxo-phytodienoic acid (cis-OPDA) (Zhang et al., 2021). Notably, in rice, AA rapidly activated the JA signaling cascade, producing physiological changes that closely resembled those induced by drought stress, suggesting a mechanism by which AA may prime plants for acclimatization and improve readiness for subsequent water deficits (Ogawa et al., 2021).

The molecular basis of AA–JA interactions appears to involve epigenetic regulation. AA can function as an acetyl group donor, and its application has been shown to enhance histone H4 acetylation (Kim et al., 2017; Kudo et al., 2023). This chromatin modification likely facilitates the transcription of genes involved in JA biosynthesis and signaling, amplifying JA-dependent stress responses. Additionally, the acetate pathway itself may contribute to the production of methyl jasmonate (MeJA), an active JA derivative (Tamogami et al., 2008). The transcription factor *MYC2*, a master regulator of JA signaling, is also believed to mediate AA's role in enhancing JA-driven stress tolerance (Samanta and Roychoudhury, 2025).

Modulation of JA signaling by AA has been linked to improved plant resistance against biotic stressors. For instance, in tomato, AA pre-treatment significantly enhanced resistance to the chewing herbivore *Spodoptera litura* by boosting JA-mediated transcriptional responses and defense-related enzymatic activities (Chen et al., 2019). More broadly, AA-mediated priming of JA pathways is increasingly viewed as a key mechanism through which AA contributes to plant resilience against both abiotic and biotic challenges (Roychowdhury et al., 2025).

8.5.1.7 Salicylic Acid

Salicylic acid is a phenolic phytohormone that plays a central role in plant defense against a broad spectrum of pathogens and also contributes to responses against various abiotic stresses (Song et al., 2023). SA is particularly well known for establishing systemic acquired resistance (SAR), a durable and broad-spectrum defense mechanism in plants (Wani et al., 2017).

SA biosynthesis occurs mainly via two routes: the isochorismate pathway, mediated by isochorismate synthase (ICS), and the phenylalanine pathway, mediated by phenylalanine ammonia-lyase (PAL) (Zhong et al., 2021). SA signaling relies on key regulatory proteins such as Nonexpressor of *Pathogenesis-Related Genes 1* (*NPR1*) and *WRKY* transcription factors, which together orchestrate the expression of defense-related genes (Tian et al., 2025).

The influence of AA on SA dynamics appears to be species- and context-dependent. In perennial ryegrass under salt stress, AA treatment did not significantly alter SA content (Zhang et al., 2021). By contrast, in grapevine, AA treatment increased SA accumulation in plant tissues (Fig. 8.3), suggesting a stimulatory effect on SA biosynthesis in certain species. These contrasting outcomes highlight the complexity and specificity of hormonal responses to AA application.

The molecular mechanisms underlying AA–SA interactions remain to be fully elucidated. One plausible explanation involves AA-induced histone acetylation, which could alter chromatin accessibility and influence transcription of genes involved in SA biosynthesis (*ICS*, *PAL*) and signaling (*NPR1*, *WRKYs*) (Hu et al., 2019). Given the well-established crosstalk between SA and other phytohormones, particularly jasmonic acid, it is likely that AA's modulation of SA signaling is mediated, at least in part, by its impact on these hormonal networks.

NPR1, a master regulator of SA-dependent defense responses, may also be indirectly influenced by AA through its effects on upstream signaling pathways or through interactions with other hormones that control *NPR1* activity and expression. Thus, while direct evidence remains limited, AA appears capable of shaping SA-mediated immunity both through chromatin-level regulation and through integration within broader phytohormonal dialogues.

8.5.2 Influence on Soil Microbial Communities

Soil microbial communities are a cornerstone of terrestrial ecosystems, underpinning both ecosystem health and agricultural productivity. Among the most abundant and diverse biological assemblages in nature, they encompass bacteria, fungi, archaea, viruses, algae, and protozoa (Kiprotich et al., 2025). Their diversity and composition directly affect key ecological processes, including nutrient cycling, organic matter decomposition, disease suppression, and plant growth promotion (Wang et al., 2024). By maintaining soil fertility, recycling essential nutrients, degrading complex compounds, and protecting plants against pathogens, soil microbes provide critical functions for sustainable agriculture (Sahu et al., 2017). Conversely, disruption of these communities can severely impair ecosystem stability and food production.

AA occurs naturally in soil as a byproduct of microbial metabolism. Acetic acid bacteria (AAB), notably the genera Acetobacter and Gluconobacter, convert ethanol into AA through a two-step oxidative process: ethanol is first oxidized to acetaldehyde, which is then converted to AA (Deppenmeier and Ehrenreich, 2009; Han et al., 2024). This reaction typically occurs under high-oxygen conditions, such as in well-aerated soils (Gomes et al., 2018). Beyond natural production, AA can enter soils through anthropogenic inputs, including residues from vinegar industries, AA-rich wood vinegar from biomass pyrolysis, and its agricultural use as a herbicide (He et al., 2020; Gu et al., 2024). These inputs can increase AA concentrations above natural levels, in some cases exceeding thresholds tolerated by sensitive microbial communities (Iacomino et al., 2024).

Once in soil, AA is rapidly biodegraded by microorganisms, making its effects highly dependent on application frequency, dose, and local environmental conditions (Ogram et al., 1985). Soil pH strongly predicts microbial community composition (Jiao and Lu, 2020), and AA's influence on soil pH is therefore central to understanding its ecological impact. Depending on concentration and buffering capacity, AA can either acidify soil or be rapidly degraded with minimal pH alteration (Fig. 8.3).

AA, particularly in the form of wood vinegar, exerts notable effects on soil microbial assemblages. At low doses, it stimulates beneficial groups such as *Actinobacteriota*, *Firmicutes*, *Azotobacter*, *Pseudomonas*, and *Bacillus*—taxa often associated with PGPR (Macias-Benitez et al., 2020; Kong et al., 2022; Fig. 8.3). For example, vinegar diluted 200-fold increased Actinobacteriota by +41% and Firmicutes by +127% (Gu et al., 2024), while a 3% solution increased overall bacterial density (Koç et al., 2019). At higher concentrations, however, AA reduces microbial biomass and alpha diversity, often through acidification (Wang et al., 2022a).

AA also exhibits broad-spectrum antimicrobial activity. It is effective against soilborne pathogens such as *Rhizoctonia solani* and *Fusarium oxysporum* (Devi et al., 2024). Wood vinegar has inhibited *Salmonella*, *Bacillus subtilis*, *Colletotrichum capsici*, and *Phytophthora palmivora* (Iacomino et al., 2024). Efficacy is strongly dose-dependent: AA vapors at 4 μL/L inhibit *R. solani* growth, while 6 μL/L are required for *Fusarium* spp., *Sclerotium rolfsii*, and *Didymella melonis*. Complete inhibition of spore germination was observed at 4 μL/L (Abd-El-Kareem, 2009). At concentrations above 2.5%, AA becomes fungicidal (Trzaska et al., 2015).

At high levels, AA is directly toxic to soil microorganisms. It disrupts membranes, causes desiccation and cell death, and inhibits spore viability of multiple fungal species (Webber III et al., 2018). Its antifungal effects involve both intracellular acidification and pH-independent toxicity (Abd-El-Kareem, 2009). In addition, AA can modulate microbial gene expression. For example, in Bacillus, the regulatory protein Catabolite Control Protein A (CcpA) negatively regulates AA production (Zhang et al., 2023), while in *Pichia kudriavzevii*, tolerance to AA is associated with coordinated activation of detoxification and resistance pathways (Dolpatcha et al., 2025). Some soil bacteria also metabolize IAA as a carbon source, highlighting the complexity of interactions between AA and microbial metabolic networks (Donoso et al., 2017).

8.6 Molecular and Cellular Mechanisms of Acetic Acid-Mediated Stress Tolerance

Acetic acid has emerged as a central regulator of plant adaptation to environmental stressors, exerting protective effects through molecular, physiological, and epigenetic mechanisms. Its role in stress tolerance is mediated by the modulation of hormone signaling pathways, activation of stress-responsive genes, reinforcement of antioxidant defense systems, and maintenance of cellular homeostasis (Rahman et al., 2024).

One of the primary mechanisms by which AA confers tolerance to (a)biotic stress is through stimulation of the ABA signaling pathway. Upon AA treatment, several ABA-related genes—including *ABI2*, *OST1*, *AHG1*, and *AFP2*—are upregulated (Finkelstein and Somerville, 1990; Mustilli et al., 2002; Nishimura et al., 2007; Lynch et al., 2017). This genetic activation is accompanied by increased endogenous ABA levels, reduced stomatal conductance, and decreased transpiration (Kim et al., 2017; Rasheed et al., 2018), thereby enhancing water-use efficiency under adverse conditions. These physiological adjustments highlight the involvement of ABA-dependent signaling in AA-induced stress responses.

In *Arabidopsis thaliana* and *Oryza sativa*, AA also enhances drought tolerance through activation of the JA pathway (Kim et al., 2017; Ogawa et al., 2021). However, this effect is species-dependent; in Manihot esculenta (cassava), transcriptional activation of JA-related genes following AA treatment is minimal, underscoring the variability of hormone regulation across plant taxa.

At the epigenetic level, AA promotes drought tolerance through chromatin remodeling. By influencing histone deacetylase 6 activity, AA regulates chromatin accessibility and transcription, thereby priming the genome for faster and stronger induction of stress-responsive genes (Kim et al., 2017; Bajpai et al., 2024).

Beyond ABA and JA, AA also modulates auxin pathways. It enhances auxin accumulation by upregulating *TAA1* (involved in auxin biosynthesis) and *ARF1/ARF2* (*auxin response factors*), contributing to improved root development and stress adaptation (Kim et al., 2017).

AA treatment further induces a wide suite of stress-related genes. Heat shock proteins (HSPs)—including *HSP20*, *HSP21*, *HSP70*, and *HSP90*—are upregulated, functioning as molecular chaperones that stabilize proteins and prevent denaturation under thermal and oxidative stress (Krishna and Gloor, 2001; Tang et al., 2016; Muthusamy et al., 2017). The accessory protein HOP3, associated with endoplasmic reticulum stress responses, is also induced, ensuring proper protein folding under adverse conditions (Fernández-Bautista et al., 2017a, b).

AA strengthens the antioxidant defense system by upregulating *CAT1* and *CAT2*, encoding catalase enzymes that detoxify H_2O_2 and reduce oxidative stress. This ROS regulation is tightly coordinated with ABA signaling to maintain cellular redox balance (Zou et al., 2015). In parallel, AA induces *MBF1c*, a transcriptional co-activator that enhances abiotic stress tolerance when overexpressed in *Arabidopsis* (Alavilli et al., 2017), and TSPO, a multifunctional protein responsive to environmental stress in both roots and shoots (Guillaumot et al., 2009; Rasheed et al., 2018).

AA also contributes to osmotic adjustment. Under polyethylene glycol–induced osmotic stress, AA significantly increases expression of *P5CS1* (*Δ^1-pyrroline-5-carboxylate synthetase 1*), a key enzyme in proline biosynthesis. Proline acts as an osmoprotectant, stabilizing proteins and membranes and maintaining cellular osmotic balance, thereby protecting against both abiotic and biotic stress.

Another critical mechanism of AA-mediated tolerance is regulation of ion homeostasis, particularly under salinity. AA promotes potassium uptake while restricting sodium accumulation, thereby increasing the K^+/Na^+ ratio—a key determinant of salt tolerance (Zhang et al., 2021). This effect is associated with upregulation of the sodium transporter gene HKT1;1, which mediates Na^+ exclusion (Zhang et al., 2021). Interestingly, in the presence of AA, *HKT1;1* is specifically induced in leaf tissue but not in roots, suggesting tissue-specific transcriptional regulation possibly linked to acetylation-mediated control (Rasheed et al., 2018).

AA functions as a multifunctional regulator of stress tolerance. By orchestrating hormone signaling pathways (ABA, JA, IAA), activating stress-responsive genes, enhancing antioxidant activity, promoting osmolyte accumulation, and maintaining ion balance, AA enables plants to mount efficient defenses against environmental challenges. Its actions reflect an integrated strategy that strengthens resilience through coordinated genetic, hormonal, and physiological responses (Fig. 8.4).

8.7 Practical Applications of Acetic Acid in Agriculture

Acetic acid, an organic compound naturally present in vinegar, has diverse applications in agriculture as a herbicide, antifungal agent, biostimulant, and soil management tool (Table 8.2).

- Herbicidal activity. AA acts as a non-selective contact herbicide, disrupting plant cell membranes and causing rapid desiccation (Webber III et al., 2018). At 5% (household vinegar), it provides variable control of young annual weeds depending on species and developmental stage. At 20%, it achieves 80–100% efficacy on species such as *Mollugo verticillata* and Oenothera laciniata (Webber III et al., 2018). At 30%, AA's efficacy approaches that of glyphosate for initial weed suppression, though repeated treatments are often required (Domenghini,

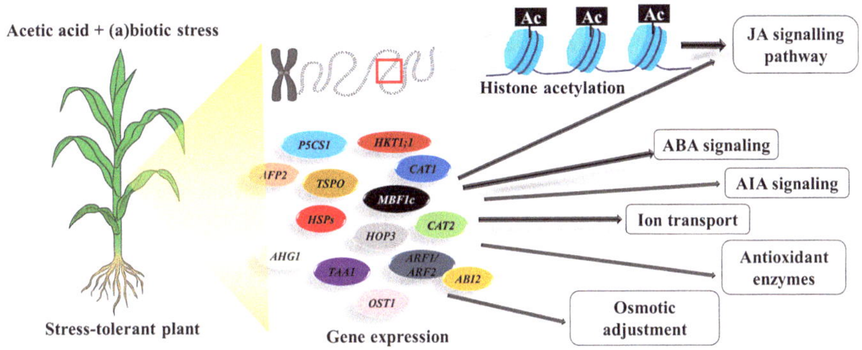

Fig. 8.4 Molecular mechanisms underlying acetic acid (AA)-mediated stress tolerance. Exogenous or endogenous AA primes plants for resilience against (a)biotic stresses by modulating multiple molecular pathways. At the epigenetic level, AA promotes histone acetylation, thereby enhancing accessibility of stress-responsive genes. Key hormonal signaling cascades, including ABA, JA, and auxin (IAA), are activated, leading to stomatal regulation, metabolic adjustment, and growth modulation. AA also induces expression of specific stress-related genes (e.g., P5CS1, TSPO, HSPs, CAT1, CAT2, HKT1;1, ARF1/2, ABI2, OST1) that collectively strengthen antioxidant defenses, maintain ion transport and homeostasis, promote osmotic adjustment, and enhance cellular protection. Through this coordinated regulation of gene expression, hormonal crosstalk, and antioxidant activity, AA functions as a central hub enabling plants to mount integrated adaptive responses and achieve enhanced stress tolerance. *Abbreviations*: ABA, abscisic acid; JA, jasmonic acid; IAA, indole-3-acetic acid; CAT, catalase; HSP, heat shock protein; P5CS1, Δ^1-pyrroline-5-carboxylate synthetase 1; TSPO, tryptophan-rich sensory protein; HKT1;1, high-affinity potassium transporter 1; ARF, auxin response factor; ABI2, abscisic acid insensitive 2; OST1, open stomata 1

2020). Because AA leaves no residual effect in the soil, applications must be targeted to avoid injury to neighboring crops. Its effectiveness depends primarily on the application volume rather than on the use of adjuvants (Webber III et al., 2018).

- Antifungal properties. AA is also recognized for its antifungal activity. It inhibits the growth of plant pathogens, including *Fusarium* spp., through penetration of fungal cells and disruption of intracellular pH (Nakashima et al., 2025). Wood vinegar, rich in AA, has been applied effectively as both a foliar and root fungicide (Narendranath et al., 2001). Post-harvest applications, such as fumigation or immersion in dilute solutions, reduce rots caused by Botrytis cinerea and *P. expansum* in fruits (Rogawansamy et al., 2015). For seed treatment, low concentrations (0.5–2%) decrease contamination by Alternaria and other pathogens without impairing germination, making AA a useful tool in organic farming (Dorna et al., 2021).
- Biostimulant effects. Beyond its pesticidal uses, AA has demonstrated biostimulant activity, particularly under water stress conditions (Table 8.2). By modulating hormone signaling (ABA, JA), enhancing proline metabolism, activating antioxidant defenses, and inducing epigenetic modifications, AA improves plant resilience (Allen and Allen, 2020; Mahmud et al., 2023; Rahman et al., 2024). Applied via seed priming, root dipping, or foliar spraying, AA increases

Table 8.2 Case studies of acetic acid (AA) application in different crops and stress scenarios

Crop/ species	Stress	Application method	Main observed effects	Reference
Cassava	Drought	External application (10 mM)	Enhanced drought avoidance; higher leaf relative water content, chlorophyll, and carotenoid levels; increased leaf temperature and ABA content; upregulation of ABA signaling and stress response genes	Utsumi et al. (2019)
Maize	Drought	Foliar spray Various concentrations	Inconsistent results; no significant interaction with deficit irrigation, some concentrations reduced transpiration without improving performance; Root growth inhibition at low pH	Allen and Allen (2020)
Begonia	Drought	6-day treatment (0, 10, 20, and 38 mM)	Reduced tray evapotranspiration (dose-dependent); 10 mM increased survival; 20 mM reduced water use without affecting flowering; 38 mM reduced flower number	Allen and Allen (2021)
Alfalfa	Drought	Pretreatment (15 mM)	Increased fresh weight and RWC; decreased H_2O_2 and proline content	Kim et al. (2024)
Canola	Salinity + lithium	Pretreatment (16 mM)	Promoted growth; re-established sodium and potassium homeostasis; enhanced antioxidant defense system; more effective than nitric oxide	Dawood et al. (2023)
Ryegrass	Salinity	Pretreatment (20 mM)	Increased K^+ content; suppressed Na^+ accumulation (higher K^+/Na^+ ratio); reduced H_2O_2 levels; enhanced SOD and CAT activity; altered phytohormone content (increased JA, IAA, CK; decreased ABA)	Zhang et al. (2021)
Mung bean	Salinity	Foliar application (20 mM)	Mitigated salt toxicity; increased uptake of Ca^{2+} and Mg^{2+}; reduced Na^+ accumulation; improved water use efficiency; enhanced accumulation of compatible solutes; activated CAT activity	Rahman et al. (2019)
Wheat	Salinity	Foliar application (15 and 30 mM)	Reduced malondialdehyde (MDA) and H_2O_2 content	Khan et al. (2022)
Grapevines	Alkali $NaHCO_3$	Exogenous application (25 mM)	Enhanced tolerance; increased acetic acid levels in vines; increased biosynthesis of salicylic acid (SA)	Lv et al. (2021)
"Mexican" lime	Cold stress	Foliar application (15 mM)	Increased tolerance to cold at 0 °C; improving enzymatic and non-enzymatic antioxidant systems; reducing electrolyte leakage and malondialdehyde levels	Sanie Khatam et al. (2023)

AA acetic acid, *SA* salicylic acid, *CK* cytokinins, *ABA* abscisic acid, *IAA* indole-3-acetic acid, *CAT* catalase, *SOD* superoxide dismutase, *MDA* malondialdehyde, H_2O_2 hydrogen peroxide photosynthesis, water retention, and drought tolerance in crops such as maize, wheat, cassava, and *Arabidopsis*. However, AA's effects are dose- and pH-dependent, and high concentrations may inhibit root growth (Allen and Allen, 2020; Mahmud et al., 2023).

Soil health and remediation. AA also plays a role in soil management. By transiently lowering soil pH, it increases the availability of certain nutrients and can facilitate phytoremediation of heavy metals such as cadmium (Yang et al., 2023). At moderate concentrations, AA stimulates beneficial microbial enzymatic activities (e.g., saccharase, catalase), but excessive levels can negatively impact beneficial fungi (LeBlanc et al., 2012). Thus, careful dose optimization is required to maximize benefits while minimizing risks to soil biodiversity.

8.8 Conclusion and Future Perspectives

Acetic acid has emerged as a central regulatory metabolite in plant stress physiology, orchestrating responses to both abiotic and biotic stressors. Its multifunctional role is reflected in its capacity to activate key hormonal signaling pathways—particularly those mediated by ABA, JA, and AIA—while simultaneously enhancing antioxidant defenses, stabilizing cellular homeostasis, and modulating stress-responsive gene expression through epigenetic mechanisms such as histone acetylation. In addition, AA contributes to root system plasticity, stomatal regulation, ion balance, and beneficial interactions with rhizosphere microbes, forming an integrated network that enhances plant adaptability under adverse environmental conditions.

These attributes highlight AA's strong potential as a low-cost, eco-friendly biostimulant for sustainable agriculture, particularly in regions facing drought, salinity, or soil degradation. Its dual capacity to improve plant resilience while supporting microbial health positions it as a promising tool in climate-smart crop management.

Nevertheless, critical research gaps remain. Future studies should aim to:

(i) Identify specific sensors or receptors that perceive AA in plants
(ii) Characterize its interactions with other phytohormones and signaling networks under combinatorial stress
(iii) Optimize application strategies—including dose, timing, and formulation—for field deployment
(iv) Evaluate long-term ecological and agronomic impacts, including effects on soil microbiota, nutrient cycling, and yield stability

Addressing these questions will be essential for harnessing acetic acid's full potential and integrating it into next-generation, climate-resilient agricultural systems.

References

Abd-El-Kareem F (2009) Effect of acetic acid fumigation on soil-borne fungi and cucumber root rot disease under greenhouse conditions. Arch Phytopathol Plant Protect 42:213–220. https://doi.org/10.1080/03235400600999380

Adedayo AA, Musser R, Aanaenson M, Babalola OO (2025) The biochemical and molecular mechanisms of plants: a review on insect herbivory. Plant Signal Behav 20:2439248. https://doi.org/10.1080/15592324.2024.2439248

Adhikari TS, Banerjee S, Sarkar K, Sil PC (2023) Crosstalk of hormones, second messengers, and MAPK in plant defense. In: Hormonal cross-talk, plant defense and development. Elsevier, pp 335–351. https://doi.org/10.1016/B978-0-323-95375-7.00016-1

Akhtar SS, Mekureyaw MF, Pandey C, Roitsch T (2020) Role of cytokinins for interactions of plants with microbial pathogens and pest insects. Front Plant Sci 10:1777. https://doi.org/10.3389/fpls.2019.01777

Alavilli H, Lee H, Park M, Lee B (2017) Antarctic moss multiprotein bridging factor 1c overexpression in Arabidopsis resulted in enhanced tolerance to salt stress. Front Plant Sci 8:1206. https://doi.org/10.3389/fpls.2017.01206

Al-Baldawy MSM, Matloob AAAH, Almammory MKN (2023) The importance of nitrogen-fixing bacteria Azotobacter chroococcum in biological control to root rot pathogens (review). IOP Conf Ser Earth Environ Sci 1259:012110. https://doi.org/10.1088/1755-1315/1259/1/012110

Allen MM, Allen DJ (2020) Biostimulant potential of acetic acid under drought stress is confounded by pH-dependent root growth inhibition. Front Plant Sci 11:647. https://doi.org/10.3389/fpls.2020.00647

Allen MM, Allen DJ (2021) Acetic acid is a low cost antitranspirant that increases begonia survival under drought stress. Sci Hortic 287:110257. https://doi.org/10.1016/j.scienta.2021.110257

Anli M, Boutasknit A, Ait-El-Mokhtar M, Ben-Laouane R, Abdouroihamane H, Meddich A, Baslam M (2026) Plant extracts for boosting plant stress tolerance. In: Abdel Latef AAH (eds) Using Stimulants to Improve Plant Health: From Stress to Strength. Elsevier. https://doi.org/10.1016/B978-0-443-45576-6.00009-1

Anwaar MA, Salem MMH, Ayat IE-K (2024) Efficiency of gibberellic acid on feeding, biological and biochemical parameters of Spodoptera littoralis (Lepidoptera: Noctuidae) on tomato Solanum lycopersicum. Egypt J Plant Prot Res Inst 7:288–294. https://doi.org/10.4314/ejppri.v7i3.2

Bajguz A, Piotrowska A (2009) Conjugates of auxin and cytokinin. Phytochemistry 70:957–969. https://doi.org/10.1016/j.phytochem.2009.05.006

Bajpai SK, Nisha, Pandita S, Bahadur A, Verma PC (2024) Recent advancements in the role of histone acetylation dynamics to improve stress responses in plants. Mol Biol Rep 51:413. https://doi.org/10.1007/s11033-024-09300-3

Bharath P, Gahir S, Raghavendra AS (2021) Abscisic acid-induced stomatal closure: an important component of plant defense against abiotic and biotic stress. Front Plant Sci 12:615114. https://doi.org/10.3389/fpls.2021.615114

Cao FY, Yoshioka K, Desveaux D (2011) The roles of ABA in plant–pathogen interactions. J Plant Res 124:489–499. https://doi.org/10.1007/s10265-011-0409-y

Castro-Camba R, Sánchez C, Vidal N, Vielba JM (2022) Plant development and crop yield: the role of gibberellins. Plants 11:2650. https://doi.org/10.3390/plants11192650

Chen D, Shao M, Sun S, Liu T, Zhang H, Qin N et al (2019) Enhancement of jasmonate-mediated antiherbivore defense responses in tomato by acetic acid, a potent inducer for plant protection. Front Plant Sci 10:764. https://doi.org/10.3389/fpls.2019.00764

Chen H, Bullock DA, Alonso JM, Stepanova AN (2021) To fight or to grow: the balancing role of ethylene in plant abiotic stress responses. Plants 11:33. https://doi.org/10.3390/plants11010033

Chowardhara B, Saha B, Borgohain P, Awasthi JP, Kityania S, Panda SK (2022) Effect of ethanol, putresciene and acetic acid on cadmium accumulation and toxicity in Indian mustard. S Afr J Bot 147:42–52. https://doi.org/10.1016/j.sajb.2021.12.019

Cui J, Zhang L, Li Q, Qi Y, Ma J, Guo D et al (2024) Seed dressing containing gibberellic acid, indole-3-acetic acid, and brassinolide improves maize seed germination and seedling growth under cold stress. Agronomy 14:2933. https://doi.org/10.3390/agronomy14122933
Dawood MFA, Tahjib-Ul-Arif M, Sohag AAM, Abdel Latef AAH (2023) Role of acetic acid and nitric oxide against salinity and lithium stress in canola (Brassica napus L.). Plants 13:51. https://doi.org/10.3390/plants13010051
Dello Ioio R, Linhares FS, Sabatini S (2008) Emerging role of cytokinin as a regulator of cellular differentiation. Curr Opin Plant Biol 11:23–27. https://doi.org/10.1016/j.pbi.2007.10.006
Deppenmeier U, Ehrenreich A (2009) Physiology of acetic acid bacteria in light of the genome sequence of Gluconobacter oxydans. Microb Physiol 16:69–80. https://doi.org/10.1159/000142895
Devi VV, Rani KCME, Asaph RS, Suresh P, Gomathinayagam S, Shanmugaiah V (2024) Prevalent plant growth hormone indole-3-acetic acid produced by Streptomyces sp. VSMKU1027 and its potential antifungal activity against phytofungal pathogens. J Pure Appl Microbiol 18:2721–2733. https://doi.org/10.22207/JPAM.18.4.45
Dolpatcha S, Phong HX, Thanonkeo S, Klanrit P, Boonchot N, Yamada M et al (2025) Transcriptional regulation mechanisms in adaptively evolved Pichia kudriavzevii under acetic acid stress. J Fungi 11:177. https://doi.org/10.3390/jof11030177
Domenghini JC (2020) Comparison of acetic acid to glyphosate for weed suppression in the garden. Horttechnology 30:82–87. https://doi.org/10.21273/HORTTECH04453-19
Dong H, Bai L, Zhang Y, Zhang G, Mao Y, Min L et al (2018) Modulation of guard cell turgor and drought tolerance by a peroxisomal acetate–malate shunt. Mol Plant 11:1278–1291. https://doi.org/10.1016/j.molp.2018.07.008
Donoso R, Leiva-Novoa P, Zúñiga A, Timmermann T, Recabarren-Gajardo G, González B (2017) Biochemical and genetic bases of indole-3-acetic acid (auxin phytohormone) degradation by the plant-growth-promoting rhizobacterium Paraburkholderia phytofirmans PsJN. Appl Environ Microbiol 83:e01991-16. https://doi.org/10.1128/AEM.01991-16
Dorna H, Rosińska A, Szopińska D (2021) The effect of acetic acid treatments on the quality of stored carrot (Daucus carota L.) seeds. Agronomy 11:1176. https://doi.org/10.3390/agronomy11061176
El Sabagh A, Islam MS, Hossain A, Iqbal MA, Mubeen M, Waleed M et al (2022) Phytohormones as growth regulators during abiotic stress tolerance in plants. Front Agron 4:765068
El-rahman A, El-mogy SMM, Al-Sharnouby SFS, Abo-Kora HA (2019) An attempt towards reducing salt stress on Flame Seedless grape rootings through the inoculation of Arbuscular Mycorrhizae fungi and Pseudomonas fluorescens. Middle East J Appl Sci 09:575–590
El-Sappah AH, Rather SA (2021) Genomics approaches to study abiotic stress tolerance in plants. In: Plant abiotic stress physiology. Apple Academic Press, Boca Raton, pp 25–46. https://doi.org/10.1201/9781003180579-2
Fernández-Bautista N, Fernández-Calvino L, Muñoz A, Castellano MM (2017a) HOP3 a new regulator of the ER stress response in Arabidopsis with possible implications in plant development and response to biotic and abiotic stresses. Plant Signal Behav 12:e1317421. https://doi.org/10.1080/15592324.2017.1317421
Fernández-Bautista N, Fernández-Calvino L, Muñoz A, Castellano MM (2017b) HOP3, a member of the HOP family in Arabidopsis, interacts with BiP and plays a major role in the ER stress response. Plant Cell Environ 40:1341–1355. https://doi.org/10.1111/pce.12927
Fichman Y, Mittler R (2020) Rapid systemic signaling during abiotic and biotic stresses: is the ROS wave master of all trades? Plant J 102:887–896. https://doi.org/10.1111/tpj.14685
Finkelstein RR, Somerville CR (1990) Three classes of abscisic acid (ABA)-insensitive mutations of Arabidopsis define genes that control overlapping subsets of ABA responses. Plant Physiol 94:1172–1179. https://doi.org/10.1104/pp.94.3.1172
Gomes RJ, Borges M d F, Rosa M d F, Castro-Gómez RJH, Spinosa WA (2018) Acetic acid bacteria in the food industry: systematics, characteristics and applications. Food Technol Biotechnol 56:139–151. https://doi.org/10.17113/ftb.56.02.18.5593

Gu S, Zhu W, Ren L, Sun B, Ren Y, Niu Y et al (2024) New insights into the impact of wood vinegar on the growth and rhizosphere microorganisms of cherry radish (Raphanus sativus L.). PeerJ 12:e18505. https://doi.org/10.7717/peerj.18505

Guillaumot D, Guillon S, Morsomme P, Batoko H (2009) ABA, porphyrins and plant TSPO-related protein. Plant Signal Behav 4:1087–1090. https://doi.org/10.4161/psb.4.11.9796

Han D, Yang Y, Guo Z, Dai S, Jiang M, Zhu Y et al (2024) A review on the interaction of acetic acid bacteria and microbes in food fermentation: a microbial ecology perspective. Foods 13: 2534. https://doi.org/10.3390/foods13162534

Hawrylak-Nowak B, Dresler S, Matraszek R (2015) Exogenous malic and acetic acids reduce cadmium phytotoxicity and enhance cadmium accumulation in roots of sunflower plants. Plant Physiol Biochem 94:225–234. https://doi.org/10.1016/j.plaphy.2015.06.012

He M, Shah Jahan M, Wang Y, Sun J, Shu S, Guo S (2020) Compost amendments based on vinegar residue promote tomato growth and suppress bacterial wilt caused by Ralstonia solanacearum. Pathogens 9:227. https://doi.org/10.3390/pathogens9030227

He Y, Guo Z, Liu R, Li Q, He Y, Feng J et al (2024) Enhancing temperature resilience in dioecious Salix myrtillacea: the role of acetic acid in eco-physiological and microbial adaptations. Ind Crop Prod 218:118888. https://doi.org/10.1016/j.indcrop.2024.118888

Hossain MS, Hasanuzzaman M, Sohag MMH, Bhuyan MHMB, Fujita M (2019) Acetate-induced modulation of ascorbate: glutathione cycle and restriction of sodium accumulation in shoot confer salt tolerance in Lens culinaris Medik. Physiol Mol Biol Plants 25:443–455. https://doi.org/10.1007/s12298-018-00640-6

Hossain MZ, Bahar MM, Sarkar B, Donne SW, Ok YS, Palansooriya KN et al (2020) Biochar and its importance on nutrient dynamics in soil and plant. Biochar 2:379–420

Hossain MS, Abdelrahman M, Tran CD, Nguyen KH, Chu HD, Watanabe Y et al (2022) Modulation of osmoprotection and antioxidant defense by exogenously applied acetate enhances cadmium stress tolerance in lentil seedlings. Environ Pollut 308:119687. https://doi.org/10.1016/j.envpol.2022.119687

Howe GA, Major IT, Koo AJ (2018) Modularity in jasmonate signaling for multistress resilience. Annu Rev Plant Biol 69:387–415. https://doi.org/10.1146/annurev-arplant-042817-040047

Hu Y, Lu Y, Zhao Y, Zhou D-X (2019) Histone acetylation dynamics integrates metabolic activity to regulate plant response to stress. Front Plant Sci 10:1236. https://doi.org/10.3389/fpls.2019.01236

Hu Q, Cui H, Ma C, Li Y, Yang C, Wang K et al (2021) Lipidomic metabolism associated with acetic acid priming-induced salt tolerance in Carex rigescens. Plant Physiol Biochem 167:665–677. https://doi.org/10.1016/j.plaphy.2021.08.045

Husain T, Fatima A, Suhel M, Singh S, Prasad SM, Singh VP (2020) A brief appraisal of ethylene signaling under abiotic stress in plants. Plant Signal Behav 15:1782051. https://doi.org/10.1080/15592324.2020.1782051

Iacomino G, Idbella M, Staropoli A, Nanni B, Bertoli T, Vinale F et al (2024) Exploring the potential of wood vinegar: chemical composition and biological effects on crops and pests. Agronomy 14:114. https://doi.org/10.3390/agronomy14010114

Ihtisham M, Hasanuzzaman M, El-Sappah AH, Zaman F, Khan N, Raza A et al (2023) Primary plant nutrients modulate the reactive oxygen species metabolism and mitigate the impact of cold stress in overseeded perennial ryegrass. Front Plant Sci 14:1149832. https://doi.org/10.3389/fpls.2023.1149832

In Y, Kim J, Kim H, Oh S (2013) Antimicrobial activities of acetic acid, citric acid and lactic acid against Shigella species. J Food Saf 33:79–85. https://doi.org/10.1111/jfs.12025

Ivany JA (2010) Acetic acid for weed control in potato (Solanum tuberosum L.). Can J Plant Sci 90: 537–542. https://doi.org/10.4141/CJPS09026

Janah I, Elhasnaoui A, Abouloifa H, Ait-El-Mokhtar M, Ben Laouane R (2025) Hormonal priming to increase germination of Stevia rebaudiana Bertoni seeds in saline environments. Int J Plant Biol 16:2. https://doi.org/10.3390/ijpb16010002

Jasrotia S, Jasrotia R (2022) Role of ethylene in combating biotic stress. In: Ethylene in plant biology. Wiley, pp 388–397. https://doi.org/10.1002/9781119744719.ch18

Jiao S, Lu Y (2020) Soil pH and temperature regulate assembly processes of abundant and rare bacterial communities in agricultural ecosystems. Environ Microbiol 22:1052–1065. https://doi.org/10.1111/1462-2920.14815

Jin G, Qi J, Zu H, Liu S, Gershenzon J, Lou Y et al (2023) Jasmonate-mediated gibberellin catabolism constrains growth during herbivore attack in rice. Plant Cell 35:3828–3844. https://doi.org/10.1093/plcell/koad191

Jones B, Gunnerås SA, Petersson SV, Tarkowski P, Graham N, May S et al (2010) Cytokinin regulation of auxin synthesis in Arabidopsis involves a homeostatic feedback loop regulated via auxin and cytokinin signal transduction. Plant Cell 22:2956–2969. https://doi.org/10.1105/tpc.110.074856

Kappachery S, AlHosani M, Khan TA, AlKharoossi SN, AlMansoori N, AlShehhi SAS et al (2024) Modulation of antioxidant defense and PSII components by exogenously applied acetate mitigates salinity stress in Avena sativa. Sci Rep 14:620. https://doi.org/10.1038/s41598-024-51302-5

Khan A, Bilal S, Khan AL, Imran M, Shahzad R, Al-Harrasi A et al (2020) Silicon and gibberellins: synergistic function in harnessing ABA signaling and heat stress tolerance in date palm (Phoenix dactylifera L.). Plants 9:620. https://doi.org/10.3390/plants9050620

Khan I, Iqbal M, Mahmod A, Magbool R, Ali M, Aslam MT et al (2022) Foliar applied proline and acetic acid improves growth and yield of wheat under salinity stress by improving photosynthetic pigments, physiological traits, antioxidant activities and nutrient uptake. Not Bot Horti Agrobot Cluj Napoca 50:12820. https://doi.org/10.15835/nbha50312820

Khan I, Hussan S, Chattha MU, Maqbool R, Mahmood A, Ali M et al (2023) Acetic acid mitigates salinity-induced toxic effects in wheat by maintaining photosynthetic efficiency, antioxidant activities, ionic homeostasis, and synthesis of stress-protection hormones and osmolytes. Gesunde Pflanz 75:979–992. https://doi.org/10.1007/s10343-022-00759-3

Kim J-M, To TK, Matsui A, Tanoi K, Kobayashi NI, Matsuda F et al (2017) Acetate-mediated novel survival strategy against drought in plants. Nat Plants 3:17097. https://doi.org/10.1038/nplants.2017.97

Kim H, Seomun S, Yoon Y, Jang G (2021) Jasmonic acid in plant abiotic stress tolerance and interaction with abscisic acid. Agronomy 11:1886. https://doi.org/10.3390/agronomy11091886

Kim M-J, Kim M-J, Yoon I-K, Lee B-H (2024) Effect of acetic acid pretreatment on drought stressed alfalfa plants. J Korean Soc Grassl Forage Sci 44:210–214. https://doi.org/10.5333/KGFS.2024.44.3.210

Kiprotich K, Muema E, Wekesa C, Ndombi T, Muoma J, Omayio D et al (2025) Unveiling the roles, mechanisms and prospects of soil microbial communities in sustainable agriculture. Discov Soil 2:10. https://doi.org/10.1007/s44378-025-00037-4

Koç İ, Öğün E, Namli A, Mendeş M, Kutlu E, Yardim EN (2019) The effects of wood vinegar on some soil microorganisms. Appl Ecol Environ Res 17:2437–2447. https://doi.org/10.15666/aeer/1702_24372447

Kong X, Guo Z, Yao Y, Xia L, Liu R, Song H et al (2022) Acetic acid alters rhizosphere microbes and metabolic composition to improve willows drought resistance. Sci Total Environ 844:157132. https://doi.org/10.1016/j.scitotenv.2022.157132

Krishna P, Gloor G (2001) The Hsp90 family of proteins in Arabidopsis thaliana. Cell Stress Chaperones 6:238. https://doi.org/10.1379/1466-1268(2001)006<0238:THFOPI>2.0.CO;2

Kudo T, To TK, Kim J-M (2023) Simple and universal function of acetic acid to overcome the drought crisis. Stress Biol 3:15. https://doi.org/10.1007/s44154-023-00094-1

Lang D, Lyu D, Zhu Z, Qin S (2019) Exogenous glucose mediates the regulation of root morphology and carbon–nitrogen metabolism by indole-3-acetic acid (IAA) in Malus baccata (L.) Borkh. in soil with low organic carbon content. J Plant Growth Regul 38:1598–1615. https://doi.org/10.1007/s00344-019-10005-2

Lazebnik J, Frago E, Dicke M, van Loon JJA (2014) Phytohormone mediation of interactions between herbivores and plant pathogens. J Chem Ecol 40:730–741. https://doi.org/10.1007/s10886-014-0480-7

LeBlanc L, Paré A, Jean-François J, Hébert M, Surette M, Touaibia M (2012) Synthesis and antiradical/antioxidant activities of caffeic acid phenethyl ester and its related propionic, acetic, and benzoic acid analoguesc. Molecules 17:14637–14650. https://doi.org/10.3390/molecules171214637

Li B, Wang R, Wang S, Zhang J, Chang L (2021a) Diversified regulation of cytokinin levels and signaling during Botrytis cinerea infection in Arabidopsis. Front Plant Sci 12:584042. https://doi.org/10.3389/fpls.2021.584042

Li C, Kong X, Luo Z, Li W, Tang W, Zhang D et al (2021b) Exogenous application of acetic acid improves the survival rate of cotton by increasing abscisic acid and jasmonic acid contents under drought stress. Acta Physiol Plant 43:32. https://doi.org/10.1007/s11738-021-03202-8

Li S, Liu S, Zhang Q, Cui M, Zhao M, Li N et al (2022) The interaction of ABA and ROS in plant growth and stress resistances. Front Plant Sci 13:1050132. https://doi.org/10.3389/fpls.2022.1050132

Lv X, Lv J, Gao S, Xiang G, Yao Y (2021) Acetic acid enhances the tolerance of grapevines to NaHCO3 stress by increasing SA production. Sci Hortic 288:110338. https://doi.org/10.1016/j.scienta.2021.110338

Lynch TJ, Erickson BJ, Miller DR, Finkelstein RR (2017) ABI5-binding proteins (AFPs) alter transcription of ABA-induced genes via a variety of interactions with chromatin modifiers. Plant Mol Biol 93:403–418. https://doi.org/10.1007/s11103-016-0569-1

Macias-Benitez S, Garcia-Martinez AM, Caballero Jimenez P, Gonzalez JM, Tejada Moral M, Parrado Rubio J (2020) Rhizospheric organic acids as biostimulants: monitoring feedbacks on soil microorganisms and biochemical properties. Front Plant Sci 11:633. https://doi.org/10.3389/fpls.2020.00633

Mahmud S, Kamruzzaman M, Bhattacharyya S, Alharbi K, Abd El Moneim D, Mostofa MG (2023) Acetic acid positively modulates proline metabolism for mitigating PEG-mediated drought stress in maize and Arabidopsis. Front Plant Sci 14:1167238. https://doi.org/10.3389/fpls.2023.1167238

Medrano HA, Dungque J, Espulgar HR, Chavez LV, Bautista LS, Sevilla T et al (2024) Optimization of acetic acid—infused insecticide from avocado (persea americana) seed extract using microwave—assisted extraction method. IOP Conf Ser Earth Environ Sci 1372:012076. https://doi.org/10.1088/1755-1315/1372/1/012076

Mine A (2019) Multifaceted involvement of abscisic acid in plant interactions with pathogenic and mutualistic microbes. Adv Bot Res 92:219–253. https://doi.org/10.1016/bs.abr.2019.06.002

Mirfattahi Z, Eshghi S (2023) Acetic acid alleviates salinity damage and improves fruit yield in strawberry by mediating hormones and antioxidant activity. Erwerbs-Obstbau 65:1403–1412. https://doi.org/10.1007/s10341-023-00840-9

Mustilli A-C, Merlot S, Vavasseur A, Fenzi F, Giraudat J (2002) Arabidopsis OST1 protein kinase mediates the regulation of stomatal aperture by abscisic acid and acts upstream of reactive oxygen species production. Plant Cell 14:3089–3099. https://doi.org/10.1105/tpc.007906

Muthusamy SK, Dalal M, Chinnusamy V, Bansal KC (2017) Genome-wide identification and analysis of biotic and abiotic stress regulation of small heat shock protein (HSP20) family genes in bread wheat. J Plant Physiol 211:100–113. https://doi.org/10.1016/j.jplph.2017.01.004

Nagar S, Singh VP, Arora A, Dhakar R, Singh N, Singh GP et al (2021) Understanding the role of gibberellic acid and paclobutrazol in terminal heat stress tolerance in wheat. Front Plant Sci 12:692252. https://doi.org/10.3389/fpls.2021.692252

Nakashima Y, Yamauchi N, Murano H (2025) Suppressive effects of Levilactobacillus brevis on Fusarium-plant disease and its potential contribution to environmentally friendly agriculture. Biol Control 205:105758. https://doi.org/10.1016/j.biocontrol.2025.105758

Narendranath NV, Thomas KC, Ingledew WM (2001) Effects of acetic acid and lactic acid on the growth of Saccharomyces cerevisiae in a minimal medium. J Ind Microbiol Biotechnol 26:171–177. https://doi.org/10.1038/sj.jim.7000090

Nassar AMK, Adss IAA (2016) 2,4-Dichlorophenoxy acetic acid, abscisic acid, and hydrogen peroxide induced resistance-related components against potato early blight (Alternaria solani, Sorauer). Ann Agric Sci 61:15–23. https://doi.org/10.1016/j.aoas.2016.04.005

Nawaz M, Li L, Azeem F, Shabbir S, Zohaib A, Ashraf U et al (2021) Insight of transcriptional regulators reveals the tolerance mechanism of carpet-grass (Axonopus compressus) against drought. BMC Plant Biol 21:71. https://doi.org/10.1186/s12870-021-02844-7

Nishimura N, Yoshida T, Kitahata N, Asami T, Shinozaki K, Hirayama T (2007) ABA-hypersensitive germination1 encodes a protein phosphatase 2C, an essential component of abscisic acid signaling in Arabidopsis seed. Plant J 50:935–949. https://doi.org/10.1111/j.1365-313X.2007.03107.x

Ogawa D, Suzuki Y, Yokoo T, Katoh E, Teruya M, Muramatsu M et al (2021) Acetic-acid-induced jasmonate signaling in root enhances drought avoidance in rice. Sci Rep 11:1–13. https://doi.org/10.1038/s41598-021-85355-7

Ogram AV, Jessup RE, Ou LT, Rao PS (1985) Effects of sorption on biological degradation rates of (2,4-dichlorophenoxy) acetic acid in soils. Appl Environ Microbiol 49:582–587. https://doi.org/10.1128/aem.49.3.582-587.1985

Orek C (2023) A review of the functions of transcription factors and related genes involved in cassava (Manihot esculenta Crantz) response to drought stress. Trop Plants 2:14. https://doi.org/10.48130/TP-2023-0014

Panchal P, Miller AJ, Giri J (2021) Organic acids: versatile stress-response roles in plants. J Exp Bot 72:4038–4052. https://doi.org/10.1093/jxb/erab019

Rahman MM, Mostofa MG, Rahman MA, Islam MR, Keya SS, Das AK et al (2019) Acetic acid: a cost-effective agent for mitigation of seawater-induced salt toxicity in mung bean. Sci Rep 9:15186. https://doi.org/10.1038/s41598-019-51178-w

Rahman M, Mostofa MG, Keya SS, Rahman A, Das AK, Islam R et al (2021) Acetic acid improves drought acclimation in soybean: an integrative response of photosynthesis, osmoregulation, mineral uptake and antioxidant defense. Physiol Plant 172:334–350. https://doi.org/10.1111/ppl.13191

Rahman MM, Keya SS, Sahu A, Gupta A, Dhingra A, Tran L-SP et al (2024) Acetic acid: a cheap but chief metabolic regulator for abiotic stress tolerance in plants. Stress Biol 4:34. https://doi.org/10.1007/s44154-024-00167-9

Rasheed S, Bashir K, Kim J-M, Ando M, Tanaka M, Seki M (2018) The modulation of acetic acid pathway genes in Arabidopsis improves survival under drought stress. Sci Rep 8:7831. https://doi.org/10.1038/s41598-018-26103-2

Rehman A, Azhar MT, Hinze L, Qayyum A, Li H, Peng Z et al (2021) Insight into abscisic acid perception and signaling to increase plant tolerance to abiotic stress. J Plant Interact 16:222–237. https://doi.org/10.1080/17429145.2021.1925759

Riyazuddin R, Verma R, Singh K, Nisha N, Keisham M, Bhati KK et al (2020) Ethylene: a master regulator of salinity stress tolerance in plants. Biomolecules 10:959. https://doi.org/10.3390/biom10060959

Rogawansamy S, Gaskin S, Taylor M, Pisaniello D (2015) An evaluation of antifungal agents for the treatment of fungal contamination in indoor air environments. Int J Environ Res Public Health 12:6319–6332. https://doi.org/10.3390/ijerph120606319

Roychowdhury R, Hada A, Biswas S, Mishra S, Prusty MR, Das SP et al (2025) Jasmonic acid (JA) in plant immune response: unravelling complex molecular mechanisms and networking of defence signalling against pathogens. J Plant Growth Regul 44:89–114. https://doi.org/10.1007/s00344-024-11264-4

Sahu N, Vasu D, Sahu A, Lal N, Singh SK (2017) Strength of microbes in nutrient cycling: a key to soil health. In: Agriculturally important microbes for sustainable agriculture. Springer Singapore, Singapore, pp 69–86. https://doi.org/10.1007/978-981-10-5589-8_4

Samanta S, Roychoudhury A (2025) Molecular crosstalk of jasmonate with major phytohormones and plant growth regulators during diverse stress responses. J Plant Growth Regul 44:62–88. https://doi.org/10.1007/s00344-024-11412-w

Sanie Khatam A, Rastegar S, Aboutalebi Jahromi A, Hassanzadeh Khankahdani H, Bagherian SAA (2023) Physio-biochemical mechanisms of acetic acid involved in cold and freezing tolerance in Mexican lime plants (Citrus aurantifolia Swingle). Erwerbs-Obstbau 65:1693–1704. https://doi.org/10.1007/s10341-023-00885-w

Santoyo G, Hernández-Pacheco C, Hernández-Salmerón J, Hernández-León R (2017) The role of abiotic factors modulating the plant-microbe-soil interactions: toward sustainable agriculture. A review. Span J Agric Res 15:1–15. https://doi.org/10.5424/sjar/2017151-9990

Shu K, Zhou W, Chen F, Luo X, Yang W (2018) Abscisic acid and gibberellins antagonistically mediate plant development and abiotic stress responses. Front Plant Sci 9:416. https://doi.org/10.3389/fpls.2018.00416

Singh A, Roychoudhury A (2023) Abscisic acid in plants under abiotic stress: crosstalk with major phytohormones. Plant Cell Rep 42:961–974. https://doi.org/10.1007/s00299-023-03013-w

Singh D, Dhiman VK, Pandey H, Dhiman VK, Pandey D (2022) Crosstalk between salicylic acid and auxins, cytokinins and gibberellins under biotic stress. In: Auxins, cytokinins and gibberellins signaling in plants. Springer, Cham, pp 249–262. https://doi.org/10.1007/978-3-031-05427-3_11

Sinha R, Fritschi FB, Zandalinas SI, Mittler R (2021) The impact of stress combination on reproductive processes in crops. Plant Sci 311:111007. https://doi.org/10.1016/j.plantsci.2021.111007

Song W, Shao H, Zheng A, Zhao L, Xu Y (2023) Advances in roles of salicylic acid in plant tolerance responses to biotic and abiotic stresses. Plants 12:1–23. https://doi.org/10.3390/plants12193475

Sosnowski J, Truba M, Vasileva V (2023) The impact of auxin and cytokinin on the growth and development of selected crops. Agriculture 13:724. https://doi.org/10.3390/agriculture13030724

Sun T, Zhang J, Zhang Q, Li X, Li M, Yang Y et al (2022) Exogenous application of acetic acid enhances drought tolerance by influencing the MAPK signaling pathway induced by ABA and JA in apple plants. Tree Physiol 42:1827–1840. https://doi.org/10.1093/treephys/tpac034

Tadege M, Dupuis I, Kuhlemeier C, Tadege M, Dupuis I, Kuhlemeier C (1999) Ethanolic fermentation: new functions for an old pathway. Trends Plant Sci 4:320–325. https://doi.org/10.1016/S1360-1385(99)01450-8

Tamogami S, Rakwal R, Agrawal GK (2008) Interplant communication: airborne methyl jasmonate is essentially converted into JA and JA-Ile activating jasmonate signaling pathway and VOCs emission. Biochem Biophys Res Commun 376:723–727. https://doi.org/10.1016/j.bbrc.2008.09.069

Tang T, Yu A, Li P, Yang H, Liu G, Liu L (2016) Sequence analysis of the Hsp70 family in moss and evaluation of their functions in abiotic stress responses. Sci Rep 6:33650. https://doi.org/10.1038/srep33650

Tang N, Cao Z, Wu P, Liu Y, Lou J, Hu Y et al (2023) Comparative transcriptome analysis reveals hormone, transcriptional and epigenetic regulation involved in prickle formation in Zanthoxylum armatum. Gene 871:147434. https://doi.org/10.1016/j.gene.2023.147434

Theerawitaya C, Tisarum R, Samphumphuang T, Singh HP, Takabe T, Cha-um S (2020) Expression levels of vacuolar ion homeostasis-related genes, Na+ enrichment, and their physiological responses to salt stress in sugarcane genotypes. Protoplasma 257:525–536. https://doi.org/10.1007/s00709-019-01450-7

Tian H, Xu L, Li X, Zhang Y (2025) Salicylic acid: the roles in plant immunity and crosstalk with other hormones. J Integr Plant Biol 67:773–785. https://doi.org/10.1111/jipb.13820

Trzaska WJ, Correia JN, Villegas MT, May RC, Voelz K (2015) pH manipulation as a novel strategy for treating mucormycosis. Antimicrob Agents Chemother 59:6968–6974. https://doi.org/10.1128/AAC.01366-15

Utsumi Y, Utsumi C, Tanaka M, Van Ha C, Takahashi S, Matsui A et al (2019) Acetic acid treatment enhances drought avoidance in cassava (Manihot esculenta Crantz). Front Plant Sci 10:521. https://doi.org/10.3389/fpls.2019.00521

Vetterlein D, Carminati A, Kögel-Knabner I, Bienert GP, Smalla K, Oburger E et al (2020) Rhizosphere spatiotemporal organization–a key to rhizosphere functions. Front Agron 2:1–22. https://doi.org/10.3389/fagro.2020.00008

Wang J, Song L, Gong X, Xu J, Li M (2020) Functions of jasmonic acid in plant regulation and response to abiotic stress. Int J Mol Sci 21:1446

Wang Y, Mostafa S, Zeng W, Jin B (2021) Function and mechanism of jasmonic acid in plant responses to abiotic and biotic stresses. Int J Mol Sci 22:8568. https://doi.org/10.3390/ijms22168568

Wang T, Cao X, Chen M, Lou Y, Wang H, Yang Q et al (2022a) Effects of soil acidification on bacterial and fungal communities in the Jiaodong Peninsula, Northern China. Agronomy 12: 927. https://doi.org/10.3390/agronomy12040927

Wang Y, Qiao D, Han Y, Zhang D (2022b) Optimizing acetic acid application strategy can effectively promote the remediation performance of oilseed sunflower on Cd-contaminated soils. Minerals 12:1139. https://doi.org/10.3390/min12091139

Wang G, Wang Z, Zhao F, Chen J-X, Chen Y (2023) Calcium acetate enhances both drought tolerance and arsenic accumulation in Pteris vittata. Environ Pollut Bioavailab 35:2161947. https://doi.org/10.1080/26395940.2022.2161947

Wang X, Chi Y, Song S (2024) Important soil microbiota's effects on plants and soils: a comprehensive 30-year systematic literature review. Front Microbiol 15:1347745. https://doi.org/10.3389/fmicb.2024.1347745

Wani AB, Chadar H, Wani AH, Singh S, Upadhyay N (2017) Salicylic acid to decrease plant stress. Environ Chem Lett 15:101–123. https://doi.org/10.1007/s10311-016-0584-0

Waqas M, Khan AL, Kamran M, Hamayun M, Kang S-M, Kim Y-H et al (2012) Endophytic fungi produce gibberellins and indoleacetic acid and promotes host-plant growth during stress. Molecules 17:10754–10773. https://doi.org/10.3390/molecules170910754

Waqas M, Khan AL, Hamayun M, Shahzad R, Kim Y-H, Choi K-S et al (2015) Endophytic infection alleviates biotic stress in sunflower through regulation of defence hormones, antioxidants and functional amino acids. Eur J Plant Pathol 141:803–824. https://doi.org/10.1007/s10658-014-0581-8

Webber CL III, White PM Jr, Shrefler JW, Spaunhorst DJ (2018) Impact of acetic acid concentration, application volume, and adjuvants on weed control efficacy. J Agric Sci 10:1. https://doi.org/10.5539/jas.v10n8p1

Yang J, Duan G, Li C, Liu L, Han G, Zhang Y et al (2019) The crosstalks between jasmonic acid and other plant hormone signaling highlight the involvement of jasmonic acid as a core component in plant response to biotic and abiotic stresses. Front Plant Sci 10:1–12. https://doi.org/10.3389/fpls.2019.01349

Yang K, Zhu C, Li J, Meng B, Zhong K, Huang W et al (2023) Electric field-assisted leaching of valuable metals from spent lithium-ion batteries in a mixture of acetic acid and ascorbic acid. Hydrometallurgy 221:106152. https://doi.org/10.1016/j.hydromet.2023.106152

Zandalinas SI, Fritschi FB, Mittler R (2021) Global warming, climate change, and environmental pollution: recipe for a multifactorial stress combination disaster. Trends Plant Sci 26:588–599. https://doi.org/10.1016/j.tplants.2021.02.011

Zhang J, Zhang Q, Xing J, Li H, Miao J, Xu B (2021) Acetic acid mitigated salt stress by alleviating ionic and oxidative damages and regulating hormone metabolism in perennial ryegrass (Lolium perenne L.). Grass Res 1:1–10. https://doi.org/10.48130/GR-2021-0003

Zhang Y, Xiao F, Zhang L, Ding Z, Shi G, Li Y (2023) A new mechanism of carbon metabolism and acetic acid balance regulated by CcpA. Microorganisms 11:2303. https://doi.org/10.3390/microorganisms11092303

Zhong Q, Hu H, Fan B, Zhu C, Chen Z (2021) Biosynthesis and roles of salicylic acid in balancing stress response and growth in plants. Int J Mol Sci 22:11672. https://doi.org/10.3390/ijms222111672

Zhu L, Guo J, Sun Y, Wang S, Zhou C (2021) Acetic acid-producing endophyte Lysinibacillus fusiformis orchestrates jasmonic acid signaling and contributes to repression of cadmium uptake in tomato plants. Front Plant Sci 12:1–16. https://doi.org/10.3389/fpls.2021.670216

Zou J-J, Li X-D, Ratnasekera D, Wang C, Liu W-X, Song L-F et al (2015) Arabidopsis CALCIUM-DEPENDENT PROTEIN KINASE8 and CATALASE3 function in abscisic acid-mediated signaling and H2O2 homeostasis in stomatal guard cells under drought stress. Plant Cell 27:1445–1460. https://doi.org/10.1105/tpc.15.00144

Chapter 9
Benzoic Acid and Its Derivatives: Protectors Against Plant Stress

Mona Gergis Dawood

Abstract Benzoic acid and related substances are phenolic compounds that help plants resist the harmful effects of salinity (soil or irrigation water), water stress (flooding or drought), temperature extremes (heating or chilling), and heavy metal stress by enhancing various physiological processes and strengthening the plant's immune system. Phenolic compounds are a type of non-enzymatic antioxidant that helps reduce lipid peroxidation and the formation of reactive oxygen species (ROS), maintaining an optimal balance between plant growth and stress defense. Additionally, benzoic acid and related substances show antimicrobial properties by inhibiting the growth of molds, yeasts, and bacteria, especially in acidic environments. Furthermore, benzoic acid is considered a plant allelopathic substance that accumulates in soil year after year due to its poor biodegradability. Using benzoic acid or its derivatives, such as salicylic acid, p-hydroxybenzoic acid, gallic acid, and sodium benzoate, in small amounts has great potential as a management tool to help crops achieve higher yields under various biotic and abiotic stresses. This chapter focuses on the role of benzoic acid and its derivatives in protecting different kinds of plants from biotic and abiotic stress due to their role in the formation of secondary metabolites, growth promoters, and acting as antitranspirants and antimicrobials. However, the physiological and biochemical mechanisms underlying these roles of benzoic acid and its derivatives are not well understood.

Keywords Phenolic acid · Benzoic acid · Abiotic stress · Biotic stress · Antioxidant defense system · Plant tolerance

M. G. Dawood (✉)
Botany Department, Agriculture and Biological Institute, National Research Centre, Giza, Egypt

A. A. H. Abdel Latef (ed.), *Harnessing Vitamins and Organic Acids for Plant Stress Resilience*, Sustainability Sciences in Asia and Africa,
https://doi.org/10.1007/978-981-95-8593-9_9

9.1 Introduction

It is well established that biotic and abiotic stresses, along with developmental activities such as industrialization and urbanization, pose potential threats to the productivity of various crops. Abiotic stresses include water stress (flooding or drought), salinity stress (soil or irrigated water), heavy metal stress, temperature stress (heating or chilling), and nutrient stress (deficiency or excess), all of which significantly impact the global agricultural sector (Koche et al. 2021; Nasir and Toth 2022; Yang et al. 2023; Abdel Latef et al. 2026).

The main environmental obstacle to crop development is water scarcity (Sirisuntornlak et al. 2019). Water shortages may affect up to 50% of agricultural regions by 2050 (Hasanuzzaman et al. 2019). Additionally, soil salinity has increased due to human activities such as fertilizer application, groundwater intrusion, seawater backflow, and climate change (Arani et al. 2023; Azooz et al. 2026). Extreme drought and salinity often occur together in arid and semi-arid environments (Slama et al. 2008; Nazari et al. 2023), damaging plant productivity (FAO 2017). Extreme temperatures, both heating and chilling, adversely affect plants and reduce crop yields (Yang et al. 2023; Tahjib-Ul-Arif et al. 2024). Over recent decades, high temperatures have become a major stress factor due to global warming (Lippmann et al. 2019). The global air temperature is projected to rise by about 0.2 °C every decade (Koche et al. 2021; Yang et al. 2023). Historically, Ruelland and Zachowski (2010) reported that freezing and chilling cause membrane damage, chlorosis, necrosis, altered enzyme activity, increased cytoplasm viscosity, and cell rupture (Orvar et al. 2000).

Plants generally require certain nutrients, such as copper, manganese, molybdenum, zinc, and cobalt, in very small amounts for growth, development, and physiological functions (Shahid et al. 2015, 2017). However, when these nutrients surpass optimal levels, along with other toxic heavy metals like arsenic, beryllium, aluminum, lead, cadmium, and mercury, they can drastically decrease agricultural yields (Pierart et al. 2015). Heavy metal toxicity has become a more serious concern than pesticides and other pollutants for ecosystems worldwide (Chen et al. 2007). Heavy metals are released into the air, water, and land from various sources, including human activities such as the overuse of chemical fertilizers, pesticides, industrialization, and inadequate waste management, leading to soil degradation and reduced crop productivity (Tchounwou et al. 2012; Shahid et al. 2015, 2017). Excess heavy metals in the soil increase plant absorption, which disrupts metabolic processes, causing reductions in chlorophyll, mineral uptake, and resulting in chlorosis (Ullah et al. 2020; Adnan et al 2026).

Globally, both biotic and abiotic stressors induce excessive reactive oxygen species (ROS), leading to cellular responses such as oxidative or osmotic stress. These responses can damage organelles, disturb redox balance, cause lipid peroxidation, destabilize membranes, and interfere with ROS signaling pathways, ultimately limiting agricultural productivity (Noctor et al. 2018). Regardless of a plant's developmental stage, abiotic stresses can alter its biochemical, physiological, and molecular mechanisms, impacting roughly half of its productivity (Yadav et al. 2020).

Furthermore, continuous use of pesticides, herbicides, fungicides, and antibacterial agents adversely affects ecosystems and accumulates in the food chain. These chemicals influence plant biological functions by impairing defense mechanisms, affecting physiological traits, and posing health risks to humans (Ahmed et al. 2018; Jan et al. 2020; Kumar et al. 2023). There is an urgent need to mitigate the toxicity caused by biotic and abiotic stresses using safe materials such as phenolic compounds. Plants' defense systems include enzymatic and non-enzymatic antioxidants. The latter comprises hydrophilic metabolites like phenolics and flavonoids, and lipophilic antioxidants such as carotenoids (Gill and Tuteja 2010). Phenolic compounds are regarded as some of the most potent defenses against various biotic and abiotic stresses that elevate oxidative stress levels. These antioxidant molecules serve as a secondary defense line against ROS (Gill and Tuteja 2010; Dumanović et al. 2021). Earlier, Fu et al. (2011) noted that total phenolic content in higher plants correlates with increased cellular antioxidant capacity, stimulates redox reactions, and depends on electron donation (Symes et al. 2018).

9.1.1 Phenolic Compounds

Phenolic compounds are a group of secondary metabolites that contain at least one phenol unit with one or more hydroxyl groups (Dar et al. 2017) and are derived from L-phenylalanine via the shikimic acid and phenylpropanoid pathways (De la Rosa et al. 2019). These phytochemicals are ubiquitously distributed in most plant tissues and modulate the plant responses to biotic and abiotic stresses (Ancillotti et al. 2015; Smirnov et al. 2015; De la Rosa et al. 2019; Sharma et al. 2019; Naikoo et al. 2019; Wang et al. 2019a, b; Pratyusha 2022; Khawula et al. 2024). Phenolic compounds are synthesized under normal and stressful conditions to regulate different metabolic and physiological processes in higher plants (Kumar et al. 2019), like seed germination, cell division, biomass accumulation, membrane permeability (Einhellig and Leather 1988; Tanase et al. 2019), hormonal regulation, photosynthetic activity, nutrient uptake, and reproduction (Senaratna et al. 2000; El-Awadi et al. 2016). Under stressful conditions, plants can synthesize more phenolic compounds relative to those grown under normal conditions (Sharma et al. 2019). They are known as potent antioxidants, plant growth promoters, plant defense molecules, allelochemicals, and bioremediation substances (Bujor et al. 2015; Cheng and Cheng 2015; Bistgani et al. 2019; Chen et al. 2019; Radi et al. 2021). These non-enzymatic antioxidants can scavenge free radicals, decrease lipid peroxidation, and protect plant cells from the harmful effects of oxidative stress (Gawlik-Dziki et al. 2012; Chen et al. 2019; Sharma et al. 2019; Farghaly et al. 2021) and decrease the activity of oxidative enzymes (Amarowicz and Weidner 2009). Furthermore, under stress conditions, phenolic compounds have a great ability to chelate elements

(Sharma et al. 2019) and take part in membrane stability, osmolyte maintenance, and the signal transduction pathways (Zagoskina et al. 2023; Yang et al. 2024; Saini et al. 2024; Rao and Zheng 2025). It is worth mentioning that exposure of plants to unsuitable stressful conditions activates the non-enzymatic antioxidant defense system through stimulation of the phenylpropanoid biosynthetic pathway and induces the synthesis of phenolic acids (Bistgani et al. 2019; Sharma et al. 2019; Wang et al. 2019a, b).

9.1.2 Phenolic Acids

Generally, phenolic acids are a type of phenolic compound that contain one COOH group, either alone or combined with one or more OH groups, which donate the H atom and provide antioxidant properties in plants (Radi et al. 2021). Plants accumulate phenolic acids in response to biotic or abiotic stimuli, and they have been shown to increase plant stress tolerance (Williams et al. 2003; Anjum et al. 2013; Valdez-Sepúlveda et al. 2015; Metwally et al. 2018; Hussain et al. 2021). Phenolic acids can be classified into different subclasses based on their chemical structures, including hydroxycinnamic acids (e.g., p-coumaric, ferulic, sinapinic acids) and hydroxybenzoic acids (e.g., protocatechuic, gallic acid, vanillic acid). They are used in various applications, such as bioremediation and the production of antioxidants and allelochemicals (Bujor et al. 2015). The primary role of phenolic acids in plant defense is their antioxidant activity (Karamać et al. 2005). Accordingly, Chen et al. (2020) showed that the presence and position of hydroxyl groups in phenolic acids are crucial for their antioxidant activity. Earlier, Pannala et al. (2001) noted that the presence of more hydroxyl groups increases phenolic acids' ability to scavenge hydroxyl radicals and slow membrane peroxidation (Siquet et al. 2006). So far, phenolic acids can lower soil pH and form metal complexes, which enhance their absorption in soils that tend to chemically fix metal ions, thereby increasing yield components (Lopez-Bucio et al. 2000). Additionally, phenolic acids may be necessary to produce auxin and/or cytokinin (Gharib 2006). On the other hand, trans-cinnamic acid, coumarin, p-hydroxybenzoic acid, and benzoic acid are examples of phenolic acids that can be phytotoxic if they accumulate in large quantities, reducing germination and seedling growth by interfering with cellular enzymes and impairing cell division (Baleroni et al. 2000).

There is a dire need to enhance plant tolerance to biotic and abiotic stresses using efficient, economical, and safe approaches, such as benzoic acid and its derivatives, and focusing on their levels.

9.2 Benzoic Acid

Benzoic acid is an aromatic acid with a carboxyl group attached to a phenolic ring and naturally produced in plants to enhance metabolism.

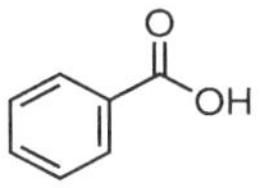

Chemical structure of benzoic acid

Benzoic acid is a crucial substance in several primary and secondary metabolites and natural products that perform essential functions in plants, including growth regulators, cofactors, defense compounds, pollinator attractors, and pest repellent compounds (Williams et al. 2003; Anjum et al. 2013; Widhalm and Dudareva 2015; Valdez-Sepúlveda et al. 2015). Furthermore, benzoic acid is the precursor of a plant hormone called salicylic acid (2-hydroxybenzoic acid) and has been considered to be one of the endogenous signals involved in activation of the defense system (Widhalm and Dudareva 2015). Another group of plant hormones derived from benzoic acids is the aromatic cytokinins, consisting of compounds such as meta-topolin, abenzyl–purine conjugate, to regulate root and shoot formation, and retard senescence of leaves in different crops (Strnad 1997; Tarkowská et al. 2003; Mutui et al. 2012; Widhalm and Dudareva 2015). Moreover, benzoic acid treatments regulated the exchange of gases, increased CO_2 fixation, stomata conductance, net photosynthetic rate, and transpiration rate in cotton, as mentioned by Shakoor et al. (2017). Likewise, the application of benzoic acid on plants exerts bio-regulatory effects on various physiological and biochemical processes regulating plant growth and development, enhancing photosynthetic activity, and ion absorption and transportation (Barros and Dixon 2020). Lopez-Bellido et al. (2010) stated that application of benzoic acid and other plant growth regulators elevated the soil's organic carbon content and capacity for cation exchange, which improved the soil's ability to absorb and retain vital micro and macronutrients like calcium, magnesium, and potassium. In addition, benzoic acid acts as an elicitor for increasing the level of some bioactive compounds, and biosynthesis of phenylpropanoid compounds, which are precursors to an extensive spectrum of secondary metabolites as flavonoids and phenolics in plants (Marchiosi et al. 2020). It also stimulated the biosynthesis of plant defense enzymes, increased activity of antioxidant enzymes, and phytoalexin content (Valdez-Sepúlveda et al. 2015; Sharma et al. 2019; Liu et al. 2021; Sachdev et al. 2021).

Regarding the application of benzoic acid on different plant species, Anjum et al. (2013), Sadak et al. (2013), Talaat et al. (2014), and Valdez-Sepúlveda et al. (2015) stated that benzoic acid treatment significantly increased growth parameters of soybean, geranium plants, as well as tomatillo, tomato, and pepper seedlings, as well and significantly increased the content of phenolic acids in leaf tissues of

Catharanthus roseus (Chipley 2020). Likewise, Zamir et al. (2023) showed that exogenous application of benzoic acid (200 ml/L) is the most pronounced treatment that increased growth and yield of wheat (flag leaf area, tiller number/m^2, length of spike, grain number/spike, and dry weight of plant/m^2). In addition, benzoic acid is released into the soil through root exudates, accumulates, and interferes with the growth of competing organisms because of its allelopathic effect (Kaur et al. 2005; Valdez-Sepúlveda et al. 2015) and accumulates year after year in the same field or soil due to its limited biodegradability (Wang et al. 2023).

9.2.1 Impact of the Level of Benzoic Acid on Different Plant Species

Application of benzoic acid at a low concentration range (less than one millimolar) acted as a stress signaling compound (Benavides-Mendoza et al. 2012; Abd Allah et al. 2015). It increased photosynthetic pigments, gas exchange attributes, stomatal conductance, transpiration rate, and photosynthetic rate (Anjum et al. 2013). Similarly, low levels of benzoic acid triggered metabolic components that play essential roles in plants (Del Mondo et al. 2021), leading to significant boosts in the dry biomass, bioactive compounds, and nutritional quality of lentil sprouts.

On the other hand, high concentrations of benzoic acid showed a negative effect (Gaucin-Delgado et al. 2024). Since numerous studies have indicated that the use of benzoic acid at elevated levels resulted in a reduction of fresh biomass accumulation in *Lactuca sativa* and *Solanum lycopersicum*, as mentioned by Prado et al. (2012) and Valdez-Sepúlveda et al. (2015), respectively, and caused a reduction in dry matter accumulation in wheat sprouts (Salas-Pérez et al. 2016). Likewise, benzoic acid at high levels alters the growth, stress tolerance, anatomy, and morphology of plant species because the elicitors at high levels disrupt normal cellular function, in addition to damaging various macromolecules, causing irreversible lipid damage (Yoo et al. 2013). Yu et al. (2003) mentioned that root exudates of cucumber containing benzoic acid showed an inhibitory impact on the net photosynthesis of cucumber plants and changed the structure of the chloroplasts of leaves, lowered chlorophyll content, and photosynthetic rate. Chen et al. (2021) showed that adding 1.17 mg kg^{-1} of benzoic acid to the soil had toxic effects on *Malus hupehensis* Rehd seedlings and significantly decreased the seedlings' root biomass, root morphology index, net photosynthetic rate, protective enzyme activity, and soil enzyme activity. Additionally, a number of studies demonstrated that the exogenous administration of benzoic acid at high levels led to oxidative stress in several plant species, reduced biosynthesis of antioxidant substances (Deng and Lu 2017), and significantly produced reactive oxygen species via the shikimate/chorismate pathway and phenylalanine (Valdez-Sepúlveda et al. 2015), which, in turn, triggered lipid peroxidation processes and increased malondialdehyde and hydrogen peroxide levels (Wildermuth 2006; Yadav and Singh 2013; Dias et al. 2021). Furthermore, high doses of benzoic acid impacted plant metabolism through unbalancing the flow of electrons in the electron transport chain, resulting in the production of reactive oxygen species as superoxide radicals and singlet oxygen (Benincasa et al. 2019).

9.2.2 *Impact of Benzoic Acid in Reducing the Deleterious Effects of Abiotic Stress*

Benzoic acid is included in the self-signaling of plants, which regulates the defensive reaction of plants against different biotic and abiotic stresses (Senaratna et al. 2003; Williams et al. 2003; Anjum et al. 2013; Pan-Pan et al. 2020; Farghaly et al. 2021; Nehela et al. 2021). Benzoic acid at appropriate levels acts as a cellular antioxidant that reduces reactive oxygen species (Marchiosi et al. 2020), alleviates the damage induced by oxidative stress and the deterioration of the antioxidant regulatory system (Ahmed et al. 2022), and resists the development of degenerative diseases (Koza et al. 2022). It was proven that benzoic acid can resist sandy soil conditions (Dawood et al. 2019), water stress (Anjum et al. 2013), salinity stress (Hussain et al. 2021), temperature extremes (Senaratna et al. 2003), and heavy metals toxicity (Pan-Pan et al. 2020; Farghaly et al. 2021).

Plants growing under sandy soil conditions are exposed to a combination of various abiotic stresses, including nutrient deprivation, low water availability, high irradiance, temperature fluctuations, and soil salinity. In this regard, El-Awadi et al. (2016) demonstrated that benzoic acid (30 mg/L) is the most promising treatment in improving the quality and quantity of lupine plants grown under sandy soil conditions. Likewise, Dawood et al. (2019) found that benzoic acid boosted growth parameters, seed yield quality, and quantity through regulating various biochemical processes of flax plants. In addition, exogenous application of benzoic acid (400 mg/L) on soybean plant—grown under sandy soil conditions—was the most pronounced treatments that caused marked increases in various biochemical constituents of the yielded seeds (protein, oil, carbohydrate, phenolic content and flavonoid) and improved oil quality as it increased total unsaturated fatty acids (TUFA) and total essential fatty acids (TEFA) accompanied by decreases in total saturated fatty acids (TSFA) as mentioned by Sadak et al. (2013).

Regarding water stress, it was noted by Van Hees et al. (2000) that one of the antitranspirants that plants synthesize naturally in trace amounts is benzoic acid. Earlier, it has been observed that using antitranspirants could increase the yield of crops that experience water stress during growth, and it was proven that exogenous administration of antitranspirants at particular stages of plant development may reduce transpiration without negatively impacting plant growth and yields (Hayat et al. 2008; Thomson et al. 2010). Hassanein et al. (2012) stated that benzoic acid acts as a plant growth regulator and antitranspirant that regulates the growth of wheat plants grown under drought conditions. Accordingly, Anjum et al. (2013) mentioned that foliar application of benzoic acid significantly boosted chlorophyll contents, transpiration rate, stomatal conductance, water use efficiency, and ambient CO_2 ratio, which ultimately translated to stimulate vegetative growth attributes, biological yield, grain yield, and harvest index in water-stressed soybean. In addition, Abd Allah et al. (2015) also reported that foliar application of benzoic acid at 400 mg L^{-1} is the most pronounced treatment in reducing the deleterious effect of moisture stress on quinoa plants. Similarly, benzoic acid treatment significantly increased some vegetative aspects of sweet potato grown under water stress because of its role as an antitranspirant to minimize transpiration rate without reducing the growth and yields of plants (Yahaya and Muhammad 2018).

Whereas, under salinity stress, it was noted by Hussain et al. (2021) that foliar application of benzoic acid is recognized as one of the effective methods to alleviate the deleterious impacts of salinity stress on maize plants. In vitro, Khairy and Roh (2016) concluded that benzoic acid treatments recovered the damage caused by high salinity levels and improved salt tolerance of tobacco through increases in chlorophyll, proline, and vitamin C.

Under heating stress, it was mentioned by Shakoor et al. (2017) that application of benzoic acid at 0.75 mM reduced the negative influence of heat stress on cotton plants and caused earliness of all parameters to occur earlier than the control and improved cotton plants' ability to absorb various nutrients (nitrogen, phosphorous, potassium, zinc) and gave maximum net income and benefit-cost ratio.

With respect to the deleterious effect of the accumulation of heavy metals, it was noted that application of either benzoic acid at a higher level (1.5 mM) or cadmium (7 mg L^{-1}) exhibited an inhibitory effect on the growth and metabolism of wheat seedlings. Whereas application of benzoic acid at a lower level (0.5 mM) reduced the toxic effect of cadmium on wheat seedlings (Yadav and Singh 2013). Farghaly et al. (2021) and Radi et al. (2021) mentioned that phenolic acids (benzoic acid, gallic acid, and salicylic acid) reduced the deleterious impact of oxidative damage resulting from an excess of boron. Obviously, benzoic acid (1 μM) was the most pronounced phenolic acid, as it mitigated the deleterious impact of excess boron on the growth of callus by reducing the boron uptake, production of malondialdehyde, damage to plasma membranes and increasing free phenolics, flavonoids, ascorbate, and non-protein thiol. In addition, Radi et al. (2021) added that the most pronounced phenolic acid was benzoic acid > gallic acid > salicylic acid in decreasing the accumulation of different boron forms in tomato calli.

However, the effect of benzoic acid on the growth and yield of different plant species grown under various abiotic stresses has not been widely studied until now.

9.3 Derivatives of Benzoic Acid

The derivatives of benzoic acid include salicylic acid (2-hydroxybenzoic acid or -ortho-hydroxybenzoic acid), *p*-hydroxybenzoic acid (4-hydroxybenzoic acid), protocatechuic acid (3,4-dihydroxybenzoic acid), gallic acid (3,4,5-trihydroxybenzoic acid), and sodium benzoate (Heleno et al. 2015; Rashmi and Negi 2020).

It is worth mentioning that when benzoic acid is applied to plants during stress, the enzyme benzoic acid-2-hydroxylase converts benzoic acid into salicylic acid (Kumar et al. 2015). Earlier, Senaratna et al. (2003) mentioned that benzoic acid is a functional group of salicylic acid, which is responsible for increasing abiotic stress tolerance.

9.3.1 Salicylic Acid

Salicyclic acid is a simple phenolic compound ($C_7H_6O_3$) with an aromatic ring to which one carboxylic group and one hydroxyl group are attached and is distributed in different species of plants. The influence of salicylic acid on various plants depends on its level, duration of application, plant species, and growth stage of the plant (Shi et al. 2009).

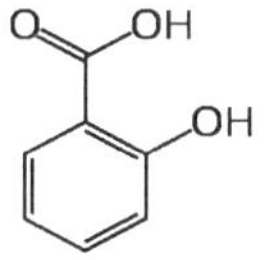

Chemical structure of salicylic acid

Earlier, Blokhina et al. (2003) illustrated that the stimulating effect of salicylic acid and its precursor benzoic acid might be attributed to their bio-regulator impacts on different physiological and biochemical processes in plants, such as the photosynthetic process, cell elongation, cell differentiation, cell division, ion uptake, enzyme activity, protein synthesis, and antioxidant defense system of plants. Additionally, salicylic acid acts as an endogenous growth regulator that is essential for the regulation of various processes, including seed germination, thermogenesis, respiration, transpiration, photosynthesis, stomatal closure, ion transport, and accumulation (Liu et al. 2015; Barman et al. 2016; Li et al. 2022), membrane permeability (Simaei et al. 2011), glycolysis, flowering, and production of fruit (Yang et al. 2023). Salicylic acid and its precursor, benzoic acid, are linked to changes in the cell wall that boost their hardness, enhance the synthesis of antioxidants, and increase mitochondrial activity (Ding et al. 2007), and improve the functionality of plants' photosynthetic apparatus (Khairy and Roh 2016). Since salicylic acid is a crucial modulator of the photosynthesis process because of its impact on leaf and chloroplast structure, carbon dioxide assimilation, chlorophyll and carotenoid levels, and the activity of RUBISCO and carbonic anhydrase ribulose-1,5-bisphosphate carboxylase/oxygenase activity (Pancheva et al. 1996; Khodary 2004; Feng et al. 2020; Wang et al. 2021). In addition, salicylic acid improves the utilization of carbon sources, which significantly increases the quantity of polysaccharides (Wang et al.

2010). It was noted that the application of salicylic acid on plants increased the concentration of non-enzymatic antioxidant substances like flavonoids and phenolic compounds, as mentioned by Zaghlool et al. (2001). It has been stated that spraying salicylic acid (1 μM) during vegetative growth on the canopy of various maize cultivars resulted in significant improvement in root/shoot dry weight, and the amount of the yielded grains. Additionally, they stated that salicylic acid increased root volume, which enhanced nutrient absorption and accumulation and improved grain quality (Tucuch-Haas et al. 2021).

9.3.1.1 Impact of the Level of Salicylic Acid on Different Plant Species

Low levels of salicylic acid (0.5 mM) tend to be beneficial, but high levels are potentially harmful (Damalas and Koutroubas 2021). Hence, a low level of salicylic acid (10 μM) boosted photosynthetic CO_2 assimilation in *Brassica juncea* seedlings (Fariduddin et al. 2003). It was hypothesized that the advantageous effects of a low dose of salicylic acid may have been connected to the blocking of auxin oxidation, thereby increasing the auxin levels (Ahmad et al. 2001). Meanwhile, application of high salicylic acid levels (1–5 mM) resulted in a reduction in photosynthetic rate and RUBISCO activity in barley plants (Pancheva et al. 1996), and chlorophyll contents (Arif et al. 2020). Yuan and Lin (2008) claimed that moderate or optimum concentrations of salicylic acid (0.01–0.05 mM) may be beneficial for abiotic stress tolerance by inducing antioxidant responses, while extremely high concentrations of salicylic acid induced reactive oxygen species overproduction and inhibited antioxidant enzymes, thereby increasing plants' susceptibility to abiotic stress. This behavior is caused by a self-amplifying feedback loop between salicylic acid and hydrogen peroxide, meaning that high levels of salicylic acid induce the production of hydrogen peroxide (Hara et al. 2012; López-González et al. 2023).

9.3.1.2 Impact of Salicylic Acid in Alleviating the Deleterious Effects of Abiotic Stress

Salicylic acid at appropriate levels plays a beneficial role in plants under either normal or stressed conditions, as mentioned by Hernández-Ruiz and Arnao (2018). It was reported that salicylic acid is an endogenous signaling molecule that is regarded as a powerful plant hormone and non-enzymatic antioxidant because it regulates plant growth and developmental processes throughout a plant's lifetime (Khan et al. 2015; Barman et al. 2016; Khan et al. 2019; Zhang et al. 2020). The level of endogenous salicylic acid is changed under abiotic stress conditions (Pavlović et al. 2018; Linić et al. 2019). It has a beneficial role in various physiological and biochemical processes (Kaya et al. 2020), boosting flowering and fruit set (Pacheco et al. 2013), reducing fruit ripening (Srivastava and Dwivedi 2000), controlling glycinebetaine formation, and proline metabolism (Miura and Tada 2014), decreasing transpiration (Gorni and Pacheco 2016), reversing the effect of ABA (Davies

2010), and maintaining gravitropism (El Tayeb and Ahmed 2010). Several researchers mentioned that either endogenous or exogenous salicylic acid modulates plant responses to different types of abiotic stresses, including salinity, heat shock, chilling, drought, UV light, heavy metal toxicity, oxidative stress, and osmotic stress (Janda et al. 1999; Metwally et al. 2003; Horváth et al. 2007a, b; Farooq et al. 2008; Saruhan and Kadioglu 2012; Khan et al. 2015; Barman et al. 2016; Cingoz and Gurel 2016; Ghazi 2017; Khan et al. 2019; Abdelhameed et al. 2021; Desoky et al. 2021; Rhaman et al. 2021; Dawood et al. 2022; Shiade et al. 2022; Roozkhosh et al. 2024; Jameel et al. 2026). Moreover, salicylic acid has been shown to alter plant hormone levels and activate plant defense mechanisms when utilized during adverse conditions (Torun et al. 2020). Besides, salicylic acid promoted genes linked to defense system expression, modifying cellular redox equilibrium, and changing transcription element activities to regulate the plant's resistance to abiotic stress (Kang et al. 2012; Koche et al. 2021). Salicylic acid administration minimizes abiotic stress by triggering transcriptional modifications that regulate mRNA levels and produce amounts of proteins involved in metabolic and signaling functions (Jan et al. 2020). Khan et al. (2015) and Hernández-Ruiz and Arnao (2018) stated that the application of salicylic acid protects plants from abiotic stress through regulating an array of plant metabolic processes and modulating the generation of different osmolytes (glycinebetaine, proline, soluble sugars, and amines) to maintain osmotic homeostasis, scavenging reactive oxygen species, boosting the synthesis of secondary metabolites (terpenes, phenolics, alkaloids, glutathione, glucosinolates), and regulating various hormone pathways (Khan et al. 2020; Koo et al. 2020). Similarly, application of salicylic acid to plants under abiotic stress decreased oxidative stress (Li et al. 2019; Koche et al. 2021) and improved the plant defenses by promoting the activity of antioxidant enzymes (Jogawat et al. 2021; Rai et al. 2024) and, in turn, decreased cellular lipid peroxidation and hydrogen peroxide accumulation (Farghaly et al. 2021; Yang et al. 2023). Additionally, under abiotic stress, salicylic acid significantly increased the amounts of carotenoids, anthocyanins, and chlorophylls (Naz et al. 2021; Azeem et al. 2023). It has previously been shown that micromolar doses of salicylic acid or acetylsalicylic acid (also known as acetyloxybenzoic acid or aspirin), either soaking the seeds or applied as a soil drench, boosted bean and tomato plants' resistance to a range of stressors, including heat, cold, and drought (Senaratna et al. 2000).

Regarding the role of salicylic acid in attenuating drought stress, salicylic acid accelerated root differentiation and enhanced its hollowness, which is reflected in the boosting of water and nutrient absorption by the roots (Youssef and Saeed 2023). Salicylic acid and acetylsalicylic acid significantly increased the drought tolerance of tomato and bean plants (Senaratna et al. 2000). Likewise, Németh et al. (2002) stated that soaking maize grains with 0.5 mM salicylic acid for 1 day enhanced the drought tolerance of maize plants by increasing their polyamine content. In addition, salicylic acid maintained the level of moisture in the internal tissues of the plant and increased the activity of the enzymes responsible for the photosynthesis process and chlorophyll synthesis (Hayat and Ahmad 2007), regulated antioxidant enzymes and scavenging active oxygen species (Fathi et al. 2024), accumulated dry biomass (Aziz et al. 2015), stabilized crop production (Sharma et al. 2017), and increased plant tolerance to stress (Singh et al. 2016). Salicylic acid suppressed the action of ethylene and the ABA-induced stomata closure (Rai et al. 1986), due to the increase in the enzyme β-glycosidase, which controls the opening and closing of stomata (Popova et al. 1997). In field conditions, Liza et al. (2016) and Yahaya and Muhammad (2018) reported that acetylsalicylic acid and benzoic acid caused significant increases in foliage weight of sweet potato and highlighted how the antitranspirants can reduce transpiration without harming the growth, development, and yields of plants (Thomson et al. 2010). Saheri et al. (2020) observed that exogenous application of salicylic acid (0, 0.5, and 1 mM) on *Portulaca oleracea* L. grown under water stress conditions (90%, 60%, and 30% field capacity) elevated the biomass accumulation, content of photosynthetic pigments, gas exchange attributes, fatty acid contents, osmolytes, and secondary metabolites, and lowered drought-induced oxidative stress biomarkers. In the same direction, salicylic acid application reduced the deleterious effect of water deficits on grape tomatoes (Chakma et al. 2021). The combined effect of salicylic acid (0.5 and 1.0 mM) and glycinebetaine (1.0 and 2.0 mM) boosted proline, ascorbic acid, and total sugar content in potato plants grown under water stress (Alhoshan et al. 2018). Likewise, applications of salicylic acid can reduce the deleterious impact of oxidative stress and osmotic stress brought on by water stress via producing osmolytes (soluble sugars and proline), elevating the activity of antioxidant enzymes, and scavenging hydrogen peroxide along with decreasing lipid peroxidation, lipoxygenase activity, and sustaining the water status of plants (Kabiri et al. 2012; Urmi et al. 2023). In addition, salicylic acid maintained the total chlorophyll content and preserved the photosynthetic process (Damalas and Koutroubas 2021), promoted the phloem-loading of sucrose, which preserved leaf turgor, osmotic potential, and delayed senescence in Arabidopsis under salinity (Lee and Park 2010), and upregulated antioxidant enzyme activity in wheat and rice under water deficit (Sharma et al. 2017; Sohag et al. 2020). It was concluded by El-Samad et al. (2019) that salicylic acid takes part in the regulation of the most important physiological processes, like nitrogen metabolism, photosynthesis, production of glycinebetaine, proline metabolism, and enhancement of the antioxidant defense system, thereby increasing plant tolerance against drought. Ahmad et al. (2019) noted that the application of salicylic acid on *Carthamus tinctorius* L. increased the phenylalanine lyase enzyme activity and anthocyanin contents of plants during drought stress.

Regarding salinity stress, salicylic acid is a naturally occurring molecule that protects plant growth and productivity and boosts plant tolerance against salinity stress, as mentioned by Horváth et al. (2007a). Earlier, Shakirova et al. (2003) mentioned that treatment with salicylic acid increased proline and defensive proteins of wheat under salinity stress. Moreover, Fahad and Bano (2012) illustrated that the foliar application of salicylic acid (10^{-5} M) boosted production of osmolytes, endogenous content of abscisic acid and indole acetic acid, and increased the activity of antioxidant enzymes, thereby ameliorating the deleterious effects of salinity stress (EC 4.2 ds/m) on *Zea mays*. Similarly, foliar spraying of salicylic acid (0.5 mM) on mung beans induced the accumulation of glycinebetaine and suppressed ethylene formation under salinity stress (Khan et al. 2014). In the same manner, Miao et al.

(2020) mentioned that salicylic acid (0.3 mM) resulted in a significant increment in relative growth rate, photosynthetic attributes, and architecture of the root system (root length, root surface area, number, and total lengths of primary and secondary roots) of cucumber plants receiving NaCl (50 mM). Furthermore, Hussain et al. (2020) stated that the application of salicylic acid (0.5 mM) and sulphur (2.0 mM) enhanced the antioxidant defense system, activity of nitrate reductase, glutathione reductase, ascorbate peroxidase, scavenged reactive oxygen species, and reduced the accumulation of malondialdehyde thereby maintaining membrane integrity and protecting the cells from oxidative damage in mung bean plants grown under salinity stress (50 mM NaCl). Exogenous application of salicylic acid at 1 mM boosted salt tolerance of *Brassica nigra* by diminishing sodium ion content, lipid peroxidation, and up-regulating antioxidant activity, secondary metabolites, and sugars under various levels of salinity (Ghassemi-Golezani et al. 2020). Mainly, salicylic acid application decreased the uptake and buildup of Na^+ and Cl^- in plant tissues, improved the absorption of essential nutrients, and consequently increased salt tolerance of sorghum (Dehnavi et al. 2019).

Concerning high temperature stress, Wen et al. (2008) stated that the application of salicylic acid can catalyze the accumulation of PAL-mRNA, biosynthesis of new peptidoglycan-associated lipoprotein (PAL) protein, and enhance PAL activity under high temperature stress. It was reported by Khan et al. (2013) that supplementing wheat plants with salicylic acid at 0.5 mM reduced the effects of heat stress (40 °C for 6 h) by modulating proline metabolism through increased γ-glutamyl kinase activity and decreased proline oxidase activity, which preserved the ideal osmotic and water potential and limited the synthesis of ethylene to an ideal range. In another related work, spraying salicylic acid minimizes the oxidative stress damage caused by heat (38 °C) in wheat plants via regulating the expression of heat-stable genes and proteins (Kumar et al. 2015) or through enhanced gas-exchange parameters and effectively alleviates the adverse effects on intercellular CO_2 concentration, net photosynthesis, transpiration rate, and stomatal conductance (Jahan et al. 2019). Exogenous salicylic acid increased thermo tolerance by regulating enzymatic activity, photosynthetic rate, and cell membrane stability in maize (Waqas et al. 2021) and sustaining antioxidant production of pepper plants (Zhang et al. 2020). Further, salicylic acid applications regulated the antioxidant enzymes activity, lowered the production of reactive oxygen species, and enhanced the photosynthetic activity by improving both the structure and function of PSII during high temperature stress, as mentioned by Khavari et al. (2021) and Mihaljević et al. (2024). Application of salicylic acid (1 mM) decreased the deleterious impacts of high temperature stress on *Festuca arundinacea* through boosting the content of photosynthetic pigments, proline accumulation, and increasing non-enzyme and enzymatic antioxidant activities, which reflected on increasing forage yield and heat tolerance (Pirnajmedin et al. 2020). Salicylic acid plays a significant role in increasing plant tolerance to both high- and low-temperature stress through activation of antioxidative enzymes and minimizing the oxidative damage (Arif et al. 2020).

Regarding chilling stress, Janda et al. (1999) stated that treating maize seedlings with salicylic acid using a hydroponic solution increased their resistance to chilling. Exogenous salicylic acid treatment increased photosynthetic activity and antioxidant content in chilling-stressed wheat plants (Yordanova and Popova 2007) and watermelon (Sayyari et al. 2013). Lower levels of either salicylic acid or acetylsalicylic acid concentrations (0.1 and 0.5 mM) showed high effectiveness against low-temperature stress in tomato (Orabi et al. 2015). In addition, exogenous supplementation of salicylic acid at 0.1 mM for 7 days enhanced plant tolerance to cold stress by reducing the production of ROS and improving the activity of antioxidative enzymes in both tolerant and sensitive cultivars of barley (Mutlu et al. 2016). In the same direction, Pan et al. (2020) stated that salicylic acid acts as an upstream signaling molecule under chilling tolerance of cucumber plants by boosting their antioxidant defense system and modifying the expression of genes that respond to chilling stress.

Regarding heavy metal stress, it was reported that external application of salicylic acid increased shoot and root growth of barley plants (Metwally et al. 2003) and maize plants (Krantev et al. 2008) grown under the toxicity of cadmium. Zaid et al. (2020) investigate the effectiveness of salicylic acid supplementation at 10^{-6} M in two menthol mint cultivars, Kosi and Kushal, cultivated under cadmium stress conditions (50 μM). Salicylic acid reduced cadmium toxicity in both cultivars; however, Kushal's cultivar response to salicylic acid was significantly stronger and superior to the Kosi cultivar. In addition, El Dakak and Hassan (2020) mentioned that the application of salicylic acid (1.0 mM) through soil drenching had an additive effect on the activities of antioxidant enzymes (catalase, superoxide dismutase, ascorbate peroxidase, and guaiacol peroxidase), the chlorophyll content, chlorophyll fluorescence attributes, and photosynthetic process in maize grown under the toxicity of cadmium (5 and 10 μM). On the other hand, dealing with aluminum toxicity, it was mentioned by Yalcin and Vardar (2016) that pretreatment of *Hordeum vulgare* with 10 μM salicylic acid might alleviate aluminum-induced toxicity via inhibition of aluminum absorption and activation of the antioxidant enzyme system. Likewise, foliar application of salicylic acid (100 μM) to *Zea mays* increased carbohydrate metabolism, photosynthetic rate, proline accumulation, membrane stability, and activity of antioxidant enzymes under chromium stress (Islam et al. 2016). Regarding the nickel effect, application of salicylic acid (10^{-5} M) lowered the toxicity of nickel (50, 100 and 150 μM) by promoting the activity of antioxidant enzymes and glyoxalase systems in Indian mustard plants (Zaid et al. 2019). Furthermore, Es-sbihi et al. (2020) concluded that salicylic acid application (0.5 mM) significantly boosted some physiological attributes, the content of calcium, phosphorus, potassium, essential oil content, and the density of peltate glands in *Salvia officinalis* L. grown under copper contamination (40 mM).

9.3.2 Other Derivatives of Benzoic Acid (p-Hydroxybenzoic Acid, Protocatechuic Acid, Gallic Acid, Sodium Benzoate)

It was stated by Chen et al. (2020) that the location and presence of the OH group in phenolic acids are crucial factors in the antioxidant activity. The derivatives of benzoic acid (hydroxybenzoic) act as signaling molecules that regulate the expression of antioxidant genes and contribute to the preservation of cellular redox homeostasis in plants by controlling reactive oxygen species toxicity (Xue et al. 2024). Regarding p-hydroxybenzoic acid, Horváth et al. (2007b) reported that p-hydroxybenzoic acid increased abiotic stress tolerance of winter wheat and increased the cell wall's impermeability, which strengthened the plant's defenses against invasion by pathogens. Lu et al. (2014) stated that p-aminobenzoic acid boosted the thermotolerance of mushrooms by eliminating hydrogen peroxide and increasing chitinase and defense heat shock proteins. It was mentioned earlier by Dana et al. (2006) that heat shock proteins are crucial for protecting organisms from high temperatures, and chitinase improved and assisted tobacco resistance against biotic and abiotic stress.

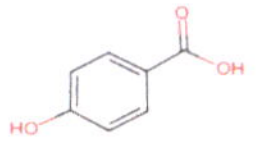

Chemical structure of p-hydroxybenzoic acid (4-hydroxybenzoic acid)

Similarly, protocatechuic acid (3, 4-dihydroxybenzoic acid) is a derivative of p-hydroxybenzoic acid and can scavenge free radicals (Yip et al. 2006) and exhibit antioxidant properties (Hur et al. 2003).

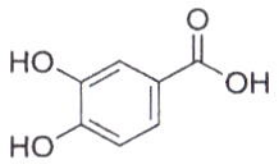

Chemical structure of protocatechuic acid (3,4-dihydroxybenzoic acid)

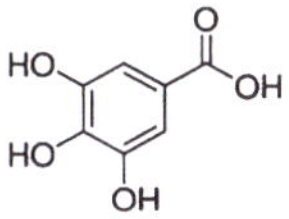

Chemical structure of gallic acid (3,4,5-trihydroxybenzoic acid)

Likewise, gallic acid was the most active phenolic acid in plants and has the largest number of hydroxyl groups of the phenolic acids (Radi et al. 2021). It also plays an essential part in defense against abiotic stress; it lowers the impacts of ozone stress (Rudolphi-Skórska and Sieprawska 2016), cold stress (Ozfidan-Konakci et al.

2019), cadmium toxicity (Ozfidan-Konakci and Kabakci 2020), and boron toxicity (Farghaly et al. 2021).

Regarding sodium benzoate, it was mentioned that sodium benzoate at a low concentration (3 g L^{-1}) reduced the oxidative damage, maintained a comparatively high photosynthetic efficiency, and delayed the loss and degradation of chlorophyll, all of which helped maize plant growth and development (Hu et al. 2014).

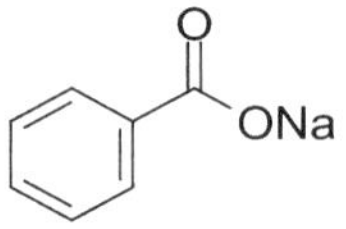

Chemical structure of sodium benzoate

Exogenous sodium benzoate significantly enhanced the contents of free proline, soluble protein, and reduced malondialdehyde, thereby increasing wheat stress tolerance against heavy metal (Zhao et al. 2018). Pan-Pan et al. (2020) mentioned that sodium benzoate at lower concentrations (2–4 g L^{-1}) can alleviate the toxicity of heavy metals to wheat seedlings and enhance their resistance to heavy metals. It significantly increased the ability to uptake and transport nutrients, osmotic protection (soluble protein and free proline), chlorophyll content and chlorophyll fluorescence parameters, activity of superoxide dismutase and reduced malondialdehyde content, accumulation of reactive oxygen species, and maintained the stability and integrity of the cell membrane. On the other hand, higher concentrations of sodium benzoate solution (>6 g L^{-1}) inhibited the growth of wheat seedlings and even caused damage to seedlings. Likewise, several investigations suggested that sodium benzoate hurts crop growth. For example, Liu (2001) mentioned that sodium benzoate has a significant inhibitory impact on the germination of garlic and peas. Chen et al. (2003) reported that sodium benzoate had obvious genetic toxicity to the root tip cells of *Vicia faba* L. seedlings, potentially increasing the rate of micronucleus formation, number of chromosomal, and chromosome structural abnormalities.

9.4 Impact of Benzoic Acid and Its Derivatives in Alleviating the Deleterious Effects of Biotic Stress

The continuous application of pesticides, herbicides, antifungal, antibacterial, etc., leads to several harmful impacts on the ecosystem and accumulates in the food chain. It is direly needed to reduce the toxicity of these substances by applying non-toxic and inexpensive materials. One of these materials includes the application of benzoic acid and its derivatives at appropriate levels. Over the years, several researches revealed that benzoic acid is an organic aromatic monocarboxylic acid that is released by plants to deal with soil-borne pathogens as a signal molecule (Windisch et al. 2021(or an antimicrobial agent (Peters et al. 2016) and possesses potential antifungal activities (Senaratna et al. 2003; Williams et al. 2003; Windisch et al. 2021), antibacterial activities (Agamy et al. 2013), and antioxidant activity (Gao et al. 2018). Benzoic acid significantly impacted the evenness of rhizosphere microorganisms, resulting in changes in microbial community structure and composition (Ding et al. 2023). Also, benzoic acid is a regulator of stress responses (Wildermuth 2006) by promoting numerous secondary metabolites associated with plant–herbivore or plant–pathogen interactions, such as salicin, a major phenolic glycoside in willow (*Salix*) (Ruuhola and Julkunen-Titto 2003); taxol in *Taxus cuspidate* (Walker and Croteau 2000); xanthones in *Hypericum androsaemum* (Abd El-Mawla and Beerhues 2002); cocaine in *Erythroxylum coca* (Bjorklund and Leete 1992); and glucosinolate esters in *Arabidopsis thaliana* (Graser et al. 2001). Recently, Barasker et al. (2024) mentioned that benzoic acid, as a phenolic carboxylic acid, is considered a part of a plant chemical defense mechanism against microbial attacks and acts as a signaling molecule in plant-microbe interactions. Benzoic acid is often present in topical antifungal preparations because it is stable and suppresses the growth of fungi. They also mentioned that benzoic acid might strengthen plant cell walls and cuticles and increase their resistance against fungus penetration. Additionally, it can encourage the deposition of structural components like lignin, which can provide physical barriers to avoid fungal infection. Furthermore, benzoic acid exerts its antimicrobial effects by disrupting the pH balance within microbial cells because it is more effective in acidic conditions. It enters the cell and lowers the intracellular pH, causing acidification of the cytoplasm. This acidification interferes with the normal metabolic processes of microorganisms, ultimately inhibiting their growth and survival. In vitro conditions, benzoic acid lowered the growth rate of *Alternaria alternata* and *Alternaria citri*, the common causes of stem-end rot and internal core rot in citrus fruits (Embaby et al. 2013). In addition, benzoic acid acted as a preservative through undissociated molecules, because undissociated benzoic acid has high lipophilicity and may rapidly enter the cells through the cell membrane and disrupt the permeability of cell membranes in mold and bacteria (Pan-Pan et al. 2020). After entering microbial cells, benzoic acid molecules can acidify the alkalinity in the cell, inhibit the activity of respiratory enzymes, and prevent the condensation reaction of acetyl-CoA, thereby resulting in an antiseptic effect (Hou 2004). Likewise, under field conditions, benzoic acid reduced the severity of *Alternaria blight* on Indian mustard leaves and pods, which is caused by *Alternaria brassicae* and *Alternaria brassicicola* (Singh et al. 2014). Exogenous application of benzoic acid significantly lowered the disease severity of rice brown spot disease, caused by the *Cochliobolus miyabeanus* fungus (Shabana et al. 2008), and the cocoa vascular streak dieback, produced due to the *Oncobasidium theobromae* fungus (Zakariyya et al. 2018). Also, several soil-borne fungi, like Fusarium sp., as well as seed-borne pathogens, like *Aspergillus flavus*, *Penicillium citrinum*, and *Alternaria alternata*, are inhibited by benzoic acid (Liu et al. 2017; Heflish et al. 2020). Regarding its level, it has been reported that low doses of elicitors such as benzoic acid induced plant tolerance to pathogens by activating signals that boosted the production of secondary metabolites (Marchiosi et al. 2020) and included them in developmental signaling cascades and control growth (Abdul et al. 2020; Cherepanov and Zhuravleva 2021).

It was mentioned earlier that the chemical structure of derivatives of benzoic acid provides them with their antifungal action (Berne et al. 2015). In this regard, benzoic acid's antifungal activity against *Aspergillus flavus* and *Aspergillus fumigatus* was increased by stirring the methyl, methoxy, or chloro group on the phenyl ring in the direction ortho > meta > para (Kim et al. 2010). Similarly, the antifungal efficacy of benzoic acid derivatives against *Pleurotus ostreatus*, *Aspergillus niger*, and *Cochliobolus lunatus* was significantly diminished by substitution with two or more methyl groups (Berne et al. 2015).

Regarding benzoic acid derivatives, one of the safest and simplest ways to lessen pesticide toxicity and accumulation in plants is to apply salicylic acid (O-hydroxybenzoic acid), which aids in pesticide detoxification (Kumar et al. 2023). It is worth mentioning that the salicylic acid was previously reported to inhibit the germination of *Penicillium expansum* conidia (da Rocha Neto et al. 2015). Yüzbaşıoğlu and Dalyan (2019) reported that the salicylic acid (1 mM) pretreatment lessened the content of hydrogen peroxide and malondialdehyde formation in thiram-treated *Solanum lycopersicum Mill.* It significantly boosted the photosynthetic pigments, increased the activity of antioxidative enzymes, and increased the expression of pesticide detoxification enzymes to regulate thiram toxicity. It has been observed by Kumar et al. (2023) that salicylic acid (50 μM to 1 mM) may act as a potential tool in reducing the deleterious impact of pesticides. It affected metabolic functioning and secondary metabolite production to regulate the level of toxicity under pesticide stress in most crops. The mechanism behind the protection of different plant species from microbes and the reduced impact of pesticides through the salicylic acid application involved the development of stress tolerance and hastening growth recovery via controlling signal transduction, specific information pathways, and intra- or extracellular signals that are responsible for regulating pesticide stress (Jan et al. 2020). Salicylic acid acts as an essential signal molecule in controlling a plant's defense response to different pathogens, and endogenous salicylic acid activates local immune responses at the infection site (Liu et al. 2015; Barman et al. 2016; Li et al. 2022; Kumar et al. 2023).

Regarding the p-hydroxybenzoic acid effect, Manuja et al. (2013) mentioned that p-hydroxybenzoic acid has various biological impacts as an antialgal, antimicrobial, antimutagenic, nematicidal, antiviral, and antioxidant. It has been reported by Liu et al. (2019) that the exogenous application of p-hydroxybenzoic acid significantly promoted root border cells of grapevine seedlings to produce more phenolic acids. Li et al. (2010) discovered that as the number of consecutive peanut cropping years increased, the amount of p-hydroxybenzoic acid in the soil increased. This resulted in a notable decrease in the number of bacteria and actinomycetes in the rhizosphere soil and an increase in the number of fungi. Likewise, protocatechuic acid (3, 4-dihydroxybenzoic acid) exhibited antifungal activity, antioxidant, and free radical scavenger properties as reported by Sroka and Cisowski (2003) and Hur et al. (2003). In vitro conditions, benzoic acid and its hydroxylated compounds can inhibit the growth of fungi and reduce the growth of *A. solani*, depending on the concentration used. This effectiveness is comparable to that of the most used fungicide-based control strategy for managing tomato early blight disease (Mandal et al. 2019). In vitro studies, Nehela et al. (2021) showed that benzoic acid and two of its hydroxylated derivatives (p-hydroxybenzoic acid and protocatechuic acid) exhibited a fungistatic activity against *A. solani* tomato leaves. The high concentration (100 ppm) of benzoic acid effectively inhibited the mycelial growth of *A. solani* through activation of the enzymatic antioxidant defense machinery and decreased oxidative stress. They mentioned that their effects were almost like a positive control (difenoconazole fungicide), without any significant difference between them.

9.5 Conclusion

Application of benzoic acid and its derivatives, such as salicylic acid, p-hydroxybenzoic acid, gallic acid, and sodium benzoate, at appropriate levels plays an essential role in inducing the immunity defense system of different plant species toward biotic and abiotic stresses. These promotions are mainly due to their roles in inducing photosynthetic rate, transpiration rate, production of growth promoters, secondary metabolites, and osmolytes, increasing activity of antioxidant enzymes, water uptake, nutrients absorption, and transportation, as well as reducing the accumulation of reactive oxygen species, lipid peroxidation, and electrolyte leakages.

In addition, benzoic acid and its derivatives exert a chemical defense mechanism against microbial attacks by acting as an antimicrobial and essential signal molecule in controlling a plant's defense response to different pathogens. Moreover, benzoic acid promotes numerous secondary metabolites associated with plant-herbivore or plant-pathogen interactions, such as salicin, taxol, xanthones, cocaine, and glucosinolate esters.

References

Abd Allah MMS, El-Bassiouny HMS, Elewa TA, El-Sebai TN (2015) Effect of salicylic acid and benzoic acid on growth, yield and some biochemical aspects of quinoa plant grown in sandy soil. Int J ChemTech Res 8(12):216–225

Abdelhameed RE, Abdel Latef AAH, Shehata RS (2021) Physiological responses of salinized fenugreek (Trigonella foenum-graecum L.) plants to foliar application of salicylic acid. Plants 10:657

Abd El-Mawla AM, Beerhues L (2002) Benzoic acid biosynthesis in cell cultures of *hypericum and rosaemum*. Planta 214(5):727–733. https://doi.org/10.1007/s004250100657

Abdul NA, Kumar IS, Nadarajah K (2020) Eliciting and receptor molecules: orchestrators of plant defense and immunity. Int J Mol Sci 21(3):963–972. https://doi.org/10.3390/ijms21030963

Abdel Latef AAH, Hameed M, Khan Z, Khan S, Imran A, Fatima N (2026) Cereals and phytohormones under temperature str foliar stimulants: A sustainable strategy for enhancing plant resilience to abiotic and biotic stresses. Physiologia Plantarum 178, no. 1: e70741

Adnan M (2026) Heavy metals stress mitigation by chitosan nanoparticles. In: Abdel Latef, A.A.H. (eds) Chitosan Nanoparticles for Plant Resilience Under Abiotic and Biotic Stressors. Smart Nanomaterials Technology. Springer, Singapore. https://doi.org/10.1007/978-981-95-6422-4_7

Agamy R, Alamri S, Moustafa MFM, Hashem M (2013) Management of tomato leaf spot caused by *Alternaria tenuissima* Wiltshire using salicylic acid and agrileen. Int J Agric Biol 15:266–272

Ahmad A, Hayat S, Fariduddin Q, Ahmad I (2001) Photosynthetic efficiency of plants of *Brassica juncea* treated with chlorosubstituted auxins. Photosynthetica 39:565–568

Ahmad F, Singh A, Kamal A (2019) Chapter 23: Salicylic acid-mediated defense mechanisms to abiotic stress tolerance. In: MIR K, Reddy PS, Ferrante A, Khan NA (eds) Plant signaling molecules. Woodhead Publishing, Sawston, pp 355–369. https://doi.org/10.1016/B978-0-12-816451-8.00022-8

Ahmed A, Shamsi A, Bano B (2018) Deciphering the toxic effects of iprodione, a fungicide and malathion, an insecticide on thiol protease inhibitor isolated from yellow Indian mustard seeds. Environ Toxicol Pharmacol 61:52–60. https://doi.org/10.1016/j.etap.2018.05.019

Ahmed OS, Tardif CC, Rouger VC, Atanasova V, Richard OF, Waffo-Téguo P (2022) Natural phenolic compounds as promising antimycotoxin agents: where are we now? Compr Rev Food Sci Food Saf 21(2):1161–1197. https://doi.org/10.1111/1541-4337.12891

Alhoshan M, Zahedi M, Ramin AA, Sabzalian MR (2018) Exogenous application of salicylic acid and glycine betaine as tools to enhance biomass and tolerance of potato cultivars. Gesunde Pflanzen 71(1):25–35. https://doi.org/10.1007/s10343-018-00438-2

Amarowicz R, Weidner S (2009) Biological activity of grapevine phenolic compounds. In: Roubelakis-Angelakis KA (ed) Grapevine molecular physiology and biotechnology. Springer Netherlands, Dordrecht, pp 389–405. https://doi.org/10.1007/978-90-481-2305-6

Ancillotti C, Bogani P, Biricolti S, Calistri E, Checchini L, Ciofi L, Gonnelli C, Del Bubba M (2015) Changes in polyphenol and sugar concentrations in wild type and genetically modified *Nicotiana langsdori* Weinmann in response to water and heat stress. Plant Physiol Biochem 97: 52–61. https://doi.org/10.1016/j.plaphy.2015.09.012

Anjum SA, Ehsanullah L, Xue L, Wang L, Saleem MF, Huang C (2013) Exogenous benzoic acid (BZA) treatment can induce drought tolerance in soybean plants by improving gas-exchange and chlorophyll contents. Aust J Crop Sci 7:555–560

Arani EA, Hosseini SA, Javadi S, Ghazavi R (2023) Field studies, ion analysis, and modeling of Kashan plain aquifer to predict the source of salinization. Environ Monit Assess 195(1):23. https://doi.org/10.1007/s10661-022-10644-1

Arif Y, Sami F, Siddiqui H, Bajguz A, Hayat S (2020) Salicylic acid in relation to other phytohormones in plant: a study towards physiology and signal transduction under challenging environment. Environ Exp Bot 175:104040. https://doi.org/10.1016/j.envexpbot.2020.104040

Azeem M, Sultana R, Mahmood A, Qasim M, Siddiqui ZS, Mumtaz S et al (2023) Ascorbic and salicylic acids vitalized growth, biochemical responses, antioxidant enzymes, photosynthetic efficiency, and ionic regulation to alleviate salinity stress in *Sorghum bicolor*. J Plant Growth Regul 42(8):5266–5279. https://doi.org/10.1007/s00344-023-10907-2

Azooz MM, Abdel Latef AAH, Abdelmajeed WG (2026) Biochar ameliorates salt stress in maize by modulating physiological responses and expression of SOS1 and NHX2 genes. Journal of Soil Science and Plant Nutrition. https://doi.org/10.1007/s42729-026-03123-w

Aziz NK, Saeed AKA, Edan K, Ameen SKM (2015) Response of narcissus plant (*Narcissus poeticus*) to foliar applications of plant growth regulators salicylic acid and KT-30. Diyala Agric Sci J 7(1):111–120

Baleroni CRS, Ferrarese MLL, Souza NE, Ferrarese-Filho O (2000) Lipid accumulation during canola seed germination in response to cinnamic acid derivatives. Biol Plant 43(2):313–316

Barasker K, Sharma P, Jain N (2024) A comprehensive study on benzoic acid and its derivatives. Int J Creat Res Thoughts 12(8):e173–e183

Barman K, Sharma S, Kumari P, Siddiqui MW (2016) Salicylic acid. In: Siddiqui MW, Zavala JFA, Hwang CA (eds) Postharvest management approaches for maintaining quality of fresh produce. Springer International Publishing, Cham, pp 51–68. https://doi.org/10.1007/978-3-319-23582-0

Barros J, Dixon RA (2020) Plant phenylalanine/tyrosine ammonia-lyases. Trends Plant Sci 25(1): 66–79. https://doi.org/10.1016/j.tplants.2019.09.011

Benavides-Mendoza A, Burgos-Limón D, Ramírez H, Robledo-Torres V, Sandoval-Rangel A (2012) Benzoic acid effect in the growth and yield of tomato in calcareous soil. Acta Hortic 938:251–256. https://doi.org/10.17660/ActaHortic.2012.938.32

Benincasa PV, Falcinelli BN, Lutts SA, Stagnari FA, Galieni AJ (2019) Sprouted grains: a comprehensive review. Nutrients 11(2):421–435. https://doi.org/10.3390/nu11020421

Berne S, KovačičL SM, Kraševec N, Gobec S, Križaj I, Komel R (2015) Benzoic acid derivatives with improved antifungal activity: design, synthesis, structure-activity relationship (SAR) and CYP_{53} docking studies. Bioorg Med Chem 23(15):4264–4276. https://doi.org/10.1016/j.bmc.2015.06.042

Bistgani ZE, Hashemi M, Dacosta M, Craker L, Maggi F, Morshedloo MR (2019) Effect of salinity stress on the physiological characteristics, phenolic compounds and antioxidant activity of *Thymus vulgaris* L. and *thymus daenensis* Celak. Indus crop. PRO 135(1):311–320. https://doi.org/10.1016/j.indcrop.2019.04.055

Bjorklund JA, Leete E (1992) Biosynthesis of the benzoyl moiety of cocaine from cinnamic acid via (*R*)-(+)-3-hydroxy-3-phenylpropanoic acid. Phytochem 31(11):3883–3887. https://doi.org/10.1016/S0031-9422(00)97546-0

Blokhina O, Virolainen E, Fagerstedt KV (2003) Antioxidants, oxidative damage and oxygen deprivations stress: a review. Ann Bot 91(2):179–194. https://doi.org/10.1093/aob/mcf118

Bujor OC, Talmaciu IA, Volf I, Popa VI (2015) Biorefining to recover aromatic compounds with biological properties. Bioact Compd TAPPI J 14(3):187–193

Chakma R, Biswas A, Saekong P, Ullah H, Datta A (2021) Foliar application and seed priming of salicylic acid affect growth, fruit yield, and quality of grape tomato under drought stress. Sci Hortic 280(5):109904. https://doi.org/10.1016/j.scienta.2021.109904

Chen G, Gao J, Han Y, Geng QS (2003) Study on the mutagenicity of antiseptic sodium benzoate. Carcinogen Teratogen Mutagen 15:178–179

Chen TM, Gokhale J, Shofer S, Kuschner WG (2007) Outdoor air pollution: nitrogen dioxide, sulfur dioxide, and carbon monoxide health effects. Am J Med Sci 333(4):249–256. https://doi.org/10.1097/MAJ.0b013e31803b900f

Chen S, Wang Q, Lu H, Li J, Yang D, Liu J, Yan C (2019) Phenolic metabolism and related heavy metal tolerance mechanism in *Kandelia obovata* under cd and Zn stress. Ecotoxicol Environ Saf 169:134–143. https://doi.org/10.1016/j.ecoenv.2018.11.004

Chen S, Lin R, Lu H, Wang Q, Yang J, Liu J, Yan C (2020) Effects of phenolic acids on free radical scavenging and heavy metal bioavailability in *Kandelia obovata* under cadmium and zinc stress. Chemosphere 249:126341. https://doi.org/10.1016/j.chemosphere.2020.126341

Chen R, Jiang W, Liu Y, Wang Y, Fan H, Chen X, Shen X, Yin C, Mao Z (2021) Amygdalin and benzoic acid on the influences of the soil environment and growth of *Malus hupehensis* Rehd seedlings. ACS Omega 6:12522–12529. https://doi.org/10.1021/acsomega.1c00206

Cheng F, Cheng Z (2015) Research progress on the use of plant allelopathy in agriculture and the physiological and ecological mechanisms of allelopathy. Front Plant Sci 6:1020–1036. https://doi.org/10.3389/fpls.2015.01020

Cherepanov IS, Zhuravleva AA (2021) Formation of primary metabolites and chlorophyll in *Cucumis sativus* L. plants under the influence of the conjugate of L-rhamnose with ma-aminobenzoic acid. Proc Univ Appl Chem Biotechnol 11(1):90–98. https://doi.org/10.21285/2227-2925-2021-11-1-90-98

Chipley JR (2020) Sodium benzoate and benzoic acid. In: Antimicrobials in food, 4th edn. CRC Press, pp 41–88

Cingoz GS, Gurel E (2016) Effects of salicylic acid on thermotolerance and cardenolide accumulation under high temperature stress in *Digitalis trojana* Ivanina. Plant Physiol Biochem 105: 145–149. https://doi.org/10.1016/j.plaphy.2016.04.023

da Rocha Neto AC, Maraschin M, Di Piero RM (2015) Antifungal activity of salicylic acid against *Penicillium expansum* and its possible mechanisms of action. Int J Food Microbiol 215:64–70. https://doi.org/10.1016/j.ijfoodmicro.2015.08.018

Damalas CA, Koutroubas SD (2021) Chapter 5: Foliar applications of salicylic acid for improving crop tolerance to drought stress a review. In: Hayat S, Siddiqui H, Damalas CA (eds) Salicylic acid – a versatile plant growth regulator. Springer, pp 65–76. https://doi.org/10.1007/978-3-030-79229-9

Dana MM, Pintor J, Cubero B (2006) Transgenic tobacco plants overexpressing chitinases of fungal origin show enhanced resistance to biotic and abiotic stress agents. Plant Physiol 142(2): 722–730. https://doi.org/10.1104/pp.106.086140

Dar SA, Wani AB, Mushtaq GA, Kandoo AA, Wani MY (2017) Resistance against insect pests by phenolics and their derivative compounds. J Crop Weed 13(2):187–193

Davies PJ (2010) Plant hormones: their nature, occurrence and functions. In: Davies PJ (ed) Plant hormones: biosynthesis, signal transduction, action, 3rd edn. Springer, pp 1–15

Dawood MG, Sadak MS, Bakry BA, El Karamany MF (2019) Comparative studies on the role of benzoic, t-cinnamic, and salicylic acids on growth, some biochemical aspects, and yield of three flax cultivars grown under sandy soil conditions. Bull Natl Res Cent 43:112. https://doi.org/10.1186/s42269-019-0152-4

Dawood MFA, Zaid A, Abdel Latef AAH (2022) Salicylic Acid Spraying-Induced Resilience Strategies Against the Damaging Impacts of Drought and/or Salinity Stress in Two Varieties of Vicia faba L. Seedlings. Journal of Plant Growth Regulation 41:1919–1942

de la Rosa LA, Moreno-Escamilla JO, Rodrigo-García J, Alvarez-Parrilla E (2019) Chapter 12: Phenolic compounds. In: Yahia E, Carrillo-Lopez A (eds) Postharvest physiology and biochemistry of fruits and vegetables. Woodhead Publishing, Elsevier, Sawston, pp 253–271. https://doi.org/10.1016/B978-0-12-813278-4.00012-9

Dehnavi AR, Zahedi M, Razmjoo J, Eshghizadeh H (2019) Effect of exogenous application of salicylic acid on salt-stressed sorghum growth and nutrient contents. J Plant Nutr 42(1):1–17. https://doi.org/10.1080/01904167.2019.1617307

Del Mondo A, Smerilli A, Ambrosino L, Albini A, Noonan DM, Sansone C, Brunet C (2021) Insights into phenolic compounds from microalgae: structural variety and complex beneficial activities from health to nutraceuticals. Crit Rev Biotechnol 41(2):155–171. https://doi.org/10.1080/07388551.2021.1874284

Deng Y, Lu S (2017) Biosynthesis and regulation of phenylpropanoids in plants. Crit Rev Plant Sci 36(1):257–290. https://doi.org/10.1080/07352689.2017.1402852

Desoky E-SM, Selem E, Abo El-Maati MF, Hassn AASA, Belal HEE, Rady MM et al (2021) Foliar supplementation of clove fruit extract and salicylic acid maintains the performance and antioxidant defense system of *Solanum tuberosum* L. under deficient irrigation regimes. Horticulturae 7(11):435. https://doi.org/10.3390/horticulturae7110435

Dias MC, Pinto DC, Silva AM (2021) Plant flavonoids: chemical characteristics and biological activity. Molecules 26(17):5377–5382. https://doi.org/10.3390/molecules26175377

Ding Z-S, Tian P, Zheng X, Zhou Z-W, Xu Y (2007) Responses of reactive oxygen metabolism and quality in mango fruit to exogenous oxalic acid or salicylic acid under chilling temperature stress. Physiol Plant 130(1):112–121. https://doi.org/10.1111/j.1399-3054.2007.00893.x

Ding Y, Ren H, Hao X, Zhang R, Hao J, Liu J, Pan H, Wang Y (2023) Enhanced phytoremediation of PCBs-contaminated soil by co-expressing tfdB and bphC in Arabidopsis aiding in metabolism and improving toxicity tolerance. Environ Exp Bot 217(5):105548. https://doi.org/10.1016/j.envexpbot.2023.105548

Dumanović J, Nepovimova E, Natić M, KučaK JV (2021) The significance of reactive oxygen species and antioxidant defense system in plants: a concise overview. Front Plant Sci 11:2106. https://doi.org/10.3389/fpls.2020.552969

Einhellig FA, Leather GR (1988) Potentials for exploiting allelopathy to enhance crop production. J Chem Ecol 14(10):1892–1842. https://doi.org/10.1007/BF01013480

El Dakak RA, Hassan IA (2020) The alleviative effects of salicylic acid on physiological indices and defense mechanisms of maize (*Zea mays* L. Giza 2) stressed with cadmium. Envi Processes 7:873–884. https://doi.org/10.1007/s40710-020-00448-1

El Tayeb MA, Ahmed NL (2010) Response of wheat cultivars to drought and salicylic acid. Am Eurasian J Agron 3(1):1–7

El-Awadi ME, Abdel-Baky YR, Sadak MS, Amin AA, Dawood MG (2016) Physiological response of *Lupinus termis* to trans-cinammic acid and benzoic acid treatments under sandy soil conditions. Res J Pharm Biol Chem Sci 7(2):120–129

El-Samad HMA, Shaddad MAK, Ragaey MM (2019) Drought strategy tolerance of four barley cultivars and combined effect with salicylic acid application. Am J Plant Sci 10(4):512–535. https://doi.org/10.4236/ajps.2019.104037

Embaby EM, Hazaa M, Hagag LF, Ibrahim TE, El-Azem FSA (2013) Decay of some citrus fruit quality caused by fungi and their control: II-control Alternaria rot or core rot decay by using some alternative fungicides. J Appl Sci Res 9(8):5671–5678

Es-sbihi FZ, Hazzoumi Z, Benhima R, Joutei KA (2020) Effects of salicylic acid on growth, mineral nutrition, glandular hairs distribution and essential oil composition in *Salvia officinalis* L. grown under copper stress. Environ Sustain 3:199–208. https://doi.org/10.1007/s42398-020-00109-x

Fahad S, Bano A (2012) Effect of salicylic acid on physiological and biochemical characterization of maize grown in saline area. Pak J Bot 44(4):1433–1438

FAO (2017) The future of food and agriculture-trends and challenges. FAO, Rome

Farghaly FA, Salam HK, Hamada AM, Radi AA (2021) The role of benzoic acid, gallic acid and salicylic acid in protecting tomato callus cells from excessive boron stress. Sci Hortic 278: 109867. https://doi.org/10.1016/j.scienta.2020.109867

Fariduddin Q, Hayat S, Ahmad A (2003) Salicylic acid influences net photosynthetic rate, carboxylation efficiency, nitrate reductase activity, and seed yield in *Brassica juncea*. Photosynthetica 41:281–284

Farooq M, Aziz T, Basra SMA, Cheema MA, Rehman H (2008) Chilling tolerance in hybrid maize induced by seed priming with salicylic acid. J Agron Crop Sci 194(2):161–168. https://doi.org/10.1111/j.1439-037X.2008.00300.x

Fathi A, Maleki A, Naseri R (2024) A review of the effects of drought stress on plants and some effective strategies in crop management. J Plant Environ Physiol 19(1):121–145

Feng BH, Li GY, Islam M, Fu WM, Zhou YQ, Chen TT et al (2020) Strengthened antioxidant capacity improves photosynthesis by regulating stomatal aperture and ribulose-1, 5-bisphosphate carboxylase/oxygenase activity. Plant Sci 290:110245. https://doi.org/10.1016/j.plantsci.2019.110245

Fu L, Xu B-T, Xu XR, Gan R-Y, Zhang Y, Xia E-Q, Li H-B (2011) Antioxidant capacities and total phenolic contents of 62 fruits. Food Chem 129(2):345–350. https://doi.org/10.1016/j.foodchem.2011.04.079

Gao Y, Liu W, Wang X, Yang L, Han S, Chen S, Strasser RJ, Valverde BE, Qiang S (2018) Comparative phytotoxicity of usnic acid, salicylic acid, cinnamic acid and benzoic acid on photosynthetic apparatus of *Chlamydomonas reinhardtii*. Plant Physiol Biochem 128:1–12. https://doi.org/10.1016/j.plaphy.2018.04.037

Gaucin-Delgado A, Márquez-Guerrero SY, Méndez FJC, Moreno-Reséndez A, Espinosa-Palomeque B, Gaucin-Delgado JM (2024) Benzoic acid improves nutritional quality in lentil sprouts. Rev Mex Cienc Agric 15(2). https://doi.org/10.29312/remexca.v15i2.3638

Gawlik-Dziki U, Swieca M, Dziki D (2012) Comparison of phenolic acids profile and antioxidant potential of six varieties of spelt (*Triticum spelta* L.). J Agric Food Chem 60(18):4603–4612. https://doi.org/10.1021/jf3011239

Gharib FA (2006) Effect of salicylic acid on the growth, metabolic activities and oil content of basil and marjoram. Int J Agric Biol 8(4):485–492

Ghassemi-Golezani K, Hassanzadeh N, Shakiba MR, Esmaeilpour B (2020) Exogenous salicylic acid and 24-epi-brassinolide improve antioxidant capacity and secondary metabolites of *Brassica nigra*. Biocatal Agric Biotechnol 26:101636. https://doi.org/10.1016/j.bcab.2020.101636
Ghazi DA (2017) Impact of drought stress on maize (*Zea mays*) plant in presence or absence of salicylic acid spraying. J Soil Sci Agric Eng 8(6):223–229. https://doi.org/10.21608/jssae.2017.37382
Gill SS, Tuteja N (2010) Reactive oxygen species and antioxidant machinery in abiotic stress tolerance in crop plants. Plant Physiol Biochem 48(12):909–930. https://doi.org/10.1016/j.plaphy.2010.08.016
Gorni PH, Pacheco AC (2016) Growth promotion and elicitor activity of salicylic acid in *Achillea millefolium* L. Afr J Biotechnol 15(16):657–665. https://doi.org/10.5897/AJB2016.15320
Graser G, Oldham NJ, Brown PD, Temp U, Gershenzon J (2001) The biosynthesis of benzoic acid glucosinolate esters in *Arabidopsis thaliana*. Phytochem 57(1):23–32. https://doi.org/10.1016/s0031-9422(00)00501-x
Hara M, Furukawa J, Sato A, Mizoguchi T, Miura K (2012) Chapter 13: Abiotic stress and role of salicylic acid in plants. In: Ahmad P, Prasad MNV (eds) Abiotic stress responses in plants. Springer, New York, pp 235–251. https://doi.org/10.1007/978-1-4614-0634-1_13
Hasanuzzaman M, Fugita M, Oku H, Islam MT (2019) Plant tolerance to environmental stress: role of Phytoprotectants, 1st edn. CRC Press, Boca Raton, p 488. https://doi.org/10.1201/9780203705315
Hassanein RA, Amiradelkader AFAB, Ali H, Amin AAE, Rashad EM (2012) Grain priming and foliar pretreatment enhanced stress defense in wheat (*Triticum aestivum* var. Gimaza 9) plants cultivated in drought land. Aust J Crop Sci 6:121–129
Hayat S, Ahmad A (2007) Salicylic acid: a plant hormone. Springer, Netherlands. https://doi.org/10.1007/1-4020-5184-0
Hayat S, Hasan SA, Fariduddin Q, Ahmad A (2008) Growth of tomato (*Lycopersicon esculentum*) in response to salicylic acid under water stress. J Plant Interact 3(4):297–304. https://doi.org/10.1080/17429140802320797
Heflish A, El Samra I, Youssef N (2020) Occurrence and control of *Alternaria alternata*, *Penicillium citrinum* and *Aspergillus flavus* mycotoxins in broad bean seeds by benzoic and sorbic acids. Egypt Acad J Biol Sci 12(2):75–87. https://doi.org/10.21608/EAJBSG.2020.134701
Heleno SA, Martins A, Queiroz MJRP, Ferreira ICFR (2015) Bioactivity of phenolic acids: metabolites versus parent compounds: a review. Food Chem 173:501–513. https://doi.org/10.1016/j.foodchem.2014.10.057
Hernández-Ruiz J, Arnao MB (2018) Relationship of melatonin and salicylic acid in biotic/abiotic plant stress responses. Agronomy 8(4):33. https://doi.org/10.3390/agronomy8040033
Horváth E, Szalai G, Janda T (2007a) Induction of abiotic stress tolerance by salicylic acid signaling. J Plant Growth Regul 26(3):290–300. https://doi.org/10.1007/s00344-007-9017-4
Horváth E, Pál M, Szalai G, Páldi E, Janda T (2007b) Exogenous 4-hydroxybenzoic acid and salicylic acid modulate the effect of short-term drought and freezing stress on wheat plants. Biol Plant 51(3):480–487
Hou ZJ (2004) Food additives and application technology. Chemical Industry Publishing, Beijing, pp 126–127
Hu C, Liu ZH, Guo J, Ji XE, Dang KK (2014) Influence of soaking seeds in sodium benzoate and potassium sorbate on growth and development of maize seedlings. J Jilin Agric Sci 39:11–13
Hur JM, Park JG, Yang KH, Park JC, Park JR, Chun SS, Choi JS, Choi JW (2003) Effect of methanol extract of *Zanthoxylum piperitum* leaves and of its compound, protocatechuic acid on hepatic drug metabolizing enzymes and lipid peroxidation in rats. Biosci Biotechnol Biochem 67(5):945–950. https://doi.org/10.1271/bbb.67.945
Hussain SJ, Khan NA, Anjum NA, Masood A, Khan MIR (2020) Mechanistic elucidation of salicylic acid and Sulphur-induced defence systems, nitrogen metabolism, photosynthetic, and

growth potential of mungbean (*Vigna radiata*) under salt stress. J Plant Growth Regul 40:1000–1016. https://doi.org/10.1007/s00344-020-10159-4

Hussain A, Noureen A, Shoukat W, Hussain S, Ali F, Ujjan JA, Memon FS, Bapar NA (2021) Mitigating the effect of salinity stress through foliar application of benzoic acid in spring maize (*Zea mays* L). Ilkogretim Online Elementary Educ Online 20(5):7707–7712. https://doi.org/10.17051/ilkonline.2021.05.874

Islam F, Yasmeen T, Arif MS, Riaz M, Shahzad SM, Imran Q, Ali I (2016) Combined ability of chromium (Cr) tolerant plant growth promoting bacteria (PGPB) and salicylic acid (SA) in attenuation of chromium stress in maize plants. Plant Physiol Biochem 108:456–467. https://doi.org/10.1016/j.plaphy.2016.08.014

Jahan MS, Wang Y, Shu S et al (2019) Exogenous salicylic acid increases the heat tolerance in tomato (*Solanum lycopersicum* L.) by enhancing photosynthesis efficiency and improving antioxidant defense system through scavenging of reactive oxygen species. Sci Hortic Amsterdam 247:421–429. https://doi.org/10.1016/j.scienta.2018.12.047

Jameel A, Akram NA, Bashir R, Ashraf M, Abdel Latef AAH (2026) Salicylic acid enhances yield and alleviates drought-induced oxidative damage in two carrot cultivars with contrasting tolerance. Russian Journal of Plant Physiology 73,23

Jan S, Singh R, Bhardwaj R, Ahmad P, Kapoor D (2020) Plant growth regulators: a sustainable approach to combat pesticide toxicity. 3Biotech 10(11):466. https://doi.org/10.1007/s13205-020-02454-4

Janda T, Szalai G, Tari T, Paldi E (1999) Hydroponic treatment with salicylic acid decreases the effects of chilling injury in maize (*Zea mays* L) plants. Planta 208:175–180

Jogawat A, Yadav B, Chhaya, Lakra N, Singh AK, Narayan OP (2021) Crosstalk between phytohormones and secondary metabolites in the drought stress tolerance of crop plants: a review. Physiol Plant 172(2):1106–1132. https://doi.org/10.1111/ppl.13328

Kabiri R, Farahbakhsh H, Nasibi F (2012) Salicylic acid ameliorates the effects of oxidative stress induced by water deficit in hydroponic culture of *Nigella sativa*. J Stress Physiol Biochem 8(3): 13–22

Kang G, Li G, Xu W et al (2012) Proteomics reveals the effects of salicylic acid on growth and tolerance to subsequent drought stress in wheat. J Proteome Res 11(12):6066–6079. https://doi.org/10.1021/pr300728y

Karamać M, Kosińska A, Pegg RB (2005) Comparison of radical-scavenging activities for selected phenolic acids. Pol J Food Nutr Sci 14/55(2):165–169

Kaur H, Inderjit KS, Kaushik S (2005) Cellular evidence of allelopathic interference of benzoic acid to mustard (*Brassica juncea* L.) seedling growth. Plant Physiol Biochem 43(1):77–81. https://doi.org/10.1016/j.plaphy.2004.12.007

Kaya C, Ashraf M, Alyemeni MN, Corpas FJ, Ahmad P (2020) Salicylic acid-induced nitric oxide enhances arsenic toxicity tolerance in maize plants by upregulating the ascorbate-glutathione cycle and glyoxalase system. J Hazard Mater 399:123020. https://doi.org/10.1016/j.jhazmat.2020.123020

Khairy AIH, Roh KS (2016) Effect of salicylic acid, benzoic acid, and *p*-coumaric acid on growth, chlorophyll, proline, and vitamin C of salinity-stressed tobacco (*Nicotiana tabacum*). Inter J Plant Soil Sci 9(4):1–10. https://doi.org/10.9734/IJPSS/2016/22396

Khan MIR, Iqbal N, Masood A, Per TS, Khan NA (2013) Salicylic acid alleviates adverse effects of heat stress on photosynthesis through changes in proline production and ethylene formation. Plant Signal Behav 8(11):e26374. https://doi.org/10.4161/psb.26374

Khan MIR, Asgher M, Khan NA (2014) Alleviation of salt-induced photosynthesis and growth inhibition by salicylic acid involves glycinebetaine and ethylene in mung bean (*Vigna radiata* L.). Plant Physiol Biochem 80(3):67–74. https://doi.org/10.1016/j.plaphy.2014.03.026

Khan MIR, Fatma M, Per TS, Anjum NA, Khan NA (2015) Salicylic acid-induced abiotic stress tolerance and underlying mechanisms in plants. Front Plant Sci 6:462. https://doi.org/10.3389/fpls.2015.00462

Khan A, Kamran M, Imran M, Al-Harrasi A, Al-Rawahi A, Al-Amri I, Khan AL (2019) Silicon and salicylic acid confer high-pH stress tolerance in tomato seedlings. Sci Rep 9:19788. https://doi.org/10.1038/s41598-019-55651-4

Khan N, Bano A, Ali S, Babar MA (2020) Crosstalk amongst phytohormones from planta and PGPR under biotic and abiotic stresses. Plant Growth Regul 90:189–203. https://doi.org/10.1007/s10725-020-00571-x

Khavari M, Fatahi R, Zamani Z (2021) Salicylic acid and kaolin effects on pomological, physiological, and phytochemical characters of hazelnut (*Corylus avellana*) at warm summer condition. Sci Rep 11(1):4568. https://doi.org/10.1038/s41598-021-83790-0

Khawula S, Gokul A, Niekerk LA, Basson G, Keyster M, Badiwe M, Klein A, Nkomo M (2024) Insights into the effects of hydroxycinnamic acid and its secondary metabolites as antioxidants for oxidative stress and plant growth under environmental stresses. Curr Issues Mol Biol 46(1): 81–95. https://doi.org/10.3390/cimb46010007

Khodary SEA (2004) Effect of salicylic acid on the growth, photosynthesis and carbohydrate metabolism in salt stressed maize plants. Int J Agric Biol 6:5–8

Kim JH, Campbell BC, Mahoney N, Chan KL, Molyneux RJ, Balajee A (2010) Augmenting the activity of antifungal agents against aspergilli using structural analogues of benzoic acid as chemosensitizing agents. Fungal Biol 114(10):817–824. https://doi.org/10.1016/j.funbio.2010.07.008

Koche D, Gandhi R, Rathod S, Shirsat R (2021) An update on role of salicylic acid (SA) in abiotic stress tolerance in crop plants: a review. Agric Biol Res 37(6):201–225

Koo YM, Heo AY, Choi HW (2020) Salicylic acid as a safe plant protector and growth regulator. Plant Pathol J 36(1):1–10. https://doi.org/10.5423/PPJ.RW.12.2019.0295

Koza NA, Adedayo AA, Babalola OO, Kappo AP (2022) Microorganisms in plant growth and development: roles in abiotic stress tolerance and secretion of secondary metabolites. Microorganisms 10(8):1528–1539. https://doi.org/10.3390/microorganisms10081528

Krantev A, Yordanova R, Janda T, Szalai G, Popova L (2008) Treatment with salicylic acid decreases the effect of cadmium on photosynthesis in maize plants. J Plant Physiol 165(9): 920–931. https://doi.org/10.1016/j.jplph.2006.11.014

Kumar RR, Sharma SK, Goswami S, Verma P, Singh K, Dixit N et al (2015) Salicylic acid alleviates the heat stress-induced oxidative damage of starch biosynthesis pathway by modulating the expression of heat-stable genes and proteins in wheat (*Triticum aestivum*). Acta Physiol Plant 37(8):143. https://doi.org/10.1007/s11738-015-1861-8

Kumar V, Sharma A, Kohli SK, Bali S, Sharma M, Kumar R, Bhardwaj R, Thukral AK (2019) Differential distribution of polyphenols in plants using multivariate techniques. Biotechnol Res Innov 3(1):1–21. https://doi.org/10.1016/j.biori.2019.03.001

Kumar A, Yadav PK, Singh S, Singh A (2023) An overview on the modulation of pesticide detoxification mechanism via salicylic acid in the plants. Environ Pollut Bioavailability 35(1): 2242701. https://doi.org/10.1080/26395940.2023.2242701

Lee S, Park CM (2010) Modulation of reactive oxygen species by salicylic acid in Arabidopsis seed germination under high salinity. Plant Signal Behav 5(12):1534–1536. https://doi.org/10.4161/psb.5.12.13159

Li PD, Wang XX, Li YL, Wang HW, Liang FY, Dai CC (2010) The contents of phenolic acids in continuous cropping peanut and their allelopathy. Acta Ecol Sin 30:2128–2134

Li Q, Wang G, Wang Y, Yang D, Guan C, Ji J (2019) Foliar application of salicylic acid alleviates the cadmium toxicity by modulation the reactive oxygen species in potato. Ecotoxicol Environ Saf 172:317–325. https://doi.org/10.1016/j.ecoenv.2019.01.078

Li A, Sun X, Liu L (2022) Action of salicylic acid on plant growth. Front Plant Sci 13:878076. https://doi.org/10.3389/fpls.2022.878076

Linić I, Šamec D, Grúz J, Bok V, Strnad M, Salopek-Sondi B (2019) Involvement of phenolic acids in short-term adaptation to salinity stress is species-specific among brassicaceae. Plants 8:155. https://doi.org/10.3390/plants8060155

Lippmann R, Babben S, Menger A, Delker C, Quint M (2019) Development of wild and cultivated plants under global warming conditions. Curr Biol 29(24):R1326–R1338. https://doi.org/10.1016/j.cub.2019.10.016

Liu JX (2001) Primary study on the influence for plants germinating of sodium benzoate. J Biol 18: 28–35

Liu X, Rockett KS, Korner CJ, Pajerowska-Mukhtar KM (2015) Salicylic acid signalling: new insights and prospects at a quarter-century milestone. Essays Biochem 58:101–113. https://doi.org/10.1042/bse0580101

Liu J, Li X, Jia Z, Zhang T, Wang X (2017) Effect of benzoic acid on soil microbial communities associated with soilborne peanut diseases. Appl Soil Ecol 110:34–42. https://doi.org/10.1016/j.apsoil.2016.11.001

Liu Q, Li K, Guo X, Ma L, Guo Y, Liu Z (2019) Developmental characteristics of grapevine seedlings root border cells and their response to *r*-hydroxybenzoic acid. Plant Soil 443:199–218

Liu W, Feng Y, Yu S, Fan Z, Li X, Li J, Yin H (2021) The flavonoid biosynthesis network in plants. Int J Mol Sci 22(23):12824–12837. https://doi.org/10.3390/ijms222312824

Liza SC, Saenz HH, Torres OC (2016) Environmental effect and acetyl salicylic acid on agronomic performance of three sweet potato genotypes. Open Access Libr J 3:e3009. https://doi.org/10.4236/oalib.1103009

Lopez-Bellido RJ, Lal R, Danneberger TK, Stree JR (2010) Plant growth regulator and nitrogen fertilizer effects on soil organic carbon sequestration in creeping bent grass fairway turf. Plant Soil 332:247–255. https://doi.org/10.1007/s11104-010-0289-2

Lopez-Bucio J, Nieto-Jacobo MF, Ramírez-Rodríguez VV, Herrera-Estrella L (2000) Organic acid metabolism in plants: from adaptative physiology to transgenic varieties for cultivation in extreme soils. Plant Sci 160:1–13. https://doi.org/10.1016/s0168-9452(00)00347-2

López-González D, Ferradás Y, Araniti F, Graña E, Hermida-Ramón JM, González MV, Teijeira M, Rey M, Reigosa MJ, Sánchez-Moreiras AM (2023) Trans-cinnamaldehyde-related overproduction of benzoic acid and oxidative stress on *Arabidopsis thaliana*. Front Plant Sci 14: 1157309. https://doi.org/10.3389/fpls.2023.1157309

Lu Z, Kong X, Lu Z, Xiao M, Chen M et al (2014) Para-aminobenzoic acid (PABA) synthase enhances thermotolerance of *Mushroom Agaricus* bisporus. PLoS One 9(3):e91298. https://doi.org/10.1371/journal.pone.0091298

Mandal D, Pal R, Mohanty AK (2019) Management of early blight of tomato. Indian J Hortic 76(3): 548–551. https://doi.org/10.5958/0974-0112.2019.00088.4

Manuja R, Sachdeva S, Jain A, Chaudhary J (2013) A Comprehensive review on biological activities of p-hydroxy benzoic acid and its derivatives. Int J Pharm Sci Rev Res 22(2):109–115

Marchiosi R, Santos WD, Constantin RP, Lima RB, Soares AR, Finger-Teixeira A, Mota TR, Oliveira DM, Foletto-Felipe MP, Abrahão JB, Ferrarese-Filho O (2020) Biosynthesis and metabolic actions of simple phenolic acids in plants. Phytochem Rev 19:865–906. https://doi.org/10.1007/s11101-020-09689-2

Metwally A, Finkemeier I, Georgi M, Dietz K-J (2003) Salicylic acid alleviates the cadmium toxicity in barley seedlings. Plant Physiol 132(1):272–281. https://doi.org/10.1104/pp.102.018457

Metwally AM, Radi AA, El-Shazoly RM, Hamada AM (2018) The role of calcium, silicon and salicylic acid treatment in protection of canola plants against boron toxicity stress. J Plant Res 131(6):1015–1028. https://doi.org/10.1007/s10265-018-1008-y

Miao Y, Luo X, Gao X, Wang W, Li B, Hou L (2020) Exogenous salicylic acid alleviates salt stress by improving leaf photosynthesis and root system architecture in cucumber seedlings. Sci Hortic 272:109577. https://doi.org/10.1016/j.scienta.2020.109577

Mihaljević I, Viljevac Vuletić M, Tomaš V, Zdunić Z, Vuković D (2024) Photosynthetic responses of heat-stressed apple leaves to foliar application of salicylic and ascorbic acid. Inter J Photosynthetic Res 62(1):79–89. https://doi.org/10.32615/ps.2024.006

Miura K, Tada Y (2014) Regulation of water, salinity, and cold stress responses by salicylic acid. Front Plant Sci 5(4):1–12. https://doi.org/10.3389/fpls.2014.00004

Mutlu S, Atıcı Ö, Nalbantoğlu B, Mete E (2016) Exogenous salicylic acid alleviates cold damage by regulating antioxidative system in two barley (*Hordeum vulgare* L.) cultivars. Front Life Sci 9: 99–109

Mutui TM, Mibus H, Serek M (2012) Effect of meta-topolin on leaf senescence and rooting in Pelargonium × hortorum cuttings. Postharvest Biol Technol 63(1):107–110. https://doi.org/10.1016/j.postharvbio.2011.09.010

Naikoo MI, Dar MI, Raghib F, Jaleel H, Ahmad B, Raina A, Khan FA, Naushin F (2019) Chapter 9: Role and regulation of plants phenolics in abiotic stress tolerance: an overview. In: Plant signaling molecules, 1st edn. Elsevier, Amsterdam, pp 157–168. https://doi.org/10.1016/B978-0-12-816451-8.00009-5

Nasir MW, Toth Z (2022) Effect of drought stress on potato production: a review. Agronomy 12(3): 635. https://doi.org/10.3390/agronomy12030635

Naz R, Sarfraz A, Anwar Z, Yasmin H, Nosheen A, Keyani R, Roberts TH (2021) Combined ability of salicylic acid and spermidine to mitigate the individual and interactive effects of drought and chromium stress in maize (*Zea mays* L.). Plant Physiol Biochem 159:285–300. https://doi.org/10.1016/j.plaphy.2020.12.022

Nazari M, Ghasemi-Soloklui AA, Kordrostami M, Abdel Latef AAH (2023) Deciphering the response of medicinal plants to abiotic stressors: A focus on drought and salinity. Plant Stress 10:100255

Nehela Y, Taha NA, Elzaawely AA, Xuan TD, Amin MA, Ahmed ME, El-Nagar A (2021) Benzoic acid and its hydroxylated derivatives suppress early blight of tomato (*Alternaria solani*) via the induction of salicylic acid biosynthesis and enzymatic and non-enzymatic antioxidant defense machinery. J Fungi 7(8):663. https://doi.org/10.3390/jof7080663

Németh M, Janda T, Horváth E, Páldi E, Szalai G (2002) Exogenous salicylic acid increases polyamine content but may decrease drought tolerance in maize. Plant Sci 162(4):569–574. https://doi.org/10.1016/S0168-9452(01)00593-3

Noctor G, Reichheld J-P, Foyer CH (2018) ROS-related redox regulation and signaling in plants. Cell Dev Biol 80:3–12. https://doi.org/10.1016/j.semcdb.2017.07.013

Orabi SA, Dawood MG, Saleem SR (2015) Comparative study between the physiological role of hydrogen peroxide and salicylic acid in alleviating low temperature stress on tomato plants grown under sandponic culture. Sci Agric 9:49–59. https://doi.org/10.15192/pscp.sa.2015.1.9.4959

Orvar BL, Sangwan V, Omann F, Dhindsra RS (2000) Early steps in cold sensing by plant cells: the role of actin cytoskeleton and membrane fluidity. Plant J 23(6):785–794. https://doi.org/10.1046/j.1365-313x.2000.00845.x

Ozfidan-Konakci C, Kabakci M (2020) The impacts of gallic acid on redox state of antioxidants related to ascorbate-glutathione cycle in wheat (*Triticum aestivum*) grown under cadmium toxicity. Agric Res 9:543–553

Ozfidan-Konakci C, Yildiztugay E, Yildiztugay A, Kucukoduk M (2019) Cold stress in soybean (*Glycine max* L.) roots: exogenous gallic acid promotes water status and increases antioxidant activities. Bot Serbica 43(1):59–71. https://doi.org/10.2298/BOTSERB1901059O

Pacheco AC, Cabral CDS, Fermino ES, Aleman CC (2013) Salicylic acid-induced changes to growth, flowering and flavonoids production in marigold plants. J Med Plant Res 7(42): 3158–3163. https://doi.org/10.5897/JMPR2013.5208

Pan D-Y, Fu X, Zhang X-W, Liu F-J, Bi H-G, Ai X-Z (2020) Hydrogen sulfide is required for salicylic acid induced chilling tolerance of cucumber seedlings. Protoplasma 257(6): 1543–1557. https://doi.org/10.1007/s00709-020-01531-y

Pancheva TV, Popova LP, Uzunova AN (1996) Effect of salicylic acid on growth and photosynthesis in barley plants. J Plant Physiol 149(1–2):57–63. https://doi.org/10.1016/S0176-1617(96)80173-8

Pannala SA, Chan TS, O'Brien PJ, Rice-Evans CA (2001) Flavonoid B-ring chemistry and antioxidant activity: fast reaction kinetics. Biochem Biophys Res Commun 282(5): 1161–1168. https://doi.org/10.1006/bbrc.2001.4705

Pan-Pan L, Chen Z, Yuan L, Ji-Jia G, Yuan C, De-Hua C, Xiang Z (2020) Effects of sodium benzoate on growth and physiological characteristics of wheat seedlings under compound heavy metal stress. J Integr Agric 19(4):1010–1018. https://doi.org/10.1016/S2095-3119(19)62723-1

Pavlović I, Petřík I, Tarkowská D, Lepeduš H, Bok VV, Brkanac SR, Novák O, Salopek-Sondi B (2018) Correlations between phytohormones and drought tolerance in selected brassica crops: Chinese cabbage, white cabbage and kale. Int J Mol Sci 19(10):2866. https://doi.org/10.3390/ijms19102866

Peters BJ, Groninger AS, Fontes FL, Crick DC, Crans DC (2016) Differences in interactions of benzoic acid and benzoate with interfaces. Langmuir 32(37):9451–9459. https://doi.org/10.1021/acs.langmuir.6b02073

Pierart A, Shahid M, Séjalon-Delmas N, Dumat C (2015) Antimony bioavailability: knowledge and research perspectives for sustainable agricultures. J Hazard Mater 289:219–234. https://doi.org/10.1016/j.jhazmat.2015.02.011

Pirnajmedin F, Majidi MM, Taleb H, Maibody SAMM, Saeidi G (2020) Amelioration of high temperature stress by exogenously applied salicylic acid: genotype-specific response of physiological traits. Agron J 112(3):1573–1579. https://doi.org/10.1002/agj2.20150

Popova L, Pancheva T, Uzunova A (1997) Salicylic acid: properties, biosynthesis and physiological role. Bulg J Plant Physiol 23(1–2):85–93

Prado WE, Fonseca AÁ, Batista EL, Larramendí LR, Gómez GG, González RP (2012) Efecto de losácidossalicílico y benzoicoen la lechuga (*Lactuca sativa* L.). Centro Agrícola 39(4):85–89

Pratyusha S (2022) Phenolic compounds in the plant development and defense: an overview. In: Hasanuzzaman M, Nahar K (eds) Plant stress physiology-perspectives in agriculture. IntechOpen, Rijeka. https://doi.org/10.5772/intechopen.102873

Radi AA, Salam HK, Hamada AM, Farghaly FA (2021) Phenolic acids mediate boron excess tolerance in tomato callus tissues by regulating antioxidant enzymes and boron accumulation. Assiut Univ J Bot Microbiol 50(2):44–66

Rai VK, Sharma SS, Sharma S (1986) Reversal of ABA-induced stomatal closure by phenolic compounds. J Exp Bot 37:129–134

Rai GK, Magotra I, Khanday DM, Choudhary SM, Bhatt A, Gupta V, Rai PK, Kumar P (2024) Boosting drought tolerance in tomatoes through stimulatory action of salicylic acid imparted antioxidant defense mechanisms. Agronomy 14(6):1227. https://doi.org/10.3390/agronomy14061227

Rao MJ, Zheng B (2025) The Role of polyphenols in abiotic stress tolerance and their antioxidant properties to scavenge reactive oxygen species and free radicals. Antioxidants 14(1):74. https://doi.org/10.3390/antiox14010074

Rashmi HB, Negi PS (2020) Phenolic acids from vegetables: a review on processing stability and health benefits. Food Res Int 136:109298. https://doi.org/10.1016/j.foodres.2020.109298

Rhaman MS, Rauf F, Tania SS, Karim MM, Sagar A, Robin, AHK, Abdel Latef AAH, Murata Y (2021) Seed Priming and Exogenous Application of Salicylic Acid Enhance Growth and Productivity of Okra (Abelmoschus esculentus L.) by Regulating Photosynthetic Attributes. Journal of Experimental Biology and Agricultural Sciences 9(6):759–769

Roozkhosh M, Tahmasebi BK, Soleimani S, Meighani H, Eshraghi-Nejad M, Manesh GRA, Nia MAV (2024) Salicylic acid can enhance quality characteristics, growth and productivity of strawberry (*Fragaria* × *ananassa* CV Camarosa) under drought stress condition. J Plant Process Func 12(58):103–115. https://doi.org/10.22034/12.58.103

Rudolphi-Skórska E, Sieprawska A (2016) Adaptation of wheat cells to short-term ozone stress: the impact of α-tocopherol and gallic acid on natural and model membranes. Acta Physiol Plant 38(4):85. https://doi.org/10.1007/s11738-016-2102-1

Ruelland E, Zachowski A (2010) How plants sense temperature. Environ Exp Bot 69(3):225–232. https://doi.org/10.1016/j.envexpbot.2010.05.011

Ruuhola T, Julkunen-Titto R (2003) Trade-off between synthesis of salicylates and growth of micropropagated *Salix pentandra*. J Chem Ecol 29(7):1565–1588. https://doi.org/10.1023/A:1024266612585

Sachdev SJ, Ansari SA, Ansari MI, Fujita M, Hasanuzzaman MG (2021) Abiotic stress and reactive oxygen species: generation, signaling, and defense mechanisms. Antioxidantes 10(2):277–285. https://doi.org/10.3390/antiox10020277

Sadak MS, El-Lethy SR, Dawood MG (2013) Physiological role of benzoic acid and salicylic acid on growth, yield, some biochemical and antioxidant aspects of soybean plant. World J Agric Sci 9(6):435–442. https://doi.org/10.5829/idosi.wjas.2013.9.6.1763

Saheri F, Barzin G, Pishkar L, Boojar MMA, Babaeekhou L (2020) Foliar spray of salicylic acid induces physiological and biochemical changes in purslane (*Portulaca oleracea* L.) under drought stress. Biologia 75(2):2189–2200. https://doi.org/10.2478/s11756-020-00571-2

Saini N, Anmol A, Kumar S, Wani A, Bakshi M, Dhiman Z (2024) Exploring phenolic compounds as natural stress alleviators in plants. A comprehensive review. Physiol Mol Plant Pathol 133(24):102383. https://doi.org/10.1016/j.pmpp.2024.102383

Salas-Pérez L, Gaucín-Delgado JM, Preciado-Rangel P, Fortis-Hernández M, Valenzuela-García JR, Ayala-Garay AV (2016) Benzoic acid effect in the antioxidant capacity of wheat sprouts. Revista Mexicana de Ciencias Agrícolas 7(17):3397–3404

Saruhan NS, Kadioglu A (2012) Salicylic acid pretreatment induces drought tolerance and delays leaf rolling by inducing antioxidant systems in maize genotypes. Acta Physiol Plant 34:97–106. https://doi.org/10.4038/jas.v17i2.9747

Sayyari M, Ghanbari F, Fatahi S, Bavandpour F (2013) Chilling tolerance improving of watermelon seedling by salicylic acid seed and foliar application. Not Sci Biol 5(1):67–73. https://doi.org/10.15835/nsb518293

Senaratna T, Touchell D, Bunn E, Dixon K (2000) Acetylsalicylic acid (Aspirin) and salicylic acid induce multiple stress tolerance in bean and tomato plants. Plant Growth Regul 30(2):157–161. https://doi.org/10.1023/A:1006386800974

Senaratna T, Merritt D, Dixon K, Bunn E, Touchell D, Sivasithamparam K (2003) Benzoic acid may act as the functional group in salicylic acid and derivatives in the induction of multiple stress tolerance in plants. Plant Growth Regul 39(1):77–81. https://doi.org/10.1023/A:1021865029762

Shabana YM, Abdel-Fattah GM, Ismail AE, Rashad YM (2008) Control of brown spot pathogen of rice (*Bipolaris oryzae*) using some phenolic antioxidants. Braz J Microbiol 39(3):438–444. https://doi.org/10.1590/S1517-83822008000300006

Shahid M, Khalid S, Abbas G, Shahid N, Nadeem M, Sabir M et al (2015) Heavy metal stress and crop productivity. In: Hakeem KR (ed) Crop production and global environmental issues. Springer International Publishing, Cham, pp 1–25. https://doi.org/10.1007/978-3-319-23162-4_1

Shahid M, Dumat C, Khalid S, Schreck E, Xiong T, Niazi NK (2017) Foliar heavy metal uptake, toxicity and detoxification in plants: a comparison of foliar and root metal uptake. J Hazard Mater 325:36–58. https://doi.org/10.1016/j.jhazmat.2016.11.063

Shakirova FM, Sakhabutdinova AR, Bezrukova MV, Fatkhutdinova RA, Fatkhutdinova DR (2003) Changes in the hormonal status of wheat seedlings induced by salicylic acid and salinity. Plant Sci 164(3):317–322. https://doi.org/10.1016/S0168-9452(02)00415-6

Shakoor A, Saleem MF, Anjum SA, Wahid MA, Saeed MT (2017) Effect of heat stress and benzoic acid as foliar application on earliness and nutrients uptake in cotton. J Agric Res 55(1):15–28

Sharma M, Gupta SK, Majumder B, Maurya VK, Deeba F, Alam A, Pandey V (2017) Salicylic acid mediated growth, physiological and proteomic responses in two wheat varieties under drought stress. J Proteome 163:28–51. https://doi.org/10.1016/j.jprot.2017.05.011

Sharma A, Shahzad B, Rehman A, Bhardwaj R, Landi M, Zheng B (2019) Response of phenylpropanoid pathway and the role of polyphenols in plants under abiotic stress. Molecules 24(13):2452. https://doi.org/10.3390/molecules24132452

Shi GR, Cai QS, Liu QQ, Wu L (2009) Salicylic acid-mediated alleviation of cadmium toxicity in hemp plants in relation to cadmium uptake, photosynthesis, and antioxidant enzymes. Acta Physiol Plant 31(5):969–977. https://doi.org/10.1007/s11738-009-0312-5

Shiade SR, Fathi A, Ghasemkheili TF, Amiri E, Pessarakli M (2022) Plants' responses under drought stress conditions: effects of strategic management approaches—a review. J Plant Nutr 46:1–33. https://doi.org/10.1080/01904167.2022.2105720

Simaei M, Khavarinejad RA, Saadatmand S, Bernard FH, Fahimia H (2011) Interactive effects of salicylic acid and nitric oxide on soybean plants under NaCl salinity. Russ J Plant Physiol 58(5): 783–790. https://doi.org/10.1134/S1021443711050220

Singh H, Kavita SR, Maurya K (2014) Management of *Alternaria blight* of Indian mustard through resistance inducing chemicals. J Agri Search 1(2):108–111

Singh S, Singh A, Bashri G, Prasad SM (2016) Impact of Cd stress on cellular functioning and its amelioration by phytohormones: an overview on regulatory network. Plant Growth Regul 80(3): 253–263. https://doi.org/10.1007/s10725-016-0170-2

Siquet C, Paiva-Martins F, Lima JFC, Reis S, Borges F (2006) Antioxidant profile of dihydroxy-and trihydroxyphenolic acids-A structure activity relationship study. Free Radic Res 40(4):433–442. https://doi.org/10.1080/10715760500540442

Sirisuntornlak N, Ghafoori S, Datta A, Arirob W (2019) Seed priming and soil incorporation with silicon influence growth and yield of maize under water deficit stress. Arch Agron Soil Sci 65(2):197–207. https://doi.org/10.1080/03650340.2018.1492713

Slama I, Ghnaya T, Savoure A, Abdelly C (2008) Combined effects of long-term salinity and soil drying on growth, water relations, nutrient status and proline accumulation of *Sesuvium portulacastrum*. C R Biol 331:442–451. https://doi.org/10.1016/j.crvi.2008.03.006

Smirnov OE, Kosyan AM, Kosyk OI, Taran NY (2015) Response of phenolic metabolism induced by aluminium toxicity in *Fagopyrum esculentum* moench. plants. Ukr Biochem J 87(6): 129–135

Sohag AAM, Tahjib-Ul-Arif M, Brestic M, Afrin S, Sakil A, Hossain T, Hossain MA, Hossain A (2020) Exogenous salicylic acid and hydrogen peroxide attenuate drought stress in rice. Plant Soil Environ 66(1):7–13. https://doi.org/10.17221/472/2019-PSE

Srivastava MK, Dwivedi UN (2000) Delaying ripening of banana fruits by salicylic acid. Plant Sci 158(1–2):87–96. https://doi.org/10.1016/S0168-9452(00)00304-6

Sroka Z, Cisowski (2003) Hydrogen peroxide scavenging, antioxidant and anti-radical activity of some phenolic acids. Food Chem Toxicol 41(6):753–758. https://doi.org/10.1016/s0278-6915(02)00329-0

Strnad M (1997) The aromatic cytokinins. Physiol Plant 101(4):674–688. https://doi.org/10.1111/j.1399-3054.1997.tb01052.x

Symes A, Shavandi A, Zhang H, Ahmed IA, Al-Juhaimi FY, Bekhit A (2018) Antioxidant activities and caffeic acid content in New Zealand asparagus (*Asparagus officinalis*) roots extracts. Antioxidants 7(4):52. https://doi.org/10.3390/antiox7040052

Tahjib-Ul-Arif M, Asaduzzaman, M, Shirazy BJ, Khan MSU, Rahman AMS, Murata Y, Abdel Hamed S, Abdel Latef AAH (2024) Seed priming improves chilling stress tolerance in rice (oryza sativa L.) seedlings. Phyton-International Journal of Experimental Botany 93(11): 3013–3027

Talaat IM, Khattab HI, Ahmed AM (2014) Changes in growth, hormones levels and essential oil content of *Ammi visnaga* L. plants treated with some bioregulators. Saudi J Biol Sci 21:355–365. https://doi.org/10.1016/j.sjbs.2013.10.008

Tanase C, Bujor O-C, Popa VI (2019) Chapter 3: Phenolic natural compounds and their influence on physiological processes in plants. In: Watson RR (ed) Polyphenols in plants, 2nd edn. Academic Press, Cambridge, pp 45–58. https://doi.org/10.1016/b978-0-12-813768-0.00003-7

Tarkowská D, Dolezal K, Tarkowski P, Astot C, Holub J, Fuksová K, Schmülling T, Sandberg G, Strnad M (2003) Identification of new aromatic cytokinins in *Arabidopsis thaliana*and Populus × canadensis leaves by LC-(+)ESI-MS and capillary liquid chromatography/frit–fast atom bombardment mass spectrometry. Physiol Plant 117(4):579–590. https://doi.org/10.1034/j.1399-3054.2003.00071.x

Tchounwou PB, Yedjou CG, Patlolla AK, Sutton DJ (2012) Heavy metal toxicity and the environment. Exp Suppl 101:133–164. https://doi.org/10.1007/978-3-7643-8340-4_6

Thomson MJ, Ismail AM, McCouch SR, Mackill DJ (2010) Chapter 20: Marker assisted breeding. In: Pareek A, Sopory SK, Bohner HJ, Govindjee (eds) Abiotic stress adaptation in plants. physiological, molecular and genomic foundation. Springer, Dordrecht. https://doi.org/10.1007/978-90-481-3112-9_20

Torun H, Novak O, Mikulik J, Pencik A, Strnad M, Ayaz FA (2020) Timing-dependent effects of salicylic acid treatment on phytohormonal changes, ROS regulation, and antioxidant defense in salinized barley (*Hordeum vulgare* L.). Sci Rep 10(1):13886. https://doi.org/10.1038/s41598-020-70807-3

Tucuch-Haas CJ, Dzib-Ek MA, Vergara-Yoisura S, Larqué-Saavedra A (2021) Chapter 2: Salicylic acid increases root size that favours the absorption and accumulation of macro and micronutrients that contribute to biomass production. In: Hayat S, Siddiqui H, Damalas CA (eds) Salicylic acid – a versatile plant growth regulator. Springer, pp 17–33. https://doi.org/10.1007/978-3-030-79229-9

Ullah S, Khan M, Idress M, Ali S, Khan MN, Adnan M, Ozdemir FA (2020) Enhancement effect of hydroxyl benzoic acid (HBA) foliar application on growth performance of *Triticum aestivum* L. (Wheat) under induced chromium toxicity. Pure Appl Biol 9(4):2529–2542. https://doi.org/10.19045/bspab.2020.90269

Urmi TA, Islam M, Zumur KN, Abedin A, Haque M, Siddiqui MH, Murata Y, Hoque A (2023) Combined effect of salicylic acid and proline mitigates drought stress in rice (*Oryza sativa* L.) through the modulation of physiological attributes and antioxidant enzymes. Antioxidants 12(7):1438. https://doi.org/10.3390/antiox12071438

Valdez-Sepúlveda LV, Morales SG, Valdez-Aguilar LA, Ramírez-Godina F, Benavides-Mendoza A (2015) Effect of exogenous application of benzoic and salicylic acid on growth of tomato, tomatillo and pepper seedling. Rev Mex Cienc Agríc 6:2331–2343

Van Hees PAW, Lundstrom US, Giesler R (2000) Low molecular weight organic acids and their Al-complexes in soil solution composition, distribution and seasonal variationin three podzolized soils. Geoderma 94(2–4):173–200. https://doi.org/10.1016/S0016-7061(98)00140-2

Walker K, Croteau R (2000) Taxol biosynthesis: molecular cloning of a benzoyl-CoA: taxane 2*a*-O-benzoyltransferase cDNA from *Taxus* and functional expression in *Escherichia coli*. Proc Natl Acad Sci USA 97(25):13591–13596. https://doi.org/10.1073/pnas.250491997

Wang L, Fan L, Loescher W, Duan W, Liu G-J, Cheng J-S, Luo H-B, Li S-H (2010) Salicylic acid alleviates decreases in photosynthesis under heat stress and accelerates recovery in grapevine leaves. BMC Plant Biol 10(1):34. https://doi.org/10.1186/1471-2229-10-34

Wang J, Yuan B, Huang B (2019a) Differential heat-induced changes in phenolic acids associated with genotypic variations in heat tolerance for hard fescue. Crop Sci 59:667–674. https://doi.org/10.2135/cropsci2018.01.0063

Wang L, Shan T, Xie B, Ling C, Shao S, Jin P, Zheng Y (2019b) Glycine betaine reduces chilling injury in peach fruit by enhancing phenolic and sugar metabolisms. J Food Chem 272:530–538. https://doi.org/10.1016/j.foodchem.2018.08.085

Wang J, Shi SH, Wang DY, Sun Y, Zhu M, Li FH (2021) Exogenous salicylic acid ameliorates waterlogging stress damages and improves photosynthetic efficiency and antioxidative defense system in waxy corn. Photosynthetica Inter J Photosyn Pig 59(1):84–94. https://doi.org/10.32615/ps.2021.005

Wang D, Sun L, Yu H, Zhang C, Guan X, Wang M, Cheng R, Wang C, Xie Z (2023) Whole-genome analysis of the benzoic acid-degrading *Bacterium bacillus* halotolerans B28 to reveal its phytoprobiotic effects. Int Biodeterior Biodegrad 185:105668. https://doi.org/10.1016/j.ibiod.2023.105668

Waqas MA, Wang X, Zafar SA, Noor MA, Hussain HA, Azher Nawaz M, Farooq M (2021) Thermal stresses in maize: effects and management strategies. Plants 10:293. https://doi.org/10.3390/plants10020293

Wen PF, Chen JY, Wan SB, Kong WF, Zhang P, Wang W, Zhan JC, Pan QH, Huang WD (2008) Salicylic acid activates phenylalanine ammonia-lyase in grape berry in response to high temperature stress. Plant Growth Regul 55:1–10. https://doi.org/10.1007/s10725-007-9250-7

Widhalm JR, Dudareva N (2015) A familiar ring to it: biosynthesis of plant benzoic acids. Mol Plant 8:83–97. https://doi.org/10.1016/j.molp.2014.12.001
Wildermuth MC (2006) Variations on a theme: synthesis and modification of plant benzoic acids. Curr Opin Plant Biol 9(3):288–296. https://doi.org/10.1016/j.pbi.2006.03.006
Williams M, Senaratna T, Dixon K, Sivasithamparam K (2003) Benzoic acid induces tolerance to biotic stress caused by *Phytophthora cinnamomi*in Banksia attenuate. Plant Growth Regul 41(1):89–91. https://doi.org/10.1023/A:1027355604096
Windisch S, Walter A, Moradtalab N, Walker F, Höglinger B, El-Hasan A, Ludewig U, Neumann G, Grosch R (2021) Role of benzoic acid and lettuce in A in the defense response of lettuce against soil-borne pathogens. Plants 10:2336. https://doi.org/10.3390/plants10112336
Xue T, Fang Y, Li H, Li M, Li C (2024) The effects of exogenous benzoic acid on the physico-chemical properties, enzyme activities and microbial community structures of *Perilla frutescens* inter-root soil. Microorganisms 12(6):1190. https://doi.org/10.3390/microorganisms12061190
Yadav K, Singh NB (2013) Effects of benzoic acid and cadmium toxicity on wheat seedlings. Chil J Agric Res 73(2):168–174. https://doi.org/10.4067/S0718-58392013000200013
Yadav S, Modi P, Dave A, Vijapura A, Patel D, Patel M (2020) Effect of abiotic stress on crops. In: Hasanuzzaman M, Filho MCMT, Fujita M, Nogueira TAR (eds) Sustainable crop production. IntechOpen. https://doi.org/10.5772/intechopen.88434
Yahaya SU, Muhammad ZY (2018) Efficacy of foliar spray with benzoic acid in mitigating moisture stress in sweet potato (*Ipomoea batatas* [L.] lam) in Sudan Savanna, Nigeria. Faman J 18(1):73–85
Yalcin G, Vardar F (2016) The alleviating effects of salicylic acid application against aluminium toxicity in barley (*Hordeum vulgare*) roots. Biologia 71:1338–1344. https://doi.org/10.1515/biolog-2016-0166
Yang H, Fang R, Luo L, Yang W, Huang Q, Yang C, Hui W, Gong W, Wang J (2023) Uncovering the mechanisms of salicylic acid-mediated abiotic stress tolerance in horticultural crops. Front Plant Sci 14:1226041. https://doi.org/10.3389/fpls.2023.1226041
Yang Y, Tilman D, Jin Z, Smith P, Barrett CB, Zhu Y-G, Burney J, D'Odorico P, Fantke P, Fargione J (2024) Climate change exacerbates the environmental impacts of agriculture. Science 385(6713):eadn3747. https://doi.org/10.1126/science.adn3747
Yip EC, Chan ASL, Pang H, Tam YK, Wong YH (2006) Protocatechuic acid induces cell death in HepG2 hepatocellular carcinoma cells through a c-Jun N-terminal kinase-dependent mechanism. Cell Biol Toxicol 22(4):293–302. https://doi.org/10.1007/s10565-006-0082-4
Yoo H, Widhalm JR, Qian Y, Maeda H, Cooper BR, Jannasch AS, Gonda I, Lewinsohn E, Rhodes D, Dudareva (2013) An alternative pathway contributes to phenylalanine biosynthesis in plants via a cytosolic tyrosine: phenylpyruvate aminotransferase. Nat Commun 4:2833. https://doi.org/10.1038/ncomms3833
Yordanova R, Popova L (2007) Effect of exogenous treatment with salicylic acid on photosynthetic activity and antioxidant capacity of chilled wheat plants. Gen Appl Plant Physiol 33(3–4): 155–170
Youssef MN, Saeed AKAJM (2023) Response of *Catharanthus roseus* cv. 'Pacifica XP' to adding copper and foliar spraying with benzoic and salicylic acids. ICASE-2022. IOP Conf Ser Earth Environ Sci 1225:012016. IOP Publishing. https://doi.org/10.1088/1755-1315/1225/1/0120162
Yu JQ, Ye SF, Zhang MF, Hu WH (2003) Effects of root exudates and aqueous root extracts of cucumber (*Cucumis sativus*) and allelochemicals, on photosynthesis and antioxidant enzymes in cucumber. Biochem Syst Ecol 31(2):129–139. https://doi.org/10.1016/S0305-1978(02)00150-3
Yuan S, Lin HH (2008) Role of salicylic acid in abiotic stress tolerance. ZNaturforsch C J Biosci 63(6):313–320. https://doi.org/10.1515/znc-2008-5-601
Yüzbaşıoğlu E, Dalyan E (2019) Salicylic acid alleviates thiram toxicity by modulating antioxidant enzyme capacity and pesticide detoxification systems in the tomato (*Solanum lycopersicum* Mill.). Plant Physiol Biochem 135:322–330. https://doi.org/10.1016/j.plaphy.2018.12.023
Zaghlool AM, Ibrahim SI, Sharaf Eldeen HAM (2001) The effect of naphthaline acetic acid (NAA), salicylic acid (SA) and their combination on growth, fruit setting yield and some correlated

components in dry bean (*Phaseolus vulgaris* L.). Ann Agric Sci Ain Shams Univ Cairo 46(2): 451–463

Zagoskina NV, Zubova MY, Nechaeva TL, Kazantseva VV, Goncharuk EA, Katanskaya VM, Baranova EN, Aksenova MA (2023) Polyphenols in plants: structure, biosynthesis, abiotic stress regulation, and practical applications. Int J Mol Sci 24:13874. https://doi.org/10.3390/ijms24181387

Zaid A, Mohammad F, Wani SH, Siddique KM (2019) Salicylic acid enhances nickel stress tolerance by up-regulating antioxidant defense and glyoxalase systems in mustard plants. Ecotoxicol Environ Saf 180:575–587. https://doi.org/10.1016/j.ecoenv.2019.05.042

Zaid A, Mohammad F, Fariduddin Q (2020) Plant growth regulators improve growth, photosynthesis, mineral nutrient and antioxidant system under cadmium stress in menthol mint (*Mentha arvensis* L.). Physiol Mol Biol Plants 26(1):25–39. https://doi.org/10.1007/s12298-019-00715-y

Zakariyya F, Susilo WA, Iman Santoso T, Susilo Addy H, Pancaningtyas S (2018) Role of exogenous salicylic acid and benzoic acid applications to vascular streak dieback disease attack on cocoa seedlings. Pelita Perkebunan 34(1):33–39. https://doi.org/10.22302/iccri.jur.pelitaperkebunan.v34i1.305

Zamir H, Hussain A, Shabir M, Aslam A, Hassan M, Din Waqas SU, Akhtar MZ, Hussain S (2023) The effect of foliar application of benzoic acid on growth and production of wheat. Haya Saudi J Life Sci 8(11):260–265

Zhang Z, Lan M, Han X, Wu J, Wang-Pruski G (2020) Response of ornamental pepper to high-temperature stress and role of exogenous salicylic acid in mitigating high temperature. J Plant Growth Regul 39(10):133–146. https://doi.org/10.1007/s00344-019-09969-y

Zhao JH, Liang PP, Chai JL, Li X (2018) Effect of sodium benzoate on physiological characteristics of wheat seedlings under compound heavy metal pollution. J Zhoukou Normal Univ 35:84–88

Chapter 10
Citric Acid: Improving Plant Metabolism Under Stress

Boujemaa Fassih, Aicha Nait Douch, Abdelhamid Aouabe, Mohamed Ait-El-Mokhtar, Raja Ben-Laouane, Abdelilah Meddich, and Marouane Baslam

Abstract Plants are persistently challenged by abiotic stressors such as drought, salinity, heat, cold, and heavy metals, as well as biotic threats including bacteria, fungi, viruses, and insect pests. These factors impair essential physiological and biochemical processes, causing oxidative damage, disruption in nutrient balance, and reduced productivity. Citric acid (CA), a key intermediate of the tricarboxylic acid (TCA) cycle, plays a central role in maintaining metabolic homeostasis during stress conditions. Beyond its metabolic function, CA acts as a signaling molecule,

B. Fassih · A. N. Douch · A. Aouabe
Centre d'Agrobiotechnologie et Bioingénierie, Unité de Recherche Labellisée CNRST (Centre AgroBiotech-URL-CNRST-05), Cadi Ayyad University, Marrakesh, Morocco

M. Ait-El-Mokhtar
Centre d'Agrobiotechnologie et Bioingénierie, Unité de Recherche Labellisée CNRST (Centre AgroBiotech-URL-CNRST-05), Cadi Ayyad University, Marrakesh, Morocco

Laboratory of Biotechnology, Agri-food, Materials, and Environment (LBAME), Department of Biology, Faculty of Science and Techniques Mohammedia, Hassan II University of Casablanca, Mohammedia, Morocco

R. Ben-Laouane
Centre d'Agrobiotechnologie et Bioingénierie, Unité de Recherche Labellisée CNRST (Centre AgroBiotech-URL-CNRST-05), Cadi Ayyad University, Marrakesh, Morocco

Laboratory of Environment and Health, Department of Biology, Faculty of Science and Techniques, Errachidia, Morocco

A. Meddich
Centre d'Agrobiotechnologie et Bioingénierie, Unité de Recherche Labellisée CNRST (Centre AgroBiotech-URL-CNRST-05), Cadi Ayyad University, Marrakesh, Morocco

African Sustainable Agriculture Research Institute (ASARI), University Mohamed VI Polytechnic (UM6P), Laayoune, Morocco

M. Baslam (✉)
Centre d'Agrobiotechnologie et Bioingénierie, Unité de Recherche Labellisée CNRST (Centre AgroBiotech-URL-CNRST-05), Cadi Ayyad University, Marrakesh, Morocco

GrowSmart, Amsterdam, The Netherlands

A. A. H. Abdel Latef (ed.), *Harnessing Vitamins and Organic Acids for Plant Stress Resilience*, Sustainability Sciences in Asia and Africa,
https://doi.org/10.1007/978-981-95-8593-9_10

chelating agent, and precursor for multiple stress-associated compounds. It enhances stress resilience by modulating antioxidant defense, aiding in heavy metal detoxification, improving nutrient uptake, and participating in the synthesis of phytohormones and osmolytes. Moreover, CA influences defense gene activation and strengthens cell walls, aiding in biotic stress mitigation. This chapter offers a comprehensive overview of CA's biosynthesis, regulation, and multidimensional role in modulating plant responses to abiotic and biotic stress. By dissecting its molecular and physiological effects, we highlight the potential of CA as a metabolic regulator and biostimulant in sustainable agriculture.

Keywords Citric acid · Stress metabolism · Antioxidant defense · Plant immunity · TCA cycle · Biotic stress · Abiotic stress · Chelation · Signaling molecules

10.1 Introduction

Global agricultural productivity is increasingly threatened by a wide array of environmental constraints. Crop stress resistance represents the cornerstone of reliable food production and, ultimately, human survival. However, both abiotic and biotic stresses—amplified by the accelerating impacts of climate change—pose severe challenges to sustainable agriculture (Palmgren and Shabala 2024). Abiotic stress factors, including drought, soil salinity, extreme temperatures, flooding, nutrient deficiencies, and heavy metal contamination, remain the leading causes of yield losses worldwide (Ait-El-Mokhtar et al. 2022; Anli et al. 2022; Raja et al. 2022; Villalobos-López et al. 2022). These stresses disrupt essential physiological processes, altering plant growth, development, and reproduction (Boutasknit et al. 2024; Lal et al. 2023; Ait-El-Mokhtar et al. 2019). Concurrently, biotic stresses—caused by fungal and bacterial pathogens, insect pests, and weeds—inflict further damage and reduce both yield and quality (Teshome et al. 2020; Afzal et al. 2026). Importantly, abiotic and biotic stresses often interact synergistically, intensifying crop damage beyond the sum of their individual effects (Teshome et al. 2020).

Horticultural crops, valued for their nutritional and economic importance, are particularly vulnerable to these combined stressors. Critical phases such as seed germination, flowering, and fruit development are highly sensitive to environmental fluctuations (Lal et al. 2023). Global assessments estimate that a large proportion of cultivated land is already affected by multiple stress factors, underscoring the urgent need to enhance crop resilience to safeguard food security for a growing population (Palei et al. 2024). The Mediterranean basin exemplifies this challenge, as climate change projections predict severe impacts from rising temperatures and prolonged drought. Developing crop varieties with improved tolerance to these conditions is therefore a priority (González Guzmán et al. 2022). Achieving this requires innovative management strategies that reinforce tolerance to both abiotic and biotic stressors, supported by a deep understanding of plant defense mechanisms spanning morphological, biochemical, and molecular levels (Ben-Laouane et al. 2025; Fassih et al. 2024; Ikan et al. 2024; Gimenez et al. 2018; Abdel Latef et al. 2026; Anli et al. 2026).

Plants possess a remarkable ability to synthesize diverse organic acids that are integral to stress responses. These compounds participate in respiration, photosynthesis, nutrient uptake, and, critically, the tricarboxylic acid (TCA) cycle, which serves as the central hub of energy metabolism. Beyond these primary roles, organic acids are increasingly recognized for their contribution to stress tolerance. Their concentration and composition vary with species, tissue type, developmental stage, and environmental context. Among the most abundant are malate and citrate, which frequently accumulate in significant quantities.

Low-molecular-weight organic acids—including salicylic acid (SA), citric acid (CA), and malic acid (MA)—have attracted attention for their ability to mitigate stress-induced damage. Their biodegradability and low toxicity render them environmentally compatible candidates for sustainable crop management. SA, for instance, is a critical phytohormone that enhances yield and growth under drought by modulating stomatal function, photosynthetic efficiency, and membrane stability. Similar benefits are observed under cold stress and heavy metal toxicity (Kleiber et al. 2024). Root exudation of organic acids such as CA and MA modifies rhizosphere pH, influencing nutrient availability (Yang et al. 2019). Under drought, CA enhances photosynthetic rates via effects on chloroplast structure, whereas MA promotes biomass accumulation by stimulating antioxidant activity (Kleiber et al. 2024). Other organic acids, such as jasmonic acid (JA) and derivatives, also function as key regulators of stress responses (Wang et al. 2021).

Among these CA has gained prominence as a particularly versatile regulator of plant responses to environmental stresses. Exogenous CA application has been shown to enhance growth and yield under adverse conditions, with effects frequently linked to increased photosynthesis, reduced reactive oxygen species (ROS) accumulation, and improved osmoregulation. CA stimulates endogenous antioxidant systems, promotes chlorophyll accumulation, and modulates secondary metabolic pathways that collectively mitigate growth limitations imposed by stress. Notably, CA reduces heavy metal toxicity through chelation and sequestration of metal ions, facilitating their precipitation or immobilization in plant tissues or soil (Zahan et al. 2021).

Empirical studies confirm CA's effectiveness across diverse stress types—including drought, salinity, temperature extremes, and heavy metal contamination (Zahan et al. 2021). For example, CA seed priming enhances germination and seedling growth under saline conditions (Yan et al. 2022), while exogenous application stimulates root development, improving water acquisition under drought (Burns 2023). In black gram, CA alleviates salt stress by enhancing growth, yield, and nutritional quality (Hossain et al. 2024). CA also protects plants against copper toxicity through proteomic adjustments (Ju et al. 2021) and facilitates nickel removal from contaminated soils while reducing oxidative stress (Ijaz et al. 2022).

The centrality of CA in metabolism—connecting the TCA cycle with the biosynthesis of phytohormones, amino acids, signaling molecules, and secondary metabolites—likely underpins its broad stress-mitigating functions (Zahan et al. 2021). A deeper understanding of these interactions is essential for harnessing CA's full potential as a regulator of stress resilience (Fig. 10.1).

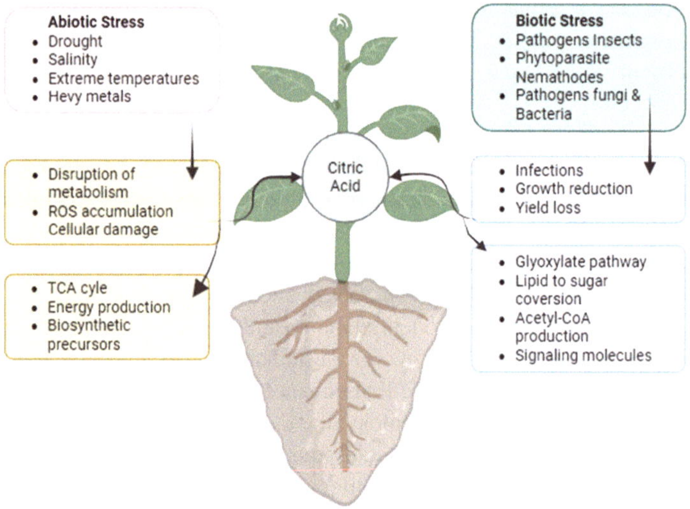

Fig. 10.1 Global abiotic and biotic stressors affecting plant health and the central role of citric acid (CA). Abiotic stress factors such as drought, salinity, extreme temperatures, and heavy metals disrupt plant metabolism, trigger excessive accumulation of reactive oxygen species (ROS), and cause cellular damage. Within the tricarboxylic acid (TCA) cycle, CA serves as a central metabolite that fuels energy production and provides precursors for biosynthetic pathways. Under biotic stresses—including insects, phytoparasitic nematodes, and pathogenic fungi and bacteria—CA contributes to tolerance mechanisms by supporting glyoxylate pathway activity, lipid-to-sugar conversion, acetyl-CoA production, and the synthesis of signaling molecules. Together, these functions highlight CA as a metabolic hub orchestrating plant resilience to diverse stressors

10.2 Biosynthesis and Metabolic Pathways of Citric Acid in Plants

The tricarboxylic acid cycle, a central metabolic pathway in plant cells, represents the primary route for citric acid biosynthesis (Nishio and Mizushima 2020). Located in the mitochondrial matrix, the TCA cycle plays a pivotal role in cellular respiration by oxidizing acetyl-CoA—derived from carbohydrates, lipids, and proteins—into reducing equivalents, thereby driving ATP generation. The first and irreversible step of this cycle is catalyzed by citrate synthase (CS), which condenses oxaloacetate (OAA; a four-carbon dicarboxylic acid) with acetyl-CoA (a two-carbon thioester), producing CA, a six-carbon tricarboxylic acid, and coenzyme A (Fig. 10.2). This step represents a critical regulatory point in carbon metabolism, controlling the entry of two-carbon units into the TCA cycle.

The mitochondrial TCA cycle operates in close coordination with chloroplastic photosynthetic metabolism to maintain a balanced supply of energy and carbon skeletons for diverse biosynthetic demands. Its regulation is multifaceted, involving both metabolic feedback and environmental cues. The redox state ($NADH/NAD^+$ ratio), cellular energy status (ATP/ADP ratio), and the thioredoxin (TRX) system—a

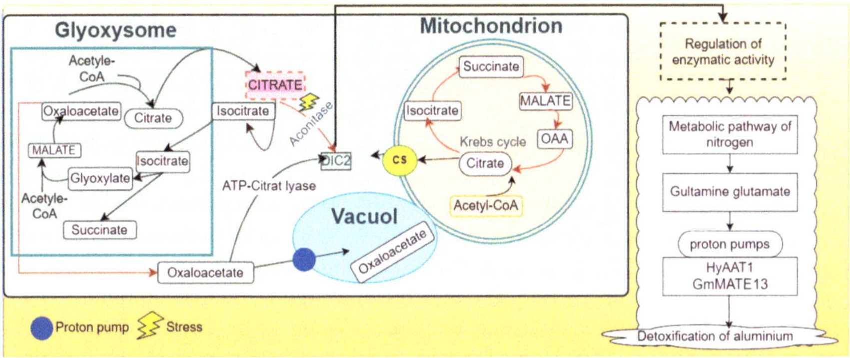

Fig. 10.2 Biochemical pathways of citric acid (CA) synthesis, transport, and functional interconnections with stress signaling. CA is synthesized in the mitochondrial matrix by citrate synthase (CS) via the condensation of oxaloacetate (OAA) and acetyl-CoA. It can be metabolized to isocitrate by aconitase, fueling the TCA cycle, or exported to the cytosol through the *DICARBOXYLATE CARRIER 2* (DIC2) in exchange for malate. In the cytosol, CA is cleaved by ATP-citrate lyase into OAA and acetyl-CoA, providing precursors for fatty acid and secondary metabolite biosynthesis. Vacuolar sequestration, mediated by proton pumps, enables pH regulation and carbon storage. CA fluxes extend beyond mitochondria, linking with the glyoxylate cycle in peroxisomes during seed germination. In roots, MATE-family transporters such as *HvAACT1* (barley) and *GmMATE13* (soybean) mediate CA efflux, conferring aluminum detoxification. Beyond central metabolism, CA integrates into nitrogen assimilation (via α-ketoglutarate/glutamate) and stress signaling networks, highlighting its role as a metabolic hub at the interface of energy production, nutrient homeostasis, and abiotic stress resilience

master regulator of redox homeostasis—each influence CS activity and other TCA enzymes. The evolutionary conservation of CS across taxa underscores its central role in primary metabolism (Nishio and Mizushima 2020). In addition, CS contributes to the glyoxylate cycle in peroxisomes, a pathway essential for early seedling growth and for converting stored lipids into carbohydrates during germination (Eckardt 2005). Beyond respiration, CA synthesized by CS also serves as a precursor for numerous metabolic processes both within and beyond the mitochondria (Nishio and Mizushima 2020).

Following its production in mitochondria, CA is subject to complex transport and metabolic partitioning across cellular compartments, notably the mitochondria, cytoplasm, and vacuole. Within mitochondria, CA may be further metabolized to isocitrate via aconitase, continuing the TCA cycle (Han et al. 2025). However, CA is also pivotal for carbon exchange between organelles. In *Arabidopsis thaliana* leaves, the dicarboxylate carrier 2 (DIC2) exports CA to the cytosol while simultaneously importing malate into mitochondria (Lee et al. 2021). This exchange integrates mitochondrial energy metabolism with cytosolic biosynthetic pathways. In the cytoplasm, CA is cleaved by ATP-citrate lyase (ACL) to generate acetyl-CoA and OAA (Zahan et al. 2021). Acetyl-CoA fuels the synthesis of fatty acids, isoprenoids, and flavonoids, whereas OAA either re-enters mitochondrial metabolism or participates in other biosynthetic routes (Lee et al. 2021).

A further fate of CA involves vacuolar storage, where it accumulates at high concentrations. This storage not only contributes to cytoplasmic pH buffering but also provides a carbon reserve mobilized under stress or nutrient limitation (Zahan et al. 2021). Proton pumps actively transport CA across the tonoplast against concentration gradients, highlighting the energy dependence of vacuolar sequestration (Han et al. 2025).

Multiple transporter families mediate CA flux across membranes. The Multidrug and Toxic Compound Extrusion (MATE) transporters play crucial roles in root citrate efflux, particularly under metal stress. For instance, *HvAACT1* in barley and *GmMATE13* in soybean secrete CA into the rhizosphere under aluminum (Al^{3+}) stress, conferring detoxification and tolerance (Wang et al. 2022). Other MATE transporters, including FRD3 in *Arabidopsis* and *OsFRDL1* in rice, facilitate long-distance iron transport by stabilizing Fe–CA complexes during root-to-shoot translocation (Yokosho et al. 2009).

CA metabolism is deeply interconnected with carbon–nitrogen (C–N) integration in plants (Baslam et al. 2021). As an early TCA intermediate, CA anchors cross-talk between C and N metabolism (Foyer et al. 2011). Notably, α-ketoglutarate, derived from CA metabolism, provides the essential carbon skeleton for glutamate and glutamine biosynthesis, which underpin N assimilation (Zahan et al. 2021). This relationship is reciprocal: nitrogen availability modulates TCA fluxes to balance energy supply and carbon skeletons, while carbon status directly influences N uptake and assimilation efficiency (Goel et al. 2016; Xin et al. 2019). Citrate export to the cytoplasm and its subsequent conversion shape precursor availability for biosynthetic pathways that couple C and N metabolism (Lee et al. 2021). Furthermore, CA pool sizes and fluxes are dynamically tuned by the nitrogen status of the plant, reflecting the tight regulation of these pathways (Xin et al. 2019). The coordinated transcriptional regulation of C and N metabolic genes highlights the importance of maintaining an optimal C/N ratio for growth, development, and secondary metabolism, particularly under stress (Goel et al. 2016; Jian et al. 2024).

In summary, citric acid is far more than a TCA intermediate. Its biosynthesis, transport, compartmentation, and metabolic fates position it as a metabolic hub linking respiration, biosynthesis, carbon–nitrogen balance, and stress adaptation. This centrality underscores the importance of CA as a key regulator of plant resilience under fluctuating environmental conditions (Fig. 10.2).

10.3 Role of Citric Acid in Plant Response to Abiotic Stress

Citric acid plays a critical role in plant adaptation to abiotic stresses through diverse physiological and biochemical mechanisms (Table 10.1). Exogenous application of CA has been shown to improve plant water status under drought, as reflected in higher relative water content (RWC) in leaves (El-Tohamy et al. 2013). This indicates a role in osmotic adjustment, a key mechanism enabling plants to maintain cell turgor during water deficit. Under drought conditions, CA enhances

Table 10.1 Citric acid-mediated effects on plant tolerance under major abiotic stress conditions

Abiotic stress	Crop/ species	Application method	Main results	Reference
Drought	Lettuce	Foliar spray	Increased yield; enhanced K^+ uptake; reduced superoxide anion and MDA; increased proline and phenolic compound accumulation	Kleiber et al. (2024)
Drought	Spotted gum	Cotyledon root	Enhanced root system development	Burns (2023)
Drought	Chili	Seed priming	Increased germination, growth, photosynthetic pigments content, decreased H_2O_2 and MDA levels	Maphalaphathwa and Nciizah (2025)
Drought	Bean	Foliar spray, 0.5–2 g/L (optimum 1.5 g/L)	Increased RWC, chlorophyll, growth, productivity, and quality	El-Tohamy et al. (2013)
Salinity	Okra	Seed priming	Increased germination percentage, seedling growth, RWC, and root and shoot length and weight	Chakrobortty et al. (2022)
Salinity	Black gram	Foliar spray	Increased plant height, leaf number, root DW, chlorophyll, RWC, WRC, grain yield, stover yield, biological yield, N, and P content	Hossain et al. (2024)
Salinity	Brinjal	Foliar spray/ 200 ppm	Increased fruit number and weight	
Salinity	Chili	Seed priming/ 0.5–1 mM	Improved germination, vigor, photosynthetic pigments; reduced water loss	
Salinity	Soybean	Seed priming/ 1–2 mM	Improved germination, vigor, RWC, growth traits, pigments; reduced H_2O_2 and MDA	
Salinity	Moringa	CA/100 mM	Increased growth, pigments, proteins; reduced H_2O_2 and MDA	Hajaji et al. (2025)
Heavy metal (Cd)	Iris	Soil treatment	Altered mineral element content, increased antioxidant enzyme, and Cd binding to root cell walls	Yu et al. (2024)
Heavy metal (Cd)	Brown mustard	Foliar spray, 0.6 mM + SA 0.01 mM	Improved growth, photosynthesis, stomatal activity, and membrane stability; reduced ROS and oxidative damage	

(continued)

Table 10.1 (continued)

Abiotic stress	Crop/ species	Application method	Main results	Reference
Heavy metal (Cd)	Willow	Soil amendment, 20 mM/kg soil	Increased Cd solubility and phytoextraction; antioxidant activity; improved photosynthesis; reduced oxidative stress	
Heavy metal (Cu)	Rapeseed	Soil treatment	Altered proteome, increased enzymatic activities (POD, SOD, CAT)	Ju et al. (2021)
Heavy metal (Ni)	Sunflower	Soil treatment	Reduced MDA and H_2O_2 and, enhanced expression of *NRAMP* genes	Ijaz et al. (2022)
Heavy metal (Pb)	Olga Bay larch	Soil irrigation, 0.2–10.0 mmol/L	Reduced Pb uptake; improved growth; enhanced SOD/POD activity, pigments, proline; reduced oxidative damage	
Heavy metal (Cr)	Castor bean	Hydroponic CA/2.5–5 mM	Improved growth, biomass, photosynthesis, and antioxidant activity; reduced H_2O_2, MDA, and electrolyte leakage; CA + microbes further mitigated Cr toxicity	
Heat	Tall fescue	Foliar spray/ 0.2 mM	Decreased electrolyte leakage and MDA, improved plant growth, chlorophyll content, Fv/Fm, root activity, antioxidant enzyme activities, HSP expression	Hu et al. (2016)
P deficiency	Tomato	Seed priming/ 7.5 mmol/L	Enhanced biomass and antioxidant; reduced MDA and ROS	Xiu et al. (2021)
P deficiency	Cucumber	CA + *Enterobacter* organic acids	Enhanced P utilization efficiency	
Fe deficiency	Tomato	Combined CA + HA substances	Upregulated Fe uptake genes; improved yield and fruit quality	

RWC relative water content, *DW* dry weight, *MDA* malondialdehyde, H_2O_2 hydrogen peroxide, *SOD* superoxide dismutase, *CAT* catalase, *POD* peroxidase, *APX* ascorbate peroxidase, *NRAMP* natural resistance-associated macrophage protein genes, *HSP* heat shock proteins, *Fv/Fm* maximum quantum efficiency of photosystem II, *CA* citric acid, *HA* humic acid, *SA* salicylic acid, *ROS* reactive oxygen species, *Cd* cadmium, *Pb* lead, *Cr* chromium, *Fe* iron, *P* phosphorus, *ppm* parts per million, *mM* millimolar photosynthetic efficiency and biomass accumulation, partly by increasing chloroplast volume and stimulating antioxidant activity. It may also contribute to osmotic adjustment through the accumulation of compatible solutes, though its effectiveness is relative to other organic acids, such as salicylic acid appears species- and context-dependent (El-Tohamy et al. 2013). In addition, CA confers tolerance to high thermal stress which is often concomitant with drought (combined hydro-thermal stress). The acid intervenes in the mitigation of thermal stress by stabilizing cell membranes and enhancing the activity of antioxidant enzymes, thereby trapping excess ROS produced by heat (Hu et al. 2016). In other species, pre-treatment at CA has been linked to better water use efficiency (WUE) and maintenance of the integrity of the photosynthetic apparatus under severe temperatures and water deficits (Xie et al. 2022; Youssef et al. 2025).

Exogenous application of CA has emerged as an effective strategy to enhance plant resilience under adverse environmental conditions while promoting optimal crop growth and productivity. Numerous studies demonstrate that CA, when applied at precise concentrations, confers multiple benefits to both plants and soil health. For example, foliar application of 200 ppm CA in brinjal under salt stress significantly increased fruit number and weight. Similarly, treatment of tall fescue with 0.2 mM CA via foliar spray improved heat tolerance by mitigating oxidative damage and preserving membrane integrity (Hu et al. 2016). In tomatoes, seed priming with 7.5 mmol/L CA markedly enhanced seedling growth, biomass accumulation, and antioxidant enzyme activity, while reducing levels of MDA and ROS (Xiu et al. 2021).

Comparable effects have been observed in chili, where seed priming with 0.5–1 mM CA alleviated salinity-induced inhibition of germination and early growth under 100 mM NaCl. Treated seeds displayed improved germination rates, greater seedling vigor, elevated photosynthetic pigment levels, and reduced water loss, collectively enhancing salt stress tolerance. Combining CA with complementary biostimulants further amplifies these benefits. In tomato, co-application of CA and humic substances improved the expression of genes involved in iron (Fe) uptake, leading to higher yields and improved fruit quality. Similarly, in cucumber, CA in conjunction with organic acids secreted by *Enterobacter* species enhanced phosphorus utilization efficiency.

Applications of CA have also proven beneficial in perennial and multipurpose crops. In *Moringa oleifera*, exogenous application of 100 mM CA under saline irrigation restored shoot growth, increased pigment and soluble protein levels, and reduced oxidative stress markers such as H_2O_2 and MDA (Hajaji et al. 2025). As a soil amendment, CA improved the performance of willow plants grown in Cd-contaminated soils by sustaining photosynthesis and gas exchange capacity, and by modestly enhancing growth and biomass depending on Cd concentration and harvest stage.

Collectively, these studies highlight CA's potential as a versatile and environmentally friendly biostimulant capable of enhancing stress resilience, nutrient utilization, and crop performance across diverse agroecosystems. Its integration into sustainable crop management strategies offers a promising avenue for improving yield stability under challenging environmental conditions.

Under salinity stress, CA has demonstrated the ability to improve seed germination and early seedling growth in several crops (Chakrobortty et al. 2022). It alleviates salt-induced damage by stabilizing photosynthetic pigments, improving plant water status, and enhancing yield and nutritional quality (Hossain et al. 2024). Ion homeostasis is essential for salinity tolerance, as excessive sodium (Na^+) accumulation disrupts cellular metabolism (Xiao and Zhou 2023). CA contributes to ionic balance by facilitating the uptake of essential nutrients, including calcium (Ca^{2+}), magnesium (Mg^{2+}), and potassium (K^+), which support osmotic regulation and counteract Na^+ toxicity (Hossain et al. 2024). Additionally, citrate application has been associated with reductions in stress-related metabolites such as free amino acids, soluble sugars, and proline under salinity, suggesting enhanced metabolic efficiency (Hajaji et al. 2025). While direct evidence on water use efficiency is limited, improvements in RWC and overall plant health strongly suggest a positive effect of CA on water uptake capacity (Chakrobortty et al. 2022). In citrus, salinity tolerance depends on maintaining high water use efficiency and limiting chloride (Cl^-) accumulation in leaves, with rootstocks exerting significant influence on this trait (Syvertsen et al. 2010; Ziogas et al. 2021).

CA also plays an important role in mitigating heavy metal (HM) toxicity through several interconnected mechanisms (Zahan et al. 2021). Its tricarboxylic structure confers strong chelating capacity, enabling CA to bind positively charged metal ions and thereby alter their solubility and bioavailability in soils (Tusei 2019). Depending on concentration and context, this action can reduce metal uptake into roots—protecting plants from toxicity—or increase solubility and uptake, an effect exploited in phytoremediation strategies (Mallhi et al. 2019). Within plants, CA facilitates HM sequestration by binding Cd and other ions to root cell walls, restricting their translocation to shoots, where metabolic processes are more vulnerable (Yu et al. 2024).

Furthermore, CA can influence HM speciation by altering soil pH and forming stable complexes, converting metals into less toxic or more mobile forms depending on the element and species involved (Chen et al. 2003). In addition, CA enhances plant defense against HM stress by activating antioxidant enzymes that counteract the oxidative stress typically induced by metal exposure (Ju et al. 2021). It may also act synergistically with plant growth-promoting bacteria to enhance HM removal from contaminated soils, further improving phytoremediation efficiency (Wu et al. 2020).

Plants subjected to temperature extremes experience severe disruptions to cellular integrity, including enzyme inactivation and membrane destabilization. Citric acid alleviates heat-induced damage by stabilizing membranes and reducing lipid peroxidation, as shown by lower malondialdehyde (MDA) levels and electrolyte leakage. It also supports the function of antioxidant enzymes such as superoxide dismutase (SOD), peroxidase (POD), and catalase (CAT), which are critical for reactive oxygen species (ROS) detoxification during thermal stress (Hu et al. 2016). Furthermore, CA has been shown to induce the expression of *heat shock proteins* (*HSPs*), molecular chaperones that maintain protein stability under high temperature. Endogenous CA accumulation under heat stress further supports antioxidant defense and respiratory metabolism, suggesting a dual protective role (Hu et al. 2016). Although direct evidence under cold stress is limited, broader roles in lipid remodeling and membrane stabilization (Sharma et al. 2023; Liu et al. 2022) suggest that CA contributes to resilience across temperature extremes.

Oxidative stress, caused by excessive ROS accumulation, is a common consequence of multiple abiotic stressors. Citric acid mitigates oxidative stress both through direct scavenging and by activating antioxidant defense systems (Kleiber et al. 2024). Exogenous CA application consistently enhances enzymatic defenses, including SOD, POD, CAT, and APX, thereby reducing ROS-mediated cellular injury (Zahan et al. 2021). In addition, CA may act directly as an antioxidant, lowering ROS indicators such as superoxide and hydrogen peroxide (Kleiber et al. 2024). It also modulates levels of non-enzymatic antioxidants, including proline, and induces alternative oxidase (AOX), a mitochondrial enzyme that helps regulate ROS balance (Zahan et al. 2021). These mechanisms collectively strengthen oxidative stress tolerance across diverse conditions, including salinity, heavy metal exposure, and extreme temperatures (Ju et al. 2021). The central position of CA in cellular metabolism, coupled with its integration into signaling pathways, underscores its importance in coordinating ROS regulation and redox homeostasis (Zahan et al. 2021) (Fig. 10.3).

10.4 Role of Citric Acid in Biotic Stress Responses

Plants are continually challenged by a wide spectrum of biotic stressors—including bacteria, fungi, viruses, and insect herbivores—that can severely compromise growth, yield, and overall fitness (Khan et al. 2025). Pathogens such as bacteria, fungi, and viruses represent the primary agents of disease, often inflicting direct cellular damage. For instance, fungal pathogens infiltrate host tissues with their mycelia or hijack core metabolic processes, thereby disrupting cellular function and inhibiting reproduction (Luo et al. 2024). To counteract such threats, plants employ multifaceted defense strategies encompassing structural and chemical barriers, hypersensitive responses, and systemic acquired resistance (SAR) (Luo et al. 2024) (Table 10.2).

Citric acid contributes significantly to plant resistance against bacterial pathogens. In kiwifruit, CA destabilizes the outer membrane of *Pseudomonas syringae pv. actinidiae* (Psa), a Gram-negative phytopathogen, through acidification that compromises membrane integrity and enhances susceptibility to treatments (Sisson et al. 2024). Similarly, CA inhibits biofilm formation in *Paracidovorax citrulli*, the causal agent of fruit blotch, by reducing extracellular polysaccharide, protein, and DNA content (Yang et al. 2024). Beyond direct antimicrobial activity, CA primes plant immunity by enhancing defense responses, although its effects are context- and species-dependent. In *Arabidopsis thaliana*, CA priming enhanced resistance to

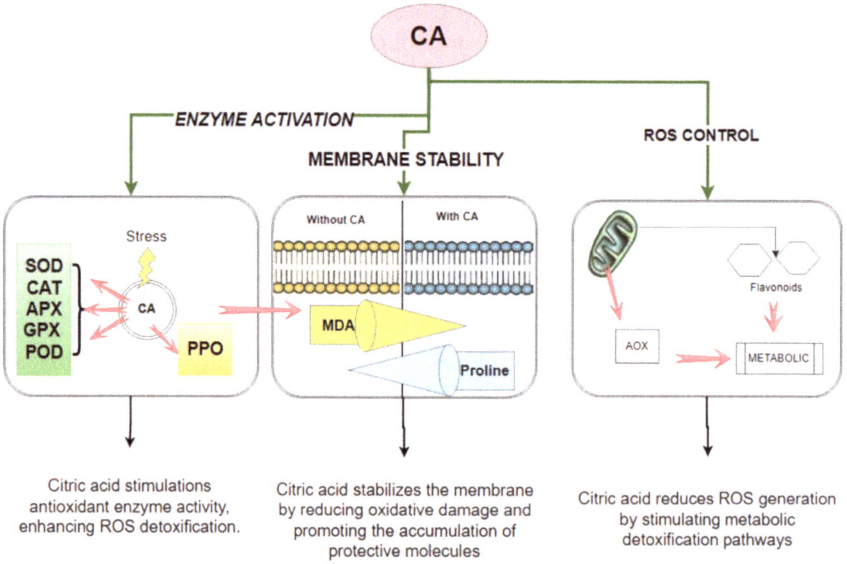

Fig. 10.3 Citric acid-mediated protective mechanisms under abiotic stress. Citric acid (CA) enhances stress tolerance through three major mechanisms: (i) enzyme activation, stimulating antioxidant enzymes (SOD, CAT, APX, GPX, POD, PPO) to improve reactive oxygen species (ROS) detoxification; (ii) membrane stability, reducing oxidative damage (as indicated by lower MDA levels) and promoting the accumulation of protective molecules such as proline; and (iii) ROS control, limiting ROS generation by stimulating metabolic detoxification pathways, including activation of AOX and related mitochondrial processes. *Abbreviations*: SOD, superoxide dismutase; CAT, catalase; APX, ascorbate peroxidase; GPX, glutathione peroxidase; POD, peroxidase; PPO, polyphenol oxidase; MDA, malondialdehyde; ROS, reactive oxygen species; AOX, alternative oxidase

Pseudomonas syringae pv. tomato DC3000 by elevating the accumulation of defense-related metabolites such as SA and camalexin during infection (Balmer et al. 2018). In addition to priming, CA modulates immunity by influencing the expression of pathogenesis-related (PR) genes within the dual-layered innate immune system: pathogen-associated molecular pattern-triggered immunity (PTI) and effector-triggered immunity (ETI). These immune responses rely on interconnected signaling networks involving Ca^{2+}, ROS, nitric oxide, and defense hormones such as SA, JA, and ethylene (ET), which collectively coordinate defense gene expression and pathogen resistance (Hu et al. 2025).

Recent studies further link CA metabolism to immune regulation. The transcription factor CitPH4, which enhances CA accumulation in citrus fruits, simultaneously promotes disease resistance by upregulating defense metabolites, including quercetin, pipecolic acid (Pip), and N-hydroxypipecolic acid (NHP), all of which inhibit bacterial and fungal proliferation. Overexpression of *CitPH4* in model plants significantly improved fungal resistance, underscoring CA's role in defense gene activation (Lu et al. 2025). Moreover, CA interacts with the SA pathway, a central regulator of SAR (Fig. 10.4). SA accumulation induces PR genes (e.g., *PR1*, *PR2*,

Table 10.2 Role of citric acid in mitigating biotic stress in various crop systems

Crop	Biotic stress	Role of citric acid	Mechanism/effect	Reference
Sugar beet	*Cercospora beticola*	Direct antifungal activity reducing disease severity	Lowers fungal cytoplasmic pH, disrupts metabolism, increases membrane permeability leading to pathogen inhibition	Hassan et al. (2015)
Maize	*Fusarium oxysporum*	Enhances efficacy of biocontrol bacteria	Chelates micronutrients, improving bioavailability for beneficial microbes and boosting antagonistic activity	Asaturova et al. (2020)
Kiwifruit	*Pseudomonas syringae pv. actinidiae (Psa)*	Direct antibacterial effect	Destabilizes outer membrane of Psa via acidification	Sisson et al. (2024)
Cucumber	*Paracidovorax citrulli*	Inhibits biofilm formation	Reduces extracellular polysaccharide, protein, and DNA content	Yang et al. (2024)
Tobacco	Tobacco mosaic virus (TMV)	Enhances viral resistance	Induces systemic resistance via signaling pathways regulated by ethylene, JA, and SA	Basit et al. (2021)
Maize	*Fusarium oxysporum*	Promotes nutrient cycling and pathogen suppression	Enhances Fe solubilization, supports soil health, and attracts siderophilic bacteria that suppress pathogens	Wang et al. (2024)
Banana	*Fusarium oxysporum*	Promotes biocontrol potential	Enhances rhizosphere colonization by Bacillus velezensis LG14-3, improving its antagonistic potential	Yang et al. (2023)
Tomato	*Meloidogyne javanica*	Reduces nematode infestations	Induces systemic resistance by elevating JA biosynthesis in roots, activating proteinase inhibitors and polyamine oxidases toxic to nematodes	Shaimaa et al. (2021)

SA salicylic acid, *JA* jasmonic acid

PR5), whose encoded proteins restrict pathogen spread. Overexpression of SA-associated regulators, particularly *NPR1*, enhances resistance to diseases such as citrus canker by sustaining PR gene expression and priming defense responses (Boccardo et al. 2019; Nascimento et al. 2022). Beneficial microbes also reinforce this connection by activating SA-dependent PR gene expression and ROS production, further integrating CA metabolism with immune signaling (Riera et al. 2018).

CA also exerts direct antifungal activity against key phytopathogens. In sugar beet, foliar CA application markedly reduced the severity of cercospora leaf spot caused by *Cercospora beticola*. In vitro assays revealed a dose-dependent inhibition of fungal mycelial growth, reaching 83.7% suppression at 50 mM CA. The proposed mechanism involves cytoplasmic acidification, disruption of metabolic function, and enhanced membrane permeability, leading to fungal cell death. These effects

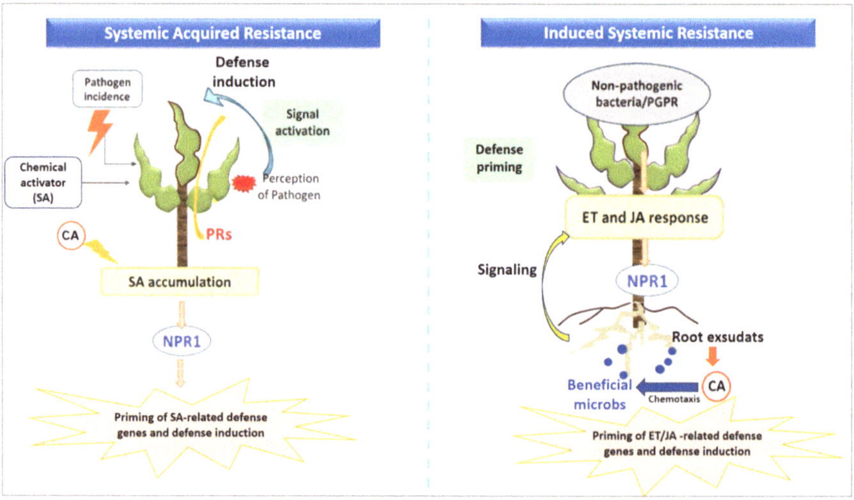

Fig. 10.4 Citric acid-mediated regulation of systemic plant immunity. Citric acid (CA) influences plant defense by modulating both systemic acquired resistance (SAR) and induced systemic resistance (ISR). In SAR, pathogen infection or chemical activators trigger SA accumulation, leading to the activation of NPR1 and priming of SA-related defense genes, including pathogenesis-related (PR) proteins. In ISR, non-pathogenic bacteria or PGPR activate ethylene (ET) and jasmonic acid (JA) signaling pathways, also requiring NPR1, thereby priming ET/JA-dependent defense genes. Root exudates enriched in CA enhance beneficial microbe recruitment, reinforcing defense priming and strengthening immune signaling. *Abbreviation*: CA, citric acid; SA, salicylic acid; ET, ethylene; JA, jasmonic acid; PRs, pathogenesis related genes; PGPR, plant growth–promoting rhizobacteria; NPR1, non-expresser of PR genes

translated in vivo to reduced disease severity, improved yield, and enhanced crop quality (Hassan et al. 2015). In maize, CA amplified the antifungal efficacy of the biocontrol bacterium *Bacillus subtilis*, enhancing its antagonism against *Fusarium oxysporum* by more than threefold. This synergy was attributed to CA-mediated micronutrient chelation, which improved nutrient availability for beneficial microbes and indirectly suppressed pathogenic fungi (Asaturova et al. 2020). Additionally, CA inhibited the growth of *Aspergillus flavus* and *Penicillium purpurogenum* through interactions of its multi-carboxyl structure with fungal cellular components, impairing membrane integrity and enzymatic activity (Hassan et al. 2015).

Secondary metabolites (SMs) also play a critical role in plant chemical defense against herbivores and pathogens, and their biosynthesis is influenced by CA (Al-Khayri et al. 2023). SMs such as alkaloids, phenolics, flavonoids, and terpenoids exhibit diverse antimicrobial, antifungal, and antiviral activities (Ashraf et al. 2018). These compounds are key to systemic resistance induction, including defense against Tobacco mosaic virus (TMV), via signaling pathways regulated by ET, JA, and SA (Basit et al. 2021). Phytoalexins, a specialized class of SMs with potent antimicrobial activity, are central to pathogen defense. Camalexin, the major phytoalexin in Arabidopsis, accumulates during pathogen invasion under the regulation of phytohormones, effectively inhibiting virulent strains of Pseudomonas (Balmer et al. 2018). Terpenoids similarly mediate plant–herbivore–pathogen interactions. For instance, monoterpenes act as phytoalexins, influence insect oviposition, and deter herbivory, while functioning as kairomones that attract natural enemies to damaged plants. In *Pinus sylvestris*, oviposition by the sawfly *Diprion pini* induced systemic production of monoterpenes and sesquiterpenes (Hijaz et al. 2013). Furthermore, terpenes such as farnesol, menthol, camphor, pyrethrins, and artemisinins exhibit broad-spectrum antimicrobial activity. Capsaicin, a characteristic alkaloid of chili pepper, demonstrates potent antibacterial properties (Al-Khayri et al. 2023).

The rhizosphere represents a key interface for plant–microbe interactions. Root exudates—comprising a diverse mixture of organic and inorganic compounds—regulate microbial density and activity in this zone (Feng et al. 2021). Citric acid, as a major root exudate component, plays a pivotal role in microbial recruitment, colonization, and activation of plant defenses. In rice, CA significantly enhanced colonization by *Bacillus altitudinis* LZP02 by upregulating chemotaxis-related genes (Jiao et al. 2022). It also promoted biofilm formation in beneficial bacteria, reinforcing rhizosphere colonization and plant–microbe interactions (Jiao et al. 2022). Similarly, CA induced chemotactic responses in *Bacillus velezensis* B26 while upregulating biofilm-related genes (Sharma et al. 2020). Additional evidence shows that CA enhanced recruitment of *Bacillus amyloliquefaciens* T-5, leading to increased rhizobacterial populations (Tan et al. 2013). In banana, CA improved colonization capacity of *Bacillus velezensis* LG14-3 in the rhizosphere, boosting its biocontrol potential against Fusarium wilt (Yang et al. 2023).

Beyond recruitment, CA also fuels microbial-driven soil processes. Acting as a carbon source, it sustains microbial metabolism and supports biotransformation reactions, such as the detoxification of arsenic in rice systems (Zou et al. 2024). The interplay between CA and rhizobacteria is reciprocal: beneficial microbes stimulate CA secretion, generating a positive feedback loop that enhances bacterial colonization and activity. This dynamic fosters a resilient rhizosphere microbiome critical for nutrient cycling, pathogen suppression, and soil health (Jiao et al. 2022). For instance, CA released from maize roots enhances Fe solubilization, improving Fe availability for siderophilic bacteria while indirectly suppressing pathogenic bacteria (Yang et al. 2025).

Beneficial microbes recruited or activated by CA-enriched root exudates also contribute to induced systemic resistance (ISR) (Fig. 10.4), a primed immune state that enhances plant responsiveness to subsequent pathogen or herbivore attack (Li et al. 2021; Pieterse et al. 2014). Once established, these microbes produce signaling molecules—including cyclic lipopeptides, volatile organic compounds, and phytohormones—that interact with host immune networks to trigger ISR (Kour et al. 2024; Zhou and Wang 2018). Microbial signals stimulate JA/ET-mediated defense pathways, leading to physiological responses such as ROS bursts, callose deposition, and upregulation of defense-related genes. While ISR is largely SA-independent, effective signaling still requires NPR1 as a central regulator (Choudhary and Prakash 2007; Zhou and Wang 2018). In rice (Oryza sativa), CA-enriched root exudates acted as chemoattractants for plant growth-promoting rhizobacteria (PGPR). In particular, CA enhanced root colonization by *Bacillus altitudinis* LZP02, thereby reinforcing ISR against sheath blight pathogens (Jiao et al. 2022).

10.5 Citric Acid as a Precursor and Regulator of Defense Compounds

The carboxylic acids generated in the TCA cycle act as central nodes linking broader metabolic networks (Sweetlove et al. 2010). Citric acid, in particular, occupies a pivotal position within these networks, contributing to the biosynthesis of SMs, osmolytes, and phytohormones that collectively enhance stress tolerance in plants (Eidyan et al. 2014; Singh et al. 2019; Sweetlove et al. 2010). During the TCA cycle, the degradation products of CA (e.g., SA salts) feed into the biosynthesis of γ-aminobutyric acid (GABA) (Fig. 10.5) (Cercós et al. 2006), as well as isoprenoids, flavonoids, fatty acids, sugars, and plant hormones (Sweetlove et al. 2010; Zahan et al. 2021).

Citric acid plays a central role in amino acid biosynthesis and GABA metabolism by acting as a hub that channels intermediates toward glutamate (Glu) production (Cercós et al. 2006). This process is mediated through the aconitase (ACO)–GABA pathway, involving aconitase (ACO), isocitrate dehydrogenase (IDH), glutamate synthase (GS), and glutamate decarboxylase (GAD) (Hussain et al. 2017). The CA cycle intermediates α-ketoglutarate (α-KG) and 2-oxoglutarate (2-OG) represent direct entry points into this pathway. Specifically, α-KG undergoes oxidative deamination catalyzed by Glu dehydrogenase to generate Glu, which subsequently branches into two major routes: conversion into glutamine or processing through the GABA shunt (Cercós et al. 2006; Hussain et al. 2017). Glu also serves as a precursor for a wide range of amino acids and derivatives—including proline, arginine, ornithine, thiamine, lysine (Zahan et al. 2021), and glutamine (Kumar and Dubey 2019). Studies in Citrus limon callus and fruit tissues have demonstrated that elevated cytosolic CA enhances the activity of enzymes in the GABA shunt pathway (Cercós et al. 2006; Hussain et al. 2017). Beyond its metabolic functions, GABA contributes to stress tolerance by regulating organic metabolites, notably water-soluble sugars such as trehalose (Choudhary et al. 2025).

A second major route of CA metabolism is the acetyl-CoA pathway (Fig. 10.5), also referred to as the ATP-citrate lyase (ACL) pathway. In this process, CA is cleaved to yield acetyl-CoA, a key precursor for the biosynthesis of diverse SMs through both the mevalonate (MVA) and non-mevalonate (non-MVA) pathways (Li et al. 2020). Acetyl-CoA generated via this route is indispensable for isoprenoid biosynthesis. In the MVA pathway, acetyl-CoA is converted into isopentenyl diphosphate (IPP), the universal precursor of isoprenoids. IPP serves as the substrate to produce phytohormones—including GA, ABA, IAA, carotenoids, strigolactones, brassinosteroids, and tocopherols (Zahan et al. 2021)—as well as volatile defense

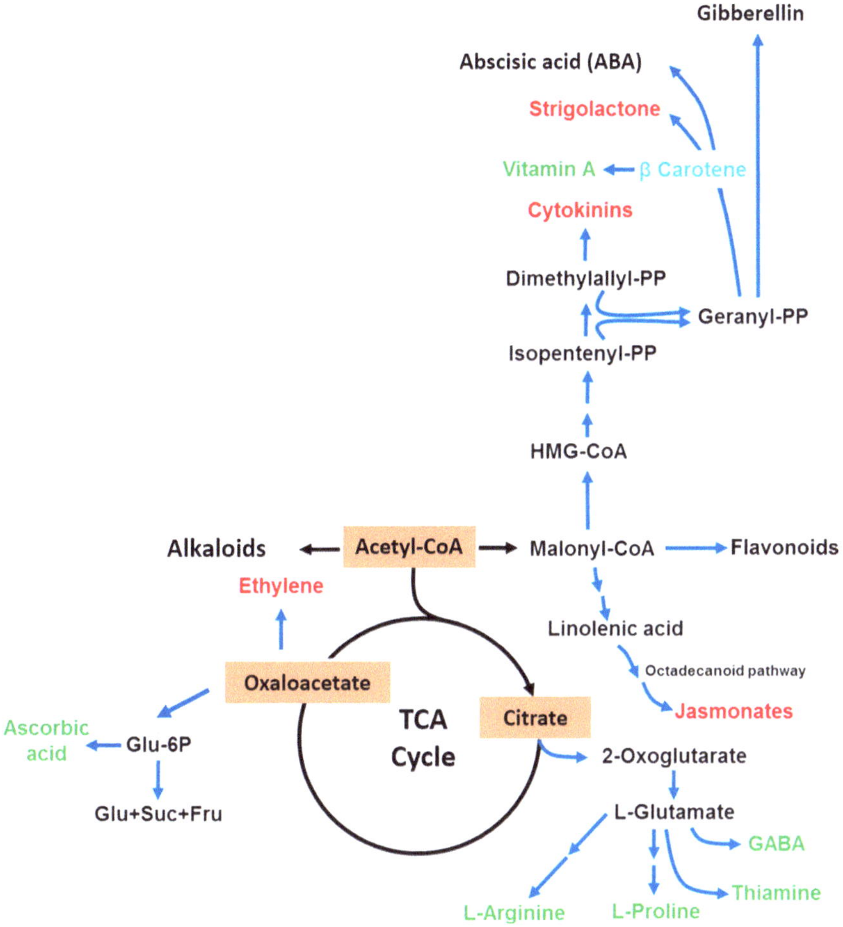

Fig. 10.5 Citric acid–derived metabolic cascades leading to protective molecule biosynthesis. Citric acid, as a central intermediate of the tricarboxylic acid (TCA) cycle, fuels multiple metabolic routes that generate protective molecules under stress. Through the aconitase–glutamate pathway, it contributes to γ-aminobutyric acid (GABA) and amino acids (proline, arginine, thiamine), while oxaloacetate (OAA) supports ascorbate production and gluconeogenesis (glucose, fructose, sucrose). Acetyl-CoA derived from citrate enters the mevalonate (MVA) pathway to yield isoprenoid precursors (isopentenyl-PP, geranyl-PP, dimethylallyl-PP) driving phytohormone biosynthesis (gibberellins, abscisic acid, strigolactones, cytokinins) and carotenoids (e.g., β-carotene, vitamin A). Malonyl-CoA generated via ATP-citrate lyase (ACL) supports flavonoid production, while acetyl-CoA elongation produces oxylipins including jasmonates through the octadecanoid pathway. Additionally, citrate-derived acetyl-CoA and OAA support alkaloid and ethylene biosynthesis. Collectively, these interconnected cascades highlight the pivotal role of citrate in linking central metabolism to secondary metabolite and phytohormone networks essential for plant defense and stress tolerance

compounds such as isoprenes, monoterpenes, and sesquiterpenes. These metabolites play key roles in mediating resistance against pathogens and herbivores. Additionally, IPP contributes to the synthesis of dolichols, which regulate diverse aspects of plant growth and development (Liao et al. 2016). Exogenous CA application has also been reported to stimulate phytohormones such as auxins, GAs, and CKs, while activating ROS cascades under stress conditions (Burns 2023).

Beyond the MVA route, acetyl-CoA also serves as a substrate for fatty acid elongation, ultimately giving rise to oxylipins, including jasmonates, via the octadecanoid (non-MVA) pathway (Zahan et al. 2021). Furthermore, the acetyl-CoA (or acetate) pathway supports the biosynthesis of N-containing SMs such as alkaloids. For example, it contributes to the formation of piperidine alkaloids and anthranilic acid, the latter being further converted into CoA esters such as anthraniloyl-CoA, which serve as intermediates in specific alkaloid biosynthetic pathways (Seneca 2007). L-amino acids—including tyrosine, phenylalanine, tryptophan, lysine, and arginine—derived from CA cycle intermediates, also function as precursors for a broad spectrum of alkaloids. These amino acids may be integrated into structural classes such as steroids, secoiridoids, and other ligands, enabling the synthesis of complex alkaloid structures (Kumar et al. 2018).

The biosynthesis of citrus flavonoids is another pathway tightly linked to CA metabolism (Fig. 10.5). Flavonoid production proceeds via the general phenylpropanoid pathway, beginning with the conversion of phenylalanine to p-coumaroyl-CoA through the sequential activities of phenylalanine ammonia-lyase (PAL, EC 4.3.1.5), cinnamate-4-hydroxylase (C4H), and 4-coumarate-CoA ligase (4CL). Concurrently, cytosolic acetyl-CoA is carboxylated by acetyl-CoA carboxylase (ACCase) to form malonyl-CoA, the extender unit used by chalcone synthase (CHS)—the first committed and rate-limiting enzyme in flavonoid biosynthesis (Fatland et al. 2002; Liu et al. 2021). CHS catalyzes the condensation of one molecule of p-coumaroyl-CoA with three molecules of malonyl-CoA to form chalcone, which is subsequently converted into naringenin. Naringenin acts as a central intermediate that gives rise to various flavonoid subclasses, including flavanones, flavonols, and anthocyanins (Zhao et al. 2020).

Finally, oxaloacetate (OAA)—another product of ACL catalysis and a key TCA cycle intermediate—plays essential roles in several metabolic pathways. OAA contributes to gluconeogenesis, generating soluble sugars such as sucrose, fructose, and glucose (Luo et al. 2021), while also serving as a precursor for organic acids, including ascorbate and aspartate (Guo et al. 2016; Zahan et al. 2021). Moreover, OAA is required for ethylene biosynthesis (Zahan et al. 2021).

10.6 Molecular Mechanisms and Gene Regulation

Citric acid is increasingly recognized as a multifunctional molecule that contributes not only to plant primary metabolism but also to the regulation of gene expression underlying stress tolerance. Beyond its central metabolic functions, CA acts as a

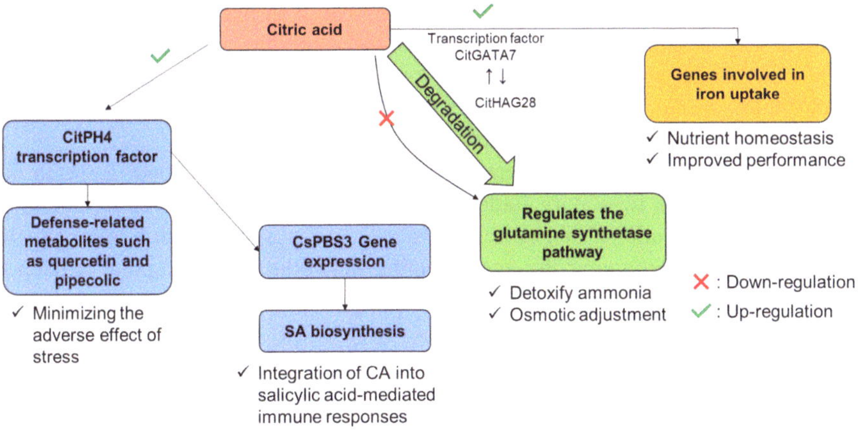

Fig. 10.6 Citric acid-responsive gene networks and transcriptional regulators under stress. CA accumulation is modulated by transcription factors such as *CitPH4*, which enhances defense-related metabolites (e.g., quercetin, pipecolic acid) and influences *CsPBS3*-mediated salicylic acid (SA) biosynthesis. Conversely, CA degradation is promoted by *CitGATA7* in coordination with the histone acetyltransferase *CitHAG28*, regulating the glutamine synthetase pathway. In parallel, CA interacts with vacuolar transport processes via *CitERF13–CitVHA-c4* complexes and contributes to nutrient acquisition by modulating the expression of genes involved in iron uptake (Fig. 10.6). Collectively, these networks illustrate CA's dual role as a metabolic intermediate and regulatory molecule linking stress responses, epigenetic regulation, and nutrient homeostasis. *Abbreviations*: *CitPH4*: *Citrus PH4* transcription factor; *CsPBS3*: *Citrus sinensis PBS3* gene; SA: Salicylic Acid

signaling mediator that modulates transcriptional, epigenetic, and protein-level processes linked to plant defense responses.

In citrus, the transcription factor *CitPH4* has been identified as a central regulator of CA accumulation. Overexpression of *CitPH4* enhances the levels of defense-associated metabolites such as quercetin and pipecolic acid (Pip), thereby mitigating stress-induced damage (Fig. 10.6). Furthermore, *CitPH4* influences SA biosynthesis through the modulation of *CsPBS3* expression, integrating CA metabolism with SA-dependent immune pathways (Hu et al. 2025; Lu et al. 2025). Another transcription factor, *CitGATA7*, promotes CA degradation by interacting with the histone acetyltransferase *CitHAG28*, which regulates the glutamine synthetase pathway. This highlights the role of epigenetic machinery in orchestrating CA turnover and its downstream effects on metabolic and stress-response networks. Similarly, *CitERF13* has been shown to regulate CA accumulation via protein–protein interaction with the vacuolar proton pump *CitVHA-c4*, emphasizing the complexity of CA homeostasis mediated by dynamic protein-level interactions.

In rice, the *OsAlR3* gene enhances tolerance to aluminum toxicity by stimulating the secretion of organic acids, including CA, while simultaneously upregulating antioxidant-related genes to alleviate oxidative stress (Tahjib-Ul-Arif et al. 2021). In soybean, *GmMATE13* encodes a CA transporter whose expression is induced under aluminum stress, facilitating CA efflux into the rhizosphere and conferring aluminum resistance (Wang et al. 2022). Moreover, CA plays a regulatory role in nutrient acquisition. In tomato, for example, the combined application of CA with humic substances modulates the expression of genes associated with iron uptake, contributing to nutrient homeostasis, enhanced vigor, and improved performance under suboptimal conditions (Tahjib-Ul-Arif et al. 2021).

Together, these findings underscore the dual role of CA as both a metabolic intermediate and a regulatory molecule. Through transcriptional reprogramming, epigenetic modifications, transporter activation, and cross-talk with hormone pathways, CA integrates metabolic and signaling networks to fortify plant resilience against environmental stresses.

10.7 Conclusion and Future Perspectives

Citric acid emerges as a pivotal and multifunctional metabolite in plant physiology, serving essential roles in both primary metabolism and in orchestrating adaptive responses to diverse environmental stresses. Its central position within the tricarboxylic acid cycle underscores its importance in carbon metabolism and energy production, while its capacity for transport across multiple cellular compartments highlights its involvement in a wide spectrum of biosynthetic pathways.

As a stress regulator, CA demonstrates remarkable versatility. It contributes to osmotic adjustment and ion balance under drought and salinity stress, mediates the chelation, sequestration, and detoxification of heavy metals, stabilizes enzyme function and preserves membrane integrity under thermal extremes, and plays a central role in mitigating oxidative stress through both direct ROS scavenging and the activation of antioxidant enzyme systems. These multifaceted protective effects underscore the strong potential of CA as a biostimulant for enhancing crop resilience under the increasingly challenging environmental conditions of modern agriculture.

The diverse mechanisms through which CA confers tolerance highlight its promise for practical applications aimed at improving plant performance, yield stability, and agricultural sustainability. Nevertheless, several key research directions remain. Advancing our understanding of the molecular and genetic regulatory networks that govern CA metabolism, elucidating its interactions with other phytohormones and signaling compounds, and optimizing application strategies in field settings will be critical to fully harness its potential. Moreover, the development of crops genetically engineered for enhanced CA biosynthesis, as well as the exploration of CA in phytoremediation strategies, represents promising avenues for innovation.

Taken together, these perspectives position CA not only as a central metabolic hub but also as a strategic tool for designing sustainable, climate-resilient agricultural systems capable of withstanding the combined pressures of abiotic and biotic stresses in the era of global environmental change.

References

Abdel Latef AAH, Hameed M, Khan Z, Khan S, Imran A, Fatima N (2026) Foliar stimulants: A sustainable strategy for enhancing plant resilience to abiotic and biotic stresses. Physiologia Plantarum. 178, no. 1: e70741

Afzal A, Ali B, Samarasinghe KGBA, Khan MN, Javed MA, Saleem A (2026) Overview of biotic stress and its agricultural impact. In: Abdel Latef, A.A.H. (eds) Chitosan Nanoparticles for Plant Resilience Under Abiotic and Biotic Stressors. Smart Nanomaterials Technology. Springer, Singapore. https://doi.org/10.1007/978-981-95-6422-4_3

Ait-El-Mokhtar M, Ben Laouane R, Anli M, Boutasknit A, Wahbi S, Meddich A (2019) Use of mycorrhizal fungi in improving tolerance of the date palm (Phoenix dactylifera L.) seedlings to salt stress. Sci Hortic 253:429–438. https://doi.org/10.1016/j.scienta.2019.04.066

Ait-El-Mokhtar M et al. (2022) Cereals and Phytohormones Under Drought Stress. In: Abdel Latef, A.A.H. (eds) Sustainable Remedies for Abiotic Stress in Cereals. Springer, Singapore. https://doi.org/10.1007/978-981-19-5121-3_13

Al-Khayri JM, Rashmi R, Toppo V, Chole PB, Banadka A, Sudheer WN, Nagella P, Shehata WF, Al-Mssallem MQ, Alessa FM, Almaghasla MI, Rezk AAS (2023) Plant secondary metabolites: the weapons for biotic stress management. Metabolites 13(6):1–37. https://doi.org/10.3390/metabo13060716

Anli M et al. (2022) Use of biostimulants to improve drought tolerance in cereals. In: Abdel Latef, A.A.H. (eds) Sustainable Remedies for Abiotic Stress in Cereals. Springer, Singapore. https://doi.org/10.1007/978-981-19-5121-3_20

Anli M, Boutasknit A, Ait-El-Mokhtar M, Ben-Laouane R, Abdouroihamane H, Meddich A, Baslam M (2026) Plant extracts for boosting plant stress tolerance. In: Abdel Latef AAH (eds) Using Stimulants to Improve Plant Health: From Stress to Strength. Elsevier. https://doi.org/10.1016/B978-0-443-45576-6.00009-1

Asaturova A, Gyrnets E, Allakhverdian V, Astakhov M, Saenko K (2020) The study of the antifungal activity of the Bacillus subtilis BZR 336g strain under the conditions of periodic cultivation with the addition of citric acid, corn extract and some microelements. BIO Web Conf 21:00015. https://doi.org/10.1051/bioconf/20202100015

Ashraf MA, Riaz M, Arif MS (2018) Environmental stress and secondary metabolites in plants: an overview. In: Plant metabolites and regulation under environmental stress. Academic Press

Balmer A, Pastor V, Glauser G, Mauch-Mani B (2018) Tricarboxylates induce defense priming against bacteria in Arabidopsis thaliana. Front Plant Sci 9:1–15. https://doi.org/10.3389/fpls.2018.01221

Basit A, Farhan M, Di Mo W, Ding HX, Ikram M, Farooq T, Ahmed S, Yang ZF, Wang Y, Hashem M, Alamri S, Bashir MA, El-Zohri M (2021) Enhancement of resistance by poultry manure and plant hormones (salicylic acid & citric acid) against tobacco mosaic virus. Saudi J Biol Sci 28(6):3526–3533. https://doi.org/10.1016/j.sjbs.2021.03.025

Baslam M, Mitsui T, Sueyoshi K, Ohyama T (2021) Recent advances in carbon and nitrogen metabolism in C3 plants. Int J Mol Sci 22(1):1–39. https://doi.org/10.3390/ijms22010318

Ben-Laouane R, Ait-el-mokhtar M, Anli M, Boutasknit A (2025) Microbial biotechnologies for salt tolerance in alfalfa: agro-nutritional comparison between local and imported varieties. Nitrogen 6:27. https://doi.org/10.3390/nitrogen6020027

Boccardo NA, Segretin ME, Hernandez I, Mirkin FG, Chacón O, Lopez Y, Borrás-Hidalgo O, Bravo-Almonacid FF (2019) Expression of pathogenesis-related proteins in transplastomic tobacco plants confers resistance to filamentous pathogens under field trials. Sci Rep 9(1): 2791. https://doi.org/10.1038/s41598-019-39568-6

Boutasknit A, Ait-El-Mokhtar M, Fassih B, Ben-Laouane R, Wahbi S, Meddich A (2024) Effect of arbuscular mycorrhizal fungi and rock phosphate on growth, physiology, and biochemistry of carob under water stress and after rehydration in vermicompost-amended soil. Metabolites 14(4):202. https://doi.org/10.3390/metabo14040202

Burns M (2023) The effect of the interaction of citric acid and drought on the growth of spotted gum (Corymbia maculata) seedlings. J Biotechnol Biomed 06(3):347–358. https://doi.org/10.26502/jbb.2642-91280097

Cercós M, Soler G, Iglesias DJ, Gadea J, Forment J, Talón M (2006) Global analysis of gene expression during development and ripening of citrus fruit flesh. A proposed mechanism for citric acid utilization. Plant Mol Biol 62(4–5):513–527. https://doi.org/10.1007/s11103-006-9037-7

Chakrobortty J, Shahin Imran M, Mahamud A, Sarker P, Paul NC (2022) Effect of citric acid (CA) priming and exogenous application on germination and early seedling growth of okra (Abelmoschus esculentus L.) plants under salinity stress condition. Arch Agric Environ Sci 7(3):318–326. https://doi.org/10.26832/24566632.2022.070303

Chen YX, Lin Q, Luo YM, He YF, Zhen SJ, Yu YL, Tian GM, Wong MH (2003) The role of citric acid on the phytoremediation of heavy metal contaminated soil. Chemosphere 50(6):807–811. https://doi.org/10.1016/S0045-6535(02)00223-0

Choudhary DK, Prakash A (2007) Induced systemic resistance (ISR) in plants: mechanism of action. Indian J Microbiol 47:289–297

Choudhary A, Nepovimova E, Rajput VD, Malik T, Choudhary M, Bhardwaj N, Peter L, Puri S, Kimta N (2025) Mechanistic understanding of GABA and trehalose in modulating plant response to drought stress. Plant Stress 16:100838. https://doi.org/10.1016/j.stress.2025.100838

Eckardt NA (2005) Peroxisomal citrate synthase provides exit route from fatty acid metabolism in oilseeds. Plant Cell 17(7):1863–1865. https://doi.org/10.1105/tpc.105.034843

Eidyan B, Hadavi E, Moalemi N (2014) Pre-harvest foliar application of iron sulfate and citric acid combined with urea fertigation affects growth and vase life of tuberose (Polianthes tuberosa L.) "Por-Par". Hortic Environ Biotechnol 55(1):9–13. https://doi.org/10.1007/s13580-014-0061-2

El-Tohamy WA, El-Abagy HM, Badr MA, Gruda N (2013) Drought tolerance and water status of bean plants (Phaseolus vulgaris L.) as affected by citric acid application. J Appl Bot Food Qual 86(1):212–216. https://doi.org/10.5073/JABFQ.2013.086.029

Fassih B, Ait-el-mokhtar M, Douch AN, Boutasknit A, Ben-Laouane R, Aganchich B, Wahbi S (2024) Combined effect of subsurface water retention technology and arbuscular mycorrhizal fungi on growth, physiology and biochemistry of argan seedlings under field conditions. Plants 13(15):2098. https://doi.org/10.3390/plants13152098

Fatland BL, Ke J, Anderson MD, Mentzen WI, Li WC, Allred CC, Johnston JL, Nikolau BJ, Wurtele ES (2002) Molecular characterization of a heteromeric ATP-citrate lyase that generates cytosolic acetyl-coenzyme A in Arabidopsis. Plant Physiol 130(2):740–756. https://doi.org/10.1104/pp.008110

Feng H, Fu R, Hou X, Lv Y, Zhang N, Liu Y, Xu Z, Miao Y, Krell T, Shen Q, Zhang R (2021) Chemotaxis of beneficial rhizobacteria to root exudates: the first step towards root–microbe rhizosphere interactions. Int J Mol Sci 22(13):6655. https://doi.org/10.3390/ijms22136655

Foyer CH, Noctor G, Hodges M (2011) Respiration and nitrogen assimilation: targeting mitochondria-associated metabolism as a means to enhance nitrogen use efficiency. J Exp Bot 62(4):1467–1482. https://doi.org/10.1093/jxb/erq453

Gimenez E, Salinas M, Manzano-Agugliaro F (2018) Worldwide research on plant defense against biotic stresses as improvement for sustainable agriculture. Sustainability 10(2):1–19. https://doi.org/10.3390/su10020391

Goel P, Bhuria M, Kaushal M, Singh AK (2016) Carbon: nitrogen interaction regulates expression of genes involved in N-uptake and assimilation in Brassica juncea L. PLoS One 11(9):1–19. https://doi.org/10.1371/journal.pone.0163061

González Guzmán M, Cellini F, Fotopoulos V, Balestrini R, Arbona V (2022) New approaches to improve crop tolerance to biotic and abiotic stresses. Physiol Plant 174(1):e13547. https://doi.org/10.1111/ppl.13547

Guo LX, Shi CY, Liu X, Ning DY, Jing LF, Yang H, Liu YZ (2016) Citrate accumulation-related gene expression and/or enzyme activity analysis combined with metabolomics provide a novel insight for an orange mutant. Sci Rep 6:1–12. https://doi.org/10.1038/srep29343

Hajaji AN, Maaroufi-Dguimi H, Ammari Y (2025) Exogenous application of citric acid mitigates salt-induced oxidative stress in Moringa oleifera seedlings. J Umm Al-Qura Univ Appl Sci 11: 699–711. https://doi.org/10.1007/s43994-024-00169-3

Han H, Li YJ, Alam SM, Zhou T, Khan MA, Thu AM, Liu YZ (2025) AP2 transcription factor CsAIL6 negatively regulates citric acid accumulation in citrus fruits by interacting with a WD40 protein CsAN11. Hortic Res 12(4):1–10. https://doi.org/10.1093/hr/uhaf002

Hassan R, El-kadi S, Sand M (2015) Effect of some organic acids on some fungal. Int J Adv Biol 2(1):1–11

Hijaz F, El-Shesheny I, Killiny N (2013) Herbivory by the insect diaphorina citri induces greater change in citrus plant volatile profile than does infection by the bacterium, Candidatus Liberibacter asiaticus. Plant Signal Behav 8(10):1–10. https://doi.org/10.4161/psb.25677

Hossain MA, Khatun MS, Hosen M, Sayed ZI, Islam MR, Chowdhury MK, Iqbal MA, Al-Ashkar I, Erden Z, Toprak CC, El Sabagh A, Islam MS (2024) Citric acid alleviated salt stress by modulating photosynthetic pigments, plant water status, yield and nutritional quality of black gram [Vigna mungo (L.) Hepper]. Legum Res 47:2068–2076. https://doi.org/10.18805/lrf-820

Hu L, Zhang Z, Xiang Z, Yang Z (2016) Exogenous application of citric acid ameliorates the adverse effect of heat stress in tall fescue (Lolium arundinaceum). Front Plant Sci 7:1–11. https://doi.org/10.3389/fpls.2016.00179

Hu B, Yuan T, Lu Z, Huang R, He J, Yang K, Wu Q, Ai W, Zhang W, Zheng W, Wu X, Wang X, Xu Y, Deng X, Xu Q (2025) "Candidatus Liberibacter asiaticus" infection induces citric acid accumulation and immune responses mediated by the transcription factor CitPH4. Mol Plant Pathol 26(2):e70062. https://doi.org/10.1111/mpp.70062

Hussain SB, Shi CY, Guo LX, Kamran HM, Sadka A, Liu YZ (2017) Recent advances in the regulation of citric acid metabolism in citrus fruit. Crit Rev Plant Sci 36(4):241–256. https://doi.org/10.1080/07352689.2017.1402850

Ijaz M, Ansari MUR, Alafari HA, Iqbal M, Alshaya DS, Fiaz S, Ahmad HM, Zubair M, Ramzani PMA, Iqbal J, Abushady AM, Attia K (2022) Citric acid assisted phytoextraction of nickle from soil helps to tolerate oxidative stress and expression profile of NRAMP genes in sunflower at different growth stages. Front Plant Sci 13:1–14. https://doi.org/10.3389/fpls.2022.1072671

Ikan C, Soussani FE, Ouhaddou R, Ech-Chatir L, Errouh F, Boutasknit A, Assouguem A, Ali EA, Ullah R, Ait Barka E, Lahlali R, Meddich A (2024) Use of biofertilizers as an effective management strategy to improve the photosynthetic apparatus, yield, and tolerance to drought stress of drip-irrigated wheat in semi-arid environments. Agronomy 14(6):1316. https://doi.org/10.3390/agronomy14061316

Jian SF, Huang YF, Wan S, Chen DL, Lin Y, Liao Q, Zhong C (2024) Sulfur redirects carbon metabolism to optimize nitrogen utilization and promote andrographolide biosynthesis in Andrographis paniculata seedlings. Curr Plant Biol 40:100422. https://doi.org/10.1016/j.cpb.2024.100422

Jiao H, Weihui X, Yunlong H, Tian R, Wang Z (2022) Citric acid in rice root exudates enhanced the colonization and plant growth-promoting ability of Bacillus altitudinis LZP02. Microbiol Spectrum 10(6):1–15. https://doi.org/10.1128/spectrum.01002-22

Ju YH, Roy SK, Choudhury AR, Kwon SJ, Choi JY, Rahman MA, Katsube-Tanaka T, Shiraiwa T, Lee MS, Cho K, Woo SH (2021) Proteome changes reveal the protective roles of exogenous citric acid in alleviating Cu toxicity in Brassica napus L. Int J Mol Sci 22(11):1–22. https://doi.org/10.3390/ijms22115879

Khan A, Kanwal F, Ullah S, Fahad M, Tariq L, Altaf MT, Riaz A, Zhang G (2025) Plant secondary metabolites—central regulators against abiotic and biotic stresses. Metabolites 15:1–27

Kleiber T, Chadzinikolau T, Formela-Luboińska M, Lartey JL, Kosiada T (2024) Enhancing lettuce drought tolerance: the role of organic acids in photosynthesis and oxidative defense. Appl Sci (Switz) 14(12):5119. https://doi.org/10.3390/app14125119

Kour D, Negi R, Sharief S, Kumar S, Kaur S, Kaur T, Sharma B, Dasila H, Kour H, Ramniwas S, Kumar K, Kumar D, Nath A (2024) Microbes mediated induced systemic response in plants: a review. Plant Stress 11:100334. https://doi.org/10.1016/j.stress.2023.100334

Kumar P, Dubey KK (2019) Citric acid cycle regulation: back bone for secondary metabolite production. In: New and future developments in microbial biotechnology and bioengineering. Elsevier, pp 165–181

Kumar SR, Shilpashree HB, Nagegowda DA (2018) Terpene moiety enhancement by overexpression of geranyl(geranyl) diphosphate synthase and geraniol synthase elevates monomeric and dimeric monoterpene indole alkaloids in transgenic Catharanthus roseus. Front Plant Sci 9:1–16. https://doi.org/10.3389/fpls.2018.00942

Lal MK, Tiwari RK, Altaf MA, Kumar A, Kumar R (2023) Editorial: abiotic and biotic stress in horticultural crops: insight into recent advances in the underlying tolerance mechanism. Front Plant Sci 14:1212982

Lee CP, Elsässer M, Fuchs P, Fenske R, Schwarzländer M, Harvey Millar A (2021) The versatility of plant organic acid metabolism in leaves is underpinned by mitochondrial malate-citrate exchange. Plant Cell 33(12):3700–3720. https://doi.org/10.1093/plcell/koab223

Li SJ, Wang WL, Ma YC, Liu SC, Grierson D, Yin XR, Chen KS (2020) Citrus CitERF6 contributes to citric acid degradation via upregulation of CitAclα1, encoding ATP-citrate lyase subunit α. J Agric Food Chem 68(37):10081–10087. https://doi.org/10.1021/acs.jafc.0c03669

Li J, Wang C, Liang W, Liu S (2021) Rhizosphere microbiome: the emerging barrier in plant-pathogen interactions. Front Microbiol 12:772420. https://doi.org/10.3389/fmicb.2021.772420

Liao P, Hemmerlin A, Bach TJ, Chye ML (2016) The potential of the mevalonate pathway for enhanced isoprenoid production. Biotechnol Adv 34(5):697–713. https://doi.org/10.1016/j.biotechadv.2016.03.005

Liu W, Feng Y, Suhang Y, Fan Z, Li X, Li J, Yin H (2021) The flavonoid biosynthesis network in plants. Int J Mol Sci 22(23):1–18. https://doi.org/10.3390/ijms222312824

Liu X, Tang Y, Ning W, Bao Y, Luo T, Wang J (2022) Responses of Issatchenkia terricola WJL-G4 upon citric acid stress. Molecules 27(9):2664. https://doi.org/10.3390/molecules27092664

Lu Z, Hu B, He J, Yuan T, Wu Q, Yang K, Zheng W, Huang Y, Xu Y, Wang X, Xu Q (2025) The transcription factor CitPH4 regulates plant defense-related metabolite biosynthesis in citrus. Plant Physiol 197(2):kiaf027. https://doi.org/10.1093/plphys/kiaf027

Luo Z, Yu S, Zeng W, Zhou J (2021) Comparative analysis of the chemical and biochemical synthesis of keto acids. Biotechnol Adv 47:107706. https://doi.org/10.1016/j.biotechadv.2021.107706

Luo C, Bashir NH, Li Z, Liu C, Shi Y, Chu H (2024) Plant microRNAs regulate the defense response against pathogens. Front Microbiol 15:1–10. https://doi.org/10.3389/fmicb.2024.1434798

Mallhi ZI, Rizwan M, Mansha A, Ali Q, Asim S, Ali S, Hussain A, Alrokayan SH, Khan HA, Alam P, Ahmad P (2019) Citric acid enhances plant growth, photosynthesis, and phytoextraction of lead by alleviating the oxidative stress in castor beans. Plants 8(11):525. https://doi.org/10.3390/plants8110525

Maphalaphathwa M, Nciizah AD (2025) Nutrient seed priming effects on water stress tolerance and nutrient uptake of chilies (Capsicum annum L.). Agronomy 15:1–20

Nascimento CA, Teixeira-Silva NS, Caserta R, Marques MOM, Takita MA, de Souza AA (2022) Overexpression of CsSAMT in Citrus sinensis induces defense response and increases resistance to Xanthomonas citri subsp. citri. Front Plant Sci 13:836582. https://doi.org/10.3389/fpls.2022.836582

Nishio K, Mizushima T (2020) Structural and biochemical characterization of mitochondrial citrate synthase 4 from Arabidopsis thaliana. Acta Crystallogr F Struct Biol Commun 76:109–115. https://doi.org/10.1107/S2053230X20001521

Palei M, Das Mohapatra M, Pradhan M, Sahoo RK (2024) Occurrence of abiotic and biotic stress tolerance in rice: a multigene approach. Indian J Agric Res 58:737–743. https://doi.org/10.18805/ijare.a-6243

Palmgren M, Shabala S (2024) Adapting crops for climate change: regaining lost abiotic stress tolerance in crops. Front Sci 2:1–23. https://doi.org/10.3389/fsci.2024.1416023

Pieterse MJ, Zamioudis C, Berendsen RL, Weller DM, Van Wees SCM, Bakker PAHM (2014) Induced systemic resistance by beneficial microbes. Annu Rev Phytopathol 52:1–29. https://doi.org/10.1146/annurev-phyto-082712-102340
Raja BL et al. (2022) Use of biostimulants to improve salinity tolerance in cereals. In: Abdel Latef, A.A.H. (eds) Sustainable Remedies for Abiotic Stress in Cereals. Springer, Singapore. https://doi.org/10.1007/978-981-19-5121-3_19
Riera N, Wang H, Li Y, Li J, Pelz-Stelinski K, Wang N (2018) Induced systemic resistance against citrus canker disease by rhizobacteria. Phytopathology 108(9):1038–1045. https://doi.org/10.1094/PHYTO-07-17-0244-R
Seneca (2007) Alkaloid chemistry. In: Alkaloids – secrets of life. Elsevier, pp 61–139
Shaimaa N, El-Zawahry AMI, El Aref HM, Hanaa S (2021) Induction of systemic resistance in tomato by some abiotic com-pounds against Meloidogyne javanica. Assiut J Agric Sci 52(1): 74–89. https://doi.org/10.21608/ajas.2021.171570
Sharma M, Saleh D, Charron JB, Jabaji S (2020) A crosstalk between Brachypodium root exudates, organic acids, and Bacillus velezensis B26, a growth promoting bacterium. Front Microbiol 11: 1–18. https://doi.org/10.3389/fmicb.2020.575578
Sharma P, Lakra N, Goyal A, Ahlawat YK, Zaid A, Siddique KHM (2023) Drought and heat stress mediated activation of lipid signaling in plants: a critical review. Front Plant Sci 14:1–19. https://doi.org/10.3389/fpls.2023.1216835
Singh SP, Upadhyay SK, Pandey A, Kumar S (2019) Molecular approaches in plant biology and environmental challenges, Energy, environment, and sustainability. Springer, pp 1–5
Sisson HM, Fagerlund RD, Jackson SA, Briers Y, Warring SL, Fineran PC (2024) Antibacterial synergy between a phage endolysin and citric acid against the Gram-negative kiwifruit pathogen Pseudomonas syringae pv. actinidiae. Appl Environ Microbiol 90:e0184623. https://doi.org/10.1128/aem.01846-23
Sweetlove LJ, Beard KFM, Nunes-Nesi A, Fernie AR, Ratcliffe RG (2010) Not just a circle: flux modes in the plant TCA cycle. Trends Plant Sci 15(8):462–470. https://doi.org/10.1016/j.tplants.2010.05.006
Syvertsen JP, Melgar JC, García-Sánchez F (2010) Salinity tolerance and leaf water use efficiency in citrus. J Am Soc Hortic Sci 135(1):33–39. https://doi.org/10.21273/jashs.135.1.33
Tahjib-Ul-Arif M, Zahan MI, Karim MM, Imran S, Hunter CT, Islam MS, Mia MA, Hannan MA, Rhaman MS, Hossain MA, Brestic M, Skalicky M, Murata Y (2021) Citric acid-mediated abiotic stress tolerance in plants. Int J Mol Sci 22(13):7235. https://doi.org/10.3390/ijms22137235
Tan S, Yang C, Mei X, Shen S, Raza W, Shen Q, Xu Y (2013) The effect of organic acids from tomato root exudates on rhizosphere colonization of Bacillus amyloliquefaciens T-5. Appl Soil Ecol 64:15–22. https://doi.org/10.1016/j.apsoil.2012.10.011
Teshome DT, Zharare GE, Naidoo S (2020) The threat of the combined effect of biotic and abiotic stress factors in forestry under a changing climate. Front Plant Sci 11:601009. https://doi.org/10.3389/fpls.2020.601009
Tusei C (2019) The effects of citric acid on PH and nutrient uptake in wheatgrass (Triticum aestivum). IdeaFest 3:7
Villalobos-López MA, Arroyo-Becerra A, Quintero-Jiménez A, Iturriaga G (2022) Biotechnological advances to improve abiotic stress tolerance in crops. Int J Mol Sci 23(19):12053. https://doi.org/10.3390/ijms231912053
Wang Y, Mostafa S, Zeng W, Jin B (2021) Function and mechanism of jasmonic acid in plant responses to abiotic and biotic stresses. Int J Mol Sci 22(16):8568
Wang Z, Liu Y, Cui W, Gong L, He Y, Zhang Q, Meng X, Yang Z, You J (2022) Characterization of GmMATE13 in its contribution of citrate efflux and aluminum resistance in soybeans. Front Plant Sci 13:1–11. https://doi.org/10.3389/fpls.2022.1027560
Wang N, Wang T, Chen Y, Wang M, Lu Q, Wang K, Dou Z, Chi Z, Qiu W, Dai J, Niu L, Cui J, Wei Z, Zhang F, Kümmerli R, Zuo Y (2024) Microbiome convergence enables siderophore-secreting-rhizobacteria to improve iron nutrition and yield of peanut intercropped with maize. Nat Commun 15(1):1–17. https://doi.org/10.1038/s41467-024-45207-0

Wu Y, Ma L, Liu Q, Vestergård M, Topalovic O, Wang Q, Zhou Q, Huang L, Yang X, Feng Y (2020) The plant-growth promoting bacteria promote cadmium uptake by inducing a hormonal crosstalk and lateral root formation in a hyperaccumulator plant Sedum alfredii. J Hazard Mater 395(866):122661. https://doi.org/10.1016/j.jhazmat.2020.122661

Xiao F, Zhou H (2023) Plant salt response: perception, signaling, and tolerance. Front Plant Sci 13: 1–16. https://doi.org/10.3389/fpls.2022.1053699

Xie H, Bai G, Lu P, Li H, Fei M, Xiao BG, Chen XJ, Tong ZJ, Wang ZY, Yang DH (2022) Exogenous citric acid enhances drought tolerance in tobacco (Nicotiana tabacum). Plant Biol 24:333–343. https://doi.org/10.1111/plb.13371

Xin W, Zhang L, Zhang W, Gao J, Yi J, Zhen X, Li Z, Zhao Y, Peng C, Zhao C (2019) An integrated analysis of the rice transcriptome and metabolome reveals differential regulation of carbon and nitrogen metabolism in response to nitrogen availability. Int J Mol Sci 20(9):2349. https://doi.org/10.3390/ijms20092349

Xiu J, Haoting C, Yu S, Longqiang B, Leiping H, Yi Z (2021) Effect of citric acid seed priming on the growth and physiological characteristics of tomato seedlings under low phosphorus stress. Chin J Eco-Agric 29(7):1159–1170

Yan L, Riaz M, Liu J, Min Y, Cuncang J (2022) The aluminum tolerance and detoxification mechanisms in plants; recent advances and prospects. Crit Rev Environ Sci Technol 52(9): 1491–1527. https://doi.org/10.1080/10643389.2020.1859306

Yang Y, Yang Z, Shizhou Y, Chen H (2019) Organic acids exuded from roots increase the available potassium content in the rhizosphere soil: a rhizobag experiment in Nicotiana tabacum. HortScience 54(1):23–27. https://doi.org/10.21273/HORTSCI13569-18

Yang L, Zhou Y, Guo L, Yang L, Wang J, Liang C, Huang J (2023) The effect of banana rhizosphere chemotaxis and chemoattractants on Bacillus velezensis LG14-3 root colonization and suppression of banana Fusarium wilt disease. Sustainability 15:351

Yang J, Mao L, Gulfam Y, Zeeshan M, Wang X, Fan T (2024) Effect of acetic acid on biofilm formation in Paracidovorax citrulli, causal agent of bacterial fruit blotch. J Basic Microbiol 64: 1–9. https://doi.org/10.1002/jobm.202400188

Yang C-X, Chen S-J, Hong X-Y, Wang L-Z, Wu H-M, Tang Y-Y, Gao Y-Y (2025) Plant exudates-driven microbiome recruitment and assembly facilitates plant health management. FEMS Microbiol Rev 49:fuaf008. https://doi.org/10.1093/femsre/fuaf008

Yokosho K, Yamaji N, Ueno D, Mitani N, Jian FM (2009) OsFRDL1 is a citrate transporter required for efficient translocation of iron in rice. Plant Physiol 149(1):297–305. https://doi.org/10.1104/pp.108.128132

Youssef EA, Abdelbaset MM, El-Shafie AF (2025) Citric acid foliar application impact on growth and yield parameters of maize (Zea mays) under drought stress conditions. Egypt J Agron 47: 145–155. https://doi.org/10.21608/AGRO.2025.319414.1510

Yu XF, Wang ZW, Chen FL, Chen ML, Zhang XY, Zhang JR, Ji WY, Xiao X, Yang LJ, Lei T, Li X, Jiang MY (2024) The mitigation of citric acid on cadmium toxicity in Iris tectorum and its effects on the composition of cell walls. Int Biodeterior Biodegrad 193:105857. https://doi.org/10.1016/j.ibiod.2024.105857

Zahan MI, Karim M, Imran S, Hunter CT, Islam S, Mia A, Hannan A, Rhaman MS (2021) Citric acid-mediated abiotic stress tolerance in plants. Int J Mol Sci 22(13):7235

Zhao C, Wang F, Lian Y, Xiao H, Zheng J (2020) Biosynthesis of citrus flavonoids and their health effects. Crit Rev Food Sci Nutr 60(4):566–583. https://doi.org/10.1080/10408398.2018.1544885

Zhou M, Wang W (2018) Recent advances in synthetic chemical inducers of plant immunity. Front Plant Sci 9:1–10. https://doi.org/10.3389/fpls.2018.01613

Ziogas V, Tanou G, Morianou G, Kourgialas N (2021) Drought and salinity in citriculture: optimal practices to alleviate salinity and water stress. Agronomy 11(7):1–16. https://doi.org/10.3390/agronomy11071283

Zou L, Jiang O, Zhang S, Duan G, Gustave W, An X, Tang X (2024) Effects of citric acid on arsenic transformation and microbial communities in different paddy soils. Environ Res 249:118421. https://doi.org/10.1016/j.envres.2024.118421

Chapter 11
Ferulic Acid: A Phenolic Compound for Plant Stress Resistance

Btissam Zoubi, Abderrahim Boutasknit, Farid Errouh, Nizar El Mazouni, Raja Ben-Laouane, Abdelilah Meddich, and Marouane Baslam

Abstract Ferulic acid (FA), a hydroxycinnamic acid derivative abundant in plant cell walls, is a key metabolite within the phenylpropanoid pathway. It is increasingly recognized as a multifunctional compound modulating a wide spectrum of

---

B. Zoubi
Laboratory of Biotechnology, Environment, Agri-Food, and Health, Faculty of Sciences Dhar El Mahraz, Sidi Mohammed Ben Abdellah University, Fez, Morocco

Center of Agrobiotechnology and Bioengineering, Research Unit Labeled CNRST, Cadi Ayyad University, Marrakech, Morocco

A. Boutasknit
Center of Agrobiotechnology and Bioengineering, Research Unit labelled CNRST (Centre AgroBiotech-URL-7 CNRST-05), Cadi Ayyad University, Marrakesh, Morocco

Laboratory of Biology, Geosciences, Physics and Environment (OLMAN-LBGPE), Pluridisciplinary Faculty of Nador, University Mohamed Premier, Nador, Morocco

F. Errouh · N. El Mazouni
Center of Agrobiotechnology and Bioengineering, Research Unit labelled CNRST (Centre AgroBiotech-URL-7 CNRST-05), Cadi Ayyad University, Marrakesh, Morocco

R. Ben-Laouane
Center of Agrobiotechnology and Bioengineering, Research Unit labelled CNRST (Centre AgroBiotech-URL-7 CNRST-05), Cadi Ayyad University, Marrakesh, Morocco

FSTE-FSM Joint Laboratory: NRHE-UMI, Bioresources, Environment and Health Research Team, Moulay Ismail University, Meknes, Morocco

A. Meddich
Center of Agrobiotechnology and Bioengineering, Research Unit labelled CNRST (Centre AgroBiotech-URL-7 CNRST-05), Cadi Ayyad University, Marrakesh, Morocco

African Sustainable Agriculture Research Institute (ASARI), University Mohamed VI Polytechnic (UM6P), Laayoune, Morocco

M. Baslam (✉)
Center of Agrobiotechnology and Bioengineering, Research Unit labelled CNRST (Centre AgroBiotech-URL-7 CNRST-05), Cadi Ayyad University, Marrakesh, Morocco

GrowSmart, Amsterdam, The Netherlands

A. A. H. Abdel Latef (ed.), *Harnessing Vitamins and Organic Acids for Plant Stress Resilience*, Sustainability Sciences in Asia and Africa,
https://doi.org/10.1007/978-981-95-8593-9_11

physiological responses to both abiotic (e.g., drought, salinity, extreme temperatures, elevated CO_2, ozone, heavy metals) and biotic (e.g., bacterial and fungal pathogens, insects) stresses. Apart from its classical roles in lignification and cell wall reinforcement, FA participates in redox balance, signal transduction, and stress gene regulation. It contributes to plant immunity via antimicrobial activity, regulation of pathogenesis-related genes, and hormone signaling interactions. Under abiotic stress, FA acts as an antioxidant, regulates ion transport, and maintains membrane stability.

This chapter examines FA biosynthesis, transport, molecular functions, and its role as a bioactive metabolite in plant defense and stress reduction. By incorporating recent discoveries and practical insights, we offer a systems-level understanding of FA's role in plant resilience and its potential as a stimulant in stress-smart agriculture.

Keywords Ferulic acid · Phenolic compounds · Plant immunity · Abiotic stress · Biotic stress · Phenylpropanoid pathway · Antioxidant system · Defense signaling

11.1 Introduction

Global agriculture is increasingly challenged by the combined impacts of abiotic and biotic stresses, exacerbated by climate change and anthropogenic activities (Hanaka et al. 2022; Anand et al. 2023; Lal et al. 2023). These overlapping pressures have become major limiting factors for crop quality, yield, and productivity, thereby posing a serious threat to global food security (Devi et al. 2022; Fu et al. 2022; Mahto et al. 2022). Abiotic stresses such as drought, salinity, extreme temperature fluctuations, and heavy metal exposure impose severe constraints on plant metabolism (Godoy et al. 2021; Du et al. 2024). These conditions disrupt germination, growth, flowering, fruiting, and seed formation, often culminating in irreversible physiological decline (Medda et al. 2022; Nowicka 2022). Concurrently, biotic stresses—including those induced by fungi, bacteria, viruses, nematodes, and insect herbivores—further compromise crop vigor and exacerbate yield losses (Iqbal et al. 2021; Yıldız et al. 2024). Soil degradation, declining fertility, and reduced microbial diversity add to this burden by destabilizing ecosystem functions critical for sustainable agriculture (Rahman et al. 2021; Bogati et al. 2023). According to FAO (2023), up to 40% of annual global agricultural production losses are attributable to such combined stresses, underscoring the urgency of integrated strategies to strengthen crop resilience.

To address these challenges, a range of intervention strategies has been developed to mitigate the deleterious impacts of environmental stresses on crops. These include the application of chemical fertilizers, biostimulants, phytohormones, elicitors, and newly identified bioactive metabolites, all of which have demonstrated the ability to activate defense pathways, enhance physiological tolerance, and improve yields (Boutasknit et al. 2021, 2024; Dukare and Paul 2021; Godoy et al. 2021; Ikan et al. 2023; Ouhaddou et al. 2023; Soussani et al. 2023; Abdel Latef et al.

2026; Anli et al. 2026; Boutasknit et al. 2026; Janah et al. 2026a, b). Among these, phytochemicals, including both primary and secondary metabolites—are gaining prominence for their roles in alleviating environmental stress (Arbona et al. 2013; Godoy et al. 2021; Yaqoob et al. 2022; Bano et al. 2023). They contribute to key processes such as osmotic adjustment, reactive oxygen species (ROS) neutralization, signaling cascades, and the maintenance of cellular integrity (Arbona et al. 2013; Jogawat et al. 2021; Yaqoob et al. 2022; Ikram et al. 2025).

Polyphenolic compounds are particularly significant in plant adaptive responses (Gharibi et al. 2019). Frequently regarded as biochemical markers of stress, these metabolites influence multiple physiological and biochemical processes, thereby enhancing resilience under adverse conditions (Pratyusha 2022). Among them, ferulic acid (FA)—chemically defined as [E]-3-[4-hydroxy-3-methoxy-phenyl] prop-2-enoic acid or 4-hydroxy-3-methoxycinnamic acid—has drawn considerable attention for its potent antioxidant activity and protective capacity against free radicals generated during stress (a)biotic episodes. FA has been reported to enhance tolerance by promoting vegetative growth, stimulating photosynthesis, increasing chlorophyll content, and activating antioxidant enzymes (Peng et al. 2021). Additionally, recent studies indicate that FA modulates the expression of stress-responsive genes, participates in ROS-dependent signaling pathways, and contributes to long-term metabolic adjustments (Chi et al. 2013; Peng et al. 2021; Shu et al. 2022).

Beyond its role as a signaling and antioxidant molecule, FA enhances the rigidity and resilience of plant cell walls through cross-linking of polysaccharides and lignin, thereby strengthening mechanical barriers against pathogens and environmental stressors (Khosravi et al. 2015). These multifunctional roles highlight FA as a promising candidate for sustainable crop management strategies aimed at resilience, productivity, and reduced dependence on chemical inputs (Yildiztugay et al. 2019; Chen et al. 2020; Shu et al. 2022, 2023).

Keyword co-occurrence clustering analyses from 2024 to 2025 (Fig. 11.1) further illustrate emerging research directions on FA and plant stress responses. Three major clusters were identified: Cluster #0 (red) focuses on FA, gallic acid, and flavonoids, emphasizing their antioxidant functions and association with plant extracts; Cluster #1 (green) highlights polyphenols, chemistry, and superoxide dismutase (SOD), reflecting interest in oxidative stress regulation; and Cluster #2 (pink) groups ascorbic acid and cell protection, underscoring research on defense mechanisms. The central positioning of FA across these clusters demonstrates its pivotal role in linking phytochemical metabolism to stress mitigation.

In this context, the study of FA has gained strategic relevance in plant science research, not only because of its multifunctional properties, but also due to its potential applications in sustainable agriculture. FA plays a key role in orchestrating plant responses to adverse environmental conditions by simultaneously modulating physiological, biochemical, and molecular mechanisms (Fig. 11.2) (Mughal et al. 2024; Saini et al. 2024). At the physiological level, FA contributes to improving water-use efficiency, preserving photosynthetic activity, and reinforcing the structural integrity of plant tissues (Peng et al. 2021). Biochemically, its powerful

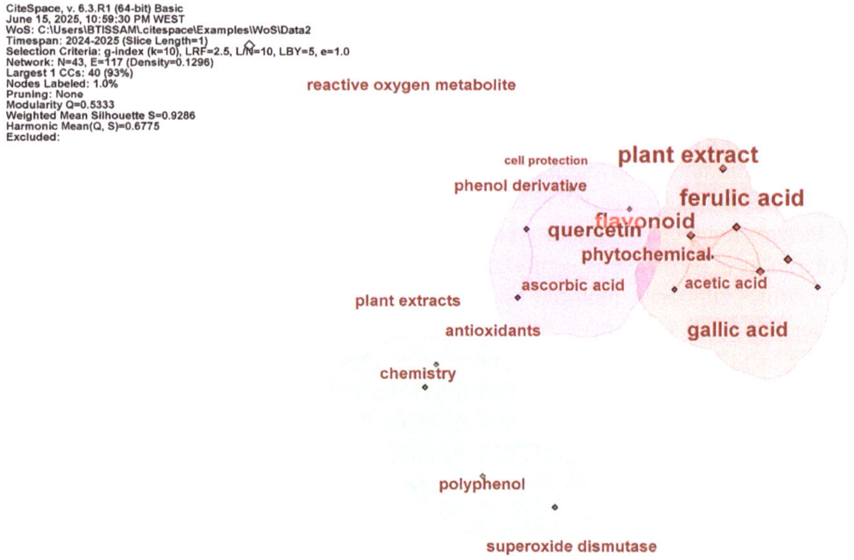

Fig. 11.1 Co-cited references of keyword co-occurrence clustering analysis in studies of ferulic acid and plant stress (2024–2025). The analysis highlights three primary clusters: Cluster #0 (red), centered on ferulic acid, gallic acid, and flavonoids, emphasizes their antioxidant functions and association with plant extracts; Cluster #1 (green) focuses on polyphenols, chemistry, and superoxide dismutase (SOD), underscoring research on oxidative stress regulation by natural compounds; Cluster #2 (pink) links ascorbic acid and cell protection, pointing to emerging work on defense mechanisms. The central positioning of ferulic acid across clusters indicates its pivotal role as a bridge between phytochemicals and stress mitigation

antioxidant capacity enables it to neutralize excessive ROS and to enhance the activity of defense-related enzymes such as SOD, catalase (CAT), and peroxidases (POD) (Zhang et al. 2022). On the molecular scale, FA can induce the expression of stress-responsive genes and regulate signaling pathways associated with hormone responses and redox homeostasis (Salami et al. 2024).

Beyond its biological effects, FA is also at the heart of interdisciplinary research efforts, linking plant physiology, biochemistry, and molecular biology (Fig. 11.2). The presence of FA in lignocellulosic biomass and its potential valorization in the development of agricultural inputs or biosourced protective agents make it a promising biotechnological tool for crop management (Sarangi et al. 2023). In addition, the integration of FA into environmentally friendly stimulant formulations could reduce dependence on synthetic agrochemicals, in line with global objectives for climate-resilient, low-input agriculture. Thus, a better understanding of the mode of action of FA represents a key step in the design of innovative strategies to improve plant resilience, productivity, and sustainability in the face of escalating environmental pressures.

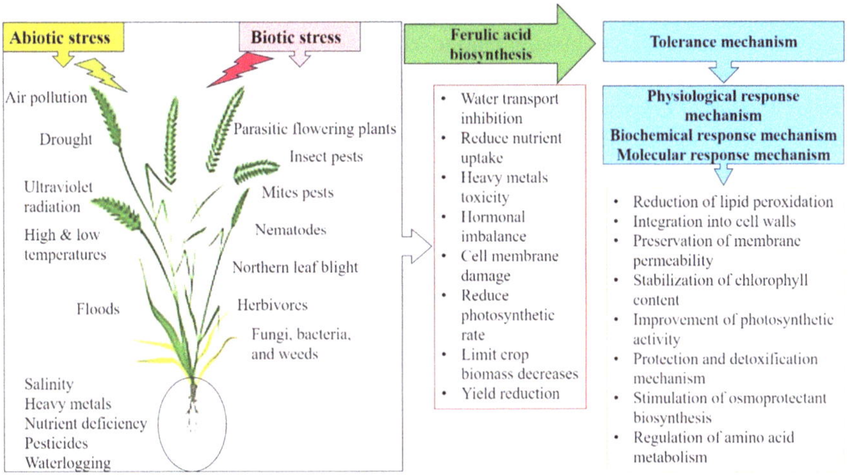

Fig. 11.2 Integrated roles of ferulic acid in plant–environment interactions. Ferulic acid (FA) biosynthesis is activated in response to diverse abiotic stressors (e.g., drought, salinity, nutrient deficiencies, heavy metals, ultraviolet radiation, extreme temperatures, flooding, and air pollution) and biotic stressors (e.g., parasitic plants, insect pests, nematodes, fungi, bacteria, and weeds). FA contributes to stress tolerance through coordinated physiological, biochemical, and molecular mechanisms. These include improved water relations, stabilization of chlorophyll and photosynthesis, reduced lipid peroxidation and membrane damage, enhanced activity of antioxidant and detoxification enzymes, regulation of amino acid metabolism, and modulation of stress-responsive gene expression. Overall, FA functions as a multifunctional metabolic and defense hub, reinforcing plant resilience under adverse environmental conditions

11.2 Ferulic Acid Biosynthesis and Metabolism

11.2.1 The Phenylpropanoid Pathway: From Phenylalanine to Ferulic Acid

The phenylpropanoid pathway, a branch of the shikimate pathway, is the central route for the biosynthesis of ferulic acid (FA) and other phenolic compounds (dos Santos et al. 2008). Phenolics are derived primarily from two biosynthetic sources: the shikimate pathway, which supplies approximately 60% of the carbon, and the malonate pathway, which contributes about 40% (Vanholme et al. 2019). The phenylpropanoid pathway is initiated by the deamination of the aromatic amino acids phenylalanine or tyrosine, catalyzed by phenylalanine ammonia-lyase (PAL) or tyrosine ammonia-lyase (TAL), respectively. This reaction produces cinnamic acid (CA) and subsequently p-coumaric acid.

Cinnamic acid, the first committed intermediate of the pathway, undergoes hydroxylation to form p-coumaric acid, which is then further transformed via oxidation to caffeic acid and O-methylation by S-adenosylmethionine (SAM)-dependent O-methyltransferases, ultimately yielding FA. Chemically, FA

([E]-3-[4-hydroxy-3-methoxy-phenyl]prop-2-enoic acid) is a hydroxycinnamic acid derivative characterized by a phenolic core, a para-substituted unsaturated side chain, and an ortho-methoxy group. This configuration permits cis–trans isomerization and confers FA with versatile biological functions, particularly in plant defense and cell wall reinforcement (Alifah et al. 2024).

Key enzymes of the phenylpropanoid pathway, including PAL, cinnamate 4-hydroxylase (C4H), and 4-coumarate-CoA ligase (4CL), are ubiquitous across land plants and direct metabolic intermediates toward the biosynthesis of structurally and functionally important molecules such as lignins, flavonoids, and UV-protective pigments. In parallel, tyrosine aminotransferase (TAT) contributes to the production of p-hydroxyphenylpyruvate (pHPP), a critical precursor for tocopherol and plastoquinone biosynthesis (Xu et al. 2018). Interestingly, enzymatic processes analogous to those in the phenylpropanoid pathway, such as the NAD(P)H-dependent reduction of hydroxypyruvate to glycerate during photoacclimation, have also been implicated in tocopherol and plastoquinone synthesis, underscoring the metabolic integration of these pathways.

11.2.2 FA Conjugation, Transport, and Integration into Lignin

Lignin is a principal structural component of secondary cell walls in higher terrestrial plants, conferring mechanical strength and rigidity, facilitating efficient water transport through vascular tissues, and serving as a physical barrier against pathogens. Its biosynthesis is driven by monolignols, the primary building blocks of lignin, with FA, a phenolic derivative of the phenylpropanoid pathway, playing a pivotal role. Beyond contributing to monolignol biosynthesis, FA functions as a cross-linking agent, bridging lignin polymers and polysaccharides to reinforce the structural integrity of the cell wall.

The pathway begins with the deamination of phenylalanine into cinnamic acid via phenylalanine ammonia-lyase (PAL), followed by hydroxylation and methylation steps that yield caffeic acid and, subsequently, FA through the activity of caffeate O-methyltransferase (COMT). FA is then activated into feruloyl-CoA by 4-coumarate: CoA ligase (4CL), a key intermediate not only in lignification but also in other secondary metabolic branches (Vanholme et al. 2019). In grasses, FA is ester-linked to arabinoxylans in secondary cell walls, forming covalent bridges between polysaccharides and lignin polymers that strengthen the three-dimensional architecture of the wall (Ralph et al. 2004; de Buanafina 2009). The enzyme feruloyl monolignol transferase (FMT) catalyzes the acylation of monolignols, such as coniferyl alcohol, with FA to generate feruloylated monolignols, which are subsequently integrated into lignin via radical polymerization (Liu et al. 2011).

These monolignols are synthesized in the cytoplasm and transported to the cell wall through mechanisms that likely include both passive diffusion and active transport mediated by ATP-binding cassette (ABC) transporters or specialized membrane proteins, particularly in the case of conjugated forms (Liu et al. 2011; Sun et al. 2024). Microscopy has also revealed vesicle-mediated trafficking of these monomers, suggesting that vesicular transport contributes to their delivery in certain plant species (Ragauskas et al. 2014). Once localized to the apoplast, polymerization proceeds via oxidative enzymes such as peroxidases and laccases, which generate phenoxyl radicals that facilitate the incorporation of FA and its esters into the lignin network. Unlike p-coumaric acids, which often remain uncoupled, ferulates actively form cross-links, introducing branching points that enhance interconnectivity within lignified matrices (Quideau and Ralph 1997; De Oliveira et al. 2015).

The integration of FA into lignin imparts multiple functional attributes, including increased rigidity, resistance to enzymatic hydrolysis, and reduced biomass digestibility. These characteristics, while advantageous for plant defense and structural resilience, present challenges for industrial applications such as biofuel production and paper pulping, where ferulate cross-links hinder depolymerization (Cesarino et al. 2012; Scheller and Ulvskov 2010). Genetic studies have further demonstrated that downregulation of COMT or FMT reduces ferulate incorporation into lignin, resulting in cell walls with enhanced degradability—an outcome of significant interest for biomass utilization (Smith et al. 2022; Chandrakanth et al. 2023; Gao et al. 2024). Moreover, emerging evidence suggests that phenolic compounds such as ferulates may integrate into lignin post-mortem, contributing to atypical lignin architectures in some plant systems (de Buanafina and Morris 2022).

11.2.3 Modulation Under Stress Conditions

As sessile organisms, plants are continuously exposed to a wide range of environmental constraints throughout their life cycle, which can severely impact growth, productivity, and survival (Khan et al. 2024). These stresses, whether abiotic or biotic, initiate complex metabolic responses designed to restore cellular homeostasis and preserve physiological integrity (Khan et al. 2024). Within this adaptive network, secondary metabolites, and particularly phenolic compounds, emerge as central regulators of stress resilience (Ahlawat et al. 2024). Among these, FA stands out due to its multifunctionality. As a structural component of cell walls, FA contributes to lignification and cross-linking of polysaccharides, enhancing wall rigidity and creating physical barriers that restrict pathogen ingress (Fig. 11.3). In parallel, its phenolic structure confers high antioxidant potential, enabling FA to neutralize ROS generated in excess under stress. This antioxidant activity is complemented by the upregulation of both expression and activity of key enzymatic antioxidants, including superoxide dismutase (SOD), catalase (CAT), and ascorbate peroxidase (APX), thereby strengthening cellular defenses against oxidative stress (Khan et al. 2024).

The role of FA in salinity tolerance has been supported by multiple experimental studies. Linić et al. (2021) demonstrated that foliar application of FA to salt-stressed

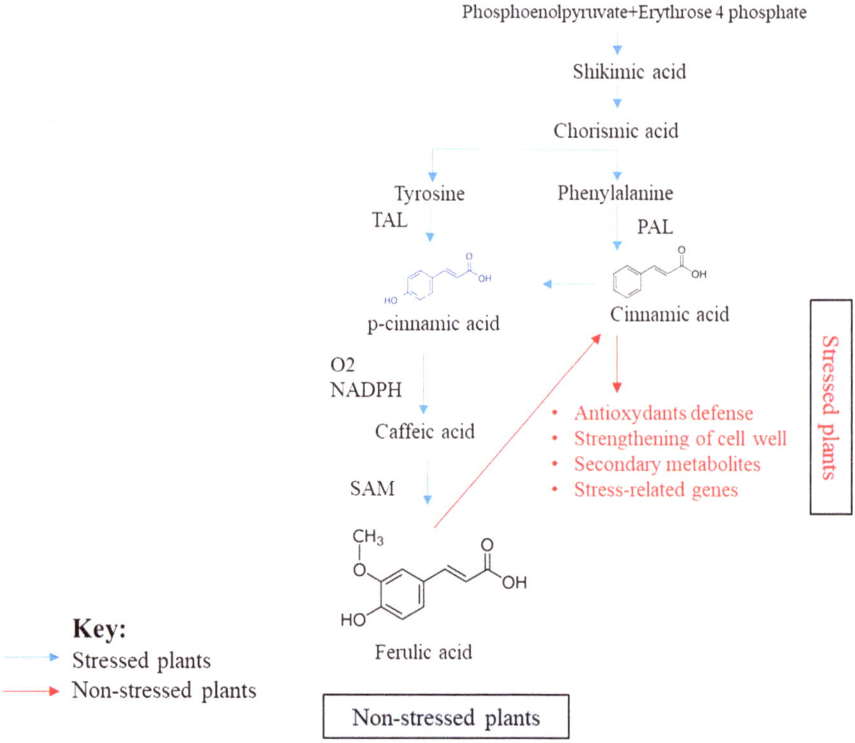

Fig. 11.3 Biosynthetic and metabolic dynamics of ferulic acid in stressed versus non-stressed plants. Schematic representation of ferulic acid (FA) biosynthesis and its differential allocation under normal and stress conditions. In non-stressed plants, FA is primarily directed toward basal metabolism and structural integration. Under stress conditions, FA biosynthesis and accumulation are upregulated, contributing to multiple protective roles, including antioxidant defense, reinforcement of the cell wall through cross-linking with polysaccharides and lignin, modulation of secondary metabolism, and regulation of stress-responsive gene networks. These combined actions underscore FA's dual function as both a metabolic intermediate and a signaling molecule central to plant stress adaptation

plants significantly improved photosynthetic parameters, particularly performance indices such as PI_ABS, φP0, and φE0, while simultaneously elevating endogenous levels of proline and salicylic acid (SA). FA application also enhanced the accumulation of phenolic metabolites, including sinapic acid, quercetin, and kaempferol, suggesting a coordinated activation of the secondary phenolic metabolism. These changes collectively mitigated the detrimental effects of salinity stress, positioning FA as a potent antioxidant and adaptive metabolite, with greater efficacy than SA (Linić et al. 2021).

FA also plays a critical role in plant responses to water deficit. Under drought conditions, enhanced accumulation of both free and wall-bound FA has been reported in leaf tissues, particularly during the flowering stage (Hura et al. 2007a, b). This accumulation appears to serve as an adaptive mechanism aimed at

Table 11.1 Chemical properties, plant sources, and metabolic roles of ferulic acid (FA)

Aspect	Description	References
Chemical properties	*Name:* 4-Hydroxy-3-methoxycinnamic acid; *Formula:* $C_{10}H_{10}O_4$ *Molar mass:* 194.18 g/mol; *Solubility:* low in water, good in ethylene glycol; *Functional groups*: -OH, $-OCH_3$, -COOH	
Plant sources	*Cereals* (e.g., wheat bran, rice, corn); *Fruits* (apple, orange, berries) *Vegetables* (beetroot, spinach); *Medicinal plants* (*Angelica sinensis*) *Coffee*	Boz (2015), Abadias et al. (2024), Shu et al. (2021), Lu et al. (2005), Kumar and Pruthi (2014)
Metabolic roles	*Cell wall integrity* (Cross-links polysaccharides and lignin, providing rigidity and defense); *Antioxidant defense* (neutralizes reactive oxygen species, ROS); *Precursor for lignin biosynthesis* (participates in monolignol pathways for lignin polymerization); *Allelopathy and defense* (functions as a chemical signal or inhibitor, suppressing neighboring plant growth or pathogen development)	Mnich et al. (2020), Zduńska et al. (2018), De Oliveira et al. (2015), Khattab et al. (2023)

limiting oxidative damage and enhancing photoprotection (Table 11.1). Meyer et al. (2003) observed that drought stress led to increased blue fluorescence emissions in foliar tissues, reflecting the presence of phenolic acids, including FA, and highlighting their potential involvement in non-photochemical dissipation of excess excitation energy. A comparative study of two winter triticale genotypes, Lamberto and Ticino, further underscored FA's role in drought adaptation (Hura et al. 2009). The more tolerant Lamberto genotype exhibited a pronounced accumulation of wall-bound FA, which correlated with better preservation of photosynthetic performance (measured via chlorophyll fluorescence) and more rapid post-stress recovery. These findings suggest that FA functions both as a spectral filter reducing the intensity of light reaching chloroplasts, and as a regulator of leaf development, restricting cell expansion and thereby reducing transpirational water loss (Hura et al. 2009).

11.3 Ferulic Acid in Plant Abiotic Stress Tolerance

Ferulic acid plays a pivotal role in plant adaptation to abiotic stresses, primarily due to its potent antioxidant capacity (Fig. 11.4) (Zduńska et al. 2018; Han et al. 2023). This capacity is conferred by its distinctive chemical structure, consisting of an aromatic ring substituted with a hydroxyl group and an unsaturated side chain, which together confer high efficiency as a free radical scavenger (Sebastian et al. 2009; Chen et al. 2020). Under abiotic and biotic stresses, plants accumulate excessive

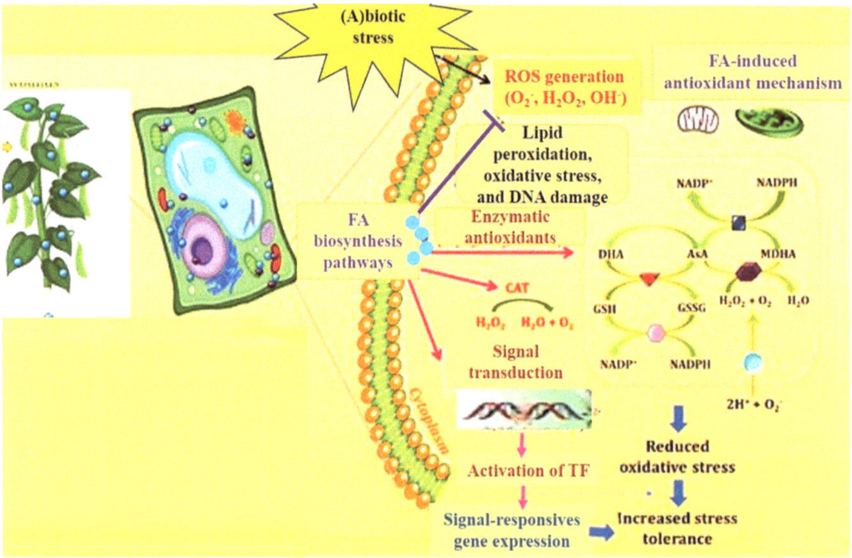

Fig. 11.4 Mechanistic model of ferulic acid (FA) function under abiotic stress. Abiotic stress triggers excessive reactive oxygen species (ROS; O_2^-, H_2O_2, OH·), leading to lipid peroxidation, oxidative damage, and genetic instability. FA biosynthesis pathways are activated in response, with FA acting both directly as an antioxidant and indirectly through modulation of enzymatic defenses. By stimulating antioxidant enzymes such as catalase (CAT), FA regulates signal transduction cascades and promotes the activation of stress-responsive transcription factors (TFs). This cascade enhances the expression of defense-related genes, reduces oxidative stress, and improves cellular homeostasis. The synergistic action of enzymatic antioxidants (e.g., APX, SOD, POD, CAT, GSH-related enzymes) and FA's radical-scavenging capacity collectively contributes to enhanced stress tolerance and improved physiological resilience in plants

ROS, which damage membrane lipids, proteins, and DNA, thereby compromising cellular integrity and metabolic function (Fig. 11.4) (Das and Roychoudhury 2014; Liu et al. 2015b; Sachdev et al. 2021). FA mitigates these effects by neutralizing ROS, preserving membrane stability, enzyme activity, and genomic stability (Zduńska et al. 2018; Yildiztugay et al. 2019; Shu et al. 2023).

Exogenous FA application has been widely reported to enhance tolerance to drought, salinity, heat, and pathogen-associated stresses (Li et al. 2013; Shu et al. 2023). Beyond direct ROS scavenging, FA also modulates key signal transduction pathways, particularly mitogen-activated protein kinase (MAPK) cascades (Shu et al. 2021, 2022). These cascades sequentially phosphorylate downstream kinases, ultimately activating transcription factors (TFs) that regulate stress-responsive gene expression (Salami et al. 2023; Khan et al. 2024). Through this mechanism, FA enhances the expression of antioxidant enzymes (SOD, CAT, POD), osmolyte biosynthesis genes, and stress-protective proteins (Li et al. 2013; Zduńska et al. 2018; Yildiztugay et al. 2019; Shu et al. 2023).

FA has also been shown to modulate transcriptional networks, upregulating genes encoding heat shock proteins (HSPs), pathogenesis-related proteins (PRs), and stress-associated TFs such as WRKY, MYB, and AP2/ERF (Li et al. 2021; Shi et al. 2022; Shu et al. 2023; Zhang et al. 2023a). These genes collectively enhance stress tolerance and facilitate recovery following stress exposure (Xiao et al. 2021; Song et al. 2022). Notably, recent studies have reported that FA strongly induces the expression of stress-inducible ZIM genes, particularly ZIM-3, which confers broad tolerance to both abiotic and biotic challenges (Peng et al. 2021; Shi et al. 2022; Khan et al. 2024; Rabeh et al. 2025).

Taken together, these findings demonstrate that moderate concentrations of FA not only directly counteract oxidative stress but also orchestrate beneficial transcriptional and signaling responses, thereby enhancing plant tolerance to diverse environmental challenges (Li et al. 2021; Shu et al. 2021, 2022, 2023).

FA possesses pronounced metal-chelating properties, which further reinforce its efficacy as a protective agent against both abiotic and biotic stresses (Amić et al. 2021; AAlikhani et al. 2022; Kumar et al. 2023). This property is particularly critical under conditions of oxidative stress, where transition metals such as copper (Cu) and iron (Fe) catalyze ROS formation through Fenton- and Haber–Weiss-type reactions (Sharma et al. 2023). By binding to these redox-active metals, FA effectively prevents their participation in such deleterious reactions, thereby limiting excessive ROS generation and mitigating oxidative injury (Chen et al. 2020; Li et al. 2021, 2025). This dual role—as both a direct ROS scavenger and a chelator of pro-oxidant metals—positions FA as an especially powerful molecule in enhancing plant stress tolerance (Guneidy et al. 2022).

Experimental studies confirm that exogenous FA application significantly improves plant tolerance to abiotic stress by reducing osmotic stress, ion toxicity, and oxidative damage (Table 11.2) (Das et al. 2023). Hasanuzzaman et al. (2020a) demonstrated that foliar FA application enhances enzymatic antioxidant activity, lowers ROS levels, and simultaneously sequesters free metals in stressed plants, thereby safeguarding membranes from oxidative degradation (Table 11.4). Similarly, under drought conditions—where ROS accumulation disrupts cellular structures (Lone et al. 2023)—FA contributes to the stabilization of membranes, protection of proteins and nucleic acids, and preservation of essential metabolic functions (Shu et al. 2023; Khan et al. 2024). In agreement, Quan and Xuan (2018) reported that FA application under drought stress improved photosynthetic efficiency, water status, and redox regulation, collectively enhancing adaptive capacity (Table 11.2). Interestingly, moderate oxidative responses elicited by FA, when tightly regulated, appear to facilitate pre-acclimation to stress by priming and strengthening endogenous defense systems (Saini et al. 2024).

At the molecular level, FA has been shown to induce the expression of various membrane transporter genes, which play crucial roles in maintaining metabolite, ion, and signal fluxes under stress conditions (Sharma et al. 2019; Hussain and Reigosa 2021). Among these, transporters of the AAAP (Amino Acid/Auxin Permease) family have attracted particular attention for their dual involvement in amino acid and auxin transport (Chi et al. 2013; Du et al. 2023; Khan et al. 2024). Upregulation

Table 11.2 Plant responses to exogenous ferulic acid (FA) under major abiotic stresses

Stress	Plant responses	Physiological and molecular responses	Plant species	References
Drought	Improve growth and yield Higher relative water content Reduced wilting, oxidative damage Protection of cell membranes Improved photosynthesis Increased endogenous hormones	Enhanced plant nutritional and water status; maintenance of water absorption; accumulation of osmotic-adjusting compounds; stimulation of antioxidant enzyme synthesis; stabilization of membranes; preservation of mesophyll and stomatal structure	*Triticum aestivum, Zea mays, Cicer arietinum, Hordeum vulgare, Thymus vulgaris, Triticosecale wittmack*	Hura et al. (2009), Hou et al. (2018), Alam (2019), Ma et al. (2022a), Kamyab et al. (2024), Mohagheghian et al. (2025)
Salinity	Improved seedling growth Maintained chlorophyll content Higher K^+/Na^+ ratio Reduced necrosis	Reduced ion toxicity Regulation of SOS1/NHX1 genes Glutathione metabolism Improve antioxidant enzyme expression	*Z. mays, Oryza sativa, Arabidopsis, Capsicum annuum, Fagopyrum tataricum, Hordeum vulgare*	Minh et al. (2016), Wu et al. (2019), Kim et al. (2019), Oliveira et al. (2020), Zamljen et al. (2022), Quoc et al. (2024), Afzal et al. (2025)
Heavy metals	Reduced root growth inhibition; increased biomass; alleviated Cd and Pb toxicity and oxidative damage	Chelation via phenolic groups; stimulation of phytochelatin and metallothionein synthesis; activation of detoxifying enzymes	*Zea mays, Kandelia obovate, Withania somnifera, S. lycopersicum, Vigna radiata*	Kısa et al. (2016), Krishnappa et al. (2017), Hasanuzzaman et al. (2020a, b), Chen et al. (2020), Narnoliya et al. (2021), Vega et al. (2022), Ma et al. (2022b)
Heat	Preserved chlorophyll; reduced lipid peroxidation; enhanced tolerance and growth	Upregulation of HSPs (HSP70, HSP90); increased activity of APX, GR, and CAT; protection of PSII and PSI centers; stabilization of membrane fluidity	*O. niloticus, V. corymbosum S. lycopersicum, T. aestivum*	Hussain et al. (2017), Cheng et al. (2018), Yildiztugay et al. (2019), Dawood et al. (2020), Hussain et al. (2017)

of AAAP transporters in response to FA may facilitate the sequestration or efflux of toxic compounds, including autotoxins (Chi et al. 2013; Wang et al. 2022). Such autotoxins, often derived from the degradation of secondary metabolites such as FA itself, can accumulate in soils and inhibit the growth of conspecific or neighboring plants, a hallmark of allelopathy (Li et al. 2010; Hussain and Reigosa 2021).

Furthermore, FA interacts with phytohormone signaling networks, modulating the balance between growth-promoting hormones and stress-related phytohormones (Gao et al. 2023; Khatun et al. 2024). This hormonal regulation optimizes physiological responses, including root architecture, stomatal closure, growth modulation, and the synthesis of defense compounds (Tripathi et al. 2022; Wahab et al. 2022; Singh et al. 2024). Importantly, FA also acts as an indirect transcriptional regulator, influencing the expression of genes involved in antioxidant metabolism and metal-stress responses, thereby extending its impact beyond direct biochemical functions.

Together, this synergy of antioxidant activity, chelating capacity, hormonal modulation, and transcriptional regulation establishes FA as a highly versatile and effective agent for enhancing plant resilience against both abiotic and biotic stresses.

11.4 Ferulic Acid in Plant Biotic Stress Resistance

11.4.1 Antifungal and Antibacterial Effects

Ferulic acid is a natural metabolite widely distributed in plant cell walls and abundantly extracted from medicinal plants such as Ferula, Asafoetida, and Angelica (Yan et al. 2023). It possesses broad-spectrum bioactivities, including antibacterial, antifungal, and anti-inflammatory properties, making it valuable not only in food preservation but also in crop protection and sustainable farming practices (Hernández-Jaime et al. 2025).

In terms of antibacterial action, FA is notably effective against human and plant pathogens such as *Escherichia coli*, *Staphylococcus aureus*, *Pseudomonas aeruginosa* (Borges et al. 2013), *Candida albicans*, *Candida tropicalis* (de Morais et al. 2021), and *Ralstonia solanacearum* (Tu et al. 2022). The latter is a devastating soilborne bacterium that infects more than 450 plant species across 54 botanical families, posing a major threat to global agriculture. A major limitation of FA is its high polarity and short alkyl chain, which reduce its penetration into bacterial biofilms. To overcome this, researchers have synthesized alkylated ferulic acid esters (FCs) using fatty alcohols, significantly improving antibacterial potency. Among these derivatives, hexyl ferulate (FC6) displayed remarkable activity against *Pseudomonas aeruginosa*, with a minimum inhibitory concentration (MIC) of just 0.4 mg/ml (Song et al. 2023). Mechanistic studies revealed that FCs disrupt bacterial membranes, damage cell walls, inhibit biofilm formation, induce membrane hyperpolarization, and cause leakage of intracellular contents, ultimately leading to cell death (Borges et al. 2013; Song et al. 2023).

FA also exhibits strong antifungal activity. For instance, it is effective against *Fusarium graminearum*, a major pathogen of cereals. Yan et al. (2023) reported EC50 and EC90 values of 20 μg/mL and 100 μg/mL, respectively, for FA against *F. graminearum*. Microscopic observations revealed severe structural damage to fungal hyphae, including roughened surfaces, blurred edges, and cytoplasmic leakage after FA treatment (Yan et al. 2023). FA also inhibits ergosterol biosynthesis—an essential component of fungal cell membranes—thereby functioning as a natural biofungicide. In addition, FA-rich extracts, such as those from maize seed pericarp, effectively suppress *Fusarium verticillioides* growth and significantly reduce fumonisin B1 (FB1) accumulation, providing both food safety benefits and crop protection (Martínez-Fraca et al. 2022).

Beyond its direct antimicrobial effects, FA interferes with microbial physiology in multiple ways: disrupting membranes, altering hydrophobicity and surface charge, impairing enzymatic systems critical for survival, and inducing localized ruptures that cause leakage of essential intracellular metabolites (Borges et al. 2013; Yan et al. 2023). It also impairs bacterial motility, thereby reducing colonization and infection potential. Recent work by Shirai et al. (2022) showed that combining FA with UV-A exposure drastically enhances its antimicrobial effect by depleting intracellular ATP and altering cytosolic proteins, while notably avoiding genotoxic effects.

In agricultural contexts, FA and its derivatives are increasingly recognized as eco-friendly alternatives to synthetic pesticides. Their dual antibacterial and antifungal efficacy particularly against destructive pathogens like *Ralstonia solanacearum* and *Fusarium verticillioides*, offers an effective means of safeguarding crops while minimizing chemical inputs. Modified derivatives, such as FC6, show particular promise for controlling biofilm-forming bacteria and enhancing postharvest storage quality.

However, despite these advantages, FA is not universally beneficial. Excessive concentrations can have phytotoxic effects. For example, FA has been reported to exacerbate Fusarium wilt in faba bean by inducing structural abnormalities in root tissues (Ting-ting et al. 2020; Yang et al. 2023; Z. Zhang et al. 2023b). Similarly, high doses of FA aggravated *Fusarium oxysporum* infections in strawberry seedlings and caused root tissue anomalies in other plant species.

11.4.2 Systemic Resistance

Ferulic acid has emerged as a pivotal metabolite in enhancing systemic acquired resistance (SAR), thereby strengthening plant defense systems against a broad spectrum of pathogens through multifaceted biochemical and molecular mechanisms. Foliar application of FA in combination with L-phenylalanine on pea plants significantly inhibited *Erysiphe pisi* conidial germination, while simultaneously stimulating phenylalanine ammonia-lyase (PAL) activity. This dual effect reinforced the phenylpropanoid pathway and highlighted the role of FA as a potential elicitor in pathogen resistance strategies (Bahadur et al. 2012). Similarly, Shu et al. (2021) demonstrated the postharvest efficacy of FA, showing that application of $0.80\ g \cdot L^{-1}$ FA to 'Pink Lady' apple fruits markedly reduced lesion development caused by *Botrytis cinerea*. This reduction correlated with enhanced activity of key phenylpropanoid enzymes, including PAL and polyphenol oxidase (PPO). In cucumber roots, FA further improved resistance to *B. cinerea* by enhancing PAL activity and boosting endogenous FA accumulation, underscoring its role in fortifying host defense metabolism (Shu et al. 2021).

At the signaling level, FA induces programmed cell death in the maize pathogen *Cochliobolus heterostrophus* via the activation of two mitogen-activated protein kinase (MAPK) cascades: Hog1, associated with stress signaling, and Chk1, linked to pathogenicity regulation. This dual activation elicits a coordinated stress response while simultaneously promoting the degradation of pathogenic structures (Shalaby et al. 2016). In addition, FA mitigates oxidative stress by modulating ROS dynamics, thereby maintaining cellular redox homeostasis and reinforcing defense signaling cascades (Yildiztugay et al. 2019).

FA-induced SAR is mediated through the modulation of hormone-dependent signaling pathways, particularly those governed by SA and jasmonic acid (JA) (Durrant and Dong 2004). These pathways converge to activate pathogenesis-related (PR) genes and stimulate the accumulation of secondary defense metabolites. Specifically, FA has been shown to induce PR-1, PR-2 (β-1,3-glucanase), and PR-3 (chitinase), enzymes that hydrolyze fungal cell wall components and restrict pathogen proliferation (Durrant and Dong 2004). Furthermore, FA interacts with JA-dependent signaling, which is particularly relevant for resistance against necrotrophic pathogens such as B. cinerea. For example, in tomato fruits, FA application induced the activation of chitinases and β-1,3-glucanases, demonstrating its ability to integrate into both SA- and JA-mediated defense networks (Shu et al. 2021).

Beyond its regulatory role in SAR, FA contributes directly to structural and biochemical defense mechanisms. By promoting lignification, FA reinforces cell wall rigidity, creating physical barriers to pathogen ingress. Simultaneously, its antioxidant properties alleviate oxidative stress during infection, while its capacity to induce pathogen cell death further suppresses pathogen spread. Collectively, these findings highlight FA as a promising natural elicitor capable of activating systemic defense networks, modulating hormonal signaling, and providing durable protection against diverse plant pathogens—making it a valuable tool for sustainable disease management in agriculture.

11.4.3 Synergy with Hormonal Signaling

Plant hormones are central regulators of adaptive responses to environmental stresses, coordinating physiological, biochemical, and molecular mechanisms that enable plants to survive and grow under adverse conditions. Within this context, FA plays a pivotal role in modulating direct plant defenses while also integrating into broader hormonal signaling cascades. Under water stress, activation of metabolic pathways such as the phenylpropanoid pathway strengthens cell walls through enhanced lignification, thereby reinforcing mechanical barriers against stress (Santos

Table 11.3 Effects of ferulic acid (FA) on biotic stress mitigation across different crops

Crops	Biotic stress	Effect of ferulic acid	References
Wheat (*Triticum aestivum*)	*Fusarium graminearum*	Disrupts fungal cell surface morphology and internal structure	Yan et al. (2023)
Tomato (*Solanum lycopersicum*)	*Botrytis cinerea*	Attenuates postharvest disease caused by *B. cinerea*	Shu et al. (2021)
Maize (*Zea mays*)	*Fusarium graminearum*	Inhibits fungal growth and enhances resistance to *F. graminearum*	Assabgui et al. (1993)
Apple (*Malus domestica*)	*Botrytis cinerea*	Enhances phenylpropanoid pathway enzyme activity, increasing phenol and flavonoid synthesis to strengthen defenses against postharvest infections	He et al. (2019)
Cucumber (*Cucumis sativus*)	Various fungal pathogens	Reduces leaf growth, dry weight, and water use, indicating potential phytotoxic effects at certain concentrations	Blum et al. (1985)
Bean (*Vicia faba*)	*F. oxysporum* f. sp. fabae	Inhibits mycelial growth but stimulates mycotoxin and pathogenic enzyme production, aggravating Fusarium wilt symptoms	Ting-ting et al. (2020)
Grape (*Vitis vinifera*)	*Botrytis cinerea* *Aspergillus niger* *Penicillium expansum* *Fusarium culmorum*	Inhibits the growth of multiple phytopathogenic fungi, including fungicide-resistant strains, highlighting potential as a natural pesticide	Patzke and Schieber (2018)

et al. 2015; Yadav et al. 2020; Kumar et al. 2020). While this defense mechanism generally improves resilience, it can also be exploited by pathogens (Table 11.3). For example, Khattab et al. (2023) demonstrated that the accumulation of FA in grapevines under drought stress conditions enhanced phytotoxin release from the pathogenic fungus *Neofusicoccum parvum* (Bt-67), thereby increasing virulence. Interestingly, whereas other phenolic acids, such as cinnamic and coumaric acids, alleviated phytotoxicity, FA exhibited a paradoxical effect by generating distress signals that appeared to favor pathogen aggressiveness.

This duality underscores FA's complex role in secondary metabolism and hormonal crosstalk. FA acts synergistically with SA, JA, and ET, three hormones that are critical for orchestrating plant defense responses (Verma et al. 2016). For example, FA potentiates SA-induced defenses by enhancing the accumulation of pathogenesis-related (PR) proteins during systemic acquired resistance (SAR) (Fig. 11.5). Moreover, FA stimulates SA biosynthesis, in part through the activation of PAL activity, thereby creating a positive feedback loop that further reinforces SA-dependent immunity (Shadle et al. 2003). In addition to strengthening SA-driven responses, FA also modulates oxidative bursts through the production of ROS, acting in coordination with JA to trigger rapid and localized immune responses, particularly critical for combating necrotrophic pathogens (Khattab et al. 2023).

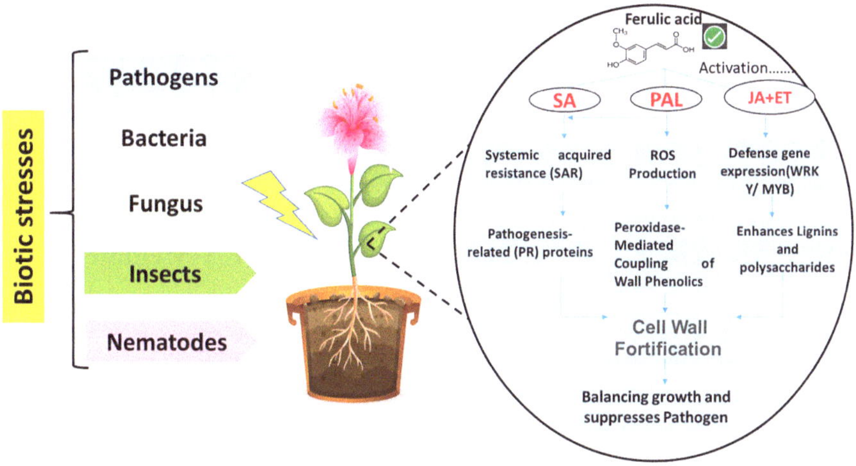

Fig. 11.5 Ferulic acid–mediated regulation of biotic stress signaling and defense activation. Ferulic acid (FA) integrates into multiple defense signaling pathways to enhance resistance against diverse biotic stressors, including pathogens, fungi, bacteria, insects, and nematodes. FA stimulates salicylic acid (SA)-dependent systemic acquired resistance (SAR) through the activation of phenylalanine ammonia-lyase (PAL) and the accumulation of pathogenesis-related (PR) proteins, while also promoting reactive oxygen species (ROS) production and peroxidase-mediated cross-linking of wall phenolics. Concurrently, FA modulates jasmonic acid (JA) and ethylene (ET) pathways, leading to the upregulation of defense-associated transcription factors (WRKY, MYB), lignin biosynthesis, and polysaccharide cross-linking. These coordinated mechanisms reinforce cell wall fortification, balance growth with defense, and collectively suppress pathogen proliferation

At the transcriptional level, FA regulates JA signaling cascades by promoting the upregulation of JA biosynthesis and downstream defense-related genes, including those encoding APETALA2/ethylene response factor (AP2/ERF), WRKY, and MYB transcription factors. Through this mechanism, FA enhances JA-dependent immunity but also contributes to growth regulation, as observed in *Oryza sativa*, where FA-mediated activation of the JA pathway resulted in root growth inhibition, consistent with JA's known role in growth repression during stress adaptation (Chi et al. 2013). Additionally, FA reinforces the cell wall by cross-linking lignins and polysaccharides, a process that integrates with ET-dependent pathways to further bolster structural defenses (Broekaert et al. 2006; de Buanafina and Morris 2022).

Several experimental studies have confirmed FA's synergistic effects with plant hormones in improving stress tolerance. In wheat seedlings, exogenous FA application enhanced oxidative stress tolerance by modulating JA and ET levels and increasing the activities of key antioxidant enzymes, including glutathione reductase (GR), polyphenol oxidase (PPO), and catalase (CAT) (Sharma et al. 2015). Similarly, Linić et al. (2021) reported that FA synergized with SA to enhance phenolic metabolism in Chinese cabbage under salinity stress, leading to improved antioxidant activity and photosynthetic efficiency. In tomato, Gorni et al. (2021) demonstrated that combined FA and SA applications significantly improved growth performance while elevating endogenous antioxidant compound levels, reinforcing the notion of FA–SA synergism in promoting plant natural defense mechanisms.

Together, these findings highlight FA as a multifunctional metabolite that not only cross-links structural defenses but also fine-tunes the interplay between SA, JA, and ET signaling pathways. Through this hormonal crosstalk, FA orchestrates a broad spectrum of defense and adaptive mechanisms, enhancing resilience to both abiotic and biotic stresses.

11.5 Molecular and Cellular Mechanisms of FA-Mediated Stress Response

11.5.1 Regulation of Antioxidant Enzyme Genes

As a cinnamic acid derivative, FA exerts multifunctional roles in plant defense, particularly under biotic and abiotic stress conditions (Shu et al. 2022). Among its most prominent functions is its strong antioxidant activity, which derives from its capacity to neutralize ROS—including superoxide anion, hydrogen peroxide, and hydroxyl radicals—that are generated both during aerobic metabolism and in response to stress-induced perturbations (Caillet et al. 2006; Wan et al. 2015). FA mitigates oxidative damage by directly scavenging free radicals, chelating redox-active metal ions such as Fe(II) and Cu(II), and enhancing the activity of antioxidant enzymes. Through these mechanisms, FA safeguards lipids, proteins, and nucleic acids from oxidative injury (Truong et al. 2020; Zduńska et al. 2018).

Exogenous application of FA has been shown to enhance plant stress tolerance by modulating both the transcription and enzymatic activity of key antioxidant systems (Cheng et al. 2018). For example, in heat-stressed blueberry seedlings, FA pretreatment upregulated genes encoding iron SOD, CAT, and glutathione peroxidase, while also elevating the accumulation of protective osmolytes such as proline, soluble sugars, ascorbate, and glutathione. These molecular and metabolic adjustments led to reduced ROS accumulation, lower lipid peroxidation, improved osmotic balance, and ultimately, enhanced stress resilience (Cheng et al. 2018). Collectively, these findings highlight FA as a versatile bioactive compound with a robust mechanism of action, integrating antioxidant, metal-chelating, and signaling functions to strengthen plant resilience.

11.5.2 Expression of Cell Wall–Modifying Genes Under Attack

The plant cell wall functions as a dynamic and responsive interface, capable of modifying its physicochemical properties in response to abiotic and biotic stressors. This dual role—as both a structural barrier and a signaling hub—renders it central to plant defense and adaptation (de Buanafina and Morris 2022). Among the cell wall reinforcements induced under stress, ferulic acid plays a particularly prominent role through its incorporation into wall polymers, a process termed feruloylation (de Buanafina and Morris 2022).

In grasses, FA is primarily ester-linked to arabinose residues at the O-5 position of arabinoxylan side chains, while in dicotyledonous plants, it also binds to arabinan and galactan side chains of pectin (de Buanafina and Morris 2022). These wall-bound ferulates are critical in cross-linking cell wall components via oxidative coupling reactions mediated by cell wall peroxidases (CW-PRX) in the presence of H_2O_2 (Wakabayashi et al. 2012; Wallace and Fry 1995). Such reactions lead to the formation of diferulates, which anchor lignin within the polysaccharide matrix, thereby strengthening the integration of lignin into the xylan–cellulose network (Kang et al. 2019; Jacquet et al. 1995). This cross-linking enhances wall rigidity, reduces enzymatic digestibility, and increases resistance to microbial degradation and mechanical stress (de Buanafina and Morris 2022; Iiyama et al. 1990; Sánchez et al. 1996).

The expression and activity of CW-PRX genes are tightly coordinated with ferulate dimerization and vary according to developmental and stress-related cues (Cho and Lim 2018). In rice shoots, for instance, elevated diferulate levels correlate with increased CW-PRX gene expression and enzymatic activity, suggesting a regulatory role in wall reinforcement and growth modulation (Cho and Lim 2018). More broadly, the ability of plants to perceive stress signals and transduce them into structural modifications is fundamental for survival and adaptation. Among the central regulators of these processes are MAPK cascades, which coordinate responses to diverse abiotic and biotic stresses (Pitzschke et al. 2009).

Beyond its structural role, FA also participates directly in stress-induced signaling pathways. Under chilling stress, exogenous FA application has been shown to trigger Ca^{2+} influx across the plasma membrane, which subsequently activates MAPK signaling cascades (Chi et al. 2013; White and Sacks 2010). These cascades play a pivotal role in transmitting cold stress signals within cells (Chu et al. 2021) and are closely linked to the production of cell wall–associated phenolic compounds such as FA. Furthermore, FA appears to enhance chilling tolerance by activating the ICE-CBF-COR signaling pathway, a canonical regulatory mechanism in plant cold stress adaptation (Shu et al. 2022). In tomato fruits, FA treatment significantly increased the expression of the cold-responsive genes *SlICE1* and *SlCBF1* compared to untreated controls (Shu et al. 2022). These findings suggest that FA not only reinforces cell wall architecture but also primes transcriptional programs associated with cold tolerance, thereby integrating structural and molecular defenses into a coordinated adaptive response.

11.5.3 Crosstalk with Transcription Factors

Ferulic acid is a multifunctional phenolic compound that not only reinforces the plant cell wall through oxidative cross-linking but also acts as a key modulator of plant signaling, defense, and transcriptional regulation (Stan et al. 2016). Its chemical structure, characterized by a phenolic hydroxyl group and electron-donating substituents, underpins its potent antioxidant capacity. FA effectively scavenges ROS such as superoxide anion radicals, suppresses lipid peroxidation, and chelates transition metals to limit hydroxyl radical generation via the Fenton reaction (Srinivasan et al. 2007). Through these redox-mediated mechanisms, FA integrates structural reinforcement with signaling functions that influence plant growth and stress adaptation.

A critical dimension of FA activity lies in its regulation of transcription factors (TFs), which—despite representing <10% of plant transcriptomes—govern expansive gene networks underlying development, metabolism, and defense. Among these, NAC TFs (NAM, ATAF, and CUC family members) orchestrate processes ranging from shoot apical meristem maintenance and inflorescence development to fruit ripening (Liu et al. 2022). MYB TFs, particularly those regulating secondary cell wall biosynthesis, play central roles in lignification and phenylpropanoid metabolism. For example, ZmMYB167 activates lignin biosynthetic and cellulose synthase genes, with overexpression in maize leading to elevated FA deposition in the cell wall, enhanced growth, and improved stress resilience (Rao and Dixon 2018; de Buanafina and Morris 2022). Similarly, *ZmMYB-IF35* promotes the accumulation of FA and chlorogenic acid, linking MYB regulation to phenylpropanoid-derived defense metabolites (Liu et al. 2015a, b).

Transcriptomic analyses further demonstrate FA's broad impact on plant regulatory networks. Short-term FA exposure (1–3 h) induces reprogramming of over 1200 genes, compared with only 176 genes under prolonged treatment (24 h), underscoring its role as a dynamic signaling cue (Chi et al. 2013). FA-responsive genes span diverse functions, including cell wall biosynthesis, detoxification, secondary metabolism, and hormone signaling. Notably, FA enhances the transcription of genes associated with ET and JA pathways—two central hormonal axes in stress and defense responses—as well as ABC transporters, auxin/amino acid permeases, and signaling receptors such as leucine-rich repeat (LRR-VIII) proteins and receptor-like cytoplasmic kinases (RLCK-VII). Moreover, FA consistently upregulates stress-related TF families, including *WRKY*, *MYB*, *AP2/ERF*, and zinc-finger proteins, thereby integrating redox, metabolic, and hormonal pathways into a coordinated defense program (Chi et al. 2013).

In tomato, exogenous FA application (100 μM) significantly upregulated PR1, NPR1, MYC2, and LoxD, activating both SA- and JA-dependent defense cascades. MYC2, a basic helix–loop–helix (*bHLH*) TF, functions as a central node in JA signaling, while *NPR1* mediates SA-induced resistance. These transcriptional shifts correlated with improved resistance to *Botrytis cinerea* and were partly dependent on nitric oxide (NO) signaling, as NO synthesis inhibition attenuated FA-induced gene expression (Shu et al. 2021).

Collectively, these findings establish FA as a dual-function molecule: a structural integrator of cell wall integrity and a regulatory signal that modulates TF-driven defense and developmental pathways. By influencing key TF families—including NAC, MYB, WRKY, and MYC2—FA coordinates transcriptional networks that couple growth, metabolic reprogramming, and immune adaptation (Fig. 11.6). This

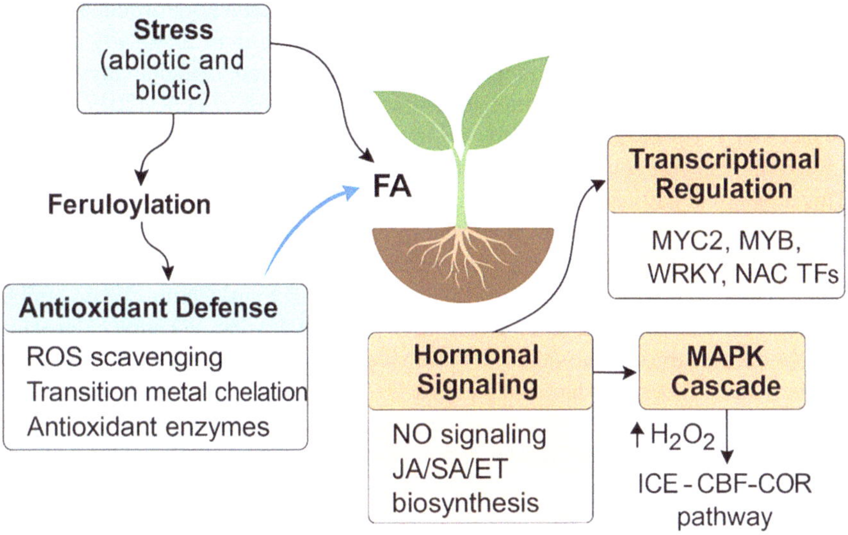

Fig. 11.6 Schematic representation of ferulic acid (FA)-mediated molecular networks under abiotic and biotic stress. FA contributes to antioxidant defense through reactive oxygen species (ROS) scavenging, transition metal chelation, and induction of antioxidant enzymes. Concurrently, FA modulates hormonal signaling, including nitric oxide (NO), jasmonic acid (JA), salicylic acid (SA), and ethylene (ET) pathways, and activates mitogen-activated protein kinase (MAPK) cascades. These processes converge on transcriptional regulation via MYC2, MYB, WRKY, and NAC transcription factors, integrating redox, hormonal, and stress-responsive pathways to reinforce cell walls and enhance stress resilience

positions FA as a central hub in the crosstalk between phenylpropanoid metabolism, redox balance, and hormone-mediated defense signaling.

11.6 Exogenous Application and Agronomic Relevance

11.6.1 Methods of Application: Seed Priming, Foliar Sprays, and Hydroponics

The agronomic effectiveness of ferulic acid as a stress mitigator is strongly influenced by the mode and timing of application. Diverse strategies—including seed priming, foliar spraying, and incorporation into hydroponic systems—have been employed to optimize FA delivery, each with distinct physiological and developmental implications (Bhardwaj et al. 2017; Caspersen et al. 1999; Linić et al. 2021). These approaches are typically tailored to enhance FA bioavailability, synchronize uptake with phenological stages, and maximize plant resilience under stress.

Seed priming with FA functions as a pre-sowing intervention to enhance seed vigor and early-stage stress tolerance. Seeds are imbibed in dilute FA solutions (commonly in the micromolar range) for a prescribed period, inducing physiological and biochemical adjustments prior to germination. Bhardwaj et al. (2017) demonstrated that FA-primed wheat seeds exhibited higher germination indices, elevated antioxidant enzyme activity, and improved tolerance to drought and oxidative stress. This strategy is particularly valuable for fortifying seedlings during establishment, where early resilience strongly determines yield outcomes.

Foliar application represents a widely adopted method that enables direct absorption of FA via the leaf surface, ensuring rapid systemic translocation. When applied at appropriate developmental stages, foliar sprays of FA can enhance chlorophyll fluorescence, activate antioxidant defense, and upregulate stress-responsive signaling pathways (Moradinezhad and Ranjbar 2024; Linić et al. 2021). In tomato, Shu et al. (2021) reported that FA sprays improved resistance to *Botrytis cinerea* by stimulating the antioxidant system and phenylpropanoid metabolism. Likewise, in Chinese cabbage (*Brassica rapa*), Linić et al. (2021) showed that foliar FA treatments (10–100 μM) mitigated short-term salt stress (150 mM NaCl for 72 h), leading to improved physiological and biochemical performance. Such evidence underscores foliar spraying as a versatile and field-relevant application mode against abiotic and biotic challenges.

Hydroponic delivery systems provide a precisely controlled environment for testing FA's physiological impacts under variable conditions. Caspersen et al. (1999), working with lettuce (*Lactuca sativa* L.) in axenic hydroponic culture, observed that FA applied at 10–100 μM inhibited root and shoot growth under low irradiance, whereas higher light intensity alleviated these inhibitory effects. The study demonstrated that FA modulates biomass accumulation, nutrient uptake, and root architecture in a light-dependent manner, emphasizing its dual potential as both a growth regulator and stress mediator in hydroponic systems (Park and Williams 2024).

Together, these application strategies highlight the context-dependent nature of FA efficacy. Seed priming offers long-term developmental advantages, foliar spraying provides rapid stress protection during active growth, and hydroponic delivery enables mechanistic insights under controlled conditions. Understanding how these methods interact with plant physiology and environmental variables is essential for translating FA applications into scalable agronomic practices.

11.6.2 Optimal Concentrations and Plant-Specific Responses

Ferulic acid exhibits a wide spectrum of physiological and biochemical activities, including antioxidant and allelopathic functions (Khan et al. 2024). However, its effects are strongly species-dependent, necessitating precise determination of optimal concentrations for targeted agronomic outcomes. Indeed, FA can exert beneficial effects at low concentrations yet display inhibitory or phytotoxic activity at higher levels, underscoring the importance of species-specific calibration.

Ikhajiagbe and Ogwu (2021) demonstrated this dual behavior in *Amaranthus* species: while moderate concentrations of FA enhanced growth and allelopathic interactions, surpassing a threshold level inhibited weed germination and reduced growth performance. Such findings emphasize that concentration-dependent responses must be aligned with desired agronomic goals, whether stimulating crop resilience or suppressing competitor species.

At low concentrations (typically 10–100 μM), FA consistently enhances plant stress tolerance through activation of antioxidant defenses and modulation of the phenylpropanoid pathway (Table 11.4). In *Brassica rapa* (Chinese cabbage), a 10 μM foliar spray increased photosynthetic efficiency and promoted phenolic accumulation under salt stress (Linić et al. 2021). Similarly, in tomato, FA treatments stimulated vegetative growth, elevated phenolic and carotenoid content, and improved resilience to environmental stressors (Gorni et al. 2021). Together, these studies illustrate that FA functions as a context-dependent modulator whose optimal efficacy relies on careful alignment of concentration, plant species, and stress environment.

11.6.3 Synergistic Use with Microbial Biostimulants and Other Phenolics

The integration of FA with microbial biostimulants and other phenolic compounds has emerged as a promising avenue for enhancing crop resilience and productivity. Beyond its established roles in reinforcing plant cell walls and modulating redox balance (Khan et al. 2024), FA contributes to synergistic interactions that amplify metabolic, physiological, and ecological responses when combined with complementary bioactive compounds.

Phenolic-rich extracts, such as those derived from spelt husk, provide a natural matrix of hydroxycinnamic acids (including p-coumaric, caffeic, and ferulic acids) and simpler phenolics such as p-hydroxybenzoic and syringic acids (Ceccarini et al. 2019). This diversity of bioactive molecules elicits broader effects than individual compounds alone, collectively inducing antioxidant defenses, strengthening immune responses, and facilitating beneficial plant–microbe symbioses. The cooperative action of FA and microbial inoculants has been shown to improve nutrient uptake, metabolic efficiency, and crop performance in species such as soybean and lettuce.

Importantly, the biochemical synergy enhances lignin biosynthesis and related structural polymers, thereby fortifying plants against pathogens while simultaneously improving yield and marketable quality (Ceccarini et al. 2019; Szparaga et al. 2018). This integrative strategy positions FA as both a direct regulator of plant defense and a functional precursor within complex phenolic networks that drive resilience and productivity under stress-prone agricultural systems.

Table 11.4 Case studies of exogenous ferulic acid (FA) application in crops exposed to abiotic and biotic stressors

Crop	Stress type	Application	Concentration	Key findings	References
Wheat (*Triticum aestivum*)	Drought stress	Seed priming	100 ppm	Enhanced germination index Increased antioxidant enzyme activity Improved early-stage tolerance to water deficit	Bhardwaj et al. (2017)
Chinese cabbage (*Brassica rapa*)	Salinity (150 mM)	Foliar spray	10–100 μM	Improved photosynthetic performance Increased phenolic compound accumulation Mitigated salt-induced physiological damage	Linić et al. (2021)
Tomato (*Solanum lycopersicum*)	Pathogen (*Botrytis cinerea*)	Foliar spray	Not specified	Enhanced resistance via phenylpropanoid metabolism Strengthened antioxidant system	Shu et al. (2021)
Lettuce (*Lactuca sativa*)	Low irradiance	Hydroponic application	10–100 μM	Growth inhibition under low light conditions Negative effects alleviated under high light intensity	Caspersen et al. (1999)
Maize (*Zea mays L.*)	Salinity (100–200 mM NaCl)	Combined with biostimulants (FA-rich extracts)	Not specified	Synergistic enhancement of structural metabolism Improved crop quality through combined phenolic activity	Ceccarini et al. (2019)

11.7 Conclusion and Future Directions

Ferulic acid, a pivotal hydroxycinnamic acid, functions as both a structural fortifier and a biochemical regulator in plants, conferring resilience against diverse abiotic and biotic stresses. Through lignification and cell-wall cross-linking, FA enhances mechanical barriers against pathogen invasion and improves tolerance to drought, salinity, and other environmental constraints. Concurrently, its potent antioxidant activity mitigates oxidative damage by scavenging ROS, thereby preserving cellular integrity and redox balance.

Beyond structural and antioxidant roles, FA acts as a versatile signaling modulator. It influences hormonal networks involving JA, SA, and ET, activates MAPK cascades, and upregulates transcriptional regulators associated with antioxidant defense, osmolyte biosynthesis, and phenylpropanoid metabolism. Collectively, these mechanisms contribute to improved water-use efficiency, nutrient uptake, and stabilized photosynthetic performance under stress conditions.

Field-level evidence highlights the agronomic value of FA. Foliar sprays and seed priming have been shown to improve germination, growth, and stress tolerance across diverse crops, including wheat, lettuce, tomato, and soybeans. FA's dual role extends further: it not only strengthens host plants but also exhibits antimicrobial properties, underscoring its bidirectional function in plant health. Synergistic integration with microbial biostimulants and other phenolic compounds offers promising avenues for bio-based formulations that enhance yield, resilience, and quality.

Future research should prioritize the following:

- Formulation innovation: development of nanoformulations and controlled-release delivery systems for precision agriculture.
- Species-specific optimization: defining concentration thresholds and crop-specific responses under varying field conditions.
- Microbiome interactions: elucidating how FA modulates rhizosphere and phyllosphere microbial communities.
- Systems-level integration: embedding FA-based strategies into circular agriculture, hydroponics, and smart-farming platforms.

As climate pressures intensify, FA emerges as a promising bioactive tool for advancing climate-smart agriculture. Its multifaceted functions—bridging structural integrity, biochemical signaling, and synergistic interactions—position FA as a cornerstone for future strategies that unite resilience, productivity, and sustainability in modern crop systems.

References

AAlikhani M, Khalili M, Jahanshahi M (2022) The natural iron chelators' ferulic acid and caffeic acid rescue mice's brains from side effects of iron overload. Front Neurol 13:951725. https://doi.org/10.3389/fneur.2022.951725

Abadias M, Bobo G, Anguera M, Ortiz-Solà J, Aguiló-Aguayo I (2024) Fortification of orange and apple juices with Ferulic acid: implications for food safety and quality. Foods 13(20):3288

Abdel Latef AAH, Hameed M, Khan Z, Khan S, Imran A, Fatima N (2026) Foliar stimulants: A sustainable strategy for enhancing plant resilience to abiotic and biotic stresses. Physiologia Plantarum 178(1):e70741

Afzal S, Hussain I, Rasheed R (2025) Foliar application of Choline chloride and Ferulic acid regulated Osmolyte accumulation, nutrient acquisition and ROS metabolism to Lessen chromium effects on Barley (Hordeum vulgare). https://www.researchgate.net/profile/Iqbal-

Ahlawat YK, Singh M, Manorama K, Lakra N, Zaid A, Zulfiqar F (2024) Plant phenolics: neglected secondary metabolites in plant stress tolerance. Rev Bras Bot 47(3):703–721. https://doi.org/10.1007/s40415-023-00949-x

Alam MA (2019) Anti-hypertensive effect of cereal antioxidant ferulic acid and its mechanism of action. Front Nutr 6:121

Alifah LHN, Jatmika C, Hayun H (2024) Exploration of Ferulic acid and its derivatives as potent anti-Tyrosinase: a systematic review. Egypt J Chem 67(3):257–271

Amić A, Marković JMD, Marković Z et al (2021) Theoretical study of radical inactivation, lox inhibition, and iron chelation: the role of ferulic acid in skin protection against uva induced oxidative stress. Antioxidants 10:1303. https://doi.org/10.3390/antiox10081303

Anand U, Pal T, Yadav N et al (2023) Current scenario and future prospects of endophytic microbes: promising candidates for abiotic and biotic stress management for agricultural and environmental sustainability. Springer New York, NY, USA

Anli M, Boutasknit A, Ait-El-Mokhtar M, Ben-Laouane R, Abdouroihamane H, Meddich A, Baslam M (2026) Plant extracts for boosting plant stress tolerance. In: Abdel Latef AAH (eds) Using Stimulants to Improve Plant Health: From Stress to Strength. Elsevier. https://doi.org/10.1016/B978-0-443-45576-6.00009-1

Arbona V, Manzi M, de Ollas C, Gómez-Cadenas A (2013) Metabolomics as a tool to investigate abiotic stress tolerance in plants. Int J Mol Sci 14:4885–4911. https://doi.org/10.3390/ijms14034885

Assabgui RA, Reid LM, Hamilton RI, Arnason JT (1993) Correlation of kernel (E)-ferulic acid content of maize with resistance to Fusarium graminearum. Phytopathology 83:949–949

Bahadur A, Singh DP, Sarma BK, Singh UP (2012) Foliar application of l-phenylalanine and ferulic acids to pea plants: induced phenylalanine ammonia lyase activity and resistance against Erysiphe pisi. Arch Phytopathol Plant Protect 45(4):398–403. https://doi.org/10.1080/03235408.2011.587963

Bano A, Qadri TA, Mahnoor KN (2023) Bioactive metabolites of plants and microbes and their role in agricultural sustainability and mitigation of plant stress. S Afr J Bot 159:98–109. https://doi.org/10.1016/j.sajb.2023.05.049

Bhardwaj RD, Kaur L, Srivastava P (2017) Comparative evaluation of different phenolic acids as priming agents for mitigating drought stress in wheat seedlings. Proc Natl Acad Sci, India, Sect B Biol Sci 87(4):1133–1142. https://doi.org/10.1007/s40011-015-0690-y

Blum U, Dalton BR, Shann JR (1985) Effects of various mixtures of ferulic acid and some of its microbial metabolic products on cucumber leaf expansion and dry matter in nutrient culture. J Chem Ecol 11(5). https://doi.org/10.1007/BF00988572

Bogati KA, Golińska P, Sewerniak P et al (2023) Deciphering the impact of induced drought in agriculture soils: changes in microbial community structure, enzymatic and metabolic diversity. Agronomy 13:1417. https://doi.org/10.3390/agronomy13051417

Borges A, Ferreira C, Saavedra MJ, Simões M (2013) Antibacterial activity and mode of action of Ferulic and Gallic acids against pathogenic bacteria. Microb Drug Resist 19(4):256–265. https://doi.org/10.1089/mdr.2012.0244

Boutasknit et al. (2026) Auxin-mediated regulation of plant stress tolerance: Mechanisms and molecular responses. In: Abdel Latef AAH (eds) Using Stimulants to Improve Plant Health: From Stress to Strength. Elsevier. https://doi.org/10.1016/B978-0-443-45576-6.00016-9

Boutasknit A, Baslam M, Ait-El-Mokhtar M et al (2021) Assemblage of indigenous arbuscular mycorrhizal fungi and green waste compost enhance drought stress tolerance in carob (Ceratonia siliqua L.) trees. Sci Rep 11:1–23. https://doi.org/10.1038/s41598-021-02018-3

Boutasknit A, Ait-El-Mokhtar M, Fassih B et al (2024) Effect of Arbuscular Mycorrhizal fungi and rock phosphate on growth, physiology, and biochemistry of carob under water stress and after rehydration in vermicompost-amended soil. Meta 14:202. https://doi.org/10.3390/metabo14040202

Boz H (2015) Ferulic acid in cereals–a review. Czech J Food Sci 33(1) https://cjfs.agriculturejournals.cz/pdfs/cjf/2015/01/02.pdf

Broekaert WF, Delauré SL, De Bolle MFC, Cammue BPA (2006) The role of ethylene in host-pathogen interactions. Annu Rev Phytopathol 44(1):393–416. https://doi.org/10.1146/annurev.phyto.44.070505.143440

Caillet S, Salmiéri S, Lacroix M (2006) Evaluation of free radical-scavenging properties of commercial grape phenol extracts by a fast colorimetric method. Food Chem 95(1):1–8. https://doi.org/10.1016/J.FOODCHEM.2004.12.011

Caspersen S, Sundin P, Munro M, Aðalsteinsson S, Hooker JE, Jensén P (1999) Interactive effects of lettuce (Lactuca sativa L.), irradiance, and ferulic acid in axenic, hydroponic culture. Plant Soil 210(1):115–126. https://doi.org/10.1023/A:1004682018888

Ceccarini C, Antognoni F, Biondi S, Fraternale A, Verardo G, Gorassini A, Scoccianti V (2019) Polyphenol-enriched spelt husk extracts improve growth and stress-related biochemical parameters under moderate salt stress in maize plants. Plant Physiol Biochem 141:95–104. https://doi.org/10.1016/j.plaphy.2019.05.016

Cesarino I, Araújo P, Domingues Júnior AP, Mazzafera P (2012) An overview of lignin metabolism and its effect on biomass recalcitrance. Rev Bras Bot 35:303–311

Chandrakanth NN, Zhang C, Freeman J, de Souza WR, Bartley LE, Mitchell RA (2023) Modification of plant cell walls with hydroxycinnamic acids by BAHD acyltransferases. Front Plant Sci 13:1088879

Chen S, Lin R, Lu H et al (2020) Effects of phenolic acids on free radical scavenging and heavy metal bioavailability in kandelia obovata under cadmium and zinc stress. Chemosphere 249:126341. https://doi.org/10.1016/j.chemosphere.2020.126341

Cheng ZY, Sun L, Wang XJ, Sun R, An YQ, An BL, Zhu MX, Zhao CF, Bai JG (2018) Ferulic acid pretreatment alleviates heat stress in blueberry seedlings by inducing antioxidant enzymes, proline, and soluble sugars. Biol Plant 62(3):534–542. https://doi.org/10.1007/S10535-018-0772-9/METRICS

Chi WC, Chen YA, Hsiung YC et al (2013) Autotoxicity mechanism of Oryza sativa: transcriptome response in rice roots exposed to ferulic acid. BMC Genomics 14:351. https://doi.org/10.1186/1471-2164-14-351

Cho DH, Lim ST (2018) Changes in phenolic acid composition and associated enzyme activity in shoot and kernel fractions of brown rice during germination. Food Chem 256:163–170. https://doi.org/10.1016/J.FOODCHEM.2018.02.040

Chu P, Wang T, Sun YR, Chu MX, Wang HY, Zheng X, Yin S (2021) Effect of cold stress on the MAPK pathway and lipidomics on muscle of Takifugu fasciatus. Aquaculture 540:736691. https://doi.org/10.1016/J.AQUACULTURE.2021.736691

Das K, Roychoudhury A (2014) Reactive oxygen species (ROS) and response of antioxidants as ROS-scavengers during environmental stress in plants. Front Environ Sci 2:1–13

Das S, Dey A, Maji S et al (2023) Attenuation of fluoride-induced hepatorenal oxidative stress by ferulic acid in vivo: An approach with in-silico analysis and interaction informatics of ferulic acid. J Trace Elem Med Biol 77:127133. https://doi.org/10.1016/j.jtemb.2023.127133

Dawood MA, Metwally AE-S, El-Sharawy ME, Ghozlan AM, Abdel-Latif HM, Van Doan H, Ali MA (2020) The influences of ferulic acid on the growth performance, haemato-immunological responses, and immune-related genes of Nile tilapia (Oreochromis niloticus) exposed to heat stress. Aquaculture 525:735320

de Buanafina MMO (2009) Feruloylation in grasses: current and future perspectives. Mol Plant 2(5):861–872

de Buanafina MMO, Morris P (2022) The impact of cell wall feruloylation on plant growth, responses to environmental stress, plant pathogens and cell wall degradability. Agronomy 12(8):1847. https://doi.org/10.3390/AGRONOMY12081847

de Morais MC, Perez-Castillo Y, Silva VR, de Santos LS, Soares MBP, Bezerra DP, de Castro RD, de Sousa DP (2021) Cytotoxic and antifungal amides derived from Ferulic acid: molecular docking and mechanism of action. Biomed Res Int 2021(1):3598000. https://doi.org/10.1155/2021/3598000

De Oliveira DM, Finger-Teixeira A, Rodrigues Mota T, Salvador VH, Moreira-Vilar FC, Correa Molinari HB, Craig Mitchell RA, Marchiosi R, Ferrarese-Filho O, Dantas Dos Santos W (2015) Ferulic acid: a key component in grass lignocellulose recalcitrance to hydrolysis. Plant Biotechnol J 13(9):1224–1232. https://doi.org/10.1111/pbi.12292

Devi B, Kumar MN, Chutia M, Bhattacharyya N (2022) Abiotic and biotic stress challenges of Cassava in changing climate and strategies to overcome: a review. Sci Hortic 305:111432. https://doi.org/10.1016/j.scienta.2022.111432

dos Santos WD, Ferrarese MLL, Ferrarese-Filho O (2008) Ferulic acid: An allelochemical trouble-maker. Funct Plant Sci Biotechnol 2(1):47–55

Du J, Du C, Ge X et al (2023) Genome-wide analysis of the AAAP gene family in Populus and functional analysis of PsAAAP21 in root growth and amino acid transport. Int J Mol Sci 24:624. https://doi.org/10.3390/ijms24010624

Du B, Haensch R, Alfarraj S, Rennenberg H (2024) Strategies of plants to overcome abiotic and biotic stresses. Biol Rev 99:1524–1536. https://doi.org/10.1111/brv.13079

Dukare A, Paul S (2021) Biological control of Fusarium wilt and growth promotion in pigeon pea (Cajanus cajan) by antagonistic rhizobacteria, displaying multiple modes of pathogen inhibition. Rhizosphere 17:100278. https://doi.org/10.1016/j.rhisph.2020.100278

Durrant WE, Dong X (2004) Systemic acquired resistance. Annu Rev Phytopathol 42(1):185–209. https://doi.org/10.1146/annurev.phyto.42.040803.140421

FAO (2023) World Food and Agriculture—Statistical Yearbook 2023

Fu P, Jaiswal D, McGrath JM et al (2022) Drought imprints on crops can reduce yield loss: nature's insights for food security. Food Energy Secur 11:1–16. https://doi.org/10.1002/fes3.332

Gao P, Qi Y, Li L et al (2023) Amorphophallus muelleri activates ferulic acid and phenylpropane biosynthesis pathways to defend against Fusarium solani infection. Front Plant Sci 14:1–14. https://doi.org/10.3389/fpls.2023.1207970

Gao Y, Wang X, Hou X, Chen J (2024) Evolution and analysis of Caffeic acid transferase (COMT) in seed plants. Biochem Genet 62(3):1953–1976. https://doi.org/10.1007/s10528-023-10525-3

Gharibi S, Sayed Tabatabaei BE, Saeidi G et al (2019) The effect of drought stress on polyphenolic compounds and expression of flavonoid biosynthesis related genes in Achillea pachycephala Rech.f. Phytochemistry 162:90–98. https://doi.org/10.1016/J.PHYTOCHEM.2019.03.004

Godoy F, Olivos-Hernández K, Stange C, Handford M (2021) Abiotic stress in crop species: improving tolerance by applying plant metabolites. Plants 10:1–19. https://doi.org/10.3390/plants10020186

Gorni PH, da Cornelissen BS, Pereira AA (2021) Exogenous salicylic acid and ferulic acid improve growth, phenolic and carotenoid content in tomato. Adv Hortic Sci 35(4):335–341. https://doi.org/10.36253/ahsc-8295

Guneidy RA, Zaki ER, Gad AAM et al (2022) Evaluation of phenolic content diversity along with antioxidant/pro-oxidant, glutathione transferase inhibition, and cytotoxic potential of selected commonly used plants. Prevent Nutr Food Sci 27:282–298. https://doi.org/10.3746/pnf.2022.27.3.282

Han H, Dye L, Mackie A (2023) The impact of processing on the release and antioxidant capacity of ferulic acid from wheat: a systematic review. Food Res Int 164:112371. https://doi.org/10.1016/j.foodres.2022.112371

Hanaka A, Majewska M, Jaroszuk-ściseł J (2022) Study of the influence of abiotic and biotic stress factors on horticultural plants. Horticulturae 8:8–13. https://doi.org/10.3390/horticulturae8010006

Hasanuzzaman M, Bhuyan MHMB, Parvin K et al (2020a) Regulation of ROS metabolism in plants under environmental stress: a review of recent experimental evidence. Int J Mol Sci 21:1–44. https://doi.org/10.3390/ijms21228695

Hasanuzzaman M, Bhuyan MB, Zulfiqar F, Raza A, Mohsin SM, Mahmud JA, Fujita M, Fotopoulos V (2020b) Reactive oxygen species and antioxidant defense in plants under abiotic stress: revisiting the crucial role of a universal defense regulator. Antioxidants 9(8):681

He J, Ma L, Wang D, Zhang M, Zhou H (2019) Ferulic acid treatment reinforces the resistance of postharvest apple fruit during gray mold infection. J Plant Pathol 101(3):503–511. https://doi.org/10.1007/s42161-018-00223-0

Hernández-Jaime AG, Castillo-Rangel F, Arévalos-Sánchez MM, Rentería-Monterrubio AL, Santellano-Estrada E, Tirado-Gallegos JM, Chávez-Martínez A (2025) Antioxidant and antimicrobial activity of Ferulic acid added to dried meat: shelf-life evaluation. Foods 14(4):Article 4. https://doi.org/10.3390/foods14040708

Hou X, Wu F, Wang X-J, Sun Z-T, Zhang Y, Yang M-T, Bai H, Li S, Bai J-G (2018) Bacillus methylotrophicus CSY-F1 alleviates drought stress in cucumber (Cucumis sativus) grown in soil with high ferulic acid levels. Plant Soil 431(1–2):89–105. https://doi.org/10.1007/s11104-018-3748-9

Hura T, Grzesiak S, Hura K, Thiemt E, Tokarz K, Wędzony M (2007a) Physiological and biochemical tools useful in drought-tolerance detection in genotypes of winter triticale: accumulation of ferulic acid correlates with drought tolerance. Ann Bot 100(4):767–775

Hura T, Hura K, Grzesiak M, Rzepka A (2007b) Effect of long-term drought stress on leaf gas exchange and fluorescence parameters in C3 and C4 plants. Acta Physiol Plant 29(2):103–113. https://doi.org/10.1007/s11738-006-0013-2

Hura T, Hura K, Grzesiak S (2009) Possible contribution of cell-wall-bound ferulic acid in drought resistance and recovery in triticale seedlings. J Plant Physiol 166(16):1720–1733. https://doi.org/10.1016/j.jplph.2009.04.012

Hussain MI, Reigosa MJ (2021) Secondary metabolites, ferulic acid and p-hydroxybenzoic acid induced toxic effects on photosynthetic process in rumex acetosa l. Biomolecules 11:1–13. https://doi.org/10.3390/biom11020233

Hussain I, Singh NB, Singh A, Singh H, Singh SC, Yadav V (2017) Exogenous application of phytosynthesized nanoceria to alleviate ferulic acid stress in Solanum lycopersicum. Sci Hortic 214:158–164

Iiyama K, Lam TBT, Stone BA (1990) Phenolic acid bridges between polysaccharides and lignin in wheat internodes. Phytochemistry 29(3):733–737. https://doi.org/10.1016/0031-9422(90)80009-6

Ikan C, Ben-Laouane R, Ouhaddou R et al (2023) Interactions between arbuscular mycorrhizal fungus and indigenous compost improve salt stress tolerance in wheat (Triticum durum). S Afr J Bot 158:417–428. https://doi.org/10.1016/j.sajb.2023.05.038

Ikhajiagbe B, Ogwu MC (2021) Growth performance and ferulic acid composition of Amaranthus species and Mitracarpus villosus in competition with weeds species. Environ Sustain 4(4):691–703. https://doi.org/10.1007/s42398-021-00163-z

Ikram M, Batool M, Ullah M et al (2025) Molecular Alchemy: converting stress into resilience via secondary metabolites and calcium signaling in rice. Rice 18:32. https://doi.org/10.1186/s12284-025-00783-7

Iqbal Z, Iqbal MS, Hashem A et al (2021) Plant Defense responses to biotic stress and its interplay with fluctuating dark/light conditions. Front Plant Sci 12:1–22. https://doi.org/10.3389/fpls.2021.631810

Jacquet G, Pollet B, Lapierre C, Mhamdi F, Rolando C (1995) New ether-linked Ferulic acid-Coniferyl alcohol dimers identified in grass straws. J Agric Food Chem 43(10):2746–2751

Janah et al. (2026a) Gibberellic acid-mediated molecular signaling in plants under environmental stresses. In: Abdel Latef AAH (eds) Using Stimulants to Improve Plant Health: From Stress to Strength. Elsevier. https://doi.org/10.1016/B978-0-443-45576-6.00012-1

Janah et al. (2026b) Unveiling the impact of salicylic acid on medicinal and aromatic plants under environmental stresses. In: Abdel Latef AAH (eds) Using Stimulants to Improve Plant Health: From Stress to Strength. Elsevier. https://doi.org/10.1016/B978-0-443-45576-6.00020-0

Jogawat A, Yadav B, Chhaya et al (2021) Crosstalk between phytohormones and secondary metabolites in the drought stress tolerance of crop plants: a review. Physiol Plant 172:1106–1132. https://doi.org/10.1111/PPL.13328

Kamyab A, Samsampour D, Ahmadinasab N, Bagheri A (2024) The combination of nanoparticles and endophytes boosts Thyme (Thymus vulgaris L.) resistance to drought stress by elevating levels of phenolic compounds, flavonoids, and essential oils. Chem Biol Technol Agric 11(1): 163. https://doi.org/10.1186/s40538-024-00682-x

Kang X, Kirui A, Dickwella Widanage MC, Mentink-Vigier F, Cosgrove DJ, Wang T (2019) Lignin-polysaccharide interactions in plant secondary cell walls revealed by solid-state NMR. Nat Commun 10(1):1–9. https://doi.org/10.1038/s41467-018-08252-0

Khan KA, Saleem MH, Afzal S et al (2024) Ferulic acid: therapeutic potential due to its antioxidant properties, role in plant growth, and stress tolerance. Plant Growth Regul 104:1329–1353. https://doi.org/10.1007/s10725-024-01243-w

Khattab IM, Fischer J, Kaźmierczak A, Thines E, Nick P (2023) Ferulic acid is a putative surrender signal to stimulate programmed cell death in grapevines after infection with *Neofusicoccum parvum*. Plant Cell Environ 46(1):339–358. https://doi.org/10.1111/pce.14468

Khatun MM, Bhuia MS, Chowdhury R et al (2024) Potential utilization of ferulic acid and its derivatives in the management of metabolic diseases and disorders: An insight into mechanisms. Cell Signal 121:111291. https://doi.org/10.1016/j.cellsig.2024.111291

Khosravi C, Benocci T, Battaglia E et al (2015) Sugar catabolism in aspergillus and other fungi related to the utilization of plant biomass. Amsterdam, The Netherlands. Elsevier Ltd

Kim NS, Kwon S-J, Cuong DM, Jeon J, Park JS, Park SU (2019) Accumulation of phenylpropanoids in tartary buckwheat (Fagopyrum tataricum) under salt stress. Agronomy 9(11):739

Kısa D, Elmastaş M, Öztürk L, Kayır Ö (2016) Responses of the phenolic compounds of Zea mays under heavy metal stress. Appl Biol Chem 59(6):813–820. https://doi.org/10.1007/s13765-016-0229-9

Krishnappa NP, Basha SA, Negi PS, Prasada Rao UJS (2017) Phenolic acid composition, antioxidant and antimicrobial activities of green gram (*vigna radiata*) exudate, husk, and germinated seed of different stages. J Food Proc Preserv 41(6):e13273. https://doi.org/10.1111/jfpp.13273

Kumar N, Pruthi V (2014) Potential applications of ferulic acid from natural sources. Biotechnol Rep 4:86–93. https://doi.org/10.1016/j.btre.2014.09.002

Kumar S, Bhushan B, Wakchaure GC, Meena KK, Kumar M, Meena NL, Rane J (2020) Plant phenolics under water-deficit conditions: biosynthesis, accumulation, and physiological roles in water stress alleviation. In: Lone R, Shuab R, Kamili AN (eds) Plant phenolics in sustainable agriculture. Springer Singapore, pp 451–465. https://doi.org/10.1007/978-981-15-4890-1_19

Kumar K, Debnath P, Singh S, Kumar N (2023) An overview of plant phenolics and their involvement in abiotic stress tolerance. Stress 3:570–585. https://doi.org/10.3390/stresses3030040

Lal MK, Tiwari RK, Altaf MA et al (2023) Editorial: abiotic and biotic stress in horticultural crops: insight into recent advances in the underlying tolerance mechanism. Front Plant Sci 14:1–6. https://doi.org/10.3389/fpls.2023.1212982

Li ZH, Wang Q, Ruan X et al (2010) Phenolics and plant allelopathy. Molecules 15:8933–8952. https://doi.org/10.3390/molecules15128933

Li DM, Nie YX, Zhang J et al (2013) Ferulic acid pretreatment enhances dehydration-stress tolerance of cucumber seedlings. Biol Plant 57:711–717. https://doi.org/10.1007/s10535-013-0326-0

Li H, Yu X, Meng F et al (2021) Ferulic acid supplementation increases lifespan and stress resistance via insulin/igf-1 signaling pathway in c. Elegans. Int J Mol Sci 22:1–19. https://doi.org/10.3390/ijms22084279

Li S, Fu W, Li B et al (2025) Insight into Cd detoxification and accumulation in wheat by foliar application of Ferulic acid. Plants 14:1436. https://doi.org/10.3390/plants14101436

Linić I, Mlinarić S, Brkljačić L, Pavlović I, Smolko A, Salopek-Sondi B (2021) Ferulic acid and salicylic acid foliar treatments reduce short-term salt stress in Chinese cabbage by increasing phenolic compounds accumulation and photosynthetic performance. Plants 10(11):2346. https://doi.org/10.3390/plants10112346

Liu C-J, Miao Y-C, Zhang K-W (2011) Sequestration and transport of lignin monomeric precursors. Molecules 16(1):710–727

Liu J, Feng L, Li J, He Z (2015a) Genetic and epigenetic control of plant heat responses. Front Plant Sci 6:1–21. https://doi.org/10.3389/fpls.2015.00267

Liu J, Osbourn A, Ma P (2015b) MYB transcription factors as regulators of Phenylpropanoid metabolism in plants. Mol Plant 8(5):689–708. https://doi.org/10.1016/J.MOLP.2015.03.012

Liu GS, Li HL, Grierson D, Fu DQ (2022) NAC transcription factor family regulation of fruit ripening and quality: a review. Cells 11(3):525. https://doi.org/10.3390/CELLS11030525

Lone R, Khan S, Al-Sadi AM (2023) Plant Phenolics in abiotic stress management. Springer Nature, Singapore

Lu G-H, Chan K, Leung K, Chan C-L, Zhao Z-Z, Jiang Z-H (2005) Assay of free ferulic acid and total ferulic acid for quality assessment of Angelica sinensis. J Chromatogr A 1068(2):209–219

Ma J, Saleem MH, Ali B, Rasheed R, Ashraf MA, Aziz H, Ercisli S, Riaz S, Elsharkawy MM, Hussain I (2022a) Impact of foliar application of syringic acid on tomato (Solanum lycopersicum L.) under heavy metal stress-insights into nutrient uptake, redox homeostasis, oxidative stress, and antioxidant defense. Front Plant Sci 13:950120

Ma R, He Y, Fang Q, Xie G, Qi M (2022b) Ferulic acid ameliorates renal injury via improving autophagy to inhibit inflammation in diabetic nephropathy mice. Biomed Pharmacother 153: 113424

Mahto RK, Ambika SC et al (2022) Chickpea biofortification for Cytokinin dehydrogenase via genome editing to enhance abiotic-biotic stress tolerance and food security. Front Genet 13:1–13. https://doi.org/10.3389/fgene.2022.900324

Martínez-Fraca J, de la Torre-Hernández ME, Meshoulam-Alamilla M, Plasencia J (2022) In search of resistance against Fusarium ear rot: Ferulic acid contents in maize pericarp are associated with antifungal activity and inhibition of Fumonisin production. Front Plant Sci 13. https://doi.org/10.3389/fpls.2022.852257

Medda S, Fadda A, Mulas M (2022) Influence of climate change on metabolism and biological characteristics in perennial Woody fruit crops in the Mediterranean environment. Horticulturae 8:273. https://doi.org/10.3390/horticulturae8040273

Meyer S, Cartelat A, Moya I, Cerovic ZG (2003) UV-induced blue-green and far-red fluorescence along wheat leaves: a potential signature of leaf ageing. J Exp Bot 54(383):757–769

Minh LT, Khang DT, Ha PT, Tuyen PT, Minh TN, Quan NV, Xuan TD (2016) Effects of salinity stress on growth and phenolics of rice (Oryza sativa L.). Int Lett Nat Sci 57. https://agro.icm.edu.pl/agro/element/bwmeta1.element.agro-b5ed5b63-bb30-4e2c-92fc-e2df553c7415

Mnich E, Bjarnholt N, Eudes A, Harholt J, Holland C, Jørgensen B, Larsen FH, Liu M, Manat R, Meyer AS (2020) Phenolic cross-links: building and de-constructing the plant cell wall. Nat Prod Rep 37(7):919–961

Mohagheghian B, Saeidi G, Arzani A (2025) Phenolic compounds, antioxidant enzymes, and oxidative stress in barley (Hordeum vulgare L.) genotypes under field drought-stress conditions. BMC Plant Biol 25(1):709. https://doi.org/10.1186/s12870-025-06750-0

Moradinezhad F, Ranjbar A (2024) Foliar application of fertilizers and plant growth regulators on pomegranate fruit yield and quality: a review. J Plant Nutr 47(5):797–821. https://doi.org/10.1080/01904167.2023.2280152

Mughal A, Jabeen N, Ashraf K, Sultan K, Farhan M, Hussain MI, Deng G, Alsudays IM, Saleh MA, Tariq S (2024) Exploring the role of caffeic acid in mitigating abiotic stresses in plants: a review. Plant Stress 12:100487

Narnoliya LK, Sangwan N, Jadaun JS, Bansal S, Sangwan RS (2021) Defining the role of a caffeic acid 3-O-methyltransferase from Azadirachta indica fruits in the biosynthesis of ferulic acid through heterologous over-expression in Ocimum species and Withania somnifera. Planta 253(1):20. https://doi.org/10.1007/s00425-020-03514-y

Nowicka B (2022) Modifications of Phytohormone metabolism aimed at stimulation of plant growth, improving their productivity and tolerance to abiotic and biotic stress factors. Plants 11:3430. https://doi.org/10.3390/plants11243430

Oliveira DM, Mota TR, Salatta FV, Sinzker RC, Končitíková R, Kopečný D, Simister R, Silva M, Goeminne G, Morreel K, Rencoret J, Gutiérrez A, Tryfona T, Marchiosi R, Dupree P, Del Río JC, Boerjan W, McQueen-Mason SJ, Gomez LD et al (2020) Cell wall remodeling under salt stress: insights into changes in polysaccharides, feruloylation, lignification, and phenolic metabolism in maize. Plant Cell Environ 43(9):2172–2191. https://doi.org/10.1111/pce.13805

Ouhaddou R, Ech-chatir L, Anli M et al (2023) Secondary metabolites, osmolytes and antioxidant activity as the main attributes enhanced by biostimulants for growth and resilience of lettuce to drought stress. Gesunde Pflanzen. https://doi.org/10.1007/s10343-022-00827-8

Park Y, Williams KA (2024) Organic hydroponics: a review. Sci Hortic 324:112604. https://doi.org/10.1016/j.scienta.2023.112604

Patzke H, Schieber A (2018) Growth-inhibitory activity of phenolic compounds applied in an emulsifiable concentrate-ferulic acid as a natural pesticide against Botrytis cinerea. Food Res Int 113:18–23

Peng T, Wang Y, Yang T et al (2021) Physiological and biochemical responses, and comparative transcriptome profiling of two Angelica sinensis cultivars under enhanced ultraviolet-B radiation. Front Plant Sci 12:1–18. https://doi.org/10.3389/fpls.2021.805407

Pitzschke A, Schikora A, Hirt H (2009) MAPK cascade signalling networks in plant defence. Curr Opin Plant Biol 12(4):421–426. https://doi.org/10.1016/J.PBI.2009.06.008

Pratyusha S (2022) Phenolic compounds in the plant development and defense: an overview. In: Plant stress physiology–perspectives in agriculture. https://doi.org/10.5772/INTECHOPEN.102873

Quan NT, Xuan TD (2018) Foliar application of vanillic and p-hydroxybenzoic acids enhanced drought tolerance and formation of phytoalexin momilactones in rice. Arch Agron Soil Sci 64:1831–1846. https://doi.org/10.1080/03650340.2018.1463520

Quideau S, Ralph J (1997) Lignin–ferulate cross-links in grasses. Part 4.1–3 incorporation of 5–5-coupled dehydrodiferulate into synthetic lignin. J Chem Soc Perkin Trans 1, 16:2351–2358

Quoc CD, Lan AB, Ngoc TN, Le Hong T, Thanh TT, Tran G-B, Hoa SP, Hung TN, Ngoc TNH, Cong HN, Le NNT, Ngoc NT (2024) Variations in some metabolic compounds in the roots of rice varieties (Oryza sativa L.) with different salinity tolerance under salinity stress during the seedling stage. Plant Physiol Rep 29(3):660–677. https://doi.org/10.1007/s40502-024-00804-5

Rabeh K, Hnini M, Oubohssaine M (2025) A comprehensive review of transcription factor-mediated regulation of secondary metabolites in plants under environmental stress. Stress Biol 5:15. https://doi.org/10.1007/s44154-024-00201-w

Ragauskas AJ, Beckham GT, Biddy MJ, Chandra R, Chen F, Davis MF, Davison BH, Dixon RA, Gilna P, Keller M, Langan P, Naskar AK, Saddler JN, Tschaplinski TJ, Tuskan GA, Wyman CE (2014) Lignin valorization: improving lignin processing in the biorefinery. Science 344(6185):1246843. https://doi.org/10.1126/science.1246843

Rahman NSNA, Hamid NWA, Nadarajah K (2021) Effects of abiotic stress on soil microbiome. Int J Mol Sci 22:9036. https://doi.org/10.3390/ijms22169036

Ralph J, Lundquist K, Brunow G, Lu F, Kim H, Schatz PF, Marita JM, Hatfield RD, Ralph SA, Christensen JH, Boerjan W (2004) Lignins: natural polymers from oxidative coupling of 4-hydroxyphenyl- propanoids. Phytochem Rev 3(1–2):29–60. https://doi.org/10.1023/B:PHYT.0000047809.65444.a4

Rao X, Dixon RA (2018) Current models for transcriptional regulation of secondary cell wall biosynthesis in grasses. Front Plant Sci 9:331750. https://doi.org/10.3389/FPLS.2018.00399/XML/NLM

Sachdev S, Ansari SA, Ansari MI et al (2021) Abiotic stress and reactive oxygen species: generation, signaling, and defense mechanisms. Antioxidants 10:1–37. https://doi.org/10.3390/antiox10020277

Saini N, Anmol A, Kumar S et al (2024) Exploring phenolic compounds as natural stress alleviators in plants- a comprehensive review. Physiol Mol Plant Pathol 133:102383. https://doi.org/10.1016/j.pmpp.2024.102383

Salami M, Heidari B, Batley J et al (2023) Integration of genome-wide association studies, metabolomics, and transcriptomics reveals phenolic acid- and flavonoid-associated genes and their regulatory elements under drought stress in rapeseed flowers. Front Plant Sci 14:1–25. https://doi.org/10.3389/fpls.2023.1249142

Salami M, Heidari B, Batley J, Wang J, Tan X-L, Richards C, Tan H (2024) Integration of genome-wide association studies, metabolomics, and transcriptomics reveals phenolic acid-and flavonoid-associated genes and their regulatory elements under drought stress in rapeseed flowers. Front Plant Sci 14:1249142

Sánchez M, Peña MJ, Revilla G, Zarra I (1996) Changes in Dehydrodiferulic acids and peroxidase activity against Ferulic acid associated with cell walls during growth of Pinus pinaster Hypocotyl. Plant Physiol 111(3):941–946. https://doi.org/10.1104/PP.111.3.941

Santos ABD, Bottcher A, Kiyota E, Mayer JLS, Vicentini R, Brito MDS, Creste S, Landell MGA, Mazzafera P (2015) Water stress alters lignin content and related gene expression in two sugarcane genotypes. J Agric Food Chem 63(19):4708–4720. https://doi.org/10.1021/jf5061858

Sarangi PK, Vivekanand V, Mohanakrishna G, Pattnaik B, Muddapur UM, Aminabhavi TM (2023) Production of bioactive phenolic compounds from agricultural by-products towards bioeconomic perspectives. J Clean Prod 414:137460

Scheller HV, Ulvskov P (2010) Hemicelluloses. Annu Rev Plant Biol 61(1):263–289. https://doi.org/10.1146/annurev-arplant-042809-112315

Sebastian S, Sundaraganesan N, Manoharan S (2009) Molecular structure, spectroscopic studies and first-order molecular hyperpolarizabilities of ferulic acid by density functional study. Spectrochimica Acta A Mol Biomol Spectrosc 74:312–323. https://doi.org/10.1016/j.saa.2009.06.011

Shadle GL, Wesley SV, Korth KL, Chen F, Lamb C, Dixon RA (2003) Phenylpropanoid compounds and disease resistance in transgenic tobacco with altered expression of L-phenylalanine ammonia-lyase. Phytochemistry 64(1):153–161

Shalaby S, Larkov O, Lamdan N-L, Goldshmidt-Tran O, Horwitz BA (2016) Plant phenolic acids induce programmed cell death of a fungal pathogen: MAPK signaling and survival of Cochliobolus heterostrophus. Environ Microbiol 18(11):4188–4199. https://doi.org/10.1111/1462-2920.13528

Sharma A, Bhardwaj RD, Gupta AK (2015) Ferulic acid: a novel inducer of antioxidant enzymes in wheat (Triticum aestivum L.) seedlings. Cereal Res Commun 43(3):394–402

Sharma A, Shahzad B, Rehman A et al (2019) Response of Phenylpropanoid pathway and the role of polyphenols in plants under abiotic stress. Molecules 24:2452. https://doi.org/10.3390/MOLECULES24132452

Sharma H, Kumari A, Raigar OP et al (2023) Strategies for improving tolerance to the combined effect of drought and salinity stress in crops. In: Salinity and drought tolerance in plants. Springer, Singapore

Shi M, Zhu R, Zhang Y et al (2022) A novel WRKY34-bZIP3 module regulates phenolic acid and tanshinone biosynthesis in Salvia miltiorrhiza. Metab Eng 73:182–191. https://doi.org/10.1016/j.ymben.2022.08.002

Shirai A, Kunimi H, Tsuchiya K (2022) Antifungal action of the combination of ferulic acid and ultraviolet-A irradiation against Saccharomyces cerevisiae. J Appl Microbiol 132(4):2957–2967. https://doi.org/10.1111/jam.15407

Shu P, Li Y, Wang X et al (2021) Exogenous ferulic acid treatment increases resistance against Botrytis cinerea in tomato fruit by regulating nitric oxide signaling pathway. Postharvest Biol Technol 182:111678. https://doi.org/10.1016/j.postharvbio.2021.111678

Shu P, Li Y, Li Z et al (2022) Ferulic acid enhances chilling tolerance in tomato fruit by up-regulating the gene expression of CBF transcriptional pathway in MAPK3-dependent manner. Postharvest Biol Technol 185:111775. https://doi.org/10.1016/j.postharvbio.2021.111775

Shu P, Li Y, Xiang L et al (2023) SlNPR1 modulates chilling stress resistance in tomato plant by alleviating oxidative damage and affecting the synthesis of ferulic acid. Sci Hortic 307:111486. https://doi.org/10.1016/j.scienta.2022.111486

Singh K, Gupta R, Shokat S et al (2024) Ascorbate, plant hormones and their interactions during plant responses to biotic stress. Physiol Plant 176:1–16. https://doi.org/10.1111/ppl.14388

Smith RA, Lu F, Muro-Villanueva F, Cusumano JC, Chapple C, Ralph J (2022) Manipulation of lignin monomer composition combined with the introduction of monolignol conjugate biosynthesis leads to synergistic changes in lignin structure. Plant Cell Physiol 63(6):744–754

Song Q, Ji K, Yu X et al (2022) Dynamic metabolic and transcriptomic profiling reveal synthetic characters and regulators of flavonoid biosynthesis in Camellia oleifera seeds. Ind Crop Prod 186:115295. https://doi.org/10.1016/j.indcrop.2022.115295

Song W, Xin J, Yu C, Xia C, Pan Y (2023) Alkyl ferulic acid esters: evaluating their structure and antibacterial properties. Front Microbiol 14:1135308

Soussani FE, Boutasknit A, Ben-laouane R, Benkirane R (2023) Arbuscular Mycorrhizal fungi and compost-based biostimulants enhance fitness, physiological responses, yield, and quality traits of drought-stressed tomato plants. Plants 12:1856

Srinivasan M, Sudheer AR, Menon VP (2007) Ferulic acid: therapeutic potential through its antioxidant property. J Clin Biochem Nutr 40(2):92. https://doi.org/10.3164/JCBN.40.92

Stan CD, Dragan M, Iacob AT, Profire L (2016) Assessment of in vitro antioxidant activity of some new Ferulic acid derivatives. Med Surg J 120(3):727–731. https://revmedchir.ro/index.php/revmedchir/article/view/234

Sun N, Wang Y, Kang J, Hao H, Liu X, Yang Y, Jiang X, Gai Y (2024) Exploring the role of the LkABCG36 transporter in lignin accumulation. Plant Sci 343:112059

Szparaga A, Kocira S, Kocira A, Czerwińska E, Świeca M, Lorencowicz E, Kornas R, Koszel M, Oniszczuk T (2018) Modification of growth, yield, and the nutraceutical and antioxidative potential of soybean through the use of synthetic biostimulants. Front Plant Sci 9. https://doi.org/10.3389/fpls.2018.01401

Ting-ting Y, Kun D, Zeng-peng GUO, Yan D (2020) Allelopathic effects of ferulic acid inducing Fusarium wilt occurrence and abnormal root tissue structure of faba bean. J Plant Nutr Fertiliz 26(5):914–923. https://doi.org/10.11674/zwyf.19388

Tripathi D, Singh M, Pandey-Rai S (2022) Crosstalk of nanoparticles and phytohormones regulate plant growth and metabolism under abiotic and biotic stress. Plant Stress 6:100107. https://doi.org/10.1016/j.stress.2022.100107

Truong DH, Nhung NTA, Dao DQ (2020) Iron ions chelation-based antioxidant potential vs. pro-oxidant risk of ferulic acid: a DFT study in aqueous phase. Comput Theor Chem 1185:112905. https://doi.org/10.1016/J.COMPTC.2020.112905

Tu Q-B, Shi H-C, Li P, Sheng S, Wu F-A (2022) Antibacterial activity of Ferulic acid Ester against Ralstonia solanacearum and its synergy with essential oils. Sustainability 14(24):Article 24. https://doi.org/10.3390/su142416348

Vanholme R, De Meester B, Ralph J, Boerjan W (2019) Lignin biosynthesis and its integration into metabolism. Curr Opin Biotechnol 56:230–239

Vega A, Delgado N, Handford M (2022) Increasing heavy metal tolerance by the exogenous application of organic acids. Int J Mol Sci 23(10):5438

Verma V, Ravindran P, Kumar PP (2016) Plant hormone-mediated regulation of stress responses. BMC Plant Biol 16(1):86. https://doi.org/10.1186/s12870-016-0771-y

Wahab A, Abdi G, Saleem MH et al (2022) Alleviate the adverse effects of drought stress: a comprehensive review. Plants 11:1620

Wakabayashi K, Soga K, Hoson T (2012) Phenylalanine ammonia-lyase and cell wall peroxidase are cooperatively involved in the extensive formation of ferulate network in cell walls of developing rice shoots. J Plant Physiol 169(3):262–267. https://doi.org/10.1016/j.jplph.2011.10.002

Wallace G, Fry SC (1995) In vitro peroxidase-catalysed oxidation of ferulic acid esters. Phytochemistry 39(6):1293–1299. https://doi.org/10.1016/0031-9422(95)00048-C

Wan YY, Zhang Y, Zhang L, Zhou ZQ, Li X, Shi Q, Wang XJ, Bai JG (2015) Caffeic acid protects cucumber against chilling stress by regulating antioxidant enzyme activity and proline and soluble sugar contents. Acta Physiol Plant 37(1):1–10. https://doi.org/10.1007/S11738-014-1706-6/METRICS

Wang D, Sun L, Yin Z et al (2022) Insights into genomic evolution and the potential genetic basis of Klebsiella variicola subsp. variicola ZH07 reveal its potential for plant growth promotion and autotoxin degradation. Microbiol Spectrum 10:e00846–e00822. https://doi.org/10.1128/spectrum.00846-22

White CD, Sacks DB (2010) Regulation of MAP kinase signaling by calcium. Methods Mol Biol (Clifton, NJ) 661:151–165. https://doi.org/10.1007/978-1-60761-795-2_9

Wu D, Cui M, Hao Y, Liu L, Zhou Y, Wang W, Xue A, Chingin K, Luo L (2019) In situ study of metabolic response of *Arabidopsis thaliana* leaves to salt stress by neutral desorption-extractive electrospray ionization mass spectrometry. J Agric Food Chem 67(46):12945–12952. https://doi.org/10.1021/acs.jafc.9b05339

Xiao R, Zhang C, Guo X et al (2021) MYB transcription factors and its regulation in secondary cell wall formation and lignin biosynthesis during xylem development. Int J Mol Sci 22:3560. https://doi.org/10.3390/ijms22073560

Xu J-J, Fang X, Li C-Y, Zhao Q, Martin C, Chen X-Y, Yang L (2018) Characterization of Arabidopsis thaliana Hydroxyphenylpyruvate reductases in the tyrosine conversion pathway. Front Plant Sci 9:1305. https://doi.org/10.3389/fpls.2018.01305

Yadav V, Wang Z, Wei C, Amo A, Ahmed B, Yang X, Zhang X (2020) Phenylpropanoid pathway engineering: An emerging approach towards plant defense. Pathogens 9(4):312

Yan H, Meng X, Lin X, Duan N, Wang Z, Wu S (2023) Antifungal activity and inhibitory mechanisms of ferulic acid against the growth of *Fusarium graminearum*. Food Biosci 52: 102414. https://doi.org/10.1016/j.fbio.2023.102414

Yang S, Zheng Y, Guo Y, Cen Z, Dong Y (2023) Allelopathic effect of phenolic acids in various extracts of wheat against Fusarium wilt in faba bean. Funct Plant Biol 50(12):1062–1072

Yaqoob U, Jan N, Raman PV et al (2022) Crosstalk between brassinosteroid signaling, ROS signaling and phenylpropanoid pathway during abiotic stress in plants: does it exist? Plant Stress 4:100075. https://doi.org/10.1016/J.STRESS.2022.100075

Yıldız Ş, Küçükrecep A, Tekdal D (2024) Physiological and molecular responses of nanoparticle application to plants grown under abiotic and biotic stress. In: Studies in natural products chemistry. Elsevier, Amsterdam, The Netherlands

Yildiztugay E, Ozfidan-Konakci C, Karahan H et al (2019) Ferulic acid confers tolerance against excess boron by regulating ROS levels and inducing antioxidant system in wheat leaves (Triticum aestivum). Environ Exp Bot 161:193–202. https://doi.org/10.1016/j.envexpbot.2018.10.029

Zamljen T, Medic A, Hudina M, Veberic R, Slatnar A (2022) Salt stress differentially affects the primary and secondary metabolism of peppers (Capsicum annuum L.) according to the genotype, fruit part, and salinity level. Plants 11(7):853

Zduńska K, Dana A, Kolodziejczak A, Rotsztejn H (2018) Antioxidant properties of ferulic acid and its possible application. Skin Pharmacol Physiol 31:332–336. https://doi.org/10.1159/000491755

Zhang Q, Li M, Yang G, Liu X, Yu Z, Peng S (2022) Protocatechuic acid, ferulic acid and relevant defense enzymes correlate closely with walnut resistance to Xanthomonas arboricola pv. Juglandis. BMC Plant Biol 22(1):598. https://doi.org/10.1186/s12870-022-03997-9

Zhang W, Jiang Y, Zhang Z (2023a) The role of different natural organic acids in postharvest fruit quality management and its mechanism. Food Front 4:1127–1143. https://doi.org/10.1002/fft2.245

Zhang Z, Yang W, Li Y, Zhao Q, Dong Y (2023b) Wheat–faba bean intercropping can control Fusarium wilt in faba bean under F. commune and ferulic acid stress as revealed by histopathological analysis. Physiol Mol Plant Pathol 124:101965

Chapter 12
Gallic Acid: Role in Plant Defense Mechanisms Against Stress

Saleha Saeed and Sami Ullah

Abstract Gallic acid, a naturally occurring phenolic compound, plays a pivotal role in enhancing plant defense mechanisms against abiotic and biotic stresses. This chapter explores the benefits of exogenous application of gallic acid to improve plant resilience, growth, and productivity under stressful conditions. Under abiotic stresses such as drought, salinity, extreme temperatures, and heavy metal toxicity, gallic acid mitigates oxidative damage, stabilizes cellular membranes, and modulates ion homeostasis. In biotic stress contexts, it exhibits antimicrobial activity, triggers induced systemic resistance (ISR), and alters secondary metabolite production to strengthen plant immunity. Mechanistic insights reveal gallic acid's influence on gene expression, hormonal signaling, and possibly epigenetic regulation. Its application in biostimulants, nanotechnology-enabled delivery, and seed priming shows promise in sustainable agriculture. Future research should optimize its practical use and explore synergistic effects with other bioactives.

Keywords Gallic acid · Abiotic stress · Biotic stress · Oxidative damage · Ion homeostasis · Secondary metabolites · ISR · Sustainable agriculture

12.1 Introduction

Abiotic stresses, including drought, salinity, extreme temperatures, heavy metal exposure, flooding, UV radiation, and nutrient deficiency, negatively affect plant growth, productivity, and survival. These stresses primarily cause oxidative damage via excessive reactive oxygen species (ROS), osmotic imbalance, disruption in ion homeostasis, and metabolic irregularities (Dikilitas et al. 2021; Imran et al. 2022; Shikari et al. 2022; Abdel Latef 2024; Kaplan et al. 2025; Abdel Latef et al. 2026; Barman et al. 2026).

S. Saeed · S. Ullah (✉)
Department of Botany, University of Peshawar, Peshawar, Pakistan
e-mail: saleha362026@uop.edu.pk; samibotany@uop.edu.pk

A. A. H. Abdel Latef (ed.), *Harnessing Vitamins and Organic Acids for Plant Stress Resilience*, Sustainability Sciences in Asia and Africa,
https://doi.org/10.1007/978-981-95-8593-9_12

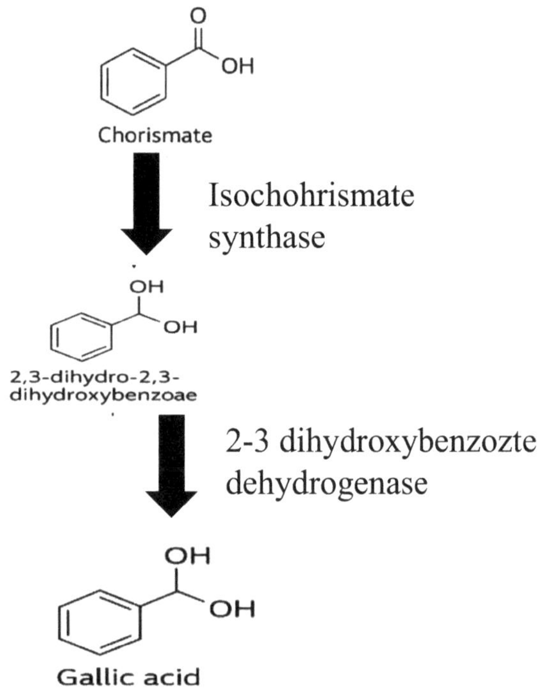

Fig. 12.1 Biosynthesis of gallic acid via the shikimate pathway in plants

Gallic acid (GA; 3,4,5-trihydroxybenzoic acid) is a plant-derived phenolic compound commonly found in fruits, vegetables, seeds, and medicinal plants. It plays key roles in antioxidant defense, growth regulation, and stress adaptation. Synthesized via the shikimate pathway (Fig. 12.1), it serves as a precursor for tannins, flavonoids, and lignins (Jalil et al. 2020). Importantly, this biosynthesis involves dehydroxyshikimate conversion by gallic acid to shikimate dehydrogenase, which also contributes to downstream hydrolysable tannin formation (Mora et al. 2022). Given the increasing global incidence of abiotic and biotic stresses, gallic acid is being investigated for its ability to bolster stress tolerance in crops (Hasanuzzaman et al. 2014).

This chapter discusses the physiological, biochemical, and molecular roles of gallic acid in plant defense, supported by species-specific studies. The aims of this chapter are as follows: (1) to explore the multifunctional role of gallic acid as a natural phenolic compound in enhancing plant defense mechanisms against abiotic and biotic stresses; (2) to analyze the biochemical, physiological, and molecular mechanisms through which gallic acid mitigates stress-induced damage in plants; (3) to explain how exogenous application of gallic acid improves plant resilience, growth, and productivity under stress conditions such as drought, salinity, extreme temperatures, and heavy metal toxicity; and (4) to evaluate gallic acid's role in plant immunity, including its antimicrobial properties, induction of systemic resistance, and modulation of secondary metabolites for pathogen and pest defense.

Gallic acid (GA), a trihydroxybenzoic acid, plays a protective role by functioning as a natural antioxidant, osmoprotectant, metal chelator, and signaling modulator, enhancing plants' physiological and biochemical stress responses.

12.1.1 Gallic Acid and Drought Stress

Drought stress significantly limits water availability, reducing cell turgor, causing osmotic imbalance, and triggering oxidative stress that results in wilting and impaired growth (Dawood et al. 2022, 2025). Exogenous application of GA alleviates drought-induced damage via multiple protective mechanisms:

- *Antioxidant Activation*

GA enhances the activities of key antioxidant enzymes. It boosts superoxide dismutase (SOD) levels—converting superoxide radicals into hydrogen peroxide—as well as catalase (CAT) and peroxidase (POD), which detoxify hydrogen peroxide and reduce lipid peroxidation, thereby preserving membrane integrity (Khan et al. 2023; El-Mahdy et al. 2025).

- *Osmolyte Accumulation*

GA promotes the accumulation of osmolytes like proline and glycine betaine, which help maintain osmotic balance, stabilize proteins and membranes, and contribute to ROS scavenging (Martínez et al. 2023).

- *Physiological Benefits*

GA treatment improves relative water content (RWC), a marker of better hydration, and enhances membrane stability, lowering electrolyte leakage. Notably, GA-treated wheat, soybean, and barley plants under drought stress showed marked improvements in RWC, photosynthetic efficiency, and antioxidant enzyme activities (Khan et al. 2023; El-Mahdy et al. 2025). Additionally, studies reveal that wild barley genotypes under drought exhibit substantial increases in gallic acid levels and antioxidant capacity, underscoring GA's intrinsic protective role (Sato et al. 2025).

12.1.2 Gallic Acid and Salinity Stress

Salinity stress results in excessive uptake of sodium (Na^+) and chloride (Cl^-), leading to ion toxicity, osmotic imbalance, and nutrient disruption (Osman et al. 2021; Tahjib-Ul-Arif et al. 2023; Dawood et al. 2024; Azooz et al. 2026). Exogenous gallic acid (GA) helps plants cope with salinity through multiple key mechanisms:

- *Ion Homeostasis*

GA maintains a favorable Na^+/K^+ balance, helping to modulate ion transporters and pumps (like H^+-ATPase and components of the salt overly sensitive (SOS) signaling pathway). In *Lepidium sativum*, GA treatment reduced Na^+ uptake and increased K^+ levels, improving the K^+/Na^+ ratio, which is critical for photosynthesis and enzyme function (Martínez et al. 2023). In *Oryza sativa* (rice), GA application enhanced ion regulation and osmoregulation under salt stress (Sato et al. 2025).

- *Antioxidant Defense*

GA strengthens antioxidant systems, reducing reactive oxygen species like hydrogen peroxide (H_2O_2) and malondialdehyde (MDA), while boosting activities of detoxifying enzymes and elevating phenolic and glutathione content (Sato et al. 2025).

- *Photosynthesis and Growth*

GA preserves chlorophyll content and supports photosynthetic efficiency under saline conditions. Treated plants showed increased biomass, photosynthetic pigments, and better growth compared to untreated controls (Sato et al. 2025).

- *Root Vigor and Nutrient Uptake*

By improving ion balance and antioxidant status, GA enhances root health, enabling more efficient water and nutrient uptake under stress (Sato et al. 2025).

12.1.3 Gallic Acid and Waterlogging/Flooding Stress

Flooding or waterlogging creates hypoxic or anoxic conditions that suppress aerobic respiration and increase reactive oxygen species (ROS), compromising plant growth and survival (Martínez et al. 2023; Tahjib-Ul-Arif et al. 2023; Bandehagh et al. 2025; Iqbal et al. 2025). Exogenous application of gallic acid (GA) helps alleviate these effects through:

- *Fermentative Metabolism*

GA enhances the activity of fermentative enzymes like alcohol dehydrogenase (ADH) and lactate dehydrogenase (LDH), thereby supporting continued ATP production under low-oxygen conditions, improving energy balance and stress resilience (Meena et al. 2024).

- *ROS Management*

GA upregulates ROS-scavenging enzymes, including ascorbate peroxidase (APX) and glutathione reductase (GR), thereby maintaining redox homeostasis and preventing cellular damage.

- *Anatomical Adaptation*

GA promotes the formation of aerenchyma tissue in roots, facilitating oxygen diffusion to submerged tissues and enhancing survival in flooded soils (Meena et al. 2024).

- *Phenolic Accumulation*

Additional evidence shows flooding-induced increases in gallic acid and total phenolic content in leaves, novel antioxidant responses that may contribute to tolerance under hypoxic stress (Liu et al. 2021).

12.1.4 Gallic Acid and Temperature Extremes

Temperature extremes disrupt cellular homeostasis by altering membrane fluidity, protein folding, and metabolic enzyme activities (Abdel Latef and Chaoxing 2011; Haque et al. 2021; Hossain et al. 2021; Bidabadi and Abdel Latef 2022; Khan et al. 2025; Kassout et al. 2025; Ahmad et al. 2026).

Cold Stress
Gallic acid induces cold-responsive (COR) genes such as CBF/DREB, triggering osmoprotectant synthesis and production of stabilizing proteins. It also promotes the accumulation of sugars and amino acids that protect cellular structures during freezing (Babaei et al. 2022). Additionally, GA levels were found to increase significantly in cold-tolerant oak species, correlating with elevated antioxidant activity and higher survival rates under low-temperature exposure (*Quercus wutaishanica* vs. *Q. acutissima*) (Zhang et al. 2023).

Heat Stress
GA stimulates heat shock proteins (HSPs), functioning as molecular chaperones to refold denatured proteins and preserve the proteome under high temperatures. It also mitigates lipid peroxidation, maintaining membrane integrity (Babaei et al. 2022). Moreover, GA enhances biosynthesis of chlorophyll and osmoprotectant proline—crucial for maintaining photosynthesis and cellular homeostasis under heat stress (Khan et al. 2023).

12.1.5 Gallic Acid and Heavy Metal Stress

Heavy metals (e.g., Cd, Pb, Zn) cause toxicity by inducing oxidative stress, interfering with essential nutrient uptake, and binding to cellular components (Abdel Latef 2013; Ahmad et al. 2016; Abdel latef et al. 2017; Hakla et al. 2021; Abeed et al. 2022; Bhardwaj and Garg 2022; Fakhar et al. 2022; Adnan et al. 2025, 2026). Gallic acid protects via:

Chelation of Metals
GA contains hydroxyl groups that bind with metal ions, reducing their bioavailability and toxicity (Khan et al. 2023).

Detoxification Enzyme Induction
Increases synthesis of: phytochelatins (PCs) and glutathione-S-transferases (GSTs), which detoxify and compartmentalize metals into vacuoles (Khan et al. 2023).

Growth Protection
Enhances chlorophyll retention, photosynthetic function, and root elongation under metal stress. Colak et al. (2021) found reduced cadmium accumulation and improved plant growth in spinach with gallic acid application.

12.1.6 Gallic Acid and UV/Ozone/Mineral Deficiency

Ozone Stress
Tropospheric ozone is a strong oxidant that induces lipid peroxidation, chlorophyll degradation, and reduced photosynthetic activity in plants (Alam et al. 2021; Hasan et al. 2022; Pazarlar 2022). In model systems, gallic acid (GA) demonstrated protective antioxidant effects by inhibiting ozone-induced oxidation of chloroplast membrane lipids such as galactolipids. When combined with α-tocopherol, GA helped regenerate tocopheroxyl radicals and stabilized lipid layers, thus potentially mitigating membrane damage (Gubina et al. 2016). Although these findings are from in vitro and cell culture systems, they suggest GA's possible role in enhancing ozone stress tolerance.

UV Radiation Stress
UV-B radiation generates reactive oxygen species (ROS) and damages DNA, proteins, and membranes in plants (Shan et al. 2022; Aktar et al. 2025). While many phenolic compounds, such as flavonoids, act as UV filters and antioxidants, GA itself appears more prone to degradation under UV exposure. Studies show that GA undergoes rapid oxidative breakdown in UV–Fe^{3+} systems via ROS-mediated reactions (Xie et al. 2014). In microbial models, UV-A exposure increased GA uptake but led to elevated ROS production and oxidative stress, indicating a potential prooxidant effect rather than protective action (Kwon et al. 2017). These findings suggest that GA may not serve as a primary sunscreen compound in plants under UV stress.

Mineral Deficiency
Deficiency of micronutrients such as Fe, Zn, and Mn disrupts chlorophyll biosynthesis, enzyme activity, and plant growth (Assad et al. 2025). Although GA is known to chelate transition metals, no direct experimental evidence demonstrates its role in improving nutrient solubility, root proliferation, or uptake efficiency in plants under mineral deficiency stress. Current literature focuses on synthetic chelators or organic

acids such as citric acid and ethylenediaminetetraacetic acid (EDTA) for improving micronutrient acquisition (Alloway 2008), leaving GA's role largely unexplored.

12.2 Role of Gallic Acid in Biotic Stress Resistance

Biotic stresses in plants are caused by pathogens (fungi, bacteria, viruses) and pests (insects, nematodes, herbivores). These stresses disrupt cellular functions, reduce photosynthesis, and compromise overall plant productivity (Afzal et al. 2026). Gallic acid enhances resistance to biotic stress by acting as a direct antimicrobial agent, priming the plant immune system, and modifying metabolic pathways associated with defense.

12.2.1 Antimicrobial Properties

Gallic acid exhibits strong antimicrobial activity against a broad spectrum of pathogens. It operates through several mechanisms: membrane disruption: gallic acid integrates into microbial membranes, causing increased permeability and leakage of cellular contents, leading to cell death. This is particularly effective against Gram-negative bacteria and certain fungi (Gantner et al. 2021). Phenolic and flavonoid induction: application of gallic acid boosts the biosynthesis of antimicrobial secondary metabolites, including phenolics and flavonoids, which inhibit pathogen colonization and spread (Mierziak et al. 2021). Enzyme activation: gallic acid stimulates key defense-related enzymes such as phenylalanine ammonia-lyase (PAL) and polyphenol oxidase (PPO), both essential for lignin formation and phenolic metabolism. These enzymes enhance structural barriers and antimicrobial compound production. *Example:* In tomato plants challenged with *Pseudomonas syringae*, exogenous gallic acid increased PAL activity by over 45%, resulting in reduced bacterial colonization (Singh et al. 2022).

12.2.2 Induced Systemic Resistance (ISR)

Gallic acid primes the plant's innate immune system, facilitating a faster and stronger response upon pathogen attack. This phenomenon, known as induced systemic resistance (ISR), involves: hormonal signaling: gallic acid modulates the levels and signaling pathways of salicylic acid (SA) and jasmonic acid (JA). SA is typically involved in resistance to biotrophic pathogens, whereas JA mediates responses against necrotrophic pathogens and herbivores. Pathogenesis-related (PR) proteins: It induces the expression of PR proteins such as chitinases and β-1,3-glucanases, which break down fungal cell walls and enhance pathogen clearance. ROS and nitric oxide (NO) burst: gallic acid facilitates controlled production of reactive oxygen species (ROS) and NO, triggering the hypersensitive response (HR)–a rapid, localized cell death that restricts pathogen spread (Colak et al. 2021). *Example:* In *Arabidopsis thaliana*, gallic acid treatment resulted in early activation of NPR1 and PR1 genes, key markers of SA-mediated ISR, within 12 h of fungal exposure.

12.2.3 Insect and Herbivore Defense

Gallic acid also strengthens plants against insect and herbivore attacks via both direct toxicity and signaling-based defenses: alkaloid and tannin biosynthesis: gallic acid upregulates biosynthetic genes involved in the production of bitter or toxic compounds (e.g., tannins, alkaloids) that deter herbivory and inhibit insect growth. Volatile organic compounds (VOCs): It enhances the emission of plant volatiles like green leaf volatiles (GLVs) and terpenes that attract natural predators of herbivorous insects (Chen et al. 2021). This indirect defense mechanism reduces pest populations through ecological interactions. Lignin and cell wall fortification: Through activation of PAL and related enzymes, gallic acid contributes to lignin accumulation, making plant tissues mechanically tougher and more resistant to chewing or piercing insects. *Example:* In maize, foliar application of gallic acid reduced *Spodoptera littoralis* larval feeding by 38% compared to untreated controls, coinciding with increased leaf lignin content and tannin levels (Ahmed et al. 2022).

12.3 Mechanistic Insights into Gallic Acid Action

Gallic acid functions not only as a direct antioxidant but also as a signaling molecule capable of regulating plant responses at multiple molecular levels. These include transcriptional regulation, hormonal signaling crosstalk, epigenetic remodeling, and proteomic restructuring. The mechanistic insights into how gallic acid acts provide a clearer understanding of its role in priming plant defense and enhancing stress tolerance (Khan et al. 2023).

12.3.1 Gene Expression Regulation

Gallic acid influences the transcription of stress-responsive genes, facilitating adaptive responses under various abiotic and biotic stresses. Transcription Factors Modulated:

WRKY: A major family of transcription factors that regulate genes involved in abiotic stress (drought, salinity) and pathogen defense.

MYB: Controls phenylpropanoid biosynthesis, ROS detoxification, and ABA-responsive pathways. These transcription factors are activated by gallic acid through redox-sensitive signaling and stress-induced cascades. Antioxidant enzyme-encoding genes: GA induces transcription of genes encoding superoxide dismutase (SOD), catalase (CAT), and peroxidases, enhancing cellular ROS-scavenging capacity. Heat shock proteins (HSPs): HSP70, HSP90, and small HSPs are upregulated under heat and oxidative stress upon GA treatment, aiding in protein folding and repair. Singh et al. (2017) showed that GA upregulated stress-related gene expression in Arabidopsis and rice, leading to enhanced heat and drought resilience.

12.3.2 Hormonal Crosstalk

Plant stress responses are tightly regulated by a network of phytohormones. Gallic acid interacts with and modulates this network to fine-tune plant defenses. Abscisic acid (ABA): Central hormone in drought and salinity responses. GA modulates ABA signaling by enhancing ABA-responsive gene expression, improving stomatal regulation, and water retention. Salicylic acid (SA): key hormone in biotic stress, especially systemic acquired resistance (SAR). GA enhances SA-dependent pathways, promoting expression of PR (pathogenesis-related) genes and bolstering pathogen resistance. Jasmonic acid (JA) and ethylene (ET): primarily involved in wounding and insect/pathogen resistance, triggering induced systemic resistance (ISR). GA potentiates JA/ET signaling, likely by affecting redox status and secondary metabolite biosynthesis. Hormonal synergy: GA acts as a modulator, balancing SA and JA/ET responses–which are often antagonistic–depending on the stress type. Sun et al. (2018) reported that GA amplified SA-triggered defenses and promoted ISR in Arabidopsis challenged with *Pseudomonas syringae.*

12.3.3 Epigenetic Regulation

Epigenetic mechanisms are heritable changes in gene expression that do not alter the DNA sequence. Gallic acid influences these mechanisms, contributing to stress memory and priming. Histone modifications: GA treatment is linked to increased histone acetylation (e.g., H3K9ac, H4K5ac) at stress-responsive loci, resulting in open chromatin and active transcription. DNA methylation: changes in cytosine methylation patterns under GA influence may regulate gene silencing or activation, especially in long-term or repeated stress exposures. Stress memory: plants exposed to GA exhibit faster and stronger responses upon recurrent stress, suggesting a form of "priming" through epigenetic imprinting. Wild et al. (2012) observed that GA-like

Table 12.1 Mechanistic layers of gallic acid action

Mechanism	Details	Outcome	References
Gene expression	WRKY, MYB, SOD, CAT, HSPs	Activation of stress defense genes	Li et al. (2025)
Hormonal crosstalk	ABA (drought), SA (SAR), JA/ET (ISR)	Coordinated defense responses	Wasternack and Song (2017)
Epigenetic regulation	Histone acetylation, DNA methylation	Long-term stress memory, priming	Lämke and Bäurle (2017)
Proteome and membrane	MAPK, Ca^{2+}, HSPs, PR proteins	Signal transduction, membrane stabilization	Zhao et al. (2014)

compounds induced stress memory via epigenetic markers, enhancing tolerance upon subsequent exposures.

12.3.4 Membrane and Proteome Responses

Gallic acid also acts through signal transduction pathways, altering protein expression and cellular responses to strengthen plant defenses. MAPK signaling cascade: GA activates mitogen-activated protein kinases (MAPKs), which relay extracellular stress signals to the nucleus, triggering transcription of defense genes. Calcium (Ca^{2+}) signaling: GA induces Ca^{2+} influx, activating calcium-binding proteins (like calmodulins) and Ca^{2}-dependent protein kinases (CDPKs), key players in ROS signaling and membrane repair. Proteomic shifts: GA modifies the abundance and post-translational modification of several stress-related proteins, including: heat shock proteins (HSPs), pathogenesis-related (PR) proteins, aquaporins (for water transport), and defense enzymes (e.g., glutathione S-transferases). Membrane integrity: GA stabilizes plasma membranes under stress conditions by preventing lipid peroxidation and maintaining protein–lipid interactions. Zhao et al. (2021) identified proteomic changes in GA-treated plants under salinity and heat, including upregulation of HSPs and ROS-detoxifying enzymes (Table 12.1).

12.4 Practical Applications and Future Prospects

Gallic acid holds significant promise as a bioactive compound in sustainable agriculture due to its antioxidant, antimicrobial, and regulatory roles in stress adaptation. Its multifunctionality opens avenues for diverse applications—from preharvest foliar treatments to postharvest preservation—while ongoing research continues to explore advanced delivery systems, genetic strategies, and long-term ecological implications.

12.4.1 Biostimulants and Foliar Sprays

Gallic acid-based biostimulants can be applied to crops via foliar sprays or soil drenches, often combined with other phytoactive compounds or micronutrients. Mechanisms: Enhances photosynthetic efficiency, chlorophyll retention, and ROS-scavenging enzyme activities under drought, heat, or salinity. Promotes osmolyte accumulation (e.g., proline) for better water retention and membrane stability. Stimulates root development and nutrient uptake, supporting plant vigor even under suboptimal conditions. Crops studied: wheat, maize, rice, and soybean have shown significant yield enhancement, improved tolerance to abiotic stress, and better reproductive fitness under foliar GA treatments. *Example:* Foliar spray of 100 μM gallic acid improved grain filling, shoot biomass, and water use efficiency in drought-stressed wheat (Khan et al. 2023).

12.4.2 Nanotechnology Delivery Systems

Nanocarriers, such as chitosan nanoparticles, liposomes, mesoporous silica nanoparticles (MSNs), and carbon nanotubes, can encapsulate gallic acid for targeted and sustained delivery. Advantages: controlled release and enhanced bioavailability at target sites (e.g., root zone, stomatal pores). Protects gallic acid from premature degradation by light or enzymes. Enables lower doses with higher efficacy, reducing environmental contamination. Research gaps: Limited in vivo studies in plants. Need for optimization of size, charge, and surface modifications for plant uptake. Evaluation of toxicity and biodegradability of nanomaterials. *Example:* Experimental studies with chitosan-GA nanocomposites showed enhanced drought tolerance and disease resistance in Arabidopsis and tomato models (Zhao et al. 2014).

12.4.3 Genetic Engineering

Boost endogenous production of gallic acid through metabolic engineering of the shikimate pathway, the biosynthetic route leading to gallic acid and other phenolics. Overexpression of key enzymes such as 3-dehydroshikimate dehydratase (converts 3-dehydroshikimate to gallic acid), Shikimate kinase and chorismate synthase. Introduction of feedback-insensitive regulatory elements to increase phenolic flux. Creation of transgenic plants capable of self-synthesizing gallic acid at elevated levels, reducing reliance on exogenous application. Potential for enhanced innate stress tolerance, antioxidant activity, and disease resistance. *Example:* Experimental rice lines overexpressing shikimate-pathway genes showed elevated phenolic content and improved drought tolerance (Zhao et al. 2014).

12.4.4 Postharvest Applications

Perishable fruits and vegetables suffer from rapid microbial spoilage and oxidative degradation, reducing shelf life and market value. Role of gallic acid: acts as a natural antimicrobial, inhibiting the growth of spoilage organisms such as *Botrytis cinerea*, *Penicillium* spp., and *Aspergillus* spp. Prevents lipid peroxidation and maintains cell membrane integrity, slowing tissue softening and discoloration. Functions as a bio-preservative, safe for human consumption, and biodegradable. Tested commodities tomatoes, grapes, strawberries, and apples have shown significantly reduced spoilage and improved postharvest quality with gallic acid treatments. Nahar et al. (2016) demonstrated that a 200 ppm GA dip reduced fungal decay and retained firmness in table grapes stored for 14 days.

12.4.5 Research Needs and Future Directions

Despite its proven efficacy, several research gaps remain that must be addressed before gallic acid can be widely adopted in agriculture:

1. Clarifying Epigenetic Roles. More work is needed to decipher how GA modulates histone marks and DNA methylation under field conditions. Is GA a true priming agent, inducing heritable "stress memory" in crops?
2. Hormonal Crosstalk in Complex Stress. Detailed investigations on how GA interacts with phytohormones (ABA, SA, JA/ET) under combined abiotic–biotic stresses. Understanding context-specific responses will improve its reliability as a biostimulant.
3. Dose Optimization Across Species. Optimal concentration ranges vary with crop, stress type, growth stage, and environmental factors. Excessive GA can have pro-oxidant effects, so standardized protocols are needed.
4. Soil and Microbiome Impact. Long-term field trials needed to assess: Effects on rhizosphere microbial communities, potential for allelochemical interference or changes in soil enzyme activities. Ensuring GA applications do not disturb soil health is critical for sustainability.

12.5 Conclusion

Gallic acid emerges as a promising multifunctional compound enhancing plant defense against abiotic and biotic stresses. Its antioxidant, antimicrobial, and regulatory roles strengthen tolerance across diverse conditions. From drought mitigation to pathogen resistance, it influences molecular, biochemical, and physiological pathways. Integrating gallic acid into modern agriculture—via foliar sprays, biostimulants, or engineered biosynthesis—could support sustainable crop protection. Further studies will help refine its utility and mechanism of action for broad agricultural adoption.

References

Abdel Latef AAH (2013) Growth and some physiological activities of pepper (Capsicum annuum L.) in response to cadmium stress and mycorrhizal symbiosis. Journal of Agricultural Science and Technology. 15:1437–1448

Abdel Latef AHA (2024) Biochar in mitigating abiotic stress in plants. Academic Press. Elsevier Inc. pp. 1–375. https://doi.org/10.1016/C2023-0-01389-9

Abdel Latef AAH, Chaoxing H (2011) Arbuscular mycorrhizal influence on growth, photosynthetic pigments, osmotic adjustment and oxidative stress in tomato plants subjected to low temperature stress. Acta Physiologia Plantarum 33:1217–1225

Abdel Latef AAH, Alhmad MF and Abdelfattah KE (2017) The possible roles of priming with ZnO nanoparticles in mitigation of salinity stress in Lupine (Lupinus termis) plants. Journal of Plant Growth Regulation. 36: 60-70

Abdel Latef AA, Hameed M, Khan Z, Khan S, Imran A, Fatima N (2026) Foliar stimulants: a sustainable strategy for enhancing plant resilience to abiotic and biotic stresses. Physiologia Plantarum 178(1):e70741

Abeed AHA, Mahdy RE, Alshehri D, Hammami I, Eissa MA, Abdel Latef AA, Mahmoud GA (2022) Induction of resilience strategies against biochemical deteriorations prompted by severe cadmium stress in sunflower plant when Trichoderma and bacterial inoculation were used as biofertilizers. Frontiers in Plant Science. 13: 1004173

Adnan M et al. (2025) Use of Biostimulants to Increase Heavy Metal Tolerance in Oilseed Crops. In: Abdel Latef, A.A.H. (eds) Oilseed Crops Under Abiotic Stress. Sustainability Sciences in Asia and Africa. Springer, Singapore. https://doi.org/10.1007/978-981-96-8346-8_6

Adnan M et al. (2026) Heavy Metals Stress Mitigation by Chitosan Nanoparticles. In: Abdel Latef, A.A.H. (eds) Chitosan Nanoparticles for Plant Resilience Under Abiotic and Biotic Stressors. Smart Nanomaterials Technology. Springer, Singapore. https://doi.org/10.1007/978-981-95-6422-4_7

Afzal A, Ali B, Samarasinghe KGBA, Khan MN, Javed MA, Saleem A (2026) Overview of Biotic Stress and Its Agricultural Impact. In: Abdel Latef, A.A.H. (eds) Chitosan Nanoparticles for Plant Resilience Under Abiotic and Biotic Stressors. Smart Nanomaterials Technology. Springer, Singapore. https://doi.org/10.1007/978-981-95-6422-4_3

Ahmad P, Abdel Latef AAH, Abd_Allah EF, Hashem A, Sarwat M, Anjum NA and Gucel S (2016) Calcium and potassium supplementation enhanced growth, osmolyte secondary metabolite production, and enzymatic antioxidant machinery in cadmium-exposed chickpea (Cicer arietinum L.). Frontiers in Plant Science 7: 513

Ahmad N, Ahmad S, Islam MS, Khan MN, Abdel Latef AAH (2026) Chitosan nanoparticles under temperature extremes. In: Abdel Latef, A.A.H. (eds) Chitosan Nanoparticles for Plant Resilience Under Abiotic and Biotic Stressors. Smart Nanomaterials Technology. Springer, Singapore. https://doi.org/10.1007/978-981-95-6422-4_6

Ahmed S, Ali M, Khan Z (2022) Foliar application of gallic acid enhances tannin–lignin accumulation and reduces *Spodoptera littoralis* feeding in maize. Entomol Exp Appl 170(2):123–135. https://doi.org/10.1111/eea.13145

Aktar SN et al. (2025) Oilseed Crops and Phytohormones Under UV Stress. In: Abdel Latef, A.A. H. (eds) Oilseed Crops Under Abiotic Stress. Sustainability Sciences in Asia and Africa. Springer, Singapore. https://doi.org/10.1007/978-981-96-8346-8_14

Alam et al. (2021) Ozone and Enzymatic and Non-Enzymatic Antioxidant Enzymes in Plants. In: Abdel Latef AAH (ed.) Organic Solutes, Oxidative Stress, and Antioxidant Enzymes under Abiotic Stressors. CRC Press/Taylor & Francis Group. https://doi.org/10.1201/9781003022879-17

Alloway BJ (2008) Zinc in soils and crop nutrition, 2nd edn. International Zinc Association & International Fertilizer Industry Association

Assad N et al. (2025) Phytohormones as Key Players: Boosting Oilseed Crop Tolerance to Nutrient Deficiency Stress. In: Abdel Latef, A.A.H. (eds) Oilseed Crops Under Abiotic Stress. Sustainability Sciences in Asia and Africa. Springer, Singapore. https://doi.org/10.1007/978-981-96-8346-8_13

Azooz MM, Abdel Latef AA, Abdelmajeed WG et al. (2026) Biochar ameliorates salt stress in maize by modulating physiological responses and expression of SOS1 and NHX2 genes. Journal of Soil Science and Plant Nutrition. https://doi.org/10.1007/s42729-026-03123-w

Babaei A, Farajpour M, Rahimian M (2022) Transcriptional upregulation of HSPs and COR genes in gallic acid–treated tomato under temperature extremes. Plant Stress 3:Article 100110. https://doi.org/10.1016/j.stress.2022.100110

Bandehagh A, Dehghanian Z, Taylor NL (2025) Phytohormones in waterlogging and flooding resilience of oilseed crops. In: Abdel Latef, A.A.H. (eds) Oilseed Crops Under Abiotic Stress. Sustainability Sciences in Asia and Africa. Springer, Singapore. https://doi.org/10.1007/978-981-96-8346-8_11

Barman M, Das S, Kundu R, Mukherjee S, Singh SK, Abdel Latef AAH (2026) Overview of abiotic stress and its agricultural impact. In: Abdel Latef, A.A.H. (eds) Chitosan Nanoparticles for Plant Resilience Under Abiotic and Biotic Stressors. Smart Nanomaterials Technology. Springer, Singapore. https://doi.org/10.1007/978-981-95-6422-4_2

Bhardwaj I, Garg N (2022) Cereals and phytohormones under heavy metal stress. In: Abdel Latef, A.A.H. (eds) Sustainable Remedies for Abiotic Stress in Cereals. Springer, Singapore. https://doi.org/10.1007/978-981-19-5121-3_15

Bidabadi SS, Abdel Latef AAH (2022) Cereals and phytohormones under temperature stress. In: Abdel Latef, A.A.H. (eds) Sustainable Remedies for Abiotic Stress in Cereals. Springer, Singapore. https://doi.org/10.1007/978-981-19-5121-3_14

Chen C, Chen H, Huang S, Jiang T, Wang C, Tao Z, He C, Tang Q, & Li P (2021) Volatile DMNT directly protects plants against Plutella xylostella by disrupting the peritrophic matrix barrier in insect midgut. eLife, 10:e63938. https://doi.org/10.7554/eLife.63938

Colak B, Demir O, Yilmaz G (2021) Gallic acid chelation reduces cadmium accumulation and upregulates phytochelatin and GST production in spinach. Environ Exp Bot 182:Article 104110. https://doi.org/10.1016/j.envexpbot.2021.104110

Dawood MFA, Zaid A, Abdel Latef AAH (2022) Salicylic acid spraying-induced resilience strategies against the damaging impacts of drought and/or salinity stress in two varieties of vicia faba L. seedlings. Journal of Plant Growth Regulation 41:1919–1942

Dawood MFA, Tahjib-Ul-Arif M, Sohag AAM, Abdel Latef AAH (2024) Role of Acetic Acid and Nitric Oxide against Salinity and Lithium Stress in Canola (Brassica napus L.). Plants 13(1):51

Dawood MFA, Tahjib-Ul-Arif MD, Shirazy BJ, Abdel Latef AAH (2025) Unraveling the role of ⍰-carrageenan on the combined effect of drought and chromium stress in wheat (Triticum aestivium L.). Plant Physiology and Biochemistry 223:J109781

Dikilitas M, Simsek E, Karakas S, Abdel Latef AAH (2021) Abiotic stresses and their interactions with each other on plant growth, development and defense mechanisms. In: Abdel Latef AA (ed.) Organic Solutes, Oxidative Stress, and Antioxidant Enzymes under Abiotic Stressors. CR100C Press/Taylor & Francis Group. https://doi.org/10.1201/9781003022879-1

El-Mahdy R et al (2025) Exogenous gallic acid improves water retention and membrane stability in drought-exposed barley by upregulating antioxidant enzymes. [Journal details pending]

Fakhar A, Gul B, Rafique M, Ortas I (2022) Use of Biostimulants to Increase Heavy Metal Tolerance in Cereals. In: Abdel Latef, A.A.H. (eds) Sustainable Remedies for Abiotic Stress in Cereals. Springer, Singapore. https://doi.org/10.1007/978-981-19-5121-3_22

Gantner J, Bauer M, Neumann M (2021) Mode of action of gallic acid on Gram-negative bacterial and fungal membranes. Int J Antimicrob Agents 57(3):106112. https://doi.org/10.1016/j.ijantimicag.2021.106112

Gubina AA, Terekhova IV, Karpov AV, Novikova GV (2016) Antioxidant protection of plant galactolipids against ozone by gallic acid and α-tocopherol. Acta Physiol Plant 38(1):1–8. https://doi.org/10.1007/s11738-016-2102-1

Hakla HR, Sharma S, Urfan M, Yadav NS, Rajput P, Kotwal D, Abdel Latef AAH, Pal S (2021) Gibberellins Target Shoot-Root Growth, Morpho-Physiological and Molecular Pathways to Induce Cadmium Tolerance in Vigna radiata L. Agronomy 11: 896

Haque MS, Husna MT, Uddin MN, Hossain MA, Sarwar AKMG, Ali OM, Abdel Latef AAH, Hossain A (2021) Heat stress at early reproductive stage differentially alters several physiological and biochemical traits of three tomato cultivars. Horticulturae 7:330

Hasan MN, Hossain MA, Burritt DJ, Ashrafuzzaman M (2022) Cereals and Phytohormones Under Ozone Stress. In: Abdel Latef, A.A.H. (eds) Sustainable Remedies for Abiotic Stress in Cereals. Springer, Singapore. https://doi.org/10.1007/978-981-19-5121-3_18

Hasanuzzaman M, Alam MM, Rahman MM (2014) Role of gallic acid and related polyphenolics in plant abiotic stress tolerance. Arch Agron Soil Sci 60(8):1031–1048. https://doi.org/10.1080/03650340.2013.805790

Hossain et al. (2021). The key roles of proline against heat, drought and salinity-induced oxidative stress in wheat (Triticum aestivum L.). In: Abdel Latef AAH (ed.) Organic Solutes, Oxidative Stress, and Antioxidant Enzymes under Abiotic Stressors. CR100C Press/Taylor & Francis Group. https://doi.org/10.1201/9781003022879-9

Imran S, Sarker P, Hoque NMd, Paul NC, Mahamud SMd, Chakrobortty J, Ul-Arif, TMd, Abdel Latef AAH, Hasanuzzaman M, Rhaman SM (2022) Biochar actions for the mitigation of plant abiotic stress. Crop and Pasture Science 74:6–20

Iqbal M et al. (2025) Use of biostimulants to enhance waterlogging and flooding tolerance in oilseed crops. In: Abdel Latef, A.A.H. (eds) Oilseed Crops Under Abiotic Stress. Sustainability Sciences in Asia and Africa. Springer, Singapore. https://doi.org/10.1007/978-981-96-8346-8_3

Jalil A, Khan MA, Akhtar N (2020) Shikimate-pathway–derived gallic acid as precursor for phenolics in stress adaptation. Plant Biochem J 45(2):55–66

Kaplan, A. et al. (2025). Integrative omics strategies for enhancing abiotic stress tolerance in oilseed crops. In: Abdel Latef, A.A.H. (eds) Oilseed Crops Under Abiotic Stress. Sustainability Sciences in Asia and Africa. Springer, Singapore. https://doi.org/10.1007/978-981-96-8346-8_15

Kassout J et al. (2025) Phytohormone-mediated protection in oilseed crops: A key strategy against temperature stress. In: Abdel Latef, A.A.H. (eds) Oilseed Crops Under Abiotic Stress. Sustainability Sciences in Asia and Africa. Springer, Singapore. https://doi.org/10.1007/978-981-96-8346-8_12

Khan IR, Khan NA, Sofo A, Siddique KHM (2023) Critical role of gallic acid in enhancing physiological and biochemical stress responses in drought-exposed wheat seedlings. Plant Soil 486(1):45–56. https://doi.org/10.1007/s11104-023-05789-2

Khan AA, Alnusaire TS, Akbar R, Iqbal B, Zeb A, Soliman MH (2025) Use of biostimulants to enhance temperature tolerance in oilseed crops. In: Abdel Latef A.A.H. (eds) Oilseed Crops Under Abiotic Stress. Sustainability Sciences in Asia and Africa. Springer, Singapore. https://doi.org/10.1007/978-981-96-8346-8_4

Kwon YI, Apostolidis E, Shetty K, Kim DO (2017) Combined UV-A and gallic acid treatment increases oxidative stress in *Escherichia coli* through enhanced uptake and reactive oxygen species formation. Photochem Photobiol 93(5):1209–1217. https://doi.org/10.1111/php.12780

Lämke J, Bäurle I (2017) Epigenetic and chromatin-based mechanisms in environmental stress adaptation and stress memory in plants. Genome Biol 18:124. https://doi.org/10.1186/s13059-017-1263-6

Li Z, Zhang S, Yu X, Zhang Y (2025) MfWRKY40 positively regulates drought tolerance in Arabidopsis by scavenging reactive oxygen species. Plant Signal Behav 20(1):213–226

Liu MM et al (2021) Effects of waterlogging stress on secondary metabolites including gallic acid and total flavonoids in herbaceous peony. Am J Plant Sci 12(4):547–550. https://doi.org/10.4236/ajps.2021.124035

Martínez D et al (2023) Gallic acid modulates ABA and proline pathways to enhance drought response in maize. [Journal details pending]

Meena RK, Verma SK, Yadav A (2024) Root aerenchyma development and fermentation enzyme activity in waterlogged maize enhanced by gallic acid. Crop Environ 28(2):78–86. https://doi.org/10.2134/cropenv.2024.02.006

Mierziak J, Kostyn K, Kulma A (2021) Gallic acid boosts phenolic and flavonoid production in pathogen-challenged plants. Phytochemistry 183:112–120. https://doi.org/10.1016/j.phytochem.2021.112120

Mora L, Salminen JP, Karoren M (2022) The role of shikimate pathway derived gallic acid hydrolysable tannin biosynthesis and plant defence. Plant Physiol Biochem 172:1–10
Nahar G, Rahman MM, Hasanuzzaman M (2016) Effect of gallic acid dipping on postharvest fungal decay and quality retention in table grapes. J Agric Food Chem 64(6):1234–1242
Osman MS, Badawy AA, Osman, AI, Abdel Latef AAH (2021) Ameliorative impact of an extract of the halophyte Arthrocnemum macrostachyum on growth and biochemical parameters of soybean under salinity stress. Journal of Plant Growth Regulation 40:1245–1256
Pazarlar S (2022) Use of Biostimulants to Improve Ozone Tolerance in Cereals. In: Abdel Latef, A. A.H. (eds) Sustainable Remedies for Abiotic Stress in Cereals. Springer, Singapore. https://doi.org/10.1007/978-981-19-5121-3_24
Sato H et al (2025) Contribution of phenolic compounds and hormones in antioxidant regulation under combined drought and heat stress in barley. Sci Rep. https://www.nature.com/articles/s41598-025-11496-8
Shan T, Gul B, Rafique M, Jabeen A, Gurmani AR, Ortas I (2022) Cereals and Phytohormones Under UV Stress. In: Abdel Latef, A.A.H. (eds) Sustainable Remedies for Abiotic Stress in Cereals. Springer, Singapore. https://doi.org/10.1007/978-981-19-5121-3_17
Shikari AB et al. (2022) Cereals under abiotic stress: An overview. In: Abdel Latef, A.A.H. (eds) Sustainable Remedies for Abiotic Stress in Cereals. Springer, Singapore. https://doi.org/10.1007/978-981-19-5121-3_1
Singh R, Sharma P, Ahmad A (2017) Gallic acid regulates WRKY and MYB transcription factors, SOD, CAT, and HSP gene expression in stress-exposed rice and *Arabidopsis*. Plant Mol Biol Report 35(4):489–502. https://doi.org/10.1007/s11105-017-1028-3
Singh TV, Sharma U, Mehta N (2022) Exogenous gallic acid elevates phenylalanine ammonia-lyase activity and reduces *Pseudomonas syringae* colonization in tomato. J Plant Interact 17(1): 45–55. https://doi.org/10.1080/17429145.2022.2034567
Sun L, Wang Z, Liu J (2018) Gallic acid promotes salicylic acid–mediated induced systemic resistance against *Pseudomonas syringae* in *Arabidopsis*. Plant Physiol Biochem 128:12–20. https://doi.org/10.1016/j.plaphy.2018.04.015
Tahjib-Ul-Arif, M, Hasan MT, Rahman MA, Nuruzzaman M, Rahman AMS, Hasanuzzaman M, Haque MR, Hossain MA, Abdel Latef AAH, Murata Y, Brestic M (2023) Plant response to combined salinity and waterlogging stress: Current research progress and future prospects. Plant Stress 7:100137
Wasternack C, Song S (2017) The crosstalk between jasmonic acid and other plant hormone signaling highlights the involvement of jasmonic acid as a key integrator in plant responses. Front Plant Sci 8:1203
Wild R, Smith JL, Thompson CR (2012) Gallic acid–mediated epigenetic priming: histone acetylation and DNA methylation improve stress memory in plants. Epigenetics 7(6):617–624. https://doi.org/10.4161/epi.20705
Xie J, Lin J, Ye J (2014) Photochemical degradation of gallic acid in the presence of Fe^{3+} under UV light. J Photochem Photobiol A Chem 287:8–15. https://doi.org/10.1016/j.jphotochem.2014.04.008
Zhang X et al (2023) Metabolomic profiling reveals elevated gallic acid levels and antioxidant responses in cold-tolerant oak species under chilling stress. Plant Metabolom J 9(2):101–110. https://doi.org/10.1234/pmj.v9i2.101
Zhao C, Nie H, Shen Q, Zhang S, Lukowitz W, Tang D (2014) EBS1 regulates plant immunity by controlling MAPK kinase 6 activity through the MKK9–MPK6 cascade in Arabidopsis. Plant Physiol 164(4):2149–2159
Zhao X, Xu Y, Li Y (2021) Proteomic responses and membrane stabilization in salinity- and heat-stressed plants treated with gallic acid. J Proteome 235:Article 104093. https://doi.org/10.1016/j.jprot.2021.104093

Chapter 13
Glutamic Acid: A Key Player in Stress-Induced Metabolism

Saeideh Alipoor, Jaber Mehdiniya Afra, Reza Afsharianzadeh, Nafiseh Jalali, Afshar Zafarani, Seyede Roghie Ghadirnezhad Shiade, and Mohamed Ait-El-Mokhtar

Abstract Glutamic acid (Glu) plays a basic and complex function in plant physiology as a central molecule of nitrogen (N) assimilation, amino acid biosynthesis, redox regulation, and stress response. Moreover, Glu is actively engaged in adapting plants to abiotic stress components such as drought, salinity, temperature, and heavy metals via antioxidant functions, reactive oxygen species (ROS) detoxification, and metabolic signaling. In addition, Glu was reported to be a key component in plant response to biotic stress. This chapter aims to describe the Glu biosynthesis pathways and highlight its function as a primary precursor molecule in the plant cell. Specific focus is laid on recent literature regarding exogenously applied Glu and its

S. Alipoor
Department of Agronomy and Plant Breeding Sciences, College of Aburaihan, University of Tehran, Pakdasht, Iran

J. M. Afra
Department of Agricultural Sciences, National University of Skills (NUS), Tehran, Iran

R. Afsharianzadeh
Department of Agronomy, Faculty of Agriculture, Science and Research Branch, Islamic Azad University, Tehran, Iran

N. Jalali
Agroecology, Shirvan Institute of Higher Education, Ministry of Science, Research and Technology, Shirvan, Iran

A. Zafarani
Department of Plant Production and Genetics, Faculty of Agriculture and Natural Resources, University of Mohaghegh Ardabili, Ardabil, Iran

S. R. G. Shiade
Department of Agronomy, Sari Agricultural Science and Natural Resources University, Sari, Iran

M. Ait-El-Mokhtar (✉)
Laboratory of Biotechnology, Agri-food, Materials, and Environment (LBAME), Department of Biology, Faculty of Science and Techniques—Mohammedia, Hassan II University of Casablanca, Mohammedia, Morocco
e-mail: mohamed.aitelmokhtar@univh2c.ma

A. A. H. Abdel Latef (ed.), *Harnessing Vitamins and Organic Acids for Plant Stress Resilience*, Sustainability Sciences in Asia and Africa,
https://doi.org/10.1007/978-981-95-8593-9_13

potential to enhance plant tolerance against environmental stress and the underlying mechanisms of its functions.

Keywords Glutamate · Biostimulant · Signaling · Plant tolerance · Environmental stress · Antioxidant enzymes

13.1 Introduction

Amino acids can potentially function as biostimulants, applied singularly or in blends, to boost plant growth, yield, and quality. Their use under stressful environmental conditions has been demonstrated to increase plant tolerance (Matysiak et al. 2020). As nitrogen (N)-containing molecules, amino acids function as the fundamental units of proteins. Furthermore, they function as precursors to various metabolites, contributing to plants' responses to the eco-physiological stimuli, including vitamins (Asensi-Fabado and Munné-Bosch 2010), various pigments (Cho et al. 2009), secondary metabolites and phytohormones (Westfall et al. 2013). These compounds can also regulate ion transport and stomatal closure as osmolytes (Franzoni et al. 2022). Besides, such molecules contribute to carbon (C) and N metabolism and assimilation in plants. Previous studies have documented that the use of amino acids via foliar spray can effectively enhance the performance of agronomic and horticultural crops, highlighting their value as an effective biostimulant strategy (Franzoni et al. 2022). Notably, the effectiveness of these applications depends on both the approaches used as well as on the form that the amino acids take upon application, whether individually or in combination with other substances (Xu et al. 2017). Global climate change poses a major threat to the survival of natural ecosystems due to its dynamic and multifaceted nature, altering environmental conditions and impacting both abiotic and biotic components worldwide (Fathi et al. 2024d). Environmental stresses such as drought, salinity, temperature, heavy metals, and nutrient deficiency can be exacerbated under this global phenomenon. Besides, biotic stresses, pathogens, and insect effects are considerably affected by climate change. Therefore, it is required to address effective measures to combat this increasing issue.

Glutamic acid (Glu) is a significant amino acid in plants since it chiefly sustains the plant osmotic balance by proline, glutathione (GSH), gamma-aminobutyric acid (GABA), Glutamine (Gln), and other N-containing compounds biosynthesis (Matysiak et al. 2020). Specifically, Glu is decarboxylated to form GABA, which mainly contributes to balancing C and N metabolism and numerous cellular processes. In addition, under the action of glutamine synthetase, Glu is converted to Gln, a compound of great importance to amino acid, nucleic acid, and vitamin synthesis (Lee et al. 2023).

Application of exogenous Glu was reported to mitigate various environmental strains. Earlier studies have noted that its exogenous application enhances leaf functionality and photosynthetic activity (Lee et al. 2023). This response can be attributed to an increase in the leaf N concentration, which is positively correlated with improved photosynthetic performance. Besides, glycine and Glu are incorporated into aminolevulinic acid biosynthesis, thereby contributing significantly to chlorophyll and key metabolites biosynthesis (Franzoni et al. 2021). Similarly, Lee et al. (2017) demonstrated that Glu foliar spray improved antioxidant enzyme activities in *Brassica rapa* subsp. pekinensis (Kimchi Cabbages) leaves exposed to cold stress, improving their tolerance to the low air temperature.

Considering its central role in N and C metabolism, antioxidant defense, and osmotic regulation, Glu emerges as a key molecule in plant responses to environmental stresses. Here, this chapter aims to explore its biosynthesis, functions, and applications, particularly its potential as a biostimulant to promote plant performance under abiotic and biotic stress conditions.

13.2 Biosynthesis and Metabolic Role of Glutamic Acid in Plants

Glu accounts for the majority of plant metabolism and acts as an important mediator for N metabolism, amino acid synthesis, and redox balance (Liao et al. 2022). It is a product of two main biosynthesis pathways in plants: (1) glutamine synthetase–glutamate synthase (GS–GOGAT) pathway that functions as the predominant N assimilation mechanism and converts inorganic ammonium (NH_4^+) to Gln, which in turn can be used by plant cells. In the GS–GOGAT pathway, NH_4^+ generated from nitrate reduction or photorespiration is initially assimilated into Gln by glutamine synthetase (GS). This Gln, in turn, transfers its amide group to 2-oxoglutarate via GOGAT, forming two glutamates. This cycle is very efficient and is the main pathway for ammonium assimilation in higher plants. (2) The glutamate dehydrogenase (GDH) pathway, which mediates the reversible reaction for both oxidative deamination of Glu and reductive amination of 2-oxoglutarate (2-OG) yielding glutamate. Alternatively, the GDH pathway allows the reversible transformation of 2-oxoglutarate into glutamate (through amination) and plays an ancillary role, particularly under conditions of stress or high NH_4^+ concentration (Liao et al. 2022) (Fig. 13.1).

13.3 Glutamic Acid in Abiotic Stress Resilience

13.3.1 Drought Stress and Glu-Mediated Metabolic Responses

Drought stress is increasingly recognized as one of the acute environmental concerns adversely affecting world agricultural management (Fathi et al. 2024c; Dawood et al. 2025). It has the potential to trigger extreme declines in grain yield and overall crop

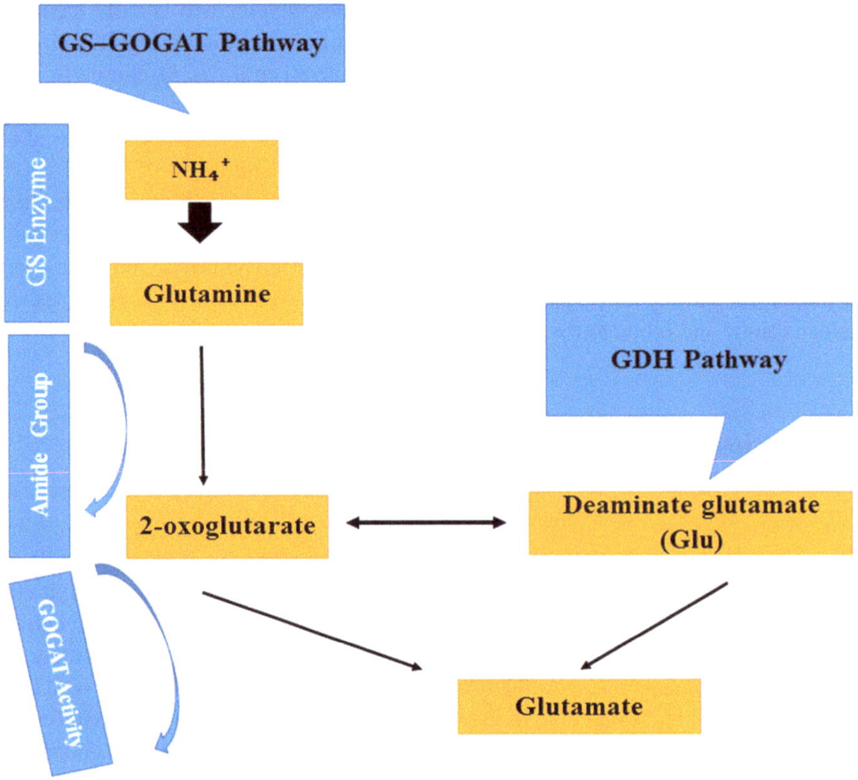

Fig. 13.1 Biosynthesis pathways and metabolic role of glutamate (Glu) over two main pathways of the glutamine synthetase–glutamate synthase (GS–GOGAT) pathway and glutamate dehydrogenase (GDH) pathway. Glutamine synthetase (GS), glutamate synthase (GOGAT), glutamate dehydrogenase (GDH), ammonium (NH_4^+)

productivity (Ghadirnezhad Shiade et al. 2024). Under drought conditions, plants exhibit concurrent reduction in photosynthetic efficiency, disrupted root growth, and oxidative cellular damage, ultimately causing growth and development inhibition (Ghadirnezhad Shiade et al. 2023a). As a result of these adverse effects, plants undergo several metabolic adjustments, namely, the production of osmoprotectants such as proline (Meena et al. 2019; Umer et al. 2026) and many phytohormones (Xu et al. 2020). Moreover, antioxidant enzyme activity, along with malondialdehyde (MDA) content, is usually utilized as a marker for assaying plant tolerance against abiotic stress (Ghadirnezhad Shiade et al. 2023b).

Enzymes such as ascorbate peroxidase (APX), peroxidase (POD), catalase (CAT), and superoxide dismutase (SOD) are significant in controlling reactive oxygen species (ROS), whereas the content of MDA indicates the extent of lipid peroxidation under stress situations (Ghadirnezhad Shiade et al. 2023b; Azooz et al. 2026). Proline synthesis is largely performed in plants by the glutamate pathway, specifically owing to abiotic stress agents like drought and salinity. Accordingly, glutamate is first converted into Δ^1-pyrroline-5-carboxylate (P5C) by the enzyme P5C synthetase (P5CS), which is considered a rate-limiting step and is transcriptionally upregulated during stress. Subsequently, P5C is converted to proline by P5C reductase (P5CR) using NAD(P) H as a cofactor. An alternative pathway, particularly active in certain species or under specific conditions, involves the conversion of ornithine to P5C through ornithine δ-aminotransferase (OAT), after which P5C again enters the final reduction step catalyzed by P5CR. While these biosynthetic pathways are induced during stress to accumulate proline, the catabolic pathway, comprising proline dehydrogenase (ProDH) and P5C dehydrogenase, becomes suppressed to prevent degradation (Meena et al. 2019). Proline not only acts as a compatible osmolyte in osmotic adjustment but also performs pivotal functions in ROS scavenging, redox homeostasis maintenance, membrane and protein stabilization, and photosynthetic machinery protection under stress (Fathi et al. 2024b). In addition, proline metabolism is interconnected with a number of metabolic pathways like nitrogen assimilation, tricarboxylic acid (TCA) cycle intermediates, and the urea cycle and thereby rendering it a focal point in plant stress physiology and cell homeostasis (Meena et al. 2019) (Table 13.1).

Besides, Xu et al. (2020) assessed the use of poly-γ-glutamic acid (γ-PGA), a natural biopolymer produced by certain strains of *Bacillus* spp. under water stress conditions. The research demonstrated a clear enhancement of drought tolerance in *Brassica napus*. Treatment with γ-PGA induced a systemic response resulting in a higher level of endogenous abscisic acid (ABA) in plants, a key phytohormone and critical to plant response to abiotic stresses. The increased ABA level was found to be strongly correlated with the expression of ABA biosynthesis genes, indicating that γ-PGA is actively involved in the regulation of hormonal signal pathways at the transcript level. These accumulated quantities provided improved stomatal control, lowered transpiration rate, and increased water holding capacity under stress. Furthermore, plants treated with γ-PGA exhibited strengthened antioxidant defense mechanisms, including higher activities of enzymes such as SOD and CAT, which collectively mitigated oxidative damage induced by drought stress (Xu et al. 2017). These findings highlight the γ-PGA contribution as an effective biostimulant that enhances drought resilience through ABA-mediated signaling and physiological adjustments. Moreover, the impact of γ-PGA on maize seedlings under 7-day irrigation deprivation (drought-stressed) conditions was also investigated (Ma et al. 2022). The γ-PGA treatment enhanced plant dry weight, osmoprotectant content, and photosynthetic performance, although the growth parameters were lower than the control, while drought stress depressed all of these. Concomitantly, response analysis and RNA-seq data demonstrated that the γ-PGA also affected the ABA biosynthetic pathway and ABA signal transduction pathways (Fig. 13.2).

Table 13.1 Effects of γ-PGA and glutamic acid (Glu)-based treatments on plant responses to abiotic stresses

Abiotic stress	Plant	Treatment dose	Effect	Reference
Drought stress (irrigation inhibition)	*Zea mays*	(0–100 mg/L) of γ-PGA	Improved photosynthesis, enhanced osmolyte accumulation, ABA content, and improved root growth	Ma et al. (2022)
Drought stress (−0.40 MPa osmotic potential using 150 g/L PEG-6000)	*Brassica napus* L.	γ-PGA at 20 mg/L, applied via hydroponic solution	Improved chlorophyll and relative water content (RWC), elevated proline accumulation, and antioxidant enzyme activities Promoted abscisic acid (ABA) accumulation through upregulation of ABA biosynthesis genes (*BnNCED3*, *BnZEP*, *BnAAO4*)	Xu et al. (2020)
Salt stress (100 mM NaCl)	*Brassica napus* L.	γ-PGA at 20 mg/L, applied 96 h before stress	Enhanced fresh weight Reduced MDA, while enhanced and proline content activation of Ca^{2+} and H_2O_2 signaling pathways	Xu et al. (2017)
Salt stress (100 mM NaCl)	*Lactuca sativa* L. (lettuce)	Foliar application of 1 mM glutamic acid	Improved proline accumulation and antioxidant enzyme activities	Franzoni et al. (2022)
Heat stress (45 °C in 16-h light/35 °C in 8-h darkness)	*Brassica rapa* L. *ssp. pekinensis*	Foliar treatment of 1 mM Glu or 1 mM of γ-PGA	Upregulation of genes involved in carotenoid biosynthesis, DNA repair, and ROS detoxification pathways	Quan et al. (2022)
Heat stress (35 °C for 14 days)	*Agrostis stolonifera*	Foliar application of 1 mM glutamate	Reduced visible leaf senescence symptoms, such as leaf yellowing, under heat stress Downregulated senescence-associated genes and upregulated genes related to chlorophyll biosynthesis and stress response pathways	Rossi et al. (2021)
Cold stress (4 °C)	*Brassica napus*	Poly(-γ-glutamic acid) (γ-PGA), 20 mg L^{-1}	Enhanced fresh weight Improved chlorophyll content Regulated calcium-dependent protein kinases	Lei et al. (2015)

(continued)

Table 13.1 (continued)

Abiotic stress	Plant	Treatment dose	Effect	Reference
Cold stress (4 °C, 48 h)	*Arabidopsis thaliana*	Glutamate receptors of AtGLR1.3 and AtGLR1.2	Enhanced cold tolerance, upregulated jasmonate signaling and *CBF/DREB1* expression	Zheng et al. (2018)
Cold stress (4 °C)	*Brassica napus* L.	γ-PGA at 20 mg/L, applied 96 h before stress	Upregulation of brassinolide and jasmonic acid biosynthesis genes (*DWF4* and *LOX2*).	Xu et al. (2017)
Heavy metal stress; Cd (1 mM), Pb (1 mM), As (2 mg/L^{-1}), and Ni (1 mM)	*Phaseolus vulgaris* L.	L-glutamic and L-aspartic acid (1.5 and 3 mM)	Reduced oxidative stress (H_2O_2 and MDA), improved antioxidant enzyme activities Enhanced photosynthetic pigment levels Increased nutrient uptake under heavy metal stress.	Yuksel (2024)
Cadmium stress (50 µM $CdCl_2$)	*Oryza sativa* L.	Exogenous glutamate at 1 mM concentration	Reduced oxidative stress markers Suppressed cadmium uptake and translocation	Jiang et al. (2020)
Silver stress (0, 100, 250, 500 µM Ag)	*Helianthus annuus* L.	Soil application of Glu	Alleviated ag-induced toxicity by enhancing antioxidant enzyme activities and reducing ROS levels Increased "Ag" accumulation in sunflower roots	Farid et al. (2018)

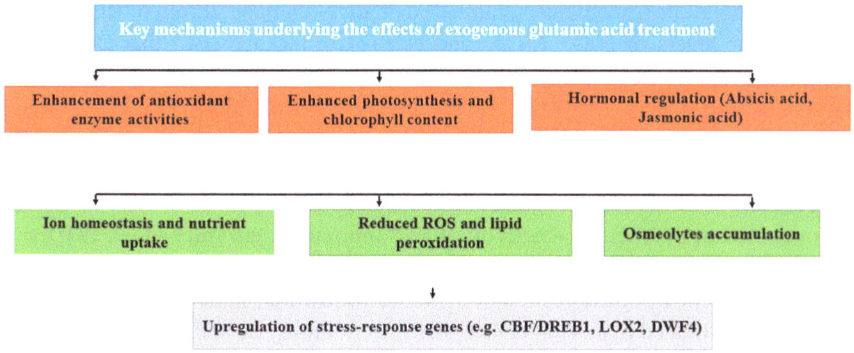

Fig. 13.2 Key mechanisms underlying the effects of exogenous glutamic acid (Glu) treatment under stress conditions

13.3.2 Salinity and Glu-Mediated Metabolic Responses

It is estimated that around 20% of global cultivable lands are affected by salinity stress, amounting to approximately 800 million hectares (ha). Moreover, it was exhibited that salinity is becoming one of the most pressing threats to agricultural productivity as it is extending and affecting around half of the current cultivable lands by the end of the twenty-first century (Ghadirnezhad Shiade et al. 2023b; Hayat et al. 2026). When plants are exposed to salinity above the threshold level of 4 dSm^{-1}, they experience osmotic stress. Besides, the accumulation of toxic ions, particularly sodium (Na^+) and chlorine (Cl^-) ions, disrupts cellular homeostasis and leads to leaf necrosis (Fathi et al. 2024d). Subsequently, this stress results in ionic imbalance, inhibition of key enzyme activities, and the excessive production of ROS, including singlet oxygen ($^1O_2$), superoxide radical ($O_2\bullet^-$), hydrogen peroxide (H_2O_2), and hydroxyl radical (·OH). These factors collectively contribute to a decline in photosynthetic efficiency, protein denaturation, and lipid peroxidation, ultimately reducing plant growth and survival (Fathi et al. 2024b).

Glutamic acid contributes to maintaining ion balance by supporting the synthesis of compatible solutes and regulating ion transport mechanisms. Accordingly, Fardus et al. (2021) investigated the external treatment of L-glutamic acid (L-Glu) and found that this compound significantly maintained ion homeostasis in salt-stressed lentil seedlings. Under 110 mM NaCl stress, untreated plants exhibited a marked increase in Na^+ accumulation and a decrease in potassium (K^+) content, resulting in a disturbed Na^+/K^+ ratio, which is a critical factor in salt toxicity. However, pretreatment with 10 mM L-Glu notably reduced Na^+ accumulation while sustaining higher K^+ levels, thereby improving the Na^+/K^+ balance, which is essential for maintaining cellular function and metabolic stability under salinity stress. In addition to its role in ion regulation, L-Glu treatment also improved photosynthetic pigment content, including chlorophylls and carotenoids, which are typically degraded under salt stress (Lei et al. 2015).

Furthermore, the treatment inhibited oxidative damage through significantly lowering the levels of H_2O_2, MDA, and electrolyte leakage (EL) that are utilized as indicators of cell damage. Besides, it has been demonstrated that Glu triggers the antioxidant enzymes activity, thereby playing a major role in the physiological and biochemical defense system of plants. Iqbal et al. (2021) exhibited that exogenous application of Glu effectively mitigated the harmful impacts of salt stress in cucumber (*Cucumis sativus* L.) by enhancing these defense responses. Remarkably, Glu treatment led to the increased activity of key antioxidant enzymes SOD, POD, and APX, which participate in ROS scavenging and prevention of oxidative damage. This was evident from the reduction in MDA content, an indicator of lipid peroxidation in Glu-treated plants under salt growth conditions. Besides, Glu can also play a role in osmotic adjustment through proline biosynthesis stimulation, thereby enhancing water retention and cell turgor under osmotic stress conditions. All these observations indicate the potential of exogenous Glu as an effective strategy for salinity tolerance improvement (Fardus et al. 2021). Moreover, Lei et al. (2016)

subjected rape (*Brassica napus*) seedlings to salt stress (100 mM NaCl) for varying durations (48 to 144 h). The results indicated that γ-PGA treatment enhanced root, shoot, and total plant dry weight under salt stress conditions. Additionally, the K^+/Na^+ ratio significantly increased, highlighting the mitigating effect of γ-PGA in reducing sodium toxicity through improved ion homeostasis. Furthermore, gene expression analysis revealed that the proline biosynthesis-related genes *BnP5CS1* and *BnP5CS2* were upregulated, while the proline degradation gene *BnPDH* was downregulated, suggesting a role for γ-PGA in promoting osmotic adjustment through proline accumulation (Table 13.1).

13.3.3 Temperature Stress and Glu-Mediated Metabolic Responses

Currently, climate change is increasingly associated with extreme weather variability, including both high and low-temperature extremes (Lee et al. 2021; Fathi et al. 2024d; Ahmad et al. 2026). Heat stress, primarily driven by global warming, is linked to a global temperature escalation of around 2 °C in comparison to pre-industrial levels (Quan et al. 2022). Conversely, cold stress can occur in two distinct temperature ranges: chilling stress (0–15 °C) and freezing stress (below 0 °C) (Lee et al. 2021). These environmental stresses can cause ROS accumulation, leading to enhanced oxidative damage to cell organelles such as membranes, proteins, nucleotides, and enzymes. These effects ultimately impair photosynthetic efficiency and inhibit plant survival. Although plants develop enzymatic (e.g. CAT, POD, SOD, APX) and non-enzymatic (antioxidant metabolites such as phenolics, GSH, carotenoids, and flavonoids) ROS-scavenging systems to keep cells redox homeostasis, these defense mechanisms may not be sufficient to combat this abiotic stress (Quan et al. 2022). Hence, the Glu application has developed as a potential approach to improve plants' tolerance to temperature stress.

The role of Glu and its derivatives in enhancing plant thermotolerance has been increasingly supported by recent studies. The exogenous application of Glu and γ-PGA has been shown to significantly enhance heat stress tolerance in *Brassica rapa* L. ssp. pekinensis (*Chinese cabbage*) through multiple physiological and molecular mechanisms (Quan et al. 2022). Under elevated temperatures (45 °C in 16-h light/35 °C in 8-h dark), treated plants with Glu and γ-PGA exhibited improved biomass accumulation, as reflected by a substantial increase in dry weight compared to untreated controls, suggesting enhanced growth performance under stress conditions. Besides, at the physiological level, Glu and γ-PGA treatments led to increased chlorophyll and carotenoid contents, which are critical for maintaining photosynthetic efficiency and protecting chloroplasts against oxidative damage. Concurrently, the treatments upregulated the activities of key antioxidant enzymes, including SOD, CAT, APX, and POD. Furthermore, the applied treatment significantly reduced MDA levels and EL, indicating reduced lipid peroxidation and improved membrane stability under heat stress. Notably, transcriptomic analysis further revealed that these treatments upregulated genes associated with carotenoid biosynthesis, photosynthetic machinery, and ROS signaling pathways. Similarly, another study emphasized the potential of Glu in mitigating chilling stress in tomato plants. Lee et al. (2021) investigated the effects of exogenous Glu application (2.5 mM) applied on tomato plants subjected to chilling stress (25 °C/18 °C, day/night). The study found that this pretreatment significantly enhanced photosynthetic efficiency and strengthened the antioxidant defense system, leading to improved plant growth and yield under low-temperature conditions. The expression levels of genes encoding antioxidant enzymes and C-repeat binding factors (CBFs), the main transcription elements in cold-inducible gene expression, were also transformed in Glu-preexposed plants in comparison with the control. In another study, Jia et al. (2017) studied exogenous application of Glu, GABA, and calcium chloride ($CaCl_2$) on rice (*Oryza sativa* L.) plants under low temperature stress. Although some physiological parameters that were negatively affected by cold treatment, such as the relative water content (RWC), photosystem II (PSII) efficiency, and relative EL (REL), the combined treatment substantially ameliorated these effects. Notably, the treatment increased photosynthetic efficiency and maintained membrane integrity, which suggested its protective effects from the cold stress on the physiological status. These findings suggest that exogenous Glu application can partially suppress the adverse impacts of temperature stress by augmenting photosynthetic efficiency and activating antioxidant defense mechanisms, thereby enhancing the overall resilience of tomato plants to low-temperature stress (Table 13.1).

13.3.4 Heavy Metal Stress and Glu-Mediated Metabolic Responses

Heavy metal stress refers to the adverse physiological, biochemical, and molecular effects exerted on plants when exposed to elevated concentrations of non-essential or essential metal ions (Taghavi Ghasemkheili et al. 2023; Adnan et al. 2026) such as cadmium (Cd), lead (Pb), arsenic (As), mercury (Hg), chromium (Cr), nickel (Ni), copper (Cu), silver (Ag), and zinc (Zn) (Fathi et al. 2024a). While some metals like Cu, Zn, and Ni are required in trace amounts for normal cellular functions, excessive accumulation leads to phytotoxicity. Non-essential metals like Cd and Pb are toxic even at low concentrations (Anas et al. 2025). Recent studies have explored the role of Glu in alleviating heavy metal stress in various plant species. In a study on *Olea europaea* (olive) plants exposed to Cd stress, exogenous Glu application was found to alleviate the adverse effects of the heavy metal. Cd exposure typically reduces plant growth and chlorophyll content, while increasing oxidative stress markers (Saleem et al. 2024). However, Glu treatment significantly improved plant growth and chlorophyll levels, thereby enhancing photosynthetic efficiency. Additionally, the Glu application activated the plant's antioxidant defense system, increasing the activity of key antioxidant enzymes. These enzymes helped mitigate ROS accumulation, reducing oxidative damage (Saleem et al. 2024). Similar protective role of Glu is reported for sunflowers subjected to Ag-contaminated stressful environments. Farid et al. (2018) investigated the role of exogenous Glu in improving the phytoremediation potential of *Helianthus annuus* cultivated in silver-polluted soils. In addition to the positive effects of Glu on reduced photosynthetic efficiency and antioxidant enzyme production, the application of Glu was shown to improve the accumulation of Ag in the roots of sunflower plants, indicating an enhancement in their phytoremediation capacity. The study suggests that Glu acts as a biostimulant that not only improves plant health under heavy metal stress but also aids in the extraction and detoxification of Ag from contaminated soils. In another study (Pang et al. 2018), γ-PGA was utilized to investigate its effects on cucumber seedlings under Pb and Cd pollutions. This treatment considerably decreased the growth and development-inhibitory impact of the mentioned pollutants. Furthermore, MDA content decreased while the proline concentration increased. Notably, infrared spectral analysis indicated that Cd^{2+} and Pb^{2+} ions interact with the free carboxyl groups of γ-PGA. The chelation capacity of γ-PGA was moderate for Pb and relatively lower for Cd. Moreover, γ-PGA enhanced the alleviation of the detrimental impacts of Cd and Pb on soil microbial populations. These results indicate that γ-PGA has a noteworthy potential to control heavy metal accumulation in crops resulting from contaminated farmland, thus underlining its importance for sustainable agriculture (Table 13.1).

13.3.5 Mineral Deficiency Stress and Glu-Mediated Metabolic Responses

The use of Glu in plants has demonstrated great potential in reducing nutrient-deficiency stress, particularly for essential nutrients (Svietlova et al. 2024). This amino acid is also involved in signal transduction, contributes mainly to nutrient loading in plants and contributes to greater stress tolerance. Besides, Glu serves as a signal molecule for the plant to sense nutrient deficiency, especially phosphorus (P), and mediate rhizosphere acidification, which contributes to P uptake from soil. For example, the growth of *Cardamine pratensis* cell suspensions was stimulated by the addition of casein hydrolysate and Gln, both at the same concentration, in the medium (Bister-Miel et al. 1985). In fact, the positive effect of Glu only occurred in *C. pratensis* and *Salix alba cell* suspensions under P-deficient conditions. This indicates that the effectiveness of casein hydrolysate may be attributed to its high Gln concentration, which in turn could compensate for the decreased synthesis of this amino acid when P is in low supply. Besides, Cui et al. (2020) reported that expression of the *glutamate synthase 1* (*GLU1*) gene was up-regulated by iron (Fe) deficiency in *Arabidopsis thaliana*. Correlated with this, knockout of *GLU1* results in a substantial decrease in the expression of many Fe-deficiency-upregulated genes in both roots and shoots. The *GLU1* mutant had reduced leaf Fe burden compared with wild-type plants and displayed severe leaf chlorosis when grown under iron-deficient conditions. Moreover, it has been reported that Glu (*GLU1*'s metabolic product) can chelate iron in plant organs and transport it out. Hence, these results give novel insight into the functions of *GLU1* in Fe homeostasis in Arabidopsis, particularly in controlling Glu biosynthesis under Fe deficiency. Plants convert inorganic N into Glc, the most plentiful amino acid in various species (Svietlova et al. 2024). Glu functions as the primary N donor for all N-containing combinations in the cell and is the initial organic N molecule synthesized from the inorganic N absorbed via the roots. Beyond its nutritional contribution, Gln is believed to act as a signaling factor for the regulation of N metabolism (Svietlova et al. 2024). Svietlova et al. (2024) examined the expression status of the nitrate transporter NRT2.4. This is an inducible factor under N deficiency in *Arabidopsis thaliana* cultivated under various N starvation conditions. It aimed to compare nitrate and Gln as the individual N sources. The results displayed that the regulation of *NRT2.4* expression is contingent on the overall availability of N, such as Gln, rather than specifically on nitrate. Unlike high nitrate concentrations, high Gln concentrations caused an accumulation of amino acids with more than one N atom, indicating a shift toward N storage metabolism. Furthermore, the *nrt2.4–2* knockout line exhibited unexpected changes in *NRT2.5* gene expression and amino acid profiles in the shoots under high Gln supply, rather than the wild-type of Arabidopsis and the *nrt2.4–1* line, signifying non-*NRT2.4*-related metabolic effects in this knockout line (Table 13.1).

13.4 Glutamic Acid in Biotic Stress Resilience

13.4.1 Microbial Pathogens and Glu-Mediated Metabolic Responses

Microbial pathogens, including fungi, bacteria, and viruses, represent a significant biotic hazard to global agriculture, capable of reducing crop yield and quality through tissue degradation, toxin production, and suppression of host defenses (Appu et al. 2021; Abdolmaleki et al. 2023). In response to such infections, plants activate a multipronged array of defense machinery, including ROS accumulation, the synthesis of pathogenesis-related proteins, and the reinforcement of structural barriers (Appu et al. 2021).

Recent studies have highlighted the potential of exogenous Glu application to enhance plant resistance against microbial pathogens. Glu can contribute to pathogen defense through multiple physiological and biochemical pathways. Exogenous application of L-glutamate (L-Glu) significantly enhanced resistance toward *Penicillium expansum* infection in pear fruit during postharvest storage at both room temperature (25 °C) and cold circumstances (4 °C) (Jin et al. 2019). The treatment suppressed the germination of *Penicillium expansum* spores both in vitro and in the fruit wound sites after 24 h, indicating a direct antifungal effect. Furthermore, L-Glu application induced a strong biochemical defense response, evidenced by increased activities of defense-related enzymes, namely, β-1,3-glucanase, chitinase, phenylalanine ammonia-lyase (PAL), POD, and polyphenol oxidase (PPO) (Jin et al. 2019).

Additionally, Glu may modulate defensive gene expression and induce local or systemic resistance pathways, making it a promising candidate for integrative plant protection strategies against microbial pathogens. Recent findings by Feng et al. (2021) exhibited that Glu or glycine (Gly) exogenous application considerably improved tomato resistance toward the necrotrophic fungus *Botrytis cinerea*. This enhanced resistance was shown to be mediated by Glu receptor-like channels, particularly GLR3.3 and GLR3.5. Knockout mutants lacking these genes, especially *glr3.3*, exhibited increased susceptibility to infection and impaired defense responses. Furthermore, treatment with Glu or Gly induced rapid membrane depolarization, an electrical signaling event associated with early defense responses. However, this signaling response was diminished in *glr3.3* mutants, confirming the essential role of *GLR3.3* in mediating Glu-triggered electrical signals. Application of the Glu receptor (GLR) inhibitor DNQX also suppressed resistance, further supporting the importance of GLR-mediated electrical signaling in fungal defense. Besides, it was shown that Glu can modulate the plant microbial community in the rhizosphere and phyllosphere to enrich beneficial microbiota associated with disease suppression. In a study by Kim et al. (2021), L-Glu treatment significantly enriched beneficial microbial taxa, particularly *Streptomyces* species, which are recognized for their antagonistic activity against fungal pathogens. This shift in microbiota composition was associated with enhanced suppression of *Fusarium oxysporum* and *B. cinerea* in tomato and strawberry plants, respectively. The findings suggest that L-Glu functions not only as a metabolic regulator in plant defense but also as a selective microbial nutrient, promoting the growth of protective microbes and thereby contributing to microbiota-mediated disease resistance. Moreover, Kim and Kwak (2023), revealed that population build-up of bacteria can be significantly enhanced upon L-Glu treatment. Application of three amino acids (L-proline, L-Glu, and L-asparagine) to tomato provoked typical responses: although L-asparagine and L-proline negatively affected growth and morphology, the response to L-Glu was particularly noticeable, particularly when tomato growth was promoted, and the density of bacterial cells was augmented. The beneficial bacteria established in both the rhizosphere and root endosphere, and L-Glu played a role in this colonization. Notably, *Streptomyces* sp. showed the ability to promote plant growth and increase resistance to abiotic stress. L-Glu was significantly effective in alleviating the adverse effects of salinity stress, while it had slight effects on drought conditions. These findings indicate that L-Glu is involved in the improvement of salinity tolerance through the regulation of core-microbiota and shed new light on its application as a potential prebiotic candidate in sustainable agriculture (Table 13.2).

Table 13.2 Effects of γ-PGA and glutamic acid (Glu)-based treatments on plant responses to biotic stresses

Biotic stress	Plant	Treatment dose	Effect	Reference
Fungal (*Botrytis cinerea*)	Tomato	100 ppm L-Glu spray	Enhanced gray mold disease resistance by inducing Glu receptor expression and increasing amino acid accumulation; reduced lesion size and decay	Sun et al. (2019)
Fungal (*Penicillium expansum*)	*Pear fruit*	L-glutamate (NA)	Induced resistance through activation of defense-related proteins and increased amino acid levels; reduced fungal decay	Jin et al. (2019)
Fungal (*Alternaria alternata*)	Tomato	L-glutamate (NA)	Inhibited fungal growth by inducing host resistance; involved enhanced activities of defense enzymes and amino acid metabolism	Yang et al. (2017)
Bacterial (*Pseudomonas syringae*)	Tomato	Not a direct glutamate treatment	Glutathione and glutathione-metabolizing enzymes, along with ascorbate, contribute significantly to tomato resistance by modulating oxidative stress and enhancing defense responses	Kuźniak and Skłodowska (2004)
Viral (*tobacco mosaic* virus, TMV)	Tobacco	Glu supplementation (endogenous levels)	Glu can compensate for SA deficiency in tobacco, maintaining resistance against TMV by modulating redox status and defense signaling pathways	Künstler et al. (2019)

13.4.2 Insect Herbivory and Glu-Mediated Metabolic Responses

Herbivorous insects impose great injury on farm crops by feeding on plant tissues and creating interference in essential metabolic processes (Xue et al. 2022). Plants, in turn, trigger defense mechanisms by producing secondary metabolites such as alkaloids, phenolics, terpenoids, and glucosinolates that repel herbivores or make them less digestible (Divekar et al. 2022). They also release volatile organic compounds (VOCs), such as green leaf volatiles, terpenes, and jasmonate (JA)-induced volatiles; such releases may attract natural enemies of herbivores or prime nearby plants for defense responses (Divekar et al. 2022). Physical defenses, such as are represented in the thickening of the cell wall via lignification, the deposition of callose, or the development of trichomes and spines, function to deter or prevent insect feeding activity (Abebe 2021). The hormone-signaling pathways that most frequently regulate these defense responses typically include JA, salicylic acid (SA), and ethylene (Divekar et al. 2022). Emerging evidence notes that Glu may indirectly, yet importantly, contribute to modulating herbivory responses.

Glu also acts as a metabolic precursor and as a signaling molecule in plant systems (Toyota et al. 2018). During insect herbivory, the high levels of Glu in the wounded sites may activate ion channels of the GLUTAMATE RECEPTOR–LIKE family, thus gating intracellular calcium fluxes and ROS bursts, respectively, which represent key elements in early defense signal transduction cascades (Toyota et al. 2018). Moreover, Li et al. (2025) have found GSH to be an essential signaling molecule of the plant's systemic defense processes. The research illustrated that GSH triggers long-distance calcium-mediated signaling among leaves in a manner independent of its function as an antioxidant. This type of signaling pathway is reliant on the Glu receptor-like channel GLR3.3, causing the activation of JA biosynthesis as well as defense-related gene induction in unwounded but distal leaves. Such observations highlight the role of GSH in facilitating rapid systemic communication within plants, inducing resistance to herbivory along with pathogen attack.

Besides, recent advances in plant signaling have revealed that Glu and its perception through glutamate receptor-like channels (GLRs) play pivotal roles in modulating early defense responses to herbivore attacks. Xue et al. (2022) investigated the specific roles of GLR3.3 and GLR3.6 in *Arabidopsis thaliana* for insect feeding and demonstrated that these GLRs are central to the plant's systemic resistance in the plant through their effects on calcium signaling and by controlling essential metabolic and transcriptional responses to insect predation. Through transcriptomic and metabolomic approaches, it was shown that GLR3.3 and GLR3.6 control the genes responsible for JA biosynthesis, phenylpropanoid metabolism, and biosynthesis of secondary metabolites, processes critical for deterring herbivores and for repairing tissues. Moreover, *glr3.3* and *glr3.6* mutants showed compromised resistance, confirming the functional role of these channels in coordinating systemic defense. These findings underline the potential of Glu signaling in activating broad-spectrum insect resistance mechanisms in plants (Xue et al. 2022).

13.5 Future Perspectives and Research Directions

Although exogenous Glu applications are promising in the development of plant stress resilience, several aspects are yet to be investigated. Further biochemical and molecular investigations on enzymes and genes associated with this molecule will continue to shed more light on Glu metabolic regulation. Research in the future also needs to be directed toward signaling pathways, gene expression profiles, and interactions with plant hormones to fully exploit the potential of Glu. Moreover, the application of multi-omics approaches (e.g., transcriptomics, metabolomics, and proteomics) would be of great assistance in identifying the regulatory networks targeted by this compound.

The integration of Glu with other biostimulants, e.g., humic substances, microbial inoculants, or seaweed extracts, holds the potential for synergistic advantages through complementary modes of action, leading to improved resilience against stresses. Such combinatorial approaches might enable the next generation of biostimulant products. Notably, one promising direction for future research can be the formulation of Glu-based products with the aid of sophisticated delivery systems, e.g., encapsulation, slow-release carriers, or nano-formulations. This will enhance their stability, effectiveness, and ease of application, rendering them both environmentally friendly and economical.

13.6 Conclusion

The application of exogenous Glu has been reported as a highly effective method for alleviating several abiotic stresses to which plants are subjected, including drought, salinity, cold, heavy metal toxicity, and nutrient deficiency. It can also be utilized for the alleviation of biotic stress. Glu regulates a variety of physiological and biochemical functions, including improving antioxidant activity, osmotic regulation, nutrient accumulation, and gene expression of stress-activated genes, thereby improving plant growth as well as tolerance to stress. Glu specifically participates in dual action as both a signaling compound and a metabolic precursor, contributing as an essential factor in plant stress response. Plant stress control using Glu is an environmentally friendly chemical input option that is sustainable in application within agricultural operations.

Declaration The authors declare that state that the AI tool (Chat GPT) was used to edit the text.

References

Abdolmaleki AK, Pirdashti H, Yaghoubian Y et al (2023) Endophytic fungi improve growth and yield of wheat (*Triticum aestivum* L.) under limited light conditions. Gesunde Pflanz 75:1517–1529. https://doi.org/10.1007/s10343-022-00816-x

Abebe W (2021) Review on plant defense mechanisms against insect pests. Int J Nov Res Interdiscip Stud 8:15–39

Adnan M. et al. (2026) Heavy metals stress mitigation by chitosan nanoparticles. In: Abdel Latef, A.A.H. (eds) Chitosan Nanoparticles for Plant Resilience Under Abiotic and Biotic Stressors. Smart Nanomaterials Technology. Springer, Singapore. https://doi.org/10.1007/978-981-95-6422-4_7

Ahmad N, Ahmad S, Islam MS, Khan MN, Abdel Latef AAH (2026) Chitosan nanoparticles under temperature extremes. In: Abdel Latef, A.A.H. (eds) Chitosan Nanoparticles for Plant Resilience Under Abiotic and Biotic Stressors. Smart Nanomaterials Technology. Springer, Singapore. https://doi.org/10.1007/978-981-95-6422-4_6

Anas M, Falak A, Hassan S et al (2025) Microbial interactions and bacterial responses to metal stress in plants: mechanisms, adaptations, and applications for sustainable agriculture. J Crop Health 77:36. https://doi.org/10.1007/s10343-025-01108-w

Appu M, Ramalingam P, Sathiyanarayanan A, Huang J (2021) An overview of plant defense-related enzymes responses to biotic stresses. Plant Gene 27:100302. https://doi.org/10.1016/j.plgene.2021.100302
Azooz MM, Abdel Latef AAH, Abdelmajeed WG et al. (2026) Biochar ameliorates salt stress in maize by modulating physiological responses and expression of SOS1 and NHX2 genes. Journal of Soil Science and Plant Nutrition. https://doi.org/10.1007/s42729-026-03123-w
Asensi-Fabado MA, Munné-Bosch S (2010) Vitamins in plants: occurrence, biosynthesis and antioxidant function. Trends Plant Sci 15:582–592
Bister-Miel F, Guignard JL, Bury M, Agier C (1985) Glutamine as an active component of casein hydrolysate: Its balancing effect on plant cells cultured in phosphorus deficient medium. Plant Cell Rep 4:161–163. https://doi.org/10.1007/BF00571307
Cho J, Lee EJ, Yoo KS et al (2009) Identification of candidate amino acids involved in the formation of blue pigments in crushed garlic cloves (*Allium sativum* L.). J Food Sci 74:C11–C16. https://doi.org/10.1111/j.1750-3841.2008.00986.x
Cui M, Gu M, Lu Y et al (2020) Glutamate synthase 1 is involved in iron-deficiency response and long-distance transportation in Arabidopsis. J Integr Plant Biol 62:1925–1941. https://doi.org/10.1111/jipb.12985
Dawood MFA, Tahjib-Ul-Arif MD, Shirazy BJ, Abdel Latef AAH (2025) Unraveling the role of κ-carrageenan on the combined effect of drought and chromium stress in wheat (Triticum aestivium L.). Plant Physiology and Biochemistry 223,J109781
Divekar PA, Narayana S, Divekar BA et al (2022) Plant secondary metabolites as defense tools against herbivores for sustainable crop protection. Int J Mol Sci 23:2690. https://doi.org/10.3390/ijms23052690
Fardus J, Hossain MS, Fujita M (2021) Modulation of the antioxidant defense system by exogenous l-glutamic acid application enhances salt tolerance in lentil (*Lens culinaris* Medik.). Biomolecules 11:587. https://doi.org/10.3390/biom11040587
Farid M, Ali S, Zubair M et al (2018) Glutamic acid assisted phyto-management of silver-contaminated soils through sunflower; physiological and biochemical response. Environ Sci Pollut Res 25:25390–25400. https://doi.org/10.1007/s11356-018-2508-y
Fathi A, Ghadirnezhad Shiade SR, Parmoon G et al (2024a) Bioremediation of heavy metals contaminated soils using nanotechnology. In: Husen A, Iqbal M, Ditta A et al (eds) Bio-organic amendments for heavy metal remediation. Elsevier, pp 611–628
Fathi A, Shiade SRG, Ali B, Zeidali E (2024b) Plant growth, development, and photosynthesis in cereals under salt stress. In: Pessarakli M (ed) Handbook of photosynthesis, 4th edn. CRC Press, Boca Raton
Fathi A, Shiade SRG, Kianersi F et al (2024c) Photosynthesis in cereals under drought stress. In: Pessarakli M (ed) Handbook of photosynthesis, 4th edn. CRC Press, pp 568–576
Fathi A, Shiade SRG, Mohamed Ait-El-Mokhtar VDR (2024d) Crop photosynthesis under climate change. In: Pessarakli M (ed) Handbook of photosynthesis, 4th edn. CRC Press, Boca Raton, pp 203–210
Feng S, Pan C, Ding S et al (2021) The glutamate receptor plays a role in defense against *Botrytis cinerea* through electrical signaling in tomato. Appl Sci 11:11217
Franzoni G, Cocetta G, Ferrante A (2021) Effect of glutamic acid foliar applications on lettuce under water stress. Physiol Mol Biol Plants 27:1059–1072. https://doi.org/10.1007/s12298-021-00984-6
Franzoni G, Cocetta G, Trivellini A et al (2022) Effect of exogenous application of salt stress and glutamic acid on lettuce (*Lactuca sativa* L.). Sci Hortic (Amsterdam) 299:111027. https://doi.org/10.1016/j.scienta.2022.111027
Ghadirnezhad Shiade SR, Fathi A, Taghavi Ghasemkheili F et al (2023a) Plants' responses under drought stress conditions: effects of strategic management approaches—a review. J Plant Nutr 46:2198–2230. https://doi.org/10.1080/01904167.2022.2105720
Ghadirnezhad Shiade SR, Pirdashti H, Esmaeili MA, Nematzade GA (2023b) Biochemical and physiological characteristics of mutant genotypes in rice (*Oryza sativa* L.) contributing to salinity tolerance indices. Gesunde Pflanz 75:303–315. https://doi.org/10.1007/s10343-022-00701-7

Ghadirnezhad Shiade SR, Rahimi R, Zand-Silakhoor A et al (2024) Enhancing seed germination under abiotic stress: exploring the potential of nano-fertilization. J Soil Sci Plant Nutr 24:5319–5341. https://doi.org/10.1007/s42729-024-01910-x

Hayat K et al. (2026) Salinity stress management using chitosan nanoparticles. In: Abdel Latef, A.A.H. (eds) Chitosan Nanoparticles for Plant Resilience Under Abiotic and Biotic Stressors. Smart Nanomaterials Technology. Springer, Singapore. https://doi.org/10.1007/978-981-95-6422-4_4

Iqbal P, Ghani MA, Ali B et al (2021) Exogenous application of glutamic acid promotes cucumber (*Cucumis sativus* L.) growth under salt stress conditions. Emirates J Food Agric 33:407. https://doi.org/10.9755/ejfa.2021.v33.i5.2699

Jia Y, Zou D, Wang J et al (2017) Effects of γ-aminobutyric acid, glutamic acid, and calcium chloride on rice (*Oryza sativa* L.) under cold stress during the early vegetative stage. J Plant Growth Regul 36:240–253. https://doi.org/10.1007/s00344-016-9634-x

Jiang M, Jiang J, Li S et al (2020) Glutamate alleviates cadmium toxicity in rice via suppressing cadmium uptake and translocation. J Hazard Mater 384:121319. https://doi.org/10.1016/j.jhazmat.2019.121319

Jin L, Cai Y, Sun C et al (2019) Exogenous l-glutamate treatment could induce resistance against *Penicillium expansum* in pear fruit by activating defense-related proteins and amino acids metabolism. Postharvest Biol Technol 150:148–157. https://doi.org/10.1016/j.postharvbio.2018.11.009

Kim D-R, Kwak Y-S (2023) Endophytic Streptomyces population induced by L-glutamic acid enhances plant resilience to abiotic stresses in tomato. Front Microbiol 14:1180538. https://doi.org/10.3389/fmicb.2023.1180538

Kim D-R, Jeon C-W, Cho G et al (2021) Glutamic acid reshapes the plant microbiota to protect plants against pathogens. Microbiome 9:244. https://doi.org/10.1186/s40168-021-01186-8

Künstler A, Király L, Kátay G et al (2019) Glutathione can compensate for salicylic acid deficiency in tobacco to maintain resistance to tobacco mosaic virus. Front Plant Sci 10:1115

Kuźniak E, Skłodowska M (2004) Differential implication of glutathione, glutathione-metabolizing enzymes and ascorbate in tomato resistance to *Pseudomonas syringae*. J Phytopathol 152:529–536

Lee HJ, Kim JS, Lee SG et al (2017) Glutamic acid foliar application enhances antioxidant enzyme activities in Kimchi Cabbages leaves treated with low air temperature. Hortic Sci Technol 35:700–706. https://doi.org/10.12972/kjhst.20170074

Lee HJ, Lee JH, Wi S et al (2021) Exogenously applied glutamic acid confers improved yield through increased photosynthesis efficiency and antioxidant defense system under chilling stress condition in *Solanum lycopersicum* L. cv. Dotaerang Dia. Sci Hortic (Amsterdam) 277:109817. https://doi.org/10.1016/j.scienta.2020.109817

Lee K-T, Liao H-S, Hsieh M-H (2023) Glutamine metabolism, sensing and signaling in plants. Plant Cell Physiol 64:1466–1481. https://doi.org/10.1093/pcp/pcad054

Lei P, Xu Z, Ding Y et al (2015) Effect of poly (γ-glutamic acid) on the physiological responses and calcium signaling of rape seedlings (*Brassica napus* L.) under cold stress. J Agric Food Chem 63:10399–10406

Lei P, Xu Z, Liang J et al (2016) Poly(γ-glutamic acid) enhanced tolerance to salt stress by promoting proline accumulation in *Brassica napus* L. Plant Growth Regul 78:233–241. https://doi.org/10.1007/s10725-015-0088-0

Li R, Yang Y, Lou H et al (2025) Glutathione triggers leaf-to-leaf, calcium-based plant defense signaling. Nat Commun 16:1915. https://doi.org/10.1093/jxb/erac399

Liao HS, Chung YH, Hsieh MH (2022) Glutamate: a multifunctional amino acid in plants. Plant Sci 318:111238. https://doi.org/10.1016/j.plantsci.2022.111238

Ma H, Li P, Liu X et al (2022) Poly-γ-glutamic acid enhanced the drought resistance of maize by improving photosynthesis and affecting the rhizosphere microbial community. BMC Plant Biol 22:11. https://doi.org/10.1186/s12870-021-03392-w

Matysiak K, Kierzek R, Siatkowski I et al (2020) Effect of exogenous application of amino acids L-arginine and glycine on maize under temperature stress. Agronomy 10:769. https://doi.org/10.3390/agronomy10060769

Meena M, Divyanshu K, Kumar S et al (2019) Regulation of L-proline biosynthesis, signal transduction, transport, accumulation and its vital role in plants during variable environmental conditions. Heliyon 5:e02952. https://doi.org/10.1016/j.heliyon.2019.e02952

Pang X, Lei P, Feng X et al (2018) Poly-γ-glutamic acid, a bio-chelator, alleviates the toxicity of Cd and Pb in the soil and promotes the establishment of healthy *Cucumis sativus* L. seedling. Environ Sci Pollut Res 25:19975–19988. https://doi.org/10.1007/s11356-018-1890-9

Quan J, Zheng W, Tan J et al (2022) Glutamic acid and poly-γ-glutamic acid enhanced the heat resistance of Chinese Cabbage (*Brassica rapa* L. ssp. pekinensis) by improving carotenoid biosynthesis, photosynthesis, and ROS signaling. Int J Mol Sci 23:11671. https://doi.org/10.3390/ijms231911671

Rossi S, Chapman C, Yuan B, Huang B (2021) Glutamate acts as a repressor for heat-induced leaf senescence involving chlorophyll degradation and amino acid metabolism in creeping bentgrass. Grass Res 1:1–10

Saleem MH, Zafar S, Javed S et al (2024) Modulatory effects of glutamic acid on growth, photosynthetic pigments, and stress responses in olive plants subjected to cadmium stress. J King Saud Univ Sci 36:103540. https://doi.org/10.1016/j.jksus.2024.103540

Sun C, Jin L, Cai Y et al (2019) L-Glutamate treatment enhances disease resistance of tomato fruit by inducing the expression of glutamate receptors and the accumulation of amino acids. Food Chem 293:263–270

Svietlova N, Zhyr L, Reichelt M et al (2024) Glutamine as sole nitrogen source prevents induction of nitrate transporter gene NRT2. 4 and affects amino acid metabolism in Arabidopsis. Front Plant Sci 15:1369543

Taghavi Ghasemkheili F, Jenabiyan M, Ghadirnezhad Shiade SR et al (2023) Screening of some endophytic fungi strains for zinc biofortification in wheat (*Triticum aestivum* L.). J Soil Sci Plant Nutr 23:5196–5206. https://doi.org/10.1007/s42729-023-01392-3

Toyota M, Spencer D, Sawai-toyota S et al (2018) Glutamate triggers long-distance, calcium-based plant defense signaling. Science 6:1112–1115

Umer S, Akram NA, Ashraf M, Abdel Latef AAH (2026) Pre-sowing application of methionine and proline mitigates drought stress in two canola (Brassica napus L.) Cultivars. Russian Journal of Plant Physiology 73:48

Westfall CS, Muehler AM, Jez JM (2013) Enzyme action in the regulation of plant hormone responses. J Biol Chem 288(19304–19):311

Xu Z, Lei P, Pang X et al (2017) Exogenous application of poly-γ-glutamic acid enhances stress defense in *Brassica napus* L. seedlings by inducing cross-talks between Ca2+, H2O2, brassinolide, and jasmonic acid in leaves. Plant Physiol Biochem 118:460–470. https://doi.org/10.1016/j.plaphy.2017.07.015

Xu Z, Ma J, Lei P et al (2020) Poly-γ-glutamic acid induces system tolerance to drought stress by promoting abscisic acid accumulation in *Brassica napus* L. Sci Rep 10:252. https://doi.org/10.1038/s41598-019-57190-4

Xue N, Zhan C, Song J et al (2022) The glutamate receptor-like 3.3 and 3.6 mediate systemic resistance to insect herbivores in Arabidopsis. J Exp Bot 73:7611–7627. https://doi.org/10.1093/jxb/erac399

Yang J, Sun C, Fu D, Yu T (2017) Test for l-glutamate inhibition of growth of Alternaria alternata by inducing resistance in tomato fruit. Food Chem 230:145–153

Yuksel EA (2024) L-glutamic acid and L-aspartic acid supplementation mitigate heavy metal-induced stress in *Phaseolus vulgaris* L. Phyton 93:2189

Zheng Y, Luo L, Wei J et al (2018) The glutamate receptors AtGLR1.2 and AtGLR1.3 increase cold tolerance by regulating jasmonate signaling in *Arabidopsis thaliana*. Biochem Biophys Res Commun 506:895–900. https://doi.org/10.1016/j.bbrc.2018.10.153

Chapter 14
Humic Acid: Boosting Soil and Plant Resilience Against Stress

A. K. M. Golam Sarwar, Md. Riyadh Arefin, and Mahmud Hossain Sumon

Abstract Humic acid (HA), a decomposition product of organic matter, is the key component of humic substances (HSs). It plays a pivotal role in enhancing soil structure, nutrient availability, and plant growth. The complex molecular composition of HA enables it to chelate nutrients, buffer the soil pH, and improve water retention. HA acts as a biostimulant, stimulating root architecture, enzyme activity, and hormone signalling pathways, thereby improving nutrient acquisition and stress tolerance in plants, and therefore, necessitates its application in crop production. HAs can be applied in different methods—foliar sprays, seed treatments/priming, and soil incorporation, while their efficacy depends on crop types, soil conditions, and environmental stresses (salinity, drought, extreme temperature, waterlogging, and heavy metal contamination). Additionally, HAs enhance plant defence against biotic stresses (insects and diseases) by altering microbial communities and inducing resistance-related genes. HAs also show promise in environmental and industrial applications. Overall, the multifunctional properties of HA make it a sustainable tool for improving crop productivity and ecological resilience.

Keywords Humic acid · Biotic stress · Drought · Waterlogging · Temperature · Salinity · Heavy metal

A. K. M. G. Sarwar (✉)
Department of Crop Botany, Bangladesh Agricultural University, Mymensingh, Bangladesh
e-mail: drsarwar@bau.edu.bd

M. R. Arefin
Department of Crop Botany, Bangladesh Agricultural University, Mymensingh, Bangladesh

Botany Division, Bangladesh Tea Research Institute, Moulvibazar, Bangladesh
e-mail: riyadh@btri.gov.bd

M. H. Sumon
Department of Soil Science, Bangladesh Agricultural University, Mymensingh, Bangladesh
e-mail: mahmud.ss@bau.edu.bd

A. A. H. Abdel Latef (ed.), *Harnessing Vitamins and Organic Acids for Plant Stress Resilience*, Sustainability Sciences in Asia and Africa,
https://doi.org/10.1007/978-981-95-8593-9_14

Abbreviations

ABA	Abscisic acid
APX	Ascorbate peroxidase
CAT	Catalase
EL	Electrolyte leakage
FA	Fulvic acid
GR	Glutathione reductase
GS	Glutamine synthetase
Gs	Stomatal conductance
H_2O_2	Hydrogen peroxide
HA	Humic acid
HMs	Heavy metals
HSP	Heat shock protein
HSs	Humic substances
Hu	Humin
MDA	Malondialdehyde
NR	Nitrate reductase
PAL	Phenylalanine ammonia-lyase
Pn	Photosynthetic rate
POD	Peroxidase
ROS	Reactive oxygen species
RWC	Relative water content
SA	Salicylic acid
SOD	Superoxide dismutase
TCA	Tricarboxylic acid
Tr	Transpiration rate

14.1 Introduction

The Latin word "*Humus*" means "Earth" or "ground". Humic substances (HSs) are dissolved organic compounds released during the decomposition of plant and animal remains and are characterized by their yellow to brown colour, high molecular mass, and refractory nature.

These acids are mixtures of acids that can be separated into humic acid (HA), fulvic acid (FA), and humin (Hu) fractions based on variations in their solubility, along with ulmic acid and a few microelements (Fig. 14.1) (Goel and Dhingra 2021). HA, the main constituent of HSs, is often the most complex structure with a large molecular mass, making it extremely stable and resistant to microbial breakdown (Rathor et al. 2023). Unique chemical structures cannot be described by these structures and are operationally defined by extraction and fractionation techniques. HAs are macromolecular fractions of HSs found in terrestrial soil, water, and

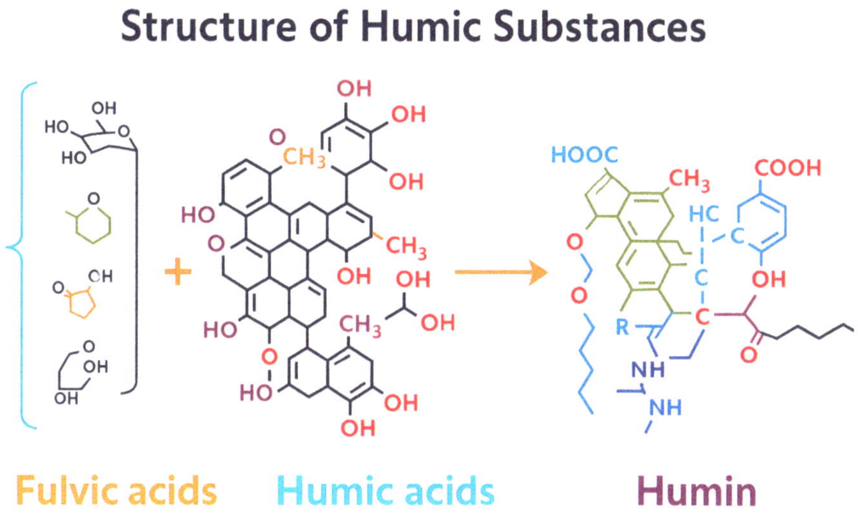

Fig. 14.1 Simplified structure of humic substances (AI modification from Yang and Antonietti 2020)

sediment (Maccarthy 2001) and are soluble in dilute bases but precipitate via acidification of alkaline soil extracts. HAs have varying chemical compositions due to their geographical origin, age, climate, and biological conditions; however, the average characteristics of the extracted HAs are strikingly comparable and do not degrade further (Yang and Antonietti 2020). The molecular weights range from 2.0 to 1300 kDa, and various functional groups, including phenolic, carboxylic acid, enolic, quinone, sugars, and peptides, are characterized as both hydrophilic (e.g. hydroxyl groups) and hydrophobic groups (such as aliphatic chains and aromatic rings) (de Melo et al. 2016). Their functional groups significantly influence key properties, such as ion exchange capacity, water retention, dispersion behaviour, cohesion energy, and their ability to form complexes (Yang and Antonietti 2020). On average, HAs—regardless of their source, including commercial products—consist of approximately 50% carbon (C), 35% oxygen, and 5% hydrogen (H), with nitrogen (N) and sulphur (S) comprising the remainder (de Melo et al. 2016). Notably, HAs rich in acidic and phenolic hydroxyl groups are more effective at mitigating environmental stresses, particularly salinity stress (Meng et al. 2023).

HAs enhance plant nutrition by upregulating genes that encode root plasma membrane proton pumps (PM H^+-ATPases), thereby facilitating ion transport into plant tissues (de Moura et al. 2023). They also modulate the expression of genes involved in primary nutrient transporters in roots and increase the activity of key enzymes related to nutrient utilization, such as nitrate reductase (NR), glutamine synthetase (GS), glutamate synthase, and phosphoenolpyruvate carboxylase (de Moura et al. 2023). The importance of HSs is increasingly acknowledged due to their multifunctional roles—as microbial carriers, plant protective agents under stress conditions, and effective biostimulants (de Melo et al. 2016). As a major component of soil organic matter, HAs play a vital role in maintaining and improving soil physicochemical and biological properties (Meganid et al. 2015). They contribute to soil quality and productivity by breaking down clay particles, increasing micronutrient availability, improving water retention, and promoting seed germination, plant growth, and nutrient uptake. Furthermore, HA application stimulates soil microbial activity, notably increasing the enzymatic functions of microorganisms, such as dehydrogenase and alkaline phosphatase activities (Fathy et al. 2023). It is used in industrial fields such as concrete control, natural indigo dyeing, and leather tanning. HA removes toxic metals, organic chemicals, and pollutants in environmental applications. The immune-modulating effects, detoxifying and gastrointestinal health-supporting properties, and antioxidant qualities of HAs make them beneficial for human health (Dauletbay et al. 2024). Owing to their structural diversity, they have potential for use in pharmaceutical applications, including drug delivery, wound healing, antioxidant treatments, antibacterial activities, estrogenic properties, and destruction of bacterial biofilms. Studies have shown that HA can protect against cancer and cancer-causing viruses, with some therapies reversing deadly cancers (Rupiasih and Vidyasagar 2005).

HAs accelerate physiological and molecular responses through phytohormonal regulation, signal transduction, innate phytohormone biosynthesis, root structural modifications, plant metabolism alterations, and increased nutrient use efficiency (Nunes et al. 2019; Rathor et al. 2023; Santos et al. 2025). This chapter explores the potential of HA as an organic biostimulant for plant protection and performance. This study examined the role of HA on soil properties, plant growth and physiology, and how it can increase plant resilience to abiotic and biotic stressors.

14.2 Methods for the Application of HAs in Agriculture

Depending on the type of soil and crop, growth stage, and available equipment, HA can be applied to plants or soil via foliar sprays, seed treatments, root drenching, or soil incorporation techniques (Garcia-Caparros et al. 2025). Seeds treated with 50% HS increased the germination percentage and reduced the time required for germination in rice (Adetumbi et al. 2019). In contrast, maize seeds soaked in 1% HA for 3 min achieved 100% germination (Nandhini et al. 2018). Waqas et al. (2014) assessed HA application methods—soil application, foliar spray, and seed priming—in a triplicate field experiment and reported a dose-dependent increase in the yield components and grain yield of mung bean. Across all three HA applications, soil application was found to be the most effective. Oktem and Oktem (2020) reported that the combination of seed, soil, and foliar application of HA produced a relatively high yield of maize, although the increased yield may result from relatively high HA levels. In potato, three HA application methods were evaluated, including soil application via drip irrigation, foliar spray, and a combined method of spray and drip (50% foliar spray +50% drip method). The combined spray and drip method was the most effective, followed by the sprayer method (El Komy et al.

2021). Kutlu and Gulmezoglu (2023) evaluated three application methods—soil, foliar, and a combination of soil and foliar applications of HA—to eliminate increasing salt stresses (0, 50, 100, and 150 mM) on bean plants. They reported that a combination of soil and foliar application of HA was more effective than other methods. de Moura et al. (2023) noted that factors such as the structural complexity and origin of the humic material, dosage, phenological stage, number of applications, characteristics of plant species, and local experimental and environmental conditions lead to variations in their effects, even when they are applied to the same plant species. Both the soil and foliar fertilization methods increase production via distinct mechanisms. Some studies suggest that soil application is more efficient, whereas others suggest that foliar spraying is better and that there is a complementary action between them (de Moura et al. 2023).

14.3 Effects of HA on Soil Properties and Plant Growth

HA in soil originates from the natural breakdown of soil organic materials and plays a critical role in enhancing soil functionality, particularly in terms of water retention and structural integrity. The role of HAs in regulating moisture dynamics by promoting soil aggregate formation and increasing the water retention capacity in soil contributes to soil formation. With the development of stable soil aggregates, HA increases the surface area and infiltration while minimizing runoff and compaction in the root zone. Olk et al. (2018) further demonstrated that HA application can increase soil aggregate stability by up to 20%, which contributes to better porosity, air circulation, and hydraulic conductivity—factors essential for deeper root access to water and nutrients. These properties are especially beneficial in areas with unpredictable weather patterns, where soil resilience is critical.

HA also influences soil chemistry through elemental complexation and pH buffering. Its functional groups, hydroxyl/phenolic and carboxylic moieties, are responsible for binding essential cations such as Fe^{2+}, Zn^{2+}, Ca^{2+}, and Mg^{2+}. This chelation enhances nutrient availability by preventing fixation, precipitation, and leaching. Furthermore, the interaction of HA with exchangeable soil cations stabilizes the pH, creating favourable conditions for microbial activity and nutrient cycling (Xiong et al. 2023). Commercially, more than 63% of humic products are derived from peat, which is known for its high chelating efficacy, especially in powdered formulations suitable for precision agriculture. In saline and contaminated soils, HA can sequester excess sodium (Na^{+}) and chelate heavy metals such as cadmium (Cd), thus reducing their phytotoxicity and promoting healthier plant growth (Wang et al. 2024). However, while HA improves the availability of micronutrients in the soil matrix, plant uptake may not always align due to the stability of HA–metal complexes, emphasising the need to match HA type, by molecular weight and composition, with the intended agronomic outcomes. Adding HA to sandy soil can enhance the plant growth-promoting bacterial community, reduce drought stress, promote plant development, and increase C-sequestration capacity (Hoefle et al. 2024). HSs and microorganisms combinedly aid in the biodegradation of soil organic pollutants and limit metal bioavailability by inhibiting electron transfer to CO_2, hence reducing CH_4 generation in anoxic environments (Kulikova and Perminova 2021).

On the nutrient front, HA significantly enhances N retention by stabilizing ammonium (NH_4^+) in the soil, reducing losses through volatilization and leaching from the soil. A meta-analysis by Ma et al. (2024) revealed substantial gains in N-use efficiency (27%), N uptake (17%), and crop yield (12%) when HA was used in combination with moderate N fertilization. Phosphorus (P) availability is also enhanced through dual mechanisms: HA disrupts its sorption onto aluminium and iron (Fe) oxides and stimulates phosphatase enzyme activity. High-temperature HA–phosphate formulations have been shown to increase plant-available P by up to 47.7%, with notable benefits in crops such as tomato and sugarcane (Hartina et al. 2025). These improvements extend to the soil microbiome; HA stimulates beneficial microbes, including P-solubilizing bacteria, thereby improving nutrient turnover and soil fertility. In saline-sodic conditions, HA application decreased soil bulk density, pH, electrical conductivity, and total alkalinity, while increasing soil organic matter, total N, and accessible potassium (K) levels (Zhaoyang et al. 2025). Additionally, the average 35.5% yield increase was reported in maize crops.

In terms of plant development, HA acts not only as a soil conditioner but also as a potent biostimulant that directly affects metabolic processes (Fig. 14.2). It modulates both C and N metabolism by increasing the activity of enzymes such as NR, GS, and glutamate dehydrogenase (GDH), supporting the efficient assimilation of N into amino acids (Canellas et al. 2013; Nardi et al. 2016). HA also promotes glycolytic and tricarboxylic acid (TCA) cycle activity, fuelling energy production and biomass accumulation (Nardi et al. 2007). These systemic effects extend to foliar

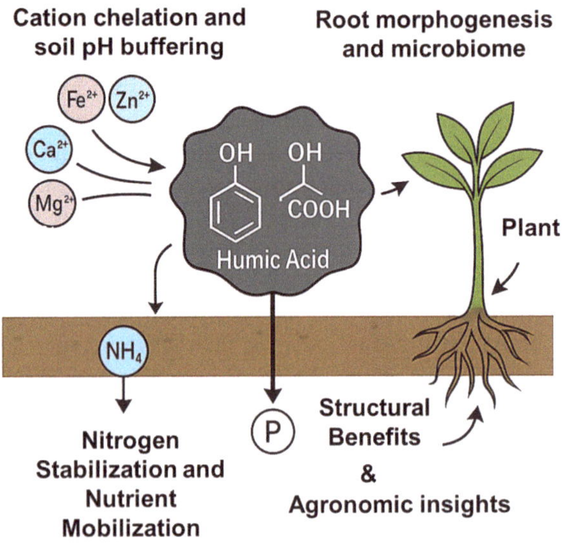

Fig. 14.2 Effects of humic acids on soil physicochemical properties and plant growth

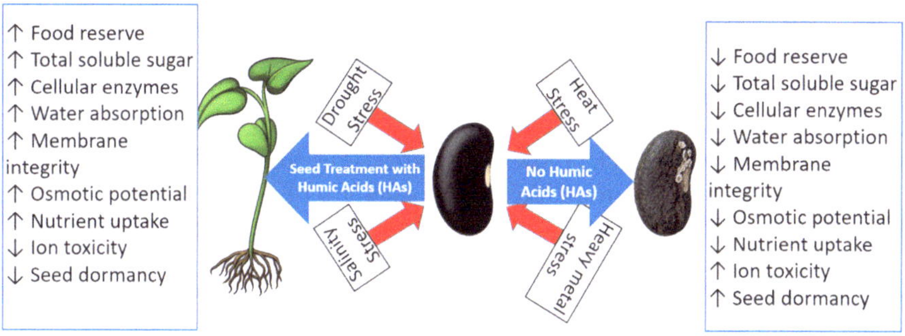

Fig. 14.3 Physiological effects of priming seeds with humic acids under stress conditions

applications, where HA influences the biosynthesis of cytokinin-like compounds such as trans-zeatin, reorganizing the plant's internal C and N pathways to optimize growth and nutrient distribution. Seed priming or treatment with HA helps prevent oxidative damage and induces biochemical changes such as the activation of cellular enzymes, the disruption of seed dormancy, improved water absorption, the stabilization of membrane integrity, the restoration of osmotic potential, and the accumulation of essential nutrients (Fig. 14.3). These modifications facilitate faster and more successful germination under stress (Bahrani and Pourreza 2012).

HA's influence on root morphogenesis is closely tied to nitric oxide (NO) signalling and auxin-like hormonal activity. Maize research has shown that HA-induced NO accumulation enhances lateral root emergence, root hair development, and overall root biomass (Mora et al. 2012; Zandonadi et al. 2010). This is mediated through increased activity of plasma membrane H^+-ATPases, which regulate nutrient uptake by creating electrochemical gradients. Proteomic analyses have confirmed that HA treatment upregulates proteins related to energy metabolism, cytoskeletal reorganization, and ion transport while also activating reactive oxygen species (ROS) pathways that signal root cell division and elongation (Gilroy et al. 2014; Pizzeghello et al. 2020). These physiological adjustments result in expanded root systems with increased absorptive capacity, which is especially valuable in nutrient-poor or compacted soils.

Finally, HA contributes to stress mitigation and nutrient uptake synergy through hormone regulation and microbial interactions. It enhances ion transport by stimulating H^+-ATPase activity and facilitates the uptake of key nutrients such as K^+, Ca^{2+}, NO_3^-, and phosphate. HA mimics plant hormones by influencing auxin transport (e.g. via PIN proteins) and amplifying cytokinin signalling, which not only supports root and shoot growth but also enhances photosynthetic efficiency and water use. Furthermore, HA combines with phosphate-solubilizing bacteria and mycorrhizal fungi, increasing P efficiency even when the total P content remains unchanged (Jindo et al. 2020). Its antioxidant effects, evident through upregulated superoxide dismutase (SOD) and catalase (CAT) activity, support plant resilience against different plant stresses.

14.4 Stresses—Types and Remediation with HAs

Plant stress refers to unfavourable conditions that hinder a plant's metabolism, growth, or development, resulting from biotic (pathogens, parasites, herbivores, and weeds) and abiotic (high or low temperature, waterlogging or drought, salinity, and toxins) factors. This stress can lead to significant crop yield loss, potentially threatening global food security, and affects various systems across the plant body. Plant species respond to stress conditions in their own ways, and their mitigation strategies are also different and species-dependent. Global environmental changes are increasing the frequency and complexity of the effects of multiple types of stress on plants, particularly heat stress, drought, and freshwater resource limitations, and may worsen soil salinization. These stresses can negatively impact crop productivity individually but can significantly influence plant growth and health when combined. Organic amendments (/biostimulants) improve soil fertility and ameliorate the deleterious effects of stresses for sustainable agricultural production (Ennab et al. 2023).

HAs influence plant growth through their biologically active components, which are organized within a supermolecular structure held together by weak H and covalent bonds. Organic acids can disrupt these bonds, releasing small bioactive molecules. These molecules can modulate physiological processes and metabolic pathways, increasing plant tolerance to both biotic and abiotic stresses and ultimately promoting improved growth (Canellas and Olivares 2014; Rathor et al. 2023). Glycolysis and the tricarboxylic acid (TCA) cycle are fundamental metabolic pathways that provide energy and key intermediates necessary for normal plant growth and development. Studies have shown that HA treatment in maize leads to the upregulation of genes associated with glycolysis and the TCA cycle, indicating increased metabolic activity (Nardi et al. 2007).

HAs, like short-chain organic acids, lower the intracellular pH, as indicated by the increased fluorescence of the BCECF probe (2,7-bis(2-carboxyethyl)-5(and 6)-carboxyfluorescein, acetoxymethyl ester) (Baía et al. 2020). This reduction in cytosolic pH is persistent and is accompanied by elevated levels of protein kinases involved in cell energy sensing and transcription factors linked to stress signalling. HAs can be considered chemical priming agents that activate plant responses to abiotic stress and improve tolerance to such stresses at appropriate concentrations.

14.4.1 Abiotic Stresses

Abiotic stress refers to the negative impact of non-living elements, such as temperature, water, light, salinity, toxins, and others, on living organisms, requiring the non-living variable to alter the environment beyond its typical range to significantly affect population performance or individual physiology.

14.4.1.1 Temperature Stresses

Cardinal temperatures are the minimum, optimum, and maximum temperatures a plant can withstand, defining its growth and development boundaries. These temperatures can change depending on developmental stage and plant species (https://glossary.ametsoc.org/wiki/Cardinal_temperatures). Temperature stress can negatively impact plant growth and development when temperatures drop below or rise above a plant's minimum or maximum temperature (Fig. 14.4).

Low-temperature stress is a major abiotic factor that limits plant geographic distribution and crop productivity by increasing ROS levels and disrupting N metabolism. HAs increased cold tolerance by increasing proline and abscisic acid (ABA) levels, increasing polyphenol oxidase activity, and reducing the transpiration rate, stomatal conductance, and net photosynthetic rate (Li et al. 2023; Zhu et al. 2025). Li et al. (2023) demonstrated that HA application to zucchini seedlings mitigated oxidative damage and increased antioxidant enzyme activity under low-temperature stress. Transcriptomic analysis revealed 17 commonly induced differentially expressed genes (DEGs) in HA-treated leaves, suggesting that HA supplementation reprograms plant metabolism and modulates the expression of

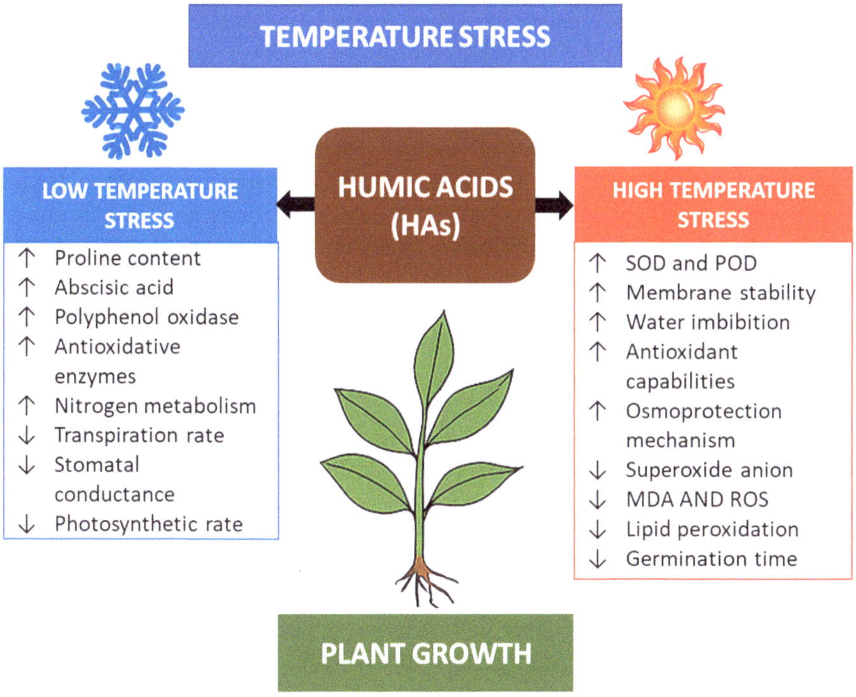

Fig. 14.4 Physiological effects of temperature extremes in plants and the mitigating role of humic acids. SOD, superoxide dismutase; POD, peroxidase; MDA, malondialdehyde; ROS, reactive oxygen species

genes related to ROS detoxification, phytohormone signalling, and stress response pathways. Similarly, Zhu et al. (2025) reported that HA treatment in melon significantly improved plant height, stem diameter, biomass, chlorophyll content, photochemical efficiency, root development, and the activity of antioxidative enzymes such as SOD and CAT. HA also increased nitrate degradation, ammonium-N synthesis, and overall N metabolism by upregulating genes associated with the N and proline pathways.

HAs enhance crop resilience to high-temperature stress by stimulating antioxidant defence mechanisms—particularly by increasing SOD and peroxidase (POD) activities—while reducing the accumulation of superoxide anions and hydrogen peroxide (H_2O_2), thus improving overall crop performance (Abdellatif et al. 2017; Choi et al. 2024; Poomani et al. 2023). High temperatures negatively impact seed germination and early development in lentils, mainly through increased membrane leakage and decreased chlorophyll content (Poomani et al. 2023). HA seed priming has been shown to mitigate these effects by increasing germination rates, seedling vigour, and the membrane stability index while decreasing the mean germination time and H_2O_2, malondialdehyde (MDA), and carbohydrate accumulation. HA promotes rapid water uptake during germination, enhances antioxidant activity, supports osmoprotection mechanisms, and reduces ROS levels and lipid peroxidation, thereby stabilizing cellular membranes. In spinach, HA improved heat tolerance through early root development, increased transpiration, and enhanced water use efficiency under daily heat stress (Choi et al. 2024). However, under combined daily heat and drought stress, HA supplementation had limited protective effects, suggesting that it may not fully alleviate prolonged, compound stress conditions. Additionally, Abdellatif et al. (2017) reported that HA application at 14.4 kg ha^{-1} significantly enhanced tomato vegetative growth, flowering, and yield under hot continental summer conditions. While improvements were observed in most growth and productivity parameters, the effects on fruit number per plant, vitamin C content, and total soluble solids were minimal compared with those of the control.

HAs boost plant stress tolerance by reducing ROS overaccumulation and the MDA content, increasing the chlorophyll content and thylakoid structure, and promoting photosynthetic efficiency (Li et al. 2023). They also promote growth and nutrient assimilation by inducing the expression of genes and genes encoding enzymes, including nitrate absorption and assimilation genes, in maize roots (García et al. 2012; Quaggiotti et al. 2004). HA upregulates genes involved in the stress response, particularly heat response. Heat stress induces heat shock protein (HSP)-encoding genes, including *HSP101*, *HSP81.1*, *HSP26.5*, *HSP23.6*, and *HSP17.6A*, which protect against thermal denaturation and facilitate denatured substrate refolding (Cha et al. 2020). Wild-type plants grown in HA were heat-tolerant, whereas the *Arabidopsis HSP101* null mutant was insensitive. HA accelerates the transcription of HSPs, suggesting that *HSP101* is a molecular target of HA that promotes heat stress tolerance in *Arabidopsis*. This information contributes to understanding the genetic and agronomic traits acquired by HA, which confers tolerance to environmental stresses in plants (Cha et al. 2020).

14.4.1.2 Water Stress

Water stress, caused by too much or too little water, significantly reduces crop productivity and global food security. Human activities lead to increasingly extreme climate events, such as prolonged dry spells, intense rains, and flash floods (Pitann et al. 2024).

14.4.1.3 Waterlogging Stress

Waterlogging, even for a few days, can cause growth depression in maize plants due to root anoxia or hypoxia, leading to ROS formation, membrane oxidation, and chloroplast degradation (Fig. 14.5; Kawałko et al. 2023; Pitann et al. 2024). Pitann et al. (2024) reported that foliar application of HAs, depending on their composition and application rate, can mitigate waterlogging stress and maintain plant productivity during temporary submergence. This effect is due to the disappearance of anthocyanin discolouration and an increase in Zn status, which is significantly reduced under waterlogged conditions (Pitann et al. 2024). Kawałko et al. (2023) noted that HAs in periodically flooded soils are more aromatic and lignin-like than those in unflooded Fluvisols, possessing fewer C and H atoms, which results in a less aliphatic or aromatic structure. The higher H/C ratio of HAs in Fluvisols after drainage also contributes to their aromatic character. HAs do not lose as many -OCH3 groups, leading to a lignin-like aromatic structure in flooded areas with periodically reductive conditions (Kawałko et al. 2023). The aromatic structures of HAs enhance plant development under stress and stimulate root growth.

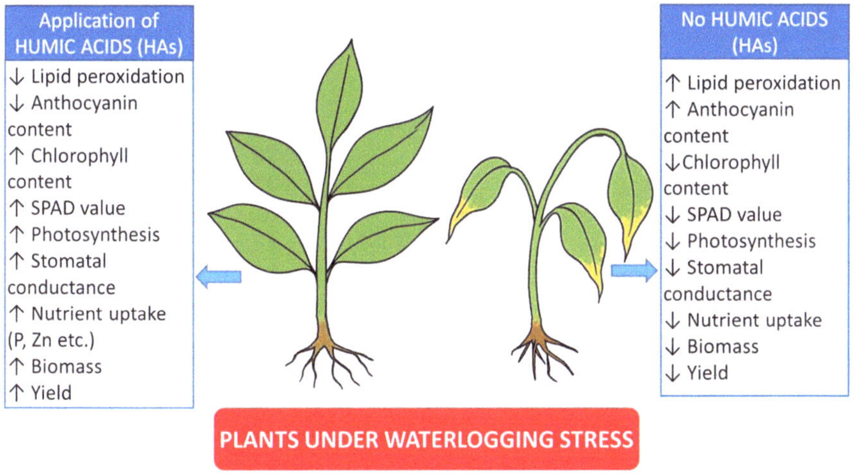

Fig. 14.5 Effects of waterlogging stress on the physiology, growth, and yield of plants and the ameliorative role of humic acids. SPAD, soil plant analysis development

14.4.1.4 Drought Stress

Drought causes a decrease in water potential, which impacts plant growth by causing transpiration to exceed the amount of absorbed water. This affects the photosynthetic process and crop yield (Fig. 14.6). Most plants adapt to this water deficiency through physiological, anatomical, and cellular changes, favouring greater water entry and efficient use (Abu-Ria et al. 2024; Aytaç et al. 2024; Fathy et al. 2023; Hernández-Campos et al. 2021; Ibrahim et al. 2024; Kaya et al. 2020; Roy et al. 2025). HAs also enhanced plant growth, stress resilience, and yield in agriculture by altering gene expression (Garcia-Caparros et al. 2025). This includes stress-responsive, nutrient uptake, hormone-related, and defense-related genes, and gene expression networks as well. Nazi et al. (2014) reported relatively high germination percentages in wheat seeds primed with HA under polyethylene glycol (PEG)-mediated drought stress. Drought impacts water content and fresh weight, soil nutrient transport, and nutrient availability. Nutrient diffusion and the mass flow of water-soluble nutrients (SO_4^{2-}, NO_3^-, Ca, Mg, and Si) are strongly disturbed and therefore affect plant growth. Drought and P deficiency reduce the biomass, grain yield, chlorophyll content, Fv/Fm, relative water content (RWC), and antioxidant activity of maize plants (Kaya et al. 2020). HA enhances leaf P and yield under stress conditions, and the greatest increase in yield and plant traits was observed under combined stress. Fathy et al. (2023) reported that HA combined with plant growth-promoting *Rhizobacteria* inocula can reduce fertilizer requirements, minimize environmental pollution, and reduce crop water requirements to 75% evapotranspiration without significantly affecting faba bean yield. It promotes plant growth, increases crop production, and reduces water requirements and production costs under drought stress by increasing the availability and uptake of NPK.

HA application to drought-stressed plants increased the chlorophyll content, water content, shoot and root length, and fresh–dry weight and decreased the

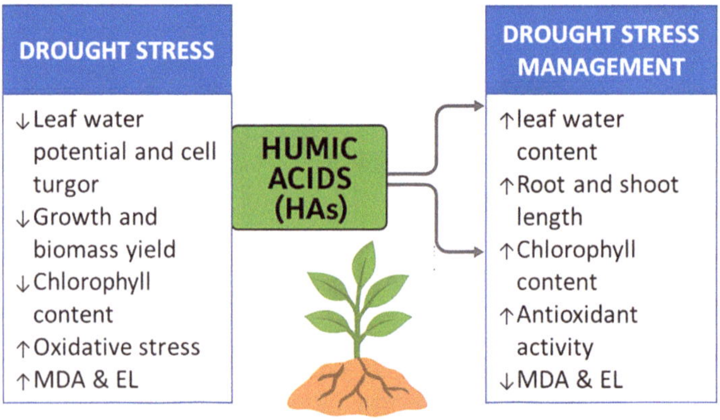

Fig. 14.6 Effects of drought stress on the physiology, growth, and yield of plants and the ameliorative role of humic acids. MDA, malondialdehyde; EL, electrolyte leakage

MDA, ion leakage, and proline levels in tomatoes (Aytaç et al. 2024). It also increased CAT gene expression and reduced glutathione reductase (GR) gene expression, but there was no significant change in the SOD and ascorbate peroxidase (APX) gene expression levels. HA positively affects the physiological and metabolic responses of tomato plants by reducing the adverse effects of drought stress (Aytaç et al. 2024). HA stimulated plant growth in maize populations, increasing height, root number, and biomass production. It attenuated water deficit effects, with the response varying with population and HA concentration. HA reduces water stress effects by increasing growth parameters and protein synthesis (Hernández-Campos et al. 2021).

Water stress negatively affects plant physiology and productivity, but HSs, particularly HA, have shown promising mitigation potential. Ibrahim et al. (2024) reported that water-stressed broccoli plants presented reduced growth, yield, leaf chlorophyll content, and nutrient content, alongside increases in leaf proline, curd dry matter, total soluble solids, and water use efficiency. However, HA soil application effectively alleviated these adverse effects, presenting a sustainable strategy for enhancing broccoli yield in arid and semiarid regions. Drought stress also significantly reduced biomass accumulation in maize and sorghum, impairing growth, RWC, photosynthetic pigment levels, and CO_2 assimilation. Moreover, it triggered increases in root traits, H_2O_2, MDA, and electrolyte leakage (EL). HA treatment improved plant performance under these conditions, with maize responding more positively than sorghum did. HA application increased the total chlorophyll content, stomatal conductance, RWC, sucrose content, total soluble sugars, carbohydrate, proline, and total soluble protein contents (Abu-Ria et al. 2024).

Lengrand et al. (2024) investigated the effects of HS and osmotic stress—individually and in combination—on the endophytic bacterial communities of tomato plants. Osmotic stress alone reduced tomato growth, whereas HS application promoted growth and increased the abundance of beneficial bacterial genera such as *Frigoribacterium*, *Roseateles*, and *Hymenobacter*. Under nonstress conditions, HS did not significantly affect plant growth but altered the endophytic profile, increasing *Hymenobacter* and reducing *Sphingomonas*, suggesting that HS may act as a biostimulant by modulating microbial communities. Roy et al. (2025) assessed the effects of four growing conditions on two wheat varieties and reported that water deficit stress significantly reduced morphological, physiological, and yield traits. Notably, foliar application of HA was more effective than soil application in alleviating drought stress, leading to a yield increase of up to 25.19% compared with the 13.86% achieved through soil treatment.

HA enhances plant growth under drought stress by improving osmotic adjustment, antioxidant capacity, and photosynthesis. This could improve soil hydrophysical properties and increase plant drought tolerance (Aytaç et al. 2024). HAs stimulated the antioxidant enzymatic system in plants, converting ROS into harmless ones. They change root cell membrane functionality, reduce hydraulic conductivity and promote foliar organ growth, transpiration, and water stress resistance through colloidal stress, a mechanism of action (Hernández-Campos et al. 2021). HA reduces drought-induced oxidative stress through the induction of antioxidants (Abu-Ria et al. 2024). HA treatment reduced the expression of P5CS and its products, as well as stress defence hormones such as ABA, SA, jasmonic acid, and polyamines. Drought stress and HS can significantly influence the endophytic communities in tomato seeds and roots. The effects can be attributed to several hypotheses, including the composition of HSs favouring certain genera capable of using compounds, influencing microbial community dynamics through the release of metabolites, indirectly altering plant metabolism, such as root Fe homeostasis, favouring specific phyla, such as *Actinobacteria* with the *Frigoribacterium* genus, activating plant defence mechanisms through the recruitment of certain microbes, and the effects of HSs being independent of endophytic composition, leading to a parallel benefit for plants and a change in their microbiota communities (Lengrand et al. 2024).

14.4.1.5 Salinity Stress

Soil salinity, characterized by the accumulation of water-soluble salts such as Na^+, potassium (K^+), chloride (Cl^-), and sulfate (SO_4^{2-}), poses a significant abiotic stress that severely limits agricultural productivity, particularly in arid and semiarid regions. Among these nutrients, Na^+ and Cl^- are especially detrimental, as they are not considered essential plant nutrients. Salinity levels are influenced by natural factors such as climate, lithology, topography, and paedology, as well as human activities, especially irrigated agriculture (Kutlu and Gulmezoglu 2023; Rao et al. 2023).

Salinity adversely affects plant physiology by reducing the leaf RWC and increasing turgor loss, electrolyte leakage (EL), and MDA levels (Fig. 14.7). In beans, the application of HA—particularly through combined soil and foliar treatments—was found to mitigate these effects by maintaining RWC and turgor pressure, decreasing EL and MDA, increasing the activity of antioxidant enzymes (e.g. SOD and GR), and increasing micronutrient uptake in aerial parts (Kutlu and Gulmezoglu 2023). Seed priming with HA also improved lentil performance under salt stress by increasing germination, growth, and nutrient uptake and increasing antioxidant enzyme activity, thereby reducing oxidative damage and lipid peroxidation (Rao et al. 2023), in maize (Yang et al. 2023), and in *Brassica juncea* (Aly and Abdelsalam 2025). The combined use of HA and silicic acid has synergistic benefits, improving osmotic regulation, membrane stability, and antioxidant responses (Rao et al. 2023).

In rice, HA treatment confers greater salt stress tolerance by promoting antioxidant enzyme activity, reducing MDA accumulation, maintaining ion homeostasis, and preserving membrane integrity (Abu-Ria et al. 2023). Similarly, Ennab et al. (2023) reported that HA improved soil nutrient availability (N, P, K, Ca, Mg, Fe, Mn, Zn, Cu, and B) and microbial activity, leading to increased tree growth, leaf chlorophyll content, and fruit yield in drip-irrigated Mexican lime cultivated on saline clay soils. Combined soil and foliar HA applications yielded the best

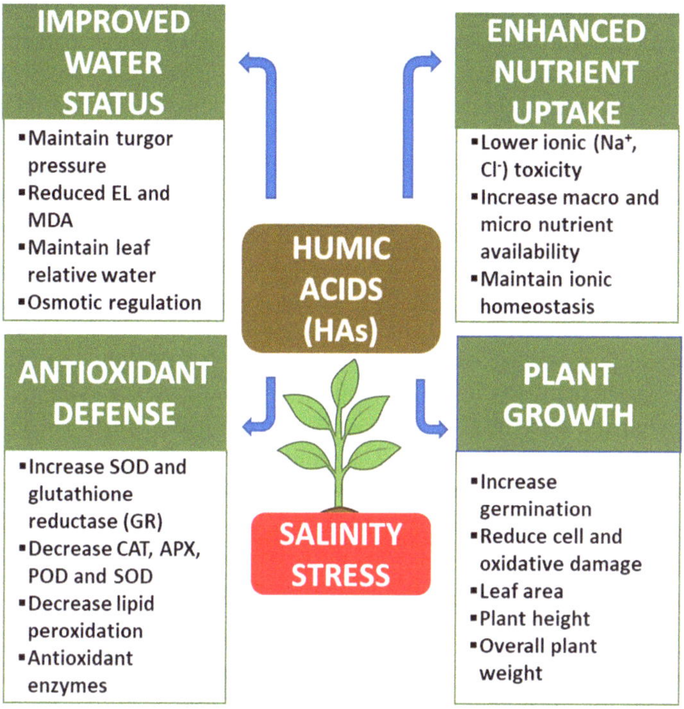

Fig. 14.7 Impact of salinity stress on plant function and the alleviating effects of humic acids. EL, electrolyte leakage; MDA, malondialdehyde; SOD, superoxide dismutase; CAT, catalase; APX, ascorbate peroxidase; POD, peroxidase

outcomes in terms of productivity and stress mitigation. In perennial ryegrass, HA application reduced oxidative damage, proline content, and intercellular CO_2 while increasing the photosynthetic rate (Pn), transpiration (Tr), and stomatal conductance (Gs). HA also promotes the synthesis of anti-senescence hormones, such as indole-3-acetic acid and brassinosteroids and restores the activities of antioxidant enzymes, including APX, CAT, POD, and SOD (Meng et al. 2023).

Spinach was tolerant to low and moderate salinity levels under high temperatures but experienced severe stress or mortality beyond 100 mM NaCl. HA-treated plants survive even under the highest salinity levels, although root growth is reduced (Choi et al. 2024). In cucumbers, foliar application of HA under 10 dS m^{-1} NaCl increased antioxidant enzyme activity and phenolic compound production, suggesting improved stress resilience (Amerian et al. 2024). The application of 200 mg/L HA increased the root strength and yield. The use of fig-leaf gourd as a rootstock also increased water and nutrient uptake under saline conditions. In barley, combining HA with P fertilizers significantly improved nutrient availability and uptake, enhanced growth traits, and increased grain yield by up to 22.30%. The integration of HA with a reduced P fertilizer dose offers a sustainable strategy for improving productivity in saline-affected soils (Alsudays et al. 2024).

HA enhances plant tolerance to salinity and alkalinity by modulating physiological and biochemical responses. It reduces the levels of stress markers such as proline, ABA, and MDA, increases the glutathione content, and improves membrane permeability (Kutlu and Gulmezoglu 2023). HA also stimulates water relations and stomatal conductance and activates antioxidant defence mechanisms to mitigate ROS damage. In addition, HA improves nutrient uptake, reduces lipid peroxidation, and increases total protein content, especially under marginal soil conditions. Canellas et al. (2020) demonstrated that HA priming in maize seedlings modulates stress markers (e.g. proline and CAT activity) and upregulates genes related to phytohormone signalling, cell signalling, and stress responses. Hormonal pathways involving ABA, gibberellins, and auxins were positively regulated, which alleviated the negative impacts of stressors such as salinity, drought, and chromium (Cr) toxicity. HA activates both enzymatic antioxidants (SOD, POD, CAT, APX) and nonenzymatic scavengers (ascorbate, tocopherol, phenolics) and induces the production of compatible solutes such as proline to protect against oxidative stress. Salinity stress typically hampers plant growth by causing Na^+ accumulation; increasing electrolyte leakage (EL); increasing the levels of MDA, soluble sugars, sucrose, glucose, and phenolics; and reducing the K^+ content, K^+/Na^+ ratio, carbohydrate reserves, and enzyme activities. HA mitigates these effects by reducing Na^+ toxicity, restoring the ion balance, regulating osmolyte concentrations, and enhancing antioxidant systems under both saline and non-saline conditions (Shukry et al. 2023). Furthermore, it improves salt stress tolerance in perennial ryegrass by modulating *P5CS* gene expression and increasing antioxidant and antisenescence hormone activity (Meng et al. 2023). These HA-driven hormonal interactions help preserve photosynthetic function and promote osmotic balance under stress.

The application of HA stimulated the expression of the SnRK2.2 protein kinase (serine/threonine kinase), which may be a mechanism for the increased abiotic stress tolerance observed in plants treated with HSs (Zubo and Schaller 2020). The application of HA to HIGH-AFFINITY K^+ TRANSPORTER 1 (*HKT1*)-overexpressing transgenic plants (*HKT1-OX*) was found to rescue salt-induced root growth inhibition, block salt-induced degradation of the HKT1 protein, increase Na^+ distribution in roots, and increase Na^+ reabsorption by xylem and parenchyma cells (Khaleda et al. 2017). These results suggest that HA could increase plant growth and salt stress tolerance by controlling the *HKT1* transporter gene under saline conditions.

14.4.1.6 Heavy Metal Stress

Heavy metal (HM) contamination in soils—caused by industrial waste, fertilizers, pesticides, smelting, sewage sludge, and groundwater irrigation—poses a significant environmental threat. Common HMs include arsenic (As), Cd, lead (Pb), Cr, cobalt (Co), zinc (Zn), Fe, manganese (Mn), copper (Cu), nickel (Ni), and mercury (Hg). These nonbiodegradable elements can leach into groundwater or accumulate on soil surfaces, disrupting plant growth, physiology, and metabolism (Ghori et al. 2019).

While some metals serve as essential micronutrients in small quantities, their excessive accumulation leads to oxidative stress, impaired photosynthesis, hormone disruption, and premature senescence, particularly in sensitive crops such as tea (Xu et al. 2018).

HA plays a dual role in promoting plant growth and alleviating heavy metal toxicity. Through complexation, chelation, and redox interactions, HA binds heavy metals to soil particles, reducing their solubility, mobility, accumulation, and uptake by plant roots; mitigating oxidative stress; and improving overall plant growth and nutrient content (Canal et al. 2022, 2023; Duan et al. 2020; Huang et al. 2025; Oliynyk et al. 2025; Özkay et al. 2016; Xu et al. 2018). For example, HA enhances Pb translocation from roots to shoots in tea plants, thereby reducing Pb accumulation in roots—a potential strategy for mitigating root toxicity. However, it may pose health risks because of Pb accumulation in leaves used for consumption (Xu et al. 2018). The amount of metal ions bound to HAs, especially Cu, Co, and Ni, increases linearly with pH and starting concentration (Oliynyk et al. 2025). At low pH values and low concentrations of Fe, Al, and Cr ions, the quantity and concentration of metal ions bound to HAs in aqueous solutions are directly correlated.

Both natural and artificial HAs improve wheat seed germination under Cu^{2+} stress, with increases of 11.2% and 8.2%, respectively (Anatoliivna et al. 2020). Under As stress, rice seed germination improved from 65% to 75% with potassium humate priming (Mridha et al. 2021). Duan et al. (2020) demonstrated that HA reduced Pb toxicity by increasing pectin and pectic acid contents in cell walls and decreasing arabinose (Ara) and galactose (Gal) levels. This improved the Pb adsorption capacity, promoted cell elongation, and stabilized protein and S metabolism. HA also strengthened the antioxidant defence system and supported cell wall biosynthesis under metal stress. In tomato, the combined or individual application of *Fusarium solani* IK-105 and HA significantly mitigated Pb-induced stress by increasing the leaf area, water retention, and membrane stability and increasing the levels of photosynthetic pigments, protein, sugars, and starch. These treatments also reduce both enzymatic and nonenzymatic oxidative stress markers, decrease ABA, and limit Pb uptake while increasing Ca and Mg levels (Khan et al. 2025).

In Cd-stressed plants, HA foliar application increased the chlorophyll content and photosynthetic efficiency. In forage turnip, chelated HA outperformed EDTA in reducing H_2O_2, MDA, and antioxidative enzyme activity while also promoting growth, making it a promising tool for phytoremediation (Canal et al. 2023; Song et al. 2020). Similar Cd-mitigating effects have been observed in rapeseed (Canal et al. 2022), garden cress (Yildirim et al. 2021), and strawberry, where HA, either alone or combined with silicon (Si), improved physiological traits and reduced oxidative stress, proline levels, and membrane permeability (Dogan et al. 2022).

In sweet orange, excessive Cu exposure increased the levels of oxidative stress markers, including superoxide anions, H_2O_2, MDA, and methylglyoxal. HA treatment counteracts these effects by enhancing the antioxidant and glyoxalase systems, reducing oxidative injury in both roots and leaves (Huang et al. 2025).

However, HA application may not always be beneficial. Adhikari et al. (2023) reported that HA increased As, Cd, and Pb accumulation in rice, leading to elevated lipid peroxidation. Nevertheless, co-application of Si and biochar mitigated toxicity by lowering ABA, increasing salicylic acid (SA), and increasing the uptake of Ca^{2+} and Si. These treatments modulated the expression of ion transporters (*Lsi1*, *Lsi2*, *and OsNHX1*) and stress-related genes (*OsMTP1 and OsNramp*) while reducing melatonin overproduction by 50%, thus helping the plant adapt to combined salinity and drought stress. Nevertheless, caution is advised when HA is used for HM remediation. In the case of As toxicity, rice plants presented increased lipid peroxidation, EL, proline, and H_2O_2, along with decreased antioxidant enzyme activity. HA application effectively reversed these symptoms, improved plant physiological traits, and enhanced As stress tolerance (Hasanuzzaman et al. 2024).

HA decreases the mobility, bioavailability, and phytotoxicity of HMs and their uptake by plants and strengthens the endogenous plant defence systems in HM-exposed plants (Huang et al. 2025). As a result, HA has been applied to plants to counteract HM toxicity. Pb stress downregulated genes involved in ABA (*Solyc11g071600*) and JA (*Solyc04g079730* and *Solyc02g085730*) signalling, but inoculation with *F. solani* IK-105 upregulated these genes, suggesting an increase in stress signalling pathways. HA downregulated the *Solyc11g071600* and *Solyc02g085730* genes while upregulating *Solyc04g079730*, suggesting different microbial inoculation effects. *F. solani* IK-105 and HA downregulated genes related to DREB and antioxidant defences, suggesting the suppression of stress response pathways, possibly due to increased microbial interactions. However, combined with IK-105 and HA, genes related to HM stress (*Solyc02g067620* and *Solyc07g009130*) were upregulated, suggesting that these biostimulants enhance HM detoxification processes through mechanisms such as metal sequestration and transport (Khan et al. 2025).

14.4.2 Biotic Stresses

14.4.2.1 Diseases

HAs have dual effects on fungal populations, as demonstrated by Loffredo et al. (2008). In growth media, HA inhibited the mycelial growth of *Sclerotinia sclerotiorum* while simultaneously stimulating sclerotia formation, leading to an earlier onset and increased number of sclerotia. Interestingly, the same HA treatments did not hinder the development of beneficial fungi such as *Trichoderma* spp.; in fact, *T. viride* showed significant growth stimulation in response to HA. The inhibitory effects on *S. sclerotiorum*, including the number of sclerotia formed, are strongly correlated with specific chemical and functional properties of HSs, such as total acidity, carboxyl group content, and elemental composition (Loffredo et al. 2008). Furthermore, HA significantly influences soil microbial communities, particularly the structure and complexity of bacterial and fungal networks. Jiang et al. (2022) reported that HA application increased the relative abundance of bacterial phyla such as *Firmicutes*, *Planctomycetes*, and *Chloroflexi* and increased fungal network complexity in the rhizosphere. These microbial shifts contributed to a reduced *Fusarium* wilt disease index and improved cucumber growth.

In tomato plants, HA foliar applications were effective in controlling bacterial spot disease caused by *Xanthomonas hortorum* pv. *gardneri*. The lowest disease severity was observed in plants treated with HA (Faccin and Di Piero 2022). Additionally, HA-treated plants presented a threefold increase in defense-related enzyme activity and improved physiological performance, including the highest NR activity and increased root biomass. HA individually or in combination with chitosan and *Bacillus subtilis* (HBC) significantly increased the height and fresh weight of tomato plants and reduced the incidence of plant diseases up to 45.1% (Qiu et al. 2025). HBC treatment increased the diversity of fungal and bacterial communities, increasing the abundance of *Bacillus*, *Gemmatimonas*, *Neobacillus*, *Acinetobacter*, and *Humicola*. However, it also decreased the abundance of *Sphingomonas*, particularly the soil-borne pathogens *Fusarium* and *Ralstonia*. The content of phenolic acids, which are allelochemicals related to continuous cropping disorder, also decreases. In contrast, Eman et al. (2017) reported the deleterious effects of HA application on the growth, height, and dry weight of cotton seedlings growing in *F. oxysporum*-contaminated soil.

Silva and Canellas (2022) reported that organic matter effectively inhibited disease and pests, with pest control efficacy of 67% and fungal control efficacy of 75%. Vermicompost is the most commonly reported source of organic matter. Compared with both fungal and bacterial disease control, humic compounds, which are commonly HAs, showed the highest efficacy (74%), with the highest concentration (500 mg L^{-1}) used for soil application (Silva and Canellas 2022). Low doses (150 and 1 mg L^{-1}) of HS had the highest efficacy (84% and 78%, respectively) in treating *Xanthomonas arbicola* and *Fusarium* spp. in peach and cucumber plants. HS increased the activities of POD, PAL, and β-1,3-glucanase while also positively regulating the *CsPR-7*, *CsPR-3*, and *CsPR-11* genes (da Silva et al. 2023) and triggered plant immune responses, potentially reducing Huanglongbing disease symptoms and promoting more resilient citrus cultivation. The application of HA (as a humate at a rate of 12 ml m^{-2}) did not alter the diversity of the soil bacterial communities or mineral profile; it altered the soil microbiota–mineral interactions and introduced subtle changes that could aid pasture dieback recovery (Whitton et al. 2023).

HSs can enhance microorganisms, potentially impacting plant growth and defence against pathogens. HAs can increase the activity of phenylalanine ammonia-lyase (PAL) and total phenolic compounds in maize leaves, which are crucial for plant defence against biotic and abiotic stress (Silva and Canellas 2022). The PAL pathway in the cytoplasm and the isochorismate pathway in the chloroplast lead to endogenous SA synthesis from chorismate. Increased SA activates a plant defence response, but the effect of HA on this pathway remains unknown in diverse plant species. This study suggested that the synergistic effect of HBC on tomato growth and disease resistance is related to its regulation of the microbial community and root exudates (Qiu et al. 2025).

14.4.2.2 Insects

The defence of plants against herbivorous insects can be influenced by biostimulants through enhancing plant health and resilience (Nasab et al. 2019). El-Sayed et al. (2021) used foliar spraying and soaking seeds of FA, HA, imidacloprid (Gaucho), and their combination to control three insect pests of cotton, namely, *Thrips tabaci*, *Empoasca lybica* and *Tetranychus urticae*. HAs, either individually or in combination with chemical pesticides, significantly reduce insect populations. It acts as a repellent and also enhances the effectiveness and reduces toxicity of traditional pesticides by acting as a synergist and reducing leaching. HA amendment increases the levels of metabolites linked to the shikimic acid pathway, such as flavonoids and benzoic acids, due to the increased activity of phenylalanine ammonia-lyase and primary metabolism induction (Pereira et al. 2021). Despite the role of phenolics in plant resistance against insect pests, few suggestions in the literature discuss the protective function of HA against biotic stress, highlighting the potential to enhance crop resistance to pests.

Nasab et al. (2019) investigated the effects of HA, plant growth-promoting *Rhizobacteria* (PGPR), and their combined application on *Brevicoryne brassicae* L. life table parameters and canola tolerance. The net reproductive rate under the HA, PGPR, and HA + PGPR treatments was lower than that under the control. The plants treated with PGPR alone or in combination with HA presented the greatest tolerance to *B. brassicae*. The highest total phenol, flavonoid, and glucosinolate contents were detected in the PGPR and HA + PGPR treatments (Nasab et al. 2019). HA acts as a potent and toxic agent that effectively controls plant-parasitic nematodes such as *Meloidogyne* spp., *Rotylenchulus reniformis*, *Radopholus similis*, and *Helicotylenchus multicinctus* by killing juvenile nematodes, inhibiting their hatching, and reducing their ability to infect plants (Nagachandrabose and Baidoo 2021). Its effective dose ranges from 0.04% to 2.0%. HA is effective in various methods, such as soil application, drenching, seedling root dip treatment, and foliar spray on leaves. HA is the most effective method (100%) for controlling root-knot infestations in bananas among HS studies (Silva and Canellas 2022). HA accelerates cotton growth, promotes callus induction, and enhances resistance against cotton bollworms (Ghaffor et al. 2025).

14.4.2.3 Other Stresses

The weed control agent, which is made from HA and other natural substances, has a strong herbicidal effect and contains essential nutrients for crop growth and fruiting, thereby influencing its effectiveness (https://patents.google.com/patent/KR100791980B1/en). Compared with the control treatment, the application of HA and/or EM1 to grapevines significantly reduced weed growth, increased yield, and improved fruit quality (Ahmed et al. 2013). Compared with herbicides alone, herbicides (e.g. Grodil Maxi) and HA preparations (e.g. Humifeld) effectively control weeds, including perennial species, by 23.6% (Korotkova et al. 2021). Moreover, applying various herbicides in a mixture with the HA preparation (4R foliar concentrate) has been reported to have the opposite effect. HAs could be used for the removal of herbicides in water as reducing agents. The maximum reducing ability was 72.72% for thifensulfuron methyl and tribenuron methyl and 37.52% for imazamox (Vrantsi et al. 2021). HAs showed superior binding ability than did FAs.

14.5 Conclusions and Future Research Directions

HA has excellent potential as a natural biostimulant in modern agriculture. The diverse functional groups of HAs give them a unique ability to improve the soil structure, buffer the soil pH, enhance nutrient retention, and increase the water-holding capacity. In addition to soil conditioning, HA plays a critical role in promoting plant growth and resilience by stimulating enzyme activity, nutrient uptake, hormone signalling, and root development. It effectively mitigates abiotic stresses such as drought, salinity, temperature extremes, waterlogging, and heavy metal toxicity while also increasing plant defences against biotic stresses, including pathogens, pests, and diseases. Furthermore, HA positively influences the soil and plant microbiomes, promoting beneficial microbial communities that contribute to plant health and productivity.

Despite these promising outcomes, further research is essential to maximize HA's potential and ensure safe, effective applications. Future studies should focus on elucidating the molecular mechanisms of HA action via advanced tools such as transcriptomics, metabolomics, and proteomics to understand its role in plant stress physiology. Field-based, long-term studies are also needed to validate HA efficacy under diverse agroclimatic conditions, particularly under combined stress scenarios. Additionally, exploring the structure–function relationships of HA will help in the design of crop- and stress-specific formulations. Understanding its interactions with microbial inoculants and root exudates can improve the understanding of synergy in soil ecosystems. Finally, standardizing HA sources, extraction methods, and application protocols is necessary to ensure consistent performance across products and regions. Research in these areas will enable the integration of HA into climate-smart, sustainable agricultural practices, supporting global food security while reducing dependence on chemical inputs.

References

Abdellatif IMY, Abdel-Ati YY, Abdel-Mageed YT, Hassan MA-MM (2017) Effect of humic acid on growth and productivity of tomato plants under heat stress. J Hortic Res 25(2):59–66. https://doi.org/10.1515/johr-2017-0022

Abu-Ria M, Shukry W, Abo-Hamed S, Albaqami M, Almuqadam L, Ibraheem F (2023) Humic acid modulates ionic homeostasis, osmolytes content, and antioxidant defense to improve salt tolerance in rice. Plants 12:1834. https://doi.org/10.3390/plants12091834

Abu-Ria ME, Elghareeb EM, Shukry WM, Abo-Hamed SA, Ibraheem F (2024) Mitigation of drought stress in maize and sorghum by humic acid: differential growth and physiological responses. BMC Plant Biol 24:514. https://doi.org/10.1186/s12870-024-05184-4

Adetumbi JA, Orimadegun IO, Akinyosoye ST, Akintayo OT, Agbeleye OA (2019) Enhancing planting value of rice seed through priming with humic substance. J Exp Agric Int 29:1–8. https://doi.org/10.9734/jeai/2019/46194

Adhikari A, Aneefi AG, Sisuvanh H, Singkham S, Pius MV, Akter F, Kwon E-H, Kang S-M, Woo Y-J, Yun B-W, Lee I-J (2023) Dynamics of humic acid, silicon, and biochar under heavy metal, drought, and salinity with special reference to phytohormones, antioxidants, and melatonin synthesis in rice. Int J Mol Sci 24:17369. https://doi.org/10.3390/ijms242417369

Ahmed HMA, Ahmed FF, Ebrahiem ME, El-Kareem AMA (2013) The beneficial effects of some humic acid, EM1 and weed control treatments on fruiting of superior grapevines. Stem Cell 4(3):25–32

Alsudays IM, Alshammary FH, Alabdallah NM, Alatawi A, Alotaibi MM, Alwutayd KM, Alharbi M, Alghanem SMS, Alzuaibr FM, Gharib HS, Awad-Allah MMA (2024) Applications of humic and fulvic acid under saline soil conditions to improve growth and yield in barley. BMC Plant Biol 24:191. https://doi.org/10.1186/s12870-024-04863-6

Aly H, Abdelsalam A (2025) Effect of humic acid and selenium on seed germination, seedling growth, and priming with mitigate salt stress effect on mustard (*Brassica juncea*) seeds. Russ J Plant Physiol 72:72. https://doi.org/10.1134/s1021443724609364

Amerian M, Palangi A, Gohari G, Ntatsi G (2024) Humic acid and grafting as sustainable agronomic practices for increased growth and secondary metabolism in cucumber subjected to salt stress. Sci Rep 14:15883. https://doi.org/10.1038/s41598-024-66677-8

Anatoliivna LV, Ivanovych DS, Mykolaivna PL, Abi NR (2020) Effects of humic substances on seed germination of wheat under the influence of heavy metal. Cherk Univ Bull Biol Sci Ser 2020(1):42–52. https://doi.org/10.31651/2076-5835-2018-1-2020-1-42-52

Aytaç E, Ünlü HÖ, Turan T, Erkan İE, Akçay UÇ (2024) Humic acid mitigates drought stress in tomato. Bilge Int J Sci Technol Res 8(1):27–37. https://doi.org/10.30516/bilgesci.1421304

Bahrani A, Pourreza J (2012) Gibberellic acid and salicylic acid effects on seed germination and seedlings growth of wheat (*Triticum aestivum* L.) under salt stress condition. World Appl Sci J 18:633–641. https://doi.org/10.5829/idosi.wasj.2012.18.05.1372

Baía DC, Olivares FL, Zandonadi DB, CDP S, Spaccini R, Canellas LP (2020) Humic acids trigger the weak acids stress response in maize seedlings. Chem Biol Technol Agric 7:31. https://doi.org/10.1186/s40538-020-00193-5

Canal SB, Bozkurt MA, Yilmaz H (2022) Effects of humic acid and EDTA on phytoremediation, growth and antioxidant activity in rapeseed (*Brassica napus* L.) grown under heavy metal stress. Pol J Environ Stud 31:4051–4060. https://doi.org/10.15244/pjoes/148120

Canal SB, Bozkurt MA, Yilmaz H (2023) Humic acid ameliorates phytoremediation, plant growth and antioxidative enzymes in forage turnip (*Brassica rapa* L.). Plant Soil Environ 69:567–576. https://doi.org/10.17221/394/2023-PSE

Canellas LP, Olivares FL (2014) Physiological responses to humic substances as plant growth promoter. Chem Biol Technol Agric 1:3. https://doi.org/10.1186/2196-5641-1-3

Canellas LP, Balmori DM, Médici LO, Aguiar NO, Campostrini E, Rosa RC, Façanha AR, Olivares FL (2013) A combination of humic substances and *Herbaspirillum seropedicae* inoculation enhances the growth of maize (*Zea mays* L.). Plant Soil 366:119–132. https://doi.org/10.1007/s11104-012-1382-5

Canellas LP, Canellas NOA, da S. Irineu LES, Olivares FL, Piccolo A (2020) Plant chemical priming by humic acids. Chem Biol Technol Agric 7:12. https://doi.org/10.1186/s40538-020-00178-4

Cha J-Y, Kang S-H, Ali I, Lee SC, Ji MG, Jeong SY, Shin G-I, Kim MG, Jeon J-R, Kim W-Y (2020) Humic acid enhances heat stress tolerance via transcriptional activation of heat-shock proteins in *Arabidopsis*. Sci Rep 10:15042. https://doi.org/10.1038/s41598-020-71701-8

Choi S, Harvey JT, Leskovar DI (2024) Solid humic substance enhanced spinach abiotic stress tolerance under combined drought, salinity, and daily heat stress. Plant Stress 13:100544. https://doi.org/10.1016/j.stress.2024.100544

da Silva RM, Canellas NA, Olivares FL, Piccolo A, Canellas LP (2023) Humic substances trigger plant immune responses. Chem Biol Technol Agric 10:123. https://doi.org/10.1186/s40538-023-00468-7

Dauletbay A, Hanzheng D, Ongalbek AN, Tursynbolat S, Dalbanbay A (2024) Humic acids: properties, structure, and application. Bull Shakarim Univ Tech Sci 3(15):321–340. https://doi.org/10.53360/2788-7995-2024-3(15)-41

de Melo BAG, Motta FL, Santana MHA (2016) Humic acids: structural properties and multiple functionalities for novel technological developments. Mater Sci Eng C 62:967–974. https://doi.org/10.1016/j.msec.2015.12.001

de Moura OVT, Berbara RLL, Torchia DFO, Da Silva HFO, van Tol de Castro TA, Tavares OCH, Rodrigues NF, Zonta E, Santos Leandro A, García AC (2023) Humic foliar application as sustainable technology for improving the growth, yield, and abiotic stress protection of agricultural crops. A review. J Saudi Soc Agric Sci 22:493–513. https://doi.org/10.1016/j.jssas.2023.05.001

Dogan M, Bolat I, Karakas S, Dikilitas M, Gutiérrez-Gamboa G, Kaya O (2022) Remediation of cadmium stress in strawberry plants using humic acid and silicon applications. Life 12:1962. https://doi.org/10.3390/life12121962

Duan D, Tong J, Xu Q, Dai L, Ye J, Wu H, Xu J, Shi J (2020) Regulation mechanisms of humic acid on Pb stress in tea plant (*Camellia sinensis* L.). Environ Pollut 267:115546. https://doi.org/10.1016/j.envpol.2020.115546

El Komy BY, Hussein EAA, Tantawy MF, El Sayed EMN (2021) Evaluation effect of application methods of different sources of humic substances on potatoes plant growth, productivity and quality in newly reclaimed sandy soil under different levels of mineral fertilizers. Menoufia J Soil Sci 6:227–252. https://doi.org/10.21608/mjss.2021.198703

El-Sayed GH, Ahmed IA, Amira EM (2021) Effect of seed soaking and foliar spray of humic acid and fulvic acid on cotton plant pests and yield, compared with some recommended pesticides. Egypt Acad J Biol Sci (B Zool) 13(1):63–75. https://doi.org/10.21608/EAJBSZ.2021.147916

Eman AMO, Maggie EH, Howida AM, Heba Y (2017) Effect of humic acid on the efficacy of some biocontrol agents in controlling damping-off of cotton seedlings caused by *Fusarium oxysporum*. Egypt J Phytopathol 45(2):53–66. https://doi.org/10.21608/ejp.2017.88579

Ennab HA, Mohamed AH, El-Hoseiny HM, Omar AA, Hassan IF, Gaballah MS, Khalil SE, Mira AM, Abd El-Khalek AF, Alam-Eldein SM (2023) Humic acid improves the resilience to salinity stress of drip-irrigated Mexican lime trees in saline clay soils. Agronomy 13:1680. https://doi.org/10.3390/agronomy13071680

Faccin D, Di Piero RM (2022) Extracts and fractions of humic substances reduce bacterial spot severity in tomato plants, improve primary metabolism and activate the plant defense system. Physiol Mol Plant Path 121:101877. https://doi.org/10.1016/j.pmpp.2022.101877

Fathy N, Tewfike TA, Abdel-Rahman HM, Zaghloul RA (2023) Alleviation of drought stress in faba bean using humic acid and inoculation with plant growth promoting *Rhizobacteria*. Ann Agric Sci Moshtohor 61(1):219–230. https://doi.org/10.21608/ASSJM.2023.312214

García AC, Santos LA, Izquierdo FG, Sperandio MVL, Castro RN, Berbara RLL (2012) Vermicompost humic acids as an ecological pathway to protect rice plant against oxidative stress. Ecol Eng 47:203–208. https://doi.org/10.1016/j.ecoleng.2012.06.011

Garcia-Caparros P, Ciriello M, Rouphael Y, Giordano M (2025) The role of organic extracts and inorganic compounds as alleviators of drought stress in plants. Horticulturae 11:91. https://doi.org/10.3390/horticulturae11010091

Ghaffor A, Qadir F, Hafeez A, Chang MS, Khaskheli MA, Khaskheli MA, Suthar V, Kolhi ML, Alam R, Shehzad S, Jaweria W, Aslam P, Misbah F, Haleem M, Afzal A, Shoukat A, Rasool G, Zafar MM, Razzaq A (2025) Humic acid-mediated resistance in cotton (*Gossypium hirsutum* L.) against bollworm (*Helicoverpa armigera*). Int J Agric Biosci 14(3):447–454. https://doi.org/10.47278/journal.ijab/2025.042

Ghori NH, Ghori T, Hayat MQ, Imadi SR, Gul A, Altay V, Ozturk M (2019) Heavy metal stress and responses in plants. Int J Environ Sci Technol 16:1807–1828. https://doi.org/10.1007/s13762-019-02215-8

Gilroy S, Suzuki N, Miller G, Choi WG, Toyota M, Devireddy AR, Mittler R (2014) A tidal wave of signals: calcium and ROS at the forefront of rapid systemic signaling. Trends Plant Sci 19(10):623–630. https://doi.org/10.1016/j.tplants.2014.06.013

Goel P, Dhingra M (2021) Humic substances: prospects for use in agriculture and medicine. In: Makan A (ed) Humic substances. IntechOpen. https://doi.org/10.5772/intechopen.99651

Hartina MT, Vityakon P, Sukitprapanon T-S (2025) Co-application of humic acid and gypsum affects soil chemical properties, rice yield, and phosphorus use efficiency in acidic paddy soils. Sci Rep 15:4350. https://doi.org/10.1038/s41598-025-89132-8

Hasanuzzaman M, Nowroz F, Raihan MRH, Siddika A, Alam MM, Prasad PVV (2024) Application of biochar and humic acid improves the physiological and biochemical processes of rice (*Oryza sativa* L.) in conferring plant tolerance to arsenic-induced oxidative stress. Environ Sci Pollut Res 31:1562–1575. https://doi.org/10.1007/s11356-023-31119-x

Hernández-Campos R, Robles C, García AC (2021) Humic acids effects on plant growth and protection against water stress in selected native maize populations from Mexico. Rev Fitotec Mex 44(4):561–569. https://doi.org/10.35196/rfm.2021.4.561

Hoefle D, Sperber S, Marzban N, Antonietti M, Hoffmann T, Wicaksono WA, Abdelfattah A, Berg G (2024) Artificial humic acid diminishes the effect of drought on the soil microbiome. J Sustain Agric Environ 3:e70034. https://doi.org/10.1002/sae2.70034

Huang W-T, Chen X-F, Huang W-L, Shen Q, Lu F, Lai N-W, Guo J, Yang L-T, Ye X, Chen L-S (2025) Humic acid enhances antioxidant and glyoxalase systems to combat copper toxicity in citrus. Agronomy 15:99. https://doi.org/10.3390/agronomy15010099

Ibrahim EA, Ebrahim NES, Mohamed GZ (2024) Mitigation of water stress in broccoli by soil application of humic acid. Sci Rep 14:2765. https://doi.org/10.1038/s41598-024-53012-4

Jiang N, Wu M, Li G, Petropoulos E, Sun F, Wang X, Liu J, Liu M, Li Z (2022) Humic substances suppress *Fusarium oxysporum* by regulating soil microbial community in the rhizosphere of cucumber (*Cucumis sativus* L.). Appl Soil Ecol 174:104389. https://doi.org/10.1016/j.apsoil.2022.104389

Jindo K, Canellas LP, Albacete A, dos Santos LF, Rocha RLF, Baia DC, Canellas NOA, Goron TL, Olivares FL (2020) Interaction between humic substances and plant hormones for phosphorus acquisition. Agronomy 10(5):640. https://doi.org/10.3390/agronomy10050640

Kawałko D, Jamroz E, Jerzykiewicz M, Ćwieląg-Piasecka I (2023) Characteristics of humic acids in drained floodplain soils in temperate climates: a spectroscopic study. Sustainability 15:11417. https://doi.org/10.3390/su151411417

Kaya C, Şenbayram M, Akram NA, Ashraf M, Alyemeni MN, Ahmad P (2020) Sulfur-enriched leonardite and humic acid soil amendments enhance tolerance to drought and phosphorus deficiency stress in maize (*Zea mays* L.). Sci Rep 10:6432. https://doi.org/10.1038/s41598-020-62669-6

Khaleda L, Park HJ, Yun D-J, Jeon J-R, Kim MG, Joon-Yung Cha J-Y, Kim W-Y (2017) Humic acid confers HIGH-AFFINITY K+ TRANSPORTER 1-mediated salinity stress tolerance in *Arabidopsis*. Mol Cells 40(12):966–975. https://doi.org/10.14348/molcells.2017.0229

Khan I, Asaf S, Kang S-M, Lubna, Methela NJ, Back MY, Yun B-W, Lee I-J (2025) Synergistic role of *Fusarium solani* IK-105 and humic acid in combating lead stress in tomato plants through physiological, biochemical and molecular modulation. Plant Stress 15:100780. https://doi.org/10.1016/j.stress.2025.100780

Korotkova I, Marenych M, Hanhur V, Laslo O, Chetveryk O, Liashenko V (2021) Weed control and winter wheat crop yield with the application of herbicides, nitrogen fertilizers, and their mixtures with humic growth regulators. Acta Agrobot 74:748. https://doi.org/10.5586/aa.748

Kulikova NA, Perminova IV (2021) Interactions between humic substances and microorganisms and their implications for nature-like bioremediation technologies. Molecules 26:2706. https://doi.org/10.3390/molecules26092706

Kutlu I, Gulmezoglu N (2023) Suitable humic acid application methods to maintain physiological and enzymatic properties of bean plants under salt stress. Gesunde Pflanzen 75:1075–1086. https://doi.org/10.1007/s10343-022-00766-4

Lengrand S, Dubois B, Pesenti L, Debode F, Legrève A (2024) Humic substances increase tomato tolerance to osmotic stress while modulating vertically transmitted endophytic bacterial communities. Front Plant Sci 15:1488671. https://doi.org/10.3389/fpls.2024.1488671

Li H, Kong F, Tang T, Luo Y, Gao H, Xu J, Xing G, Li L (2023) Physiological and transcriptomic analyses revealed that humic acids improve low-temperature stress tolerance in zucchini (*Cucurbita pepo* L.) seedlings. Plants 12:548. https://doi.org/10.3390/plants12030548

Loffredo E, Berloco M, Senesi N (2008) The role of humic fractions from soil and compost in controlling the growth *in vitro* of phytopathogenic and antagonistic soil-borne fungi. Ecotoxicol Environ Saf 69:350–357. https://doi.org/10.1016/j.ecoenv.2007.11.005

Ma Y, Cheng X, Zhang Y (2024) The impact of humic acid fertilizers on crop yield and nitrogen use efficiency: a meta-analysis. Agronomy 14(12):2763. https://doi.org/10.3390/agronomy14122763

Maccarthy P (2001) The principles of humic substances: an introduction to the first principle. Soil Sci 166:738–751. https://doi.org/10.1039/9781847551085-00019

Meganid AS, Hassan S, Al-Zahrani, EL-Metwally, Selim M (2015) Effect of humic acid application on growth and chlorophyll contents of common bean plants (*Phaseolus vulgaris* L.) under salinity stress conditions. Int J Innovative Res Sci 4(5):2651–2660. https://doi.org/10.15680/IJIRSET.2015.0405001

Meng Q, Yan M, Zhang J, Zhang Q, Zhang X, Yang Z, Luo Y, Wu W (2023) Humic acids enhance salt stress tolerance associated with pyrroline 5-carboxylate synthetase gene expression and hormonal alteration in perennial ryegrass (*Lolium perenne* L.). Front. Plant Sci 14:1272987. https://doi.org/10.3389/fpls.2023.1272987

Mora V, Baigorri R, Bacaicoa E, Zamarreno AM, García-Mina JM (2012) The humic acid-induced changes in the root concentration of nitric oxide, IAA and ethylene do not explain the changes in root architecture caused by humic acid in cucumber. Environ Exp Bot 76:24–32. https://doi.org/10.1016/j.envexpbot.2011.10.001

Mridha D, Paul I, De A, Ray I, Das A, Joardar M, Chowdhury NR, Bhadoria PBS, Roychowdhury T (2021) Rice seed (IR64) priming with potassium humate for improvement of seed germination, seedling growth and antioxidant defense system under arsenic stress. Ecotoxicol Environ Saf 219:112313. https://doi.org/10.1016/j.ecoenv.2021.112313

Nagachandrabose S, Baidoo R (2021) Humic acid—a potential bioresource for nematode control. Nematology 24(1):1–10. https://doi.org/10.1163/15685411-bja10116

Nandhini RS, Shelishiyah R, Prakash P (2018) Effect of humic acid on seed germination of *Zea mays*. Indian J Environ Prot 38(10):862–866

Nardi S, Muscolo A, Vaccaro S, Baiano S, Spaccini R, Piccolo A (2007) Relationship between molecular characteristics of soil humic fractions and glycolytic pathway and Krebs cycle in maize seedlings. Soil Biol Biochem 39:3138–3146. https://doi.org/10.1016/j.soilbio.2007.07.006

Nardi S, Ertani A, Francioso O (2016) Soil–root cross-talking: the role of humic substances. J Plant Nutr Soil Sci 180:5–13. https://doi.org/10.1002/jpln.201600348

Nasab RS, Yali MP, Bozorg-Amirkalaee M (2019) Effects of humic acid and plant growth-promoting *Rhizobacteria* (PGPR) on induced resistance of canola to *Brevicoryne brassicae* L. Bull Entomol Res 109:479–489. https://doi.org/10.1017/S0007485318000779

Nazi F, Reza H, Rahman R (2014) Effect of humic fertilizer on germination of wheat seeds under drought stress. Adv Biores 5(4):98–102. https://doi.org/10.15515/abr.0976-4585.5.4.98102

Nunes RO, Domiciano GA, Alves WS, Melo ACA, Nogueira FCS, Canellas LP et al (2019) Evaluation of the effects of humic acids on maize root architecture by label-free proteomics analysis. Sci Rep 9:12019. https://doi.org/10.1038/s41598-019-48509-2

Oktem AG, Oktem A (2020) Effect of humic acid application methods on yield and some yield characteristics of corn plant (*Zea mays* L. indentata). J Appl Life Sci Int 23(11):31–37. https://doi.org/10.9734/JALSI/2020/v23i1130196

Oliynyk L, Popadiuk I, Fedoryshyn O, Yaremkevych O, Kohut A (2025) Characterization and interaction of humic acids with metal ions as a function of solution p^H. Chem Chem Technol 19(2):214–220. https://doi.org/10.23939/chcht19.02.214

Olk DC, Dinnes DL, Scoresby JR, Callaway CR, Darlington JW (2018) Humic products in agriculture: potential benefits and research challenges—a review. J Soils Sediments 18:2881–2891. https://doi.org/10.1007/s11368-018-1916-4

Özkay F, Kıran S, Kuşvuran Ş, Ellialtıoğlu ŞŞ (2016) The effects of humic acid applications on heavy metal stress in lettuce. Turk J Agric-Food Sci Technol 4(6):431–437. (in Turkish with English abstract)

Pereira RV, Filgueiras CC, Dória J, Peñaflor MFGV, Willett DS (2021) The effects of biostimulants on induced plant defense. Front Agron 3:630596. https://doi.org/10.3389/fagro.2021.630596

Pitann B, Khan K, Mühling KH (2024) Does humic acid foliar application affect growth and nutrient status of water-stressed maize? Plant-Environ Interact 5:e10156. https://doi.org/10.1002/pei3.1015

Pizzeghello D, Schiavon M, Francioso O, Dalla Vecchia FA, Nardi S (2020) Bioactivity of size-fractionated and unfractionated humic substances from two forest soils and comparative effects on N and S metabolism, nutrition, and root anatomy of *Allium sativum* L. Front Plant Sci 11:1203. https://doi.org/10.3389/fpls.2020.01203

Poomani S, Yadav S, Choudhary R, Singh D, Dahuja A, Yadav SK (2023) Seed priming with humic acid modifies seedling vigor and biochemical response of lentil under heat stress conditions. Turk J Agric For 47:20. https://doi.org/10.55730/1300-011X.3147

Qiu C, Bao Y, Lü D, Yan M, Li G, Liu K, Wei S, Wu M, Li Z (2025) The synergistic effects of humic acid, chitosan and *Bacillus subtilis* on tomato growth and against plant diseases. Front Microbiol 16:1574765. https://doi.org/10.3389/fmicb.2025.1574765

Quaggiotti S, Ruperti B, Pizzeghello D, Francioso O, Tugnoli V, Nardi S (2004) Effect of low molecular size humic substances on nitrate uptake and expression of genes involved in nitrate transport in maize (*Zea mays* L.). J Exp Bot 55:803–813. https://doi.org/10.1093/jxb/erh085

Rao D, Yadav S, Choudhary R, Singh D, Bhardwaj R, Barthakur S, Yadav SK (2023) Silicic and humic acid priming improves micro- and macronutrient uptake, salinity stress tolerance, seed quality, and physio-biochemical parameters in lentil (*Lens culinaris* spp. *culinaris*). Plants 12:3539. https://doi.org/10.3390/plants12203539

Rathor P, Gorim LY, Thilakarathna MS (2023) Plant physiological and molecular responses triggered by humic based biostimulants—a way forward to sustainable agriculture. Plant Soil 492:31–60. https://doi.org/10.1007/s11104-023-06156-7

Roy D, Sayed MZ, Mondal D, Bandhan BS, Bahadur MM, Islam MR, Gaber A, Kabir MP, Hossain A, Pramanik SK (2025) Humic acid mediates drought tolerance in wheat through the modulation of morphophysiological traits, leading to improve grain yield in wheat. Phyton-Int J Exp Bot 94(3):763–779. https://doi.org/10.32604/phyton.2025.062717

Rupiasih NN, Vidyasagar PB (2005) A review: compositions, structures, properties and applications of humic substances. J Adv Sci Technol 8:16–25

Santos MP, Zupunski M, Monda H et al (2025) Humic acids modify root architecture in Arabidopsis through H+-ATPase-dependent target of rapamycin activation in concert with Ca2+ and ROS signaling. Chem Biol Technol Agric 12:47. https://doi.org/10.1186/s40538-025-00764-4

Shukry WM, Abu-Ria ME, Abo-Hamed SA, Anis GB, Ibraheem F (2023) The efficiency of humic acid for improving salinity tolerance in salt sensitive rice (*Oryza sativa*): growth responses and physiological mechanisms. Gesunde Pflanzen 75:2639–2653. https://doi.org/10.1007/s10343-023-00885-6

Silva RM, Canellas LP (2022) Organic matter in the pest and plant disease control: a meta-analysis. Chem Biol Technol Agric 9:70. https://doi.org/10.1186/s40538-022-00332-0

Song X, Chen M, Chen W, Jiang H, Yue X (2020) Foliar application of humic acid decreased hazard of cadmium toxicity on the growth of hybrid *Pennisetum*. Acta Physiol Plant 42(8):1–11. https://doi.org/10.1007/s11738-020-03118-9

Vrantsi E, Lakka A, Bozinou E, Athanasiadis V, Papadaki ES, Dourtoglou VG, Lalas SI (2021) Humic and fulvic acids as specific sorbents of herbicides in water. Clean—soil, air, Water 49: 2000467. https://doi.org/10.1002/clen.202000467

Wang M, Song G, Zheng Z, Mi X, Song Z (2024) Exploring the impact of fulvic acid and humic acid on heavy metal availability to alfalfa in molybdenum contaminated soil. Sci Rep 14:32037. https://doi.org/10.1038/s41598-024-83813-6

Waqas M, Ahmad B, Arif M, Munsif F, Latif Khan A, Amin M, Kang S, Kim Y, Lee I (2014) Evaluation of humic acid application methods for yield and yield components of mungbean. Am J Plant Sci 5:2269–2276. https://doi.org/10.4236/ajps.2014.515241

Whitton MM, Ren X, Yu SJ, Irving AD, Trotter T, Bajagai YS, Stanley D (2023) Humate application alters microbiota–mineral interactions and assists in pasture dieback recovery. Heliyon 9(2):e13327. https://doi.org/10.1016/j.heliyon.2023.e13327

Xiong Q, Wang S, Lu X, Xu Y, Zhang L, Chen X, Xu G, Tian D, Zhang L, Jing J, Ye X (2023) The effective combination of humic acid phosphate fertilizer regulating the form transformation of phosphorus and the chemical and microbial mechanism of its phosphorus availability. Agronomy 13(6):1581. https://doi.org/10.3390/agronomy13061581

Xu Q, Duan D, Cai Q, Shi J (2018) Influence of humic acid on Pb uptake and accumulation in tea plants. J Agric Food Chem 66:12327–12334. https://doi.org/10.1021/acs.jafc.8b03556

Yang F, Antonietti M (2020) Artificial humic acids: sustainable materials against climate change. Adv Sci 7:1902992. https://doi.org/10.1002/advs.201902992

Yang F, Yuan Y, Liu Q, Zhang X, Gai S, Jin Y, Cheng K (2023) Artificial humic acid promotes growth of maize seedling under alkali conditions. Envr Pollut 327:121588. https://doi.org/10.1016/j.envpol.2023.121588

Yildirim E, Ekinci M, Turan M, Agar G, Dursun A, Kul R, Alim Z, Argin S (2021) Humic + fulvic acid mitigated Cd adverse effects on plant growth, physiology and biochemical properties of garden cress. Sci Rep 11:8040. https://doi.org/10.1038/s41598-021-86991-9

Zandonadi DB, Santos MP, Dobbss LB, Olivares FL, Canellas LP, Binzel ML, Okorokova-Façanha AL, Façanha AR (2010) Nitric oxide mediates humic acids-induced root development and plasma membrane H+-ATPase activation. Planta 231(5):1025–1036. https://doi.org/10.1007/s00425-010-1106-0

Zhaoyang N, Lu Z, Tieyi Z, Liangliang G, Jie Z, Fenghua A, Hongyuan M, Zhichun W, Fan Y (2025) Effects of lignite humic acid and lignite humic acid-based combined amendment on soil quality in saline-sodic farmlands in the West Liaohe Plain, China. Chin Geogra Sci 35(2): 401–414. https://doi.org/10.1007/s11769-025-1491-8

Zhu L, Liu H, Zhang Y, Cao Y, Hu Y, Wang Y, Zheng H, Liu M (2025) Humic acid alleviates low-temperature stress by regulating nitrogen metabolism and proline synthesis in melon (*Cucumis melo* L.) seedlings. Horticulturae 11:16. https://doi.org/10.3390/horticulturae11010016

Zubo YO, Schaller GE (2020) Role of the cytokinin-activated type-B response regulators in hormone crosstalk. Plants 9:166. https://doi.org/10.3390/plants9020166

Chapter 15
Malic Acid: Improving Photosynthesis and Stress Tolerance

Sina Siavash Moghaddam, Latifeh Pourakbar, Sevda Alizadeh, and Zahra Alizadeh

Abstract Plants are constantly exposed to various types of biotic stresses, including pathogen attacks and pest infestations, as well as abiotic stressors such as drought, salinity, temperature extremes, and environmental pollution like heavy metals. These stresses cause significant and damaging effects on plant growth, development, and their physiological, biochemical, and molecular processes. Their impacts include harm to the photosynthetic system, metabolic disruption, impaired uptake of essential nutrients, and decreased crop yields. To combat these stresses, plants activate both enzymatic and non-enzymatic antioxidant defense systems; however, additional methods are necessary to further enhance stress tolerance. In this context, applying organic acids, such as malic acid, to leaves has proven to be an effective way to boost stress tolerance in vulnerable plants. Malic acid (MA), a vital part of the Krebs cycle, is essential for ATP production. When applied externally, it can enhance photosynthesis, preserve chlorophyll, and stabilize cell membranes. It also functions as an osmolyte, aiding in water regulation in plants under salinity and drought stress conditions. Research has shown increased antioxidant enzyme activity, reduced levels of stress-related metabolic markers, and improved mineral absorption under various stress scenarios. Therefore, applying MA externally is an economical and eco-friendly strategy for supporting sustainable agriculture and increasing crop yields. Nonetheless, existing research on its external use is limited, and further studies are needed to determine optimal application rates, plant-specific responses, stress conditions, and underlying molecular mechanisms.

S. S. Moghaddam (✉)
Department of Plant Production and Genetics, Faculty of Agricultural Sciences, University of Guilan, Rasht, Iran
e-mail: ss.moghaddam@guilan.ac.ir

L. Pourakbar · S. Alizadeh
Department of Biology, Faculty of Science, Urmia University, Urmia, Iran
e-mail: la.pourakbar@urmia.ac.ir; s.alizadeh@urmia.ac.ir

Z. Alizadeh
Department of Plant, Cell and Molecular Biology, Faculty of Natural Sciences, University of Tabriz, Tabriz, Iran

A. A. H. Abdel Latef (ed.), *Harnessing Vitamins and Organic Acids for Plant Stress Resilience*, Sustainability Sciences in Asia and Africa,
https://doi.org/10.1007/978-981-95-8593-9_15

Keywords Organic acid · Drought · Heavy metal · Salinity · Antioxidants

15.1 Introduction

Malic acid (MA, malate, 2-hydroxybutanedioic acid) is a naturally occurring four-carbon dicarboxylic acid that plays a key role in multiple metabolic pathways in plant cells. Its biosynthesis happens through various routes, reflecting its metabolic flexibility. In glyoxysomes, malic acid is produced from isocitrate, a product of fatty acid breakdown, and is then converted to succinate and other metabolites (Ma et al. 2016). The photosynthetic mechanisms of C_4 and Crassulacean acid metabolism (CAM) plants show different ways of producing and using malate. In C_4 plants, malate forms in mesophyll cells during CO_2 fixation catalyzed by phosphoenolpyruvate (PEP) carboxylase (Lundgren et al. 2014; Calace et al. 2021), while in CAM plants, night-time CO_2 fixation causes malate buildup in vacuoles (Cockburn 1983; Winter and Smith 2022). Additionally, glycolytic pyruvate can be enzymatically turned into malate via malic enzyme activity (Zelle et al. 2008; Gao et al. 2022). Mitochondrial metabolism also plays a major part in malate production, where it functions as an important intermediate in the tricarboxylic acid (TCA) cycle. The process of converting fumarate to malate in plant mitochondria is a key metabolic step (Stutz and Burris 1951; Doireau et al. 2024), with malate dehydrogenase enabling this process, a plant-specific enzyme activity (El-Shanhorey and Barakat 2020). The fate of malate extends beyond energy production, as mitochondrial malate can be oxidized to oxaloacetate (OAA), which then enters amino acid biosynthesis pathways. Through transamination reactions catalyzed by glutamate oxaloacetate transaminase (GOT), OAA is converted to aspartate, a precursor for amino acids like glutamine, lysine, and threonine (Galili et al. 2016; Zhao et al. 2020; Zamani-Nour et al. 2021). Malic acid's multifunctionality, characterized by its non-toxicity, natural abundance, and low cost, makes it suitable for various uses, including as a cross-linking agent (Sethi et al. 2022). Physiologically, malate plays crucial roles in cellular homeostasis, especially in pH regulation and osmotic balance within vacuoles (Saavedra et al. 2022). Its role in modulating cytoplasmic pH and maintaining cell turgor has been well documented across different plant species (Lu et al. 2019). These various metabolic and physiological functions emphasize the fundamental importance of malic acid in plant biochemistry and cell activity. Plants constantly face threats from biotic stressors like fungi, bacteria, viruses, and pests such as insects, parasitic plants, and weeds, as well as abiotic stresses like extreme temperatures (cold and heat), water stress (drought and flooding), salinity, and heavy metals. While plants have internal mechanisms to fight and tolerate these stresses, these defenses may not always be enough, especially in crops. Applying organic acids like malic acid externally has become an effective way to boost stress tolerance and protect plant yields (El-Shanhorey and Barakat 2020; Ahmed et al. 2022; Khan et al. 2023).

15.2 Effect of Abiotic Stress on the MA Content of Plants

Malate serves as a crucial metabolic intermediate in the photosynthetic pathways of C_4 and CAM plants. It functions as a vacuolar osmolyte, particularly in guard cells, where it is primarily synthesized via phosphoenolpyruvate carboxylase (PEPCase) (Calace et al. 2021). Elevated PEPCase activity suggests that the higher malate content observed in cold-tolerant winter rye leaves is generated through this enzymatic pathway. Additionally, the increased activities of key malate-metabolizing enzymes—including NADP-malate dehydrogenase, NAD-malate dehydrogenase, PEP carboxylase, and NADP-malic enzyme—further support enhanced malate production and turnover under these conditions (Crecelius et al. 2003). Malic acid plays a dual role in plant stress responses by modulating reactive oxygen species (ROS) levels under low-temperature conditions. While it can exacerbate photooxidative damage, this mechanism may serve as a protective strategy against cold injury. Additionally, MA acts as a supplementary carbon reservoir, storing carbon dioxide to mitigate photosystem damage induced by chilling stress (Crecelius et al. 2003). Similarly, under salt stress, MA contributes to ionic homeostasis by accumulating alongside anions such as Cl^- to counterbalance excessive sodium uptake. This response is accompanied by increased expression of malic enzyme (ME) genes and elevated ME protein levels, suggesting a critical role for ME in regulating malate metabolism during salinity adaptation (Shao et al. 2011). Metabolomic analysis revealed that waterlogging stress induced higher MA accumulation in the resistant pakchoi cultivar Heiyoudong (HYD), suggesting its potential role in enhancing hypoxia tolerance. The enriched metabolic pathways—particularly the TCA cycle (citrate) and pyruvate metabolism—were directly associated with malic acid biosynthesis. Notably, malic acid was identified as a critical metabolite in carbohydrate metabolism and ATP generation under stress conditions (Xi et al. 2025). Similarly, drought stress has been shown to trigger MA and malate accumulation in winter wheat, where malate functions as a vital carbon reserve during water deficit (Marček et al. 2019).

15.3 Effects of Exogenous Application of MA on Plant Stresses

15.3.1 Abiotic Stress

15.3.1.1 Heavy Metals

“Heavy metals” refer to a group of metals and metalloids characterized by an atomic number greater than 20 and a relative density exceeding 5 g/cm^3 (Zhang et al. 2019). Common heavy metals include lead (Pb), cadmium (Cd), copper (Cu), zinc (Zn), chromium (Cr), arsenic (As), mercury (Hg), and nickel (Ni). Their prevalence in food products has shown an increasing trend, as reported in multiple studies (Peng et al. 2019; Zhang et al. 2020b; Ahmed et al. 2022). Plants absorb heavy metals primarily through their roots, with most of these metals becoming immobilized in the cell wall and root tissues. Consequently, their concentration in aerial parts—particularly in seeds—tends to be lower than in roots (Wang et al. 2021; Liang et al. 2018; Xue et al. 2022). In conventional agriculture, sources of heavy metal contamination include fertilizers, pesticides, manure, and sewage (Li et al. 2014). The toxicity of heavy metals in plants is influenced by multiple factors, including plant species, metal type, concentration, chemical form, and soil properties (such as composition and pH). Notably, some heavy metals, such as Cu and Zn, are essential for plant growth, functioning as cofactors or activators in enzymatic reactions (e.g., in enzyme-substrate metal complexes) (Chen et al. 2022; Bastakoti 2023). Additionally, they serve as prosthetic groups in metalloproteins, participating in redox reactions, electron transport, and nucleic acid metabolism. Conversely, non-essential metals like cadmium, mercury, and arsenic exhibit high toxicity, disrupting metal-sensitive enzymes and leading to growth inhibition, impaired photosynthetic activity, or organism death (Alizadeh et al. 2025; Gill 2014; Malaie et al. 2024). As a mitigation strategy, the external application of organic acids (e.g., malic acid) has demonstrated efficacy in chelating metals and alleviating heavy metal toxicity. This approach has been widely validated as an effective countermeasure in numerous studies (Vega et al. 2022). Exogenous application of 100 μM MA to *Salix variegata* seedlings improved biomass, reduced Cd accumulation, reduced malondialdehyde (MDA) levels, and increased enzymatic activity. It also increased root glutathione content and improved phytochelatin biosynthesis potential (Zhang et al. 2020a). In *Miscanthus sacchariflorus* 200 μg MA increased plant growth, increased the net photosynthesis rate, and reduced Cd uptake. Malic acid application led to tolerance to Cd stress through the regulation of enzymatic and non-enzymatic antioxidants, thereby reducing oxidative damage (Guo et al. 2017). Concentrations of 250 or 500 μM MA in a Cd-containing culture medium caused root accumulation of Cd and reduced cadmium transfer to aerial parts in *Helianthus annuus*. In addition, MA increased growth parameters, increased the content of photosynthetic pigments such as chlorophylls and anthocyanins, reduced H_2O_2 accumulation, and improved root activity, thereby reducing cadmium toxicity. The expression of stress tolerance genes, including chlorophyll synthase, CHLG (a gene key to the formation of chlorophyll), OsNramp1, superoxide dismutase (SOD), and peroxidase (POD), increased under the influence of MA (Hawrylak-Nowak et al. 2015). In corn plants under Ni stress, external application of 0.1 mM MA per kg increased plant biomass, improved the activity of glutathione reductase (GR) and SOD enzymes, and increased total plant biomass by increasing photosynthetic parameters and upregulating the expression of stress tolerance genes (Ghasemi-Fasaei 2012). In okra (*Abelmoschus esculentus* L.), MA at two levels (0.5 and 1.0 mM) improved the dry weight and length of roots and shoots that had been reduced by Cd. It also reduced the content of Cd and MDA. The content of coumaric acid and chlorogenic acid increased under Cd and MA (Mousavi et al. 2021). It has been reported that in alfalfa (*Medicago sativa* L.) under aluminum (Al) stress, foliar application of

100 μM MA increased plant growth (shoot and root weight and root length), improved mineral nutrient concentrations, reduced lipid peroxidation, and reduced Al content in plant roots. Additionally, external application of malic acid increased the transcription of malate dehydrogenase (MDH) and phosphoenol pyruvate carboxylase (PEPC) genes in roots, leading to improved plant resistance to heavy metal stress (An et al. 2014). Reports indicate that in Cr-exposed Indian mustard (*Brassica juncea*), 0.25 mM MA increased Cr retention in roots but suppressed its translocation to shoots. MA boosted non-enzymatic antioxidants (ascorbate, glutathione) and activated enzymes including ascorbate peroxidase (APX), dehydroascorbate reductase (DHAR), GR, glutathione peroxidase (GPX), SOD, catalase (CAT), reducing oxidative stress and restoring water status, chlorophyll content, and biomass (Al Mahmud et al. 2017). In Japanese rice plants (*Oryza sativa* cultivars NPB and the brittle stem mutant fc8), foliar spraying with MA at the pollination stage significantly reduced the Cd, Pb, and Ni contents by 27.9%–41.0%, while the Ca and glutamate contents increased significantly in both NPB and fc8. MA also upregulated GLR genes (OsGLR3.1–3.5) and enhanced glutamate synthesis via glutamic-oxalate transaminase (Zhang et al. 2024). Yao et al. (2020) reported that in *Pinus massoniana* seedlings, exogenous MA at concentrations (0.00, 0.01, 0.02, 0.04, 0.08, 0.16 mmol/L) increased the activity of antioxidant enzymes in leaves, reduced H_2O_2, $O_2^{\bullet-}$, MDA levels, osmotic regulators and alleviated Al toxicity. It is documented that in Cd-stressed *Salix variegata*, MA application significantly improved biomass by preserving chlorophyll and protecting photosynthetic structures (Chen et al. 2020). Malic acid is effective in increasing the growth potential of *S. variegata* and improving its resistance under Cd stress (Liu et al. 2016). The external application of MA at 15 mmol/kg to *Bidens pilosa* L. in Cd-contaminated soil resulted in increased values of F_0 (basal fluorescence) and Fm (maximum fluorescence) of leaves, but the ratios F_v/Fm (photosystem II quantum yield) and F_v/F_0 (photochemical performance index) decreased. In the investigation of rapid light response curves, MA at low concentration (1 mmol/kg) improved ETR (electron transfer rate), Y(II) (actual photosystem II quantum yield), and Y(NPQ) (non-chemical energy loss yield). However, at higher concentrations (15 mmol/kg), MA increased Y(NO) (non-photochemical passive loss yield) to an extent that indicated a decrease in photosynthetic efficiency at high doses. A significant correlation was observed between growth and photosynthesis in these plants (Yang et al. 2022). Literature supports that in Pb-stressed centipedegrass, 400 μM MA decreased membrane lipid peroxidation, increased biomass, activated POD, CAT, and APX, and elevated proline, ascorbic acid, and phytochelatins. MA also restricted Pb translocation by sequestering it in roots (Cai et al. 2022). Studies report that in Ni-stressed *Salvia splendens*, 500 ppm MA reduced Ni toxicity, improved chlorophyll and carbohydrate content. Moreover, MA significantly reduced Ni accumulation in plant organs. MA prevented Ni accumulation in shoots by reducing its roots to shoots transfer factor (TF). These results indicated that MA increases plant resistance by reducing Ni uptake and improving photosynthetic status (Nader and Galal 2018). In Co(II)-exposed barley (*Hordeum vulgare*), MA combined with Co

Table 15.1 Effects of exogenous malic acid application on heavy metal stress in plants

Metal	Plant species	Key findings/mechanisms of action	References
Cd	*Salix variegata*	Improved biomass, reduced Cd accumulation and MDA levels, increased enzymatic activity and root glutathione content	Zhang et al. (2020a)
Cd	*Miscanthus sacchariflorus*	Increased plant growth and net photosynthesis rate, reduced Cd uptake	Guo et al. (2017)
Cd	*Helianthus annuus*	Cd accumulation in roots, reduced translocation to shoots; increased photosynthetic pigments, reduced H_2O_2, improved root activity; upregulation of stress genes (SOD, POD, CHLG)	Hawrylak-Nowak et al. (2015)
Ni	*Zea mays* (Corn)	Increased plant biomass, improved activity of GR and SOD enzymes, upregulation of stress tolerance genes	Ghasemi-Fasaei (2012)
Cd	*Abelmoschus esculentus* (Okra)	Improved dry weight and length of roots and shoots; reduced Cd and MDA content; increased coumaric and chlorogenic acid	Mousavi et al. (2021)
Cd	*Bidens pilosa* L.	Low concentration improved ETR, Y(II); high concentration decreased F_v/F_m, F_v/F_0 and increased Y (NO), indicating a dose-dependent effect on photosynthesis	Yang et al. (2022)
Ni	*Salvia splendens*	Reduced Ni toxicity, improved chlorophyll and carbohydrate content; reduced Ni accumulation and transfer factor (TF)	Nader and Galal (2018)
Al	*Medicago sativa* (alfalfa)	Increased plant growth, improved mineral nutrients, reduced lipid peroxidation and Al content in roots; increased transcription of *MDH* and *PEPC* genes	An et al. (2014)
Al	*Pinus massoniana*	Increased activity of antioxidant enzymes in leaves, reduced H_2O_2, $O_2^{\bullet-}$, MDA levels; alleviated Al toxicity	Yao et al. (2020)
Cr	*Brassica juncea*	Increased Cr retention in roots; boosted enzymatic (APX, DHAR, GR, GPX, SOD, CAT) and non-enzymatic antioxidants; reduced oxidative stress, restored water status and chlorophyll	Al Mahmud et al. (2017)
Pb	Centipedegrass	Decreased membrane lipid peroxidation, increased biomass; activated POD, CAT, APX; elevated proline, ascorbic acid, phytochelatins; restricted Pb translocation	Cai et al. (2022)
Multi	*Oryza sativa* (Rice)	Significantly reduced Cd, Pb, Ni contents in grains; increased Ca and glutamate; upregulated *GLR* genes	Zhang et al. (2024)
Co	*Hordeum vulgare* (barley)	Enhanced root/shoot length; lowered Co-induced MDA levels	Mahey et al. (2019)

(II) enhanced root/shoot length. While Co(II) or MA alone increased SOD, CAT, GPX, APX, and GR activity, their combination reduced enzyme activity. MA also lowered Co-induced MDA levels (Mahey et al. 2019) (Table 15.1).

15.3.1.2 Drought Stress

Among abiotic stresses, drought represents the most significant obstacle to plant growth, yield, and development, with its impacts expected to intensify due to climate change driven by human activities (Dawood et al. 2025; Fatima et al. 2026). Moreover, systems for monitoring drought, which assess its direct effects on crop productivity, have heightened concerns regarding the fragility of global food security (Liu et al. 2016). Drought severely disrupts plants' morphological, physiological, biochemical, and molecular attributes (Salehi-Lisar and Bakhshayeshan-Agdam 2016; Shafiq et al. 2021; Dawood et al. 2022; Kosar et al. 2022; Akram et al. 2024; Jameel et al. 2026; Roychoudhury et al. 2026; Umer et al. 2026). Consequently, there is an urgent need to develop eco-friendly, efficient, and cost-effective approaches to alleviate its effects on food production to sustain the increasing global population. Malic acid, a crucial organic acid in plant metabolic pathways, significantly contributes to drought tolerance by aiding osmotic adjustment, bolstering antioxidant defenses, and enhancing photosynthetic performance. Drought tolerance in plants is often assessed by analyzing shifts in organic compound levels under environmental stress (Khan et al. 2020). Among these compounds, malic acid—a key organic acid—exhibits the most significant increase during prolonged drought conditions. For instance, Gao et al. (2012) found that malic acid concentrations doubled in drought-stressed plants compared to well-watered controls. The role of malic acid in enhancing drought resilience has been documented across multiple plant species, including wheat and cotton (Guo et al. 2018; Kang et al. 2019; Bai et al. 2025). The foliar application of MA on lettuce under drought stress positively influenced fresh and dry weight yields, as well as dry matter production, promoting biomass accumulation, particularly under moderate to severe drought conditions. Additionally, MA improved the absorption of nitrogen, phosphorus, and potassium in drought-stressed plants. It positively impacted trends in maximum fluorescence (FM), while the FV/F0 ratio, reflecting the efficiency of primary photochemical reactions, showed improvement. The performance index (PI_ABS) at 5.0% polyethylene glycol (PEG) positively correlated with crop yield trends following MA application. At 7.5% PEG, MA significantly enhanced absorbed energy flux per reaction center (ABS/RC) and trapped photon flux per reaction center (TRo/RC). MA application notably reduced superoxide anion ($O_2^{\bullet-}$) and MDA levels in both stressed and control plants, while increasing proline accumulation. Carotenoid levels, which decreased under stress, were elevated with MA treatment (Kleiber et al. 2024) (Table 15.2). These results align with studies demonstrating that foliar MA enhances rooting and biomass in stressed rose plants (Igamberdiev and Bykova 2018). Drought typically suppresses nutrient uptake by inhibiting enzymes like nitrate reductase (Han et al. 2022), but MA likely mitigates this by stabilizing cellular pH and supporting energy metabolism via the tricarboxylic acid cycle (Kleiber et al. 2024).

15.3.1.3 Chilling

Low temperature stress, including cold (0–15 °C) and freezing, inhibits plant growth and physiological activity. To reduce the effects of this stress, plants activate physiological and cellular responses such as adjustments in membrane structure, photosynthesis, and metabolic processes (Abdel Latef and Chaoxing et al. 2011; Theocharis et al. 2012; Tahjib-Ul-Arif et al. 2024; Ahmad et al. 2026). One of the important effects of low temperatures is the disruption of photosynthesis. Cold stress causes damage to chloroplasts and pigments, disrupting the efficiency of photosynthesis and carbon metabolism (Ambavaram et al. 2014). When *Larix olgensis* seedlings were subjected to low-temperature stress (4 °C), foliar application of MA reduced electrolyte leakage and decreased MDA levels, indicating mitigated membrane damage. Additionally, MA treatment enhanced the activity of antioxidant enzymes, including SOD, POD, APX, and CAT, while also increasing proline and soluble protein concentrations. Chlorophyll content was also elevated in response to MA application. These findings suggest that exogenous organic acids, such as MA, can strengthen cold resistance in *L. olgensis* seedlings by boosting protective enzyme activity, reducing oxidative membrane damage, and improving osmoregulation and ROS scavenging capacity (Song et al. 2012) (Table 15.2).

15.3.1.4 Flooding

Exogenous MA, when combined with uric acid, notably elevated the levels of dissolved organic carbon (DOC), NH_4^+/NO_3^-–N, and dissolved inorganic

Table 15.2 Effects of exogenous malic acid application on other abiotic stresses

Stress type	Plant species	Key findings/mechanisms of action	References
Drought	*Lactuca sativa* (lettuce)	Improved fresh/dry weight and nutrient (N, P, K) uptake; improved chlorophyll fluorescence parameters (FV/F0, PI_ABS); reduced $O_2^{\bullet-}$ and MDA; increased proline and carotenoids	Kleiber et al. (2024)
Chilling	*Larix olgensis*	Reduced electrolyte leakage and MDA; enhanced activity of SOD, POD, APX, CAT; increased proline, soluble protein, and chlorophyll content	Song et al. (2012)
Salinity	*Solanum lycopersicum* (tomato)	Enhanced growth parameters, photosynthetic pigments, osmolytes (sugars, proteins, proline); increased enzymatic & non-enzymatic antioxidants; reduced MDA, H_2O_2, and Na^+ accumulation	Attia et al. (2021)
Flooding	*Kandelia obovata* (mangrove)	Stimulated release of root exudates (organic acids); reduced iron plaque formation; increased beneficial rhizosphere bacteria (Firmicutes, Actinobacteria)	Meng et al. (2022)
Elevated CO_2	*Oryza sativa* (rice)	Reduction in endogenous MA content, potentially influencing stomatal behavior and TCA cycle activity	Qi et al. (2021)

phosphorus (DIP) in *Kandelia obovata* root exudates. This effect was more pronounced at higher uric acid concentrations and extended waterlogging durations (10 h). Additionally, MA stimulated the release of formic, acetic, citric, oxalic, and lactic acids in root exudates, with secretion peaking at a moderate malic acid concentration (2.5 mg/L) after 5 h of waterlogging. The presence of exogenous malic acid also led to a significant decrease in iron plaque formation on *K. obovata* roots, particularly under conditions of no waterlogging or short-term waterlogging (5 h) (Meng et al. 2022) (Table 15.2). This reduction is linked to rhizosphere acidification and the enhanced solubilization of Fe(III) through organic acid-mediated complexation, including MA, which binds with hydroxyl groups to facilitate iron release (Chen et al. 2016). Furthermore, malic acid and related organic acids act as chemoattractants for iron-oxidizing bacteria (FeOB), promoting their activity near the root surface and supporting plant growth by influencing the iron redox cycle (Hedrich et al. 2011). When applied in mixtures, exogenous malic acid amplified the rhizosphere effect, significantly boosting bacterial diversity in both the rhizosphere and rhizoplane. Notably, it increased the relative abundance of Firmicutes (2–14 times higher) and Actinobacteria (2–6 times higher) in the rhizoplane compared to control conditions. As a key intermediate in the TCA cycle, malic acid likely serves as a metabolic signal, attracting beneficial rhizosphere bacteria such as Bacillus species to the root zone and enhancing microbial diversity (Feng et al. 2018).

15.3.1.5 Salinity

Soil salinity is a major environmental challenge worldwide, significantly threatening food security and agricultural sustainability, especially in arid and semi-arid regions. High salt concentrations in soil lead to disruption of physiological processes such as photosynthesis, nutrient imbalance, ionic toxicity, and osmotic stress in plants. These adverse effects impair plant growth and reduce crop yields (Ahmed et al. 2023; Su et al. 2022; Azooz et al. 2026). In tomato plants (*Solanum lycopersicum* L.) subjected to salt stress (100 mM NaCl), foliar application of a combined malic acid and chitosan solution enhanced growth parameters, photosynthetic pigment content, and the accumulation of osmolytes such as soluble sugars, proteins, and proline. Additionally, this treatment boosted both enzymatic antioxidants (including POD, CAT, polyphenol oxidase (PPO), and SOD) and non-enzymatic antioxidants (such as phenols and abscisic acid), improving the plant's stress tolerance. The application also mitigated salinity-induced damage by lowering oxidative stress markers like MDA and hydrogen peroxide (H_2O_2), as well as reducing Na^+ accumulation. Under NaCl stress, certain protein bands with varying molecular weights disappeared, but foliar treatment with malic acid and chitosan restored their expression, indicating a recovery in protein synthesis and stress adaptation mechanisms (Attia et al. 2021) (Table 15.2).

15.3.1.6 Elevated CO_2

Malic acid is a key metabolic intermediate involved in multiple physiological processes in plants (Santelia and Lawson 2016). One of its critical roles is in regulating stomatal aperture, where it typically suppresses stomatal activity (Lee et al. 2008; Qi et al. 2021). It has been reported that elevated CO_2 levels led to a reduction in MA content in rice seedlings, potentially influencing stomatal behavior. Additionally, the observed decline in malic acid, along with other TCA cycle intermediates such as fumaric acid, suggests a slowdown in TCA cycle activity (Qi et al. 2021) (Table 15.2). Alternatively, the decrease in malic acid may be attributed to enhanced activity of malic enzymes in chloroplasts, which decarboxylate malate to generate CO_2 and NADPH, thereby supporting the C_3 cycle and boosting photosynthetic efficiency (Rangel et al. 2010). These findings highlight the dual role of malic acid in both respiratory metabolism and photosynthetic enhancement under elevated CO_2 conditions.

15.3.2 Biotic Stress

Plants face numerous biotic stresses triggered by living organisms, including fungi, viruses, bacteria, nematodes, and insects. These pathogens and pests cause infections, diseases, and physical damage to crops, ultimately reducing agricultural productivity (Fahad et al. 2017; Afzal et al. 2026). *Sclerotinia sclerotiorum*, a widely distributed necrotrophic fungal pathogen, infects a diverse range of host plants, causing significant crop losses. Fei et al. (2023) demonstrated that MA, at a concentration of 2 mg/mL, effectively reduced the pathogenicity of *S. sclerotiorum* by suppressing mycelial growth, inhibiting sclerotia formation, and downregulating key pathogenicity factors. Additionally, treating detached leaves with MA significantly decreased lesion diameters at wound sites, underscoring its potential for disease control. Increasing MA under treatment with dissolved organic matter and selenium improves the growth inhibition of *S. sclerotiorum* (Jia et al. 2019). Yeon et al. (2022) reported that MA demonstrated antifungal effects against various fungi and oomycetes, with minimum inhibitory concentrations (MIC) ranging from 312.5 to approximately 2500 μg/mL. The wettable powder formulation of MA and chitosan (CS) mixture demonstrated strong protective efficacy in pot trials, achieving 73% control of *Phytophthora blight* on red pepper, 91% suppression of cucumber damping-off, and 84% reduction in *Pythium blight* on creeping bentgrass. These findings indicate that the combination of MA and CS exhibits a synergistic effect, offering a sustainable and effective alternative for managing destructive oomycete diseases (Yeon et al. 2022) (Table 15.3). Malic acid exerts antifungal effects by lowering the cytoplasmic pH of fungi, thereby suppressing their reproductive capacity while simultaneously promoting plant growth. In a study on *Fusarium oxysporum*, malic acid inhibited mycelial growth by 49.6% and significantly reduced

Table 15.3 Effects of exogenous malic acid application on biotic stresses

Pathogen	Disease / host plant	Key findings/mechanisms of action	References
Sclerotinia sclerotiorum	Sclerotinia stem rot	Suppressed mycelial growth, inhibited sclerotia formation, downregulated key pathogenicity factors, reduced lesion diameters on detached leaves	Fei et al. (2023)
Fusarium oxysporum	Fusarium wilt (cucumber)	Inhibited mycelial growth by 49.6%, significantly reduced disease incidence, led to higher plant survival rates, promoted plant growth	Eskander and Emara (2025)
Phytophthora capsici	Phytophthora blight (pepper)	Achieved 73% control of the disease in pot trials	Yeon et al. (2022)
Pythium spp.	Damping-off (cucumber)	Achieved 91% suppression of the disease in pot trials	Yeon et al. (2022)
Pythium spp.	Pythium blight (Bentgrass)	Achieved 84% reduction of the disease in pot trials	Yeon et al. (2022)

Fusarium wilt incidence in *Cucumis sativus* (cucumber), leading to higher plant survival rates. Additionally, at an optimal concentration of 4000 ppm, malic acid demonstrated the greatest enhancement in plant growth (Eskander and Emara 2025).

15.4 Role of MA Application in Plants Under Biotic and Abiotic Stress

Figure 15.1 summarizes the protective effects of externally applied MA on plants under biotic and abiotic stress. The figure illustrates that MA enhances biomass growth and overall plant performance through multiple key mechanisms. These include metal detoxification via chelation and sequestration in roots, antioxidant enhancement through upregulation of the defense system, protection of the photosynthetic apparatus, osmotic regulation to maintain cellular turgor and homeostasis, rhizosphere modulation, a direct antifungal effect against pathogens, and an overall improvement in stress tolerance.

15.5 Conclusions and Future Strategies

Malic acid plays a vital role in plant stress responses, demonstrating remarkable adaptability in mitigating both abiotic and biotic stresses. As a key intermediate in C_4 and CAM photosynthesis, MA serves multiple protective functions, acting as an osmolyte, ROS regulator, and carbon sink during exposure to cold, drought, salinity, and heavy metals. Studies highlight that its biosynthesis occurs primarily through

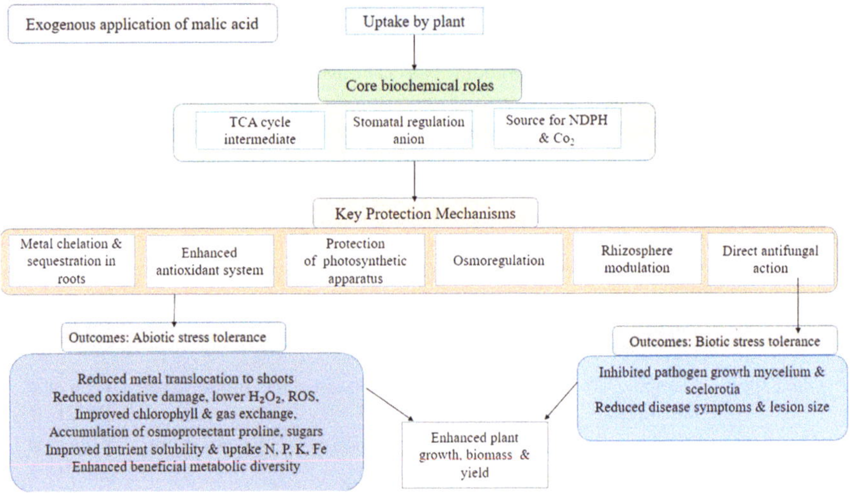

Fig. 15.1 Malic acid (MA) application and its role in plants under biotic and abiotic stress

PEPCase and malate-metabolizing enzymes. In plants such as winter rye and pakchoi cultivars, MA accumulation increases under cold and waterlogging stress, enhancing stress tolerance. Under heavy metal stress, MA helps maintain ionic homeostasis, restores photosynthetic efficiency, and strengthens the antioxidant defense system. During drought and salinity stress, it enhances water retention and nutrient uptake while reducing sodium ion accumulation. In cold stress, MA stabilizes cell membranes and boosts antioxidant enzyme activity. Additionally, its demonstrated antifungal properties against pathogens like Sclerotinia and Fusarium present promising applications in integrated pest management. To fully exploit MA's potential, further genetic studies and long-term field trials on its exogenous application are essential. Such research will deepen our understanding of MA's mechanisms and advance its use in sustainable agriculture.

References

Abdel Latef AAH, Chaoxing H (2011) Arbuscular mycorrhizal influence on growth, photosynthetic pigments, osmotic adjustment and oxidative stress in tomato plants subjected to low temperature stress. Acta Physiologia Plantarum 33:1217–1225

Ahmed N, Ehsan A, Danish S, Ali MA, Fahad S, Dawar K, Taban S, Akça H, Shah AA, Ansari MJ, Babur E (2022) Mitigation of lead (Pb) toxicity in rice cultivated with either ground water or wastewater by application of acidified carbon. J Environ Manag 307:114521

Ahmed R, Zia-ur-Rehman M, Sabir M, Usman M, Rizwan M, Ahmad Z, Alharby HF, Al-Zahrani HS, Alsamadany H, Aldhebiani AY, Alzahrani YM, Bamagoos AA (2023) Differential response of nano zinc sulphate with other conventional sources of Zn in mitigating salinity stress in rice grown on saline-sodic soil. Chemosphere 327:138479

Ahmad N, Ahmad S, Islam MS, Khan MN, Abdel Latef AAH (2026) Chitosan Nanoparticles Under Temperature Extremes. In: Abdel Latef, A.A.H. (eds) Chitosan Nanoparticles for Plant Resilience Under Abiotic and Biotic Stressors. Smart Nanomaterials Technology. Springer, Singapore. https://doi.org/10.1007/978-981-95-6422-4_6

Afzal A, Ali B, Samarasinghe KGBA, Khan MN, Javed MA, Saleem A (2026) Overview of Biotic Stress and Its Agricultural Impact. In: Abdel Latef, A.A.H. (eds) Chitosan Nanoparticles for Plant Resilience Under Abiotic and Biotic Stressors. Smart Nanomaterials Technology. Springer, Singapore. https://doi.org/10.1007/978-981-95-6422-4_3

Al Mahmud J, Hasanuzzaman M, Nahar K, Rahman A, Hossain MS, Fujita M (2017) Maleic acid assisted improvement of metal chelation and antioxidant metabolism confers chromium tolerance in *Brassica juncea* L. Ecotoxicol Environ Saf 144:216–226

Alizadeh S, Pourakbar L, Moghaddam SS (2025) Green synthesis of magnetic silver/zinc/iron nanocomposite mitigates detrimental effects of polymethyl methacrylate nanoplastics and Arsenic and ameliorates biochemical compositions in *Triticum aestivum* L. Plant Stress 17: 100984

Akram NA, Fatima K, Kong H, Zafar N, Mahmood S, Ashraf M, Abdel Latef AAH (2024) Interactive effect of drought stress and l-methionine on the growth and physio-biochemical changes in broccoli (Brassica oleracea L. var. italica): Leaf and Head. Journal of Plant Growth Regulation 43:1954–1966

Ambavaram MM, Basu S, Krishnan A, Ramegowda V, Batlang U, Rahman L, Baisakh N, Pereira A (2014) Coordinated regulation of photosynthesis in rice increases yield and tolerance to environmental stress. Nat Commun 5:5302

An Y, Zhou P, Xiao Q, Shi D (2014) Effects of foliar application of organic acids on alleviation of aluminum toxicity in alfalfa. J Plant Nutr Soil Sci 177(3):421–430

Attia MS, Osman MS, Mohamed AS, Mahgoub HA, Garada MO, Abdelmouty ES, Abdel Latef AAH (2021) Impact of foliar application of chitosan dissolved in different organic acids on isozymes, protein patterns and physio-biochemical characteristics of tomato grown under salinity stress. Plants 10:388

Azooz MM, Abdel Latef AAH, Abdelmajeed WG, et al. (2026) Biochar ameliorates salt stress in maize by modulating physiological responses and expression of SOS1 and NHX2 genes. Journal of Soil Science and Plant Nutrition. https://doi.org/10.1007/s42729-026-03123-w

Bai S, Zhao P, Su F, Wang J, Wang S, Li Y, Mohamed IA, Yin Z (2025) Silencing proline dehydrogenases improves salt and drought tolerance in *Gossypium hirsutum*. bioRxiv 2025-07

Bastakoti S (2023) Role of zinc in management of plant diseases: a review. Cogent Food Agric 9(1): 2194483

Cai X, Fu J, Li X, Peng L, Yang L, Liang Y, Jiang M, Ma J, Sun L, Guo B, Yu X (2022) Low-molecular-weight organic acid-mediated tolerance and Pb accumulation in centipedegrass under Pb stress. Ecotoxicol Environ Saf 241:113755

Calace P, Tonetti T, Margarit E, Figueroa CM, Lobertti C, Andreo CS, Gerrard Wheeler MC, Saigo M (2021) The C4 cycle and beyond: diverse metabolic adaptations accompany dual-cell photosynthetic functions in Setaria. J Exp Bot 72(22):7876–7890

Chen Z, Liu F, Bu T, Liu Y, Zhu J (2016) Effects of organic acids on dissolution of Fe and Mn from weathering coal gangue. Acta Geochim 35:316–328

Chen HC, Zhang SL, Wu KJ, Li R, He XR, He DN, Huang C, Wei H (2020) The effects of exogenous organic acids on the growth, photosynthesis and cellular ultrastructure of *Salix variegata* Franch. under Cd stress. Ecotoxicol Environ Saf 187:109790

Chen G, Li J, Han H, Du R, Wang X (2022) Physiological and molecular mechanisms of plant responses to copper stress. Int J Mol Sci 23(21):12950

Cockburn W (1983) Stomatal mechanism as the basis of the evolution of CAM and C4 photosynthesis. Plant Cell Environ 6(4):275–279

Crecelius F, Streb P, Feierabend J (2003) Malate metabolism and reactions of oxidoreduction in cold-hardened winter rye (*Secale cereale* L.) leaves. J Exp Bot 54(384):1075–1083

Dawood MFA, Zaid A, Abdel Latef AAH (2022) Salicylic acid spraying-induced resilience strategies against the damaging impacts of drought and/or salinity stress in two varieties of vicia faba L. seedlings. Journal of Plant Growth Regulation 41:1919–1942

Doireau R, Jaślan J, Cubero-Font P, Demes-Causse E, Bertaux K, Cassan C, Pétriacq P, De Angeli A (2024) At ALMT 5 mediates vacuolar fumarate import and regulates the malate/fumarate balance in Arabidopsis. New Phytol 244(3):811–824

Dawood MFA, Tahjib-Ul-Arif M, Shirazy BJ, Abdel Latef AAH (2025) Unraveling the role of κ-carrageenan on the combined effect of drought and chromium stress in wheat (Triticum aestivium L.). Plant Physiol Biochem 223:J109781

El-Shanhorey N, Barakat A (2020) Effect of lead and cadmium in irrigation water and foliar applied malic acid on vegetative growth, flowering and chemical composition of salvia Splendens plants (B) effect of cadmium. Sci J Flowers Ornam Plants 7:483–499

Eskander MA, Emara AR (2025) Study the efficacy of formulated malic acid alone and its mixture with Thiourea against *Fusarium oxysporium* in *Cucumis sativus* plants. J Plant Prot Pathol 16(2):111–118

Fahad S, Bajwa AA, Nazir U, Anjum SA, Farooq A, Zohaib A et al (2017) Crop production under drought and heat stress: plant responses and management options. Front Plant Sci 29:01147

Fatima A, Akram NA, Talha M, Younis M, Ashraf, M, Abdel Latef AAH (2026) Comparative Effects of Externally Applied Glycinebetaine and Neem (Azadirachta indica) Leaf Extract on Plant Growth, Physio-Biochemical Characteristics and Yield Production of Canola (Brassica napus L.) Subjected to Drought Stress. Russian Journal of Plant Physiology 73:18

Fei YC, Cheng Q, Zhang H, Han C, Wang X, Li YF, Li SQ, Zhao XH (2023) Maleic acid and malonic acid reduced the pathogenicity of *Sclerotinia sclerotiorum* by inhibiting mycelial growth, sclerotia formation and virulence factors. Stress Biol 3:45

Feng H, Zhang N, Du W, Zhang H, Liu Y, Fu R, Shao J, Zhang G, Shen Q, Zhang R (2018) Identification of chemotaxis compounds in root exudates and their sensing chemoreceptors in plant growth-promoting rhizobacteria *Bacillus amyloliquefaciens* SQR9. Mol Plant-Microbe Interact 31:995–1005

Galili G, Amir R, Fernie AR (2016) The regulation of essential amino acid synthesis and accumulation in plants. Annu Rev Plant Biol 67:153–178

Gao QH, Wu CS, Wang M, Xu BN, Du LJ (2012) Effect of drying of jujubes (*Ziziphus jujuba* mill.) on the contents of sugars, organic acids, α-tocopherol, β-carotene, and phenolic compounds. J Agric Food Chem 60(38):9642–9648

Gao M, Zhao H, Zheng L, Zhang L, Peng Y, Ma W, Tian R, Yuan Y, Ma F, Li M, Ma B (2022) Overexpression of apple Ma12, a mitochondrial pyrophosphatase pump gene, leads to malic acid accumulation and the upregulation of malate dehydrogenase in tomato and apple calli. Hortic Res 9:uhab053

Ghasemi-Fasaei R (2012) Malic acid and phosphorus influences on nickel phytoremediation efficiency and metal nutrients relationships in a Ni-polluted calcareous soil. Int Res J Appl Basic Sci 3:2805–2808

Gill M (2014) Heavy metal stress in plants: a review. Int J Adv Res 2(6):1043–1055

Guo H, Chen H, Hong C, Jiang D, Zheng B (2017) Exogenous malic acid alleviates cadmium toxicity in *Miscanthus sacchariflorus* through enhancing photosynthetic capacity and restraining ROS accumulation. Ecotoxicol Environ Saf 141:119–128

Guo R, Shi L, Jiao Y, Li M, Zhong X, Gu F, Liu Q, Xia X, Li H (2018) Metabolic responses to drought stress in the tissues of drought-tolerant and drought-sensitive wheat genotype seedlings. AoB Plants 10(2):ply016

Han ML, Lv QY, Zhang J, Wang T, Zhang CX, Tan RJ, Wang YL, Zhong LY, Gao YQ, Chao ZF, Li QQ (2022) Decreasing nitrogen assimilation under drought stress by suppressing DST-mediated activation of nitrate reductase 1.2 in rice. Mol Plant 15(1):167–178

Hawrylak-Nowak B, Dresler S, Matraszek R (2015) Exogenous malic and acetic acids reduce cadmium phytotoxicity and enhance cadmium accumulation in roots of sunflower plants. Plant Physiol Biochem 94:225–234

Hedrich S, Schlomann M, Johnson DB (2011) The iron-oxidizing proteobacteria. Microbiology 157:1551–1564

Igamberdiev AU, Bykova NV (2018) Role of organic acids in the integration of cellular redox metabolism and mediation of redox signalling in photosynthetic tissues of higher plants. Free Radic Biol Med 22:74–85

Jameel A, Akram NA, Bashir R, Ashraf M, Abdel Latef AAH (2026) Salicylic Acid Enhances Yield and Alleviates Drought-Induced Oxidative Damage in Two Carrot Cultivars with Contrasting Tolerance. Russian Journal of Plant Physiology 73:23

Jia W, Hu C, Xu J, Ming J, Zhao Y, Cai M, Sun X, Liu X, Zhao X (2019) Dissolved organic matter derived from rape straw pretreated with selenium in soil improves the inhibition of *Sclerotinia sclerotiorum* growth. J Hazard Mater 369:601–610

Kang Z, Babar MA, Khan N, Guo J, Khan J, Islam S, Shrestha S, Shahi D (2019) Comparative metabolomic profiling in the roots and leaves in contrasting genotypes reveals complex mechanisms involved in post-anthesis drought tolerance in wheat. PLoS One 14(3):e0213502

Khan N, Ali S, Zandi P, Mehmood A, Ullah S, Ikram M, Ismail I, Babar MA (2020) Role of sugars, amino acids and organic acids in improving plant abiotic stress tolerance. Pak J Bot 52(2): 355–363

Khan EA, Yadav S, Yadav M, Ahmed HM (2023) Role of low molecular organic acids in regulating physiological and molecular aspects of plants under abiotic stress. In: The role of growth regulators and Phytohormones in overcoming environmental stress. Academic Press, pp 289–315

Kleiber T, Chadzinikolau T, Formela-Luboińska M, Lartey JL, Kosiada T (2024) Enhancing lettuce drought tolerance: the role of organic acids in photosynthesis and oxidative defense. Appl Sci 14(12):5119

Kosar F, Alshallash KS, Akram NA, Sadiq M, Ashraf M, Alkhalifah DHM, Abdel Latef AAH, Elkelish A (2022) Trehalose-Induced Regulations in Nutrient Status and Secondary Metabolites of Drought-Stressed Sunflower (Helianthus annuus L.) Plants 11:2780

Lee M, Choi Y, Burla B, Kim YY, Jeon B, Maeshima M, Yoo JY, Martionoia E, Lee Y (2008) The ABC transporter AtABCB14 is a malate importer and modulates stomatal response to CO2. Nat Cell Biol 10:1217–1223

Li Z, Ma Z, van der Kuijp TJ, Yuan Z, Huang L (2014) A review of soil heavy metal pollution from mines in China: pollution and health risk assessment. Sci Total Environ 468-469:843–853

Liang C, Xiao H, Hu Z, Zhang X, Peng L, Geng Y, Li H, Yu S, Yang Y (2018) Uptake, transportation, and accumulation of C60 fullerene and heavy metal ions (Cd, Cu, and Pb) in rice plants grown in an agricultural soil. Environ Pollut 235:330–338

Liu X, Zhu X, Pan Y, Li S, Liu Y, Ma Y (2016) Agricultural drought monitoring: progress, challenges, and prospects. J Geogr Sci 26:750–767

Lu J, Liu J, Li J, Li X, Li Y, Yang Y (2019) Molecular cloning and functional characterization of the aluminum-activated malate transporter gene MdALMT14. Sci Hortic 244:208–217

Lundgren MR, Osborne CP, Christin PA (2014) Deconstructing Kranz anatomy to understand C4 evolution. J Exp Bot 65:3357–3369

Ma Z, Marsolais F, Bernards MA, Sumarah MW, Bykova NV, Igamberdiev AU (2016) Glyoxylate cycle and metabolism of organic acids in the scutellum of barley seeds during germination. Plant Sci 248:37–44

Mahey S, Kumar R, Kumar V, Sharma R, Sharma M, Bhardwaj R, Thukral AK (2019) Maleic acid and EDTA mediated extenuation of Co (II) stress in Hordeum vulgare seedlings. Biotechnol Res Innov 3(2):221–230

Malaie S, Pourakbar L, Siavash Moghaddam S, Xiao J, Khezrnejad N (2024) Phytoremediation of mercury-contaminated soil by *Vigna radiata* L. plant in companion with bacterial and fungal biofertilizers. Environ Sci Pollut Res 31(43):55549–55561

Marček T, Hamow KA, Vegh B, Janda T, Darko E (2019) Metabolic response to drought in six winter wheat genotypes. PLoS One 14(2):e0212411

Meng H, Yan Z, Li X (2022) Effects of exogenous organic acids and flooding on root exudates, rhizosphere bacterial community structure, and iron plaque formation in *Kandelia obovata* seedlings. Sci Total Environ 830:154695

Mousavi A, Pourakbar L, Moghaddam SS, Popović-Djordjević J (2021) The effect of the exogenous application of EDTA and maleic acid on tolerance, phenolic compounds, and cadmium phytoremediation by okra (*Abelmoschus esculentus* L.) exposed to cd stress. J Environ Chem Eng 9(4):105456

Nader A, Galal AS (2018) Effect of nickel in the irrigation water and foliar applied malic acid on vegetative growth, flowering and chemical composition of *Salvia splendens* plant. Alex J Agric Sci 63(6):327–337

Peng H, Chen Y, Weng L, Ma J, Ma Y, Li Y, Islam MS (2019) Comparisons of heavy metal input inventory in agricultural soils in North and South China: a review. Sci Total Environ 660:776–786

Qi XF, Wang XH, Wang Q, Li M, Ma LJ, Li YY, Li XM, Wang LL (2021) Photosynthesis, stomatal conductance, endogenous hormones and organic acid synergistic regulation in leaves of rice (*Oryza sativa* L.) under elevated CO2. Appl Ecol Environ Res 19(5):3773

Rangel AF, Rao IM, Braun HP, Horst WJ (2010) Aluminum resistance in common bean (*Phaseolus vulgaris*) involves induction and maintenance of citrate exudation from root apices. Physiol Plant 138(2):176–190

Roychoudhury A, Chatterjee U, Bhattacharya A, Dutta S (2026) Using Chitosan Nanoparticles in the Alleviation of Drought Stress in Plants. In: Abdel Latef, A.A.H. (eds) Chitosan Nanoparticles for Plant Resilience Under Abiotic and Biotic Stressors. Smart Nanomaterials Technology. Springer, Singapore. https://doi.org/10.1007/978-981-95-6422-4_5

Saavedra T, Almeida AM, Vaz M et al (2022) Effects of foliar application of organic acids on strawberry plants. Plant Physiol Biochem 188:12–20

Salehi-Lisar SY, Bakhshayeshan-Agdam H (2016) Drought stress in plants: causes, consequences, and tolerance. In: Hossain M, Wani S, Bhattacharjee S, Burritt D, Tran LS (eds) Drought stress tolerance in plants, vol 1. Springer, Cham

Santelia D, Lawson T (2016) Rethinking guard cell metabolism. Plant Physiol 172(3):1371–1392

Sethi S, Saruchi, Kaith BS, Kaur M, Sharma N, Kumar V (2022) Malic acid cross-linked chitosan based hydrogel for highly effective removal of chromium (VI) ions from aqueous environment. React Funct Polym 177:105318

Shafiq S, Akram NA, Ashraf M, García-Caparrós P, Ali OM, Abdel Latef AAH (2021) Influence of glycine betaine (natural and synthetic) on growth, metabolism and yield production of drought-stressed maize (zea mays L.) plants. Plants 10(11):2540

Song JF, Li F, Wang YJ, Gao HS, Cui XY, Peng HM (2012) Effects of exogenous organic acids on physiological and biochemical characteristics of *Larix olgensis* seedlings under low temperature stress. CABI Digital Library

Stutz RE, Burris RH (1951) Photosynthesis and metabolism of organic acids in higher plants. Plant Physiol 26(2):226

Su X, Wang Y, Wang G, Zhang Y, Gong X, Yu J, Gou F, Lyu H (2022) Assessment and prediction of coastal saline soil improvement effects combining substrate amendments and salt barrier materials in typical region of the Yangtze River Delta. Soil Tillage Res 223:105483

Shao HB, Liu ZH, Zhang ZB, Chen QJ, Chu LY, Brestic M (2011) Biological roles of crop NADP-malic enzymes and molecular mechanisms involved in abiotic stress. Afr J Biotechnol 10:4947–4953

Tahjib-Ul-Arif M, Asaduzzaman, M, Shirazy BJ, Khan MSU, Rahman AMS, Murata Y, Abdel Hamed S, Abdel Latef AA (2024) Seed priming improves chilling stress tolerance in rice (oryza sativa L.) seedlings. Phyton-International Journal of Experimental Botany 93(11):3013–3027

Theocharis A, Clement C, Barka EA (2012) Physiological and molecular changes in plants grown at low temperatures. Planta 235:1091–1105

Umer S, Akram NA, Ashraf M, Abdel Latef AAH (2026) Pre-Sowing Application of Methionine and Proline Mitigates Drought Stress in Two Canola (Brassica napus L.) Cultivars. Russian Journal of Plant Physiology 73:48

Vega A, Delgado N, Handford M (2022) Increasing heavy metal tolerance by the exogenous application of organic acids. Int J Mol Sci 23(10):5438

Wang C, Huang Y, Zhang C, Liu Z, Zhang X, Yang J, Wang Y, Peng L (2021) Inhibition effects of long-term calcium-magnesia phosphate fertilizer application on cd uptake in rice: regulation of the iron-nitrogen coupling cycle driven by the soil microbial community. J Hazard Mater 416: 125916

Winter K, Smith JA (2022) CAM photosynthesis: the acid test. New Phytol 233(2):599–609

Xi DD, Gao L, Ge LA, Zhu HF, Miao LM, Zhang DY, Zhang CW, Li Y, Dong YX, Hou XL, Zhu YY (2025) Combined transcriptome and metabolome analysis provides insight into the ERF073-malic acid network in pakchoi under submergence stress. Plant Stress 15:100708

Xue W, Zhang C, Huang Y, Liu Z, Wang C, Wang Y, Peng L, Zhang X (2022) Rice organs concentrate cadmium by chelation of amino acids containing dicarboxyl groups and enhance risks to human and environmental health in Cd-contaminated areas. J Hazard Mater 426:128130

Yang Q, Xie J, Liu H, Fang Z (2022) The addition of exogenous low-molecular-weight organic acids improved phytoremediation by *Bidens pilosa* L. in Cd-contaminated soil. Environ Sci Pollut Res 29(51):76766–76781

Yao H, Zhang S, Zhou W, Liu Y, Liu Y, Wu Y (2020) The effects of exogenous malic acid in relieving aluminum toxicity in *Pinus massoniana*. Int J Phytoremediation 22(6):669–678

Yeon J, Park AR, Nguyen HTT, Gwak H, Kim J, Sang MK, Kim JC (2022) Inhibition of oomycetes by the mixture of maleic acid and copper sulfate. Plant Dis 106(3):960–965

Zamani-Nour S, Lin HC, Walker BJ et al (2021) Overexpression of the chloroplastic 2-oxoglutarate/malate transporter disturbs carbon and nitrogen homeostasis in rice. J Exp Bot 72(1):137–152. https://doi.org/10.1093/jxb/eraa343

Zelle RM, de Hulster E, van Winden WA, de Waard P, Dijkema C, Winkler AA, Geertman JM, van Dijken JP, Pronk JT, van Maris AJ (2008) Malic acid production by *Saccharomyces cerevisiae*: engineering of pyruvate carboxylation, oxaloacetate reduction, and malate export. Appl Environ Microbiol 74:2766–2777

Zhang X, Yan L, Liu J, Zhang Z, Tan C (2019) Removal of different kinds of heavy metals by novel PPG-nZVI beads and their application in simulated stormwater infiltration facility. Appl Sci 9: 4213

Zhang S, Chen H, He D, He X, Yan Y, Wu K, Wei H (2020a) Effects of exogenous organic acids on Cd tolerance mechanism of *Salix variegata* Franch. Under Cd Stress. Front Plant Sci 11:594352

Zhang X, Yan Y, Wadood SA, Sun Q, Guo B (2020b) Source apportionment of cadmium pollution in agricultural soil based on cadmium isotope ratio analysis. Appl Geochem 123:104776

Zhang X, Xue W, Qi L, Zhang C, Wang C, Huang Y, Wang Y, Peng L, Liu Z (2024) Malic acid inhibits accumulation of cadmium, lead, nickel and chromium by down-regulation of OsCESA and up-regulation of OsGLR3 in rice plant. Environ Pollut 341:122934

Zhao Y, Yu H, Zhou JM, Smith SM, Li J (2020) Malate circulation: linking chloroplast metabolism to mitochondrial ROS. Trends Plant Sci 25(5):446–454

Chapter 16
Oxalic Acid: A Dual Role in Plant Stress Response and Detoxification

Mona F. A. Dawood and Yara T. AbdElaziz

Abstract Oxalic acid (OA) is a dicarboxylic organic acid that serves a dual function in plants: acting as a protective agent against environmental and biotic stressors and, paradoxically, posing potential toxicity risks when accumulated excessively. This multifaceted molecule is involved in a wide range of physiological and biochemical responses, including chelation of heavy metals, reactive oxygen species regulation, interaction with plant hormones, and genetic modulation under stress conditions. However, despite the growing body of research, substantial gaps remain in our understanding of its regulation, biosynthesis, signaling functions, and environmental implications. This chapter explores both the protective and toxicological dimensions of OA, highlighting its potential in phytoremediation and sustainable agriculture, industry, and biotechnological applications while proposing future research directions to bridge current knowledge gaps.

Keywords Organic acids · Oxalic acid · Abiotic stress · Fruit ripening · Mineralization · Biotic stress · Virulence factor

16.1 Introduction

The external application of organic acids has been extensively recognized as an effective strategy for strengthening plant tolerance to various stresses that impact plants, while also influencing fruit ripening and postharvest senescence (Walker and Famiani 2018). Oxalic acid (OA) occurs naturally in many biological systems; it is produced in fungal cultures and is also detected in human urine (Munir et al. 2001). OA has several ecological and biochemical multifunctions, such as sharing in the breakdown of organic pollutants, lignin degradation, and detoxification of copper-based wood preservatives through the activity of wood-decaying fungi. It also

M. F. A. Dawood (✉) · Y. T. AbdElaziz
Botany and Microbiology Department, Faculty of Science, Assiut University, Assiut, Egypt
e-mail: Mo_fa87@aun.edu.eg

A. A. H. Abdel Latef (ed.), *Harnessing Vitamins and Organic Acids for Plant Stress Resilience*, Sustainability Sciences in Asia and Africa,
https://doi.org/10.1007/978-981-95-8593-9_16

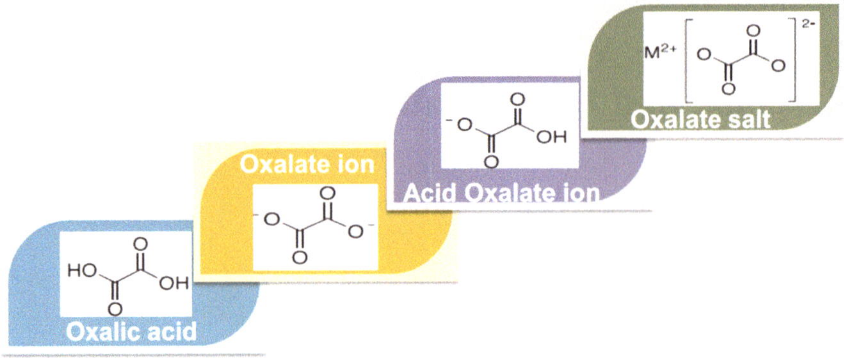

Fig. 16.1 Different forms of oxalate

improves aluminum toxicity and contributes to nitrogen fixation in legumes by serving as an electron donor in symbiotic rhizobia (Munir et al. 2001). In higher plants and their fruits, OA plays multiple roles: it facilitates calcium storage, allows protection against herbivores, improves metal-ion toxicity, and contributes to pH regulation closely connected to nitrate assimilation. It can also serve as a precursor for hydrogen peroxide formation, which is involved in defense signaling, fungal resistance, and the formation of extracellular polymer cross-links through peroxidase activity (Walker and Famiani 2018). Due to its antioxidant activity, OA participates in redox homeostasis, up-regulates programmed cell death, boosts systemic acquired resistance, and helps delay postharvest senescence in fruits and vegetables (Ding et al. 2007).

As a ubiquitous low-molecular-weight compound in higher plants, OA and its salts (Fig. 16.1) indirectly participate in various environmental and biochemical processes, including pollutant degradation, wood biodeterioration, and mineral weathering (Janusz et al. 2017; Gadd 2007). Their strong chelating property allows oxalates to bind and detoxify heavy metals (Vlasov et al. 2023; Grąz et al. 2009, 2011), participate in nutrient cycling, and affect nutrient availability in soils (Gadd 2007). Additionally, oxalates are involved in fungal pathogenicity (McCaghey et al. 2019). OA is particularly reactive with calcium, forming calcium oxalate crystals, which are structures commonly present in plants and fungi (Gadd et al. 2014); however, problematic in humans, causing renal disorders (Demoulin et al. 2022).

Increasing evidence highlights that OA is a crucial component of plant responses to stresses and plays an influential role in plant adaptation and developmental processes. This chapter describes the existing knowledge about the role of oxalates in plants, focusing on their functions, biosynthesis, and interactions with environmental factors. By mentioning the key features and gaps in current understanding, this chapter presents the significance of oxalates in plant biology and their applications in agriculture and ecosystem management.

16.2 History and Properties of Oxalic Acid

The scientific study of crystals began in the seventeenth century. In 1675, Leeuwenhoek was the first to report the occurrence of calcium oxalate crystals in plant tissues. Later, McNazir documented the presence of ergastic crystals across 215 different plant groups. Based on Webb et al. (1995), the crystals of Ca-oxalate are widely reported as biominerals in plants. Calcium oxalate in plants has been detected in plant tissues at various stages of plant development, from germination to vegetative and then reproductive. In addition to its detection in various organs, including roots, leaves, stems, seeds, floral organs, root nodules, and anthers of plants (Hussain et al. 2024).

Oxalic acid $(COOH)_2$, systematically named ethanedioic acid, is a common simple organic acid in the biosphere. Oxalate is widely distributed across biological systems, including plants, animals, fungi, bacteria, as well as in soils and mineral deposits (Kumar et al. 2019). Like many members of the Amaranthaceae, red orach accumulates antinutritional compounds such as nitrate (NO_3^-) and oxalate (Sharma and Dietz 2006). In plants, OA concentrations vary remarkably from 3% to 80% of dry weight depending on species, tissue type, and environmental conditions (Li et al. 2022a). It exists either in the free acid form ($H_2C_2O_4$) or as oxalate salts, where it forms insoluble complexes with Ca^{2+} and Mg^{2+} or soluble salts with Na^+ and K^+. The term oxalate denotes the deprotonated anion ($C_2O_4^{2-}$), a strong ligand that readily binds metal cations, whereas *oxalic acid* refers to the protonated molecule. Chemically, oxalic acid is a highly acidic, low-molecular-weight dicarboxylic acid ($pKa_1 \approx 1.27$), much stronger than acetic acid.

In nature, it occurs in two forms as free acid or mineral salts of divalent metals such as calcium or copper, while also forming soluble complexes with sodium and potassium. Owing to its strong chelating ability, oxalic acid binds efficiently to metals such as Mn^{2+} and Fe^{3+} and serves as a source of protons and electrons (Li et al. 2022a). This underpins its industrial uses in bleaching and wood restoration (Prasad and Shivay 2017). Biologically, oxalate plays essential roles in Ca-regulation, carbon cycling, detoxification of heavy metals, insect defense, and storage processes (Dauer and Perakis 2014; Franceschi and Nakata 2005; Korth et al. 2006; Li et al. 2022b; Ruiz et al. 2002; Tooulakou et al. 2016b). However, excessive levels can be detrimental, promoting stomatal opening, weakening the cell wall, activating pectin-degrading enzymes under acidic conditions, and triggering programmed cell death (Guimarães and Stotz 2004; Kim et al. 2008; Marciano et al. 1983). Thus, plants rely on tightly regulated oxalate biosynthesis and degradation pathways to maintain cellular homeostasis (Weijuan et al. 2025).

16.3 Occurrence of Oxalate

After scientists recognized the importance of OA in plant metabolism, many studies focused on measuring its levels in different plant tissues and understanding its roles (Bhandari and Kawabata 2004; Ergonül and Nergiz 2010; Nguyễn and Savage 2013a; Pereira et al. 2013). OA in fruits and vegetables is commonly measured using high-performance liquid chromatography or ultra-fast liquid chromatography with photodiode array detection (Nguyễn and Savage 2013a, b; Pereira et al. 2013). Calcium oxalate crystals in plants are among the earliest biological structures ever observed with a light microscope (Leeuwenhoek 1975). Since then, these crystals have been found in many environments, including rocks, soil, and organisms across all five biological kingdoms. In every case, the crystals form when calcium from the environment combines with oxalate produced by living organisms (Hodgkinson 1977).

In plants, Ca-oxalate crystals are widespread. More than 215 plant families are known to store these crystals within their tissues (McNair 1932). High concentrations of OA have been reported in several crops, such as beet greens, poppy seeds, purslane, rhubarb, sweet peppers, sorrel, Swiss chard, and spinach (Zheng and Brecht 2017). Plants that naturally produce oxalate may accumulate it at levels ranging from 3% to 80% of their dry weight, and up to 90% of the plant's total calcium may exist in the form of calcium oxalate (Zindler-Frank 1976; Libert and Franceschi 1987; Gallaher 1975; Horner and Wagner 1995). These oxalate crystals are typically stored inside vacuoles of specialized cells known as crystal idioblasts (Horner and Wagner 1995; Nakata 2003). It has been reported that calcium oxalate is present in plants at every stage of development, from embryonic through vegetative to reproductive stages. Crystals have been found in various plant parts, including roots, leaves, stems, seeds, floral organs, root nodules, and anthers (Malik et al. 2024). Some plants like rhubarb, oxalate levels generally rise as the plants grow older. However, for other plants, such as spinach, beet leaves and bananas, oxalate content significantly increases during the early stages of growth and then decreases as the plants mature. The rhizome of the water lily is considered safe for consumption, even though it has been reported to contain an oxalate concentration of 8.76 mg/100 g (Abelt et al. 2024). The formation, size, distribution, and accumulation of calcium oxalate crystals from oxalic acid in some vegetables can be influenced by conditions like temperature, light exposure, pressure, harvest timing, pH, ion concentration, and the ratios of macro and micronutrients in the soil (Santoso 2013). Oxalate can be detected as soluble or insoluble oxalates in the leaves, forming the total oxalates. The soluble oxalates can link to Na^+, K^+, and NH_4^+, while insoluble oxalates are only bound to Ca^{2+}, Mg^{2+}, and Fe^{2+} (Bong et al. 2017). The Ca-oxalates have been detected in the sap of vacuoles (Nakata 2012).

Although oxalates are classified as antinutrients, they can be beneficial to some extent, especially at low concentrations (Egbuna and Ifemeje 2015). In this regard, the leaves of *Corchorus olitorus* and *Hibiscus sabdariffa* contain a low amount of oxalate and are rich in essential nutrients and polyphenols that are crucial for human health (Dappah et al. 2018). Certain plants high in oxalates, e.g., *Medicago sativa*, also have a high calcium content, which makes them safe to feed (Rahman et al. 2013). Abbey et al. (2006) reported that although most indigenous leafy vegetables are high in anti-nutritional compounds like oxalate, they are also very nutritious and adapt very well to the tropical environments compared with foreign vegetables.

16.4 Biosynthesis of Oxalic Acid in Plants

Oxalic acid is one of the most widely investigated organic acids in both plants and animals, and several biosynthetic routes have been proposed for its formation. Historically, OA was first isolated from *Oxalis acetosella* L., a rhizomatous species identified in Switzerland by Francois Pierre Savary around 1773 (Prasad and Shivay 2017). In plants, OA can be produced through the oxaloacetate pathway, the ascorbic acid pathway, and the glycolate/glyoxylate pathway. Thus, there are three precursors of OA: ascorbate, oxaloacetate, and glyoxylate/glycolate, which are produced during photorespiration as well as the glyoxylate cycle in plants due to glyoxylate/glycolate oxidation (Goldsmith et al. 2022; Yu et al. 2010). In this sense, the oxidation reaction causes the production of oxalate and acetate (Zayed et al. 2025). Li et al. (2022a) explained that plants produce oxalates through three main metabolic routes: the glycolate/glyoxylate pathway, the oxaloacetate pathway, and the ascorbic acid pathway. Earlier studies also suggested that oxaloacetate and glyoxylate (both products of photorespiration) can serve as crucial starting molecules for oxalate biosynthesis (Franceschi and Loewus 1995; Yu et al. 2010). Oxalate can also be generated from ascorbate or iso-citrate. Plants use several enzymes to break down oxalate, including oxalate decarboxylase, oxalate oxidase, and oxalyl-CoA synthetase. The contribution of oxalate decarboxylase, however, is considered very small in most plants (Franceschi and Loewus 1995). Research on lettuce leaves showed that lactate dehydrogenase in the cytosol and glycolate oxidase in the peroxisome participate in oxalate formation (Davies and Asker 1983). Many plant tissues contain specialized cells called idioblasts, which are responsible for producing and storing calcium oxalate crystals (Franceschi and Horner 1980). In these idioblasts, oxalate is mainly synthesized from ASA, while its catabolism occurs through the oxalyl-CoA synthetase pathway (Foster et al. 2012).

In detail, the glyoxalic acid/glycolic acid pathway was first proposed in relation to photorespiration, where isotope tracing demonstrated the conversion of glycolate or glyoxylate to oxalate via glycolate oxidase (GLO) (Richardson and Tolbert 1961; Igamberdiev and Eprintsev 2016). GLO, localized in peroxisomes, catalyzes glyoxylate oxidation to oxalate and regulates photosynthesis and H_2O_2 homeostasis through its interaction with catalase (Xu et al. 2009; Yu et al. 2010). Although photorespiration contributes to oxalate synthesis, glyoxylate can also arise from the glyoxylate and TCA cycles, suggesting broader metabolic integration (Ji and Peng 2005; Lu et al. 2014).

In the ascorbic acid pathway, degradation of ascorbic acid through dehydroascorbic acid and oxalyl-threonate intermediates leads to oxalic acid formation, likely within vacuoles (Parsons and Fry 2012). Isotopic and mutant studies provide evidence that ascorbate can serve as an oxalate precursor, although results differ among species (Nakata and McConn 2007; Truffault et al. 2017; Mirahmadi et al. 2022). The pathway appears to be species- and condition-dependent, with further genetic confirmation still required (Li et al. 2022a).

A third, less characterized route involves 3-oxaloacetate acetylhydrolase (OXAC), detected in beet and spinach extracts (Chang and Beevers 1968). Transcriptomic data link elevated oxalate levels in spinach to genes associated with oxaloacetate degradation, including OXAC, malate synthase (MLS), and malate dehydrogenase (MDH) (Cai et al. 2018). However, direct enzymatic and molecular evidence for this pathway remains limited. These findings indicate that oxalic acid biosynthesis in plants is multifactorial, with overlapping pathways linked to photorespiration, ascorbate catabolism, and organic acid metabolism, reflecting both metabolic flexibility and environmental regulation.

16.5 Oxalate Degradation in Plants

Maintaining oxalate balance in cells of various organisms has included many oxalate catabolism routes, including oxidation, decarboxylation, and acetylation.

16.5.1 Oxalic Acid Degradation by Oxidation Route

Oxalate oxidase (OXO) is an enzyme that needs manganese ions (Mn^{2+}) to catalyze the conversion of oxalate into CO_2 and H_2O_2 (Whittaker and Whittaker 2002). In plants, OXO is part of the germin family, a group of cell wall-associated proteins. The enzyme was first identified in wheat, where it served as a marker for seed germination and was therefore named "germin" (Lane et al. 1993). Current evidence suggests that true OXO activity is found only in monocotyledonous plants. Although dicotyledonous plants also contain germin-related proteins, none of them have been shown to exhibit oxalate oxidase activity.

16.5.2 Oxalic Acid Degradation by the Decarboxylation Route

Oxalate-degrading enzymes, particularly oxalate decarboxylase (OXDC), show strong specificity for oxalate. This enzyme requires an appropriate pH and manganese ions (Mn^{2+}) to function and converts oxalate into formate and carbon dioxide

(Tanner et al. 2001; Reinhardt et al. 2003). OXDC activity was first reported in the fungus *Flammulina velutipes* (Shimazono 1955) and is mainly found in fungi, with some occurrence in bacteria (Kumar et al. 2019). Introducing the *F. velutipes* OXDC gene (FvOXDC) into crops such as tobacco, tomato, lettuce, and soybean significantly improved resistance to *Sclerotinia sclerotiorum* (Kesarwani et al. 2000; Dias et al. 2006; Cunha et al. 2010; Ghosh et al. 2016). In tomato, fruit-specific expression of FvOXDC reduced oxalate levels and helped maintain stable pH and organic acid balance (Chakraborty et al. 2013; Narula et al. 2016). Moreover, constitutive and seed-specific expression of FvOXDC in soybean and grass pea lowered seed oxalate content while increasing essential minerals such as calcium, iron, and zinc (Kumar et al. 2016). Therefore, upregulating oxalate metabolism via OXDC is a good way to enhance the nutritional status and disease resistance of plants. Studies have shown that oxalate decarboxylase is not naturally present in any known plant species, which means that it cannot be regulated or modified using the gene-editing tools currently available (Kumar et al. 2016).

16.5.3 Oxalic Acid Degradation by Acetylation Pathway

This route is commonly found in bacteria, fungi, and plants and degrades oxalate through an acetylation process that depends on CoA and ATP. It has been suggested that the pathway proceeds through a series of four enzymatic steps, ultimately converting oxalate into carbon dioxide. In the first step, oxalate is activated by oxalyl-CoA synthase to form oxalyl-CoA. This intermediate is then converted into formyl-CoA by oxalyl-CoA decarboxylase (Lou et al. 2016a, b).

16.6 Oxalic Acid Roles in Plants

Oxalates are produced by most plants, including vegetables (Yankey et al. 2015). Oxalate is a molecule originating from plants and acts as a final toxic metabolite with no identified physiological role in humans (Bargagli et al. 2020). These molecules are stored in various plant tissues, and their concentration can vary greatly between different species and under different environmental conditions (Franceschi and Loewus 2020).

Oxalates play various roles in vegetables, including seed germination, Ca-storage and regulation, ion balance, detoxification, structural strength, and insect repulsion (Yankey et al. 2015). Oxalate shares in the carbon cycle (Tooulakou et al. 2016b), acts as a defending molecule against herbivores and pathogens (Paiva 2021). Rahman and Kawamura (2011) reported that the sensitivity of herbivores to oxalate poisoning is also dependent on certain factors, including the type of oxalate chemical complexes formed and the adaptability to oxalate forage. Oxalates in plants are involved in maintaining ion balance, defending against threats, supporting tissue structure, detoxifying harmful substances, and managing light absorption and reflection. The phenomenon of light scattering by biominerals such as calcium oxalate complexes in leaves might be widespread, aiding in the redistribution of light patterns and improving the leaf's efficiency in utilizing incoming light (Minocha and Long 2023). These complexes have functions that encompass detoxifying heavy metals, defending against herbivores, and providing protection from both environmental and biological stressors.

Controlling calcium levels through the formation of oxalates has important consequences for plant growth and development. Calcium is essential for cell wall stability, signal transduction, and enzyme activation. By regulating the availability of calcium, oxalates have an indirect effect on these critical functions (Pandey and Sanyal 2021). Calcium oxalate crystals help mitigate calcium toxicity by sequestering excess calcium (He et al. 2024).

Oxalates have been reported to play a vital role in regulating the pH of the cell content of plants (Salgado et al. 2023; Kumar et al. 2021). The pH level of the vacuole content of these plants, where oxalates are primarily located, influences the chemical forms of oxalates. If the vacuole content pH is 7, oxalic acid also loses protons and appears as the oxalate ion. However, at acidic pH, like pH 2, the content is mainly acid oxalate. Also, at a pH of around 6, the oxalate ion becomes the predominant form (Salgado et al. 2023).

Another study also reported that pH levels increase in the soil environment as leaf litter on the surface deposits oxalic acid, which is then absorbed and oxidized by microorganisms (Cowan et al. 2024). Oxalates help plants manage aluminum toxicity in acidic soils and are key in phytoremediation, which cleanses soils polluted with heavy metals like cadmium, lead, and zinc. Calcium oxalate complex has been reported to play a crucial role in the regulation of cadmium distribution in cacao (Hester et al. 2024). They form insoluble calcium oxalates that assist plants in controlling the levels of calcium and oxalic acid, both of which can be harmful if present in high amounts.

Beyond its functions in plants, oxalates in the form of oxalic acid significantly influence the bioavailability of aluminum, calcium, iron, and phosphorus in soils, where they exist alongside other organic acids (Allison et al. 2020). It also plays a role in decreasing soil minerals and the formation of insoluble metal oxalates, which is vital for the survival of fungi and other organisms that thrive in environments with potentially toxic metal compounds. Although the nutritional and toxicological conditions of terrestrial ecosystems can be profoundly influenced by aerobic or anaerobic microbial processes that affect oxalate-metal interactions or the availability of oxalates. These processes also have a significant impact on plant physiology (Allison et al. 2020). Moreover, oxalic acid acts as a toxin for various plant pathogens; it is released into plants during infection, aiding in the degradation of plant cells and either being reused by the pathogens or eventually forming calcium oxalate crystals (Baran 2020).

Calcium oxalate complexes formed have been known to serve as an anticoagulant and serve as an internal source of carbon in plants when the stomata are closed (Tooulakou et al. 2016b). The results of a study in rhubarb showed that calcium oxalate crystals have the tendency to accumulate around the vascular bundles of the vegetable (Mohd-Ishfaq et al. 2024). Calcium oxalate complex in plant materials has also been reported to play a role in enhancing the stability of hydrochars (Luo et al. 2024). A powdered analysis of *Hibiscus hirtus* Linn, which included a microscopic examination of calcium oxalate crystals, revealed that these crystals are dispersed throughout the cortical tissue, including the mesophyll and phloem (Radhika et al. 2019).

By forming insoluble salts with calcium and other minerals like iron and magnesium (Kaushal et al. 2022), oxalates reduce the availability of these nutrients to potential pests and cause variations in the extractable iron content in the vegetables (Amoah et al. 2024). Oxalates in the form of oxalic acid make water-soluble salts with Na^+, K^+, and NH_4^+ ions, and bind with Ca^{2+}, Fe^{2+}, and Mg^{2+} to form insoluble salts, reducing the availability of these minerals (Savage and Klunklin 2018). Calcium oxalate complex formed in plants has been reported to cause irritation or proteolytic poisons that can hurt foraging herbivores (Malik et al. 2024). The nutritional value of vegetables is dependent on their nutrient composition, including oxalate levels (Borquaye et al. 2017). Anti-nutrients (oxalate) to nutrients ratios can be calculated and used to predict the bioavailability of calcium and iron in vegetables, as reported in a study on some leafy vegetables of Côte d'Ivoire (Koua et al. 2015). The overall effects of oxalic acid on plants can be summarized in Table 16.1. Considering the effects of oxalates on human health and agriculture is crucial for enhancing plant-based diets and refining crop management practices (Barkla et al. 2024).

Table 16.1 Functions of oxalic acid in plants

Function	Details	References
Calcium regulation and storage	Forms insoluble calcium oxalate crystals to regulate calcium levels, prevent toxicity, and maintain calcium homeostasis essential for cell wall stability and signal transduction	Pandey and Sanyal (2021), He et al. (2024)
pH regulation	Oxalates regulate cellular pH by existing in different ionic forms depending on vacuolar pH; oxalic acid also affects soil pH when released through litter decomposition	Salgado et al. (2023), Kumar et al. (2021), Cowan et al. (2024)
Defense against herbivores and pathogens	Acts as a chemical defense; oxalate toxicity deters herbivory and resists pathogen attack by forming crystals or being released during infection	Paiva (2021), Rahman and Kawamura (2011), Baran (2020), Malik et al. (2024)
Detoxification of heavy metals	Binds and immobilizes toxic metals (e.g., Cd, Pb, Zn), contributing to phytoremediation and reduced metal toxicity in plant tissues	Hester et al. (2024)

(continued)

Table 16.1 (continued)

Function	Details	References
Structural support	Calcium oxalate crystals reinforce plant tissues, particularly around vascular bundles, improving rigidity and mechanical strength	Mohd-Ishfaq et al. (2024), Radhika et al. (2019)
Light absorption and reflection	Calcium oxalate biominerals scatter and redistribute light within leaves, enhancing photosynthetic efficiency under varying light conditions	Minocha and Long (2023)
Carbon storage and recycling	Serves as an internal carbon source when stomata are closed, participating in the plant's carbon cycle	Tooulakou et al. (2016b)
Ion balance maintenance	Maintains ion homeostasis by binding with various cations (Ca^{2+}, Mg^{2+}, Fe^{2+}, Na^+, K^+, NH_4^+), forming soluble and insoluble oxalate salts	Savage and Klunklin (2018), Kaushal et al. (2022)
Seed germination and growth	Promotes or regulates early developmental processes such as seed germination and initial calcium mobilization	Yankey et al. (2015)
Influence on nutrient bioavailability	Reduces mineral availability (e.g., Ca, Fe, Mg) by forming insoluble oxalates, affecting nutritional quality of vegetables	Amoah et al. (2024), Borquaye et al. (2017), Koua et al. (2015)
Soil interactions and metal cycling	Affects bioavailability of metals (Al, Fe, Ca, P) and contributes to soil metal complexation, influencing microbial and plant processes	Allison et al. (2020)
Enhancing hydrochar stability	Calcium oxalate complexes improve structural and chemical stability of hydrochar derived from plant biomass	Xin et al. (2024)

16.7 Role of Oxalic Acid against Against Abiotic and Biotic Stress

Plants are vulnerable to stressful conditions (abiotic/biotic) that negatively affect their growth and limit agricultural crop productivity, hence causing loss of important crop plants globally (Dawood et al. 2024, 2025; Abdelrahiem et al. 2024; Abdelmoneim et al. 2024, 2025). The exogenous supplementation of OAs critically manages resistance against biotic and abiotic stresses and mediates ripening and senescence of fruits (Walker and Famiani 2018).

16.7.1 Oxalic Acid and Abiotic Stress

16.7.1.1 Oxalic Acid and Salinity Stress

Saline soil negatively impacts plant growth and productivity by inhibiting water uptake, creating a chemical drought, and causing ion toxicity from excess salts like sodium and chloride. This leads to stunted growth, reduced photosynthesis, and lower crop yields through a combination of osmotic stress, ionic imbalances, oxidative stress, metabolic disorders, and damage to plant tissues (Dawood et al. 2022; Azooz et al. 2026). The application of moderating agents such as OA stimulated plant development under salinity stress (Table 16.2). Organic acids play an important role in improving plant growth under stress conditions. In rice exposed to salinity, organic acids enhance growth by increasing intercellular CO_2 concentration (Ci) and net photosynthetic rate (Pn), while reducing excessive H_2O_2

Table 16.2 The role of oxalic acid in plant responses to different abiotic stresses

Plant	Stress	Effects/plant responses	References
Oryza sativa (Rice)	Salinity	OA and other organic acids improved growth by increasing intercellular CO_2 concentration (Ci), photosynthesis rate (Pn), and reducing H_2O_2 accumulation	Das et al. (2019)
Medicago sativa (Alfalfa)	Saline–alkali	OA enhanced growth, photosynthesis, and antioxidant activity; reduced ROS accumulation and upregulated photosynthesis and antioxidant-related genes	Chen et al. (2024)
Fragaria × ananassa (Strawberry)	Salinity	OA (10 mM) reduced soil salinity, enhanced leaf cell expansion, xylem development, and increased ascorbic and malic acids, improving growth and fruit quality	ARAS (2025)
Triticum aestivum (Wheat)	Drought	Foliar application of humic acid + OA improved growth, grain, and protein yield under drought stress	El-Shabrawi et al. (2015)
Cicer arietinum (Chickpea)	Cd toxicity	OA reduced Cd accumulation, improved root and shoot growth, restored redox balance (GSH/GSSG, NAD(P)H/NAD(P)), and enhanced antioxidant enzymes	Sakouhi et al. (2022)
Zea mays (Maize)	Pb stress	OA + salicylic acid enhanced protein, amino acids, antioxidant enzymes (CAT, GPX), and GSH content, mitigating Pb toxicity	Gupta et al. (2024)
Larix olgensis (Changbai larch)	Pb stress	OA mitigated Pb-induced oxidative injury, reduced lipid peroxidation, and upregulated antioxidant defenses	Song et al. (2018)
Various species	Heavy metals (Pb, Cd)	OA formed metal complexes, reducing metal mobility, availability, and toxicity; acted as pH/osmoregulator and antioxidant	Liu et al. (2015), Grąz (2024)
Triticum aestivum (Wheat)	Heat stress	OA (and citric acid) improved growth, yield quality, and antioxidant enzyme activity (CAT, SOD); increased IAA, GA_3, CKs, DNA/RNA; decreased ABA, MDA, H_2O_2; induced heat-responsive proteins	Sadak and Orabi (2015)
Triticum aestivum (Wheat)	Heat stress	Foliar OA + SA improved plant height, dry matter, and antioxidant content under heat stress	Pandey and Sanyal (2021)

accumulation (Das et al. 2019). Oxalic acid is one such organic acid and occurs in plants in both soluble and insoluble oxalate forms (Ghanati et al. 2024). In plants, oxalate production is closely linked to maintaining ionic balance between major cations (K^+, Na^+, NH_4^+, Ca^{2+}, Mg^{2+}) and anions (NO_3^-, Cl^-, $H_2PO_4^-$, SO_4^{2-}). Accumulation of oxalate, particularly in vacuoles, contributes to improved tolerance to abiotic stresses such as salinity by supporting ion homeostasis. Chen et al. (2024) stated that under saline–alkali stress, oxalic acid plays a key protective role in enhancing alfalfa (*Medicago sativa* L.) tolerance. They stated that OA mitigates the negative effects of saline–alkali conditions by promoting plant growth, enhancing photosynthetic efficiency, and reducing oxidative stress through lowering reactive oxygen species accumulation. It also stimulates antioxidant enzyme activities and increases osmoregulatory substances, contributing to cellular protection. Additionally, OA upregulates genes related to photosynthesis and the antioxidant defense system, highlighting its crucial role in root-mediated adaptive mechanisms under saline–alkaline environments. Recently, ARAS (2025) reported that oxalic acid effectively mitigates the adverse effects of salinity on strawberry plants by improving growth and fruit quality. Application of 10 mM OA reduced soil salinity, enhanced leaf cell expansion and xylem development, and increased levels of L-ascorbic and malic acids, indicating improved stress tolerance. Thus, OA serves as a promising treatment to alleviate salt-induced damage and support strawberry production under saline conditions.

16.7.1.2 Oxalic Acid and Drought

Drought is one of the most unpredictable and unavoidable abiotic stresses affecting many regions worldwide, and its severity is expected to increase in the upcoming years (Dawood et al. 2025; Ahmed et al. 2022; Sallam et al. 2019; Fatima et al. 2026; Jameel et al. 2026; Umer et al. 2026). Organic acids released from plant roots play an important role in supporting plant growth by improving photosynthetic activity and limiting oxidative damage (Kleiber et al. 2024). The metabolism of OA is also closely linked to plant tolerance mechanisms under both abiotic and biotic stress conditions. It contributes to the regulation of reactive oxygen species (ROS), detoxification of heavy metals, production of secondary metabolites, interactions with plant hormones, and enhanced defense against pathogens (Li et al. 2022a). Gómez-Espinoza et al. (2020) stated that conditions of CO_2 deficiency due to drought stress reduced calcium oxalate levels due to decomposition, which acts as a plant response to provide carbon for photosynthesis when atmospheric CO_2 in the leaf mesophyll is limited by drought stress. El-Shabrawi et al. (2015) showed that foliar application of humic acid and oxalic acid improved growth and productivity in two wheat cultivars. Plants treated with these acids exhibited better overall performance and significantly higher yields. The greatest increases in grain yield and protein yield per hectare were recorded when wheat plants were sprayed with a combined treatment of 17 mg/L humic acid and 300 mg/L oxalic acid (Table 16.2).

16.7.1.3 Oxalic Acid and Heavy Metals

As previously mentioned, oxalic acid is a common natural metabolite in plants and occurs in several forms, including calcium oxalate crystals, the free acid, and soluble oxalate salts (Kumar et al. 2017). It plays a key role in metal chelation across wide range of organisms, such as bacteria, fungi, and plants, mainly through metabolic routes, namely, such as the tricarboxylic acid (TCA) cycle and the glyoxylate pathway, which help in metal complex formation (Grąz 2024). In fungi and plants, OA can also function as an electron acceptor in the manganese peroxidase catalytic cycle, helping regulate reactive oxygen species through Fenton-related reactions, which is crucial for stress survival. OA additionally serves as a pH and osmotic regulator; its low pKa values (1.23 and 4.19) allow efficient proton donation. Under stress conditions, plants release organic acids, including OA, that improve tolerance by limiting metal mobility and promoting sequestration into vacuoles (Sakouhi et al. 2022). For example, Liu et al. (2015) highlighted OA's strong chelating ability in decreasing soil Pb availability, resulting in lower Pb uptake by plants. Similarly, exogenous OA reduced Cd accumulation in chickpea (*Cicer arietinum* L.) due to its affinity for metal ions, forming complexes that reduce Cd bioavailability. Other studies also report the protective effects of OA against heavy metal stress. In *Larix olgensis* seedlings, OA application mitigated Pb toxicity by lowering oxidative damage, reducing lipid peroxidation, and enhancing antioxidant defenses (Song et al. 2018). Sakouhi et al. (2022) also demonstrated that OA alleviates Cd stress in chickpea by improving root and shoot growth, limiting Cd buildup in tissues, and reducing markers of oxidative injury such as malondialdehyde and carbonyl content, thereby maintaining membrane stability. Furthermore, OA helped restore the glutathione redox balance disrupted by cadmium stress by increasing the glutathione pool and regulating the activities of glutathione peroxidase and glutathione reductase. OA also re-established the balance between oxidized [NAD(P)] and reduced [NAD(P)H] forms. In addition, Gupta et al. (2024) reported that the combined application of oxalic acid and salicylic acid (SA) in maize (*Zea mays* L.) exposed to lead stress enhanced protein and amino acid contents, antioxidant enzyme activities (guaiacol peroxidase and catalase), glutathione levels, lipid peroxidation markers, and Pb accumulation in leaves (Table 16.2).

16.7.1.4 Oxalic Acid and Temperature Stress

Organic acids play an important role in various metabolic pathways of plants (Panchal et al. 2021). Under heat stress conditions, due to molecular auto-oxidation properties of organic acids, they perform as a main cofactor for some specific enzymes, i.e., SOD, CAT and POD, which catalyze the breakdown of the toxic radicals (H_2O_2, OH, O_2^-) (Aono et al. 1993). Citric acid or OA may denote non-enzymatic antioxidants, hence act as a chelator of oxygen radicals and protect plants from injury. In addition, organic acid extends the shelf life of plant cells and enhances growth characteristics (Rao et al. 2000). Sadak and Orabi (2015) found that citric acid or OA treatment with different concentrations at the suitable and late sowing dates, where the last treatment induced heat stress increased growth traits, improved grain-yield properties, stimulated the activities of CAT and SOD enzymes, and increased endogenous IAA, GA_3 and cytokinins, with attenuation of the contents of ABA, malondialdehyde (MDA), and H_2O_2 relative to control plants. Further, the alteration in protein electrophoretic pattern of wheat shoots under various treatments indicated instigation of responsive proteins, causing enhanced thermotolerance of wheat plants. Pandey et al. (2020) displayed that salicylic acid and OA improved the growth of wheat, namely, plant height and dry matter accumulation, as well as stimulation of antioxidants relative to control plants (Table 16.2).

16.7.2 Oxalic Acid and Biotic Stress

Oxalic acid is naturally present in many plants, and some species can accumulate large amounts of this simple dicarboxylic acid. One of its most crucial chemical features is its strong ability to bind multivalent cations, which makes it an effective chelating agent. Recently, OA has gained considerable attention because of its role in inducing systemic resistance to diseases and enhancing antioxidant capacity in plants (Malencic et al. 2004). Over the last two decades, the external application of OA has been widely studied in fruits and vegetables, both before and after harvest, due to its potential to improve quality and extend storage life. Treatments with OA have been shown to increase plant tolerance to various pathogens like fungi, bacteria, and viruses, mainly by boosting antioxidant defenses (Zheng and Brecht 2017). Biological control using endophytic or rhizospheric microorganisms has also proven effective in managing plant diseases (Bagy et al. 2019; Abdelrahiem et al. 2024). Several studies have documented the diversity of oxalate-degrading (oxalotrophic) bacteria in different environments (Sahin 2003). For instance, Jin et al. (2007) demonstrated that a *Pandoraea* strain with oxalate decarboxylase activity could protect *Brassica napus* from infection by *Sclerotinia sclerotiorum*, although the protective effect was not directly linked to the enzyme. Similarly, Schoonbeek et al. (2007) found that soil bacteria capable of degrading OA helped protect *Arabidopsis* from *Botrytis cinerea*, provided that the plant's defense signaling pathways remained functional. Oxalic acid is also known as a pathogenicity factor that can trigger programmed cell death (PCD) during fungal infections. At the same time, it can act as a powerful inducer of local defense responses and systemic acquired resistance (Toal and Jones 1999). OA is frequently produced as a secondary metabolite by many plant-pathogenic fungi (Cessna et al. 2000; Nakajima and Akutsu 2014) and has also been detected in cultures of certain Bacillus species (Tahir et al. 2013). Application of OA has been shown to enhance resistance in postharvest fruits such as pear and muskmelon against Alternaria and pink rot diseases by stimulating the expression of pathogenesis-related genes and increasing the activity of defense-related enzymes (Deng et al. 2015; Tian et al. 2006). Yu et al. (2022) reported that *Bacillus* spp. are well known for their plant disease–suppressive ability, though their mechanisms of action remain incompletely understood. Recent studies identified OA, secreted by *Bacillus cereus* AR156, as a key metabolite involved in disease control. OA was shown to suppress lesions caused by *Botrytis cinerea* in tomato without direct antifungal activity, suggesting a plant-mediated defense mechanism. At low concentrations, OA inhibited cell death induced by higher OA levels and stimulated the accumulation of ROS-scavenging enzymes, thereby enhancing oxidative balance. Pretreatment with low OA concentrations activated the expression of defense-related genes through JA and ET signaling pathways, while the SA pathway was not involved. The disease-suppressive effect of OA was confirmed in wild-type tomato and SA-deficient mutants but was significantly reduced in JA-deficient mutants, confirming the dependence on JA/ET signaling. Furthermore, OA secretion was found to be widespread among Bacillus strains and tightly regulated by a pH-sensitive degradation system, maintaining low environmental concentrations essential for defense induction. Collectively, these findings highlight low-concentration OA as a signaling molecule that triggers plant resistance and suppresses gray mold disease through the JA/ET-dependent defense network. OA plays multiple roles in helping fungal pathogens colonize their host plants (Guimarães and Stotz 2004). Research suggests that OA can weaken plant cell walls by lowering the pH in the apoplast and by binding calcium ions that stabilize pectin, making the cell wall more accessible to fungal enzymes (Franceschi and Loewus 1995). During infection, OA can also suppress plant defense responses by altering the redox balance, as shown in soybean and tobacco (Cessna et al. 2000; Williams et al. 2011). In many cases, OA triggers PCD, such as responses in plant tissues, which support disease incidence and development (Williams et al. 2011; Kabbage et al. 2013). Additionally, OA may help *Sclerotinia sclerotiorum* tolerate the toxic effects of calcium ions released when the plant cell wall is degraded (Heller and Witt-Geiges 2013). Thus, several studies have shown that plants engineered to express oxalate-degrading enzymes exhibit stronger activation of defense-related genes and enhanced resistance to necrotrophic pathogens such as *S. sclerotiorum* and *Botrytis cinerea* (Kumar et al. 2011). In this context, external application of oxalic acid (OA) improved postharvest resistance of pear fruits and muskmelon to Alternaria rot and pink rot, respectively, by inducing PR gene expression and increasing the activity of defense-related enzymes (Deng et al. 2015; Tian et al. 2006). Although a beneficial effect of OA as a biocontrolling agent, it is considered a virulence factor for *Sclerotinia sclerotiorum*. The ascomycete fungus *Sclerotinia sclerotiorum* is among the most destructive necrotrophic pathogens affecting crops. Its successful colonization of host tissues depends, in part, on the secretion of OA, which functions as a non–non-host-specific toxin (Liang et al. 2015a, b). Although the role of OA in the pathogenicity of this fungus is well recognized, its mechanisms are not yet fully understood (McCaghey et al. 2019). Studies have shown that mutants that are deficient in oxalic acid production exhibit significantly reduced virulence (Rana et al. 2022; McCaghey et al. 2021; Liang et al.

2015a). Oxalic acid facilitates infection by acidifying the local environment, chelating calcium ions, and enhancing cell wall-depolymerizing enzyme activity. Moreover, OA interacts with the host ROS system, modulating redox homeostasis and thereby weakening plant defense responses (Hossain et al. 2023; McCaghey et al. 2019). It has also been proposed that OA protects fungal mycelia from the toxic effects of excess calcium in the infection zone (Heller and Witt-Geiges 2013). Plants expressing oxalate-degrading enzymes, such as oxalate oxidase or oxalate decarboxylase, display increased resistance to *S. sclerotiorum* infection (Kumar et al. 2016). These enzymes not only degrade OA but also produce hydrogen peroxide, a ROS that serves as a signaling molecule involved in activating plant defense mechanisms such as programmed cell death during the hypersensitive response (Elena-Real et al. 2021). Thus, targeting oxalic acid metabolism represents a promising strategy for reducing the virulence of necrotrophic fungi and enhancing crop resistance. Kim et al. (2008) denoted that oxalic acid secreted by *Sclerotinia sclerotiorum* acts as a key pathogenicity factor by triggering programmed cell death in host plants, characterized by ROS accumulation and apoptotic-like features such as DNA fragmentation and TUNEL-positive cells. This PCD is essential for fungal infection and tissue colonization, as mutants lacking OA fail to induce these responses and are nonpathogenic. Thus, OA functions not as a direct toxin but as a signaling molecule that manipulates the plant's defense pathways to promote disease development. Also, Marina et al. (2019) reported that oxalotrophic *Stenotrophomonas* strains (OxA and OxB), are isolated from the tomato rhizosphere, protecting plants from oxalate-producing pathogens like *Sclerotinia sclerotiorum* and *Botrytis cinerea* through multiple mechanisms. These include degradation of oxalic acid, induction of plant defense responses such as callose deposition, phenolic compound production, and PR-1 gene expression, as well as strengthening of plant cell walls, making them less susceptible to fungal penetration. Importantly, their protective effect occurs even in plants with impaired salicylic acid or jasmonate signaling, highlighting their potential as broad-spectrum biological control agents against oxalate-dependent fungal pathogens.

16.8 Oxalic Acid and Environment

16.8.1 Biomineralization

Fungi have been studied for their role in bioweathering of minerals, which occurs directly via biomechanistic action of the mycelium and indirectly through the fungal metabolites. The indirect mode of action includes fungal exudates using compounds like organic acids, oxidizing or reducing agents, or ligands such as siderophores (Muksy et al. 2023). OA is widely detected in the fungal secretome, and biomineralization and bioweathering processes are connected to OA (Sindhu et al. 2022). OA plays multiple roles in helping fungal pathogens colonize their host plants

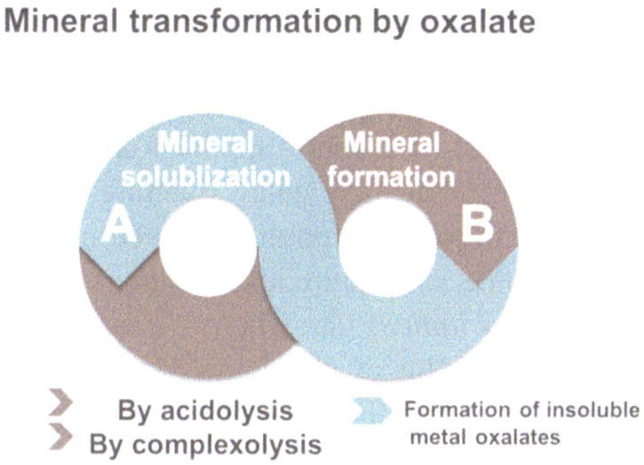

Fig. 16.2 Mineralization by oxalate

(Guimarães and Stotz 2004). Research suggests that OA can weaken plant cell walls by lowering the pH in the apoplast and by binding calcium ions that stabilize pectin, making the cell wall more accessible to fungal enzymes (Franceschi and Loewus 1995). During infection, OA can also suppress plant defense responses by altering the redox balance (Williams et al. 2011). Furthermore, OA triggers PCD-induced responses in plant tissues, which supports disease progression (Williams et al. 2011; Kabbage et al. 2013). Additionally, OA may help *Sclerotinia sclerotiorum* tolerate the toxic effects of calcium ions released when the plant cell wall is degraded (Heller and Witt-Geiges 2013). Similarly, several studies have shown that transgenic plants expressing oxalate-degrading enzymes display stronger activation of defense-related genes and increased resistance to necrotrophic pathogens, including *Sclerotinia sclerotiorum* and *Botrytis cinerea* (Kesarwani et al. 2000; Dias et al. 2006; Kumar et al. 2011). During aerobic growth, fungal mycelia release carbon dioxide, which forms carbonic acid in soils and lowers pH. This acidification promotes mineral dissolution and mobilization of divalent metals previously immobilized as oxalates (Gadd et al. 2014). Overall, oxalate accumulation is a key fungal metabolite driving geomycological processes such as bioweathering and biomineralization, as summarized in Fig. 16.2 (Gadd et al. 2014).

16.9 Role of Oxalic Acid in Fruit Ripening and Postharvest Biotechnology

Oxalic acid plays a crucial role in fruit ripening, senescence, and postharvest physiology. Its endogenous biosynthesis influences fruit internal quality by modulating ascorbate and tartaric acid metabolism, as observed in grapes (*Vitis vinifera* cv. 'Shiraz'), where OA accumulation during early berry development contributes to fruit and wine quality (Melino et al. 2009). OA levels fluctuate throughout fruit development—typically high in immature fruits and declining as ripening progresses, as seen in *Prunus mume*, persimmon, fig, and banana (*Musa sapientum*), where oxalate decarboxylation correlates with increased respiratory activity and CO_2

release during ripening (Walker and Famiani 2018). Similar developmental trends have been reported in stone fruits, litchi, and other climacteric species (Bae et al. 2014; Wang et al. 2006), highlighting the regulatory role of OA in fruit maturation.

Preharvest OA spray applications have emerged as an effective approach to improve fruit quality, delay ripening, and enhance storage performance. Studies report that low concentrations (1–10 mM) of OA enhance yield, firmness, sugar–acid balance, and antioxidant content across several fruit crops, including mango (*Mangifera indica*), strawberry (*Fragaria* × *ananassa*), plum, apricot, artichoke, and lemon (Zhu et al. 2011; Serna-Escolano et al. 2021). OA treatments also suppress ethylene production and respiration rates, delay senescence, and enhance postharvest resistance by stimulating defense-related enzymes such as chitinase, β-1,3-glucanase, and peroxidase, as well as increasing bioactive antioxidant compounds (Zhu et al. 2016; El-Zaeddi et al. 2017; Razavi and Hajilou 2016). Optimal outcomes are typically achieved using 2–5 mM OA, depending on crop species and application frequency, with multiple preharvest sprays (three or more) after fruit set showing the best effects on fruit firmness, antioxidant potential, and delayed ripening (Giménez et al. 2017; Ahmed et al. 2021; García-Pastor et al. 2020, 2021; Retamal-Salgado et al. 2023). Collectively, OA serves as a natural bioregulator, improving fruit development, enhancing postharvest quality, and extending shelf life through its influence on calcium deposition, oxidative metabolism, and defense activation.

16.10 Oxalic Acid and Industry

OA is classified as a generally recognized as safe (GRAS) compound due to its eco-friendly nature and non-toxic profile (Ali et al. 2020). It has gained commercial importance in the apiculture industry, where it is widely used to control infestations of *Varroa destructor* mites in beehives. The compound is also registered with the US Environmental Protection Agency (EPA) as a safe and approved product for this purpose (USDA 2021). Although early reports raised concerns about OA's potential antinutritional effects—mainly through binding essential minerals such as calcium and magnesium and forming insoluble calcium oxalate crystals that could contribute to kidney stone formation (Libert and Franceschi 1987)—recent studies have shown that these risks are minimal when OA occurs naturally in foods. In fact, many oxalate-rich plant foods (e.g., leafy greens, nuts, and fruits) also contain abundant minerals and organic acids that enhance mineral bioavailability, thus reducing oxalate's adverse nutritional effects. Beyond nutritional implications, OA has attracted attention as a natural antioxidant and preservative agent. It can maintain antioxidant potential, slow down metabolic activities, and reduce storage-related decay and physiological disorders in fresh produce. Postharvest application of OA has been reported to delay ripening and senescence, enhance resistance to pathogens, reduce enzymatic browning, and alleviate chilling injury in a wide range of fruits and vegetables (Zheng and Brecht 2017). Furthermore, Li et al. (2022a) have elucidated the biosynthetic pathways of oxalate in plants and highlighted its multifunctional role in plant metabolism, growth, and development, as well as its protective involvement in responses to biotic and abiotic stresses. These insights collectively underscore OA's potential as a safe, sustainable, and multifunctional compound with broad agricultural, nutritional, and industrial applications.

16.11 Conclusion and Future Prospects

Oxalic acid is a key organic acid that significantly influences plant growth, development, and metabolism. In plants, oxalate metabolism involves both biosynthetic and degradative pathways, though their regulatory mechanisms remain only partially understood. Oxalate can be produced through multiple metabolic routes, primarily originating from glycolate and ascorbate catabolism, whereas its degradation is mainly facilitated by oxalate oxidase, oxalate decarboxylase, and oxalyl-CoA synthetase, which collectively maintain oxalate homeostasis within plant tissues. Oxalic acid emerges as a multifunctional and eco-sustainable biomolecule with remarkable significance across agricultural, industrial, and environmental domains. Its GRAS status, natural occurrence, and biocompatible properties make it an exceptional alternative to synthetic chemicals for enhancing crop quality, postharvest longevity, and stress tolerance. In plants, OA plays a dual role—as a metabolic regulator influencing ripening, senescence, and antioxidant defense, and as a signaling molecule that contributes to plant resilience under biotic and abiotic stresses. OA holds tremendous potential for innovation in sustainable agriculture and food preservation. Integrating OA into precision agriculture through optimized foliar sprays, seed priming, or nanocarrier-based delivery systems could revolutionize crop management by enhancing yield, improving fruit quality, and extending storage life without environmental harm. Furthermore, genetic and metabolic engineering approaches could be employed to fine-tune endogenous OA biosynthesis, balancing its protective benefits while minimizing risks of excessive accumulation. In the broader context, OA's antioxidant and antimicrobial properties position it as a promising natural preservative in food processing and storage industries, contributing to the reduction of postharvest losses and promoting green food technologies. Future research should focus on unraveling the molecular mechanisms underlying OA-mediated stress signaling, its interaction with other plant hormones, and its synergistic roles in redox homeostasis. Overall, oxalic acid stands at the frontier of next-generation biostimulants—a sustainable, safe, and versatile compound poised to play a pivotal role in advancing climate-smart agriculture, postharvest technology, and food security in a changing world.

References

Abbey L, Bonsu KO, Glover AM, Ahenkora K (2006) Evaluation of some common leafy vegetables used in Ghana Crop Research Institute, Kumasi. Ghana J Hortic 5:23

Abdelmoneim MS, Hafez EE, Dawood MF, Hammad SF, Ghazy MA (2024) Toxicity of bisphenol A and p-nitrophenol on tomato plants: Morphophysiological, ionomic profile, and antioxidants/defense-related gene expression studies. Biomol Concepts 15(1):20220049

Abdelmoneim MS, Dawood MF, Hafez EE, Hammad SF, Ghazy MA (2025) Unveiling the role of cadaverine in mitigating salinity and/or Bisphenol A toxicity in tomato plants and reduced Bisphenol A accumulation in tomato roots. Plant Physiol Biochem 224:109799

Abdelrhim AS, Hemeda NF, Mwaheb MA, Omar MO, Dawood MF (2024) The role of Trichoderma koningii and Trichoderma harzianum in mitigating the combined stresses motivated by Sclerotiniasclerotiorum and salinity in common bean (Phaseolusvulgaris). Plant Stress 11:100370

Abelti AL, Teka TA, Bultosa G, Vermeir P (2024) Characterization of water lily (Nymphaea lotus) for nutrients, anti-nutrients, phytochemicals, and antioxidant capacities. Food Biosci 61: 104588. https://doi.org/10.1016/j.afres.2023.100264

Aguilar C, Urzúa U, Koenig C, Vicňa R (1999) Oxalate oxidase from Ceriporiopsis subvermispora: biochemical and cytochemical study. Arch Biochem Biophys 366:275–282

Ahmed M, Ullah S, Razzaq K, Rajwana IA, Akhtar G, Nazb A, Aminc M, Khalidd MS, Khalidd S (2021) Pre-harvest oxalic acid application improves fruit size at harvest, physico-chemical and sensory attributes of 'Red Flesh' apricot during fruit ripening. J Hortic Sci Technol. https://doi.org/10.46653/jhst2142048

Ahmed AA, Dawood MF, Elfarash A, Mohamed EA, Hussein MY, Börner A, Sallam A (2022) Genetic and morpho-physiological analyses of the tolerance and recovery mechanisms in seedling stage spring wheat under drought stress. Front Genet 13:1010272

Ali S, Khan AS, Anjum MA, Nawaz A, Naz S, Ejaz S, Hussain S (2020) Effect of postharvest oxalic acid application on enzymatic browning and quality of lotus (Nelumbo nucifera Gaertn.) root slices. Food Chem 312:126051. https://doi.org/10.1016/j.foodchem.2019.126051

Allison MJ, Daniel SL, Cornick NA (2020) Oxalate-degrading bacteria. In: Calcium oxalate in biological systems. CRC Press, Boca Raton, pp 131–168

Amoah J, Otchere JK, Quaye T, Ntow J (2024) A study to determine the total iron and extractable iron content available in ten indigenous vegetables of Ghana. Asian J Food Res Nutri 3(2): 211–218

Aono M, Kubo A, Saji H, Tanaka K, Kondo N (1993) Enhanced tolerance to photooxidative stress of transgenic Nicotiana tabacum with high chloroplastic glutathione reductase activity. Plant Cell Physiol 34:129–136

ARAS S (2025) Effects of Oxalic Acid on Strawberry Plant Growth and Fruit Quality Under Salinity Stress. Applied Fruit Science vol. 67, no.3

Azooz MM, Abdel Latef AAH, Abdelmajeed WG (2026) Biochar ameliorates salt stress in maize by modulating physiological responses and expression of SOS1 and NHX2 genes. Journal of Soil Science and Plant Nutrition. https://doi.org/10.1007/s42729-026-03123-w

Bae H, Yun SK, Yoon IK, Nam EY, Kwon JH, Jun JH (2014) Assessment of organic acid and sugar composition in apricot, plumcot, plum, and peach during fruit development. J Appl Bot Food Qual 87. https://doi.org/10.5073/JABFQ.2014.087.004

Bagy HMK, Hassan EA, Nafady NA, Dawood MF (2019) Efficacy of arbuscular mycorrhizal fungi and endophytic strain Epicoccum nigrum ASU11 as biocontrol agents against blackleg disease of potato caused by bacterial strain Pectobacterium carotovora subsp. atrosepticum PHY7. Biol Control 134:103–113

Baldrian P, Valaskova V (2008) Degradation of cellulose by basidiomycetous fungi. FEMS Microbiol Lett 32:501–521

Baran EJ (2020) Oxalate biosynthesis and degradation in plants and fungi. Elsa K:95–131

Bargagli M, Tio MC, Waikar SS, Ferraro PM (2020) Dietary oxalate intake and kidney outcomes. Nutrients 12(9):2673. https://doi.org/10.3390/nu12092673

Barkla BJ, Farzana T, Rose TJ (2024) Commercial cultivation of edible halophytes: The issue of oxalates and potential mitigation options. Agronomy 14(2):242

Bhandari MR, Kawabata J (2004) Organic acid, phenolic content and antioxidant activity of wild yam (Dioscorea spp.) tubers of Nepal. Food Chem 88:163–168. https://doi.org/10.1016/j.foodchem.2003.12.027

Bong WC, Vanhanen LP, Savage GP (2017) Addition of calcium compounds to reduce soluble oxalate in a high oxalate food system. Food Chem 221:54–57

Borquaye LS, Darko G, Laryea MK, Gasu EN, Amponsah NAA, Appiah EN (2017) Nutritional and anti-nutrient profiles of some Ghanaian spices. Cogent Food Agric 3:1348185

Cai XF, Ge CH, Xu CX, Wang XL, Wang S, Wang QH (2018) Expression analysis of oxalate metabolic pathway genes reveals oxalate regulation patterns in spinach. Molecules 23:1286

Cessna SG, Sears VE, Dickman MB, Low PS (2000) Oxalic acid, a pathogenicity factor for *Sclerotinia sclerotiorum*, suppresses the oxidative burst of the host plant. Plant Cell 12:2191–2199

Chakraborty N, Ghosh R, Ghosh S, Narula K, Tayal R, Datta A, Chakraborty S (2013) Reduction of oxalate levels in tomato fruit and consequent metabolic remodeling following overexpression of a fungal oxalate decarboxylase. Plant Physiol 162:364–378

Chang CC, Beevers H (1968) Biogenesis of oxalate in plant tissues. Plant Physiol 1968(43):1821–1828

Chen J, Tang L, Guo W, Wang D, Sun Y, Guo C (2024) Oxalic acid secretion alleviates saline-alkali stress in alfalfa by improving photosynthetic characteristics and antioxidant activity. Plant Physiol Biochem 208:108475

Cowan DA, Babenko D, Bird R, Botha A, Breecker DO, Clarke CE, Van Zyl L (2024) Oxalate and oxalotrophy: An environmental perspective. Sustain Microbiol 1(1):qvad004. https://doi.org/10.1093/sumbio/qvad004

Cunha WG, Tinoco MLP, Pancoti HL, Ribeiro RE, Aragão FJL (2010) High resistance to *Sclerotinia sclerotiorum* in transgenic soybean plants transformed to express an oxalate decarboxylase gene. Plant Pathol 59:654–660

Curhan GC (1997) Dietary calcium, dietary protein, and kidney stone formation. Miner Electrolyte Metab 23:261–264

Dappah DK, N'Dri DY, Kouassi NK, Amani GNG (2018) Evaluation of nutritive value of two traditional African vegetables (Corchorus olitorus and Hibiscus sabdariffa) commonly consumed in Côte d'Ivoire. Int J Biochem Res Rev 23(4):1–18

Das PR, Kim Y, Hong SJ, Eun JB (2019) Profiling of volatile and non-phenolic metabolites—Amino acids, organic acids, and sugars of green tea extracts obtained by different extraction techniques. Food Chem 296:69–77

Dauer JM, Perakis SS (2014) Calcium oxalate contribution to calcium cycling in forests of contrasting nutrient status. For Ecol Manag 334:64–73

Davies DD, Asker H (1983) Synthesis of oxalic acid by enzymes from lettuce leaves. Plant Physiol 72:134–138. https://doi.org/10.1104/pp.72.1.134

Dawood MF, Zaid A, Latef AAHA (2022) Salicylic acid spraying-induced resilience strategies against the damaging impacts of drought and/or salinity stress in two varieties of Vicia faba L. seedlings. J Plant Growth Regul 41(5):1919–1942

Dawood MF, Tahjib-Ul-Arif M, Shirazy BJ, Latef AAHA (2025) Unraveling the role of κ-carrageenan on the combined effect of drought and chromium stress in wheat (Triticum aestivium L.). Plant Physiol Biochem 223:109781

Demoulin N, Aydin S, Gillion V, Morelle J, Jadoul M (2022) Pathophysiology and management of hyperoxaluria and oxalate nephropathy: a review. Am J Kidney Dis 79(5):717–727

Deng J, Bi Y, Zhang Z, Xie D, Ge Y, Li W, Wang Y (2015) Postharvest oxalic acid treatment induces resistance against pink rot by priming in muskmelon (*Cucumis melo* L.) fruit. Postharvest Biol Technol 106:53–61

Dias BBA, Cunha WG, Morais LS, Vianna GR, Rech EL, de Capdeville G, Aragão FJL (2006) Expression of an oxalate decarboxylase gene from Flammulina sp. in transgenic lettuce (*Lactuca sativa*) plants and resistance to *Sclerotinia sclerotiorum*. Plant Pathol 55:187–193

Ding ZS, Tian SP, Zheng XL, Zhou ZW, Xu Y (2007) Responses of reactive oxygen metabolism and quality in mango fruit to exogenous oxalic acid or salicylic acid under chilling temperature stress. Physiol Plant 130:112–121

Egbuna C, Ifemeje JC (2015) Biological Function and Anti-Nutritional Effects of Phytochemicals in Living System. Journal of Pharmacy and Biological Science 10,10-19

Elena-Real CA, González-Arzola K, Pérez-Mejías G, Díaz-Quintana A, Velázquez-Campoy A, Desvoyes B, Gutiérrez C, De la Rosa MA, Díaz-Moreno I (2021) Proposed mechanism for regulation of H 2O2-induced programmed cell death in plants by binding of cytochrome c to 14-3-3 proteins. Plant J 106:74–85

El-Shabrawi HM, Bakry BA, Ahmed MA, Abou-El-Lail M (2015) Humic and oxalic acid stimulates grain yield and induces accumulation of plastidial carbohydrate metabolism enzymes in wheat grown under sandy soil conditions. Agricultural Sci 6(1):175–185. https://doi.org/10.4236/as.2015.61016

El-Zaeddi H, Calín-Sanchez A, Nowicka P, Martínez-Tome J, Noguera-Artiaga L, Burlo F, Wojdyło A, Carbonell-Barrachina AA (2017) Preharvest treatments with malic, oxalic, and acetylsalicylic acids affect the phenolic composition and antioxidant capacity of coriander, dill and parsley. Food Chem 226:179–186. https://doi.org/10.1016/j.foodchem.2017.01.067

Ergonül PG, Nergiz C (2010) Determination of organic acids in olive fruit by HPLC. Czech J Food Sci 28:202–205. https://doi.org/10.17221/1379-CJFS

Fatima A, Akram NA, Talha M, Younis M, Ashraf M, Abdel Latef AAH (2026) Comparative Effects of Externally Applied Glycinebetaine and Neem (Azadirachta indica) Leaf Extract on Plant Growth, Physio-Biochemical Characteristics and Yield Production of Canola (Brassica napus L.) Subjected to Drought Stress. Russian Journal of Plant Physiology 73:18

Foster J, Kim HU, Nakata PA, Browse J (2012) A previously unknown oxalyl-CoA synthetase is important for oxalate catabolism in Arabidopsis. Plant Cell 24:1217–1229

Franceschi VR, Horner HT (1980) Calcium oxalate crystals in plants. Bot Rev 46:361–427

Franceschi VR, Loewus FA (1995) Oxalate biosynthesis and function in plants and fungi. In: Khan S (ed) Calcium oxalate in biological systems. CRC Press, Boca Raton, pp 113–130

Franceschi VR, Loewus FA (2020) Oxalate biosynthesis and function in plants and fungi. In: Calcium oxalate in biological systems. CRC Press, Boca Raton, pp 113–130

Franceschi VR, Nakata PA (2005) Calcium oxalate in plants: formation and function. Annu Rev Plant Biol 56:4171. https://doi.org/10.1146/annurev.arplant.56.032604.144106

Gadd GM (2007) Geomycology: Biogeochemical transformations of rocks, minerals, metals and radionuclides by fungi, bioweathering and bioremediation. Mycol Res 111:3–49

Gadd GM (2021) Fungal biomineralization. Curr Biol 31:R1549–R1567

Gadd GM, Bahri-Esfahani J, Li Q, Rhee YJ, Wei Z, Fomina M, Liang X (2014) Oxalate production by fungi: significance in geomycology, biodeterioration and bioremediation. Fungal Biol Rev 28:36–55

Gallaher RN (1975) The occurrence of calcium in plant tissue as crystals of calcium oxalate. Commun Soil Sci Plant Anal 6:315

García-Pastor ME, Gimenez MJ, Valverde JM, Guillen F, Castillo S, MartínezRomero D, Serrano M, Valero D, Zapata PJ (2020) Preharvest application of oxalic acid improved pomegranate fruit yield, quality, and bioactive compounds at harvest in a concentration-dependent manner. Agronomy 10:1522. https://doi.org/10.3390/agronomy10101522

García-Pastor ME, Gimenez MJ, Serna-Escolano V, Guillen F, Valero D, Serrano M, García-Martínez S, Terry LA, Alamar MC, Zapata PJ (2021) Oxalic acid preharvest treatment improves colour and quality of seedless table grape 'Magenta' upregulating onvine abscisic acid metabolism, relative VVNCED1 gene expression, and the antioxidant system in berries. Front Plant Sci 12. https://doi.org/10.3389/fpls.2021.740240

Ghanati K, Oskoei V, Rezvani-Ghalhari M, Shavali-Gilani P, Mirzaei G, Sadighara P (2024) Oxalate in plants, amount and methods to reduce exposure; a systematic review. Toxin Rev 43(3):411–422

Ghosh S, Narula K, Sinha A, Ghosh R, Jawa P, Chakraborty N, Chakraborty S (2016) Proteometabolomic analysis of transgenic tomato overexpressing oxalate decarboxylase uncovers novel proteins potentially involved in defense mechanism against sclerotinia. J Proteome 143:242–253

Giménez M, Martínez-Espla A, Castillo-Girones S, Valverde J, Martínez-Romero D, Castillo S (2017) Effect of preharvest application of methyl salicylate, oxalic acid and methyl jasmonate on yield and quality of ''Sweet Heart' cherries. Acta Hortic 1256:501–504. https://doi.org/10.17660/ActaHortic.2019.1256.71

Goldsmith M, Barad S, Peleg Y, Albeck S, Dym O, Brandis A, Mehlman T, Reich Z (2022) The identification and characterization of an oxalyl-CoA synthetase from grass pea (*Lathyrus sativus* L.). RSC Chem Biol 3(3):320–333

Gómez-Espinoza O, González-Ramírez D, Bresta P, Karabourniotis G, Bravo LA (2020) Decomposition of calcium oxalate crystals in Colobanthus quitensis under CO2 limiting conditions. Plants 9(10):1307

Grąz M (2024) Role of oxalic acid in fungal and bacterial metabolism and its biotechnological potential. World J Microbiol Biotechnol 40(6):178. https://doi.org/10.1007/s11274-024-03973-5

Grąz M, Jarosz-Wilkołazka A, Pawlikowska-Pawlęga B (2009) Abortiporus biennis tolerance to insoluble metal oxides: Oxalate secretion, oxalate oxidase activity, and mycelial morphology. Biometals 22:401–410

Grąz M, Pawlikowska-Pawlęga B, Jarosz-Wilkołazka A (2011) Growth inhibition and intracellular distribution of Pb ions by the white-rot fungus Abortiporus biennis. Int Biodeterior Biodegrad 65:124–129

Grąz M, Rachwał K, Zan R, Jarosz-Wilkołazka A (2016) Oxalic acid degradation by a novel fungal oxalate oxidase from Abortiporus biennis. Acta Biochim Pol 63:595–600

Grąz M, Jarosz-Wilkołazka A, Janusz G, Mazur A, Wielbo J, Koper P, Żebracki K, Kubik-Komar A (2017) Transcriptome-based analysis of the saprophytic fungus Abortiporus biennis—response to oxalic acid. Microbiol Res 199:79–88

Guimarães RL, Stotz HU (2004) Oxalate production by *Sclerotinia sclerotiorum* deregulates guard cells during infection. Plant Physiol 136(3):3703–3711. https://doi.org/10.1104/pp.104.0496507

Gupta M, Kumar S, Dwivedi V, Gupta DG, Ali D, Alarifi S et al (2024) Selective synergistic effects of oxalic acid and salicylic acid in enhancing amino acid levels and alleviating lead stress in Zea mays L. Plant Signal Behav 19(1):2400451

He H, Li D, Li X, Fu L (2024) Research progress on the formation, function, and impact of calcium oxalate crystals in plants. Crystallogr Rev 30:31–60. https://doi.org/10.1080/0889311X.2024.2309486

Heller A, Witt-Geiges T (2013) Oxalic acid has an additional, detoxifying function in Sclerotinia sclerotiorum pathogenesis. PLoS One 8(8):e72292

Hester B, Hiram C-M, Giulia V, Rémi T, Jacques B, Pathmanathan U, Erik S, Géraldine S (2024) Ca-oxalate crystals are involved in cadmium storage in a high Cd accumulating cultivar of cacao. Environ Exp Bot 221:105713. https://doi.org/10.1016/j.envexpbot.2024.105713

Hodgkinson A (1977) Oxalic acid metabolism in higher plants. In: Hodgkinson A (ed) Oxalic acid biology and medicine. Academic Press, New York, pp 131–158

Horner HT, Wagner BL (1995) Calcium oxalate formation in higher plants. In: Khan SR (ed) Calcium oxalate in biological systems. CRC Press, Boca Raton, pp 53–72

Hossain MM, Sultana F, Li W, Tran L-SP, Mostofa MG (2023) Sclerotinia sclerotiorum (Lib.) de Bary: insights into the pathogenomic features of a global pathogen. Cells 12(7):1063

Hussain Z, Mir R, Dar M, Somasundaram R (2024) Crystalline chronicles: navigating the wonders and worries of plant calcium oxalate—a review. Indian J Nat Sci 15:1–12

Igamberdiev AU, Eprintsev AT (2016) Organic acids: the pools of fixed carbon involved in redox regulation and energy balance in higher plants. Front Plant Sci 7:1042
Jameel A, Akram NA, Bashir R, Ashraf M, Abdel Latef AA (2026) Salicylic Acid Enhances Yield and Alleviates Drought-Induced Oxidative Damage in Two Carrot Cultivars with Contrasting Tolerance. Russian Journal of Plant Physiology 73:23
Janusz G, Pawlik A, Sulej J, Świderska-Burek U, Jarosz-Wilkołazka A, Paszczyński A (2017) Lignin degradation: microorganisms, enzymes involved, genomes analysis and evolution. FEMS Microbiol Rev 41:941–962
Ji XM, Peng XX (2005) Oxalate accumulation as regulated by nitrogen forms and its relationship to photosynthesis in rice (*Oryza sativa* L.). J Integr Plant Biol 47:831–838
Jin ZX, Wang C, Chen W, Chen X, Li X (2007) Induction of oxalate decarboxylase by oxalate in a newly isolated Pandoraea sp. OXJ-11 and its ability to protect against Sclerotinia sclerotiorum infection. Can J Microbiol 53(12):1316–1322
Kabbage M, Williams B, Dickman MB (2013) Cell death control: the interplay of apoptosis and autophagy in the pathogenicity of *Sclerotinia sclerotiorum*. PLoS Pathog 9:e1003287. https://doi.org/10.1371/journal.ppat.1003287
Kaushal N, Sood N, Singh T, Singh A, Kaur M, Sains M (2022) Oxalates. In: Handbook of plant and animal toxins in food. CRC Press, Boca Raton, pp 97–124
Kesarwani M, Azam M, Natarajan K, Mehta A, Datta A (2000) Oxalate decarboxylase from Collybia velutipes. Molecular cloning and its overexpression to confer resistance to fungal infection in transgenic tobacco and tomato. J Biol Chem 275:7230–7238
Kim KS, Min JY, Dickman MB (2008) Oxalic acid is an elicitor of plant programmed cell death during *Sclerotinia sclerotiorum* disease development. Mol Plant-Microbe Interact 21(5): 605–612
Kleiber T, Chadzinikolau T, Formela-Luboińska M, Lartey JL, Kosiada T (2024) Enhancing Lettuce drought tolerance: the role of organic acids in photosynthesis and oxidative defense. Appl Sci 14(12):5119. https://doi.org/10.3390/app14125119
Korth KL, Doege SJ, Park SH, Goggin FL, Wang Q, Gomez SK, Liu GJ, Jia LL, Nakata PA (2006) *Medicago truncatula* mutants demonstrate the role of plant calcium oxalate crystals as an effective defense against chewing insects. Plant Physiol 141:188–195
Koua GA, Zoue LT, Ouattara S, Niamke SL (2015) Studies on nutritive value and antioxidant properties of three neglected leafy vegetables of Cote d'Ivoire. Pak J Nutr 14(12):877. https://doi.org/10.9734/ARRB/2024/v39i52080
Kumar A, Prakash A, Johri BN (2011) Bacillus as PGPR in crop ecosystem. In: Bacteria in agrobiology: crop ecosystems. Springer, Berlin, pp 37–59
Kumar V, Chattopadhyay A, Ghosh S, Irfan M, Chakraborty N, Chakraborty S, Datta A (2016) Improving nutritional quality and fungal tolerance in soya bean and grass pea by expressing an oxalate decarboxylase. Plant Biotechnol J 14(6):1394–1405
Kumar V, Khare T, Sharma M, Wani SH (2017) ROS-induced signaling and gene expression in crops under salinity stress. In: Khan MIR, Khan NA (eds) Reactive oxygen species and antioxidant systems in plants: role and regulation under abiotic stress. Springer, Singapore, pp 159–184
Kumar V, Irfan M, Datta A (2019) Manipulation of oxalate metabolism in plants for improving food quality and productivity. Phytochemistry 158:103–109
Kumar S, Natalio F, Elbaum R (2021) Protein-driven biomineralization: comparing silica formation in grass silica cells to other biomineralization processes. J Struct Biol 213(1):107665. https://doi.org/10.1016/j.jsb.2020.107665
Leeuwenhoek A (1975) Microscopical observations. Philos Trans R Soc Lond 10:380–385
Lane BG, Dunwell JM, Ray JA, Schmitt MR, Cuming AC (1993) Germin, a marker of early plant growth, is an oxalate oxidase. J Biol Chem 268:12239–12242
Li P, He Q, Jin J, Liu Y, Wen Y, Zhao K, Mao G, Fan W, Yang J (2022a) Tomato oxalyl-CoA synthetase degrades oxalate and affects fruit quality. Front Plant Sci 13:951386. https://doi.org/10.3389/fpls.2022.951386

Li P, Liu C, Luo Y, Shi H, Li Q, PinChu C, Li X, Yang JL, Fan W (2022b) Oxalate in plants: metabolism, function, regulation, and application. J Agric Food Chem 70:16037–16049. https://doi.org/10.1021/acs.jafc.2c04787

Liang X, Moomaw EW, Rollins JA (2015a) Fungal oxalate decarboxylase activity contributes to Sclerotinia sclerotiorum early infection by affecting both compound appressoria development and function. Mol Plant Pathol 16(8):825–836

Liang X, Liberti D, Li M, Kim Y-T, Hutchens A, Wilson R, Rollins JA (2015b) Oxaloacetate acetylhydrolase gene mutants of Sclerotinia sclerotiorum do not accumulate oxalic acid, but do produce limited lesions on host plants. Mol Plant Pathol 16(6):559–571

Libert B, Franceschi VR (1987) Oxalate in crop plants. J Agric Food Chem 35:926–938

Liu K, Lv J, He W et al (2015) Major factors influencing cadmium uptake from the soil into wheat plants. Ecotoxicol Environ Saf 113:207–213. https://doi.org/10.1016/j.ecoenv.2014.12.005

Lou HQ, Fan W, Xu JM, Gong YL, Jin JF, Chen WW, Liu LY, Hai MR, Yang JL, Zheng SJ (2016a) An oxalyl-CoA synthetase is involved in oxalate degradation and aluminum tolerance. Plant Physiol 172(3):1679–1690

Lou HQ, Gong YL, Fan W, Xu JM, Liu Y, Cao MJ, Wang MH, Yang JL, Zheng SJ (2016b) A formate dehydrogenase confers tolerance to aluminum and low pH. Plant Physiol 171:294–305

Lu YS, Li Y, Yang QS, Zhang ZS, Chen Y, Zhang S, Peng XX (2014) Suppression of glycolate oxidase causes glyoxylate accumulation that inhibits photosynthesis through deactivating rubisco in rice. Physiol Plant 150:463–476

Luo X, Du H, Du j, Zhang X, Xiao W, Qin L (2024) The influence of biomass type on hydrothermal carbonization: Role of calcium oxalate in enhancing carbon sequestration of hydrochar. J Environ. https://doi.org/10.1016/j.jenvman.2023.119586

Malencic DJ, Vasic D, Popovic M, Devic D (2004) Antioxidant systems in sunflower as affected by oxalic acid. Biol Plant 48:243–247

Malik Z, Mir RA, Dar MS, Somasundaram R (2024) Crystalline Chronicles: Navigating the wonders and worries of plant calcium oxalate - A review. Indian J Nat Sci 15(85). http://www.tnsroindia.org.in/

Marciano P, Di Lenna P, Magro P (1983) Oxalic acid, cell wall-degrading enzymes and pH in pathogenesis and their significance in the virulence of two *Sclerotinia sclerotiorum* isolates on sunflower. Physiol Plant Pathol 22(3):339–345. https://doi.org/10.1016/S0048-4059(83)81021-2

Marina M, Romero FM, Villarreal NM, Medina AJ, Gárriz A, Rossi FR et al (2019) Mechanisms of plant protection against two oxalate-producing fungal pathogens by oxalotrophic strains of *Stenotrophomonas* spp. Plant Mol Biol 100(6):659–674

McCaghey M, Willbur J, Smith DL, Kabbage M (2019) The complexity of the sclerotinia sclerotiorum pathosystem in soybean: Virulence factors, resistance mechanisms, and their exploitation to control Sclerotinia stem rot. Trop Plant Pathol 44:12–22

McCaghey M, Shao D, Kurcezewski J, Lindstrom A, Ranjan A, Whitham SA, Conley SP, Williams B, Smith DL, Kabbage M (2021) Host-induced gene silencing of a Sclerotinia sclerotiorum oxaloacetate acetylhydrolase using bean pod mottle virus as a vehicle reduces disease on soybean. Front Plant Sci 12:677631

McNair JB (1932) The intersection between substances in plants: essential oils and resins, cyanogen and oxalate. Am J Bot 19:255–271

Melino VJ, Soole KL, Ford CM (2009) Ascorbate metabolism and the developmental demand for tartaric and oxalic acids in ripening grape berries. BMC Plant Biol 9(1):14. https://doi.org/10.1186/1471-2229-9-145

Minocha R, Long S (2023) Seasonal changes in foliar calcium oxalate concentrations in conifer and hardwood trees: A potentially bioavailable source of cellular calcium and/or oxalate under stress. Front Forests Glob Change 6:1161088

Mirahmadi SF, Hassandokht M, Fatahi R, Naghavi MR, Rezaei K (2022) High and low oxalate content in spinach: an investigation of accumulation patterns. J Sci Food Agric 102:836–843

Mohd-Ishfaq K, Nargis B, Shahzad P, Manzoor S, Zafar R, Anwar S (2024) Rhubarb: A novel model plant to study the conundrum of calcium oxalate synthesis. Food Chem 434:137458. https://doi.org/10.1016/j.foodchem.2023.137458

Muksy R, Kolo K, Abdullah SM (2023) Bacterial and fungal-mineral interactions and their application in bioremediation—a review. J Ecol Eng 24(5):1–13

Munir E, Yoon JJ, Tokimatsu T, Hattori T, Shimada M (2001) A physiological role for oxalic acid biosynthesis in the wood-rotting basidiomycete Fomitopsis palustris. Proc Natl Acad Sci 98(20):11126–11130

Nakajima M, Akutsu K (2014) Virulence factors of *Botrytis cinerea*. J Gen Plant Pathol 80:15–23. https://doi.org/10.1007/s10327-013-0492-0

Nakata PA (2003) Advances in our understanding of calcium oxalate crystal formation and function in plants. Plant Sci 164:901–909

Nakata PA (2012) Engineering calcium oxalate crystal formation in Arabidopsis. Plant Cell Physiol 53:1275–1282

Nakata PA, McConn M (2007) Isolated *Medicago truncatula* mutants with increased calcium oxalate crystal accumulation have decreased ascorbic acid levels. Plant Physiol Biochem 45: 216–220

Narula K, Ghosh S, Aggarwal P, Sinha A, Chakraborty N, Chakraborty S (2016) Comparative proteomics of oxalate downregulated tomatoes points toward cross talk of signal components and metabolic consequences during post-harvest storage. Front Plant Sci 7:1147. https://doi.org/10.3389/fpls.2016.01147

Nguyễn HVH, Savage G (2013a) Oxalate content of New Zealand grown and imported fruits. J Food Compos Anal 31:180–184. https://doi.org/10.1016/j.jfca.2013.06.001

Nguyễn HVH, Savage GP (2013b) Total, soluble and insoluble oxalate contents of ripe green and golden kiwifruit. Foods 2:76–82. https://doi.org/10.3390/foods2010076

Paiva ÉAS (2021) Do calcium oxalate crystals protect against herbivory? Naturwissenschaften 108(3):24. https://doi.org/10.1007/s00114-021-01735-z

Panchal P, Miller AJ, Giri J (2021) Organic acids: versatile stress-response roles in plants. J Exp Bot 72(11):4038–4052

Pandey GK, Sanyal SK (2020) In Functional dissection of calcium homeostasis and transport machinery in plants

Pandey G, Sanyal S (2021) Functional dissection of calcium homeostasis and transport machinery in plants. Springer. https://doi.org/10.1007/978-3-030-58502-0

Parsons HT, Fry SC (2012) Oxidation of dehydroascorbic acid and 2, 3-diketogulonate under plant apoplastic conditions. Phytochemistry 75:41–49

Pereira C, Barros L, Carvalho AM, Ferreira IC (2013) Use of UFLC-PDA for the analysis of organic acids in thirty-five species of food and medicinal plants. Food Anal Methods 6:1337–1344. https://doi.org/10.1007/s12161-012-9548-6

Prasad R, Shivay Y (2017) Oxalic acid/oxalates in plants: from self-defence to phytoremediation. Curr Sci 112:1665–1667. https://doi.org/10.18520/cs/v112/i08/1665-1667

Radhika B, Samreen N, Ramya N (2019) Pharmacognostic and phytochemical evaluation of Hibiscus hirtus Linn. Int J Biomed Investig 2(1):117. https://doi.org/10.31531/2581

Rahman M, Kawamura O (2011) Oxalate accumulation in forage plants: some agronomic, climatic and genetic aspects. Asian Australas J Anim Sci 24:439–448. https://doi.org/10.5713/ajas.2011.10208

Rahman MM, Abdullah RB, Wan Khadijah WE (2013) A review of oxalate poisoning in domestic animals: tolerance and performance aspects. J Anim Physiol Anim Nutr 97:605–614. https://doi.org/10.1111/j.1439-0396.2012.01309.x

Rana K, Yuan J, Liao H, Banga SS, Kumar R, Qian W, Ding Y (2022) Host-induced gene silencing reveals the role of Sclerotinia sclerotiorum oxaloacetate acetylhydrolase gene in fungal oxalic acid accumulation and virulence. Microbiol Res 258:126981

Rao MV, Koch JR, Davis KR (2000) Ozone: Atool for probing programmed cell death in plants. Plant Mol Biol 44:345–358

Razavi F, Hajilou J (2016) Enhancement of postharvest nutritional quality and antioxidant capacity of peach fruits by preharvest oxalic acid treatment. Sci Hortic 200:95–101. https://doi.org/10.1016/j.scienta.2016.01.011

Reinhardt LA, Svedruzic D, Chang CH, Cleland WW, Richards NG (2003) Heavy atom isotope effects on the reaction catalyzed by the oxalate decarboxylase from *Bacillus subtilis*. J Am Chem Soc 125:1244–1252

Retamal-Salgado J, Adaos G, Cedeno-García G, Ospino-Olivella SC, VergaraRetamales R, Lopez MD, Olivares R, Hirzel J, Olivares-Soto H, Betancur M (2023) Preharvest applications of oxalic acid and salicylic acid increase fruit firmness and polyphenolic content in blueberry (Vaccinium corymbosum L.). Horticulturae 9:639. https://doi.org/10.3390/horticulturae9060639

Richardson KE, Tolbert NE (1961) Oxidation of glyoxylic acid to oxalic acid by glycolic acid oxidase. J Biol Chem 236:1280–1284

Ruiz N, Ward D, Saltz S (2002) Calcium oxalate crystals in leaves of *Pancratium sickenbergeri*: constitutive or induced defense? Funct Ecol 16:99–105

Sadak MS, Orabi SA (2015) Improving thermo tolerance of wheat plant by foliar application of citric acid or oxalic acid. Int J Chem Tech Res 8(1):111–123

Sahin N (2003) Oxalotrophic bacteria. Res Microbiol 154:399–407

Sakouhi L, Kharbech O, Massoud M, Munemasa S, Murata Y, Chaoui A (2022) Oxalic acid mitigates cadmium toxicity in *Cicer arietinum* L. germinating seeds by maintaining the cellular redox homeostasis. J Plant Growth Regul 41:697–709. https://doi.org/10.1007/s00344-021-10334-1

Salgado N, Silva MA, Figueira ME, Costa HS, Albuquerque TG (2023) Oxalate in foods: extraction conditions, analytical methods, occurrence, and health implications. Foods 12(17):3201. https://doi.org/10.3390/foods12173201

Sallam A, Alqudah AM, Dawood MF, Baenziger PS, Börner A (2019) Drought stress tolerance in wheat and barley: Advances in physiology, breeding and genetics research. Int J Mol Sci 20(13):3137

Santoso AM (2013) Distribution of calcium oxalate cristal, reduction of oxalates, and the effect of cultivation method on its formation in some vegetables. In: Proceeding biology education conference: Biology, science, environmental, and learning, pp 329–334

Savage G, Klunklin W (2018) Oxalates are found in many different European and Asian foods effects of cooking and processing. J Food Res (Canadian Center of Science and Education) 7(3):76

Schoonbeek HJ, Jacquat-Bovet AC, Mascher F, Métraux JP (2007) Oxalate-degrading bacteria can protect Arabidopsis thaliana and crop plants against botrytis cinerea. Mol Plant-Microbe Interact 20(12):1535–1544. https://doi.org/10.1094/MPMI-20-12-1535. PMID: 17990961

Serna-Escolano V, Giménez MJ, Castillo S, Valverde JM, Mar-tínez-Romero D, Guillén F, Serrano M, Valero D, Zapata PJ (2021) Preharvest treatment with oxalic acid improves post-harvest storage of lemon fruit by stimulation of the antioxidant system and phenolic content. Antioxidants 10(6):963

Sharma SS, Dietz K-J (2006) The significance of amino acids and amino acid-derived molecules in plant responses and adaptation to heavy metal stress. J Exp Bot 57:711–726. https://doi.org/10.1093/jxb/erj073

Shimazono H (1955) Oxalic acid decarboxylase, a new enzyme from the mycelium of wood destroying fungi. J Biochem 42:321–340

Sindhu SS, Sehrawat A, Glick BR (2022) The involvement of organic acids in soil fertility, plant health and environment sustainability. Arch Microbiol 204:720

Song J, Markewitz D, Wu S, Sang Y, Duan C, Cui X (2018) Exogenous oxalic acid and citric acid improve lead (Pb) tolerance of Larix olgensis A. Henry seed Forests 9(9):510

Tahir M, Mirza MS, Zaheer A, Dimitrov MR, Smidt H, Hameed S (2013) Isolation and identification of phosphate solubilizer *Azospirillum*, *Bacillus* and *Enterobacter* strains by 16S rRNA sequence analysis and their effect on growth of wheat (*Triticum aestivum* L.). Aust J Crop Sci 7:1284–1292

Tanner A, Bowater L, Fairhurst SA, Bornemann S (2001) Oxalate decarboxylase requires manganese and dioxygen for activity overexpression and characterization of *Bacillus subtilis* Yvrk and YoaN. J Biol Chem 276:43627–43634. https://doi.org/10.1074/jbc.M107202200

Tian S, Wan Y, Qin G, Xu Y (2006) Induction of defense responses against Alternaria rot by different elicitors in harvested pear fruit. Appl Microbiol Biotechnol 70:729. https://doi.org/10.1007/s00253-005-0125-4

Toal ES, Jones PW (1999) Induction of systemic resistance to *Sclerotinia sclerotiorum* by oxalic acid in oilseed rape. Plant Pathol 48:759–767

Tooulakou G, Giannopoulos A, Nikolopoulos D, Bresta P, Dotsika E, Orkoula MG, Kontoyannis CG, Fasseas C, Liakopoulos G, Klapa MI, Karabourniotis G (2016b) Reevaluation of the plant "gemstones": calcium oxalate crystals sustain photosynthesis under drought conditions. Plant Signal Behav 11:e1215793

Truffault V, Fry SC, Stevens RG, Gautier H (2017) Ascorbate degradation in tomato leads to accumulation of oxalate, threonate and oxalyl threonate. Plant J 89:996–1008

USDA (2021) Information on the upcoming amendment to the varroacide, oxalic acid (API-Bioxal™). Agricultural Research Service, USDA

Umer S, Akram NA, Ashraf M, Abdel Latef AA (2026) Pre-Sowing Application of Methionine and Proline Mitigates Drought Stress in Two Canola (Brassica napus L.) Cultivars. Russian Journal of Plant Physiology 73:48

Vlasov DY, Zelenskaya MS, Izatulina AR, Janson SY, Frank-Kamenetskaya OV (2023) Oxalate crystallization under the action of brown rot fungi. Crystals 13:432

Walker R, Famiani F (2018) Organic acids in fruits. In: Horticultural reviews, pp 371–430. https://doi.org/10.1002/9781119431077.ch8

Wang H, Huang H, Huang X, Hu Z (2006) Sugar and acid compositions in the arils of Litchi chinensis Sonn.: cultivar differences and evidence for the absence of succinic acid. J Hortic Sci Biotechnol 81:57–62. https://doi.org/10.1080/14620316.2006.11512029

Webb MA, Cavaletto JM, Carpita NC, Lopez LE, Arnott HT (1995) The intravacuolar organic matrix associated with calcium oxalate crystals in leaves of Vitis. Plant Journal 7:633–648

Whittaker MM, Whittaker JW (2002) Characterization of recombinant barley oxalate oxidase expressed by *Pichia pastoris*. J Biol Inorg Chem 7:136–145

Williams B, Kabbage M, Kim HJ, Britt R, Dickman MB (2011) Tipping the balance: *Sclerotinia sclerotiorum* secreted oxalic acid suppresses host defenses by manipulating the host redox environment. PLoS Pathog 7:e1002107. https://doi.org/10.1371/journal.ppat.1002107

Xu HW, Zhang JJ, Zeng JW, Jiang LR, Liu E, Peng CL, He ZH, Peng XX (2009) Inducible antisense suppression of glycolate oxidase reveals its strong regulation over photosynthesis in rice. J Exp Bot 60:1799–1809

Xu W, Li J, Jian S, Zhang M, Jia Y, Xia K (2025) HubHLH36 promotes oxalate degradation through HuAAE3 to enhance salt tolerance in pitaya (*Hylocereus polyrhizus*). Plant Physiol Biochem 227:110135

Yankey R, Awua A, Doe E, Achoribo E, Agbenyega S, Hayford M, Torto R (2015) Evaluation of oxalate content of some indigenous and produced spices and seasonings in Ghana. Food Sci Q Manag 39:67–73

Yu L, Jiang J, Zhang C, Jiang L, Ye N, Lu Y, Yang G, Liu E, Peng C, He Z (2010) Glyoxylate rather than ascorbate is an efficient precursor for oxalate biosynthesis in rice. J Exp Bot 61(6): 1625–1634

Yu YY, Si FJ, Wang N, Wang T, Jin Y, Zheng Y, Yang W, Luo YM, Niu DD, Guo JH, Jiang CH (2022) Bacillus-secreted oxalic acid induces tomato resistance against Gray Mold disease caused by *Botrytis cinerea* by activating the JA/ET pathway. Mol Plant-Microbe Int 35(8): 659–671. https://doi.org/10.1094/MPMI-11-21-0289-R

Zayed A, Adly G, Farag M (2025) Management strategies for the anti-nutrient oxalic acid in foods: a comprehensive overview of its dietary sources, roles, metabolism, and processing. Food Bioprocess Technol 18:4280–4300. https://doi.org/10.1007/s11947-024-03726-0

Zheng XL, Brecht JK (2017) Oxalic acid treatments. In: Pareek S (ed) Novel postharvest treatments of fresh produce. CRC Press, Boca Raton, pp 35–50. https://doi.org/10.1201/9781315370149
Zhu XF, Zheng C, Hu YT, Jiang T, Liu Y, Dong NY, Yang JL, Zheng SJ (2011) Cadmium-induced oxalate secretion from root apex is associated with cadmium exclusion and resistance in *Lycopersicon esulentum*. Plant Cell Environ 34:1055–1064
Zhu Y, Yu J, Brecht JK, Jiang TJ, Zheng XL (2016) Pre-harvest application of oxalic acid increases quality and resistance to Penicillium expansum in kiwifruit during postharvest storage. Food Chem 190:537–543. https://doi.org/10.1016/j.foodchem.2015.06.001
Zindler-Frank E (1976) Oxalate biosynthesis in relation to photosynthetic pathways and plant productivity: a survey. Z Pflanzenphysiol 80:1–13

Chapter 17
Succinic Acid: A Metabolic Enhancer for Plants Under Stress

Nizar El Mazouni, Abdelhamid Aouabe, Farid Errouh, Mohamed Ait-El-Mokhtar, Abderrahim Boutasknit, Raja Ben-Laouane, Marouane Baslam, and Abdelilah Meddich

Abstract One important metabolic signal that aids in plant adaptation to environmental stressors is succinic acid (SUA), a crucial intermediate of the tricarboxylic acid (TCA) cycle. SUA is essential for redox regulation, osmoprotection, metabolic rewiring, and stress-responsive gene expression in addition to its traditional function in cellular respiration. SUA strengthens antioxidant defenses, stabilizes membranes, controls ion balance, and primes immune responses in the face of biotic challenges like bacterial and fungal infections as well as abiotic stresses like drought, salinity,

N. El Mazouni · A. Aouabe · F. Errouh
Faculty of Sciences Semlalia Marrakech, Laboratory of Excellence in Agrobiotechnology and Bioengineering, AgroBiotech Center, CNRST-Labeled Research Unit (URL05-CNRST), (A)biotic Stress Physiology, Cadi Ayyad University, Marrakech, Morocco

M. Ait-El-Mokhtar
Laboratory of Biochemistry, Environment & Agri-Food URAC 36, Department of Biology, Faculty of Science and Techniques—Mohammedia, Hassan II University of Casablanca, Mohammedia, Morocco

A. Boutasknit
Faculty of Sciences Semlalia Marrakech, Laboratory of Excellence in Agrobiotechnology and Bioengineering, AgroBiotech Center, CNRST-Labeled Research Unit (URL05-CNRST), (A)biotic Stress Physiology, Cadi Ayyad University, Marrakech, Morocco

Department of Biology, Multidisciplinary Faculty of Nador, Mohamed First University—Oujda, Nador, Morocco

R. Ben-Laouane
Laboratory of Environment and Health, Department of Biology, Faculty of Science and Technology of Errachidia, Moulay Ismail University of Meknes, Meknes, Morocco

M. Baslam (✉)
Faculty of Sciences Semlalia Marrakech, Laboratory of Excellence in Agrobiotechnology and Bioengineering, AgroBiotech Center, CNRST-Labeled Research Unit (URL05-CNRST), (A)biotic Stress Physiology, Cadi Ayyad University, Marrakech, Morocco

GrowSmart, Amsterdam, The Netherlands

A. A. H. Abdel Latef (ed.), *Harnessing Vitamins and Organic Acids for Plant Stress Resilience*, Sustainability Sciences in Asia and Africa,
https://doi.org/10.1007/978-981-95-8593-9_17

heat, cold, and heavy metals. According to recent studies, SUA functions as a signaling molecule that interacts with phytohormones, beneficial microbes, and transcriptional networks.

This chapter gives a full picture of the biochemical, physiological, and molecular ways that SUA makes organisms more resistant to both abiotic and biotic stressors. Emphasis is placed on its exogenous application strategies, molecular targets, and relevance in sustainable agriculture.

Keywords Abiotic stress · Biotic stress · Redox homeostasis · TCA cycle · Gene regulation · Plant immunity · Signaling pathways · Metabolic priming

17.1 Introduction

Plant growth and productivity are continually challenged by biotic stresses, including bacteria, fungi, viruses, parasites, and insect pests, and abiotic stresses such as drought, flooding, extreme temperatures, salinity, and heavy metals (Mareri et al. 2022). Together, these environmental stressors account for more than half of the global crop yield losses, which is projected to worsen as climate change drives increasingly severe and unpredictable (Dos Reis et al. 2012; Mareri et al. 2022; Eckardt et al. 2023). In addition, these stressors alter essential elements of plant growth, development, and metabolism, causing even, in worse scenarios, plant death (Tuteja and Sopory 2008; Gill and Tuteja 2010).

To maintain food production, a likely option is the exogenous supplementation of protective metabolites, including organic acids that are integrated into central metabolism and regulate stress signaling cascades. Of these, succinic acid (SUA) is an especially interesting metabolite since it is both a tricarboxylic acid (TCA) cycle intermediate and a bioactive signal molecule (Zhu and Tang 2017).

SUA (butanedioic acid, or ethylene succinate) is a four-carbon dicarboxylic acid and an important part of the mitochondrial TCA cycle. Derived from glucose catabolism, pyruvate enters mitochondria to generate oxaloacetic acid (OAA) via either pyruvate carboxylation or phosphoenolpyruvate carboxylase (PEPC)-mediated carboxylation of phosphoenolpyruvate (PEP) (Zhang and Fernie 2023). Oxaloacetic acid contributes to the TCA cycle by producing a number of organic acids, such as SUA, which succinate dehydrogenase (Complex II of the electron transport chain) then oxidizes to fumarate. Energy signaling and ATP synthesis are directly linked to SUA metabolism through this reaction (Zhang and Fernie 2023).

A. Meddich (✉)
Faculty of Sciences Semlalia Marrakech, Laboratory of Excellence in Agrobiotechnology and Bioengineering, AgroBiotech Center, CNRST-Labeled Research Unit (URL05-CNRST), (A)biotic Stress Physiology, Cadi Ayyad University, Marrakech, Morocco

African Sustainable Agriculture Research Institute (ASARI), University Mohammed VI Polytechnic (UM6P), Laayoune, Morocco
e-mail: a.meddich@uca.ma

SUA can be exported into the cytoplasm and extracellular space, where it affects gene expression, takes part in hormone-like signaling, and even modifies the epigenetic landscape, in addition to mitochondrial metabolism (Pucher and Vickery 1941). SUA thus connects cellular regulation and stress adaptation to essential metabolic fluxes. Despite the paucity of research, there is growing evidence that SUA contributes to stress tolerance. For example, seed priming with SUA improved drought resistance in flowering kale (*Brassica oleracea* var. acephala) (Kiliç 2023). Drought stress was also shown to alter endogenous SUA accumulation in multiple species, suggesting its regulatory role in water-deficit responses (Ashrafi et al. 2018; Hamanishi et al. 2015). Transcriptomic analysis in *Cynanchum thesioides* revealed that drought stress controls candidate genes that are involved in SUA biosynthesis (Zhang et al. 2023). In *Leymus chinensis*, SUCLA has been reported to be a key regulator of SUA synthesis under alkaline salt stress (Wu et al. 2025). In *Medicago sativa*, cold and saline-alkali stress enhanced the accumulation of SUA and malic acid in the TCA cycle (Liu et al. 2024). Exogenous GABA treatment has also been demonstrated to promote stress tolerance by rapidly being converted into SUA and being assimilated into the TCA cycle (Hijaz and Killiny 2019). In wheat (*Triticum aestivum*), SUA and other organic acids were accumulated very much in roots under alkaline salt stress, though the reaction was less pronounced in leaves (Guo et al. 2015). Also, exogenous SUA relieved heavy metal toxicity by improving cadmium and lead tolerance in *Larix olgensis* (Song et al. 2018; Duan et al. 2019).

Together, these findings position SUA as a multifunctional metabolite that bridges primary metabolism with stress-responsive signaling. This chapter synthesizes current knowledge on the physiological, biochemical, and molecular functions of SUA in plant stress tolerance, while also exploring its crosstalk with other signaling pathways. By integrating metabolic and regulatory perspectives, we highlight SUA's emerging role as a versatile hub for improving plant resilience and adaptability under the pressures of climate change.

17.2 Succinic Acid in Plant Primary and Secondary Metabolism

17.2.1 Role in the Krebs Cycle and Carbon Fluxes

Succinic acid (succinate) is a central intermediate of the TCA cycle, or Krebs cycle, where it serves as both a substrate and a product, positioning it at the crossroads of plant and cellular metabolism (Camarasa et al. 2003). Its dual role integrates energy production, redox homeostasis, and biosynthetic processes, linking primary carbon fluxes to mitochondrial respiration and broader metabolic regulation (Ryan and O'Neill 2020).

Under aerobic conditions, succinate is generated via the oxidative branch of the TCA cycle. In this route, 2-oxoglutarate (2-OG) is first converted to succinyl-CoA by the 2-oxoglutarate dehydrogenase (OGDH) complex. Succinyl-CoA is subsequently transformed into succinate by succinyl-CoA synthetase (Tretter et al. 2016). Succinate is then oxidized to fumarate by succinate dehydrogenase (SDH), which also functions as Complex II of the mitochondrial electron transport chain, coupling the TCA cycle directly to oxidative phosphorylation through the reduction of ubiquinone to ubiquinol (Gill 2012). This oxidative branch yields high levels of reducing equivalents, namely, NADH and $FADH_2$, that drive ATP synthesis; complete metabolism of one glucose molecule through this pathway produces approximately 5 NADH and 2 $FADH_2$, highlighting succinate's importance for cellular energy supply (Camarasa et al. 2003).

Under conditions of restricted oxygen supply, metabolism is shunted towards the reductive pathway of the TCA cycle. Under this alternate pathway, fumarate is reduced to succinate by fumarate reductase, essentially reversing the SDH reaction. This adaptation maintains carbon fluxes under conditions where oxidative phosphorylation is impaired (Nghiem et al. 2017). Importantly, the reductive pathway incorporates CO_2 fixation during the conversion of phosphoenolpyruvate (PEP) to oxaloacetate (OAA), improving carbon retention efficiency. Theoretical modeling indicates that this route can yield up to 1.71 moles of succinate per mole of glucose ($\approx$1.12 g/g), with a standard free energy change (ΔG°) of -173 kJ/mol, underscoring the exergonic nature of the process:

$$7C_6H_{12}O_6 + 6CO_2 \rightarrow 12C_4H_6O_4 + 6H_2O \qquad \text{(McKinlay et al. 2007).}$$

Experimental evidence from TCA cycle mutants demonstrates that SDH is inactive under anaerobic conditions, preventing succinate oxidation and leading to its accumulation via the reductive branch (Camarasa et al. 2003). Succinate synthesis is surprisingly not dependent upon nitrogen source, although when glutamate is the sole nitrogen source present, there is a second route of metabolism that is capable of generating succinate via reverse conversion pathways. Other than its pivotal position within plant metabolism, succinate is now a bio-platform chemical due to its high theoretical yield and ease of carbon flux re-direction. Metabolic engineering has utilized this and diverted carbon from C_3 pathways (e.g., pyruvate synthesis) into C_4 pathways, with succinate formation preferred, particularly under anaerobic conditions (Zhang et al. 2019). Engineered microbial platforms are now achieving higher titers of succinate, which is the basis for its sustainable application at an industrial scale as a feedstock for pharmaceuticals, bioplastics, and green chemistry (Mitrea et al. 2024).

17.2.2 Connections with the Biosynthesis of Amino Acids, Fatty Acids, and Organic Acids

Succinic acid is at the hub of anabolic metabolism, bridging energy formation and the biosynthesis of the major macromolecules. Its roles extend across the production of amino acids, fatty acids, and C_4-dicarboxylic organic acids, enabling plants and microorganisms to adapt their metabolism under diverse environmental conditions.

Amino Acid Biosynthesis

One of the most significant anabolic functions of succinate lies in its contribution to the synthesis of amino acids in the glutamate family, glutamate, glutamine, proline, and arginine. These amino acids not only serve as building blocks for proteins but also act as critical nitrogen donors in multiple metabolic reactions (Allison and Robinson 1970). Succinate enables this pathway by serving as a precursor to α-ketoglutarate (2-oxoglutarate), the entry point for glutamate biosynthesis (Huergo and Dixon 2015). Under aerobic conditions, this conversion occurs in the oxidative branch of the Krebs cycle. Under anaerobic or fermentative conditions, some microbes use another process in which succinate or succinyl-CoA are carboxylated by reduction, catalyzed by ferredoxin-dependent enzymes and using CO_2 (Buchanan and Evans 1965). This metabolic flexibility allows amino acid biosynthesis to continue when oxidative metabolism is inhibited. Radiolabeling experiments with succinate-1,4-^{14}C have provided strong evidence for direct incorporation into amino acids, particularly glutamate and its derivatives, proline, and arginine. In vitro tests with acellular systems have also provided evidence of preferential carbon transport from succinate to glutamate, confirming its function as a carbon source for amino acid biosynthesis (Allison and Robinson 1970).

Fatty Acid and Lipid Metabolism

Succinyl-CoA, an intermediate of succinate, is equally crucial for lipid metabolism. It is involved in the catabolism of branched-chain amino acids, in which the amino acids methionine, valine, and isoleucine are broken down into propionyl-CoA, which is then converted into succinyl-CoA and recycled into the Krebs cycle (Burch et al. 2018). This metabolic diversion allows carbon skeletons from amino acid degradation to be diverted to lipid biosynthesis pathways. In turn, the β-oxidation of dicarboxylic fatty acids produces succinyl-CoA, directly linking fatty acid degradation to central carbon metabolism (Lynen and Ochoa 1953). Succinyl-CoA is also involved in fatty acid biosynthesis through CoA transfer reactions. Fatty acid activation can occur through:

- Direct acyl-CoA synthesis via ATP- and CoA-dependent enzymes.
- CoA transfers from existing acyl-CoAs, such as acetyl-CoA or succinyl-CoA.

This second mechanism enables succinyl-CoA to donate CoA moieties, producing fatty acyl-CoA intermediates that are essential precursors for lipid assembly and membrane biosynthesis (Lynen and Ochoa 1953).

Biosynthesis of C_4-Dicarboxylic Organic Acids

As an essential intermediate in the Krebs cycle, succinate serves as a branching point for other C_4 dicarboxylic acids, including fumarate, malate, and oxaloacetate. These organic acids are central to redox balance, energy turnover, and metabolic flexibility (Yin et al. 2015). Succinyl-CoA synthetase catalyzes the conversion of succinyl-CoA to succinate, which is then oxidized to fumarate by succinate dehydrogenase (SDH). Notably, SDH also functions as Complex II of the mitochondrial electron transport chain, coupling succinate oxidation with ubiquinone reduction and thus directly linking organic acid turnover with oxidative phosphorylation (Rutter et al. 2010).

Downstream, fumarase catalyzes the hydration of fumarate to malate, a reaction exploited in industrial biotechnology for malic acid production, a compound widely used in the food, beverage, and pharmaceutical industries (Katayama et al. 2019). Malate is subsequently oxidized to oxaloacetate by malate dehydrogenase (MDH) in an NAD^+-dependent reaction, which is essential both for the continuation of the TCA cycle and for maintaining cellular redox balance (Anadón et al. 2014). The structural similarity and interconvertibility of these C_4-dicarboxylic acids confer remarkable metabolic flexibility, allowing plants and microbes to reroute carbon fluxes according to environmental and energetic demands (Zhang et al. 2011; Yang et al. 2017; Cao et al. 2011).

17.2.3 Link with Secondary Metabolites Under Stress Conditions

Succinic acid, a central intermediate of the TCA cycle, plays a multifaceted role in primary metabolism but also exerts a profound influence on secondary metabolic pathways, particularly under abiotic stress conditions such as drought (Rezayian et al. 2023). Traditionally recognized as a component of energy metabolism, the TCA cycle is also a reservoir of amino acid, organic acid, and an incredibly diverse set of defense-related secondary metabolite precursors. Stress-induced metabolic reprogramming positions succinate as a significant carbon flux integrator of carbon flux, redox homeostasis, and biosynthetic activity.

Succinic Acid and Mitochondrial Energy-ROS Crosstalk
Succinate dehydrogenase (SDH) catalyzes succinate oxidation to fumarate and simultaneously acts as Complex II of the mitochondrial electron transport chain (ETC) (Oyedotun and Lemire 2004). This provides the possibility of direct crosstalk between oxidative phosphorylation and carbon metabolism. Stress-responsive and frequently genotype-dependent, SDH activity is linked to hormone-mediated drought signaling in drought-tolerant Ilex paraguariensis, where increased SDH1 expression was associated with higher SDH activity and increased accumulation of abscisic acid (ABA) (Acevedo et al. 2013). Additionally, SDH is a significant source of reactive oxygen species (ROS) in the mitochondria. Although potentially cytotoxic at high levels, ROS serve as essential signaling molecules during stress, initiating transcriptional cascades that upregulate antioxidant enzymes and secondary metabolite biosynthesis (Rezayian et al. 2023; Jardim-Messeder et al. 2015).

Exogenous Succinic Acid as a Metabolic Elicitor
The potential of exogenous succinate application to improve stress adaptation has been emphasized by recent studies. Treatments with succinate boosted the expression of TCA cycle enzymes and ETC components in tomato and other systems, which resulted in increased ATP production, mitochondrial activity, and a brief burst of reactive oxygen species (ROS) (Piechowiak and Balawejder 2025). This ROS

pulse, in turn, activated antioxidant defense systems, inducing genes encoding ROS-scavenging enzymes and stimulating the accumulation of antioxidant metabolites (Jardim-Messeder et al. 2015). Notably, treatment with 10 mM succinate increased ascorbic acid by 18%, glutathione (GSH) by 29%, and led to measurable gains in flavonoid and polyphenol content, thereby enhancing total antioxidant capacity (Piechowiak and Balawejder 2025). These findings underscore succinate's capacity to act as a metabolic elicitor, reprogramming redox and secondary metabolism under stress.

Contribution to Secondary Metabolite Biosynthesis
Beyond ROS signaling, succinate metabolism directly contributes to secondary metabolite production through succinyl-CoA, a derivative central to multiple biosynthetic pathways. Succinyl-CoA serves as a precursor for heme, chlorophyll, and diverse secondary metabolites, including phenolics, flavonoids, alkaloids, and terpenoids (Studart-Guimarães et al. 2005). Genetic studies in tomato indicate that repression of succinyl-CoA ligase (SUCoAL) caused only mild phenotypic effects, suggesting metabolic redundancy and compensatory mechanisms that sustain succinate levels (Studart-Guimarães et al. 2007). Nevertheless, succinate's role as a branching metabolite places it at the interface of primary energy metabolism and secondary defense pathways.

Physiological Implications Under Stress
The regulation of succinate flux and SDH activity exerts significant effects on stomatal conductance, photosynthetic performance, and biomass allocation under drought stress. Genetic modulation of SDH subunits in *Arabidopsis* and tomato has produced divergent physiological outcomes, suggesting that fine-tuning succinate metabolism influences how plants balance carbon allocation between temporary energy needs and secondary metabolite production (Acevedo et al. 2013).

Succinic acid is a synthetic link between mitochondrial energy metabolism, ROS-mediated signaling, and secondary metabolite biosynthesis. By modulating energy-generating and defense-eliciting pathways, succinate enhances metabolic plasticity and stress tolerance. Double-edged roles as a primary metabolite and a signaling elicitor put succinate at a pivotal metabolic node as a candidate target for strategies to render plants more stress tolerant to environmental stresses (Fig. 17.1).

17.3 Succinic Acid-Mediated Abiotic Stress Tolerance

17.3.1 Drought and Salinity

Drought and salinity are widespread abiotic stresses that limit plant productivity due to osmotic imbalance, ionic toxicity, and oxidative damage. Plant tolerance to these adverse conditions relies on significant metabolic reprogramming, such as the accumulation of soluble organic compounds such as proline, soluble sugars

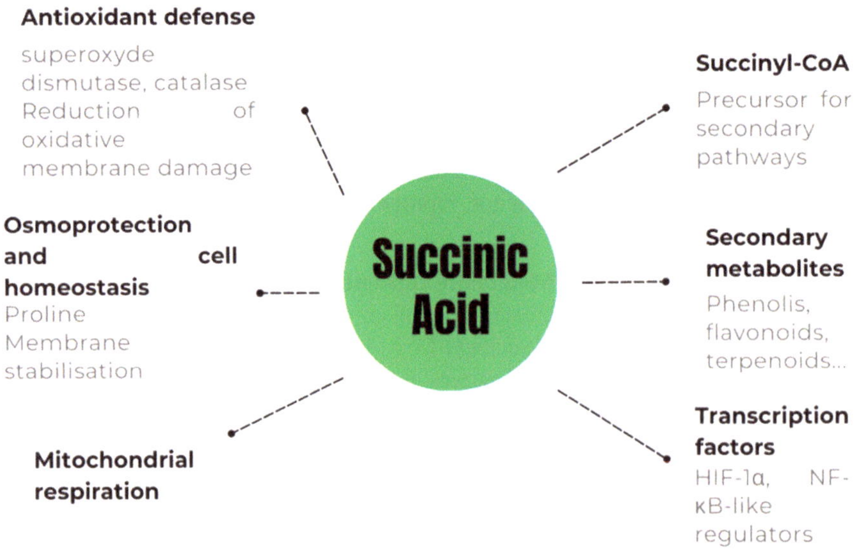

Fig. 17.1 Metabolic network of succinic acid and its role in stress adaptation. Succinic acid, as a key intermediate of the tricarboxylic acid (TCA) cycle, links mitochondrial respiration and primary and secondary metabolism. Through succinate dehydrogenase (SDH), succinate oxidation powers the electron transport chain and generates ROS activating stress-inducible transcription factors (e.g., HIF-1α, NF-κB-like regulators). This induces antioxidant enzymes

(glucose, fructose, sucrose), and organic acids (Boutasknit et al. 2021; Benaffari et al. 2022; Ikan et al. 2023; Ouhaddou et al. 2023; Soussani et al. 2023). Among these metabolites, succinic acid (succinate) occupies a central role in both energy regulation and cellular homeostasis, functioning as a hub that links mitochondrial activity, antioxidant defense, and stress signaling (Kiliç 2023; Gao et al. 2025).

Succinate Accumulation and Stress Adaptation During drought and salt stress, succinate often accumulates in plant tissues, reflecting its role as an adaptive metabolite (Ashrafi et al. 2018). This accumulation facilitates the reorganization of central metabolism in response to energy constraints, enabling plants to sustain ATP production while activating protective mechanisms (Kiliç 2023; Zhi et al. 2024; Gao et al. 2025). At the physiological level, succinate influences multiple defense pathways:

- Activation of antioxidant enzymes, including superoxide dismutase (SOD), catalase (CAT), and ascorbate peroxidase (APX).
- Regulation of stomatal behavior, contributing to improved water-use efficiency.
- Promotion of osmoprotectant synthesis, such as proline and soluble sugars.
- Modulation of gene expression, particularly of TCA- and stress-related genes.

Mitochondrial ROS and Signaling A key feature of succinate's role under stress is its link to mitochondrial redox metabolism. As a direct substrate for respiratory oxidation, succinate regulates hydrogen peroxide (H_2O_2) production in mitochondria (Eprintsev et al. 2023). Controlled ROS production is a matter of antioxidant detoxification, and also a signal transduction mechanism that stimulates transcriptional programs to enhance antioxidant protection and stress responses. The dynamic regulation of ROS levels by succinate suggests its potential use as a biomarker of stress status (Ashrafi et al. 2018).

Exogenous Application and Stress Alleviation Exogenous treatment with succinate has been seen to mitigate drought- and salt-induced oxidative injury. In different species, succinate treatments augmented the activities of antioxidant enzymes, reduced ROS accumulation, and limited cell damage (Song et al. 2018; Zhang et al. 2021; Wu et al. 2025). Succinate also mitigated Na^+ in root accumulation under salt stress, membrane stabilization, root elongation, and water maintenance (Wu et al. 2025).

GABA Shunt and Ca^{2+} Crosstalk The GABA shunt is another mandatory pathway linking succinate to stress resistance, which is a by-path of the TCA cycle. Under drought and salt stress, the concentration of GABA is extremely high (Neuroscience 2021; Dabravolski and Isayenkov 2023; Hayat et al. 2023). GABA is directly converted to succinate, which again feeds into the TCA cycle, sustaining energy metabolism as well as strengthening antioxidant defense systems (Du et al. 2012; Jardim-Messeder et al. 2015; Piechowiak and Balawejder 2025). The process is in close association with cytosolic calcium (Ca^{2+}) signaling since drought-induced Ca^{2+} influx stimulates GABA and succinate formation (Yuan et al. 2023; Ashraf et al. 2024). Through this process, succinate is involved in ROS detoxification, membrane stabilization, and maintenance of metabolic function under stress.

Genetic Regulation of Succinate Metabolism Genetically, succinate accumulation has been linked to expression of the succinyl-CoA synthetase genes (LcSUCLAα and LcSUCLAβ). Transgenic overexpression of these subunits led to higher succinate content, improved salt-alkali stress resistance, reduced cellular damage, increased root growth, and improved water retention (Wu et al. 2025). These results represent direct evidence that manipulation of succinate metabolism yields adaptive advantage under abiotic stress.

Succinate as a Metabolic Signal Succinate most likely acts as a signaling metabolite. Through the regulation of ROS-Ca^{2+} cascades and interaction with mitochondrial retrograde signaling, succinate can reprogram nuclear gene expression to orchestrate systemic responses to stress (Gleason et al. 2011; Li et al. 2021; Kiliç 2023). This dual function as a metabolite and signaling molecule highlights the integrative role of succinate in linking mitochondrial metabolism and transcriptional reprogramming, and thus in plant tolerance to water stress and salinity.

17.3.2 Extreme Temperature

Extreme temperatures, both hot and cold, impose extreme restrictions on plant metabolism, disrupting enzyme function, membrane integrity, and redox balance. New evidence implicates succinic acid (succinate) as a key metabolite in heat stress tolerance, at the nexus of energy metabolism, antioxidant defense, and stress signaling.

Succinate Accumulation under Heat and Cold Stress Succinate levels are elevated following exposure to cold and heat. In *Arabidopsis thaliana*, succinate accumulation has been observed during the development of heat and frost tolerance, demonstrating its role in heat adaptation (Al-Quraan and Al-Share 2016; Zhang et al. 2023). Through its structural integrity, succinate preserves enzyme function and membrane fluidity, which are essential for cellular homeostasis during heat stress.

Mitochondrial Respiration and ROS Balance Succinate regulates mitochondrial respiration to ensure stable ATP yield under heat stress while limiting ROS levels (Fedorin et al. 2022). As a direct substrate of the electron transport chain, succinate oxidation produces a redox signal but carries the risk of excessive ROS production. Studies have established that succinate exogenously induces the activity of primary antioxidant enzymes, such as SOD and CAT, and so increases the resistance of plants to destroy excess ROS during heat stress (Kolupaev et al. 2011, 2012; Dong et al. 2022; Piechowiak and Balawejder 2025). Redox balance prevents oxidative damage and promotes stress resistance.

Integration with the GABA Pathway Succinate is directly linked to the GABA shunt, an alternative metabolic pathway induced in stressful situations. The conversion of GABA to succinate via succinate semialdehyde dehydrogenase redirects carbon to the Krebs cycle, supporting energy production and enhancing antioxidant metabolism (Fedorin et al. 2022). Bypassing metabolism in this way provides plants with a robust mechanism for maintaining respiration during periods of heat and cold stress and mitigating oxidative damage.

Crosstalk with Hormonal Pathways Succinate can modulate stomatal movement, transpiration, and reactive physiological responses to stress (Li et al. 2017). Exogenous succinate can improve stress tolerance by increasing stomatal conductance, maintaining leaf water status, enhancing antioxidant capacity, and increasing proline content, thereby preventing oxidative damage (Kiliç 2023).

Enhancement of Secondary Metabolism Under Stress Remarkably, succinate has also been linked to changes in secondary metabolism brought on by temperature stress. For instance, by controlling gene expression and modifying ROS levels, succinate improved astaxanthin and lipid biosynthesis when combined with sodium hydrosulfide (NaHS), indirectly promoting thermal tolerance (Yu et al. 2022a). A useful method to improve photosynthetic performance under stress is the foliar application of succinate, which also increased chlorophyll content and optimized plant nutrient balance (Saavedra et al. 2022).

17.3.3 Heavy Metals and UV Stress

Heavy Metal Stress Succinic acid mainly aids in heavy metal tolerance by means of metabolic reprogramming and metal chelation. Succinate reduces the free availability of toxic metal ions and lessens cellular toxicity by forming stable complexes with them when combined with other organic acids (such as oxalic, citric, and butyric acids) (Kocaman 2023). For instance, succinate has the ability to bind manganese (Mn) and change its distribution by promoting Mn accumulation in roots and decreasing its translocation to shoots. However, growth, chlorophyll content, photosynthetic efficiency, and chloroplast structure are still compromised, so this redistribution in *Pisum sativum* does not completely reduce Mn toxicity (Doncheva et al. 2005). Additionally, there is evidence that when plants are stressed by heavy metals, they release succinate into the rhizosphere, which alters soil chemistry and lowers metal uptake. Succinate root exudation was also seen in *Ricinus communis* under severe Cu stress, likely through rhizosphere chelation inhibiting Cu translocation and toxicity (Huang et al. 2016).

In contaminated soils, increased secretion of succinate and other organic acids reduces the mobility and bioavailability of heavy metals, thereby enabling adaptation in multi-contaminated soils (Zhixin et al. 2013). Exogenously applied succinate improves the detoxification capacity of plants. Succinate alleviated lead (Pb)-induced oxidative stress in *Larix olgensis* by stimulating antioxidant enzymes, increasing biomass, and promoting Pb sequestration in fine roots, thereby reducing systemic toxicity (Song et al. 2018). In a study, succinate treatment promoted cadmium (Cd) tolerance by increasing Cd solubility and storage, antioxidant defense, gas exchange efficiency, and stress response gene expression (Mumtaz et al. 2025).

UV Stress Ultraviolet B (UV-B) radiation damages DNA, inhibits photosynthesis, and causes excessive accumulation of ROS (Chen et al. 2022). Succinic acid plays a key role in regulating UV tolerance by modulating energy metabolism, antioxidant defense pathways, and secondary metabolism. Succinate levels increase during UV-B exposure, reflecting increased Krebs cycle activity to meet higher energy and biosynthetic demands (Yu et al. 2024). In *Elymus sibiricus*, succinate was discovered by metabolomic analysis among nearly 700 metabolites associated with UV stress under the heading "organic acids and their derivatives," illustrating its function in metabolic regulation during UV stress (Zhang et al. 2024). Figure 17.2 provides an overview of the cellular and molecular responses of SUA under abiotic stress conditions.

Functionally, succinate supports UV protection in several ways:

- Antioxidant defense: Exogenous succinate enhances ascorbic acid accumulation (+18%), glutathione levels, and the activity of enzymes such as SOD and CAT, thereby neutralizing ROS generated under UV stress (Piechowiak and Balawejder 2025).

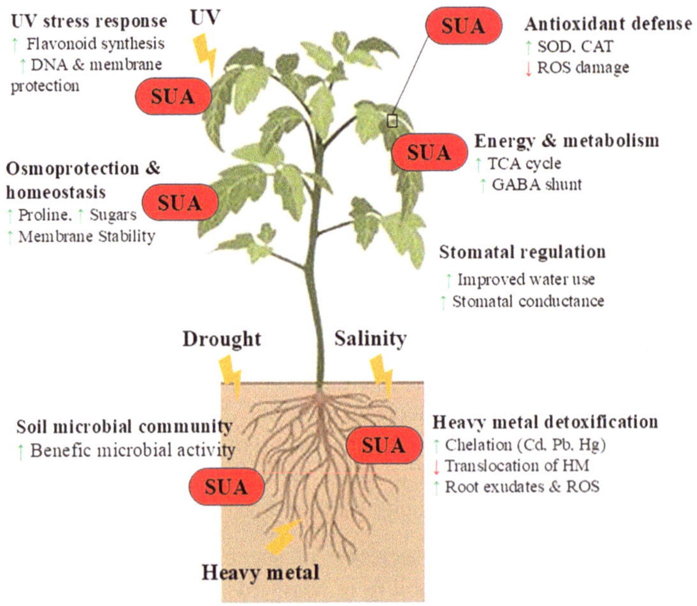

Fig. 17.2 Cellular and molecular responses to succinic acid (SUA) under abiotic stress conditions. SUA integrates multiple defense pathways to enhance plant tolerance. Under UV stress, SUA promotes flavonoid accumulation, stabilizes DNA and membranes, and enhances ROS protection. During drought and salinity, SUA contributes to osmoregulation by increasing proline and soluble sugar levels, regulates stomatal conductance, and sustains TCA cycle activity through the GABA shunt. In the antioxidant system, SUA enhances the activities of SOD and CAT, reducing ROS burden. At the root level, SUA mediates heavy metal detoxification through chelation (Cd, Pb, Hg), sequestration in roots, modulation of exudates, and ROS regulation, while also benefiting the soil microbial community. Collectively, these interconnected processes position SUA as a multifunctional metabolite bridging energy metabolism, redox homeostasis, and environmental adaptation

- Secondary metabolites: Succinate indirectly fuels the synthesis of UV-absorbing compounds, including flavonoids and polyphenols, which act as natural sunscreens and ROS scavengers (Shi and Liu 2021).
- Stomatal regulation: UV radiation is likely to disrupt stomatal conductance, yet succinate has been shown to maintain stomatal opening, thus CO_2 assimilation and photosynthetic performance in stressed plants (Gago et al. 2016; del Socorro Sánchez Correa et al. 2023).
- Amplified respiration under UV stress increases ROS production, which, when controlled, acts as a signal mechanism to induce gene encoding for antioxidants and start the biosynthesis of low-molecular-weight antioxidants, presenting an adaptive advantage (Piechowiak and Balawejder 2025).

17.4 Succinic Acid and Plant Immunity Under Biotic Stress

17.4.1 Pathogen Resistance

Succinic acid has turned into a polyfunctional metabolite involved in the regulation of plant immunity, contributing both to direct suppression of pathogens and to priming defense response in hosts (Nyzhnyk et al. 2025; Gao et al. 2025). Its role varies from transcriptional activation of defense gene expression to affecting rhizosphere microbial interactions, thereby increasing local and systemic immunity.

Priming and Activation of Defense Genes One of the most important features of plant immune priming is the induction of pathogenesis-related (PR) genes, whose proteins are fundamental components of systemic acquired resistance (SAR). Similar to salicylic acid- mediated immunity, succinate is involved in the transcriptional induction of PR proteins, including PR1, PR2 (β-1,3-glucanase), and PR3 (chitinase) (Kumar et al. 2025). These proteins degrade structural components of the cell walls of bacteria and fungi and directly halt pathogen growth (Zribi et al. 2024). Studies associate succinate accumulation with enhanced antioxidant activity and enhanced growth under stress by a pathogen, positioning it in signaling and metabolic molecule functions upon defense activation (Jin et al. 2022; Jan et al. 2024).

Direct Antimicrobial Activity Ez-Zoubi et al. (2023) showed that a biopesticide formulation with succinate and *Artemisia herba-alba* essential oils effectively suppressed *Botrytis cinerea*, a necrotrophic pathogen responsible for grey mold disease.

Modulation of Plant–Microbe Interactions Organic acids such as succinate act as chemotactic compounds, which attract beneficial microbes like plant growth-promoting rhizobacteria (PGPR), to the root zone (Sharma et al. 2020).

17.4.2 Interaction with Phytohormones

Succinic acid is implicated in modulating plant defense responses through phytohormonal signaling networks. And phytohormones are implicated in the growth-immunity trade-off so that plants can optimize resource partitioning under biotic stress (Kots et al. 2023). Exogenous succinic acid can circumvent or substitute part of SA signaling, an essential coordinator of systemic acquired resistance (SAR). For instance, while SA is known to trigger pathogenesis-related (PR) gene expression and enhance antioxidant defense (Mishra et al. 2024), succinate treatment also stimulated phenylalanine ammonia-lyase (PAL) activity and ethylene (ET) emission in resistant wheat cultivars (Nyzhnyk et al. 2025). These responses were, however, weaker in susceptible cultivars, suggesting that succinate is rather a supporting modulator than a pivotal defense signal.

Crosstalk with Ethylene Signaling Ethylene modulates defense signaling, in particular intermediating necrotrophic pathogen resistance (Shu et al. 2024). Exogenous application of 0.1 mM succinic acid substantially increased ethylene release in resistant *Triticum aestivum* (winter wheat) cultivars after infection with *Blumeria graminis* f. sp. tritici, the causal agent of powdery mildew (Nyzhnyk et al. 2025). This enhancement was associated with elevated PAL activity, underscoring succinate's role in stimulating the phenylpropanoid pathway and downstream synthesis of defense-related secondary metabolites.

Integration with ROS and Antioxidant Regulation The interplay between succinate and phytohormones extends into redox regulation. In resistant wheat lines, succinate treatment reduced excessive hydrogen peroxide (H_2O_2) accumulation and modulated the activity of antioxidant enzymes, including (SOD and CAT. This suggests that succinate not only amplifies hormone-driven defense signals but also contributes to ROS homeostasis, ensuring that oxidative bursts function as signals rather than as damaging agents (Nyzhnyk et al. 2025).

MAPK Cascades as Convergence Points At the molecular level, the Mitogen-Activated Protein Kinase (MAPK) cascades serve as central nodes integrating phytohormonal signaling. These cascades act as convergence points for SA, JA, ET, and abscisic acid (ABA) pathways, mediating both the perception and transduction of pathogen-derived signals (Jagodzik et al. 2018). Succinate's influence on ET and SA signaling suggests that it may indirectly modulate MAPK activity, thereby amplifying defense gene expression and establishing feedback loops that fine-tune immune responses.

17.4.3 Microbial Crosstalk and Induced Systemic Resistance

Succinic acid has gained recognition as an important mediator of Induced Systemic Resistance (ISR), a plant immune strategy triggered by beneficial microbes such as plant growth-promoting rhizobacteria (PGPR) (Yu et al. 2022b). Unlike systemic acquired resistance (SAR), which is largely dependent on salicylic acid signaling, ISR is typically associated with JA and ET pathways. However, recent evidence indicates that succinate participates in both immune modulation and microbial recruitment, thereby linking primary metabolism with ecological interactions in the rhizosphere (Nyzhnyk et al. 2025).

Recruitment of Beneficial Microbes Organic acids exuded by roots, including succinate, are increasingly recognized as chemoattractants that shape microbial community structure. Succinic acid can selectively enrich beneficial microbial taxa, such as *Comamonadaceae*, which are known to promote plant health (Wen et al. 2020). Through root exudation, succinate provides both a carbon source and a metabolic signal, thereby creating a selective niche that favors PGPR colonization (Yang et al. 2024). In turn, these microbes enhance nutrient uptake, suppress pathogens, and induce systemic resistance in plants (Lee et al. 2024b).

Energy Metabolism and Plant Priming At the plant level, succinic acid enhances ATP production by stimulating the mitochondrial TCA cycle, ensuring sufficient energy supply for defense-related processes under stress (Khan et al. 2019). Elevated succinate pools act as both substrates for respiration and metabolic modulators, supporting the biosynthesis of defense compounds and reinforcing antioxidant metabolism. This dual function, providing energy while signaling through phytohormonal crosstalk, primes plants for faster and stronger immune responses upon pathogen attack.

Integration into ISR Pathways Through its integration into ISR-associated signaling cascades, succinic acid connects microbial enrichment, energy metabolism, and root exudation. Its impact on ET and JA signaling points to a cooperative function in coordinating responses between systemic immune activation and local rhizosphere events. Because it primes host defense systems and cultivates advantageous microbial communities, succinic acid is positioned as a bridge metabolite.

Figure 17.3 illustrates the mechanisms of succinic acid (SUA)-induced plant immune responses under biotic stress, while Table 17.1 summarizes its effects on abiotic and biotic stress resistance in various crops.

On the other hand, the production of succinic acid by microbial fermentation has garnered attention for its potential applications in sustainable agriculture. Utilizing microbial fermentation processes to produce succinic acid from renewable resources provides a valuable chemical for industrial applications (Chen et al. 2023). For instance, species such as *Actinobacillus succinogenes* and *Escherichia coli* have been identified as potential microbial sources for succinic acid production using diverse substrates (Kaur et al. 2018).

17.5 Molecular Regulation of Succinic Acid-Mediated Stress Tolerance

Through metabolic and epigenetic regulation, succinic acid improves plant tolerance to abiotic stress. It improves the expression of succinate dehydrogenase genes and the mitochondrial energy metabolism by raising succinate levels and succinate semialdehyde dehydrogenase activity. The reduction in the methylation of the promoters of related genes indicates epigenetic control of this adaptation (Fedorin et al. 2022). The early activation of succinate semialdehyde dehydrogenase stimulates the GABA pathway, while succinate dehydrogenase activity and mitochondrial oxidation increase progressively. These responses are regulated by transcription factors such as WRKY and NAC, indicating coordinated transcriptional control of succinate metabolism (Eprintsev et al. 2025). In a study by Bilal et al. (2023), succinic acid, enhanced by the application of silicon under low temperature stress, contributes to plant tolerance by modulating gene expression. It downregulates ABA-related genes (*NCED-1*, *PyL-4*) and upregulates cold-sensitive genes (*ICE-1*, *SRC-2*), antioxidants

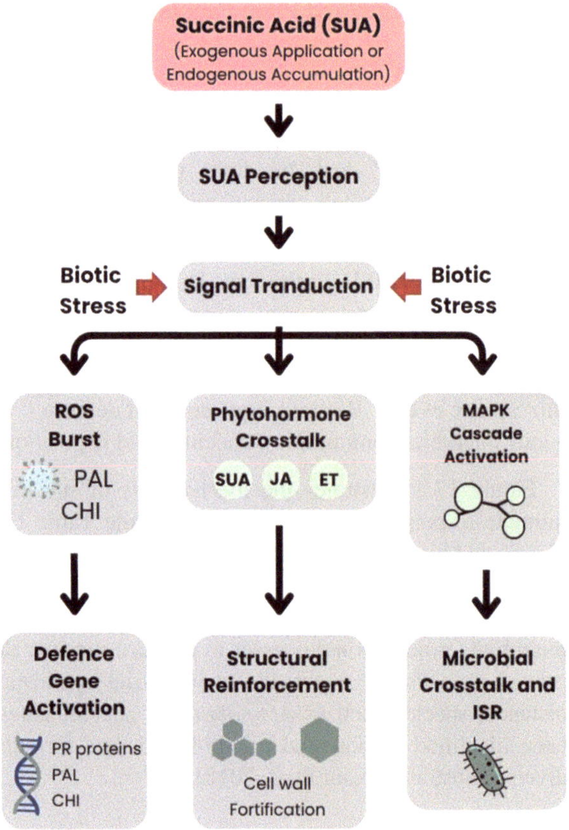

Fig. 17.3 Mechanisms of succinic acid (SUA)-induced plant immune responses under biotic stress. *SUA* succinic acid, *ROS* reactive oxygen species, *MAPKs* mitogen-activated protein kinases, *PR proteins* pathogenesis-related proteins, *PAL* phenylalanine ammonia-lyase, *CHI* chitinase, *ISR* induced systemic resistance, *JA* jasmonic acid, *ET* ethylene, *SUA* succinic acid

(SOD, CAT), and transport-related genes (*PPMA3*, *PPMA4*, *SDR*), promoting nutrient uptake, reducing oxidative stress, and boosting resistance to cold (Bilal et al. 2023). A study has shown that succinic acid—via its intermediate, succinic semialdehyde (SUSA)–contributes to plant stress tolerance through transcriptional and biochemical regulation. Under abiotic stress, SUSA was converted to γ-hydroxybutyrate (GHB) by NADPH-dependent glyoxylate reductases (*GR1* and *GR2*), whose gene expression was upregulated. This response, linked to increased GABA and NADPH/NADP$^+$ ratios, supports redox balance and stress adaptation, highlighting a generegulated pathway through which succinic acid metabolism enhances plant resilience (Allan et al. 2008). In another study on microcystin-producing species, SUA was shown to inhibit the growth of *Microcystis aeruginosa* by downregulating key photosynthesis-related genes such as *psbA*, *psaB*, *psbD*, *psbO*, and *rbcL*, resulting in reduced photosynthetic efficiency and CO_2 assimilation. SUA indirectly defends the plants against the oxidative and metabolic injury induced by such toxic cyanotoxins as it interference suppresses the production of microcystin (Chen et al. 2021).

Table 17.1 Effects of succinic acid on abiotic and biotic stress resistance in various crops

Crop	Stress type	Observed effects of SUA	References
Wheat (*Triticum aestivum*)	Biotic (Powdery mildew)	↑ Ethylene production, ↑ PAL activity, ↓ H_2O_2 accumulation, ↑ SOD and CAT activity in resistant varieties	Nyzhnyk et al. (2025)
Multiple crops	Biotic (Fungal/ bacterial)	↑ Expression of PR1, PR2 (β-1,3-glucanase), PR3 (chitinase); enhanced hydrolysis of pathogen cell walls	Kumar et al. (2024) and Zribi et al. (2024)
General (multiple plants)	Biotic (Various pathogens)	↑ PR gene expression; enhanced structural and biochemical defense priming	Jan et al. (2024)
Not crop-specific	Biotic (*Botrytis cinerea*)	Succinic acid + *Artemisia herba-alba* essential oils = strong biopesticide effect	Ez-Zoubi et al. (2023)
General/ model plants	Abiotic/ biotic	↑ Organic acids; ↑ antioxidant capacity; improved growth and pathogen tolerance	Jin et al. (2022)
Rhizosphere-associated crops	Biotic (pathogens)	Organic acid chemotaxis promotes PGPR recruitment and ISR	Sharma et al. (2020)
Model systems	Hormonal signaling	SUA modulates SUA, ET, JA, ABA crosstalk via MAPK cascade integration	Jagodzik et al. (2018)
Multiple crops	Biotic (ISR via PGPR)	Enhanced ISR; enriched *Comamonadaceae*; modified microbial recruitment and root exudation dynamics	Wen et al. (2020)
General/ stress-prone crops	Abiotic (energy limitation)	↑ ATP via mitochondrial TCA stimulation; ↑ energy availability during stress	Khan et al. (2019)

SUA succinic acid, *PAL* phenylalanine ammonia-lyase, H_2O_2 hydrogen peroxide, *SOD* superoxide dismutase, *CAT* catalase, *PR* pathogenesis-related protein, *PGPR* plant growth-promoting rhizobacteria, *ISR* induced systemic resistance, *ET* ethylene, *JA* jasmonic acid, *ABA* abscisic acid, *MAPK* mitogen-activated protein kinase, *TCA* tricarboxylic acid (cycle), *ATP* adenosine triphosphate

The bioactive succinic acid of *Cynanchum thesioides* is primarily produced through the TCA cycle and the GABA shunt, in which succinic acid is a significant intermediate. Under water stress, the transcriptome analysis revealed that genes involved in succinic acid biosynthesis, including malate dehydrogenase (MDH), succinate dehydrogenase (SDH), and glutamate decarboxylase (GAD), were differentially expressed. Succinic acid levels were positively correlated with MDH but negatively correlated with GAD and SDH, showing a close metabolic network control. The GABA shunt is a significant pathway in plant abiotic stress response through its contribution to the accumulation of GABA, a non-proteinogenic amino acid that was found to induce plant growth and stress resistance. This pathway recovers free radicals and regulates enzymatic reactions, thereby ensuring cellular homeostasis under environmental stresses such as drought, salinity, and oxidative stress (Renault et al. 2010; Zhang et al. 2023). A recent study in wheat showed that adenine methylation in the promoter region of the SSADH gene (succinic semialdehyde dehydrogenase), a key enzyme of the GABA shunt, controls its expression under salt stress (Fedorin et al. 2024). Alkaline salt stress induces a remarkable increase in succinic acid content, especially in root

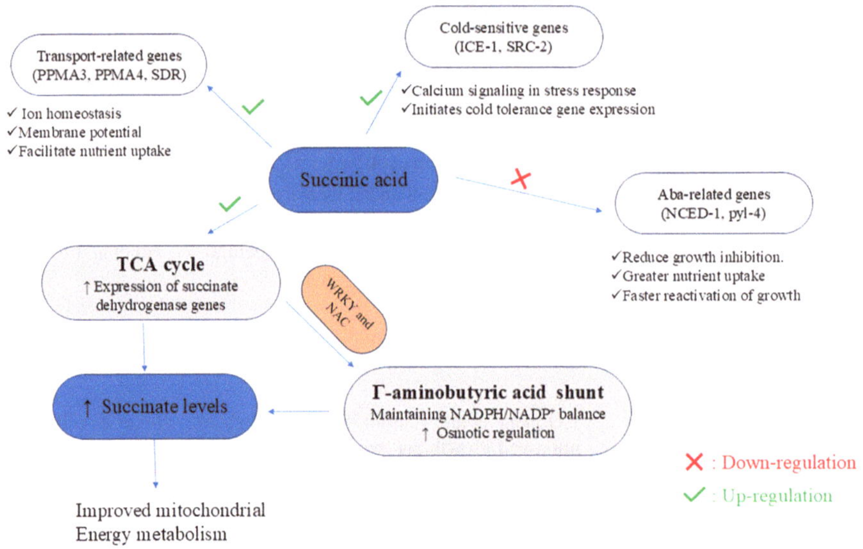

Fig. 17.4 Molecular regulatory network of succinic acid (SUA) in plant stress tolerance. In abiotic stress, succinic acid regulates several gene networks, coordinating transcription and hormone action with energy metabolism. Increased levels of SUA enhance redox balance and mitochondrial energy metabolism. They are produced by the TCA cycle and the GABA shunt. SUA strengthens ion homeostasis, nutrient uptake, and antioxidant defense by controlling genes involved with transport (*PPMA3*, *PPMA4*, *SDR*) through transcription. SUA also initiates cold-induced genes (*ICE-1*, *SRC-2*) and calcium signaling, providing improved tolerance to low temperature. SUA, however, represses ABA-related genes (*NCED-1*, *PYL-4*), suppressing ABA accumulation, arresting growth inhibition, and accelerating stress recovery

tissues, in *Leymus chinensis*. This deposition is accountable for root protection and cellular stability, allowing growth to continue under stressed conditions. LcSUCLAα and LcSUCLAβ are two mitochondrial genes that regulate succinate synthesis through the production of subunits of succinyl-CoA synthetase, an enzyme in the TCA cycle, which converts succinyl-CoA to succinate. Overexpression in rice raises succinic acid and improves $NaHCO_3$ tolerance and signifies the key function of succinic acid in stress adaptation (Wu et al. 2025). A mechanistic summary of the molecular networks by which succinic acid operates is given in Fig. 17.4, along with information on how it regulates gene expression involved in ion transport, tolerance to cold stress, mitochondrial metabolism, and osmotic adjustment.

17.6 Exogenous Application Strategies and Practical Impacts

The application of SUA, typically through foliar sprays, seed priming, or soil/hydroponic solutions, has been shown to induce physiological as well as biochemical responses, albeit varying in efficacy with plant species, stress type, and concentration.

Foliar Application and Disease Resistance Exogenous treatment of 0.1 mM SUA was tried in winter wheat (*Triticum aestivum*) against *Blumeria graminis* f. sp. tritici infection (powdery mildew). Though the treatment did not enhance the physiology of the plant or the yield significantly, it induced defense reactions. PAL and APX activities increased in flag leaves (Nyzhnyk et al. 2025).

Rooting and Vegetative Development Succinic acid treatment in quince cuttings significantly increased root number and length in pear, which is a sign of its suitability for clonal propagation and nursery cultivation (Zacepina 2023).

Hydroponic Systems and Leafy Vegetables For hydroponically cultured lettuce ('Robin'), SUA was supplied both at the time of seed inoculation and by drip irrigation throughout the entire growth phase. The treatment accelerated seedling development, shortened germination time, and enhanced photosynthetic efficiency. Leaf area was greatly improved with higher productivity, whereas dry matter accumulation rose by 17% with seed inoculation and by 50% with the addition of hydroponic solution (Knyazeva et al. 2021).

Drought Mitigation and Combined Treatments SUA activity can be enhanced when applied in conjunction with other bioactive compounds. In flower kale, SA and SUA seed priming significantly alleviated drought stress by increasing proline buildup (a plant osmoprotectant) and water-use efficiency (Kiliç 2023). This suggests that succinic acid can act synergistically with other signaling molecules to optimize stress adaptation. Table 17.2 summarizes reported case studies highlighting the effects of exogenous succinic acid application on stress-prone crops.

Table 17.2 Case studies of exogenous succinic acid application in stress-prone crops

Crop	Method	Concentration	Observed effects	References
Winter wheat (*Triticum aestivum*)	Foliar spray	0.1 mM	Increased activities of phenylalanine ammonia-lyase (PAL), ascorbate peroxidase (APX) in flag leaves; limited impact on growth, productivity under powdery mildew infection	Nyzhnyk et al. (2025)
Pear (*Pyrus communis*)	Root dipping/ cutting treatment	200 mg/L	Significant improvement in root system architecture, higher root number, root length in quince cuttings	Zacepina (2023)
Lettuce (*Lactuca sativa*)	Seed inoculation and drip irrigation	2.5 mM	Accelerated development, increased photosynthesis, leaf area; dry matter accumulation (seed inoculation +17%; hydroponic supplementation +50%)	Knyazeva et al. (2021)
Flowering kale (*Brassica oleracea*)	Seed priming (combined with salicylic acid)	0.5, 1.0, 2.0 mM	Alleviated drought stress via enhanced proline accumulation, water-use efficiency	Kiliç (2023)

17.7 Future Directions and Agricultural Potential

Succinic acid is widely recognized as a platform chemical in industrial biotechnology, yet its potential in agriculture is only beginning to be fully explored. In addition to its role as a major metabolite, succinate's physiological, ecological, and biochemical functions provide it with a strong candidacy to increase crop resistance, yield, and soil quality, important imperatives in climate-smart and sustainable agricultural systems.

Crop Stress Tolerance and Productivity Exogenously applied SA has also had beneficial effects on plant tolerance to abiotic stressors. In flowering kale, SUA treatments improved drought resilience by modulating antioxidant enzyme activity and increasing osmolyte accumulation, processes essential for maintaining cellular stability during water deficit (Kiliç 2023). Similarly, in winter wheat, SUA was linked to the activation of protective stress responses, resulting in enhanced resistance to *Blumeria graminis* (powdery mildew) (Nyzhnyk et al. 2025). In lettuce (*Lactuca sativa*), SUA application in hydroponics accelerated plant development, increased photosynthetic performance, and induced biomass growth and leaf area development, which indicated its potential to improve yield as well as quality in managed agroecosystems (Knyazeva et al. 2021).

Soil Health and Environmental Remediation As root and microbial exudates, SUA participates in heavy metal detoxification, reducing their mobility and toxicity in the rhizosphere. The activity is engaged in environmental remediation towards cleaner and healthier farming systems (Ansari et al. 2023). Through enhanced soil microbial functions and nutrient cycling, succinate creates rhizosphere conditions that enhance plant growth and soil stability.

Bio-Based Succinate Production and Circular Agriculture Incorporating succinic acid into agriculture is in line with the principles of the circular bioeconomy. Microbial biotechnology has made it possible to produce SUA from renewable biomass and crop residues and replace petroleum-based production. Engineered microbes such as *Corynebacterium glutamicum* and *Escherichia coli* have been engineered to become more proficient at lignocellulosic hydrolysate and galactose-based feedstock conversion to high-titer succinic acid. For example, Lee et al. (2024a, b) demonstrated high-efficiency bio-based succinate production from softwood by genome-edited *C. glutamicum* with petrochemical comparable yields. Zhu et al. (2020) increased the productivity of succinate in *E. coli* through targeted metabolic engineering of galactose pathway metabolism.

Value-Added By-Products for Agriculture Another opportunity lies in the succinate fermentation residual by-products. Microbial platform fermentations such as those with *Yarrowia lipolytica* yield nutrient-rich effluents that have been proved to enhance soil fertility, stimulate microbial activity, and enhance plant growth (Xiao et al. 2020).

17.8 Conclusion

Succinic acid is now well known to be an important part of plant metabolism, signaling, and stress adaptation. Positioned at the crossroads of the TCA cycle and multiple regulatory pathways, SUA fulfills a dual role: as a core metabolic intermediate sustaining energy metabolism and as a signaling molecule orchestrating physiological, molecular, and immune responses. This multifunctional character allows succinate to influence diverse processes that determine plant survival and resilience under adverse conditions.

Under abiotic and biotic stress, succinic acid enhances plant tolerance by reinforcing antioxidant defenses, stabilizing cellular membranes, maintaining ionic balance, and activating defense-related genes. It contributes to osmoprotection, supports ROS signaling for adaptive reprogramming, and interacts with phytohormonal pathways, including salicylic acid, jasmonic acid, ethylene, and ABA, thereby fine-tuning the balance between growth and defense. Beyond intrinsic metabolism, succinate also facilitates microbial crosstalk in the rhizosphere, stimulating beneficial plant–microbe associations and activating ISR. These ecological interactions further reinforce plant health and nutrient uptake.

The cumulative evidence positions succinic acid as a promising biotechnological tool for agriculture. Its capacity to promote crop growth, improve stress tolerance, and reduce reliance on synthetic inputs highlights its potential in the design of sustainable, climate-resilient farming systems. Advances in bio-based succinate production from renewable feedstocks and agricultural residues further underscore its role in the circular bioeconomy, where succinate functions both as a plant stimulant and as a platform chemical for value-added applications.

The targeted use of succinic acid, through exogenous applications, integrated microbial systems, and bio-based formulations, offers a practical pathway to environmentally friendly agriculture. By uniting metabolic efficiency with adaptive regulation, succinate exemplifies a next-generation strategy to address the dual challenges of global food security and sustainable resource management.

References

Acevedo RM, Maiale SJ, Pessino SC et al (2013) A succinate dehydrogenase flavoprotein subunit-like transcript is upregulated in Ilex paraguariensis leaves in response to water deficit and abscisic acid. Plant Physiol Biochem 65:48–54. https://doi.org/10.1016/j.plaphy.2012.12.016

Allan WL, Simpson JP, Clark SM, Shelp BJ (2008) Hydroxybutyrate accumulation in Arabidopsis and tobacco plants is a general response to abiotic stress: putative regulation by redox balance and glyoxylate reductase isoforms. J Exp Bot 59:2555–2564. https://doi.org/10.1093/jxb/ern122

Allison MJ, Robinson IM (1970) Biosynthesis of α-Ketoglutarate by the Reductive Carboxylation of Succinate in Bacteroides ruminicola. J Bacteriol 104:50–56. https://doi.org/10.1128/JB.104.1.50-56.1970

Al-Quraan NA, Al-Share AT (2016) Characterization of the γ-aminobutyric acid shunt pathway and oxidative damage in Arabidopsis thaliana pop 2 mutants under various abiotic stresses. Biol Plant 60:132–138. https://doi.org/10.1007/s10535-015-0563-5

Anadón A, Castellano V, Martínez-Larrañaga MR (2014) Biomarkers of drug toxicity. In: Biomarkers in toxicology, pp 593–607. https://doi.org/10.1016/B978-0-12-404630-6.00034-8

Ansari M, Devi BM, Sarkar A et al (2023) Microbial exudates as biostimulants: role in plant growth promotion and stress mitigation. J Xenobiot 13:572–603. https://doi.org/10.3390/jox13040037

Ashraf U, Anjum SA, Naseer S et al (2024) Gamma amino butyric acid (GABA) application modulated the morpho-physiological and yield traits of fragrant rice under well-watered and drought conditions. BMC Plant Biol 24:1–12. https://doi.org/10.1186/s12870-024-05272-5

Ashrafi M, Azimi-Moqadam M-R, Moradi P et al (2018) Effect of drought stress on metabolite adjustments in drought tolerant and sensitive thyme. Plant Physiol Biochem 132:391–399. https://doi.org/10.1016/j.plaphy.2018.09.009

Benaffari W, Boutasknit A, Anli M et al (2022) The native Arbuscular mycorrhizal fungi and vermicompost-based organic amendments enhance soil fertility, growth performance, and the drought stress tolerance of Quinoa. Plants 11:393. https://doi.org/10.3390/plants11030393

Bilal S, Khan T, Asaf S et al (2023) Silicon-induced morphological, biochemical and molecular regulation in Phoenix dactylifera L. under low-temperature stress. Int J Mol Sci 24:6036. https://doi.org/10.3390/ijms24076036

Boutasknit A, Baslam M, Ait-El-Mokhtar M et al (2021) Assemblage of indigenous arbuscular mycorrhizal fungi and green waste compost enhance drought stress tolerance in carob (Ceratonia siliqua L.) trees. Sci Rep 11:1–23. https://doi.org/10.1038/s41598-021-02018-3

Buchanan BB, Evans MC (1965) The synthesis of alpha-ketoglutarate from succinate and carbon dioxide by a subcellular preparation of a photosynthetic bacterium. Proc Natl Acad Sci USA 54: 1212–1218. https://doi.org/10.1073/PNAS.54.4.1212

Burch JS, Marcero JR, Maschek JA et al (2018) Glutamine via α-ketoglutarate dehydrogenase provides succinyl-CoA for heme synthesis during erythropoiesis. Blood 132:987–998. https://doi.org/10.1182/BLOOD-2018-01-829036

Camarasa C, Grivet JP, Dequin S (2003) Investigation by 13C-NMR and tricarboxylic acid (TCA) deletion mutant analysis of pathways of succinate formation in Saccharomyces cerevisiae during anaerobic fermentation. Microbiology (N Y) 149:2669–2678. https://doi.org/10.1099/MIC.0.26007-0/CITE/REFWORKS

Cao Y, Cao Y, Lin X (2011) Metabolically engineered Escherichia coli for biotechnological production of four-carbon 1,4-dicarboxylic acids. J Ind Microbiol Biotechnol 38:649–656. https://doi.org/10.1007/S10295-010-0913-4

Chen Y, Zhu Y, Xin J et al (2021) Succinic acid inhibits photosynthesis of Microcystis aeruginosa via damaging PSII oxygen-evolving complex and reaction center. Environ Sci Pollut Res 28: 58470–58479. https://doi.org/10.1007/s11356-021-14811-8

Chen Z, Dong Y, Huang X (2022) Plant responses to UV-B radiation: signaling, acclimation and stress tolerance. Stress Biol 2:51. https://doi.org/10.1007/s44154-022-00076-9

Chen X, Wu H, Chen Y et al (2023) Recent advancements and strategies of improving CO2 utilization efficiency in bio-succinic acid production. Fermentation 9:967. https://doi.org/10.3390/fermentation9110967

Dabravolski SA, Isayenkov SV (2023) The role of the γ-Aminobutyric Acid (GABA) in plant salt stress tolerance. Horticulturae 9:230. https://doi.org/10.3390/horticulturae9020230

del Socorro Sánchez Correa M, el Rocío Reyero Saavedra M, Antonio Estrella Parra E et al (2023) Ultraviolet radiation and its effects on plants. In: Abiotic stress in plants – adaptations to climate change. IntechOpen

Doncheva S, Georgieva K, Vassileva V et al (2005) Effects of succinate on manganese toxicity in pea plants. J Plant Nutr 28:47–62. https://doi.org/10.1081/PLN-200042161

Dong X, Yang Z, Liu Z et al (2022) Metabonomic analysis provides new insights into the response of Zhikong Scallop (Chlamys farreri) to heat stress by improving energy metabolism and antioxidant capacity. Antioxidants 11:1084. https://doi.org/10.3390/antiox11061084

Dos Reis SP, Lima AM, De Souza CRB (2012) Recent molecular advances on downstream plant responses to abiotic stress. Int J Mol Sci 13:8628–8647. https://doi.org/10.3390/ijms13078628
Du H, Wang Z, Yu W, Huang B (2012) Metabolic responses of hybrid bermudagrass to short-term and long-term drought stress. J Am Soc Hortic Sci 137:411–420. https://doi.org/10.21273/jashs.137.6.411
Duan C, Wu S, Sang Y et al (2019) Exogenous succinic acid mediates responses of *Larix olgensis* A. Henry to cadmium stress. Int J Phytoremediat 21:742–751. https://doi.org/10.1080/15226514.2018.1556593
Eckardt NA, Ainsworth EA, Bahuguna RN et al (2023) Climate change challenges, plant science solutions. Plant Cell 35:24–66. https://doi.org/10.1093/plcell/koac303
Eprintsev AT, Fedorin DN, Igamberdiev AU (2023) Light-dependent expression and promoter methylation of the genes encoding succinate dehydrogenase, fumarase, and NAD-malate dehydrogenase in maize (*Zea mays* L.) leaves. Plants 24:68. https://doi.org/10.3390/ijms241210211
Eprintsev AT, Fedorin DN, Caro OJF (2025) Functioning of succinate metabolism enzymes in wheat leaves under salt stress. Russ J Plant Physiol 72:37. https://doi.org/10.1134/S1021443724608644
Ez-Zoubi A, Ez Zoubi Y, Bentata F et al (2023) Preparation and characterization of a biopesticide based on Artemisia herba-alba essential oil encapsulated with succinic acid-modified beta-cyclodextrin. J Chem 2023:1–8. https://doi.org/10.1155/2023/3830819
Fedorin DN, Eprintsev AT, Florez Caro OJ, Igamberdiev AU (2022) Effect of salt stress on the activity, expression, and promoter methylation of succinate dehydrogenase and succinic semialdehyde dehydrogenase in maize (*Zea mays* L.) leaves. Plants 12:68. https://doi.org/10.3390/plants12010068
Fedorin DN, Novoseltseva DP, Eprintsev AT (2024) The role of the methyl status of adeniine in the regulation of the expression of the gene of succinic semialdehyde dehydrogenase in wheat leaves under salinity. J Stress Physiol Biochem 20:105–110
Gago J, de Daloso DM, Figueroa CM et al (2016) Relationships of leaf net photosynthesis, stomatal conductance, and mesophyll conductance to primary metabolism: a multispecies meta-analysis approach. Plant Physiol 171:265–279. https://doi.org/10.1104/pp.15.01660
Gao L-L, Dong Y, Cun Z et al (2025) Moderate shading elicits succinic acid accumulation aligning with simultaneous expression of genes involved in TCA cycle and photosynthetic pathway in a medicinal plant Pinellia ternata. Plant Physiol Biochem 224:109911. https://doi.org/10.1016/j.plaphy.2025.109911
Gill AJ (2012) Succinate dehydrogenase (SDH) and mitochondrial driven neoplasia. Pathology 44:285–292. https://doi.org/10.1097/PAT.0B013E3283539932
Gill SS, Tuteja N (2010) Reactive oxygen species and antioxidant machinery in abiotic stress tolerance in crop plants. Plant Physiol Biochem 48:909–930. https://doi.org/10.1016/j.plaphy.2010.08.016
Gleason C, Huang S, Thatcher LF et al (2011) Mitochondrial complex II has a key role in mitochondrial-derived reactive oxygen species influence on plant stress gene regulation and defense. Proc Natl Acad Sci USA 108:10768–10773. https://doi.org/10.1073/pnas.1016060108
Guo R, Yang Z, Li F et al (2015) Comparative metabolic responses and adaptive strategies of wheat (Triticum aestivum) to salt and alkali stress. BMC Plant Biol 15:170. https://doi.org/10.1186/s12870-015-0546-x
Hamanishi ET, Barchet GL, Dauwe R et al (2015) Poplar trees reconfigure the transcriptome and metabolome in response to drought in a genotype- and time-of-day-dependent manner. BMC Genomics 16:329. https://doi.org/10.1186/s12864-015-1535-z
Hayat F, Khan U, Li J et al (2023) γ Aminobutyric Acid (GABA): a key player in alleviating abiotic stress resistance in horticultural crops: current insights and future directions. Horticulturae 9:647. https://doi.org/10.3390/horticulturae9060647

Hijaz F, Killiny N (2019) Exogenous GABA is quickly metabolized to succinic acid and fed into the plant TCA cycle. Plant Signal Behav 14:e1573096. https://doi.org/10.1080/15592324.2019.1573096

Huang G, Guo G, Yao S et al (2016) Organic acids, amino acids compositions in the root exudates and Cu-accumulation in castor (*Ricinus communis* L.) Under Cu stress. Int J Phytoremediat 18:33–40. https://doi.org/10.1080/15226514.2015.1058333

Huergo LF, Dixon R (2015) The emergence of 2-oxoglutarate as a master regulator metabolite. Microbiol Mol Biol Rev 79:419. https://doi.org/10.1128/MMBR.00038-15

Ikan C, Ben-Laouane R, Ouhaddou R et al (2023) Interactions between arbuscular mycorrhizal fungus and indigenous compost improve salt stress tolerance in wheat (Triticum durum). S Afr J Bot 158:417–428. https://doi.org/10.1016/j.sajb.2023.05.038

Jagodzik P, Tajdel-Zielinska M, Ciesla A et al (2018) Mitogen-activated protein kinase cascades in plant hormone signaling. Front Plant Sci 9. https://doi.org/10.3389/fpls.2018.01387

Jan R, Asif S, Asaf S et al (2024) Gamma-aminobutyric acid treatment promotes resistance against Sogatella furcifera in rice. Front Plant Sci 15. https://doi.org/10.3389/fpls.2024.1419999

Jardim-Messeder D, Caverzan A, Rauber R et al (2015) Succinate dehydrogenase (mitochondrial complex II) is a source of reactive oxygen species in plants and regulates development and stress responses. New Phytol 208:776–789. https://doi.org/10.1111/nph.13515

Jin F, Hu Q, Zhao Y et al (2022) Enhancing quinoa growth under severe saline-alkali stress by phosphate solubilizing microorganism Penicillium funicuiosum P1. PLoS One 17:e0273459. https://doi.org/10.1371/journal.pone.0273459

Katayama N, Takeya M, Osanai T (2019) Biochemical characterisation of fumarase C from a unicellular cyanobacterium demonstrating its substrate affinity, altered by an amino acid substitution. Sci Rep 9:1–11. https://doi.org/10.1038/S41598-019-47025-7

Kaur R, Kaur R, Sharma A et al (2018) Microbial production of dicarboxylic acids from edible plants and milk using GC-MS. J Anal Sci Technol 9:21. https://doi.org/10.1186/s40543-018-0154-0

Khan N, Bano A, Babar MA (2019) Metabolic and physiological changes induced by plant growth regulators and plant growth promoting rhizobacteria and their impact on drought tolerance in Cicer arietinum L. PLoS One 14:e0213040. https://doi.org/10.1371/journal.pone.0213040

Kiliç T (2023) Seed treatments with salicylic and succinic acid to mitigate drought stress in flowering kale cv. "Red Pigeon F1". Sci Hortic 313:111939. https://doi.org/10.1016/j.scienta.2023.111939

Knyazeva IV, Vershinina OV, Gudimo VV et al (2021) The effect of succinic acid on the productivity of Lactuca sativa L. in artificial agroecosystems. Agron Res 19

Kocaman A (2023) Combined interactions of amino acids and organic acids in heavy metal binding in plants. Plant Signal Behav 18. https://doi.org/10.1080/15592324.2022.2064072

Kolupaev YE, Yastreb TO, Karpets YV, Miroshnichenko NN (2011) Influence of salicylic and succinic acids on antioxidant enzymes activity, heat resistance and productivity of Panicum miliaceum L. J Stress Physiol Biochem 7:154–163

Kolupaev YE, Yastreb TO, Shvidenko NV, Karpets YV (2012) Induction of heat resistance of wheat coleoptiles by salicylic and succinic acids: connection of the effect with the generation and neutralization of reactive oxygen species. Appl Biochem Microbiol 48:500–505. https://doi.org/10.1134/S0003683812050055

Kots SY, Rybachenko LI, Kukol KP et al (2023) Peculiarities of soybean-rhizobial systems subject to different levels of water supply fol-lowing treatment with succinic acid and epibrassinolide. Biosyst Divers 31:484–492. https://doi.org/10.15421/012357

Kumar P, Kumar A, Agrawal A et al (2024) Osmolytes as stress sensors in plants: acclimatizing plants under stress conditions. In: Cellular osmolytes. Springer, Singapore, pp 199–210

Kumar P, Pandey S, Pati PK (2025) Interaction between pathogenesis-related (PR) proteins and phytohormone signaling pathways in conferring disease tolerance in plants. Physiol Plant 177. https://doi.org/10.1111/ppl.70174

Lee D, Cho EJ, Nguyen D et al (2024a) Succinic acid production from softwood with genome-edited *Corynebacterium glutamicum* using the CRISPR-Cpf1 system. Biotechnol J 19. https://doi.org/10.1002/biot.202300309

Lee S-M, Thapa Magar R, Jung MK et al (2024b) Rhizobacterial syntrophy between a helper and a beneficiary promotes tomato plant health. ISME J 18. https://doi.org/10.1093/ismejo/wrae120

Li Z, Yu J, Peng Y, Huang B (2017) Metabolic pathways regulated by abscisic acid, salicylic acid and γ-aminobutyric acid in association with improved drought tolerance in creeping bentgrass (*Agrostis stolonifera*). Physiol Plant 159:42–58. https://doi.org/10.1111/ppl.12483

Li H, Tang X, Yang X, Zhang H (2021) Comprehensive transcriptome and metabolome profiling reveal metabolic mechanisms of Nitraria sibirica Pall. to salt stress. Sci Rep 11:1–19. https://doi.org/10.1038/s41598-021-92317-6

Liu L, Si L, Zhang L et al (2024) Metabolomics and transcriptomics analysis revealed the response mechanism of alfalfa to combined cold and saline-alkali stress. Plant J 119:1900–1919. https://doi.org/10.1111/tpj.16896

Lynen F, Ochoa S (1953) Enzymes of fatty acid metabolism. Biochim Biophys Acta 12:299–314. https://doi.org/10.1016/0006-3002(53)90149-8

Mareri L, Parrotta L, Cai G (2022) Environmental stress and plants. Int J Mol Sci 23:5416. https://doi.org/10.3390/ijms23105416

McKinlay JB, Vieille C, Zeikus JG (2007) Prospects for a bio-based succinate industry. Appl Microbiol Biotechnol 76:727–740. https://doi.org/10.1007/S00253-007-1057-Y/METRICS

Mishra S, Roychowdhury R, Ray S et al (2024) Salicylic acid (SA)-mediated plant immunity against biotic stresses: an insight on molecular components and signaling mechanism. Plant Stress 11:100427. https://doi.org/10.1016/j.stress.2024.100427

Mitrea L, Teleky BE, Nemes SA et al (2024) Succinic acid – a run-through of the latest perspectives of production from renewable biomass. Heliyon 10:e25551. https://doi.org/10.1016/J.HELIYON.2024.E25551

Mumtaz S, Anas M, Javed S et al (2025) Mitigating cadmium stress in rice (Oryza sativa L.) using succinic and oxalic acids with focus on cellular integrity and antioxidant responses. Plant Physiol Biochem 220:109548. https://doi.org/10.1016/j.plaphy.2025.109548

Neuroscience T (2021) Mechanisms and opportunities for. Plants 10:1939. https://doi.org/10.3390/plants10091939

Nghiem NP, Kleff S, Schwegmann S (2017) Succinic acid: technology development and commercialization. Fermentation 3(2):26. https://doi.org/10.3390/FERMENTATION3020026

Nyzhnyk T, Kiedrzyński M, Kiedrzyńska E, Kots S (2025) Salicylic and succinic acids as inducers of phytoimmunity in winter wheat for the management of powdery mildew (*Blumeria graminis* (DC) Speer f. sp. tritici). BMC Plant Biol 25:376. https://doi.org/10.1186/s12870-025-06403-2

Ouhaddou R, Ech-chatir L, Anli M et al (2023) Secondary metabolites, osmolytes and antioxidant activity as the main attributes enhanced by biostimulants for growth and resilience of lettuce to drought stress. Gesunde Pflanzen. https://doi.org/10.1007/s10343-022-00827-8

Oyedotun KS, Lemire BD (2004) The quaternary structure of the Saccharomyces cerevisiae succinate dehydrogenase: homology modeling, cofactor docking, and molecular dynamics simulation studies. J Biol Chem 279:9424–9431. https://doi.org/10.1074/JBC.M311876200/ASSET/63FECCF0-A79A-411D-83A0-2E9A11F116EA/MAIN.ASSETS/GR7.JPG

Piechowiak T, Balawejder M (2025) Succinic acid treatment enhances energy metabolism and antioxidant biosynthesis in radish sprouts. J Biotechnol 404:144–151. https://doi.org/10.1016/j.jbiotec.2025.04.017

Pucher GW, Vickery HB (1941) Succinic acid as a metabolite in plant tissues. Plant Physiol 16: 771–783. https://doi.org/10.1104/pp.16.4.771

Renault H, Roussel V, El Amrani A et al (2010) The Arabidopsis pop2-1mutant reveals the involvement of GABA transaminase in salt stress tolerance. BMC Plant Biol 10:20. https://doi.org/10.1186/1471-2229-10-20

Rezayian M, Niknam V, Arabloo M (2023) Iron nanoparticle regulate succinate dehydrogenase activity in canola plants under drought stress. Sci Rep 13:1–14. https://doi.org/10.1038/S41598-023-36105-4

Rutter J, Winge DR, Schiffman JD (2010) Succinate dehydrogenase – assembly, regulation and role in human disease. Mitochondrion 10:393. https://doi.org/10.1016/J.MITO.2010.03.001

Ryan DG, O'Neill LAJ (2020) Krebs cycle reborn in macrophage immunometabolism. Annu Rev Immunol 38:289–313. https://doi.org/10.1146/ANNUREV-IMMUNOL-081619-104850/CITE/REFWORKS

Saavedra T, Gama F, Rodrigues MA et al (2022) Effects of foliar application of organic acids on strawberry plants. Plant Physiol Biochem 188:12–20. https://doi.org/10.1016/j.plaphy.2022.08.004

Sharma M, Saleh D, Charron J-B, Jabaji S (2020) A crosstalk between brachypodium root exudates, organic acids, and Bacillus velezensis B26, a growth promoting bacterium. Front Microbiol 11. https://doi.org/10.3389/fmicb.2020.575578

Shi C, Liu H (2021) How plants protect themselves from ultraviolet-B radiation stress. Plant Physiol 187:1096–1103. https://doi.org/10.1093/plphys/kiab245

Shu P, Li Y, Sheng J, Shen L (2024) Recent advances in dissecting the function of ethylene in interaction between host and pathogen. J Agric Food Chem 72:4552–4563. https://doi.org/10.1021/acs.jafc.3c07978

Song J, Zhang H, Duan C, Cui X (2018) Exogenous application of succinic acid enhances tolerance of Larix olgensis seedling to lead stress. J For Res (Harbin) 29:1497–1505

Soussani FE, Boutasknit A, Ben-laouane R, Benkirane R (2023) Arbuscular mycorrhizal fungi and compost-based biostimulants enhance fitness, physiological responses, yield, and quality traits of drought-stressed tomato plants. Plants 12:1856

Studart-Guimarães C, Gibon Y, Frankel N et al (2005) Identification and characterisation of the α and β subunits of succinyl CoA ligase of tomato. Plant Mol Biol 59:781–791. https://doi.org/10.1007/S11103-005-1004-1/METRICS

Studart-Guimarães C, Fait A, Nunes-Nesi A et al (2007) Reduced expression of Succinyl-Coenzyme A ligase can be compensated for by up-regulation of the γ-Aminobutyrate shunt in illuminated tomato leaves. Plant Physiol 145:626. https://doi.org/10.1104/PP.107.103101

Tretter L, Patocs A, Chinopoulos C (2016) Succinate, an intermediate in metabolism, signal transduction, ROS, hypoxia, and tumorigenesis. Biochim Biophys Acta Bioenerg 1857:1086–1101. https://doi.org/10.1016/J.BBABIO.2016.03.012

Tuteja N, Sopory SK (2008) Chemical signaling under abiotic stress environment in plants. Plant Signal Behav 3:525–536. https://doi.org/10.4161/psb.3.8.6186

Wen T, Yuan J, He X et al (2020) Enrichment of beneficial cucumber rhizosphere microbes mediated by organic acid secretion. Hortic Res 7:154. https://doi.org/10.1038/s41438-020-00380-3

Wu J, Sun M, Pang A et al (2025) Succinic acid synthesis regulated by succinyl-coenzyme A ligase (SUCLA) plays an important role in root response to alkaline salt stress in Leymus chinensis. Plant Physiol Biochem 220:109485. https://doi.org/10.1016/j.plaphy.2025.109485

Xiao Y, Zhang Z, Wang Y et al (2020) Two-stage crystallization combining direct succinimide synthesis for the recovery of succinic acid from fermentation broth. Front Bioeng Biotechnol 7. https://doi.org/10.3389/fbioe.2019.00471

Yang L, Christakou E, Vang J et al (2017) Overexpression of a C4-dicarboxylate transporter is the key for rerouting citric acid to C4-dicarboxylic acid production in Aspergillus carbonarius. Microb Cell Factories 16. https://doi.org/10.1186/S12934-017-0660-6

Yang W, Li X, Yan H et al (2024) Recruitment of beneficial cucumber rhizosphere microbes mediated by amino acid secretion induced by biocontrol Bacillus subtilis isolate 1JN2. Front Microbiol 15. https://doi.org/10.3389/fmicb.2024.1379566

Yin X, Li J, Dong SH et al (2015) Metabolic engineering in the biotechnological production of organic acids in the tricarboxylic acid cycle of microorganisms: advances and prospects. Biotechnol Adv 33:830–841. https://doi.org/10.1016/J.BIOTECHADV.2015.04.006

Yu C, Wang H-P, Yu X (2022a) The associative induction of succinic acid and hydrogen sulfide for high-producing biomass, astaxanthin and lipids in Haematococcus pluvialis. Bioresour Technol 358:127397. https://doi.org/10.1016/j.biortech.2022.127397

Yu Y, Gui Y, Li Z et al (2022b) Induced systemic resistance for improving plant immunity by beneficial microbes. Plants 11:386. https://doi.org/10.3390/plants11030386

Yu W, Gong F, Zhou X et al (2024) Comparative metabolomics and transcriptome studies of two forms of rhododendron chrysanthum Pall. under UV-B stress. Biology (Basel) 13:211. https://doi.org/10.3390/biology13040211

Yuan D, Wu X, Gong B et al (2023) GABA metabolism, transport and their roles and mechanisms in the regulation of abiotic stress (hypoxia, salt, drought) resistance in plants. Metabolites 13: 347. https://doi.org/10.3390/metabo13030347

Zacepina I (2023) The effect of the plant growth stimulant (succinic acid) and cutting terms on the rooting ability of green pear and quince cuttings. Proc Komi Sci Centre Ural Div Russ Acad Sci 0:21–28. https://doi.org/10.19110/1994-5655-2023-7-21-28

Zhang Y, Fernie AR (2023) The role of TCA cycle enzymes in plants. Adv Biol 7. https://doi.org/10.1002/adbi.202200238

Zhang X, Wang X, Shanmugam KT, Ingram LO (2011) L-malate production by metabolically engineered *Escherichia coli*. Appl Environ Microbiol 77:427–434. https://doi.org/10.1128/AEM.01971-10/SUPPL_FILE/V_9_LOI_SUPPLEMENT_S1_REV.DOC

Zhang W, Yang Q, Wu M et al (2019) Metabolic regulation of organic acid biosynthesis in Actinobacillus succinogenes. Front Bioeng Biotechnol 7:485945

Zhang Y, Cui Y, Cheng Y et al (2021) Succinate accumulation contributes to oxidative stress and iron accumulation in pentylenetetrazol-induced epileptogenesis and kainic acid-induced seizure. Neurochem Int 149:105123. https://doi.org/10.1016/j.neuint.2021.105123

Zhang X, Zhang F, Li Z et al (2023) Comparative transcriptome analysis of Cynanchum thesioides under drought stress reveals candidate genes involved in succinic acid biosynthesis. J Plant Biol 66:283–295. https://doi.org/10.1007/s12374-021-09301-w

Zhang F, Sun M, Li D et al (2024) Metabolomic analysis of Elymus sibiricus exposed to UV-B radiation stress. Molecules 29:5133. https://doi.org/10.3390/molecules29215133

Zhi X, Bian X, Yu J et al (2024) Comparative metabolomics analysis of tolerant and sensitive genotypes of rapeseed (Brassica napus L.) seedlings under drought stress. Agric Water Manag 296:108797. https://doi.org/10.1016/j.agwat.2024.108797

Zhixin N, Xiaodong L, Lina S, Tieheng S (2013) Dynamics of three organic acids (malic, acetic and succinic acid) in sunflower exposed to cadmium and lead. Int J Phytoremediat 15:690–702. https://doi.org/10.1080/15226514.2012.723066

Zhu L-W, Tang Y-J (2017) Current advances of succinate biosynthesis in metabolically engineered Escherichia coli. Biotechnol Adv 35:1040–1048. https://doi.org/10.1016/j.biotechadv.2017.09.007

Zhu F, San K, Bennett GN (2020) Improved succinate production from galactose-rich feedstocks by engineered *Escherichia coli* under anaerobic conditions. Biotechnol Bioeng 117:1082–1091. https://doi.org/10.1002/bit.27254

Zribi I, Ghorbel M, Brini F (2024) Pathogenesis-related proteins and plant defense response. In: Defense-related proteins in plants. Elsevier, pp 53–97

Chapter 18
Tartaric Acid: Contributing to Plant Stress Adaptation

Affan Afzal, Muhammad Nauman Khan, Abdul Razzaq, Tabassum Yaseen, Alevcan Kaplan, Hansa Gul, Shabab Hussain, Baber Ali, Mohsin Ali, and Arafat Abdel Hamed Abdel Latef

Abstract Tartaric acid (TA), an organic acid that occurs naturally in plants, is essential for plant metabolism and stress adaptation. It is involved in several physiological functions, including antioxidant defense, osmoregulation, and signaling pathways that help plants endure biotic and abiotic stressors. Tartaric acid helps maintain redox equilibrium by scavenging reactive oxygen species (ROS) and modulating enzymatic antioxidant activity in the face of environmental stressors such as dehydration, salt, severe temperatures, and heavy metal toxicity. It also contributes to osmotic regulation by modifying cellular turgor pressure, which improves tolerance to drought and salinity. Tartaric acid also interacts with secondary metabolites and phytohormones that influence plant development and defense processes. Furthermore, its action in metal chelation helps in the detoxification of heavy metal stress, reducing their toxicity in plant tissue. The control of tartaric acid production and accumulation varies among plant species and is regulated by environmental variables, implying that it has the potential to improve crop stress tolerance through metabolic engineering or exogenous applications. This chapter emphasizes the importance of tartaric acid for plant stress tolerance, focusing on its

A. Afzal
Department of Microbiology, Faculty of Veterinary and Animal Sciences, The Islamia University of Bahawalpur, Bahawalpur, Pakistan

M. N. Khan (✉)
Department of Botany, University of Chakwal, Punjab, Pakistan
e-mail: nomiflora@uop.edu.pk

A. Razzaq
Qarshi Herb Research Center, Qarshi Industries (Pvt.) Ltd., Hattar, Haripur, Pakistan

Qarshi University, Lahore, Punjab, Pakistan

T. Yaseen
Department of Botany, Bacha Khan University Charsadda, Khyber Pakhtunkhwa, Pakistan

A. Kaplan
Department of Crop and Animal Production, Sason Vocational School, Batman University, Batman, Turkey

A. A. H. Abdel Latef (ed.), *Harnessing Vitamins and Organic Acids for Plant Stress Resilience*, Sustainability Sciences in Asia and Africa,
https://doi.org/10.1007/978-981-95-8593-9_18

multiple activities and its prospective use in sustainable agriculture and crop enhancement techniques.

Keywords Agriculture sustainability · Plant stress adaptations · ROS · Signaling pathways · Tartaric acid

18.1 Introduction

18.1.1 Importance of Plant Stress Adaptation

Plants have been continually evolving for many years as stress has increased due to human activities polluting the environment and limiting biodiversity (Nawaz et al. 2023). As plants are rooted in one place, they have to cope with daily challenges like drought, high or low temperatures, salty conditions, and attacks from pathogens that can seriously affect their health and productivity. Consequently, they have designed complex routes to recognize different factors, initiate the right signaling pathways, and respond and react appropriately by altering the levels of many genes at transcriptional, post-transcriptional, and epigenetic levels. This enables them to cope with problems such as drought, heat, nutrient deficiency, and disease (Ijaz et al. 2024). As a result of the salt content, the plants change in their shape, their functions, their chemistry, and their genes. Ion imbalance leads to ion toxicity and an increase in reactive oxygen species (ROS). It appears that ROS are part of a messaging system that enables leaves to protect themselves from salt stress (AbdElgawad et al. 2016; Wahid et al. 2025; Azooz et al. 2026). Plants need to persist in the early stages, but such mechanisms are irrelevant for achieving higher yields. Increasing the drought resistance of crops should be based on yield reliability and not just survival. So far, progress in boosting grain yield under drought has centered on root growth,

H. Gul
Department of Zoology, University of Mianwali, Mianwali, Pakistan

S. Hussain
Department of Chemical, Biological, Pharmaceutical and Environmental Sciences, University of Messina, Messina, Italy

B. Ali
School of Science, Western Sydney University, Penrith, NSW, Australia

M. Ali
State Key Laboratory of Microbial Resources, Institute of Microbiology, Chinese Academy of Sciences, Beijing, China

University of Chinese Academy of Sciences, Beijing, China

A. A. H. Abdel Latef
Department of Botany and Microbiology, Faculty of Science, Qena University, Qena, Egypt

plant water and leaf levels, osmotic adaptation and water content in young plants, all of which have little to do with final grain yield (Basu et al. 2016).

Plants that tolerate salty conditions undergo various changes in their structure, functions, and chemistry. These alterations involve expanding the root/canopy relationship, adding chlorophyll, and modifying leaf anatomy so that leaf ion toxicity is prevented, and water balance is maintained to prevent water loss and support photosynthesis (Acosta-Motos et al. 2017). Pollution and global warming, as well as other stressors, pose a threat to both biodiversity and the services offered by ecosystems.

18.1.2 Overview of Plant Responses to Biotic and Abiotic Stressors

Biotic and abiotic environmental stress can significantly limit the growth and productivity of plants and lead to yield losses. Stress in plants is caused by outside factors that can strongly influence the growth, development, or efficiency of plants. Stress often causes plants to change their DNA, cell activity, rate of development, and the amount of products produced (Fahad et al. 2021).

18.1.2.1 Biotic Stress

Viruses, bacteria, fungi, worms, insects, spiders, and weeds cause a lot of biotic stress in plants. Biotic stress causes plants to lose important nutrients, which may eventually lead to their death. Biotic stress can become important when crop losses occur before and after harvest. Although plants do not have adaptive immunity, they are able to resist biotic stress by changing and developing specific strategies. Plants respond to stress with defense mechanisms that are influenced by genes in their DNA. There are hundreds of resistant genes in the plant genome that act against biotic stress. Biotic stress is different from abiotic stress, which happens when plants are confronted with non-living factors such as high salinity, strong sunlight, high temperatures, cold weather, flooding, and drought. The air, water, and soil conditions to which a crop is exposed determine what problems the plant may have and whether it can cope with them. Photosynthesis is often affected by biotic stress, as insects that feed on leaves limit leaf area, and virus infections lead to lower photosynthetic efficiency per leaf area (Gull et al. 2019; Afzal et al. 2026). Over the years, researchers have generated various ways to deal with biotic stress. Biotic stress in plants can be managed by understanding the genetics of the causal agent. Genetically modified plants have successfully reduced the problems caused by living organisms on plants by producing resistant crops (Umar et al. 2021).

18.1.2.2 Abiotic Stresses in Plants

Crop yields are declining worldwide because certain abiotic stress factors, such as temperature fluctuations, water deprivation, and UV radiation, make it difficult for plants to thrive. The study demonstrated the influence of temperature, drought, heat, cold, and other abiotic factors on plant growth and development. At any time when plant tissue is exposed to abiotic stress, it is the non-living factors that affect their metabolism, growth, and development. The presence of these abiotic stress factors in the environment has encouraged plants to develop many ways to tolerate stress. Genes involved in coping with stress are formed because of abiotic stress factors that help plants to stay alive even under difficult conditions. For this reason, the existence of abiotic stress factors in plants usually leads to poor growth and development, but at the same time, plants produce mechanisms to cope with stress and minimize the effects (Fahad et al. 2021; Abdel Latef et al. 2026; Barman et al. 2026).

Abiotic stress can change the way the plant breaks down food, contributing to a decrease in plant productivity. This could be the biggest negative impact, as abiotic stress factors usually only lead to negative outcomes (Hasanuzzaman et al. 2020). New research shows that the molecular mechanisms for plant response to various stress factors operate at multiple levels, combining sensing, signaling, transcription, transcriptional processing, translation, and protein modifications. With this better understanding, people can work towards sustainable agriculture using genetics, chemicals, and microbes (Zhang et al. 2022).

Today's research demonstrates that DNA methylation allows plants to adapt to the challenges from their surroundings. Environmental stress can lead to methylation changes in plant DNA that alter gene expression, control transposable elements (TEs), and thus enhance the plant's ability to adapt and respond to such stresses. Most of these stress-induced changes in DNA methylation are not permanent, but some plant studies have indicated that a kind of "memory" can occur. Due to transient changes in morphology and chemical substances, plants can defend themselves against stress for a short time or a lifetime thanks to a short-term stress memory, whereas children can inherit the memory of long-term stress. Epigenetic alterations resulting from DNA methylation are handed down through generations even if the DNA sequence remains the same, and these changes in plant offspring are linked to the stress of the parents. DNA methylation affects gene expression during drought in multiple ways, implying that it can control gene activity by acting on various regulatory pathways. The methylation status of these plants was more stable during drought than that of plants without drought tolerance (Sun et al. 2022).

As a result of these stresses, the normal functions of plant physiology and biochemistry can change, leading to negative effects on plant development and growth. To build a sustainable ecosystem with plant species that resist multiple stresses, it is important to understand how environmental hazards, such as biotic and abiotic factors, influence plants.

18.2 Brief Introduction to Tartaric Acid and Its Relevance

Tartaric acid (TA) is a dicarboxylic acid with the molecular formula $C_4H_6O_6$ (Fig. 18.1) and is also known as 2,3-dihydroxysuccinic acid or tartaric acid. TA obtained from grapes is naturally colored, but the purified form is crystalline and white (Li et al. 2023).

The acidity of tartaric acid is generally 1.2 to 1.3 times higher than that of citric acid (CA) at the same concentration. Tartaric acid mixes with citric acid at matching concentrations. Tartaric acid mixes with water and ethanol. Most of the TA in grapes is L-tartaric acid, sometimes also called dextro-tartaric acid. The mirror image, the enantiomer, of tartaric acid is known as D-tartaric acid or laevotartaric acid. One type of TA, meso-tartaric acid, also known as mesotartaric acid, is optically inactive. DL-tartaric acid is an optically inactive substance that is produced by mixing the levorotatory and dextrorotatory forms in equal amounts. Organic acids are essential for the maturation of wine, and TA is the most important type of them. The sensitivity of TA content to the effects of weathering during maturation is considered to be relatively low. Better protection against the effects of climate change is associated with a higher TA content in grapes. It is worth noting that climate change can affect the content of different organic acids in grapes (Li et al. 2023).

TA is used worldwide as an excipient for medicines and as a buffering agent. It is also used as an acidifier in food and beverages. In chemistry, it is used as a starting material for the development of new forms of chiral ligands that participate in metal-assisted reactions and develop unique organocatalysts. Astrochemical researchers believe that TA could be a molecule found in the interstellar medium (ISM) (Alonso et al. 2021). Tartrates are used in science and technology for optical, ferroelectric, ferroelastic, and dielectric purposes, among others. These materials are frequently used in transducers and linear and non-linear elements. The presence of pairs of enantiomers means that these materials are used in industry, pharmacy, and medicine for the production of corrosion-resistant refrigerants, light-stabilizing plastic additives, and substances for the electrochemical and fertilizer industries. The zigzag arrangement of tartrate groups with the metallic ions in a single crystalline form allows these crystals to demonstrate notable optical non-linearity. It is also used in the extraction of potassium, germanium, and gallium from minerals (Sachdev et al. 2023).

Although tamarind, grapes, bananas, apples, and many citrus fruits (high in antioxidants) all contain TA, and TA is also found in wine, this compound has

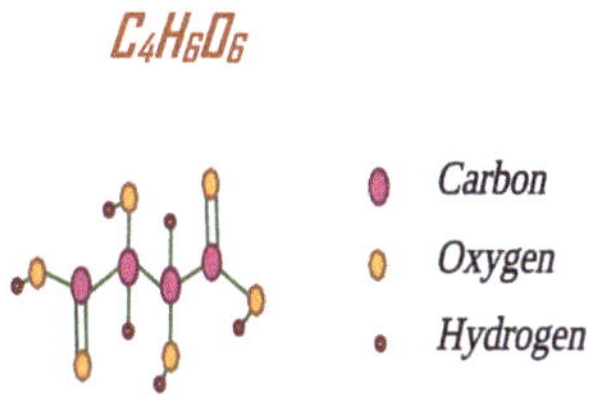

Fig. 18.1 Structure of tartaric acid. (Created by Biorender)

many unexplored uses as an antioxidant. TA helps to scavenge harmful toxins and attaches with other antioxidants to do its job. Its regular intake helps stop kidney stones and it fights cancer. Just because a substance has certain pharmacological actions does not mean we forget that we still need supervision. You should only take 30 milligrams per kilogram of body weight per day, as this can lead to toxicity and problems like acidosis and dental erosion (Jantwal et al. 2022). The acidity, which makes the wine enjoyable, but not strong, refreshing and firm, is largely due to TA. Moreover, it is not usually changed during winemaking. Because of TA, the wine develops a special flavor and at the same time is resistant to spoilage, stable against microbes and suitable for aging. While weather changes normally do not influence how much TA is formed, not enough is understood about where it is made in the grapes and where it is kept. Most of the TA present in plants and fruits is in the form of potassium salt, while only a small amount is present as free TA. While TA is mainly formed at the beginning of leaf and berry growth, biosynthesis is limited in the remaining tissues and fully mature berries (Li et al. 2023). Even though L-TA has a similar structure to several common C4-dicarboxylates, only a few species are known to accumulate it. In cultivated crops, the TA content usually remains unchanged by stress or the type of care (Burbidge et al. 2021).

This chapter deals with the question of how TAs can make plants more resilient when they are confronted with environmental stresses. Originally recognized mainly for its effects on fruit flavor, more researchers now believe that tartaric acid plays an important role in plants when they are under stress. This chapter focuses on how tartaric acid is produced in plants, how it accumulates and what role it plays in the defense against stress factors. This chapter points out the importance of tartaric acid for plant survival under drought, salinity, and reactive oxygen species stress through its effect on osmotic control, scavenging of reactive oxygen species, and control of managing cellular pH. Defense mechanisms related to pathogen attacks, which are part of biotic stress, are also investigated. In this chapter, existing knowledge in the fields of molecular biology, metabolomics, and plant physiology is used to show how tartaric acid functions in coping with stress. In addition, the possibility of stress-resistant crops created through metabolic engineering or breeding is discussed. The synthesis aims to increase our understanding of the role of tartaric acid and stimulate further work on its importance for sustainable agriculture and plant biotechnology.

18.3 Biochemical and Physiological Role of Tartaric Acid in Plants

18.3.1 Biosynthesis of TA in Plants

Grapes are classified as TA-accumulators, with these molecules being formed in both the leaf and the berry. There is very little information on the enzymes of the TA biosynthetic pathway in all plant species, but this is beginning to change. The use of advanced technologies in plant research, such as co-expression networks and association studies, will support the rapid advancement of grapevine research. Little is known about the role of organic acids in plant metabolism. In higher plants, two main pathways are known for the production of TA: through ascorbic acid (AsA) as a starting product, where it is cut at C4 or C5, and through D-gluconic acid, which is also cut at C4 or C5. In the case of grapes use ascorbic acid is mainly used as a starting product (Burbidge et al. 2021).

It is difficult for plant scientists to understand how tartaric acid is synthesized from plants and excreted via vitamin C (L-ascorbic acid). Vitamin C and L-tartaric acid, found in plants, are beneficial to humans. This is not the case with most fruits, but grapes receive L-tartaric acid in the early growth phase and store it in their berries while they are still ripening. The taste of the berries, the sensory characteristics, and aging capacity of the wine depend to a large extent on the L-tartaric acid contained in the grapes and on the substances added during production. The reactions linking L-tartaric acid to vitamin C catabolism in the Vitaceae have been found to proceed through the oxidation of L-idonic acid, which is the key step limiting the reaction rate (DeBolt et al. 2006). Several higher plants, including *Vitis vinifera* L, have a tartaric acid pathway that forms after the oxidation of L-ascorbic acid and the degradation of L-idonate to 5-keto-D-gluconate (Cholet et al. 2016). Rose Pelargonium (*Pelargonium* sp.) is popular because it is aromatic and has medicinal properties, and the substances it produces, such as oils, ascorbic acid, and tartaric acid, have important economic value. Because of their many uses, ascorbic acid and tartaric acid are valuable to humans as antioxidants and flavoring agents in food. There is no data on how the enzymes in rose geranium are structured or how they work in the production of ascorbic acid and tartaric acid (Narnoliya et al. 2019).

Fresh fruit is considered a reliable way to include ascorbic acid (AsA, vitamin C) in the diet. Still, there are some fruits, such as grapes, that do not contain particularly large amounts of AsA. Grapes, although similar to most cultivated fruits, need AsA to form both oxalic acid (OA) and TA. As can be further broken down by the body to form oxalic acid, L-threonic acid, and L-tartaric acid, and also serves as an end-product (Fig. 18.2). Wagner and Loewus found that during the digestion of geranaceous plants between carbon atoms 2 and 3, OA is formed from carbon atoms 1 and 2 and L-threonic acid, which can be further oxidized to TA. A new pathway shows how dehydroascorbic acid (DHA) is converted to oxalate or L-threonate by first becoming 4-O-oxalyl-L-threonate. In Vitaceous species, cleavage between carbon atoms 4 and 5 of 5-keto-D-gluconic acid produces TA, and the two-carbon fragments from the C5 and C6 atoms are expected to be reused in important metabolic processes. Young grapes show the ability to transform L-ascorbic acid in a different process than TA (Melino et al. 2009).

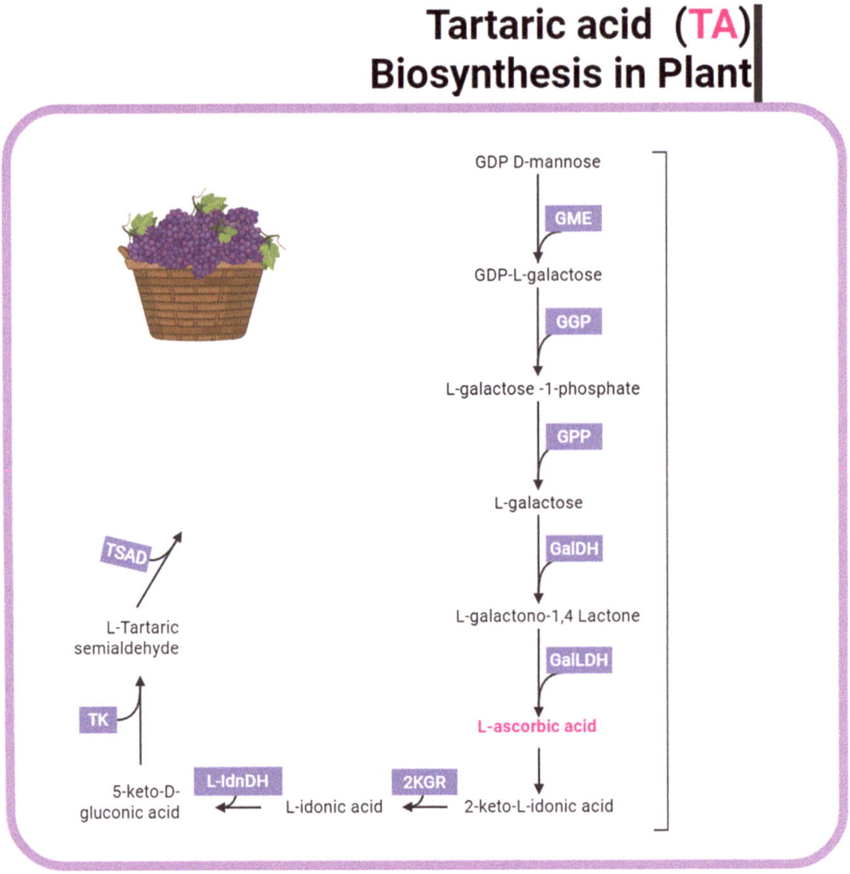

Fig. 18.2 The image illustrates that in plants, L-ascorbic acid (vitamin C) is converted into tartaric acid through a pathway that first produces GDP-D-mannose, which is then converted to tartaric acid through a group of enzymes such as GME, GGP, and GalLDH. L-ascorbic acid is then degraded to tartaric acid via 2-keto-L-idonic acid, L-idonic acid, and L-tartrate semialdehyde. The important enzymes here are 2KGR, L-IdnDH, TK, and TSAD (Cui et al. 2024). (Created by Biorender)

18.3.2 Natural Occurrence and Concentration in Different Species

Although tartaric acid is found in many plants, it is only produced in appreciable quantities in the fruits of some Vitaceae genus. The absence of the gene that controls the tartaric acid content in *Ampelopsis aconitifolia* leads to the accumulation of ascorbic acid instead. Tartaric acid is found in both grapes and leaves, mainly as a potassium salt. Tartaric acid in petrographic studies of vessels from the Middle East leads experts to believe that wine was produced thousands of years ago. Tartaric acid is found in large quantities in the fruits of several plants, and also in grapes.

Tamarind (*Tamarindus*) and hawthorn (*Crataegus*) are good examples of this kind of plant. Some yeasts are known to produce small amounts of tartaric acid (Jackson 2008). Acids measured in hawthorn fruits and plants include a range of compounds such as malic acid, citric acid, succinic acid, ascorbic acid, tartaric acid (Martinelli et al. 2021). L-tartaric acid is only found in grapes, bananas, hawthorn, and tamarind, with the highest content in grapes. Since tartaric acid is only found in grapes, no other sources are needed; therefore, the tartaric acid content in wine is directly linked to the amount in grapes. The tartaric acid content of grapes is much higher than that of table grapes and grapes intended for the production of juice (Niu et al. 2022).

Tamarind (*Tamarindus indica* L.) is an important fruit tree, valued mainly for its edible pulp. It has been selected as a prime candidate for domestication in the Sahel countries of sub-Saharan Africa. In drier areas, tamarinds are harvested with less pulp than in areas with more rain. Little real pulp can be extracted from the tamarind trees in Mali, which bear sour fruit. It was possible to identify some trees with sweeter fruit, but the sweetness was not very high (Van den Bilcke et al. 2014). In this work, a Soxhlet extraction was used to obtain tartaric acid from tamarind pulp, resulting in 17.8% of the acid being extracted from the pulp (Sachdev et al. 2023).

Since the grapes are unripe, they contain a lot of organic acid, which determines whether we can use them straightforwardly or whether they add a sour taste to other food and beverage recipes. It also acts as an antimicrobial and antioxidant and can prevent food from turning brown. Ninety percent of the total organic acids in grapes are tartaric acid and malic acid. The presence of organic acids in young black grapes was established by high-performance liquid chromatography (HPLC). The tartaric acid concentration ranged from 5.51 to 11.78 mg/mL in the three types of white unripe grapes (Shakir and Rashid 2019). *Rheum tataricum* is characterized by a lower content of oxalic and succinic acid and a significantly higher content of malic acid and tartaric acid (Golubkina et al. 2022).

The most common organic acid in apple fruit is tartaric and is associated with increased levels of enzyme activities and NADP-dependent isocitrate dehydrogenase (NADP-IDH), NADP-malic enzyme (NADP-ME), and phosphoenolpyruvate carboxylase (PEPC) (Yang et al. 2023).

18.3.3 Involvement in Cellular Metabolism and pH Regulation

TA in grapes rises considerably during fruit development and leads to lower sugar, later-ripening fruit, more flavor, improved wine stability, and better taste (Li et al. 2023). In many previous experiments, AsA was detected in the phloem of barley, peas, potatoes, tobacco, and turnips. The research found that AsA is carried in the phloem to the leaves and fruits. The location of TA biosynthesis could be the cytoplasm, as the cytoplasm contains much more AsA. Otherwise, L-Idonate dehydrogenase (L-IdnDH) is the only known enzyme of the TA biosynthesis pathway in grape berries, which is typically located in the vacuole in young berries and in the cytoplasm in older berries. At the same time, the scientists discovered that TA accumulates in the vacuoles of ripe berries. From the beginning of berry production, these TA molecules are developed and added to the berries until a few days after flowering. The TA content is therefore closely related to the ripening of the berries. While TA is formed in the cytoplasm of unripe berries, it mainly settles in the vacuole as the berries ripen. Scientists hypothesize that AsA is at the forefront of TA synthesis, that TA enters the vacuole due to AsA migrating to these areas, and that the production of AsA in the leaves and its transport to the berries provides better control over where and when TA is synthesized in the grapes (Ford 2012; Li et al. 2023). Most scientists agree that AsA, because it is present in the cell in much larger amounts than its oxidized form, is the main catalyst in the conversion of NADH to NAD^+ in the cytosol. However, the location within the cell where TA is formed has not yet been established with certainty (Burbidge et al. 2021).

The TA value remains stable during the ripening of the grapes, so that it often has no significant effect on the pH value. Because of the wide range of TA in grapes, tartaric acid plays an important role in balancing the pH of all wines taken together. They were differentiated according to the structures of their different acid compounds (malic acid (MA), TA, inorganic cations) and the sensitivity of their pH to acidity levels during the ripening phase. Nevertheless, the results have to be proven at locations with dozens of climatic conditions. The big contributor to positive flavor profiles in berry juice is partially salted organic acids. Initially, TA acts as a buffer in grape juice or wine because it has the lowest pKa1, but other acidic components probably also react. Buffering the pH helps to maintain the flavor of the wine in the initial weeks after mid-ripening. TA is important for grape pH, but its importance reduces later in the ripening period. TA has been shown to influence pH at certain stages of fruit formation and to affect the amount of acidity and quality of grape must (Plantevin et al. 2024). Three organic acids (citric acid, TA, and MA) were applied to *Salix variegata* under Cd stress to investigate their effects on biomass, pigment number, and photosynthetic performance. The addition of three organic acids significantly increased the biomass of *S. variegata* growing under Cd stress. The impact of MA or TA on shoot and total biomass production was higher than that of citric acid. The decrease in biomass could be related to the way the plants photosynthesize. Under Cd stress, each organic acid increased total photosynthetic activity compared to the values observed without organic acid treatment. Tartaric acid did not affect photosynthetic pigments, but the use of tartaric acid improved the condition of Cd-Cd-stressed plant cells (Chen et al. 2020).

18.3.4 Antioxidant Potential and ROS Scavenging Role

Research on how MA and TA supplied externally affect growth, physiology, biochemistry, and Cd in *S. oleracea* plants is preliminary. Even though MA and TA are involved in dominating physiology and metabolism, scientists do not yet

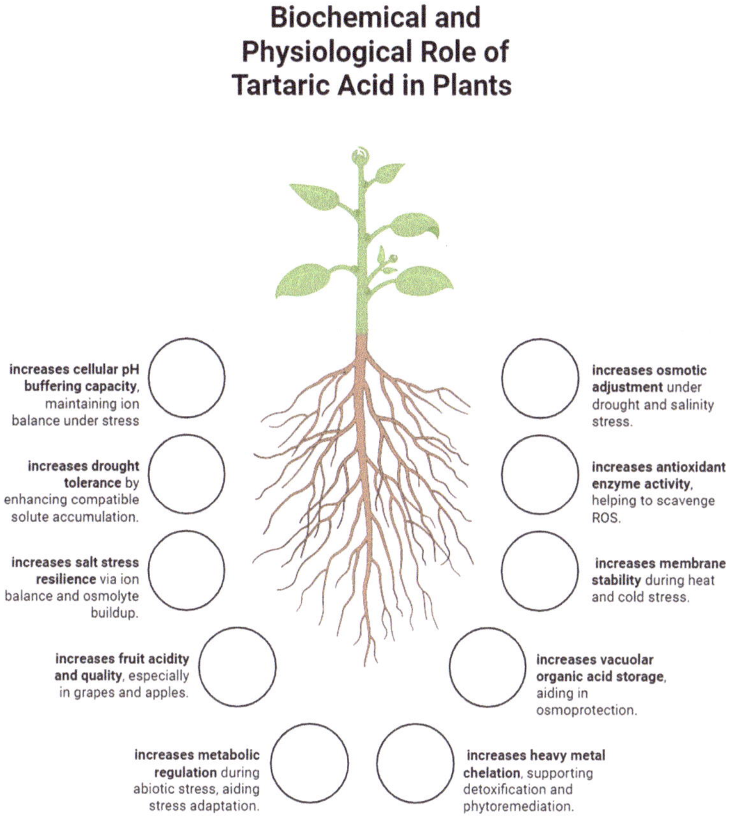

Fig. 18.3 Tartaric acid increases stress tolerance in plants through a balance of pH, regulation of ions, and osmotic adjustment in drought and salinity. Its secure cell membrane decreases oxidative injuries and helps eliminate hazardous metals. It also enhances the quality of fruits as well as sustainable agriculture. (Created by Biorender)

understand how they improve growth and reduce the uptake of heavy metals (HMs) by plants under difficult conditions. The role of MA and TA as metal chelators is to move and remove harmful metals from various plant tissues. In most cases, research to control the harmful effects of heavy metals has focused on well-known organic acids such as citric acid, oxalic acid, or ethylenediaminetetraacetic acid (EDTA). As they are known to be strong chelating agents for heavy metals, these compounds are well studied. Because MA and TA are natural organic acids, less research has been done on their use to reduce heavy metal stress. Even though research on MA and TA is still minimal, many are interested in them because plants produce them and they can help detoxify metals. Since they help to chelate and also improve the body's energy utilization, they could play an important role in further studies (Fig. 18.3). It is hypothesized that the external application of MA and TA supplements can prevent the accumulation of Cd in *S. oleracea* plants under stress (Shabbir et al. 2025).

It has been discovered that small levels of Zn and tartaric acid could safeguard *Tagetes patula* L. from the harmful effects of heavy metal stress. Among other things, zinc plays a key role in plant physiology by increasing enzymes and strengthening the defense response. Moreover, zinc and tartaric acid in medium amounts were found to help plants in the development and production of chlorophyll. Compared to other conditions, both flower development and chlorophyll content increased considerably with the combination of Zn and tartaric acid. Also, the use of small amounts of Zn and tartaric acid in the treatments seemed to reduce oxidative stress, as hydrogen peroxide and malondialdehyde (MDA) levels were reduced. The use of zinc and tartaric acid also increased both the total soluble protein levels and the relative water content in the cells. The combined approach increased the activity of antioxidant enzymes and thus reduced the damage caused by arsenic. Moreover, the combination of Zn and tartaric acid greatly improved the oxidative state and raised melatonin levels after arsenic exposure. Proline levels increased in response to the different treatments, and the greatest increase was observed with the combination of Zn and tartaric acid. The treatments differed in their effects on relative membrane permeability (RMP), but a regular pattern was observed in plants treated with both Zn and tartaric acid. The Pipeline Research Corporation tests also found that mixing targeted additives with Zn helped reduce soil contamination. For example, *Tagetes patula* plants benefited greatly when Zn was combined with tartaric acid (Jamal et al. 2023). Among the active ingredients of *H. sabdariffa*, anthocyanins, tartaric acid, and ascorbic acid act as antioxidants (Ramadhani et al. 2022).

During the change from chloroplasts to chromoplasts in the exocarp, an increase in ROS was detected in the chloroplasts. It is not certain why an oxidative burst occurs in grapes, but it is possible that the dissolution of chloroplasts, which diminishes photosynthesis, leads to a reduction in AsA and TA during ripening. Less AsA and TA may overwhelm the antioxidant system of ascorbate and glutathione, resulting in more ROS, causing a higher amount of ROS to form from respiratory or photonic activity. As a result, less AsA (or TA) and more ROS may drive ripening (Burbidge et al. 2021).

18.4 Tartaric Acid and Abiotic Stress Mitigation

18.4.1 Drought Stress: Mechanisms of Protection

During their growth and development, plants face many environmental difficulties in nature and in farms. Plant productivity can be seriously threatened by drought, which is considered one of the greatest environmental stresses. About 80–95% of the fresh plant body consists of water, which is needed for important developmental and metabolic activities in the plant (Seleiman et al. 2021; Ahmad et al. 2025). In their natural habitat, plants use different adaptation methods to survive in water-rich settings. To respond to drought, plants can be tolerant, resistant, or avoid the effects. Tolerance is obtained by changing the way the cell handles water, rearranging metabolic products, and strengthening its antioxidant system. Nevertheless, avoiding stress leads to more water being transferred to the leaves and stems. Besides, *H. salicornicum* achieves drought resistance by shortening its life cycle before the drought period begins, based on the metabolomic analysis of its shoot (Panda et al. 2021). In each species of simple leaves and roots, we identified eighteen primary metabolites. Fructose, methyl fructoside, citric acid, tartaric acid, and scylloinositol were more abundant in the leaves of drought-stressed grapevines, while erythrono-1,4-lactone and epigallocatechin were more abundant in the leaves of control grapevines (Tamayo et al. 2023).

During an experiment, researchers watched how water stress changed the tartaric acid content of Cabernet Sauvignon grape berries. Initially, the tartaric acid content in the grapes increased with different water stress treatments, then dropped rapidly and finally settled at a more stable level. At moderate water stress, tartaric acid levels decreased at 20–40 days after anthesis (DAA), but thereafter, during ripening, acidity increased to 40 to 90 DAA; all treatments showed a quick decrease in tartaric acid at fruit set. The tartaric acid concentration in each berry was generally higher in the water stress treatments. There was a rapid increase in tartaric acid in weeks 20 to 40 of development, followed by a moderate increase. Severe water stress resulted in much lower tartaric acid content in each berry compared to mild and moderate stress at the same stage (Geng et al. 2022).

18.4.2 Salinity Stress: Ionic Balance and Osmotic Adjustment

Plant production and productivity are strongly affected by the abiotic stress of salinity. Too much salt in the soil prevents plants from growing properly. Too much salt causes an imbalance in the plant cells and depresses the salt concentration, making the surrounding fluid toxic for many biochemical reactions. In plants, the salt content can be reduced by osmotic effects, that is, more salty water in the soil solution, specific ions in the solution that are toxic, and secondary effects such as oxidative stress (Zhang et al. 2013; Hayat et al. 2026).

The reaction of tomato roots to NaCl and $NaHCO_3$ salts was investigated in a hydroponic environment. Plants treated with NaCl and $NaHCO_3$ grew less rapidly than control plants, with the effect being more pronounced in plants under $NaHCO_3$ conditions. A 5-day treatment with $NaHCO_3$ led to an increase in all organic acids, while NaCl only increased oxalate, tartrate, and malate. This means that international control of organic acids could help tomatoes to become more resistant to alkali stress (Gong et al. 2014).

Crops and the global food supply are threatened by the severe effects of salt stress. While salt did not affect germination rates, alkali stress, in contrast, decreased ($p \leq 0.05$) the germination rate of *Amorpha fruticosa* seeds. Both abiotic stress factors reduced germination and radicle growth during germination ($p < 0.05$), with alkali being more problematic than salt stress. While the concentrations of K^+, Mg^{2+}, and Na^+/K^+ in the radicle and seedling hardly changed, the patterns of K^+, Mg^{2+}, NO^{3-}, $H_2PO_4^-$, and SO_4^{2-} remained stable under salt and alkali stress. Initially, our measurements indicated that greater NaCl and Na_2CO_3 levels led to an increase in tartaric acid; however, once enough stress was reached, tartaric acid began to decrease. The highest tartaric acid content was found in 100 mM salt water when under Na_2CO_3 stress, but otherwise, 20 mM salt provided the highest content under control conditions. Under salty conditions, tartaric acid became the main organic acid formed, and its content increased with the degree of stress. Hence, the performance of the plant can be enhanced when the seed is drenched with the right amount of salts (Zhan-Wu et al. 2024). Salinity has been found to increase the production and secretion of organic acids in *Mytilus*. The effects of treatment and individual or species differences were found to be significant in tartaric acid ($p \leq 0.05$). *Acacia* amplifiers secreted almost double the amount of tartaric acid than *Acacia nilotica* at both salinity levels (Abbas et al. 2015).

18.4.3 Temperature Extremes (Heat and Cold): Membrane Stability and Enzyme Protection

While increasing warming has usually helped crops in high latitudes, climate change is hurting crops in large parts of the world, and these issues are expected to intensify in the coming decades. Not only the high temperatures, but also the associated droughts and soil salinization could lead to major crop failures worldwide and food shortages in the future (Niu and Xiang 2018).

Due to the altitude, less rain falls, the temperature fluctuates more, and the plants in this ecosystem are easily affected by stress factors such as freeze and thaw, as well as cadmium. When TA is used as an acid fertilizer, it can reduce the exposure of plants to various agricultural stress factors and promote plant growth. Aquaculture tests were conducted to understand how the physiology of rye seedlings changes under the combined stress of FTC and Cd, and whether TA has a protective effect on rye (*Secale cereale* L.) seeds. Consequently, freeze–thaw condition (FTC) and Cd diminished the photosynthetic and transpiration activity of rye seedlings compared to TA, which did not perform as well. Even so, TA helped the seedlings to tolerate various stress situations by increasing the levels of peroxidase (POD), catalase (CAT) and superoxide dismutase (SOD), providing more soluble proteins and reducing MDA. Therefore, TA has been shown to provide some protection in stress situations with FTC and Cd stress, resulting in less damage to the plant. We advised farmers to avoid the effects of freeze-thaw and Cd stress on rye during the first weeks of cultivation (Ali et al. 2024). So far, there is little research on how TA affects plants at very cold temperatures, although new evidence suggests that TA keeps plant cell

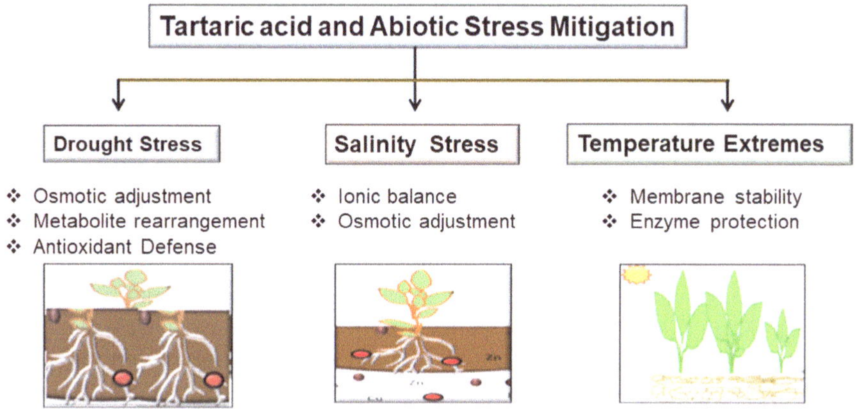

Fig. 18.4 Tartaric acid role in stress situation

enzymes and the cell membrane intact, which helps protect plants against stress (Fig. 18.4).

18.4.4 Heavy Metal Toxicity: Chelation and Detoxification

Cd, Cu, Pb, Cr, and Hg are major pollutants in the environment that occur mainly where people are present. If these metals become available to the body and are present in excessive levels, they can cause harm (Asemave 2018). Rusting metal ions can cause color, encourage spoilage, and contribute to food losing its nutrients. Such agents, including EDTA, potassium tartrate, and potassium citrate, protect against these problems by forming strong, often water-soluble bonds with the metal ions (Karimi and Goli 2021). This process is known as chelation, and the resulting compounds are called chelates. $K_2C_4H_4O_6$ stands for potassium tartrate, dipotassium tartrate, and argol. Tartaric acid is the name for potassium tartrate salt. The authors report that when tartaric acid is mixed with Rochelle salt and potassium sulfate forms potassium tartrate, which the researchers then remove, filter, purify, precipitate, and dry (Yang et al. 2016). In 2020, Goli reported that potassium tartrate and potassium citrate can help eliminate heavy metals from oilseed meal and some different types of rice grown in Iran (Goli 2020).

Up to 75% of the nutrients taken up by plant roots are expelled as exudates, where they might reduce the accessibility of heavy metals including lead, cadmium, and mercury, to plants in the soil. Pb, a toxic metal, may cause issues in the liver, kidneys, and nervous system and lead to mental disorders. In the last 50 years, at least eight million tons of Pb have been released into the environment worldwide. In addition, Pb is mostly stuck in the soil, so that it accumulates in the topsoil over time and is difficult to remove. Ten days after the application of Pb, 2.4 mM of the chelates EDTA, CA, and TA were added to the soil for the experiment. TA increased chlorophyll in all samples, while CA and EDTA increased chlorophyll only at medium concentrations of Pb. We detected that at 4.83 mM Pb stress root length became shorter, but diameter increased after 90 days. Turnip roots retained a higher lead content in the early growth stages, and the addition of EDTA increased this content when a Pb concentration of 4.83 mM was applied. Yet, under all Pb conditions, only the amendment supported ~30% transfer to shoot tissue. At both Pb concentrations of 4.83 and 9.66 mM, eddha-EX-TA greatly increased the Pb content in the roots. TA proved to be the best chelate to transport lead from the roots to the shoots, resulting in more effective growth and an increase in phytoextraction in turnip (Khan et al. 2022). The experiment showed that the use of MA and TA in sunflower plants under Cd/Pb stress increased their growth at all concentrations used. Specifically, exposure of sunflowers to MA and TA increased both Cd and Pb uptake and translocation into the stem: sunflowers receiving Cd10+MA30 had the greatest Cd uptake, and those treated with Pb100+MA30 had the greatest Pb uptake. The analysis showed a positive correlation between the uptake/displacement coefficients of Cd/Pb and the levels of these metals in the soil, but the displacement from root-to-shoot was unaffected by the acid additions. Cd in the roots promoted the movement of Cd into the shoot ($p < 0.05$) and significantly increased the Pb content in the leaves. By examining MA and TA, the study finds that they may help improve phytoextraction of Cd/Pb from contaminated soils by sunflower (Niu et al. 2025).

Shabbir et al. (2025) conducted a randomized pot experiment to evaluate the potential role of MA and TA foliar sprays to alleviate Cd stress in *Spinacia oleracea* L. plants. For Cd stress, *S. oleracea* plants were treated with $CdCl_2$ solution (100 μM). Distilled water was given to the plants as a control. One week after Cd stress, MA and TA foliar sprays were used at 100 and 150 μM concentrations for both. They reported that both low and high levels (100 and 150 μM) of MA and TA supplements effectively reversed the destructive effects of Cd stress and improved growth, photosynthesis, and defense-related traits of *S. oleracea* plants. They reported that these supplements also prevented the accumulation of excessive Cd metal ions, as indicated by decreased Cd metal contents in MA- and TA-treated plants. These findings indicated that MA and TA applications could potentially reduce Cd-induced phytotoxicity in plants by decreasing Cd uptake and increasing parameters related to photosynthesis and defense. In another study, Sharma et al. (2020) conducted a study to evaluate the effects of TA addition on the physiology of *Hordeum vulgare* L. seedlings under Cr(VI) stress. They reported that adding TA to Cr(VI) medium increased root and shoot dry weights, the activities of antioxidant enzymes (ascorbate peroxidase, guaiacol peroxidase, superoxide dismutase, catalase, and glutathione reductase), and pigment content (total chlorophyll and carotenoids). However, they reported that co-application of TA with higher concentrations of Cr(VI) reduced malondialdehyde content. They found that replacing 0.5 mM Cr(VI) with 1.5 mM TA increased Cr uptake in seedling roots by 208.7% compared to treatment with 0.5 mM Cr(VI) alone. They found that shoot and root bioconcentration factors (BCF) were improved by the application of TA to the culture medium. They confirmed that TA acts as an antagonist against Cr(VI) by increasing their antioxidant potential and improving their chromium accumulation capacity in *H. vulgare* seedlings.

The effect of tartaric acid on HMs in various fractions was compared. An attempt was made to regenerate the tartaric acid solution with HMs via chestnut shells. The experiments showed that tartaric acid helped to remove Cd, Zn, and Cu from the contaminated soil, especially from the exchangeable and carbonate fractions. The Pb and Cu contents in the Fe-Mn oxide fraction were greatly reduced after tartaric acid treatment. A solution of tartaric acid with HMs could be renewed by 10 g of chestnut shells in a batch reactor. This remediation work confirmed that tartaric acid may be suitable for the remediation of HM-contaminated soils. Further experiments are needed to make the method usable for on-site treatment of polluted sites (Ke et al. 2006).

The roots of plants near reservoirs should be able to withstand permanent changes in the water level in the area. Species such as *Salix matsudana*, which can adapt by changing their growth habit, usually survive. The study found that water availability kept the growth height under control, but did not greatly affect the diameter of *S. matsudana*. The roots of *S. matsudana* contained significantly more malic acid and tartaric acid than oxalic acid under different growth conditions. They reveal that *S. matsudana* conserves certain amounts of malic acid and tartaric acid to adapt to environmental flooding stress. If the plant is underwater for a longer period of time, part of the energy necessary for survival is stored in the roots of *S. matsudana*. This allows it to grow in the wet–dry cycle of the hydro-fluctuating part of the reservoir (Yuancai et al. 2022) (Table 18.1).

18.5 Tartaric Acid and Biotic Stress Resistance

18.5.1 Role in Enhancing Plant Immune Responses

Plants have a two-step detection response system in which pattern recognition receptors (PRR*s)* and nucleotide-binding leucine-rich repeat receptors (NLR*s)* are used to activate downstream actions following the recognition of chemicals created by pathogens. When particular conserved molecules are found, the PRRs activate PTI. When pattern-triggered immunity (PTI) does not function well due to attack by pathogenic microbes, resistance is often due to effector-triggered immunity (ETI). Host plants protected by NLRs can identify the molecules used by pathogenic organisms so that the host develops a stronger and faster protective response (Nabi et al. 2024). Fruit and beverages, such as wine, are the main sources of TA, whose ability to support a healthy immune system has only recently been considered. TA is recognized for its ability to fight antioxidant and antimicrobial damage when a person eats it in suitable amounts. However, when TA levels become higher, it can be toxic. There is evidence that TA may play a role in influencing the immune system, but exactly how is still unclear and needs to be carefully studied (Riaz et al. 2025). The accumulation of TA has been associated with stress-induced metabolic

Table 18.1 Tartaric acid application origin of TA (endogenous and exogenous) in plant stress

Stress type	Mechanism involved	Plant species/study	Observed effects	Application/origin of TA	Research approach	References
Drought stress	Osmotic adjustment, metabolic modulation	*Haloxylon salicornicum*	Enhanced accumulation of tartaric acid and other osmolytes	Endogenous	Metabolomic profiling	Panda et al. (2021)
	TA biosynthesis variation during drought	*Vitis vinifera* cv. Cabernet Sauvignon	Early-stage increase, followed by suppression during fruit ripening	Endogenous	Field study (berry development)	Geng et al. (2022)
	Drought-induced metabolite enrichment	*Vitis vinifera*	Increased TA in drought-stressed leaves	Endogenous	Leaf metabolite profiling	Tamayo et al. (2023)
Salinity stress	Ion chelation, pH buffering, osmotic adjustment	*Solanum lycopersicum*	Accumulation of oxalate, tartrate, and malate under NaCl and $NaHCO_3$	Endogenous	Organic acid profiling under salinity	Gong et al. (2014)
	Salt-stress tolerance via organic acid modulation	*Amorpha fruticosa*	Increased TA content at moderate salt levels	Endogenous	Pot experiment	Zhan-Wu et al. (2024)
	Species-specific TA secretion in saline soils	*Acacia ampliceps*, *A. nilotica*	*A. ampliceps* showed higher TA secretion and tolerance	Endogenous	Hydroponic salinity trial	Abbas et al. (2015)
Temperature stress	Antioxidant defense, membrane protection	*Secale cereale* (rye)	Improved antioxidative enzyme activity and reduced lipid peroxidation	Exogenous	Freeze-thaw + Cd stress treatment	Ali et al. (2024)
	General thermal stress mitigation (heat & cold)	Various (review-based)	Proposed stabilization of membrane and ROS control	Suggested (endogenous)	Literature-based synthesis	Niu and Xiang (2018)

Heavy metal stress	Chelation of Pb, Cd, Cu; translocation enhancement	*Brassica rapa* (turnip)	Increased Pb phytoextraction compared to EDTA or citric acid	Exogenous	Pot experiment	Khan et al. (2022)
	Phytoextraction efficiency in contaminated soils	*Helianthus annuus* (sunflower)	Better plant growth and metal uptake under TA and malic acid co-treatment	Exogenous	Cd/Pb stress trial	Niu et al. (2025)
	Soil detoxification via chelate-assisted adsorption	Biochar + tartaric acid in soil	Reduced heavy metal bioavailability	Exogenous	Soil column + bio adsorbent study	Yang et al. (2016)

changes that regularly lead to higher expression of defense genes. A recent experiment with rye showed that the use of external TA both improved plant defenses and reduced the negative effects of stress factors (Ali et al. 2024).

18.5.2 Effect on Pathogen Resistance (Bacteria, Fungi)

The efficacy of lactic, citric, and TA acids and their derivatives as antimicrobial agents has been demonstrated in the literature. TA and its salts have been developed with enhanced antibacterial qualities against multidrug-resistant diseases due to their proven safety and antimicrobial qualities (Mikaelyan et al. 2022).

Tartaric acid, a naturally occurring representative of aldaric acid, is found mainly in plants, from algae to trees, and plays a role in several of their cell biosynthesis processes. This acid and its salts (tartrates) are used as safe preservatives in the food and chemical industries as they are known to be antibacterial active chemicals.

The derivatization of natural compounds, like organic aldaric acid, to the corresponding imides and amides is one of the possible methods for the production of antimicrobial agents. Previous studies have demonstrated the efficacy of benzylimide, cyclohexylimide, cyclohexyl mono amino salt, and benzyl amino salt of tartaric acid against eight strains of Bacilli that were isolated from cave sludge and against a number of opportunistic pathogenic Pseudomonas and Stenotrophomonas. Representatives of the *P. chlororaphis* group (*P. chlororaphis* subsp. *aurantiaca, P. chlororaphis, P. chlororaphis* subsp. *chlororaphis, P. taetrolens*, and *P. fluorescens*) have also shown that these four tartaric acid derivatives are biodegradable (Babayan et al. 2020).

The efficacy of seven organic acid compounds in preventing the vegetative growth of *Monilia fructigena* conidiospores was demonstrated in vitro. Acidic substances such as acetic acid, tartaric acid, succinic acid, and malic acid have been investigated. Most of them are common food additives employed to impart a sour taste to food and are neither harmful to humans nor to the environment. In vivo phytotoxic tests were performed with two apple cultivars ("Golden Delicious" and "Jonagold"), and conidiospores of *Monilia fructigena* were extracted from naturally infested apple fruits. The experiments show that the acids are differently effective against *Monilia fructigena*. Some of them, such as acetic acid, formic acid, and succinic acid, were useless as fungicides against testicular pathogens. They can prevent the germination of conidiospores, but only in very high concentrations, which, according to the experiments, can damage treated apple plants. However, other acids such as citric acid, lactic acid, and tartaric acid were effective in extremely low concentrations (0.1–0.3%) and were therefore ideal fungicides (Ganchev 2023).

18.5.3 Influence on Insects and Pest Deterrence

Phytophagous organisms, particularly insects such as the leafminers of the genus *Liriomyza* Mik. (Diptera: Agromyzidae), are widely recognized as important agricultural pests that can significantly reduce the yield and economic value of many plant species consumed by humans. *Liriomyza cicerina* Rondani, a chickpea leaf miner, is one of the most damaging pests affecting chickpea (*Cicer arietinum* L.) cultivation in the Mediterranean region. To protect itself from pests, the chickpea secretes naturally occurring acids from its green parts, such as leaflets, stems, leaves, sepals, and pods. The susceptible and resistant genotypes showed no discernible changes in tartaric acid under field circumstances. In the chickpea genotype susceptible to leaf miner resistance, tartaric acid (0.028) showed a favorable correlation with RL scores (Chrigui et al. 2020).

The East Asian native *Drosophila suzukii* (Matsumura) (Diptera: Drosophilidae) was first identified in North America and Europe in 2008 and has since propagated to South America, where it has become an increasingly aggressive pest of fragile, thin-skinned fruit. As an exceptionally polyphagous species, *Drosophila suzukii* can germinate and/or lay eggs in a wide variety of wild and cultivated fruits. Due to its rapid development and high reproductive potential, it can lead to its rapid development and major losses in agriculture. The effects of tartaric acid concentrations on the fitness of the fly were investigated, paying particular attention to the host status of cherries, which are considered a preferred host, and grapes, which are considered a non-preferred host (Wang et al. 2019).

18.6 Exogenous Application of Tartaric Acid in Agriculture

18.6.1 Methods of Application: Foliar Spray, Soil Drenching, Seed Priming

Foliar fertilization is the application of liquid fertilizer to the leaves of the plant. Plants use their leaves to absorb important elements from the soil. The plants absorb liquids through their stomata as well as through their epidermis. When fertilizers are sprayed onto the leaves of plants, the process is called foliar spraying. This method allows you to fertilize small amounts of micronutrients. If the top layer of soil is not moist enough, you can also apply the main nutrients in this way. Instead of using foliar fertilization without soil application, both methods should be used to complement each other. In recent foliar fertilization has become standard practice in agriculture, especially for horticultural crops. Although foliar fertilization is less common in agriculture, its positive results are well established, leading to increased education for consistent results (Patil and Chetan 2018).

Many farmers rely on foliar fertilization to improve plant growth and yields. The use of foliar nutrients can improve the potential yield and important physical and biological characteristics of rice, such as *Oryza sativa* L (Mahmoodi et al. 2020).

Sedghi and Sharif (2013) found that fertilizing rice leaves with foliar spray significantly increased rice yield. A foliar fertilizer only helps the leaf and the cell if it reaches the leaf and the cytoplasm of the leaf. The nutrients must penetrate through the outer cuticle and the wall of the underlying epidermal cells. After penetration, nutrient uptake by the cells follows the same steps as absorption by the roots. Among all parts of the pathway, the cuticle offers the biggest resistance to nutrients applied through the leaf (Patil and Chetan 2018).

18.6.2 Synergistic Use with Other Biostimulants

The combination of tartaric acid with other bio-stimulants could offer new benefits for plant growth, better mineral uptake capacity, and greater stress resistance in sustainable agriculture. Tartaric acid, found in grapevines and similar plants, helps bind metal ions and mobilize phosphorus and iron in the rhizosphere (Zhang et al. 2024). When applied together with other bio-stimulants, humic acids, for example, increase root length and have a hormonal effect, while tartaric acid helps to solubilize minerals and encourages the activity of useful organisms in the surrounding soil (Canellas et al. 2015). As organic matter is a source of carbon, it also promotes the growth of various beneficial bacteria and fungi that improve soil fertility and soil-microbe relationships (Compant et al. 2010). The use of algal extracts as bio-stimulants in stressful environments can help plants improve their antioxidant system, reduce oxidative damage and improve resistance to different types of non-living stress (Shukla et al. 2018). The addition of tartaric acid to amino acid stimulants seems to help increase stress-related gene expression and stimulate photosynthesis, proving that the substance works in several ways. When tartaric acid is used together with other biostimulants, plants are healthier, yield is improved, and soil care becomes more sustainable.

18.6.3 Molecular Insights into Tartaric Acid-Induced Stress Mitigation

Until now, tartaric acid in grapevine (*Vitis vinifera*) has been considered as a factor influencing fruit acidity and fruit quality, but not as part of the plant's response to stress. While tartaric acid is important for fruit development and ripening, the way in which it affects plant stress responses is not yet fully explored. The formation of ascorbate is mostly based on ascorbic acid, a process that scientists have learned a lot about, particularly thanks to 2-keto-L-gulonic acid reductase. Nevertheless, scientists have not found convincing proof that tartaric acid changes gene expression, activates signaling pathways, or is associated with metabolism or hormone responses to stress. There is evidence that tartaric acid may be involved in the response to abiotic stress, particularly heavy metal exposure. According to one study (Wu et al. 2024), the secretion of organic acids, such as tartaric acid, increased in the roots of wheat when exposed to selenite stress (Wu et al. 2024). Despite the idea that it lowers the effects of stress in plants by trapping toxic ions or altering pH in the rhizosphere, little is known about the exact mechanisms. Tartaric acid has not been observed to affect transcriptional or protein changes associated with typical stress response pathways controlled by ABA, and SA. It follows that it is still uncertain whether tartaric acid can affect gene regulation or hormonal signaling. This story is also made more complex by the fact that there is no clear evidence that tartaric acid interacts with calcium signaling pathways, mitogen-activated protein (MAP) kinases, or transcription factors that respond to redox changes. Even though some scientists suggest that organic acids might affect the elimination of ROS or pH changes in some enzymes, there is no study showing the same for tartaric acid. Nearly all studies on tartaric acid look at its involvement in metabolism and how it is transported, but they often overlook the connection with stress (Burbidge et al. 2021).

In contrast to malate and citrate, for which there is strong evidence for stress defense in plants, there is not yet enough evidence to include tartaric acid. Tartaric acid could respond to stress by shifting metabolic processes rather than participating in the initial steps of the stress response. Analysts DeBolt et al. (2006) described the enzymes involved in tartaric acid biosynthesis but determined that they were not related to the expression of stress-inducible genes (Burbidge et al. 2021). In brief, tartaric acid is known to be important for plants in viticulture, but direct evidence for its part in stress mechanisms is still lacking. To assess whether tartaric acid regulates stress adaptation in plants, more detailed studies are needed, including transcriptomic analyses, studies of mutants or hormone interactions.

18.6.4 Environmental and Ecological Considerations

Recently, the properties of tartaric acid, lactic acid, citric acid, and oxalic acid described above have been investigated. These substances are now successfully used worldwide in the production of environmentally friendly agrochemicals. Tartaric acid, which is the most widely used organic acid in the plant world, is particularly widespread. Tartaric acid and its salts of alkali and alkaline earth metals are considered safe food additives (E334 - E337, E354), which are widely used in the world production of chelated fertilizers. Tartaric acid is also used both in free form and in amide form (Mikaelyan et al. 2021). Due to its ecological compatibility and potential use in eco-friendly agriculture, tartaric acid, which is plentiful in grapes, is catching the attention of experts. Tartaric acid degrades rapidly under aerobic conditions after 3 days and is not considered harmful to the environment. Root exudates containing tartaric acid could play a role in shaping the microbial community in the soil. Studies have shown that they have an effect on the soil bacterial community, helping to grow advantageous species that can boost healthy soil

Table 18.2 Insights into tartaric acid's role in stress signaling

Aspect	Role of tartaric acid	Evidence/findings	Mechanistic insight	References
Stress response (general)	Limited evidence for involvement in plant stress response pathways	Tartaric acid is not linked to gene expression or hormone signaling during stress	No known connection to ABA, SA, ROS, calcium signaling, or MAPKs	Burbidge et al. (2021), DeBolt et al. (2006)
Abiotic stress (heavy metals)	Possible mitigation of selenite stress	Increased secretion in wheat roots under selenite exposure	May chelate toxic ions or alter rhizosphere pH	Wu et al. (2024)
Metabolic adjustment	Involved in downstream metabolic shifts, not primary signaling	May alter metabolism during stress rather than initiate a response	No impact on transcription or protein regulation linked to stress	Burbidge et al. (2021)
Soil microbial community	May shape soil microbial structure	Promotes the growth of beneficial microbes	Acts via root exudates	Mikaelyan et al. (2021)
Environmental compatibility	Eco-friendly compound widely used in agrochemicals	Breaks down rapidly in aerobic conditions; safe food additive (E334–E337, E354)	Used in chelated fertilizers and organic farming	Mikaelyan et al. (2021)
Heavy metal uptake	Promotes metal absorption (both positive and negative implications)	Enhanced uptake of metals when applied to soil	Caution is advised in sugar industry byproduct usage	Zhang et al. (2024), Lu et al. (2013)

conditions and nutrient flow. Individual studies have confirmed that repeated applications of tartaric acid to soil promote the uptake of heavy metals into plants, which can affect plant health and food safety. For this reason, farmers must handle sugar industry waste with care to prevent unexpected negative impacts on the environment (Zhang et al. 2024; Lu et al. 2013) (Table 18.2).

18.7 Current Challenges and Future Perspectives

Plant physiologists commonly acknowledge that tartaric acid plays an important role in plant responses to stress and change, mostly in grapevines. However, the role of thick vessels in different plant species is not well understood. Although recent studies state that tartaric acid helps with antioxidation, uses its power to regulate osmoregulation and helps maintain the balance of pH in plants, there are still many obstacles to fully understanding its role in agriculture. As climate change affects traditional crop production, the role of natural plant compounds like tartaric acid in stress reduction is increasingly being discussed. However, several scientific and technical problems need to be overcome to turn this potential into usable innovations. One important problem is that we do not know enough about how tartaric acid is regulated in plant cells. Scientists know how ascorbic acid is converted into tartaric acid, but the exact regulatory steps are not yet well explored. We still need to fully understand how drought, salinity or temperatures outside the ideal range affect tartaric acid biosynthesis. So far, many studies have centered on the tartaric acid-rich grapevine, but we do not know if the same control process is found in other species as well.

Researchers may learn a lot about this organic acid by examining how it acts similarly or differently in different types of organisms. Further studies are needed to determine how tartaric acid moves and is stored in plant cells. Tartaric acid is mainly located in the vacuole, where it keeps the water content constant during growth and when the fruit is dehydrated. We still do not know the transporters that help tartaric acid pass through cell membranes. Discovering how tartaric acid is managed by the different compartments and tissues would be key to revealing its important contribution to keeping osmotic and ionic balance when the body is exposed to different stresses. Tartaric acid could also have a signaling function, but this is only one possibility so far. Not many studies have been done to show the interactions of tartaric acid with phytohormones or stress signaling pathways. It appears that citric and malic acid support plant defenses, and we can assume that tartaric acid does the same. It needs to be investigated whether tartaric acid is produced uninvolved during stress or instead plays a role by helping to control the stress response.

Switching from one species to another can also be difficult. Even though grapevine research is widespread, there are few details about tartaric acid in apples, bananas, or leafy vegetables. Consequently, we do not know exactly how to fully realize the benefits of tartaric acid in agriculture. Additional studies should be conducted on other crops in areas that are likely to be more affected by climate change, to determine whether higher levels of tartaric acid in crops can make them more resilient in unfavorable weather conditions. Another research issue is determining the most effective amount of exogenous tartaric acid for use in crops. Initial trials have found that applying tartaric acid to leaves can increase antioxidant activity and help plants cope with stress. However, the results of these applications depend on how much of the compound is ingested, when it is ingested and how it enters the stream.

Further studies are needed to establish practical guidelines for agricultural use, including crop trials and experiments to determine the best dose. With the help of new genetic engineering techniques, we can now look for better ways to control tartaric acid metabolism in plants. Using CRISPR/Cas9 and RNA interference (RNAi), it is now possible to modify specific genes in the tartaric acid metabolic pathway. The additional expression of significant enzymes could lead to more tartaric acid, and the plants could cope better with oxidative and osmotic stress. On the other hand, silencing genes could decrease the production of tartaric acid, which would help researchers to find out what it does in the body. Tartaric acid production could be triggered only under stress conditions by using stress-sensitive genes as promoters.

Nevertheless, caution is required when updating the metabolic pathways, as strengthening one part can cancel out the other parts. An increase in tartaric acid could modify the pH of the plant, interact with other organic acids, and alter both the flavor and the acceptability of the fruit to the buyer. For this reason, the effects of genetic changes on a plant should be regularly checked using biological methods such as transcriptomics and metabolomics to ensure that the plant remains healthy and productive. In addition to genetics, there is another traditional method for selecting tartaric acid-rich varieties: conventional breeding. With the help of marker-assisted selection and genomic selection, plant breeders can accelerate the development of high-performance plants without having to face the obstacles of GMO breeding. This technique is especially effective for fruit trees, as tartaric acid helps them cope with stress and is important for post-harvest flavor and quality.

Tartaric acid appears to be a promising solution for agriculture that can withstand climate change. As drought, salinization and heat waves become more common, finding ways to help crops survive them will be important for the future of food production. Farming systems would be more stable and environmentally friendly if farmers used tartaric acid-based measures along with mulching, inoculation and metered irrigation. Tartaric acid helps the body cope with antioxidant defenses and osmotic conditions, which can support the plant's defenses and improve the nutritional value of its fruit. As more and more people desire healthier and greener foods, the combined advantages of this approach are especially attractive. If plant scientists, agronomists, and biotechnologists work together, tartaric acid could play an important role in improving agriculture in the future. We still know far too little about how tartaric acid helps plants adapt to stress. To fully realize the benefits of tartaric acid, we need to resolve existing uncertainties, apply new genetic strategies, and conduct rigorous field trials. When we look at what tartaric acid research has to offer, we see hope not only for soilless plants but also for a beneficial role in agricultural practices around the world.

18.8 Conclusion

Tartaric acid, which occurs naturally in many plants and is mainly found in grapevines, now appears to have several effects on plant growth and stress response. How NO is formed, when it occurs naturally in cells, what it does and how it interacts with stress factors such as drought, salinity, and toxic heavy metals have all been investigated. New information from recent studies and mechanistic knowledge support the growing belief that tartaric acid helps plants cope with harsh weather and other conditions. This chapter reveals that tartaric acid has effects on cell activity and protects cells from different kinds of stress. TA is not only an intermediate or accumulating compound, but also plays a role in regulating osmotic balance and supporting the antioxidant defense system, and is involved in many other metabolic and signaling pathways to cope with stress. When plants suffer from drought or salt stress, resulting in both imbalanced osmotic pressure and an accumulation of harmful ROS, TA intervenes by acting osmolytically and stimulating the activity of SOD, CAT, and POD, which contribute to the elimination of ROS. The finding that TA is stored in the vacuoles underlines its vital role in controlling ions in the cell, mainly when excessive salt prevents safe ion levels. Detoxification of ions through chelation can also help the body obtain nutrients, similar to other low molecular weight organic acids. It appears that TA is also responsible for maintaining proper pH and redox balance when plants are under abiotic stress. This chapter points out that tartaric acid supports the development and ripening of grape fruit, as it is present in large amounts in grapes. Tartaric acid helps to maintain the balance of liquid and solutes during the delicate stages of development, which are sensitive to external changes. In addition, sorbitol in the vacuoles helps the fruits to regulate their water and sugar balance when they dry out or ripen. Because tartaric acid helps plants adapt to stress, it could be a valuable area for agronomic improvement. Recent studies have shown that TA, when supplied externally, improves plant growth and antioxidant levels under cadmium stress, potentially making it a reliable bio-stimulant. In addition, knowledge of how TA works could help produce plants with more antioxidants so that they can tolerate drought, high salinity, and possibly temperature fluctuations. On a global scale, our knowledge of tartaric acid in stress physiology is very useful for sustainable agriculture. As environmental problems increase due to climate change, the use of substances such as tartaric acid could reduce the use of chemicals, support plant health, and make agriculture more sustainable. As it is particularly abundant in grapes and apples, further research into its stress-responsive metabolic pathways could provide specific insights for these species that are useful for precision agriculture. Improved omics approaches will allow researchers to better explore how TA and other phytohormones and signaling molecules interact under various stress conditions. Tartaric acid is now recognized as an important factor in the response of plants to stress. As more and more people are interested in sustainability and climate-smart crops, the biochemical approach of TA could lead to useful research results and practical applications. Research into the regulation, translocation, and use of this molecule outside of plants is likely to reveal new ways to protect plants and make agriculture more environmentally friendly in the coming years.

References

Abbas G, Saqib M, Akhtar J, Murtaza G, Shahid M (2015) Effect of salinity on rhizosphere acidification and antioxidant activity of two acacia species. Can J For Res 45(1):124–129

Abdel Latef AAH, Hameed M, Khan Z, Khan S, Imran A, Fatima N (2026) Foliar Stimulants: A sustainable strategy for enhancing plant resilience to abiotic and biotic stresses. Physiologia plantarum. 178, no. 1: e70741

AbdElgawad H, Zinta G, Hegab MM, Pandey R, Asard H, Abuelsoud W (2016) High salinity induces different oxidative stress and antioxidant responses in maize seedlings organs. Front Plant Sci 7:276

Acosta-Motos JR, Ortuño MF, Bernal-Vicente A, Diaz-Vivancos P, Sanchez-Blanco MJ, Hernandez JA (2017) Plant responses to salt stress: adaptive mechanisms. Agronomy 7(1):18

Afzal A, Ali B, Samarasinghe KGBA, Khan MN, Javed MA, Saleem A (2026) Overview of biotic stress and its agricultural impact. In: Abdel Latef AAH (eds) Chitosan nanoparticles for plant resilience under abiotic and biotic stressors. Smart Nanomaterials Technology. Springer, Singapore. https://doi.org/10.1007/978-981-95-6422-4_3

Ahmad S, Ahmad N, Khan MN, Ercisli S, Iqbal R (2025) Use of biostimulants to improve drought tolerance in oilseed crops. In: Abdel Latef AAH (eds) Oilseed crops under abiotic atress. Sustainability Sciences in Asia and Africa. Springer, Singapore. https://doi.org/10.1007/978-981-96-8346-8_2

Hayat K et al (2026) Salinity stress management using chitosan nanoparticles. In: Abdel Latef AAH (eds) Chitosan nanoparticles for plant resilience under abiotic and biotic stressors. Smart nanomaterials technology. Springer, Singapore. https://doi.org/10.1007/978-981-95-6422-4_4

Ali S, Bao G, Bashir K, Chen W, Jiang Y, Fan C, Li G (2024) Tartaric acid application for improving tolerance of Rye (*Secale cereale* L.) seedlings to cadmium chloride stress and freeze-thaw condition. Russian. J Plant Physiol 71(6):206

Alonso ER, León I, Kolesniková L, Mata S, Alonso JL (2021) Unveiling five naked structures of tartaric acid. Angew Chem Int Ed 60(32):17410–17414

Asemave K (2018) Greener chelators for recovery of metals and other applications. Org Med Chem Int J 6(4):1–11

Azooz MM, Abdel Latef AAH, Abdelmajeed WG et al. (2026) Biochar ameliorates salt stress in maize by modulating physiological responses and expression of SOS1 and NHX2 genes. Journal of Soil Science and Plant Nutrition. https://doi.org/10.1007/s42729-026-03123-w

Babayan B, Mikaelyan A, Asatryan N, Soghomonyan T, Baghdasaryan A, Melkumyan M, Bagdasaryan S, Grigoryan A (2020) Tartaric acid synthetic derivatives for multi-drug resistant Phytopathogen *Pseudomonas* and *Xanthomonas* combating. Sci Technol Publ 4(5):285

Barman M, Das S, Kundu R, Mukherjee S, Singh SK, Abdel Latef AA (2026) Overview of abiotic stress and its agricultural impact. In: Abdel Latef AAH (eds) Chitosan nanoparticles for plant resilience under abiotic and biotic stressors. Smart nanomaterials technology. Springer, Singapore. https://doi.org/10.1007/978-981-95-6422-4_2

Basu S, Ramegowda V, Kumar A, Pereira A (2016) Plant adaptation to drought stress. F1000Res 5: F1000 Faculty Rev-1554

Burbidge CA, Ford CM, Melino VJ, Wong DCJ, Jia Y, Jenkins CLD, Soole KL, Castellarin SD, Darriet P, Rienth M (2021) Biosynthesis and cellular functions of tartaric acid in grapevines. Front Plant Sci 12:643024

Canellas LP, Olivares FL, Aguiar NO, Jones DL, Nebbioso A, Mazzei P, Piccolo A (2015) Humic and fulvic acids as biostimulants in horticulture. Sci Hortic 196:15–27

Chen HC, Zhang SL, Wu KJ, Li R, He XR, He DN, Huang C, Wei H (2020) The effects of exogenous organic acids on the growth, photosynthesis and cellular ultrastructure of *Salix variegata* Franch. Under Cd stress. Ecotoxicol Environ Saf 187:109790

Cholet C, Claverol S, Claisse O, Rabot A, Osowsky A, Dumot V, Ferrari G, Gény L (2016) Tartaric acid pathways in *Vitis vinifera* L.(cv. Ugni blanc): a comparative study of two vintages with contrasted climatic conditions. BMC Plant Biol 16:1–18

Chrigui N, Sari D, Sari H, Eker T, Cengiz MF, Ikten C, Toker C (2020) Introgression of resistance to leafminer (*Liriomyza cicerina* Rondani) from *Cicer reticulatum* Ladiz. to *C. arietinum* L. and relationships between potential biochemical selection criteria. Agronomy 11(1):57

Compant S, Clément C, Sessitsch A (2010) Plant growth-promoting bacteria in the rhizo-and endosphere of plants: their role, colonization, mechanisms involved and prospects for utilization. Soil Biol Biochem 42(5):669–678

Cui W, Wang X, Han S, Guo W, Meng N, Li J, Sun B, Zhang X (2024) Research progress of tartaric acid stabilization on wine characteristics. Food Chem X:101728

DeBolt S, Cook DR, Ford CM (2006) L-Tartaric acid synthesis from vitamin C in higher plants. Proc Natl Acad Sci 103(14):5608–5613

Fahad S, Saud S, Chen Y, Wu C, Wang D (2021) Abiotic stress in plants. In: BoD-books on demand. IntechOpen, London

Ford C (2012) The biochemistry of organic acids in the grape. In: The biochemistry of the grape berry. Bentham Science Publishers, Sharjah, pp 67–88

Ganchev D (2023) Antifungal effect of several organic acids towards conidiospores of *Monilia fructigena* in the in vitro conditions. Knowl Int J 59(3):213–217

Geng K, Zhang Y, Lv D, Li D, Wang Z (2022) Effects of water stress on the sugar accumulation and organic acid changes in Cabernet Sauvignon grape berries. Hortic Sci 49(3):164

Goli M (2020) The effect of soaking and cooking in the presence of chelating agents (potassium tartrate and citrate) on cadmium reduction of some imported rice varieties. J Food Sci Technol (Iran) 16(97):161–169

Golubkina N, Kharchenko V, Bogachuk M, Koshevarov A, Sheshnitsan S, Kosheleva O, Pirogov N, Caruso G (2022) Biochemical characteristics and elemental composition peculiarities of *Rheum tataricum* L. in semi-desert conditions and of European garden rhubarb. Int J Plant Biol 13(3):368–380

Gong B, Wen D, Bloszies S, Li X, Wei M, Yang F, Shi Q, Wang X (2014) Comparative effects of NaCl and $NaHCO_3$ stresses on respiratory metabolism, antioxidant system, nutritional status, and organic acid metabolism in tomato roots. Acta Physiol Plant 36:2167–2181

Gull A, Lone AA, Wani NUI (2019) Biotic and abiotic stresses in plants. In: Abiotic and biotic stress in plants. IntechOpen, London

Hasanuzzaman M, Bhuyan MB, Zulfiqar F, Raza A, Mohsin SM, Mahmud JA, Fujita M, Fotopoulos V (2020) Reactive oxygen species and antioxidant defense in plants under abiotic stress: revisiting the crucial role of a universal defense regulator. Antioxidants 9(8):681

Ijaz S, Iqbal J, Abbasi BA, Ullah Z, Yaseen T, Munir F, Hyder S, Kanwal S, Sahito ZA, Mahmood T (2024) From single nc-RNAs to networks: understanding the complexity of environmental adaption. In: ncRNAs: mediated regulation: the unarguable biological function to environmental constrains. Springer, Cham, pp 283–310

Jackson RS (2008) Chemical constituents of grapes and wine. In: Wine science. Academic Press, San Diego, pp 270–331

Jamal A, Yasin NA, Javad S, Ahmed S, Yasmin A, Chaudhry O, Daoud MS, Gatasheh MK (2023) Investigating the efficacy of tartaric acid and zinc-mediated endogenous melatonin induction for mitigating arsenic stress in *Tagetes patula* L. Sci Hortic 322:112399

Jantwal A, Durgapal S, Upadhyay J, Joshi T, Kumar A (2022) Tartaric acid. In: Antioxidants effects in health. Elsevier, Amsterdam, pp 485–492

Karimi Z, Goli M (2021) The effect of chelating agents including potassium tartrate and citrate on the maximum reduction of lead and cadmium during soaking and cooking from some different varieties of rice available in Iran. Food Sci Nutr 9(9):5112–5118

Ke X, Li P, Zhou QX, Zhang Y, Sun TH (2006) Removal of heavy metals from a contaminated soil using tartaric acid. J Environ Sci (China) 18(4):727–733

Khan I, Iqbal M, Raza SH, Anwar S, Ashraf M, Shafiq F (2022) Tartaric acid soil-amendment increases phytoextraction potential through root to shoot transfer of lead in turnip. Chemosphere 296:134055

Li M, Su J, Yang H, Feng L, Wang M, Xu G, Shao J, Ma C (2023) Grape tartaric acid: chemistry, function, metabolism, and regulation. Horticulturae 9(11):1173

Lu LL, Tian SK, Yang XE, Peng HY, Li TQ (2013) Improved cadmium uptake and accumulation in the hyperaccumulator *Sedum alfredii*: the impact of citric acid and tartaric acid. J Zhejiang Univ Sci B 14(2):106–114

Mahmoodi B, Moballeghi M, Eftekhari A, Neshaie-Mogadam M (2020) Effects of foliar application of liquid fertilizer on agronomical and physiological traits of rice (*Oryza sativa* L.). Acta Agrobot 73(3):7332

Martinelli F, Perrone A, Yousefi S, Papini A, Castiglione S, Guarino F, Cicatelli A, Aelaei M, Arad N, Gholami M (2021) Botanical, phytochemical, anti-microbial and pharmaceutical characteristics of hawthorn (*Crataegus monogyna* Jacq.), Rosaceae. Molecules 26(23):7266

Melino VJ, Soole KL, Ford CM (2009) Ascorbate metabolism and the developmental demand for tartaric and oxalic acids in ripening grape berries. BMC Plant Biol 9:1–14

Mikaelyan AR, Babayan BG, Nersisyan GH, Grigoryan AM (2021) The study of Tartaric acid based new complex preparation of plant growth activation. Bullet High Technol 1(15):10–20

Mikaelyan A, Babayan B, Vartanyan A, Tokmajyan H (2022) Tartaric acid synthetic derivatives effect on phytopathogenic bacteria. Agron Res 20:342

Nabi Z, Manzoor S, Nabi SU, Wani TA, Gulzar H, Farooq M, Arya VM, Baloch FS, Vlădulescu C, Popescu SM (2024) Pattern-Triggered Immunity and Effector-Triggered Immunity: crosstalk and cooperation of PRR and NLR-mediated plant defense pathways during host-pathogen interactions. Physiol Mol Biol Plants 30(4):587–604

Narnoliya LK, Jadaun JS, Singh SP (2019) The phytochemical composition, biological effects and biotechnological approaches to the production of high-value essential oil from geranium. In: Essential oil research: trends in biosynthesis, analytics, industrial applications and biotechnological production. Springer, Cham, pp 327–352

Nawaz M, Sun J, Shabbir S, Khattak WA, Ren G, Nie X, Bo Y, Javed Q, Du D, Sonne C (2023) A review of plants strategies to resist biotic and abiotic environmental stressors. Sci Total Environ 900:165832

Niu Y, Xiang Y (2018) An overview of biomembrane functions in plant responses to high-temperature stress. Front Plant Sci 9:915

Niu SY, Liu CH, Liu Q, Fan XC, Zhang Y, Sun L, Zhang XL, Jiang JF (2022) Composition and contents of organic acids in different grape germplasms. J Fruit Sci 29:768

Niu Z, Guo P, Wang Z, Wang H, Li X (2025) Study on the potential of malic acid and tartaric acid assisted in improving the Phytoextraction of cadmium and lead by sunflower (*Helianthus Annuus* L.). Soil Sediment Contam Int J 39:1–18

Panda A, Rangani J, Parida AK (2021) Physiological and metabolic adjustments in the xerohalophyte *Haloxylon salicornicum* conferring drought tolerance. Physiol Plant 172(2): 1189–1211

Patil B, Chetan H (2018) Foliar fertilization of nutrients. Marumegh 3(1):49–53

Plantevin M, Merpault Y, Lecourt J, Destrac-Irvine A, Dijsktra L, van Leeuwen C (2024) Characterization of varietal effects on the acidity and pH of grape berries for selection of varieties better adapted to climate change. Front Plant Sci 15:1439114

Ramadhani NF, Nugraha AP, Rahmadhani D, Puspitaningrum MS, Rizqianti Y, Kharisma VD, Noor TNEBTARR, Ernawati D, Nugraha A (2022) Anthocyanin, tartaric acid, ascorbic acid of roselle flower (*Hibiscus sabdariffa* L.) for immunomodulatory adjuvant therapy in oral manifestation coronavirus disease-19: an immunoinformatic approach. J Pharm Pharmacogn Res 10(3):418–428

Riaz B, Ryu H-M, Choi B, Sohn S (2025) Tartaric acid exacerbates DSS-induced colitis by promoting eosinophilic inflammation via IL-13 and IL-5Rα upregulation. Pathogens 14(4):366

Sachdev D, Jha PK, Rani R, Verma G, Kaur N, Sahu O (2023) Structural and optical investigation of highly fluorescent tartaric acid derived from the tamarind pulp. Mater Chem Phys 296: 127294

Sedghi M, Seyed Sharifi R (2013) Effects of foliar supplements of nitrogen, phosphorus and potassium on grain yield and macro element transport and adsorption efficiency of hybrid rice (*Oryza sativa* L.). Res Field Crop J 1(1):64–75

Seleiman MF, Al-Suhaibani N, Ali N, Akmal M, Alotaibi M, Refay Y, Dindaroglu T, Abdul-Wajid HH, Battaglia ML (2021) Drought stress impacts on plants and different approaches to alleviate its adverse effects. Plants 10(2):259

Shabbir A, Shah AA, Usman S, Ahmed S, Kaleem M, Shafique S, Gatasheh MK (2025) Efficacy of malic and tartaric acid in mitigation of cadmium stress in *Spinacia oleracea* L. via modulations in physiological and biochemical attributes. Sci Rep 15(1):3366

Shakir BK, Rashid RS (2019) Physiochemical and phytochemical profile of unripe black grape juice (verjuice). Ann Trop Med Public Health 22(12):46–81

Sharma M, Kumar V, Bhardwaj R et al (2020) Tartaric Acid Mediated Cr Hyperaccumulation and biochemical alterations in seedlings of *Hordeum vulgare* L. J Plant Growth Regul 39:1–14

Shukla PS, Shotton K, Norman E, Neily W, Critchley AT, Prithiviraj B (2018) Seaweed extract improve drought tolerance of soybean by regulating stress-response genes. AoB Plants 10(1): plx051

Sun M, Yang Z, Liu L, Duan L (2022) DNA methylation in plant responses and adaption to abiotic stresses. Int J Mol Sci 23(13):6910

Tamayo M, Sepúlveda L, Guequen EP, Saavedra P, Pedreschi R, Cáceres-Mella A, Alvaro JE, Cuneo IF (2023) Hydric behavior: insights into primary metabolites in leaves and roots of Cabernet Sauvignon and Grenache Grapevine varieties under Drought stress. Horticulturae 9(5): 566

Umar OB, Ranti LA, Abdulbaki AS, Bola AL, Abdulhamid AK, Biola MR, Victor KO (2021) Stresses in plants: biotic and abiotic. In: Current trends in wheat research. IntechOpen, London

Van den Bilcke N, Alaerts K, Ghaffaripour S, Simbo D, Samson R (2014) Physico-chemical properties of tamarind (*Tamarindus indica* L.) fruits from Mali: selection of elite trees for domestication. Genet Resour Crop Evol 61:537–553

Wahid N et al. (2025) Biostimulants as a sustainable strategy to augment salinity tolerance in oilseed crops. In: Abdel Latef AAH (eds) Oilseed crops under abiotic stress. Sustainability Sciences in Asia and Africa. Springer, Singapore. https://doi.org/10.1007/978-981-96-8346-8_1

Wang X, Kaçar G, Daane KM (2019) Temporal dynamics of host use by *Drosophila suzukii* in California's San Joaquin Valley: implications for area-wide pest management. Insects 10(7):206

Wu K, Wang L, Wu Z, Liu Z, Li Z, Shen J, Shi S, Liu H, Rensing C, Feng R (2024) Selenite reduced cadmium uptake, interfered signal transduction of endogenous phytohormones, and stimulated secretion of tartaric acid based on a combined analysis of non-invasive micro-test technique, transcriptome and metabolome. Plant Physiol Biochem 206:108107

Yang Y, Li H, Peng L, Chen Z, Zeng Q (2016) Assessment of Pb and Cd in seed oils and meals and methodology of their extraction. Food Chem 197:482–488

Yang GK, Xue SY, Li JZ, Li HB, Gao Y, Zhang XJ, Hao YY (2023) Analysis of organic acid components and malic acid metabolizing enzyme activity in Hongbaoshi apple fruits. J Fruit Sci 40:884

Yuancai Q, Arif M, Dong Z, Ting W, Qin Y, Bo P, Peng W, Wei H (2022) The effect of hydrological regimes on the concentrations of nonstructural carbohydrates and organic acids in the roots of Salix matsudana in the Three Gorges Reservoir, China. Ecol Indic 142:109176

Zhang M, Fang Y, Ji Y, Jiang Z, Wang L (2013) Effects of salt stress on ion content, antioxidant enzymes and protein profile in different tissues of *Broussonetia papyrifera*. S Afr J Bot 85:1–9

Zhang H, Zhu J, Gong Z, Zhu J-K (2022) Abiotic stress responses in plants. Nat Rev Genet 23(2): 104–119

Zhang X, Lu H, Liu J, Tadiyose B, Wan H, Zhong Z, Deng Y, Chi G, Zhao H (2024) Mechanism of tartaric acid mediated dissipation and biotransformation of tetrabromobisphenol A and its derivatives in soil. J Hazard Mater 471:134350

Zhan-Wu G, Yong-Guang M, Jian-Jun D, Ke-Jia D, Jia-Tong L, Xin-Ning L, Li-Jie H, Zhao-Jie W, Chun-Sheng M, Sulaiman AA (2024) The effects of salinity stress on *Amorpha fruticosa* Linn. seed germination, physiological and biochemical mechanisms. Notulae Botanicae Horti Agrobotanici Cluj-Napoca 52(1):13552

Chapter 19
Vanillic Acid: A Potential Elicitor of Plant Stress Resilience

Khursheda Parvin, Jannatul Ferdous, Hafsa Akter, Durjoy Kumar Datta, Sohel Hossen, and Md. Arfan Ali

Abstract Environmental stressors such as extreme temperatures, metal toxicity, UV radiation, salinity, and water shortages affect plant growth, metabolism, and productivity, posing continuous challenges for plants. In response, these stationary organisms have developed a sophisticated and disciplined defense system to survive such stressful conditions. One key strategy involves producing secondary metabolites, which help scavenge stress-induced reactive oxygen species (ROS). Among these secondary metabolites, phenolic compounds like vanillic acid (VA), produced via the phenylpropanoid pathway, are notable. VA has strong antioxidative properties and can influence stress-related signaling and physiological pathways, attracting increasing research interest. This chapter reviews prior research on VA's role in boosting plant resistance to abiotic stressors. It discusses how VA supports redox balance, maintains cellular homeostasis, regulates stress-responsive pathways, and activates defense genes. Evidence indicates VA's involvement in enhancing plant defenses and reducing oxidative damage, although our understanding of its in full signaling mechanisms remains limited. Incorporating VA into plant stress management strategies offers a sustainable approach to improving plant resilience amid changing climate conditions. We have compiled and critically examined existing literature on VA to advance plant tolerance through its antioxidant properties. Ultimately, using this class of phenolic compounds to protect plants from adverse growth conditions shows significant promise.

Keywords Abiotic stress · Plant tolerance · Antioxidant defense · Secondary metabolites · Phenolic compounds.

K. Parvin (✉) · J. Ferdous · H. Akter · D. K. Datta · S. Hossen · M. A. Ali
Department of Horticulture, Sher-e-Bangla Agricultural University, Dhaka, Bangladesh

A. A. H. Abdel Latef (ed.), *Harnessing Vitamins and Organic Acids for Plant Stress Resilience*, Sustainability Sciences in Asia and Africa,
https://doi.org/10.1007/978-981-95-8593-9_19

19.1 Introduction

Plants continuously face numerous environmental stressors, including salinity, heavy metal toxicity, dehydration, pathogen assaults, and extremely high or low temperatures, all of which greatly restrict agricultural productivity (Zhang et al. 2022). Therefore, plants are undergoing a threat to their existence, and their growth, metabolism, and reproduction become restricted, resulting in loss of the plant and its yield. But every population wants to survive and that's why plants develop different mechanisms to adapt, avoid, or tolerate abiotic stress. As a part of tolerance, a complex defense strategy is induced in plant cells where the production of secondary metabolites is an essential tactic for alleviating stress (Isah 2019). Among these metabolites, phenolic compounds have attracted considerable interest for their various biological functions, including antimicrobial, signaling, and especially antioxidant activities (Naik and Al-Khayri 2016). Vanillic acid (VA; 4-hydroxy-3-methoxybenzoic acid), a member of phenolic compounds, has gained recognition as a potential elicitor that strengthens plant stress tolerance by various amenable biochemical processes (Bhattacharya et al. 2010; Parvin et al. 2024). Vanillic acid is indeed a natural ubiquitous phenolic acid extensively present in various plants, and fruits, including vanilla beans, olives, and whole grains (Kumar and Pruthi 2014). Due to its antioxidant and antimicrobial effects, VA is an obvious component for plant defense protection (Kumar and Pruthi 2014). Like caffeic, ferulic, and sinapic acid, VA is also biosynthesized via the phenylpropanoid pathway. This pathway plays a significant role in plant defense, growth, and adaptation by generating diverse bioactive molecules with antioxidant and protective functions (Deng and Lu 2017). Under stress conditions, endogenous VA acts as a stress marker as its content becomes higher in synergy with the level of stress severity (Parvin et al. 2022). During stressful conditions, plants undergo oxidative damage due to extreme production of reactive oxygen species (ROS) as well as cellular dysfunction. Additionally, VA is increasingly recognized as a signaling molecule in plant defense mechanisms (Mittler 2017). From previous literature, it has been found that externally applied VA improved plant tolerance to environmental stresses through stimulation in enzymatic antioxidant activity, promoting osmolyte accumulation, and regulating stress–responsive gene expression (Agati et al. 2012). Vanillic acid mitigates secondary stress, including not only oxidative but also osmotic and ionic stress in plants upon various climatic changes, by enhancing plant development phases, including elongation and chlorophyll production.

Research has shown its effectiveness in increasing plant growth upon drought and salinity by regulating stomatal conductance, osmotic balance, and ion homeostasis (Farooq et al. 2017). Like, exogenous supplementation of VA increases photosynthesis capacity in salt-affected plants along with reduction of cellular damage (Ahanger et al. 2017; Raza et al. 2019). Additionally, VA has been found to promote the production of proline and trehalose, which contributed to increasing cellular strength under osmotic stress (Gupta and Huang 2014). In case of heavy metal toxicity, VA reduces ROS accumulation by binding to metal ions and boosting the synthesis of phytochelatins along with higher antioxidants and osmolytes responses (Sharma et al. 2019a, b; Bhuyan et al. 2020). Moreover, VA is also able to trigger systemic acquired resistance (SAR) in pathogen-affected plants and thus make them stronger (Vlot et al. 2021). This priming effect includes the enhanced appearance of pathogenesis-related (PR) proteins and the buildup of defense-related phytohormones, such as salicylic acid (SA) and jasmonic acid (JA) (Pandey et al. 2015). This chapter seeks to summarize the existing knowledge on VA as a potential inducer of plant stress resilience, examining its biosynthesis, mechanisms of action, and agricultural applications. By exploring its role in plant defense, researchers can leverage its potential to increase crop productivity under environmental stress.

19.2 Plant Responses to Environmental Stress

Environmental conditions are constantly changing due to climatic variation and human activities, leading to adverse effects on plant life. Abiotic stressors disrupt plant physiology and metabolism with consequent results in hampering growth and development (Fig. 19.1). One of the key global challenges is soil salinity, which diminishes crop production and affects roughly one-fifth of irrigated land (Qadir et al. 2014). Plants exhibit intricate and multifaceted physiological responses to salinity, complicating the design and interpretation of experiments. These responses can be categorized into two primary stages: shoot ion-independent is the first phase, occurring very quickly with the involvement of Na^+ sense and signaling (Gilroy et al. 2014; Roy et al. 2014). In this case, salinity can significantly impact water relations, leading to stomatal closure and restricted leaf enlargement (Munns and Termaat 1986). After that, in a prolonged second stage (days to weeks), excessive ion buildup in shoots—particularly older leaves—reaches toxic levels, triggering early leaf senescence and ultimate yield reduction and/or plant death (Munns and Termaat 1986). Salinity hinders plant growth through a triple threat: it causes osmotic stress due to increased phytotoxic ions, leads to ionic stress within the cytosol, and promotes oxidative stress by generating ROS (Hussain et al. 2019). Thus, in salt stress, Na^+ competes with K^+ uptake and causes such a nutrient imbalance, as well as disrupting stomatal regulation and photosynthesis (Van Zelm et al. 2020).

Plants' response to drought is not uniform; it varies significantly based on the growth phase of plants, the stress's severity, and its timespan (Levitt 1980). Drought stress induces modifications in a wide array of physiological and biochemical functions, including water relations (Silva et al. 2009b), metabolism of carbohydrates, proteins, and nutrients (Šircelj et al. 2005: Jameel et al. 2026), gas exchange capacity and photosynthesis (Pagter et al. 2005). Significant research in recent years has focused on adverse drought effects on key plant processes, including growth, gas exchange, water relations, and osmoregulation (Valliyodan and Nguyen 2006; Ashraf and Foolad 2007; Silva et al. 2009a, b). Heavy rainfall causes waterlogging in tropical and subtropical farmlands; it rapidly depletes oxygen from the soil. This

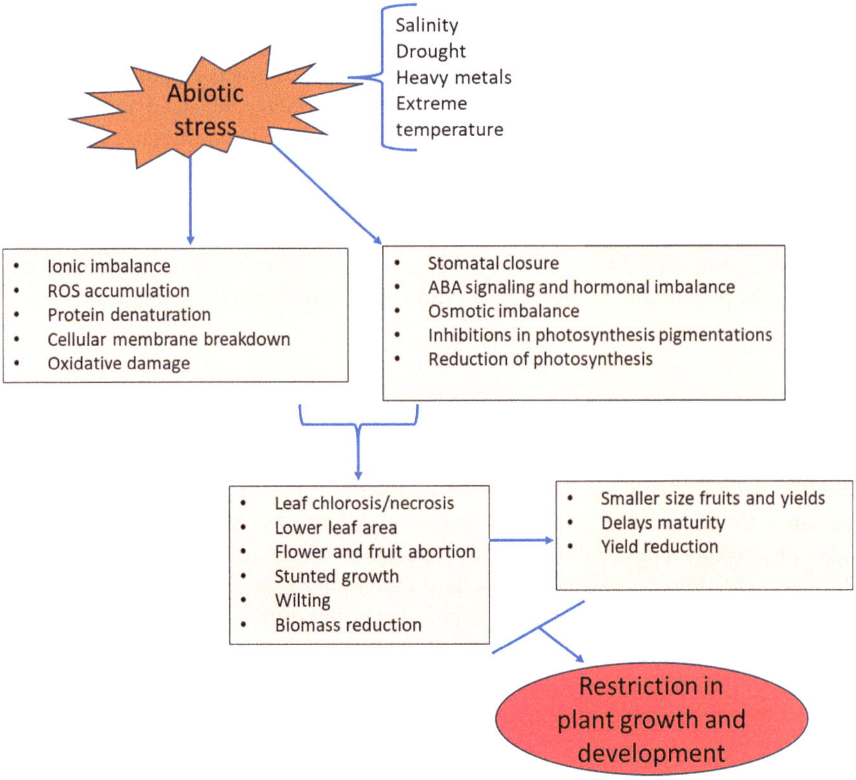

Fig. 19.1 Abiotic stress induced plant growth restriction and susceptibility through hampering various physiological, cellular, and morphological processes

hypoxic environment quickly (within 24–48 h) damages the roots' ability to respire and absorb essential nutrients (Colmer and Voesenek 2009). However, a critical adaptation to waterlogging in plants is the growth of adventitious roots. This process is driven by two main hormones: ethylene, which inhibits the growth of the original primary roots, and auxin, which stimulates cell division at the stem nodes. These new roots ultimately help the plant absorb more oxygen and nutrients (Steffens and Rasmussen 2016).

Due to rapid industrialization, environmental pollution is occurring and is ever-increasing, which results in metal and metalloid toxicity, a widespread and damaging problem for plants. Heavy metal toxicity, like excessive arsenic (As), cadmium (Cd), and lead (Pb), induces oxidative stress and disrupts essential metabolic processes (Gupta et al. 2022). As an industrial pollutant, Cd is particularly deleterious to most life forms and has a propensity to bio-accumulate in the trophic chain, primarily through plant uptake (Di Toppi and Gabbrielli 1999; Hakla et al. 2021; Abeed et al. 2022). Plants rapidly take up cadmium due to its high mobility and affinity for water, causing it to accumulate in their tissues (Besson-Bard et al. 2009). Plants poisoned by cadmium show stunted growth, yellow leaves (chlorosis), and curled leaves (epinasty). Their reproductive process is harmed as pollen fails to germinate properly. The metal also damages the internal structure of chloroplasts, which hinders photosynthesis and disrupts how the plant processes nitrogen and sulfur (Benavides et al. 2005). Lead (Pb) is a persistent and major environmental pollutant, classified as the second most hazardous material (Hengstler et al. 2003). Due to pollution, Pb levels in soil can reach thousands of mg kg^{-1}, vastly higher than the natural range of 15–40 mg kg^{-1}. When this accumulated Pb is taken up by plants, it reduces crop production. Extreme Pb levels in plant tissues also trigger the generation of ROS, which causes protein degradation, impairment in nutrient uptake, and chloroplast structures, leading to reduced membrane permeability, decreased photosynthesis and an overall decline in plant growth (Emamverdian et al. 2015; Afzaal et al. 2020). Moreover, high temperature causes restriction on plant growth through disturbing the balance of ROS by excessive production and causes the alteration of the antioxidant defense system, resulting in loss of chlorophyll pigmentation as well as photosynthesis reduction (Hossain et al. 2021).

19.3 Biosynthesis Pathway of Vanillic Acid

A large amount of plant substances comes from phenolic compounds, also known as polyphenols, which contain one or more hydroxyl substituents in an aromatic ring (Harborne and Harborne 1973). Phenolic compounds help not only in the regulation of seed germination and plant growth but also actively participate in defense response by their high antioxidant activity against environmental stresses like heavy metal, heat, and UV radiation, salinity, drought, and mechanical injury (Pratyusha 2022). Phenolic compounds consist of several groups, including phenolic acid, flavonoids, lignans, stilbenes, and tannins (Harborne and Harborne 1973).

Hence, as a member of phenolic acid, VA is produced in plant cells through the phenylpropanoid pathway (Fig. 19.2). In producing VA, the product of the shikimate pathway, named shikimate later converted to phenylalanine, which is the precursor (Vogt 2010). Thereafter, with the activities of phenylalanine ammonia-lyase (PAL), caffeic acid (CA) is produced from phenylalanine. In this pathway, as a derivative of cinnamic acid, CA originates in many plant species, which also accumulates upon stress exposure (Boerjan et al. 2003; Kısa et al. 2019).

The phenylpropanoid pathway is very stress-responsive, and, upon activation, it produces monolignols such as ferulic acid (Boerjan et al. 2003). After that, through bioconversion, ferulic acid converts into vanillin, which is then oxidized into VA (Gallage and Møller 2015). Vanillic acid is a natural antioxidant and thus capable of reducing lipid peroxidation and restoring the activity of the enzymatic antioxidant (An et al. 2019). Under stress conditions like UV-B exposure, plants produced by increasing flux of both the shikimate and phenylpropanoid pathways (Aimvijarn et al. 2023). This buildup helps plants defend by antioxidant protection.

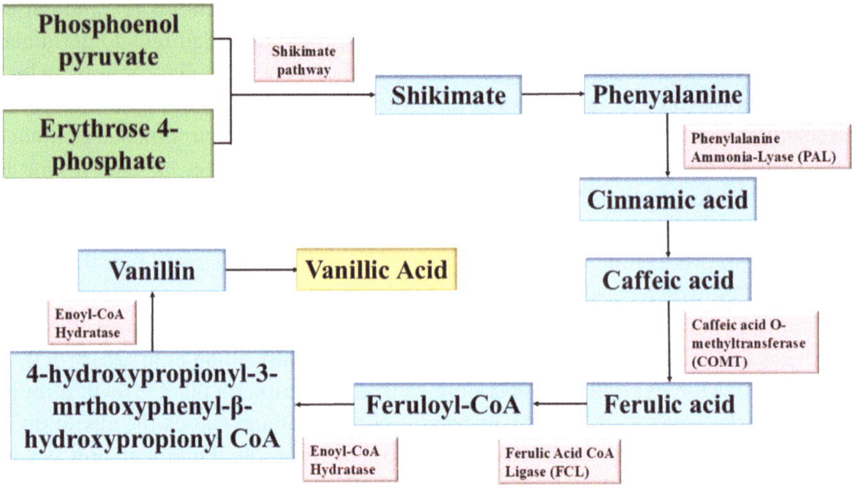

Fig. 19.2 Diagram of the biosynthesis pathway of vanillic acid in the plant cell

Low-concentrated VA application caused the protection of DNA from oxidative damage by hydrogen peroxide and UV radiation (Aimvijarn et al. 2023). Higher enzymatic activity of caffeic acid O-methyltransferase (COMT) and PAL in stress conditions causes the higher transfiguration of phenylalanine to ferulic, vanillin, and ultimately vanillic acid, contributing to the plant's defense strategy (Vogt 2010).

19.4 Vanillic Acid-Induced Regulation of Plant Tolerance to Stresses

Both abiotic (like heavy metals, salinity, and drought) and biotic (such as infections and herbivores) factors are crucial to plants for their survival. Plants are continuously subjected to undergo various climatic alterations and hence their inbuilt capacity helps them to sustain by developing various approaches for being tolerant. Accordingly, plants develop sophisticated defense mechanisms to deal with these difficulties, such as the synthesis of secondary metabolites. Of these metabolites, phenolic compounds are essential for stress reactions. Due to its participation in defense mechanisms under stressful conditions, VA has become a promising biomarker for stressed plants (Kumar et al. 2020).

Vanillic acid is a possible biomarker for determining the health and stress levels of plants since it builds up in stress response (Table 19.1). The quantification of vanillin and related phenolics, including vanillic acid, was extracted from *Vanilla planifolia* pods with a clear demonstration of excellent linearity, precision, and recovery (Proestos and Komaitis 2013). Its levels can reveal information about the degree of stress and how well a plant's defenses are working.

Table 19.1 Status of vanillic acid as a stress marker in plants under stress conditions

Environmental stress	Plant species	Status of VA accumulation	References
Drought (2 days)	*Vaccinium corymbosum* L.	Enhanced	An et al. (2019)
Lead (Pb) toxicity (1280 μM)	*Lycopersicon esculentum*	Increased	Afzaal et al. (2023)
$CdCl_2$ (1.5 mM, 3 days)	*Oryza sativa L.*	Increased	Bhuyan et al. (2020)
Drought (30% field capacity, 2 weeks)	*Pisum sativum* L.	Increased	Rahman et al. (2024a)
Cr toxicity (0.75 mM)	*Citrus nobilis* Lour *Citrus deliciosaten*	Increased	Shahid et al. (2018)
Cu, Cd, Hg, Pb toxicity (100 μM each)	*Triticum aestivum* L.	Increased	Colak et al. (2019)
Waterlogging (566 ± 216 mm and 1055 mm precipitation	*Deschampsia antarctica*	Lowered (20%) phenolic content including VA.	Park and Lee (2019).
Cold temperature (<18 °C)	*Coffea arabica* L.	The 8% rise in total phenolics, VA aids in cold stress adaptation	Acidri et al. (2020).
High temperature 38/33 °C (day/night), 21 days	*Festuca trachyphylla*	Enhanced	Wang et al. (2019)
Cd and zinc (Zn) stress (60 days)	*Kandelia obovata*	Elevated by about 45% in roots, while 127% was in shoots	Chen et al. (2019)

It has become more widely acknowledged that VA, a naturally occurring phenolic molecule that is produced when lignin and other aromatic compounds break down, plays a variety of roles in how plants react to stress. It contributes to the plant's physiological resilience in harsh environments by acting as a protective agent in addition to a biochemical marker. Its significance in plant defense and adaptation systems is highlighted by its involvement in a variety of stress conditions.

A diagrammatic representation was presented in Fig. 19.3, indicating that an increase in VA accumulation in plants exposed to various abiotic stressors resulted in the accomplishment of better stress tolerance responses. *Vaccinium corymbosum* L. actively mitigates the effects of dehydration under drought conditions, as evidenced by its increased levels of VA following 2 days of polyethylene glycol-induced dehydration stress (An et al. 2019). According to Afzaal et al. (2023), *Lycopersicon esculentum* exhibits an increase in endogenous VA in response to lead poisoning at 1280 μM, indicating its role in detoxification and oxidative stress resistance. Glutathione (GSH) levels in *Oryza sativa* are markedly elevated by 80% after exposure to $CdCl_2$ (1.5 mM for 72 h). This is strongly linked to an increase in

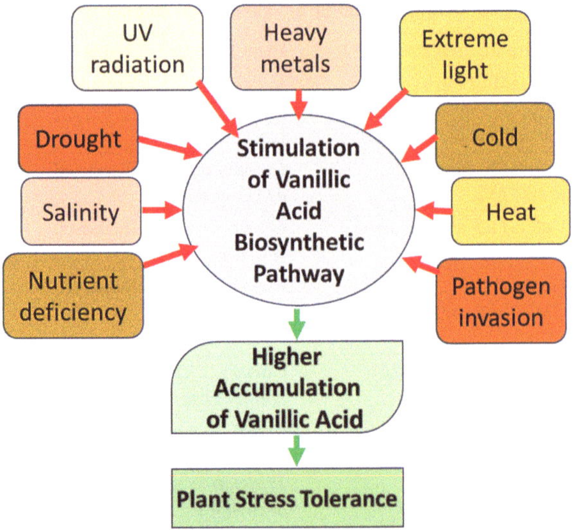

Fig. 19.3 Changes in the production status of VA are brought on by abiotic and biotic stresses. Numerous environmental stressors stimulate the biosynthesis pathway because they increase the activity of the enzymes that are responsible for accumulating more VA, which in turn increases tolerance. The red arrow is a symbolic representation of harmful effects, while the green arrow represents the beneficial effects

VA, underscoring GSH's function in improving redox balance under metal stress (Bhuyan et al. 2020). Endogenous VA content rises in *Pisum sativum* L. after 2 weeks of drought at 30% field capacity, indicating its function in osmotic adjustment and enhanced physiological response to drought (Rahman et al. 2024a). Higher VA accumulation is another effect of chromium (Cr) toxicity (0.75 mM) in Kinnow mandarin, a hybrid of *Citrus nobilis* and *Citrus deliciosa*, suggesting its defense role against metal toxicity (Shahid et al. 2018). Total phenolic compounds, including VA, rise to 45% in roots and 127% in shoots in *Kandelia obovata* upon Cd and zinc (Zn) stress, confirming their role in metal detoxification and antioxidant defense. Vanillic acid levels also increase for *Triticum aestivum* L. exposed to Cu, Cd, Hg, and Pb (each at 100 μM), indicating its wider function in reducing metal stress (Colak et al. 2019). However, in *Deschampsia antarctica*, waterlogging results in a 20% drop in total phenolic content, including VA. This could be because hypoxic circumstances inhibit phenolic production (Park and Lee 2019). The 8% rise in total phenolics that *Coffea arabica* L. exhibits in response to cold temperatures below 18 °C suggests that VA aids in cold stress adaptation (Acidri et al. 2020). Heat-exposed (38/33 °C, day/night) *Festuca trachyphylla* exhibited an increase in VA of 31–32%, showing the protection of oxidative damage (Wang et al. 2019). The collective results highlight the broad-spectrum function of VA as a protective agent and biochemical marker for better plant tolerance to stress.

From above mentioned literature, it is obvious that VA is a biological common denominator in multiple species' stress responses. Its rise in response to stress, whether from oxidative imbalance, water scarcity, temperature fluctuations, nutritional deficiencies, or pathogen invasion, suggests that it could be used as a natural, non-invasive biomarker for the diagnosis of plant stress. Vanillic acid functions as a vital stress indicator and defense mechanism in plants dealing with a variety of environmental stressors. As a trustworthy biochemical indicator of plant stress, it has been shown via considerable research that VA accumulates in response to climatic challenges, including severe temperatures, drought, salinity, and metal toxicity. Furthermore, applying VA exogenously is very successful in reducing stress-related damage in various plant species. Therefore, the modest levels of VA enhance physiological and biochemical reactions, including increasing antioxidant activity, preserving chlorophyll content, controlling ion balance, and lowering oxidative damage. These results demonstrate VA's dual function as a possible agrochemical agent for improving crop resilience in harsh conditions as well as a natural stress indicator.

19.5 VA: Elicitor of Plant Stress Tolerance

As a significant phenolic compound, VA functions as a plant growth promoter and possesses antimicrobial, antifungal, and anti-mutagenic properties. As a secondary metabolite and non-enzymatic antioxidant, it also serves as a known marker for plant stress responses (El-Beltagi et al. 2020). Exogenously applied VA increased rice tolerance against waterlogging by elevating photosynthetic properties (Xuan and Khang 2018). Treating blueberries with VA before stress reduces lipid peroxidation and alleviates dehydration damage. This protection works by boosting the plant's own VA levels, turning on genes for enzymatic antioxidants, and raising concentrations of proline and other soluble sugars (An et al. 2019). A considerable amount of various phenolic acids is found in most of the fruits and vegetables that form a significant portion of our diet. The application of VA mitigates the detrimental effects of ionic, osmotic, and oxidative stress, resulting in enhanced growth, biomass accumulation, and chlorophyll content in salt-stressed tomato seedlings (Parvin et al. 2020). Hence, the exogenous VA application up-regulated the antioxidant defense mechanism in salt-exposed tomato. This enhancement resulted in reduced ROS production, lower lipoxygenase (LOX) activity, as well as less membrane damage (Table 19.2; Parvin et al. 2020). Additionally, the VA application enhanced glyoxalase enzyme activities, with consequent reduction of toxic methylglyoxal accumulation upon salt stress. By boosting both the antioxidant defense and glyoxalase pathways, VA greatly improved plant growth and tolerance (Parvin et al. 2020). Similarly, VA suppressed the negative changes caused by Cd stress. Specifically, VA prevented Cd-induced alterations in leaf turgidity, photosynthetic pigments, and the levels of ascorbate and glutathione, under both mild and severe stress conditions (Bhuyan et al. 2020). Vanillic acid enhanced Cd tolerance by increasing phytochelatin synthesis and improving key metal accumulation metrics (accumulation factor, coefficient, and translocation factor). This suggests VA mitigates stress through improved metal chelation and sequestration, bolstered antioxidant and glyoxalase defense systems, with better nutrient balance (Bhuyan et al. 2020). The most effective VA concentrations for priming pea seeds were 2.0 and 4.0 mM. These treatments significantly improved morphological traits (length and

Table 19.2 Exogenous vanillic acid-mediated stress tolerance responses in plants

Plant species	Abiotic stress	VA application	Key effects	References
Solanum lycopersicum L.	NaCl (150 mM)	40–50 μM, 2 days	Increased glyoxalase and antioxidants; decreased Na^+/K^+ ratio; improved growth, chlorophyll (Chl) content, and RWC	Li et al. (2015)
	Drought	50 μM, 2 days	Enhanced glyoxalase and antioxidant activity; decreased oxidative damage; increased proline (Pro), and RWC	Zhang et al. (2019)
Oryza sativa L.	Cadmium toxicity (1–2 mM $CdCl_2$)	50 μM, 3 days	Enhanced photosynthesis and phytochelatins; restricted Cd absorption	Gupta et al. (2017)
Cucumis sativus L.	Drought	40 μM, 3 days	Decreased proline, ROS, and lipid peroxidation; increased RWC and Chl	Singh et al. (2021)
Triticum aestivum L.	NaCl (150 mM)	50 μM, 3 days	Increased Chl content, decreased ROS, and the Na^+/K^+ ratio	Hassan et al. (2015)
Capsicum annuum L.	Heat (38 °C, 7 days)	50 μM, 2 days	Increased growth and carotenoids (Car), and Chl; decreased MDA and ROS	Balal et al. (2020)
Helianthus annuus L.	Drought	50 μM, 3 days	Improved Chl, RWC, and growth; less damage to membranes	Yadav et al. (2020)
S. lycopersicum L.	NaCl; 150 mM	40 and 50 μM, 5 days	Reduced ROS and membrane stability with lower Na^+/K^+; higher shoot and root growth with photosynthetic pigmentation; enhanced proaccumulation with higher water status	Parvin et al. (2020)
Phaseolus vulgaris L.	Drought (PEG-induced)	40 μM, 2 days	Decreased lipid peroxidation; increased Chl and Car content	Kumar et al. (2018)
C. sativus L.	NaCl (100 mM)	50 μM, 2 days	Enhanced Chl and development; decreased oxidative stress and Na^+ toxicity	Jabeen et al. (2019)
B. oleracea L.	Heat (38 °C, 7 days)	50 μM, 2 days	Increased antioxidants; decreased ROS; increased Car and Chl content	Hossain et al. (2021)
Vitis vinifera L.	Water (drought)	40 μM, 5 days	Enhanced RWC, growth, and Chl; decreased oxidative damage	Díaz-Perales et al. (2021)

(continued)

Table 19.2 (continued)

Plant species	Abiotic stress	VA application	Key effects	References
O. sativa L.	Submergence; 3 and 6 days	1 mM, Pre-treatment, 24 h	Improved Chl *a* and *b* contents; increased shoot height and survival rate; increased endogenous VA; decreased protocatechuic acid levels while improving total phenolics and flavonoid contents	Xuan and Khang (2018)
O. sativa L.	Withholding water for 15 days	50 μM and 25 μM, pretreatment, 2 days	Increased Chl *a*, *b*, and Car content, Chl *a* and *b*; augmented flavonoids, phydroxybenzoic acid, and vanillin accumulation	Quan and Xuan (2018)
V. corymbosum L.	Drought (10% PEG, 2 days	40 μM, pretreatment, 2 days	Increased osmotic potential and RWC considerably; higher levels of endogenous VA, Pro ingredients; higher levels of soluble sugar	An et al. (2019)
S. lycopersicum L.	Drought (PEG, without water)	50 μM VA, pretreatment, 2 days	Increased growth and Chl contents; higher water content and Pro accumulation with lower leaf chlorosis	Parvin et al. (2024)

weight of shoots/roots) and elevated physiochemical metabolites, including chlorophyll, proline, phenolics, and glycine betaine (Rahman et al. 2024a). The results of this study disclosed the VA-induced betterment of growth and physio-biochemical characteristics of plants by upregulating osmoprotective mechanisms. This is particularly significant in mitigating Pb toxicity, which severely restricts plant growth, productivity, and thus poses serious risks to human health. Research shows that exogenous application of VA can enhance Pb tolerance in plants such as tomato. It achieves this by modulating growth processes and implementing a strategy to reduce Pb accumulation. Specifically, VA significantly controlled Pb uptake and distribution within plants, thereby lessening its phytotoxic effects (Afzaal et al. 2023).

Vanillic acid significantly reduced the detrimental effects of water crisis on tomato seedlings by improving height, biomass production, chlorophyll content, and osmotic status, with higher proline accumulation as well as decreased cellular membrane damage (Parvin et al. 2024). The foliar application of VA also resulted in higher proline accumulation, total phenolic content, and total soluble proteins in the cabbage (Rahman et al. 2024b). In summary, VA applied as a foliar spray improved cabbage growth during drought by increasing the accumulation of protective osmolytes and boosting metabolite levels.

19.6 Vanillic Acid-Induced Modulation of the Plant Antioxidant Defense System Under Stress

Upon environmental stress, various physiological and biochemical equilibrium in plants is disrupted. Therefore, in such stressful conditions, extreme ROS is produced, which later causes oxidative damage that compromises essential functions and cellular structures. Naturally, a complex, disciplined antioxidant defense mechanism with both non-enzymatic and enzymatic components is activated in plants to combat stress-mediated excessive ROS to be tolerant.

Phenolic chemicals like VA have garnered more attention recently due to their possible contributions to improving plant resistance to abiotic stressors, due to their antioxidant activity and regulatory roles in regulating the status of antioxidants and their redox homeostasis (Parvin et al. 2022). According to previous reports, exogenous VA application greatly increased the enzymatic antioxidant activity, where glutathione dismutase (SOD), catalase (CAT), peroxidases (POD), ascorbate peroxidase (APX), glutathione-S-transferase (GST), and glutathione peroxidase (GPX) are notable to improve ROS detoxification (Bhuyan et al. 2020; Parvin et al. 2024; Rahman et al. 2024a).

Moreover, VA treatment has been linked to enhanced membrane stability, osmolyte accumulation such as proline and glycine betaine, and redox balance maintenance through ascorbic acid (AsA) and GSH pools with the increased expression level of responsible genes (Table 19.3). *Oryza sativa*, *Brassica oleracea*, *Solanum lycopersicum*, and *Pisum sativum* are among the many crop species that benefit from the enhanced drought resistance, cadmium tolerance, and submergence resilience brought about by VA's diverse impacts (Quan and Xuan 2018; Xuan and Khang 2018; Bhuyan et al. 2020; Parvin et al. 2024; Rahman et al. 2024a).

Exogenous VA (50 μM), for example., dramatically amplified enzymatic antioxidants' activity like GPX, APX, CAT, and SOD in *O. sativa* subjected to cadmium toxicity while decreasing H_2O_2 and malondialdehyde (MDA) content. Likewise, VA treatment increased enzymatic antioxidant activity, improved redox balance, and stabilized (Bhuyan et al. 2020) membranes in *S. lycopersicum* under drought stress, all of which contributed to higher yield performance. These results imply that VA is essential for fortifying the antioxidant system, which increases plant resistance to environmental stress. It significantly boosts salinity tolerance and overall plant performance by triggering glyoxalase systems and the plant's antioxidant defense (Parvin et al. 2020).

Research demonstrates that VA performs a critical protective role in tomato seedlings subjected to drought and salinity. Its application significantly reduced the stress-induced acute ROS level and minimized cellular membrane damage. Furthermore, VA pretreatment was particularly effective in detoxifying methylglyoxal by activating glyoxalase enzyme activity. By concurrently strengthening the plant's antioxidant defense system, methylglyoxal detoxification pathway, and osmoregulation, VA effectively alleviates the damaging effects of abiotic stress (Parvin et al. 2020, 2024). Similarly, VA-induced higher antioxidant defense

Table 19.3 Regulatory role of vanillic acid in the plant antioxidant defense system for attuning plant tolerance

Plant species	Stress	Doses and application mode of vanillic acid	Status of oxidative stress and plant behavior	References
Oryza sativa L.	Drought,85% soil water content	Vanillic acid (VA); 25 and 50 μM VA	Increased antioxidant activity or DPPH radical scavenging Enhanced drought tolerance via improved ROS detoxification	Quan and Xuan (2018)
Vaccinium corymbosum L.	Dehydration (10% PEG), 2 days	VA, 40 μM	It increased gene expression of various antioxidant enzymes, like SOD, CAT, guaiacol peroxidase (GPOX), GPX, GR, and DHAR Elevated SOD and GPX activities	An et al. (2019)
O. sativa L. cv. BRRI dhan54	Cadmium toxicity (Cd; 1.0 and 2.0 mM; 72 h)	VA; 50 μM	Increased cadmium translocation factor (TF), biological concentration factor (BCF), biological accumulation coefficient (BAC) Lowered oxidative damage significantly; like H_2O_2 (27%), LOX activity (19%), MDA, and electrolyte leakage (up to 44%) Increased the ratio of both AsA:DHA and GSH:GSSG Augmented the activity of APX (24%), MDHAR (46%), DHAR (90%), and GR (26%), (SOD) by (14%), catalase (CAT) (39), GPX (22%), and GST (70%)	Bhuyan et al. (2020)
O. sativa L. cv. Koshihikari	Submergence	Protocatechuic acid (PA) and VA, 0.1–1.0 mM	Enhanced APX and SOD activities with higher expression of the responsible genes	Xuan and Khang (2018)
Brassica oleracea var. *capitata*	Drought	Foliar-applied vanillic acid (2.0 and 4.0 mM) foliar-applied vanillic acid (2.0 and 4.0 mM)	Improved the content of AsA, POD, CAT, and SOD activities Higher accumulation of proline, and total	Rahman et al. (2024b)

(continued)

Table 19.3 (continued)

Plant species	Stress	Doses and application mode of vanillic acid	Status of oxidative stress and plant behavior	References
			phenolic compounds Reduced H_2O_2 and MDA accumulation by strengthening the membrane permeability	
Solanum lycopersicum	Drought	Pretreated with 50 μM VA for 2 days	Lowered ROS scavenging enzymes activities, including SOD, CAT, GPX, and GST Enhanced MG detoxification and membrane stability Increased the ratio of ASA:DHA and GSH:GSSG Upregulated the glyoxalase enzyme activities Enhanced proline accumulation	Parvin et al. (2024)
Pisum sativum	Drought	Exogenously 4 mM,6 mM	Enhanced GB content Stimulated the activities of SOD, POD, and CAT Reduced AsA content Marked rise in other phenolic compounds. Accumulation of proline	Rahman et al. (2024a)
S. lycopersicum	NaCl; 150 mM	VA; 40 and 50 μM	Reduced ROS and membrane stability with lower Na^+/K^+ Higher shoot and root growth with photosynthetic pigmentation Enhanced proline accumulation with higher water status Higher AsA, GSH content with elevated activity of APX, MDHAR, and GR, SOD, CAT, and GST Elevated activity of glyoxalase enzymes with lower down on methylglyoxal toxicity	Parvin et al. (2020)

mechanisms of rice were reported by Bhuyan et al. (2020), where MDA, hydrogen peroxide, methylglyoxal, and LOX were significantly reduced with higher acceleration of antioxidant and glyoxalase systems. This VA-treated rice showed higher non-enzymatic antioxidant content, such as AsA and GSH, with the elevated activity of APX, monodehydroascorbate reductase (MDHAR), dehydroascorbate reductase (DHAR), glutathione reductase (GR). Besides phenolics and flavonoids, application of VA and PHBA (*p*-hydroxybenzoic acid) brought phytoalexin momilactones A (MA) and B (MB) to develop drought tolerance in rice, while the role of MB was greater than MA (Quan and Xuan 2018).

The application of exogenous VA led to an upregulation in the activities of CAT, SOD, and POD. This enhancement subsequently resulted in improved plant morphology and beneficial changes in various vital physiological as well as biochemical metabolites (Rahman et al. 2024b).

Thus, VA could be a great tool in enhancing plant antioxidant defense system to down the level of extreme ROS and consequent cellular oxidative damage in stressed conditions.

19.7 Conclusion

As a secondary metabolite, VA is vital for increasing plant growth and tolerance to environmental stress, especially abiotic stresses. As plants are sessile and are obligated to face such hazardous growth conditions, to enhance the plant growth, sustainability, and tolerance need extra stimulation for strengthening their in-built antioxidant defense system, then VA could be one of the best options for their properties. From various recent reports, it is obvious that VA plays critical roles in increasing plant growth and development, whether it is endogenously produced and/or exogenously supplemented. Excessive ROS-generated cellular damage is also well regulated by exogenous VA, with the higher involvement of the antioxidant defense system in many plants like rice, tomato, pea, and cabbage. Although its beneficial effects are promising, the molecular mechanisms behind VA-mediated stress tolerance are still not fully understood. Recent progress in omics technologies, such as transcriptomics and metabolomics, has shed light on the regulatory networks impacted by VA (Singh et al. 2019). Additional research is required to investigate its interactions with other signaling molecules and its long-term impact on various crop species. With the growing challenges posed by climate change and the demand for sustainable agriculture, VA could prove to be a beneficial resource in the development of stress-resistant crops. However, through its regulatory function in the antioxidant defense mechanism, VA can improve crop resilience, according to new research, making it a promising stimulant in sustainable agriculture. By exploring the physiological effects and mechanistic insights of VA-mediated stress amelioration in plants, this section highlights the importance of VA in enhancing plant performance in challenging environmental circumstances.

References

Abeed AHA, Mahdy RE, Alshehri D, Hammami I, Eissa MA, Abdel Latef AAH, Mahmoud GA (2022) Induction of resilience strategies against biochemical deteriorations prompted by severe cadmium stress in sunflower plant when Trichoderma and bacterial inoculation were used as biofertilizers. Frontiers in Plant Science 13: 1004173

Acidri R, Sawai Y, Sugimoto Y, Handa T, Sasagawa D, Masunaga T, Yamamoto S, Nishihara E (2020) Exogenous kinetin promotes the nonenzymatic antioxidant system and photosynthetic activity of coffee (*Coffea arabica* L.) plants under cold stress conditions. Plan Theory 9(2):281. https://doi.org/10.3390/plants9020281

Afzaal Z, Hussain I, Ashraf MA, Rasheed R, Javed MT, Ansari MUR, Iqbal M (2020) Lead induced modulation in growth, chlorophyll pigment, nutrient uptake, antioxidant enzyme regulation, gene expression and fruit quality in two tomato cultivars. Int J Agric Biol 24:1732–1744

Afzaal Z, Hussain I, Ashraf MA, Riaz S (2023) Ascorbic acid and vanillic acid application alleviate the lead toxicity in *Lycopersicon esculentum* by modulating key biochemical attributes and antioxidant defense systems. Russ J Plant Physiol 70(6):129

Agati G, Azzarello E, Pollastri S, Tattini M (2012) Flavonoids as antioxidants in plants: location and functional significance. Plant Sci 196:67–76

Ahanger MA, Tomar NS, Tittal M, Argal S, Agarwal R (2017) Plant growth under water/salt stress: ROS production; antioxidants and significance of added potassium under such conditions. Physiol Mol Biol Plants 23:731–744

Aimvijarn P, Payuhakrit W, Charoenchon N, Okada S, Suwannalert P (2023) Riceberry rice germination and UVB radiation enhance protocatechuic acid and vanillic acid to reduce cellular oxidative stress and suppress B16F10 melanogenesis relating to F-actin rearrangement. Plants 12(3):484

An L, Liu Y, Wang X, Zhang M (2019) Vanillic acid modulates drought stress response in blueberry (*Vaccinium corymbosum* L.). Plant Physiol Rep 24(3):345–354

Ashraf MFMR, Foolad MR (2007) Roles of glycine betaine and proline in improving plant abiotic stress resistance. Environ Exp Bot 59:206–216

Balal RM, Shahid MA, Iqbal Z (2020) Vanillic acid enhances heat tolerance in *Capsicum annuum*. J Therm Biol 90:102582

Benavides MP, Gallego SM, Tomaro ML (2005) Cadmium toxicity in plants. Braz J Plant Physiol 17:21–34

Bhattacharya A, Sood P, Citovsky V (2010) The roles of plant phenolics in defence and communication during *Agrobacterium* and *Rhizobium* infection. Mol Plant Pathol 11(5):705–719

Bhuyan MB, Parvin K, Mohsin SM, Mahmud JA, Hasanuzzaman M, Fujita M (2020) Modulation of cadmium tolerance in rice: insight into vanillic acid-induced upregulation of antioxidant defense and glyoxalase systems. Plants 9(2):188

Boerjan W, Ralph J, Baucher M (2003) Lignin biosynthesis. Annu Rev Plant Biol 54(1):519–546

Besson-Bard A, Gravot A, Richaud P, Auroy P, Duc C, Gaymard F, et al (2009) Nitric oxide contributes to cadmium toxicity in Arabidopsis by promoting cadmium accumulation in roots and by up-regulating genes related to iron uptake. Plant Physiol 149(3):1302–1315

Chen S, Wang Q, Lu H, Li J, Yang D, Liu J, Yan C (2019) Phenolic metabolism and related heavy metal tolerance mechanism in *Kandelia Obovata* under Cd and Zn stress. Ecotoxicol Environ Saf 169:134–143

Colak N, Torun H, Gruz J, Strnad M, Ayaz FA (2019) Exogenous N-acetylcysteine alleviates heavy metal stress by promoting phenolic acids to support antioxidant defence systems in wheat roots. Ecotoxicol Environ Saf 181:49–59

Colmer TD, Voesenek LACJ (2009) Flooding tolerance: suites of plant traits in variable environments. Funct Plant Biol 36(8):665–681. https://doi.org/10.1071/FP09144

Deng Y, Lu S (2017) Biosynthesis and regulation of phenylpropanoids in plants. Crit Rev Plant Sci 36(4):257–290

Di Toppi LS, Gabbrielli R (1999) Response to cadmium in higher plants. Environ Exp Bot 41(2): 105–130

Díaz-Perales A, Pérez JM, González M (2021) Exogenous vanillic acid boosts drought tolerance in grapevine. Plant Stress 2:100040

El-Beltagi HS, Mohamed HI, Sofy MR (2020) Role of ascorbic acid, glutathione and proline applied as singly or in sequence combination in improving chickpea plant through physiological change and antioxidant defense under different levels of irrigation intervals. Molecules 25(7): 1702

Emamverdian A, Ding Y, Mokhberdoran F, Xie Y (2015) Heavy metal stress and some mechanisms of plant defense response. Sci World J 2015(1):756120

Farooq M, Gogoi N, Barthakur S, Baroowa B, Bharadwaj N, Alghamdi SS, Siddique KH (2017) Drought stress in grain legumes during reproduction and grain filling. J Agron Crop Sci 203(2): 81–102

Gallage NJ, Møller BL (2015) Vanillin–bioconversion and bioengineering of the most popular plant flavor and its de novo biosynthesis in the vanilla orchid. Mol Plant 8(1):40–57

Gilroy S, Suzuki N, Miller G, Choi WG, Toyota M, Devireddy AR, Mittler R (2014) A tidal wave of signals: calcium and ROS at the forefront of rapid systemic signaling. Trends Plant Sci 19(10):623–630

Gupta B, Huang B (2014) Mechanism of salinity tolerance in plants: physiological, biochemical, and molecular characterization. Intl J Genomics 2014:701596

Gupta DK, Huang HG, Nicoloso FT (2017) Role of vanillic acid in cadmium detoxification in rice (*Oryza sativa*). Environ Sci Pollut Res 24(8):7390–7399

Gupta P, Mishra V, Singh N (2022) Vanillic acid mitigates heavy metal stress in plants. Environ Pollut 301:119040

Hakla HR, Sharma S, Urfan M, Yadav NS, Rajput P, Kotwal D, Abdel Latef AAH, Pal S (2021) Gibberellins target shoot-root growth, morpho-physiological and molecular pathways to induce cadmium Tolerance in Vigna radiata L. Agronomy 11: 896

Harborne JB, Harborne JB (1973) Phenolic compounds. In: Phytochemical methods: a guide to modern techniques of plant analysis, pp 33–88

Hassan FAS, Ali EF, Ramadan AA (2015) Vanillic acid modulates salt stress responses in wheat. Plant Physiol Biochem 96:279–289

Hussain S, Shaukat M, Ashraf M, Zhu C, Jin Q, Zhang J (2019) Salinity stress in arid and semi-arid climates: effects and management in field crops. In: Hossain S (ed) Climate change and agriculture. IntechOpen, London

Hengstler JG, Bolm-Audorff U, Faldum A, Janssen K, Reifenrath M, Götte W, et al (2003) Occupational exposure to heavy metals: DNA damage induction and DNA repair inhibition prove co-exposures to cadmium, cobalt and lead as more dangerous than hitherto expected. Carcinog 24(1):63–73

Hossain A, Skalicky M, Brestic M, Maitra S, Alam MA, Syed MA, Hossain J, Sarkar S, Saha S, Bhadra P. et al (2021) Consequences and mitigation strategies of abiotic stresses in wheat (Triticum aestivum L.) under the changing climate. Agronomy 11:241. https://doi.org/10.3390/agronomy11020241

Isah T (2019) Stress and defense responses in plant secondary metabolites production. Biol Res 52: 39

Jabeen N, Ahmad R, Iqbal M (2019) Ameliorative effects of vanillic acid on salt-stressed cucumber. Environ Exp Bot 162:177–185

Jameel A, Akram NA, Bashir R, Ashraf M, Abdel Latef AAH (2026) Salicylic acid enhances yield and alleviates drought-induced oxidative damage in two carrot cultivars with contrasting tolerance. Russian Journal of Plant Physiology. 73, 23

Kısa D, Kayır Ö, Sağlam N, Şahin S, Öztürk L, Elmastaş M (2019) Changes of phenolic compounds in tomato associated with the heavy metal stress. Bartın Uni Intl J Nat Appl Sci 2(1):35–43

Kumar N, Pruthi V (2014) Potential applications of ferulic acid from natural sources. Biotechnol Rep 4:86–93
Kumar V, Yadav SK, Verma S (2018) Vanillic acid boosts drought resistance in common bean. Acta Physiol Plant 40(11):214
Kumar V, Singh P, Tripathi DK (2020) Vanillic acid as a novel biomarker for plant stress. Plant Physiol Rep 25:127–134
Levitt J (1980) Responses of plants to environmental stresses, 2nd edn. Academic Press, New York
Li H, Zhu Y, Hu Y (2015) Beneficial effects of silicon in alleviating salinity stress of tomato seedlings grown under sand culture. Acta Physiol Plant 37, 71 (2015). https://doi.org/10.1007/s11738-015-1818-7
Mittler R (2017) ROS are good. Trends Plant Sci 22(1):11–19
Munns R, Termaat A (1986) Whole-plant responses to salinity. Funct Plant Biol 13(1):143–160
Naik PM, Al-Khayri JM (2016) Abiotic and biotic elicitors–role in secondary metabolites production through in vitro culture of medicinal plants. IntechOpen
Pagter M, Bragato C, Brix H (2005) Tolerance and physiological responses of *Phragmites australis* to water deficit. Aquat Bot 81(4):285–299
Pandey P, Ramegowda V, Senthil-Kumar M (2015) Shared and unique responses of plants to multiple individual stresses and stress combinations: physiological and molecular mechanisms. Front Plant Sci 6:723
Park JS, Lee EJ (2019) Waterlogging induced oxidative stress and the mortality of the Antarctic plant, *Deschampsia Antarctica*. J Ecol Environ 43(1):29. https://doi.org/10.1186/s41610-0190127-2
Parvin K, Nahar K, Hasanuzzaman M, Bhuyan MHMB, Mohsin SM, Fujita M (2020) Exogenous 1051 vanillic acid enhances salt tolerance of tomato: insight into plant antioxidant defense and 1052 glyoxalase systems. Plant Physiol Biochem 150:109–120
Parvin K, Hasanuzzaman M, Mohsin SM, Nahar K, Fujita M (2024) Vanillic acid modulates antioxidant defense and methylglyoxal detoxification systems to combat drought stress in tomato seedlings. Plants 13(22):3114. https://doi.org/10.3390/plants13223114
Pratyusha S (2022) Phenolic compounds in the plant development and defense. In: Plant stress physiology: perspectives in agriculture, vol 125, IntechOpen: London. https://doi.org/10.5772/intechopen.102873
Proestos C, Komaitis M (2013) Analysis of naturally occurring phenolic compounds in aromatic plants by RP-HPLC coupled to Diode Array Detector (DAD) and GC-MS after Silylation. Foods (Basel, Switzerland) 2(1):90–99. https://doi.org/10.3390/foods2010090
Parvin K, Nahar K, Mohsin SM, Mahmud JA, Fujita M, Hasanuzzaman M (2022) Plant phenolic compounds for abiotic stress tolerance. In: Hasanuzzaman M, Ahammed GJ, Nahar K (eds) Managing Plant Production Under Changing Environment. Springer, Singapore. p. 193–237. https://doi.org/10.1007/978-981-16-5059-8_8
Qadir M, Quillérou NV, Murtaza G, Singh M, Thomas RJ, Noble AD (2014) Economics of salt-induced land degradation and restoration. Nat Res Forum 38:282–295
Quan NT, Xuan TD (2018) Foliar application of vanillic and p-hydroxybenzoic acids enhanced drought tolerance and formation of phytoalexin momilactones in rice. Arch Agron Soil Sci 64:1831–1846
Rahman A, Akram NA, Ashraf M, Alsahli AA, Ahmad P (2024a) Deciphering the role of exogenously-applied vanillic acid in regulating drought stress tolerance in pea (*Pisum sativum* L.): key growth and physio-biochemical attributes. J King Saud Uni Sci 36(11):103544. https://doi.org/10.1016/j.jksus.2024.103544
Rahman A, Akram NA, Iqbal M (2024b) Phenolic compound vanillic acid modulates growth and defense system of cabbage (*Brassica oleracea* var. *Capitata* L.) under drought stress. Pak J Bot 56(5):1631–1644
Raza A, Razzaq A, Mehmood SS, Zou X, Zhang X, Lv Y, Xu J (2019) Impact of climate change on crops adaptation and strategies to tackle its outcome: a review. Plants 8(2):34. https://doi.org/10.3390/plants8020034

Roy SJ, Negrão S, Tester M (2014) Salt resistant crop plants. Curr Opin Biotechnol 26:115–124
Shahid MA, Balal RM, Khan N, Rossi L, Rathinasabapathi B, Liu G, Khan J, Cámara-Zapata JM, Martínez-Nicolas JJ, Garcia-Sanchez F (2018) Polyamines provide new insights into the biochemical basis of Cr-tolerance in Kinnow mandarin grafted on diploid and double-diploid rootstocks. Environ Exp Bot 156:248–260
Sharma A, Shahzad B, Rehman A, Bhardwaj R, Landi M, Zheng B (2019a) Response of phenylpropanoid pathway and the role of polyphenols in plants under abiotic stress. Molecules 24(13):2452. https://doi.org/10.3390/molecules24132452
Sharma A, Shahzad B, Kumar V, Kohli SK, Sidhu GPS, Bali AS, Handa N, Kapoor D, Bhardwaj R, Zheng B (2019b) Phytohormones regulate accumulation of osmolytes under abiotic stress. Biomolecules 9(7):285. https://doi.org/10.3390/biom9070285
Silva EC, Nogueira RJ, Valen FH, Araújo FPD, Pimenta MA (2009a) Stomatal changes induced by intermittent drought in four umbu tree genotypes. Braz J Plant Physiol 21:33–42
Silva EC, Nogueira RJ, Vale FH, Melo NFD, Araújo FPD (2009b) Water relations and organic solutes production in four umbu tree (*Spondias tuberosa*) genotypes under intermittent drought. Braz J Plant Physiol 21:43–53
Singh A, Kumar A, Yadav S, Singh IK (2019) Reactive oxygen species-mediated signaling during abiotic stress. Plant Gene 18:100173. https://doi.org/10.1016/j.plgene.2019.100173
Sabagh AE, Hossain A, Islam MS, Ahmed S, Raza A, Iqbal MA, et al (2021) Elevated CO2 in combination with heat stress influences the growth and productivity of cereals: adverse effect and adaptive mechanisms. In: Engineering tolerance in crop plants against abiotic stress. CRC Press; 125–161
Steffens B, Rasmussen A (2016) The physiology of adventitious roots. Plant Physiol 170(2): 603–17
Singh, M., Singh, P., Singh, S. et al. (2021) A global meta-analysis of yield and water productivity responses of vegetables to deficit irrigation. Sci Rep 11:22095. https://doi.org/10.1038/s41598-021-01433-w
Šircelj H, Tausz M, Grill D, Batič F (2005) Biochemical responses in leaves of two apple tree cultivars subjected to progressing drought. J Plant Physiol 162(12):1308–1318
Valliyodan B, Nguyen HT (2006) Understanding regulatory networks and engineering for enhanced drought tolerance in plants. Curr Opin Plant Biol 9(2):189–195
Van Zelm E, Zhang Y, Testerink C (2020) Salt tolerance mechanisms of plants. Annu Rev Plant Biol 71(1):403–433
Vlot AC, Sales JH, Lenk M, Bauer K, Brambilla A, Sommer A, Chen Y, Wenig M, Nayem S (2021) Systemic propagation of immunity in plants. New Phytol 229(3):1234–1250
Vogt T (2010) Phenylpropanoid biosynthesis. Mol Plant 3(1):2–20
Wang J, Yuan B, Huang B (2019) Differential heat-induced changes in phenolic acids associated with genotypic variations in heat tolerance for Hhard fescue. Crop Sci 59:667–674
Xuan T, Khang D (2018) Effects of exogenous application of protocatechuic acid and vanillic acid 1144 to chlorophylls, phenolics and antioxidant enzymes of rice (*Oryza sativa* L.) in submergence. 1145. Molecules 23(3):620. https://doi.org/10.3390/molecules23030620
Yadav AN, Rastegari AA, Yadav N, Kour D (2020) Advances in plant microbiome and sustainable Agriculture: Diversity and biotechnological applications. Springer, Singapore
Zhang H, Zhu J, Gong Z, Zhu JK (2022) Abiotic stress responses in plants. Nat Rev Genet 23(2): 104–119
Zhang Z, Cao B, Gao S et al (2019) Grafting improves tomato drought tolerance through enhancing photosynthetic capacity and reducing ROS accumulation. Protoplasma 256:1013–1024[KP7.1]. https://doi.org/10.1007/s00709-019-01357-3

MIX
Papier aus verantwortungsvollen Quellen
Paper from responsible sources
FSC® C105338

If you have any concerns about our products,
you can contact us on
ProductSafety@springernature.com

In case Publisher is established outside the EU,
the EU authorized representative is:
Springer Nature Customer Service Center GmbH
Europaplatz 3, 69115 Heidelberg, Germany

Printed by Libri Plureos GmbH
in Hamburg, Germany